Name	Structure[a]	Name ending	Example
Sulfide	C–S–C	*sulfide*	CH_3SCH_3 Dimethyl sulfide
Disulfide	C–S–S–C	*disulfide*	CH_3SSCH_3 Dimethyl disulfide
Sulfoxide	O^- above S^+; C–S^+–C	*sulfoxide*	O^- above $\overset{+}{S}$; CH_3SCH_3 Dimethyl sulfoxide
Aldehyde	O=C; R–C–H	*-al*	O=; CH_3CH Ethanal
Ketone	O=C; C–C–C	*-one*	O=; CH_3CCH_3 Propanone
Carboxylic acid	O=C; C–C–OH	*-oic acid*	O=; CH_3COH Ethanoic acid
Ester	O=C; C–C–O–C	*-oate*	O=; CH_3COCH_3 Methyl ethanoate
Thioester	O=C; C–C–S–C	*-thioate*	O=; CH_3CSCH_3 Methyl ethanethioate
Amide	O=C; C–C–N̈	*-amide*	O=; CH_3CNH_2 Ethanamide
Acid chloride	O=C; C–C–Cl	*-oyl chloride*	O=; CH_3CCl Ethanoyl chloride

Fundamentals of

Organic Chemistry

Sixth Edition

John McMurry
Cornell University

Eric Simanek
Texas A&M University

Australia • Canada • Mexico • Singapore • Spain
United Kingdom • United States

Fundamentals of Organic Chemistry, Sixth Edition
John McMurry and Eric Simanek

Publisher, Physical Sciences: David Harris
Senior Development Editor: Sandra Kiselica
Assistant Editor: Ellen Bitter
Editorial Assistant: Lauren Oliviera
Technology Project Manager: Donna Kelley
Executive Marketing Manager: Julie Conover
Marketing Assistant: Michele Colella
Marketing Communications Manager: Bryan Vann
Project Manager, Editorial Production: Lisa Weber
Creative Director: Rob Hugel
Art Director: Lee Friedman
Print Buyer: Judy Inouye
Permissions Editor: Joohee Lee
Production Service: Graphic World Inc.
Text Designer: Patrick Devine Design
Photo Researcher: Susan Kaprov
Copy Editor: Graphic World Inc.
Illustrators: 2064design and Graphic World Inc.
Cover Designer: Mende Design
Cover Image: Peter Barrett/Masterfile
Cover Printer: Phoenix Color Corp
Compositor: Graphic World Inc.
Printer: Quebecor World/Versailles

Printed in the United States of America
1 2 3 4 5 6 7 00 09 08 07 06

Library of Congress Control Number: 2006922676

Student Edition: ISBN 0-495-01203-3

Thomson Higher Education
10 Davis Drive
Belmont, CA 94002-3098
USA

Brief Contents

Contents

3

The Nature of Organic Reactions: Alkenes

4

Reactions of Alkenes and Alkynes

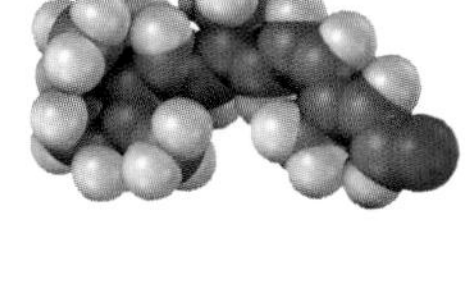

5

Aromatic Compounds

8

Alcohols, Phenols, and Ethers

9

Aldehydes and Ketones: Nucleophilic Addition Reactions

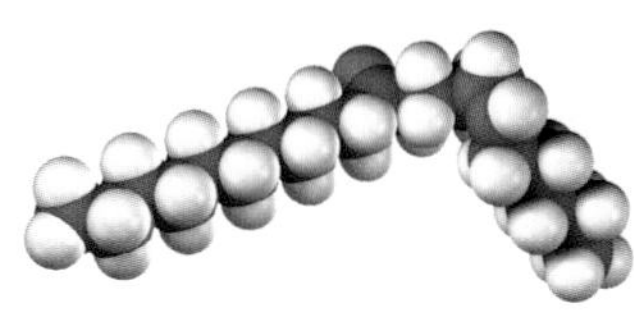

10

Carboxylic Acids and Derivatives

13

Structure Determination

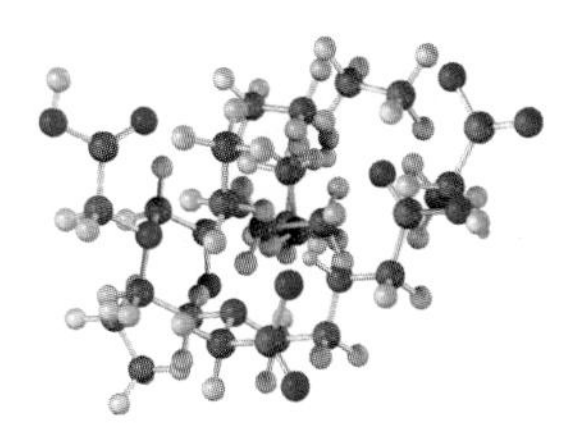

14

Biomolecules: Carbohydrates

15

Biomolecules: Amino Acids, Peptides, and Proteins

Preface

The definition of *chemistry* is changing. The role of *organic chemistry*, specifically, has evolved significantly from its humble beginnings of making soaps and dyes. *Organic chemistry* has moved to center stage in fields ranging from medicine to mechanical engineering, from agriculture to advanced communications. Chemists are learning new languages—those of molecular biology and fluid mechanics—in order to shape the world in which we live. Now more than ever before, a fundamental understanding of chemistry is critical to addressing complex, interdisciplinary problems. This sixth edition of *Fundamentals of Organic Chemistry* addresses these advances with a greater emphasis on the applications of organic chemistry, especially to medicine and agriculture. Interlude boxes appearing in each chapter are, in general, richer in the chemistry of drugs and agrochemicals. Certain topics, like triazine herbicides and drug sources and development, are woven throughout the entire text. All of the chapter openers were changed to incorporate illustrations of an application of organic chemistry, such as Augmentin to treat bacterial ear infections. New problem categories, such as "In the Field with Agrochemicals" and "In the Medicine Cabinet," were added to reinforce the focus on applications.

This book is written for the one-semester, short course in organic chemistry—where content must be thorough but to the point. We present only those subjects needed for a brief course while maintaining the important pedagogical tools commonly found in larger books. In this sixth edition, *Fundamentals of Organic Chemistry* uses clear explanations, thought-provoking examples, and the innovative vertical format for explaining reaction mechanisms. The primary organization of this book is by functional group, beginning with the simple (alkanes) and progressing to the more complex. Within the primary organization, there is also an emphasis on explaining the fundamental mechanistic similarities of reactions. Through this approach, we hope that memorization is minimized and understanding is maximized.

Some new sections on current topics have also been added. These additions do not dramatically increase the size of the book or length of the chapters: existing text, figures, and schemes have been streamlined. These sections include discussions of topics that are imperative for all students engaged at this level of inquiry. Sections dedicated to catalysis, tetrahedral intermediates and their impact on drug design, and carbon alkylations and their role in the preparation of barbiturates are new. The emerging importance of RNA in the central dogma is discussed in terms of chemistry. These topics offer both context and an opportunity to weave themes from many chapters together.

The first five editions of this text were widely regarded as the clearest and most readable treatments of introductory organic chemistry available. This edition includes a new co-author, Eric Simanek, from Texas A&M University. Simanek brings his interests in chemical education, agricultural organic chemistry, and drug delivery to this book. We hope that you find that the tradition of excellence continues with the sixth edition and that the material is interesting and fresh. We welcome all comments on these changes as well as recommendations for future editions.

Organization and Content Changes

The chapters are still organized by functional group with treatment of multifunctional biological molecules—carbohydrates, lipids, nucleic acids—and their interconversion through metabolism at the end of the book. Specific additions to chapter content are as follows:

Chapter 1: Hybridization of double and triple bonds is discussed to facilitate summations of the role of π bonding (Section 1.8). The figures and discussion surrounding hybridization have been reworked for clarity.

Chapter 2: To further add relevance, a discussion of distillation and refining is included (Section 2.4). Octane numbers are introduced as a method for defining a complex mixture based on performance rather than exact composition.

Chapter 3: Energy diagrams are more tightly correlated to reaction mechanisms (Section 3.9), and the topic of catalysis is new to this edition (Section 3.10).

Chapter 4: Biological and anti-Markovnikov alkene addition reactions are introduced as topics of contemporary interest.

Chapter 5: In addition to streamlining the text and improving the clarity of the figures, a discussion of nucleophilic aromatic substitution reactions appears in the Interlude as part of the continuing story of triazine herbicides.

Chapter 6: A new section, *The Chiral Environment* (Section 6.10), underscores the impact of diastereomeric relationships, using Pasteur, resolutions, and drugs as examples.

Chapter 7: The most substantive changes to this chapter link together a great body of chemistry in the problems sections. Questions focusing on the multistep syntheses of the antidepressant Prozac and antipsychotic Flupentixol are featured.

Chapter 8: The third in a series of Interludes focusing on the triazine herbicides introduces the mechanism of action of these molecules and highlights photosynthesis.

Chapter 9: The discussion of reduction and reducing agents is expanded and now includes the biologically relevant molecules NADH and NADPH (Section 9.6). There is a new section on the roles of acetals and hemiacetals in nature and the laboratory (Section 9.9).

Chapter 10: New Section 10.6 focuses on the tetrahedral intermediate and its role in the design of drugs, while new Section 10.15 addresses the role of enzymes in organic synthesis and reinforces the concept of the chiral environment from Chapter 6.

Chapter 11: The role of enolate alkylation chemistry is anchored in context with a discussion of barbiturates (Section 11.7) while reinforcing chemistry described in Chapter 10. Conjugate additions are highlighted in the Interlude with a detailed discussion of β-lactamase inhibitors.

Chapter 12: This chapter continues the trend of context with greater focus on bioactive amines in the body and problem sections and in the Interlude that focuses on opium and opiates.

Chapter 13: This chapter has seen significant revision and the addition of sections on X-ray crystallography (Section 13.2) and mass spectrometry (Section 13.3). The discussion of ultraviolet spectroscopy (Section 13.6) has also been expanded.

Chapter 14: A discussion (Section 14.12) of the use of carbohydrates as the basis for the energy and chemical industry is new and ties back to the use of enzymes in organic synthesis and the chiral environment that is provided by these catalysts.

Chapter 15: The discussion of synthesis in this chapter has been condensed and the roles of protecting groups and solid-phase synthesis expanded (Section 15.3). New Section 15.4 discusses the two primary routes used for peptide synthesis in nature.

Chapter 16: The chemistry of lipids and nucleic acids is expanded with new sections on the statin drugs (Section 16.6) and the catalytic activity observed in RNA (Section 16.16).

Chapter 17: This chapter benefits from condensed figures that move single metabolic pathways to a single page with transformations that are readily identified. The emphasis on context concludes with current efforts to enlist engineered microbes to aid in environmental cleanup.

Features

New to This Edition

- All new applied **chapter openers** provide students with associated illustrations relevant to the content of the chapter.
- Running application of **agricultural chemistry** builds upon a chemical theme. Students see the same compounds used in problems but with different questions.
- Reorganized **problem sets** provide students with the organizational framework in order to solve problems more effectively.
- New **applied problems** include "In the Field with Agrochemicals" and "In the Medicine Cabinet," which use compounds associated with agriculture and medicine. These problems are intended to be challenging; most tie elements from multiple chapters together under a unified theme.
- Many new brief **Interludes** are included at the end of each chapter to show interesting applications of organic chemistry to mostly biological and agricultural systems.
- In the spectroscopy chapter (Chapter 13), **X-ray crystallography** and **mass spectrometry** have been added to broaden students' knowledge of analytical techniques.
- An **enhanced art program** gives visual students more support in understanding concepts, including better orbital diagrams and more applied photographs.
- A new **Road Map of Chemical Reactions Interlude** after Chapter 12 summaries all the organic reactions students have learned in Chapters 1 through 12.
- Current **IUPAC nomenclature** rules, as updated in 1993, are used to name compounds in this text.
- Downloadable from the web, **color images** from this sixth edition are available to instructors for lecture preparation and in-class use.

Continued from the Fifth Edition

- McMurry's renowned, trademarked **vertical reaction mechanisms** give students descriptions of each step in a reaction pathway.
- Full color throughout, the text highlights the reacting parts of molecules to make it easier to focus on the main parts of a reaction.
- Nearly 100 **electrostatic potential maps** display the polarity patterns in molecules and the importance of these patterns in determining chemical reactivity.
- More than 100 **Visualizing Chemistry problems** challenge students to make the connection between typical line-bond drawings and computer-generated molecular models.
- Each chapter contains many **Practice Problems** that illustrate how problems can be solved, followed by a similar problem for the student to solve. Each worked-out problem begins with a Strategy discussion that shows how to approach the problem.
- Nearly 900 **Problems** are included both within the text and at the end of every chapter.

Book Support

Supporting instructor materials are available to qualified adopters. Please consult your local Thomson Brooks/Cole representative for details.

Visit **http://www.thomsonedu.com** to:

- Locate your local representative
- Download electronic files of text art and ancillaries
- Request a desk copy

Study Guide and Solutions Manual Written by Susan McMurry, this manual contains the answers to all the problems in the text. This indispensable tool helps students develop solid problem-solving strategies required for organic chemistry. ISBN 0-495-01932-1

OWL: Online Web-based Learning & Homework System Developed at the University of Massachusetts, Amherst; class-tested by thousands of students; and used by more than 200 institutions and 50 thousand students, OWL is a customizable and flexible cross-platform web-based homework system and assessment tool for introductory, general, allied health, liberal arts, and organic chemistry courses. The OWL Online Web-based Learning system provides students with instant analysis and feedback on homework problems, modeling questions, and animations to accompany select Thomson Brooks/Cole chemistry textbooks. With deep parameterization of chemical structures, the Organic OWL database offers more than 5000 questions as well as Marvin-Sketch, a Java applet for viewing and drawing chemical structures. The Organic OWL database also uses the MDL Chime application to assist students with viewing structures of organic compounds. This powerful system maximizes students' learning experience and, at the same time, reduces faculty workload and helps facilitate instruction. A fee-based access code is required to enter the specific OWL database selected. OWL is available for use only within North America.

Online Images For the first time, all images in the text are available on the instructor's companion website to put into your Microsoft® PowerPoint® lecture slides. Please find them at **www.thomsonedu.com/chemistry/McMurry**.

Printed and Computerized Test Bank The printed and ExamView® computerized test banks feature approximately 30 short answer questions per chapter to use as quizzes, exams, or homework assignments. ISBNs 0-495-01946-1, 0-495-01947-X

***Pushing Electrons: A Guide for Students of Organic Chemistry,* third edition** Written by Daniel P. Weeks, retired, Northwestern University, this brief text teaches a skill essential to learning organic chemistry. By working through the program, students learn to push electrons to generate resonance structures and write organic mechanisms. ISBN 0-03-020693-6

SpartanModel Electronic Modeling Kit This is a stand-alone molecular modeling program developed for chemistry students. Each copy includes the software on CD-ROM, an extensive molecular database, 3-D glasses, and a Tutorial and Users Guide with 50 pages of activities for organic chemistry. ISBN 0-495-01793-0

Organic Chemistry Laboratory Manuals Thomson Brooks/Cole is pleased to offer you a choice of organic chemistry laboratory manuals catered to fit your needs. Visit **http://www.thomsonedu.com**. Customizable laboratory manuals also can be assembled. Go to **http://cerlabs.brookscole.com** and **http://www.outernetpublishing.com** for more information.

Acknowledgments

The authors wish to extend special thanks to Sandi Kiselica, senior development editor, for her guidance on this project and to David Harris, executive editor, for his support and vision of this edition. We are grateful for the thoughtful and expert commentary of Joe Hornback (University of Denver), Kevin Minbiole (James Madison University), and Mark Erickson (Hartwick College) on content, clarity, and accuracy. We extend our thanks to Paul Martino of Flathead Valley Community College, who checked and rechecked the nomenclature as we changed to the accepted IUPAC conventions, and to Susan McMurry, who offered many helpful comments on the manuscript. We thank all the people who helped to shape this book and its message: Julie Conover, executive marketing manager, Lisa Weber, senior production project manager, Ellen Bitter, assistant editor, and Suzanne Kastner, production editor. In addition, we are grateful to colleagues who reviewed the manuscript for this book.

John McMurry
Cornell University

Eric Simanek
Texas A&M University

Manuscript Reviewers
Debbie Beard, Mississippi State University
Nick Drapela, Oregon State University
Daniel Dyer, Southern Illinois University
Peter Hanson, Wittenberg University
Steven Holmgren, Montana State University
Don Mumm, Southeast Community College
Robert S. Phillips, University of Georgia
Martin Pulver, Bronx Community College
Robert Ronald, Washington State University
Michael Toney, University of California, Davis
Gary Trammell, University of Illinois, Springfield

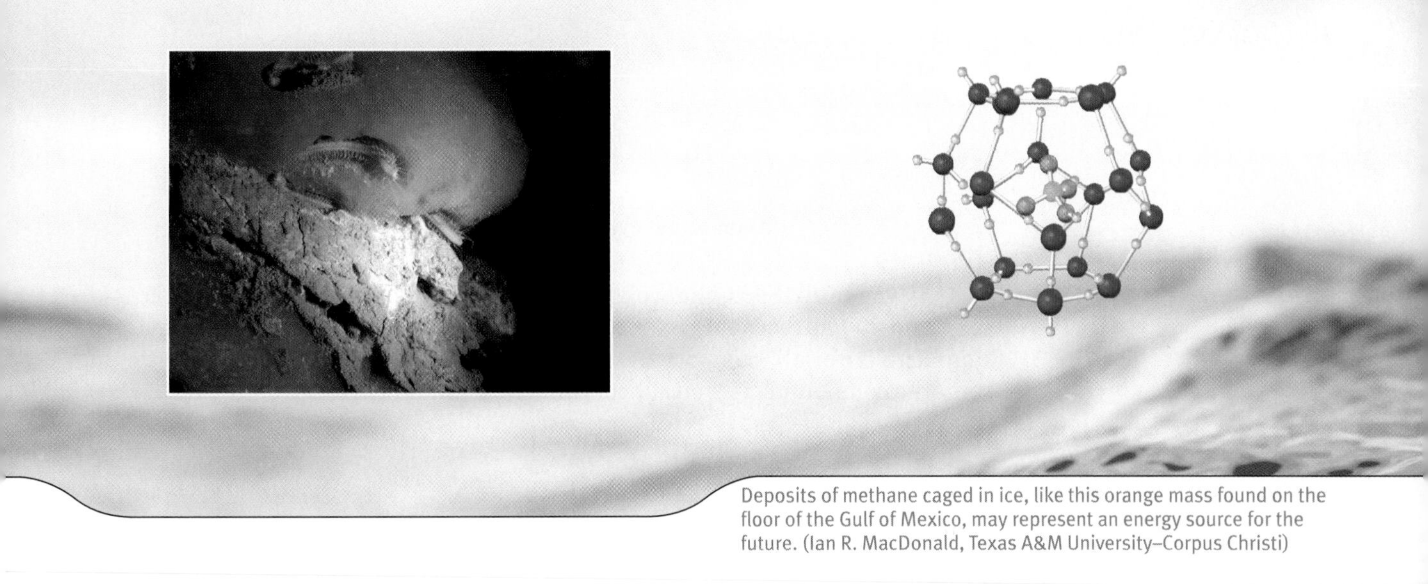

Deposits of methane caged in ice, like this orange mass found on the floor of the Gulf of Mexico, may represent an energy source for the future. (Ian R. MacDonald, Texas A&M University–Corpus Christi)

CHAPTER 1

Structure and Bonding; Acids and Bases

Organic chemistry is all around us. The reactions and interactions of organic molecules allow us to see and smell. They cause us to fight and fear. Organic chemistry is the cup that holds our take-out coffee and the caffeine that wakes us from our slumbers. Organic chemistry provides the molecules that treat illnesses, protect crops, and clean our clothes. Anyone with a curiosity about life and living things should have a basic understanding of organic chemistry.

The foundations of organic chemistry date back to the mid-1700s, when alchemists noticed unexplainable differences between compounds derived from living sources and those derived from minerals. Compounds from plants and animals were often difficult to isolate and purify. Even when pure, they were difficult to work with and tended to decompose more easily than compounds from minerals. The Swedish chemist Torbern Bergman was the first to express this difference between "organic" and "inorganic" substances in 1770, and the term *organic chemistry* soon came to mean the chemistry of compounds from living organisms.

To many chemists of the time, the only explanation for the difference in behavior between organic and inorganic compounds was that organic compounds contained an undefinable "vital force" as a result of their origin in living sources. With time, however, it became clear that organic compounds could be manipulated in the laboratory just like inorganic compounds. Friedrich Wöhler discovered in 1828, for example, that it was possible to convert the "inorganic" salt ammonium cyanate into urea, an "organic" compound isolated from urine.

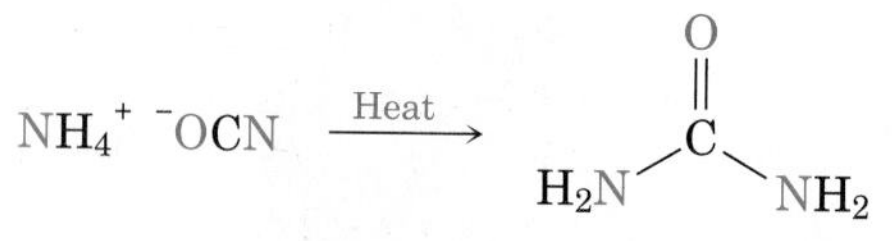

$$\underset{\textbf{Ammonium cyanate}}{NH_4^+\ ^-OCN} \xrightarrow{\text{Heat}} \underset{\textbf{Urea}}{H_2N-\overset{\overset{\displaystyle O}{\|}}{C}-NH_2}$$

By the mid-1800s, the weight of evidence was against the vitalistic theory and it had become clear that the same basic scientific principles are applicable to all compounds. The only distinguishing characteristic of organic compounds is that all contain the element carbon (Figure 1.1).

H																	He
Li	Be											B	C	N	O	F	Ne
Na	Mg											Al	Si	P	S	Cl	Ar
K	Ca	Sc	Ti	V	Cr	Mn	Fe	Co	Ni	Cu	Zn	Ga	Ge	As	Se	Br	Kr
Rb	Sr	Y	Zr	Nb	Mo	Tc	Ru	Rh	Pd	Ag	Cd	In	Sn	Sb	Te	I	Xe
Cs	Ba	La	Hf	Ta	W	Re	Os	Ir	Pt	Au	Hg	Tl	Pb	Bi	Po	At	Rn
Fr	Ra	Ac	Rf	Db	Sg	Bh	Hs	Mt	Ds	Rg	112	113	114	115	116		

FIGURE 1.1 The position of carbon in the periodic table. Other elements commonly found in organic compounds are shown in the colors typically used to represent them.

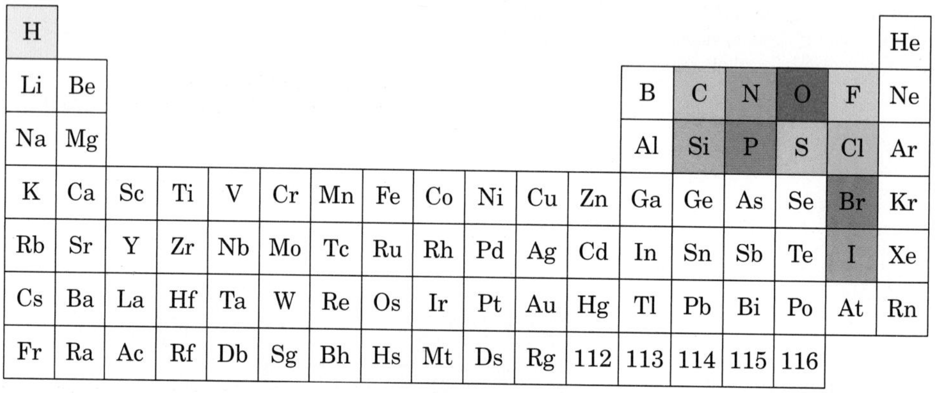

Organic chemistry, then, is the study of the compounds of carbon. But why is carbon special? The answer derives from the position of carbon in the periodic table. As a group 4A element, carbon atoms can share four valence electrons and form four strong covalent bonds. Furthermore, carbon atoms can bond to one another, forming long chains and rings. Of all elements, carbon alone is able to form an immense diversity of compounds, from simple methane, containing one carbon atom, to staggeringly complex DNA, which can contain tens of *billions* of atoms. In fact, millions of different organic compounds have been prepared just by combining carbon with hydrogen, oxygen, and nitrogen. Sometimes sulfur, phosphorus, and halogens also appear in organic molecules. Indeed, with more than 100 elements to choose from, organic chemistry focuses primarily on 10!

Not all organic compounds are derived from living organisms, of course. Modern chemists are extremely sophisticated in their ability to synthesize new organic compounds in the laboratory. Medicines, dyes, polymers, plastics, pesticides, and a host of other organic substances are all prepared in the laboratory. Organic chemistry is a subject that touches the lives of everyone. Studying it is a fascinating undertaking. Understanding organic chemistry will influence the way we think about larger, more complex assemblies of organic molecules, including cells, viruses, bacteria, and ourselves.

1.1 Atomic Structure

Let's begin our study of organic chemistry by reviewing some general ideas about atoms and bonds. Atoms consist of a dense, positively charged *nucleus*, surrounded at a relatively large distance by negatively charged *electrons*

(Figure 1.2). The nucleus consists of subatomic particles called *neutrons*, which are electrically neutral, and *protons*, which are positively charged. Although extremely small—about 10^{-14} to 10^{-15} meter (m) in diameter—the nucleus nevertheless contains essentially all the mass of the atom. Electrons have negligible mass and orbit the nucleus at a distance up to approximately 10^{-10} m. Thus, the diameter of a typical atom is about 2×10^{-10} m, or 200 picometers (pm), where 1 pm = 10^{-12} m. To give you an idea how small an atom is, a thin pencil line is about 3 *million* carbon atoms wide.

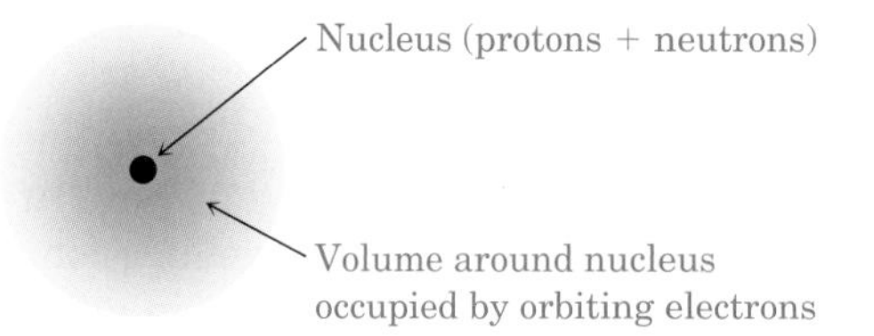

FIGURE 1.2 A schematic view of an atom. The dense, positively charged nucleus contains most of the atom's mass and is surrounded by negatively charged electrons.

An atom is described by two numbers:

The *atomic number* (Z) represents the number of protons in the nucleus.

The *mass number* (A) is the total number of protons *and* neutrons in the nucleus.

All the atoms of a given element have the same atomic number—1 for hydrogen, 6 for carbon, 17 for chlorine, and so on—but they can have different mass numbers depending on how many neutrons they contain. The term **isotopes** describes atoms that have the same number of protons in their nuclei but different numbers of neutrons. Hydrogen and deuterium are isotopes. Hydrogen has one proton and no neutrons; deuterium has one proton and one neutron. Because they are isotopes, their chemical properties are largely the same, but their masses differ.

The average mass in atomic mass units (amu) of a great many atoms of an element is called the element's *atomic weight*—1.008 amu for hydrogen, 12.011 amu for carbon, 35.453 amu for chlorine, and so on. The atomic weights for each element are given in the periodic table in the back of this book.

How are the electrons distributed in an atom? According to the *quantum mechanical model* of atomic structure, the behavior of a specific electron in an atom can be described by a mathematical expression called a *wave equation*. The solution to a wave equation is a *wave function*, or **orbital**, denoted by the Greek letter psi, ψ. An orbital can be thought of as a region of space around the nucleus where the electron can most likely be found.

Orbitals are organized into different layers, or **shells**, of successively larger size and energy. Different shells contain different numbers and kinds of orbitals. Each orbital can be occupied by two electrons. The first shell contains only a single *s* orbital, denoted 1*s*, and thus holds only two electrons. The second shell contains one *s* orbital (designated 2*s*) and three *p* orbitals (each designated 2*p*) and thus holds a total of eight electrons. The third shell contains an *s* orbital (3*s*), three *p* orbitals (3*p*), and five *d* orbitals (3*d*), for a total capacity of eighteen electrons. These orbital groupings are shown in Figure 1.3.

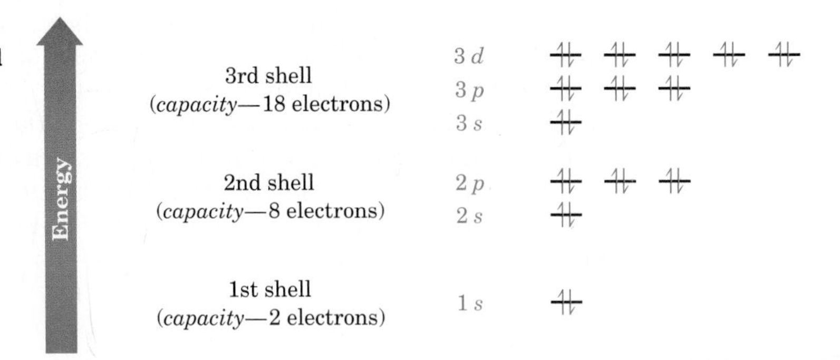

FIGURE 1.3 The energy levels of electrons in an atom. The first shell holds a maximum of two electrons in one 1*s* orbital; the second shell holds a maximum of eight electrons in one 2*s* and three 2*p* orbitals; the third shell holds a maximum of eighteen electrons in one 3*s*, three 3*p*, and five 3*d* orbitals; and so on. The two electrons in each orbital are represented by up and down arrows, ↑↓. Although not shown, the energy level of the 4*s* orbital falls between 3*p* and 3*d*.

What shapes do orbitals have? There are four different orbital shapes, denoted *s, p, d,* and *f.* Of the four, we'll be concerned only with *s* and *p* orbitals because these are the most common in organic chemistry. The *s* orbitals are spherical, with the nucleus at the center, and the *p* orbitals are dumbbell-shaped. The three different *p* orbitals within a given shell are oriented in space along mutually perpendicular directions, as shown in Figure 1.4. They are arbitrarily denoted p_x, p_y, and p_z.

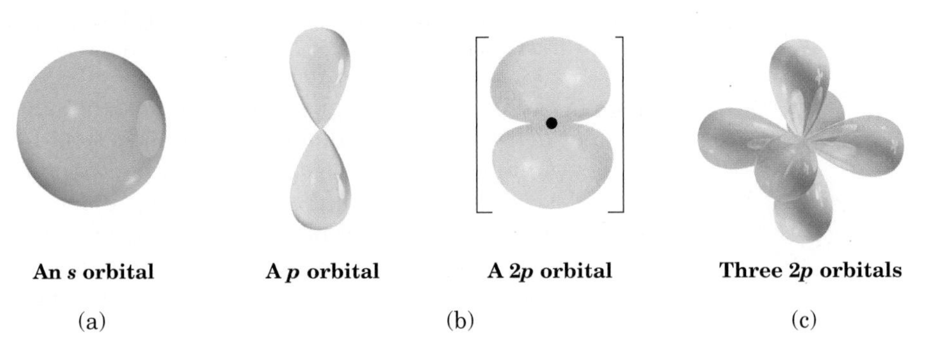

FIGURE 1.4 Representations of *s* and *p* orbitals. (a) The *s* orbitals are spherical, and (b) the *p* orbitals are dumbbell-shaped. The lobes of *p* orbitals are often drawn for convenience as "teardrops," but their true shape is more like that of a doorknob, as indicated by the computer-generated representation. (c) The three *p* orbitals in a given shell are oriented along mutually perpendicular directions.

1.2 Electron Configuration of Atoms

The lowest-energy arrangement, or **ground-state electron configuration**, of an atom is a description of the orbitals that the atom's electrons occupy. We can predict this arrangement by following three rules:

RULE 1 The orbitals of lowest energy are filled first, according to the order $1s \rightarrow 2s \rightarrow 2p \rightarrow 3s \rightarrow 3p \rightarrow 4s \rightarrow 3d$, as shown in Figure 1.3.

RULE 2 Only two electrons can occupy an orbital, and they must be of opposite *spin.* (Electrons can be thought of as spinning on an axis in much the same way that the earth spins. This spin can have two orientations, denoted as up ↑ and down ↓.)

RULE 3 If two or more empty orbitals of equal energy are available, one electron occupies each with the spins parallel until all orbitals are half-full.

Some examples of how these rules apply are shown in Table 1.1. Hydrogen, for instance, has only one electron, which must occupy the lowest-energy orbital. Thus, hydrogen has a 1*s* ground-state electron configuration. Carbon has six electrons and the ground-state electron configuration $1s^2\ 2s^2\ 2p^2$. Note that a superscript is used to represent the number of electrons in a particular orbital.

TABLE 1.1 Ground-State Electron Configuration of Some Elements

Element	Atomic number	Configuration	Element	Atomic number	Configuration
Hydrogen	1	1*s* ↑	Lithium	3	2*s* ↑ 1*s* ↑↓
Carbon	6	2*p* ↑ ↑ — 2*s* ↑↓ 1*s* ↑↓	Neon	10	2*p* ↑↓ ↑↓ ↑↓ 2*s* ↑↓ 1*s* ↑↓
Sodium	11	3*s* ↑ 2*p* ↑↓ ↑↓ ↑↓ 2*s* ↑↓ 1*s* ↑↓	Argon	18	3*p* ↑↓ ↑↓ ↑↓ 3*s* ↑↓ 2*p* ↑↓ ↑↓ ↑↓ 2*s* ↑↓ 1*s* ↑↓

PRACTICE PROBLEM 1.1

Give the ground-state electron configuration of nitrogen.

STRATEGY Find the atomic number of nitrogen to see how many electrons it has, and then apply the three rules to assign electrons into orbitals according to the energy levels given in Figure 1.3.

SOLUTION Nitrogen has atomic number 7 and thus has seven electrons. The first two electrons go into the lowest-energy orbital ($1s^2$), the next two go into the second-lowest-energy orbital ($2s^2$), and the remaining three go into the next-lowest-energy orbitals ($2p^3$), with one electron in each. Thus, the configuration of nitrogen is $1s^2\,2s^2\,2p^3$.

PROBLEM 1.1 How many electrons does each of the following elements have in its outermost electron shell?
(a) Potassium (b) Calcium (c) Aluminum

PROBLEM 1.2 Give the ground-state electron configuration of the following elements:
(a) Boron (b) Phosphorus (c) Oxygen (d) Argon

1.3 Development of Chemical Bonding Theory

By the mid-1800s, the new science of chemistry was developing rapidly and chemists had begun to probe the forces holding molecules together. In 1858, August Kekulé and Archibald Couper independently proposed that in all organic compounds, carbon always has four "affinity units." That is, carbon is *tetravalent*; it always forms four bonds when it joins other elements to form chemical compounds. (Later in the text, we'll see that a carbon atom without four bonds—one that has a positive or negative charge or a single free electron—is very reactive.) Furthermore, said Kekulé, carbon atoms can bond to one another to form extended chains, and chains can double back on themselves to form rings.

Although Kekulé was correct in describing the tetravalent nature of carbon, chemistry was still viewed in a two-dimensional way until 1874. In that year, Jacobus van't Hoff and Joseph Le Bel added a third dimension to our ideas about chemistry. They proposed that the four bonds of carbon are not randomly oriented but have a specific spatial orientation. Van't Hoff went even further and proposed that the four atoms to which a carbon atom is bonded sit at the corners of a regular tetrahedron, with carbon in the center.

A representation of a tetrahedral carbon atom is shown in Figure 1.5. Note the conventions used to show three-dimensionality: solid lines represent bonds in the plane of the paper; heavy wedged lines represent bonds coming out of the plane of the paper toward you, the viewer; and dashed lines represent bonds receding into the plane away from the viewer. Such representations are used throughout this text.

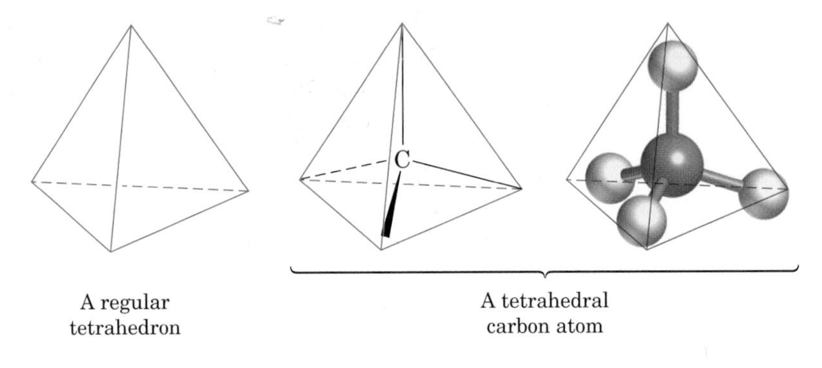

FIGURE 1.5 Representation of a tetrahedral carbon atom. The solid lines are in the plane of the paper, the heavy wedged line comes out of the plane of the paper, and the dashed line goes back into the plane.

PROBLEM 1.3 Draw a molecule of chloromethane, CH_3Cl, using solid, wedged, and dashed lines to show its tetrahedral geometry.

PROBLEM 1.4 Convert the following molecular model of ethane, C_2H_6, into a structure that uses wedged, normal, and dashed lines to represent three-dimensionality.

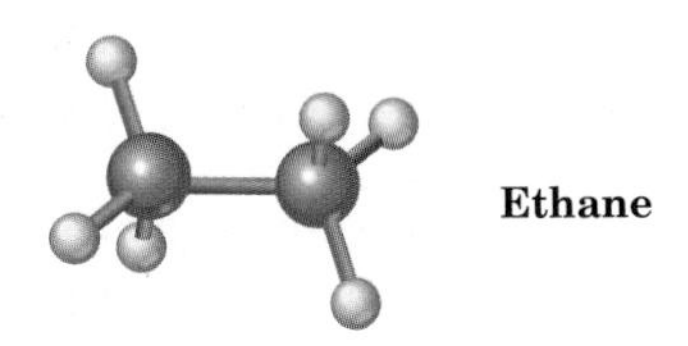

Ethane

1.4 The Nature of Chemical Bonds

Why do atoms bond together? How can bonds be described? The *why* question is relatively easy to answer: atoms bond together because the compound that results is more stable (has less energy) than the separate atoms. Energy always flows *out of* the chemical system when a bond is formed. Conversely, energy must be put *into* the system when a bond is broken. Put another way, making bonds releases energy, and breaking bonds absorbs energy. The *how* question is more difficult. To answer it, we need to know more about the properties of atoms.

We know through observation that eight electrons—an electron octet—in the outermost shell, or **valence shell**, impart special stability to the noble-gas elements in group 8A: neon (2 + 8), argon (2 + 8 + 8), krypton

(2 + 8 + 18 + 8). We also know that the chemistry of many main-group elements is governed by their tendency to take on the electron configuration of the nearest noble gas. The alkali metals in group 1A, for example, lose the single *s* electron in their outer shells to form a cation with a noble-gas configuration, while the halogens in group 7A gain an electron to form an anion with a noble-gas configuration. The resulting ions are held together in compounds like NaCl by an electrostatic attraction that we call an **ionic bond**.

How, though, do the elements near the middle of the periodic table form bonds? Look at methane, CH_4, the main constituent of natural gas, for example. The bonding in methane is not ionic because it would cost too much energy for carbon ($1s^2\ 2s^2\ 2p^2$) to either gain or lose *four* electrons to achieve a noble-gas configuration. In fact, carbon bonds to other atoms, not by losing or gaining electrons, but by *sharing* them. Such shared-electron bonds, first proposed in 1916 by G. N. Lewis, are called **covalent bonds**. The neutral collection of atoms held together by covalent bonds is called a **molecule**.

A simple shorthand way of indicating covalent bonds in molecules is to use **Lewis structures**, or *electron-dot structures*, in which an atom's valence electrons are represented by dots. Thus, hydrogen has one dot (1*s*), carbon has four dots ($2s^2\ 2p^2$), oxygen has six dots ($2s^2\ 2p^4$), and so on. A stable molecule results when a noble-gas configuration is achieved for all atoms—an octet (eight dots) for most elements or two for hydrogen—as in the following examples:

·C· + 4 H· ⟶ H:C:H (with H above and below C)

Methane (CH_4)

3 H· + ·N· ⟶ H:N:H (with H above N)

Ammonia (NH_3)

2 H· + ·O: ⟶ H:O: (with H below O)

Water (H_2O)

3 H· + ·C· + ·O: + H· ⟶ H:C:O: (with H above C, H below C, H below O)

Methanol (CH_3OH)

Bonding rules describe how many covalent bonds an atom can form. The number of bonds depends on how many valence electrons an atom has and how many additional valence electrons it needs to complete its octet. Atoms with one, two, or three valence electrons can form only one, two, or three bonds because that is the number of electrons they have available to share in bond formation. Atoms with four or more valence electrons form as many bonds as needed to fill their valence shells and reach a stable octet of electrons. Carbon, with four valence electrons, needs four more to reach an octet, so it forms four bonds, as in CH_4. Nitrogen, with five valence electrons, needs three more to reach an octet, so it forms three bonds, as in NH_3. Oxygen, with six valence electrons, needs two more, so it forms two bonds, as in H_2O. And the halogens, with seven electrons in their valence shells, form one bond, as in HCl.

H— Cl— Br— F— (One bond)

—O— (Two bonds)

—N— —B— (Three bonds)

—C— (Four bonds)

Valence electrons not used for bonding are called **nonbonding electrons**, or **lone-pair electrons**. The nitrogen atom in ammonia, for instance, shares three of its five valence electrons in three covalent bonds with hydrogen atoms; the remaining two valence electrons are a nonbonding lone pair.

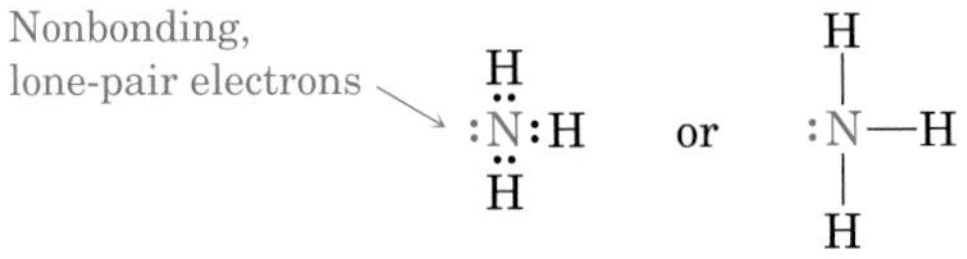

Ammonia

Lewis structures are useful because they make electron bookkeeping possible and act as reminders of the number of valence electrons present. Simpler, however, is the use of **Kekulé structures**, or **line-bond structures**, in which the two electrons of a covalent bond are indicated simply by a line between the atoms. Lone pairs of nonbonding valence electrons are often not shown when drawing line-bond structures, although it's still necessary to keep them in mind. Some examples are shown in Table 1.2.

TABLE 1.2 Lewis and Kekulé Structures of Some Simple Molecules

Name	Lewis structure	Kekulé structure	Name	Lewis structure	Kekulé structure
Water (H_2O)	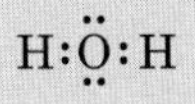		Methane (CH_4)	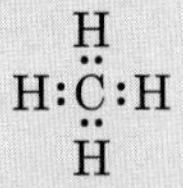	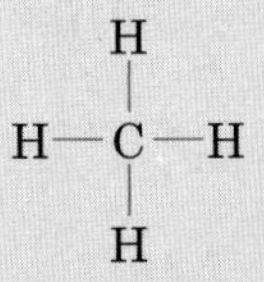
Ammonia (NH_3)	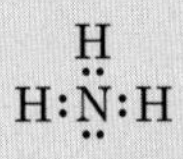	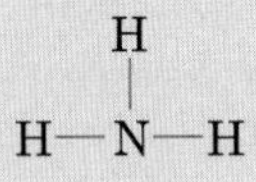	Methanol (CH_3OH)	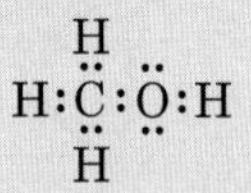	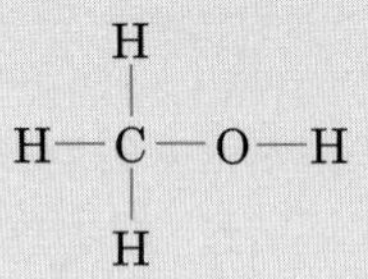

PRACTICE PROBLEM 1.2

How many hydrogen atoms does phosphorus bond to in forming phosphine, $PH_?$?

STRATEGY Identify the periodic group of phosphorus, and tell from that how many electrons (bonds) are needed to make an octet.

SOLUTION Phosphorus is in group 5A of the periodic table and has five valence electrons. It thus needs to share an additional three electrons to complete an octet. Therefore, it bonds to three hydrogen atoms, giving PH_3.

PRACTICE PROBLEM 1.3

Draw both Lewis and line-bond structures for chloromethane, CH_3Cl.

STRATEGY Remember that a bond—that is, a pair of shared electrons—is represented as a line between atoms.

SOLUTION Hydrogen has one valence electron, carbon has four valence electrons, and chlorine has seven valence electrons. Thus, chloromethane is represented as

```
  H  ..              H
H:C:Cl:              |
  H  ..          H — C — Cl      Chloromethane
                     |
                     H
```

PROBLEM 1.5 What are likely formulas for the following molecules?
(a) $CCl_?$ (b) $AlH_?$ (c) $CH_?Cl_2$ (d) $SiF_?$

PROBLEM 1.6 Draw both Lewis and line-bond structures for the following molecules, showing all nonbonded electrons:
(a) $CHCl_3$, chloroform (b) H_2S, hydrogen sulfide (c) CH_3NH_2, methylamine

PROBLEM 1.7 Why can't an organic molecule have the formula C_2H_7?

1.5 Forming Covalent Bonds: Valence Bond Theory

How does electron sharing between atoms occur? According to **valence bond theory**, a covalent bond forms when two atoms approach each other closely and a singly occupied orbital on one atom *overlaps* a singly occupied orbital on the other. The electrons are now paired in the overlapping orbitals and are attracted to the nuclei of both atoms, thereby bonding the atoms together. In the H_2 molecule, for example, the H–H bond results from the overlap of two singly occupied 1*s* orbitals.

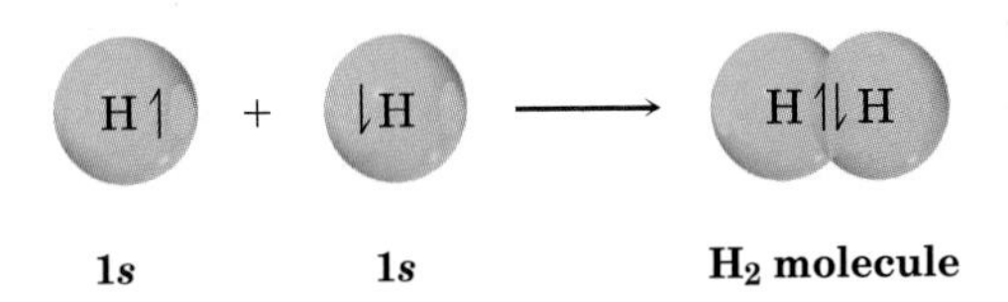

During the reaction 2 H· → H_2, 436 kJ/mol (104 kcal/mol) of energy is released. Because the product H_2 molecule has 436 kJ/mol less energy than the starting 2 H· atoms, we say that the product is *more stable* than the reactant and that the new H–H bond has a **bond strength** of 436 kJ/mol. In other words, we would have to put 436 kJ/mol of energy *into* the H–H bond to break the H_2 molecule apart into two H atoms. [For convenience, we'll generally give energies in both kilojoules (kJ) and kilocalories (kcal): 1 kJ = 0.2390 kcal; 1 kcal = 4.184 kJ.]

How close are the two nuclei in the H_2 molecule? If they are too close, they will repel each other because both are positively charged, yet if they're too far apart, they won't be able to share the bonding electrons. Thus, there is an optimum distance between nuclei that leads to maximum bond stability, a distance called the **bond length** (Figure 1.6). In the hydrogen molecule, the bond length is 74 pm. Every covalent bond has both a characteristic bond strength and bond length.

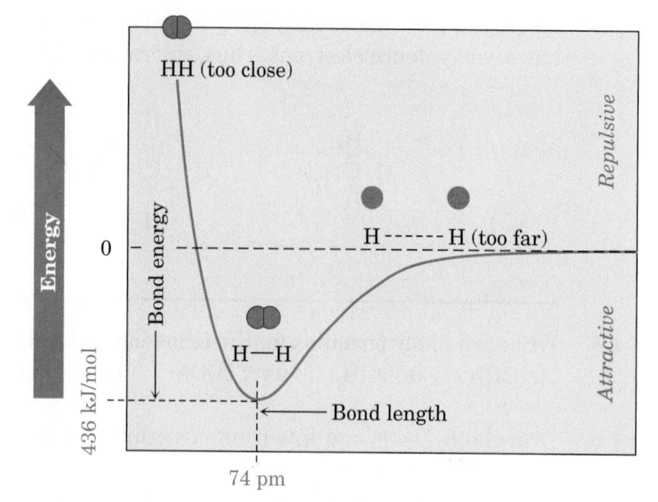

FIGURE 1.6 A plot of energy versus internuclear distance for two hydrogen atoms. The distance at the minimum energy point is the bond length, 74 pm. Breaking the H–H bond will require 436 kJ/mol of energy.

1.6 Hybridization: sp^3 Orbitals and the Structure of Methane

The bonding in the H_2 molecule is fairly straightforward, but the situation is more complex in organic molecules with tetravalent carbon atoms. Let's start with a simple case and consider methane, CH_4. Carbon has four electrons in its valence shell ($2s^2\ 2p^2$) and can form four bonds to hydrogens. What are the four C–H bonds in methane like? Because carbon uses two kinds of orbitals ($2s$ and $2p$) to form the bonds, we might expect methane to have two kinds of C–H bonds. In fact, though, all four C–H bonds in methane are identical and are spatially oriented toward the four corners of a regular tetrahedron. How can we explain this?

An answer was provided in 1931 by Linus Pauling, who proposed that an *s* orbital (shown in pink) and three *p* orbitals (blue) can combine, or *hybridize*, to form four equivalent atomic orbitals (green). These new orbitals are called **sp^3 hybrid orbitals**. The superscript 3 in sp^3 tells the fraction of each type of atomic orbital in the hybrid, not how many electrons occupy it. Thus, each sp^3 orbital is 1 part "s" and 3 parts "p," but rather than writing $s^{1/4}p^{3/4}$ or s^1p^3, we write sp^3.

The concept of hybridization explains *how* carbon forms four equivalent tetrahedral bonds but doesn't explain *why* it does so. The shape of the hybrid orbital suggests the answer. When an *s* orbital hybridizes with three *p* orbitals, the resultant hybrid orbitals are unsymmetrical about the nucleus. One of the two lobes is much larger than the other (Figure 1.7). When this larger lobe overlaps with another orbital to form a bond, the

FIGURE 1.7 To form four equivalent orbitals, the $2s$ orbital (pink) and three $2p$ orbitals (blue) hybridize to yield four sp^3 hybrid orbitals (green).

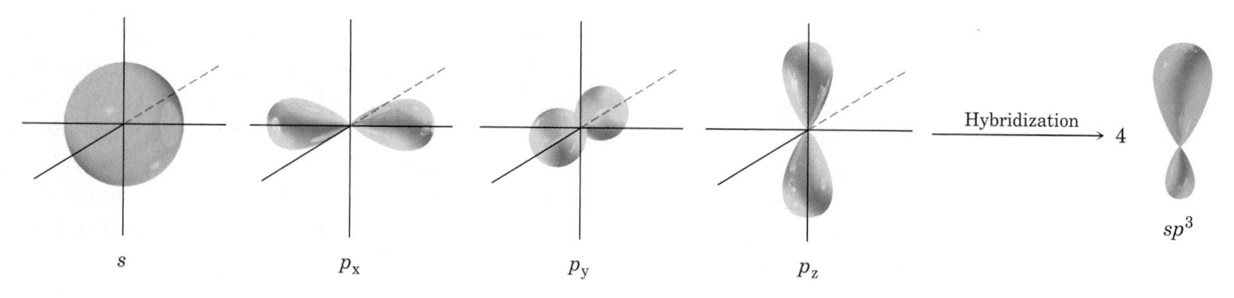

electrons are better shared between the two bonding nuclei than if unhybridized *s* or *p* orbitals were to overlap. This results in the formation of a stronger bond.

When the four identical orbitals of an sp^3-hybridized carbon atom overlap with the 1*s* orbitals of four hydrogen atoms, four identical C–H bonds are formed and methane results. Each C–H bond in methane has a strength of 438 kJ/mol (105 kcal/mol) and a length of 110 pm. Because the four bonds have a specific geometry, we can also define a property called a **bond angle**. The angle formed by each H–C–H is 109.5°, the so-called *tetrahedral angle*. The tetrahedral geometry can be depicted in a variety of ways, as shown in Figure 1.8. We call these C–H bonds **sigma (σ) bonds** to indicate that the region of orbital overlap is concentrated on the bond axis.

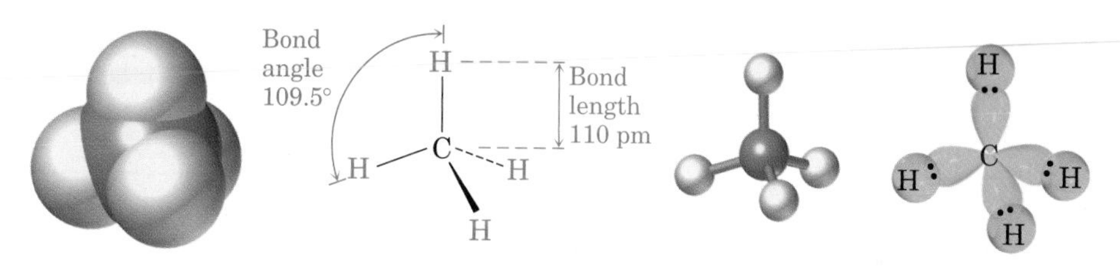

FIGURE 1.8 The structure of methane, showing its 109.5° bond angles.

PROBLEM 1.8 Draw a tetrahedral representation of tetrachloromethane, CCl_4, using the standard convention of solid, dashed, and wedged lines.

PROBLEM 1.9 Why do you think a C–H bond (110 pm) is longer than an H–H bond (74 pm)?

1.7 Hybridization: sp^3 Orbitals and the Structure of Ethane

The same kind of hybridization that explains the methane structure also explains how carbon atoms can bond together in chains and rings to make possible so many millions of compounds. Ethane, C_2H_6, is the simplest molecule containing a carbon–carbon bond:

```
 H H          H   H
 ¨ ¨          |   |
H:C:C:H    H—C—C—H     CH3CH3
 ¨ ¨          |   |
 H H          H   H
```

Some representations of ethane

We can picture the ethane molecule by imagining that the two carbon atoms bond to each other by overlap of an sp^3 hybrid orbital from each. The remaining three sp^3 hybrid orbitals on each carbon overlap with hydrogen 1*s* orbitals to form the six C–H bonds, as shown in Figure 1.9. The C–H bonds in ethane are similar to those in methane, although a bit weaker—420 kJ/mol (100 kcal/mol) for ethane versus 438 kJ/mol for methane. The C–C bond is 154 pm long and has a strength of 376 kJ/mol (90 kcal/mol). All the bond angles of ethane are near the tetrahedral value of 109.5°.

FIGURE 1.9 The structure of ethane. The carbon–carbon bond is formed by overlap of two carbon sp^3 hybrid orbitals. (For clarity, the smaller lobes of the sp^3 hybrid orbitals are not shown.)

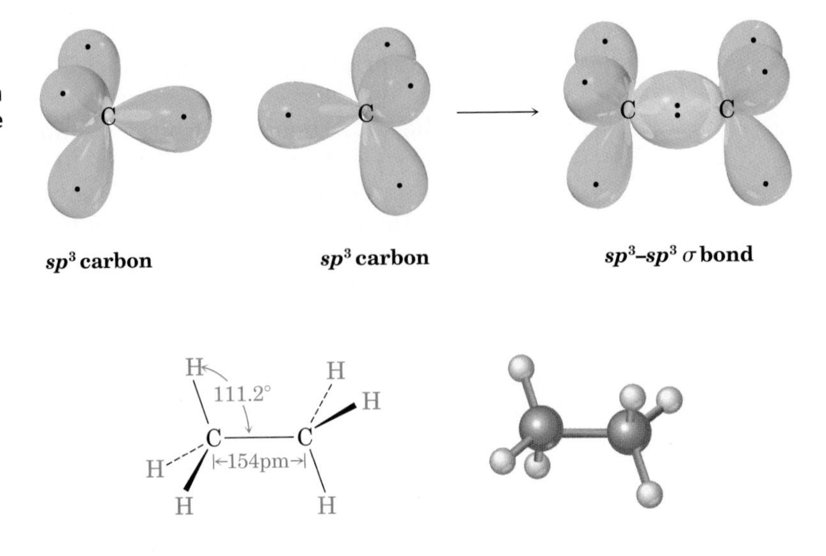

PROBLEM 1.10 Draw a line-bond structure for propane, $CH_3CH_2CH_3$; predict the value of each bond angle; and indicate the overall shape of the molecule.

1.8 Double and Triple Bonds

In addition to the single C–C bonds found in ethane, organic molecules can also have double C=C bonds and triple C≡C bonds. These bonds have a number of interesting properties:

1. Double and triple bonds are far more reactive than single bonds.
2. A C=C bond is *not* twice as strong as a C–C bond and a C≡C bond is *not* three times as strong as a single C–C bond.
3. Double bonds lead to a flat (or planar) shape. Triple bonds lead to a linear shape.

Hybridization explains these properties. When we write a C–C bond, we reinforce the idea that the two electrons, which are shared by the carbon atoms, spend time *between* both carbon atoms in a sigma bond. But, where are the four electrons of a C=C bond? Where are the six electrons of a C≡C bond? Hybridization explains where these electrons go.

Figure 1.10 summarizes the three hybridizations of carbon: sp^3, sp^2, and sp. In each case, we start with *four* orbitals and are left with *four* orbitals after hybridization. We have already seen that when all four orbitals ($s + p_x + p_y + p_z$) hybridize, four sp^3 hybrid orbitals are obtained. To form **sp^2 hybrid orbitals**, one of the p orbitals is *not* hybridized and three hybridized orbitals form. Similarly, to form **sp hybrid orbitals**, two of the p orbitals are *not* hybridized and two hybridized orbitals form. The shapes of all of the hybrid orbitals are similar, but the amount of "s" character increases from sp^3 to sp^2 to sp.

So, how do mixtures of hybridized and unhybridized orbitals organize around a single atom's nucleus? The p orbital of an sp^2-hybridized carbon (shown in blue) occupies the region above and below the plane occupied by the sp^2 hybrid orbitals. The sp^2 orbitals arrange themselves at 120° angles

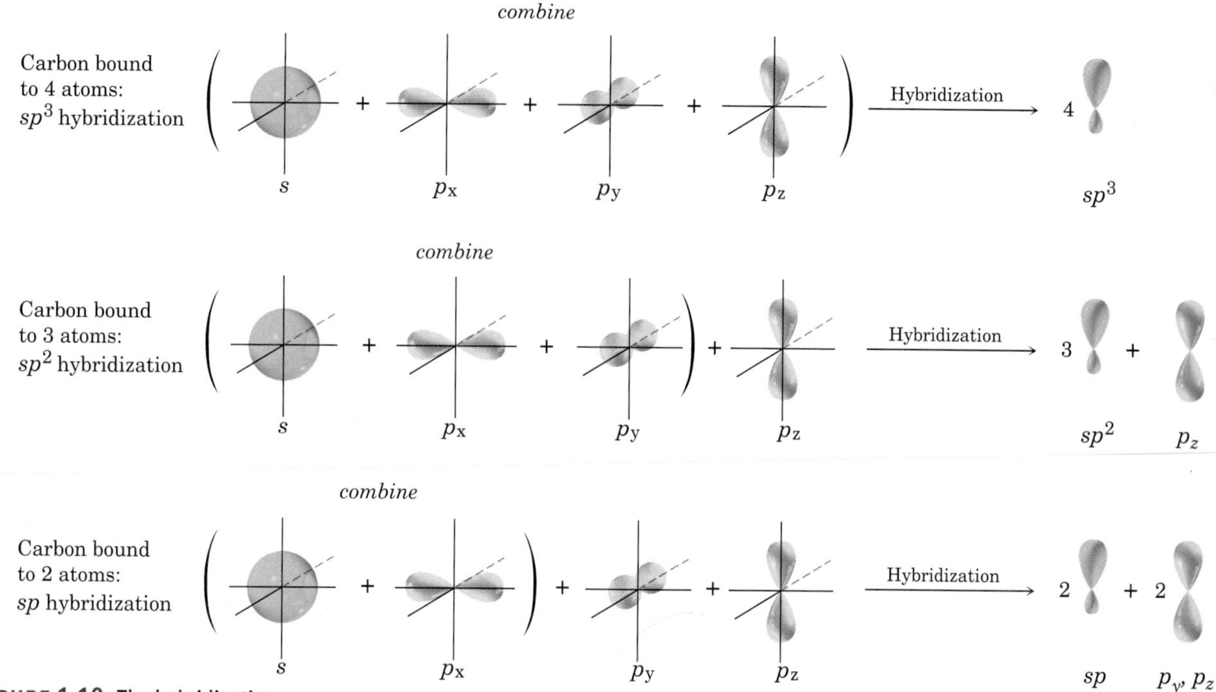

FIGURE 1.10 The hybridizations of carbon. Varying the number of 2*p* orbitals that are hybridized with the 2*s* orbital provides sp^3, sp^2, and *sp* hybrid orbitals.

in the plane. In ethylene (ethene), C_2H_4, the sp^2 hybrid orbitals of one carbon form σ bonds with either hydrogen atoms or with the other sp^2-hybridized carbon atom. The "double" part of the C=C bond is formed by overlap of the *p* orbitals of the adjacent carbon atoms. As a result, all of the atoms of ethylene lie on the same plane with 120° bond angles. The bond resulting from overlap of *p* orbitals is called a **pi** (π) **bond**. In contrast to a σ bond, the orbital overlap of a π bond is *not* concentrated on the axis between both atoms. Instead, orbital overlap occurs above and below the bond axis.

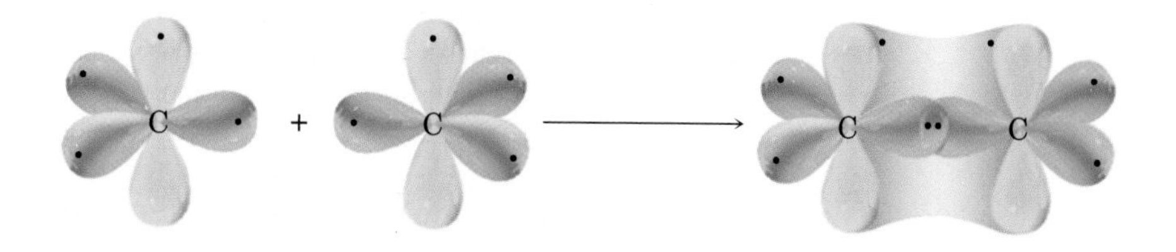

A similar situation results when *sp*-hybridized carbons form a C≡C bond. The two *p* orbitals of an sp^2-hybridized carbon share a common plane and the two *sp* orbitals occupy the perpendicular axis. In acetylene (ethyne), the two *sp* hybrid orbitals (shown in green) of one carbon form σ bonds with a hydrogen and the other *sp*-hybridized carbon atom. The overlap of the remaining *p* orbitals (shown in blue) gives the "double part" and "triple part"

of the C≡C triple bond. As a result, all the atoms are colinear, making the bond angle 180°.

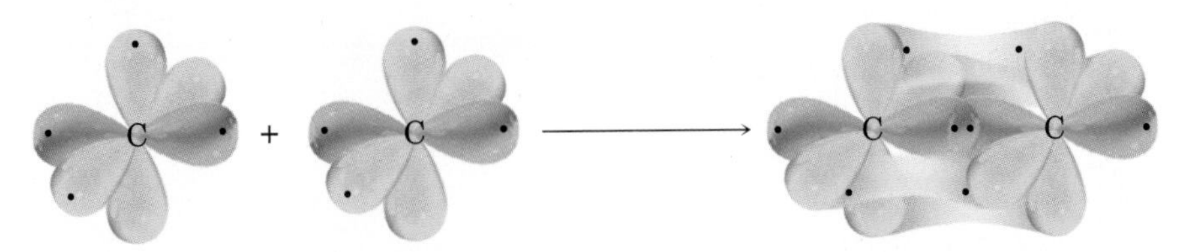

The π bonding in C=C bonds and C≡C bonds has significant implications. First, it explains how either four or six electrons can arrange themselves in the very small space around two nuclei. Instead of crowding all these electrons *between* the atoms, the π electrons are kept at a distance.

Second, π bonding explains why double and triple bonds are not two and three times stronger than single bonds, respectively. Although carbon–carbon bond energies vary somewhat from compound to compound, the average bond energy of a single bond is approximately 340 kJ/mol (81 kcal/mol). That of a double bond is approximately 610 kJ/mol (146 kcal/mol), and that of a triple bond is approximately 830 kJ/mol (196 kcal/mol). Subtracting the bond energy of a C–C single bond from that of a C=C double bond, the approximate strength of a π bond can be calculated as 270 kJ/mol (79 kcal/mol). This is significantly less than the C–C single bond energy of 340 kJ/mol. In the same manner, the bond energy of the two π bonds of a C≡C triple bond can be calculated to be 490 kJ/mol or 245 kJ/mol (59 kcal/mol) for each π bond, also significantly weaker than a C–C single bond. Because the electrons in a π bond are not as close to the nuclei as the electrons in a σ bond, they do not hold the nuclei together as tightly.

Third, molecules with double and triple bonds are more reactive than molecules with single bonds. In contrast to σ bonds, π bonds are weaker and the electrons are available for reaction instead of hidden between two nuclei.

Fourth, π bonding is consistent with observed bond lengths. Single bonds are longer than double bonds, and double bonds are longer than triple bonds. We can rationalize this in two ways. We can say that to provide the best overlap between p orbitals, the atoms squeeze closer together. Alternatively, we can say that the increased "s" character of an sp orbital makes it shorter and rounder than an sp^2 or sp^3 orbital, causing the σ portion of the bond (and therefore the entire bond) to be shorter.

Fifth, π bonding explains the differences seen in shape. The four sp^3 orbitals adopt a tetrahedral shape to maximize distance and minimize repulsion between electron pairs. However, in order to overlap p orbitals to form π bonds, a planar (or linear) geometry is required for sp^2 or sp carbons, respectively.

Table 1.3 summarizes the results of different hybridizations of carbon atoms.

PRACTICE PROBLEM 1.4

Formaldehyde, CH_2O, contains a carbon–*oxygen* double bond. Draw Lewis and line-bond structures of formaldehyde, and indicate the hybridization of the carbon atom.

STRATEGY We know that hydrogen forms one covalent bond, carbon forms four, and oxygen forms two. Trial and error, combined with intuition, must be used to fit the atoms together.

TABLE 1.3 Summary of Bonding for Carbon

Molecule	Hybridization of carbon	Bond angles	Geometry at carbon	Carbon–carbon bond length	Average bond strength, type
$H_3C—CH_3$	sp^3	109.5°	Tetrahedral	154 pm	340 kJ/mol (81 kcal/mol) σ
$H_2C═CH_2$	sp^2	120°	Planar	133 pm	610 kJ/mol (146 kcal/mol) $\sigma + \pi$
$HC≡CH$	sp	180°	Linear	120 pm	830 kJ/mol (196 kcal/mol) $\sigma + 2\pi$

SOLUTION There is only one way that two hydrogens, one carbon, and one oxygen can combine:

Lewis structure **Line-bond structure**

Like the carbon atoms in ethylene, the carbon atom in formaldehyde is sp^2-hybridized.

PROBLEM 1.11 Draw both a Lewis and a line-bond structure for acetaldehyde, CH_3CHO.

PROBLEM 1.12 Draw a line-bond structure for propene, $CH_3CH═CH_2$; indicate the hybridization of each carbon; and predict the value of each bond angle.

PROBLEM 1.13 Draw a line-bond structure for buta-1,3-diene, $H_2C═CH—CH═CH_2$; indicate the hybridization of each carbon; and predict a value for each bond angle.

PROBLEM 1.14 Convert the following molecular model of aspirin into a line-bond structure, and identify the hybridization of each carbon atom (gray = C, red = O, ivory = H):

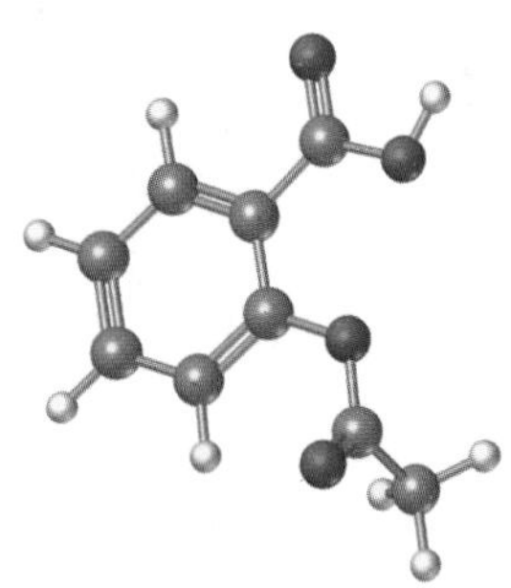

PROBLEM 1.15 Draw a line-bond structure for propyne, $CH_3C{\equiv}CH$; indicate the hybridization of each carbon; and predict a value for each bond angle.

1.9

Polar Covalent Bonds: Electronegativity

Up to this point, we've viewed chemical bonds as either ionic or covalent. The bond in sodium chloride, for instance, is ionic; an electron has been transferred from sodium to chlorine, resulting in the formation of oppositely charged Na^+ and Cl^- ions that are held together by electrostatic attractions. The carbon–carbon bond in ethane, by contrast, is electronically symmetrical and therefore covalent; the two bonding electrons are equally shared by the two equivalent carbon atoms. Many bonds, however, are neither truly ionic nor truly covalent but are somewhere between the two extremes. Such bonds are called **polar covalent bonds**, meaning that the shared electrons are attracted more strongly by one atom than by the other (Figure 1.11).

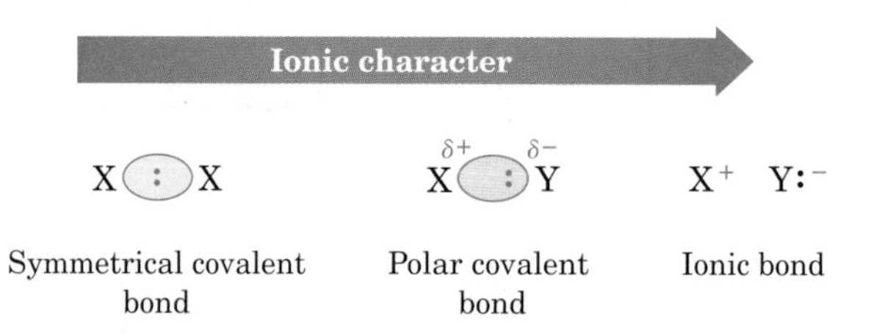

FIGURE 1.11 The bonding continuum from covalent to ionic is a result of unsymmetrical electron distribution. The symbol δ (lowercase Greek delta) means *partial* charge, either partial positive ($\delta+$) for the electron-poor atom or partial negative ($\delta-$) for the electron-rich atom.

Bond polarity is due to differences in **electronegativity**, the intrinsic ability of an atom to attract electrons in a covalent bond. As shown in Figure 1.12, metallic elements on the left side of the periodic table attract electrons weakly, whereas the halogens and other reactive nonmetal elements on the right side of the periodic table attract electrons strongly.

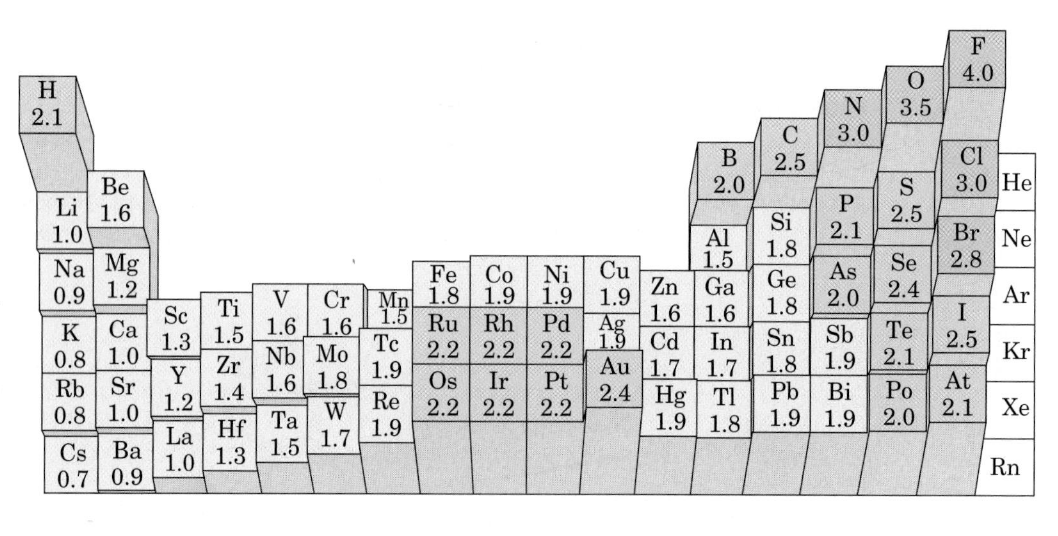

FIGURE 1.12 Electronegativity values and trends. Electronegativity generally increases from left to right across the periodic table and decreases from top to bottom. The values are on an arbitrary scale, with F = 4.0 and Cs = 0.7. Carbon has an electronegativity value of 2.5. Elements in red are the most electro-negative, those in green are medium, and those in yellow are the least electronegative.

In general, a bond between atoms with similar electronegativities is covalent, a bond between atoms whose electronegativities differ by less than 2 units is polar covalent, and a bond between atoms whose electronegativities differ by 2 units or more is largely ionic. A carbon–hydrogen bond, for instance, is relatively nonpolar because carbon and hydrogen have similar electronegativities. A bond between carbon and a *more* electronegative element such as oxygen or chlorine, however, is polar covalent. The electrons in such a bond are drawn away from carbon toward the more electronegative atom, leaving the carbon with a partial positive charge, denoted $\delta+$, and leaving the more electronegative atom with a partial negative charge, denoted $\delta-$ (δ is the lowercase Greek letter delta). An example is the C–Cl bond in chloromethane, CH_3Cl (Figure 1.13a).

A bond between carbon and a *less* electronegative element is polarized so that carbon bears a partial negative charge and the other atom bears a partial positive charge. An *organometallic* compound such as methyllithium, CH_3Li, is a good example (Figure 1.13b).

Note in the representations of chloromethane and methyllithium in Figure 1.13 that a crossed arrow $\leftarrow\!\!\!\!\!+\!\!\longrightarrow$ is sometimes used to indicate the direction of bond polarity. By convention, *electrons move in the direction of the arrow.* The tail of the arrow is electron-poor ($\delta+$), and the head of the arrow is electron-rich ($\delta-$). Note also in Figure 1.13 that charge distributions in a molecule are displayed visually with so-called *electrostatic potential maps*, which use color to indicate electron-rich (red) and electron-poor (blue) regions.

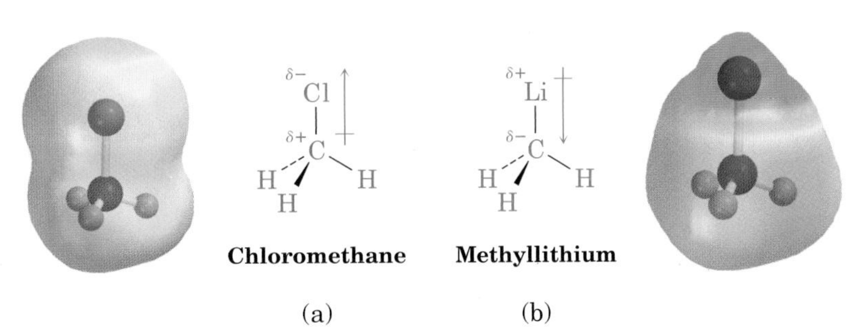

FIGURE 1.13 (a) Chloromethane, CH_3Cl, has a polar covalent C–Cl bond, and (b) methyllithium, CH_3Li, has a polar covalent C–Li bond. Called *electrostatic potential maps*, the computer-generated representations use color to show calculated charge distributions, ranging from red (electron-rich) to blue (electron-poor).

When speaking of an atom's ability to polarize a bond, we often use the term **inductive effect**. An inductive effect is simply the shifting of electrons in a bond in response to the electronegativity of nearby atoms. Metals, such as lithium and magnesium, inductively *donate* electrons, whereas electronegative elements, such as oxygen and chlorine, inductively *withdraw* electrons. Inductive effects play a major role in understanding chemical reactivity, and we'll use them many times throughout this text to explain a variety of chemical phenomena.

PRACTICE PROBLEM 1.5

Predict the extent and direction of polarization of the O–H bonds in H_2O.

STRATEGY Look at the electronegativity table in Figure 1.12 to see which atoms attract electrons more strongly.

3.5 δ−

SOLUTION Oxygen (electronegativity = 3.5) is more electronegative than hydrogen (electronegativity = 2.1) according to Figure 1.12, and it therefore attracts electrons more strongly. The difference in electronegativities (3.5 – 2.1 = 1.4) implies that an O–H bond is strongly polarized.

$$\overset{\delta+}{H}-\overset{\delta-}{O}-\overset{\delta+}{H}$$

PROBLEM 1.16 Which element in each of the following pairs is more electronegative?

(a) Li or H (b) Be or Br (c) Cl or I

PROBLEM 1.17 Use the $\delta+/\delta-$ convention to indicate the direction of expected polarity for each of the bonds shown:

(a) $Br—CH_3$ (b) $H_2N—CH_3$ (c) $H—NH_2$
(d) $HO—CH_3$ (e) $BrMg—CH_3$ (f) $F—CH_3$

PROBLEM 1.18 Order the bonds in the following compounds according to their increasing ionic character: CCl_4, $MgCl_2$, $TiCl_3$, Cl_2O.

PROBLEM 1.19 Look at the following electrostatic potential map of methyl alcohol, and indicate the direction of polarization of the C–O bond:

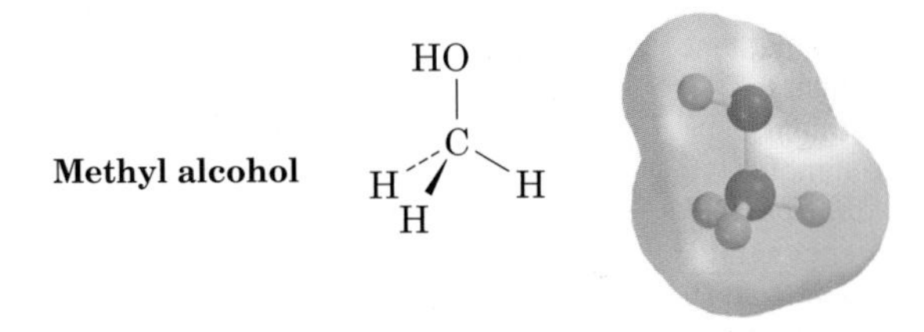

1.10 Acids and Bases: The Brønsted–Lowry Definition

Acidity and basicity are related to electronegativity and bond polarity. We'll soon see that the acid–base behavior of organic molecules helps explain much of their chemistry. You may recall from a course in general chemistry that two definitions of acidity are frequently used: the *Brønsted–Lowry definition* and the *Lewis definition*. Let's look at the Brønsted–Lowry definition first.

A **Brønsted–Lowry acid** is a substance that donates a hydrogen ion (H^+), and a **Brønsted–Lowry base** is a substance that accepts a hydrogen ion. (The name *proton* is often used as a synonym for H^+ because loss of the valence electron from a neutral hydrogen atom leaves only the hydrogen nucleus—a proton.) Acids react with bases: when gaseous hydrochloric acid dissolves in water, the acid, HCl, donates a proton to the base, a water molecule. The products of this reaction are a new acid and base. To distinguish the products from the reactants, we identify the products as a **conjugate acid** and a **conjugate base**. The hydronium ion (H_3O^+) produced is the conjugate acid because it can donate a proton, and the chloride ion (Cl^-) is the conjugate base because it can accept a proton.

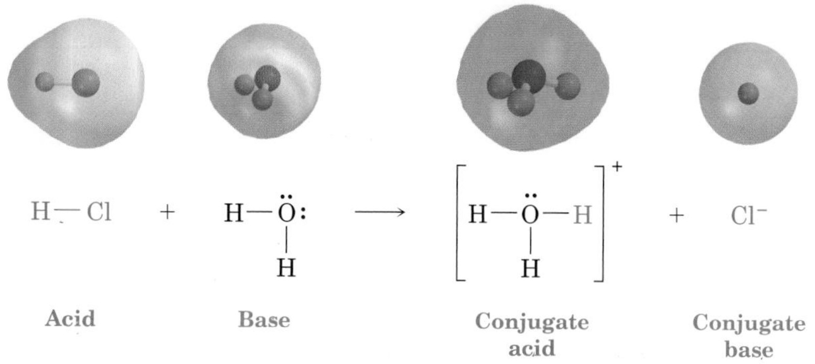

$$\text{H—Cl} + \text{H—}\ddot{\text{O}}\text{:(H)} \longrightarrow [\text{H—}\ddot{\text{O}}\text{—H (H)}]^{+} + \text{Cl}^{-}$$

Acid Base Conjugate acid Conjugate base

The ability of an acid to donate a proton depends on the acid. Stronger acids such as HCl react almost completely with water, whereas weaker acids such as acetic acid (CH_3CO_2H) react only slightly. The exact strength of a given acid in water solution can be expressed by its **acidity constant, K_a**. For the reaction of any generalized acid HA with water, the acidity constant, K_a, is

$$HA + H_2O \rightleftharpoons A^- + H_3O^+$$

$$K_a = \frac{[H_3O^+][A^-]}{[HA]}$$

where you may remember from your general chemistry course that brackets [ ] refer to the concentration of the enclosed species in molarity (M), or moles per liter (mol/L). Stronger acids have their equilibria toward the right and thus have larger acidity constants; weaker acids have their equilibria toward the left and have smaller acidity constants.

The range of K_a values for different acids is enormous, running from about 10^{15} for the strongest acids to about 10^{-60} for the weakest. The common inorganic acids such as H_2SO_4, HNO_3, and HCl have K_a's in the range 10^2 to 10^9, while many organic acids have K_a's in the range 10^{-5} to 10^{-15}. As you gain more experience, you'll develop a rough feeling for which acids are "strong" and which are "weak" (remembering that the terms are always relative).

Acid strengths are normally given using pK_a values, where the pK_a is equal to the negative common logarithm of the K_a.

$$pK_a = -\log K_a$$

A *stronger acid* (larger K_a) has a *smaller* pK_a, and a *weaker acid* (smaller K_a) has a *larger* pK_a. Table 1.4 lists the pK_a's of some common acids in order of their strength.

Notice in Table 1.4 that there is an inverse relationship between the acid strength of an acid and the base strength of its conjugate base. A *strong* acid yields a *weak* conjugate base, and a *weak* acid yields a *strong* conjugate base. To understand this inverse relationship, think about what is happening to the acidic hydrogen in a reaction. A strong acid is one that loses H^+ easily, meaning that its conjugate base does not hold the proton tightly and is therefore a weak base. A weak acid is one that loses H^+ with difficulty, meaning that its conjugate base *does* hold the proton tightly and is therefore a strong base. The fact that HCl is a strong acid, for example, means that Cl^- does not hold H^+ tightly and is thus a weak base. Water, on the other hand, is a weak acid, meaning that OH^- holds H^+ tightly and is a strong base.

In general, a proton always goes from the stronger acid to the stronger base, meaning that an acid donates a proton to the conjugate base of any acid

TABLE 1.4 Relative Strength of Some Common Acids

	Acid	Name	pK_a	Conjugate base	Name	
Weaker acid	CH_3CH_2OH	Ethanol	16.00	$CH_3CH_2O^-$	Ethoxide ion	Stronger base
	H_2O	Water	15.74	HO^-	Hydroxide ion	
	HCN	Hydrocyanic acid	9.31	CN^-	Cyanide ion	
	CH_3CO_2H	Acetic acid	4.76	$CH_3CO_2^-$	Acetate ion	
	HF	Hydrofluoric acid	3.45	F^-	Fluoride ion	
	HNO_3	Nitric acid	−1.3	NO_3^-	Nitrate ion	
Stronger acid	HCl	Hydrochloric acid	−7.0	Cl^-	Chloride ion	Weaker base

that is weaker, that is, has a larger pK_a. Conversely, the conjugate base of an acid removes a proton from any acid with a smaller pK_a. For example, the data in Table 1.4 indicate that OH^- will react with acetic acid, CH_3CO_2H, to yield acetate ion, $CH_3CO_2^-$, and H_2O. Because water ($pK_a = 15.74$) is a weaker acid than acetic acid ($pK_a = 4.76$), hydroxide ion holds a proton more tightly than acetate ion does.

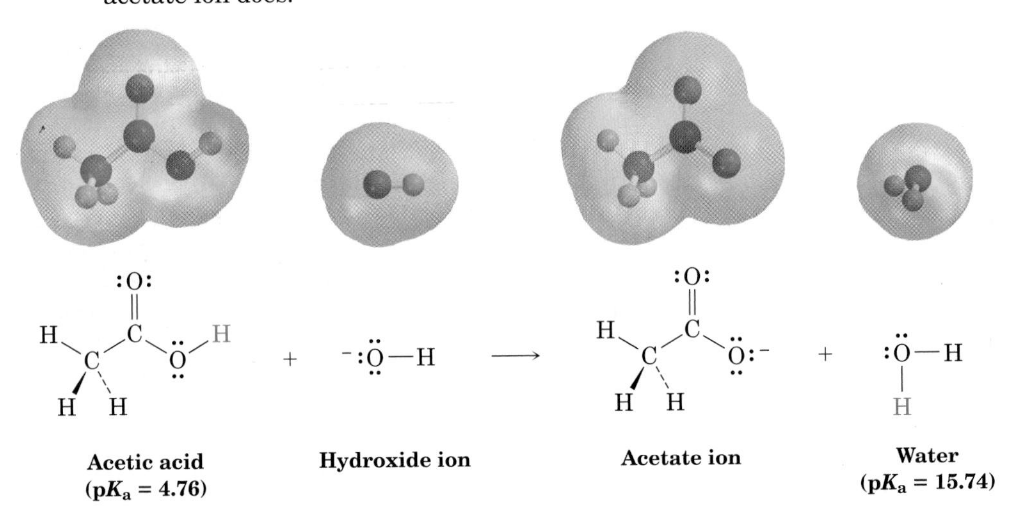

An easy way to predict acid–base reactivity is to remember that the products must be more stable than the reactants for reaction to occur. In other words, the product acid and base must be weaker and less reactive than the starting acid and base. In the reaction of acetic acid with hydroxide ion, for example, the product conjugate base (acetate ion) is weaker than the starting base (hydroxide ion) and the product conjugate acid (water) is weaker than the starting acid (acetic acid).

$$\underset{\text{Stronger acid}}{CH_3\overset{O}{\overset{\|}{C}}OH} + \underset{\text{Stronger base}}{HO^-} \rightleftarrows \underset{\text{Weaker acid}}{H\text{–}OH} + \underset{\text{Weaker base}}{CH_3\overset{O}{\overset{\|}{C}}O^-}$$

Many of the reactions we'll be seeing in future chapters involve *organic acids* and *organic bases.* Organic acids are characterized by the presence of a positively polarized hydrogen atom (blue in electrostatic potential maps). These acids can be divided into two main groups: (1) those with acidic hydrogens attached to oxygen atoms in compounds like methyl alcohol and acetic acid and (2) those where the acidic hydrogen atom is attached to a carbon atom. Because carbon is not as electronegative as oxygen, C–H hydrogens are rarely as acidic as O–H hydrogens. Later, we will see that these acidic C–H hydrogens are usually next to a C=O group (as in O=C—C—H).

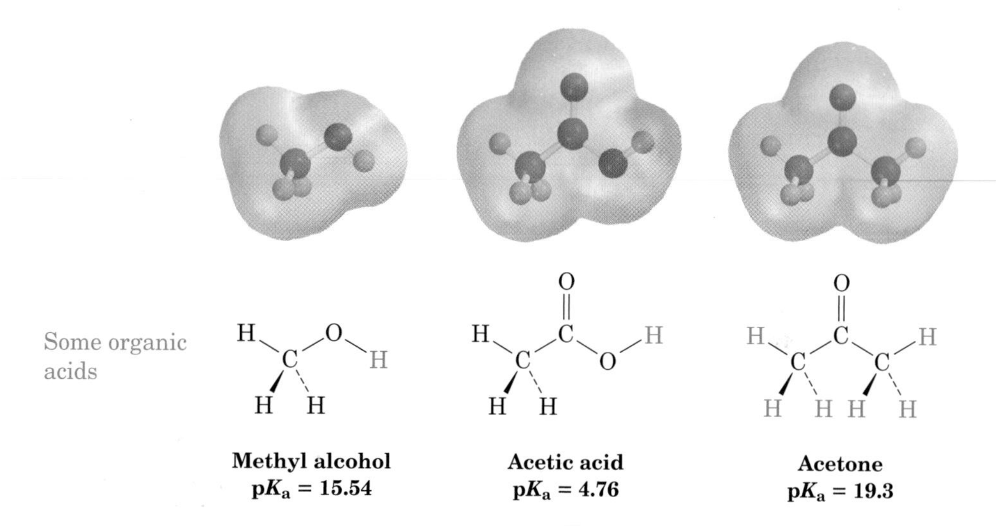

Organic bases are characterized by the presence of an atom with a lone pair of electrons (red in electrostatic potential maps) that can bond to H^+. Nitrogen-containing compounds such as trimethylamine are the most common, but oxygen-containing compounds can also act as bases when reacting with a sufficiently strong acid. Note that some oxygen-containing compounds can act as both acids and bases depending on the circumstances, just as water can. Methyl alcohol and acetone, for instance, act as *acids* when they donate a proton but act as *bases* when their oxygen atom accepts a proton.

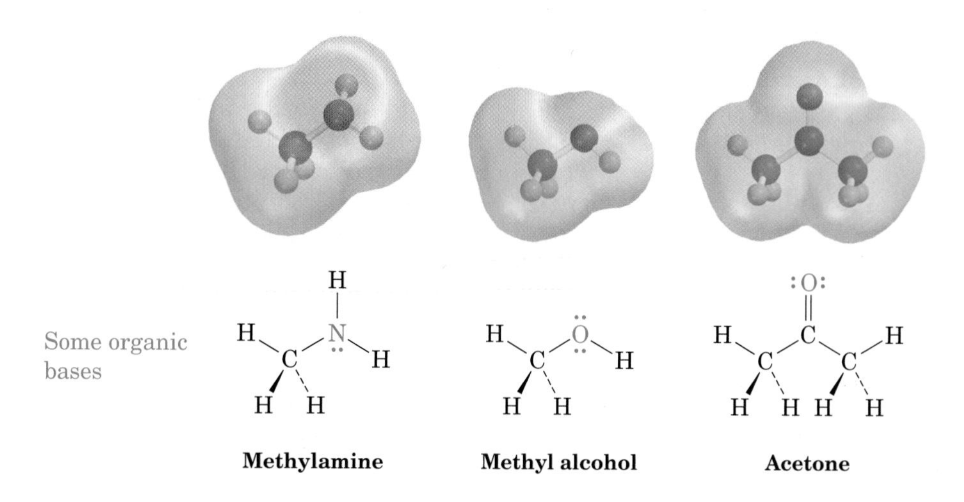

PRACTICE PROBLEM 1.6

Water has pK_a = 15.74, and acetylene has pK_a = 25. Which of the two is more acidic? Will hydroxide ion react with acetylene?

$$H{-}C{\equiv}C{-}H + H{-}O^- \xrightarrow{?} H{-}C{\equiv}C{:}^- + H{-}O{-}H$$

STRATEGY In comparing two acids, the one with the smaller pK_a is stronger. Thus, water is a stronger acid than acetylene.

SOLUTION Because water loses a proton more easily than acetylene, the HO^- ion has less affinity for a proton than the $HC{\equiv}C{:}^-$ ion. In other words, the anion of acetylene is a stronger base than hydroxide ion, and the reaction will not proceed as written.

PRACTICE PROBLEM 1.7

Butanoic acid, the substance responsible for the odor of rancid butter, has pK_a = 4.82. What is its K_a?

STRATEGY Since pK_a is the negative logarithm of K_a, it's necessary to use a calculator with an ANTILOG or INV LOG function. Enter the value of the pK_a (4.82), change the sign (−4.82), and then find the antilog (1.5×10^{-5}).

SOLUTION $K_a = 1.5 \times 10^{-5}$

PROBLEM 1.20 Formic acid, HCO_2H, has pK_a = 3.75, and picric acid, $C_6H_3N_3O_7$, has pK_a = 0.38.
(a) What is the K_a of each?
(b) Which is stronger, formic acid or picric acid?

PROBLEM 1.21 Amide ion, H_2N^-, is a stronger base than hydroxide ion, HO^-. Which is the stronger acid, $H_2N{-}H$ (ammonia) or $HO{-}H$ (water)? Explain.

PROBLEM 1.22 Is either of the following reactions likely to take place according to the pK_a data in Table 1.4?
(a) $HCN + CH_3CO_2^-\ Na^+ \longrightarrow Na^+\ ^-CN + CH_3CO_2H$
(b) $CH_3CH_2OH + Na^+\ ^-CN \longrightarrow CH_3CH_2O^-\ Na^+ + HCN$

1.11 Acids and Bases: The Lewis Definition

The Lewis definition of acids and bases differs from the Brønsted–Lowry definition in that it's not limited to substances that donate or accept protons. A **Lewis acid** is a substance that has a vacant valence orbital and can thus *accept an electron pair*; a **Lewis base** is a substance that *donates an electron pair*. The donated pair of electrons is shared between Lewis acid and base in a newly formed covalent bond.

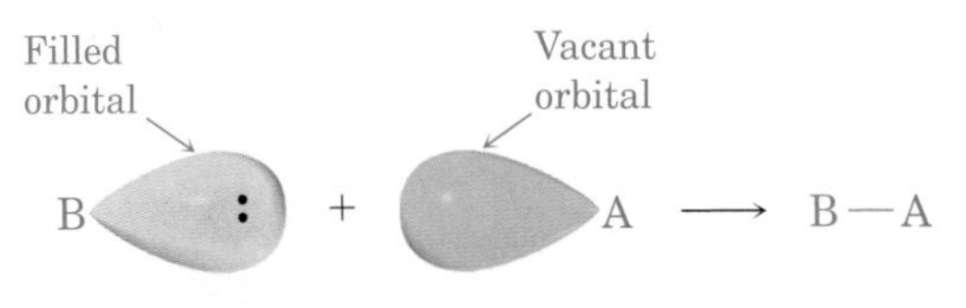

A proton (H^+) is a Lewis acid because it accepts a pair of electrons to fill its vacant 1*s* orbital when it bonds to a base. However, Lewis acids include not only proton donors but many other species as well. A compound such as $AlCl_3$ is a Lewis acid because it too accepts an electron pair from a Lewis base to fill a vacant valence orbital (Figure 1.14). The Lewis definition of basicity—a compound with a pair of nonbonding electrons that it can use in forming a bond to a Lewis acid—is similar to the Brønsted–Lowry definition.

FIGURE 1.14 The reactions of some Lewis acids with some Lewis bases. The Lewis acids accept an electron pair; the Lewis bases donate a pair of nonbonding electrons. Note how the flow of electrons from the Lewis base to the Lewis acid is indicated by the curved arrows.

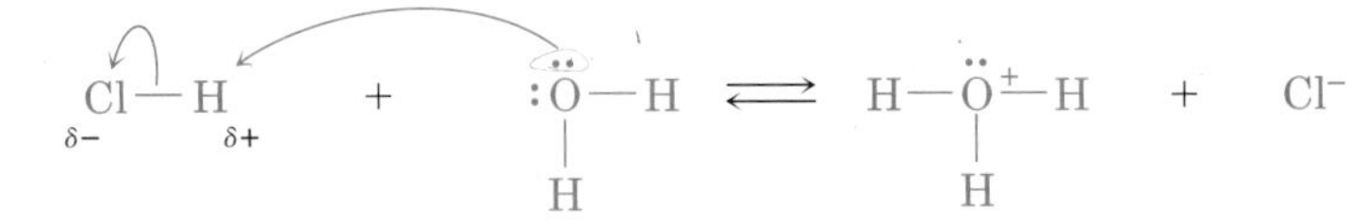

$$\overset{\delta-}{Cl}-\overset{\delta+}{H} + :\ddot{O}H_2 \rightleftharpoons H-\ddot{O}^+(H)-H + Cl^-$$

Hydrogen chloride (a Lewis acid) — **Water** (a Lewis base) — **Hydronium ion**

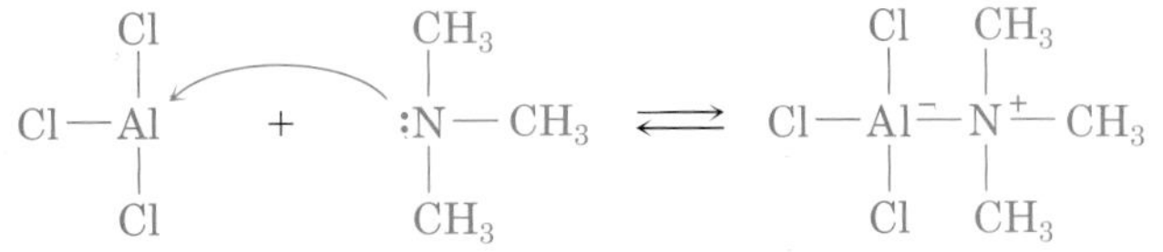

$$AlCl_3 + :N(CH_3)_3 \rightleftharpoons Cl_3Al^- - N^+(CH_3)_3$$

Aluminum trichloride (a Lewis acid) — **Trimethylamine** (a Lewis base)

Look closely at the acid–base reactions in Figure 1.14, and note how they are shown. In the first reaction, the Lewis base water uses an electron pair to abstract H^+ from the polar HCl molecule. In the second reaction, a Lewis base donates an electron pair to a vacant valence orbital of an aluminum atom. In both reactions, the direction of electron-pair flow from the electron-rich Lewis base to the electron-poor Lewis acid is shown using curved arrows. *A curved arrow always means that a pair of electrons moves from the atom at the tail of the arrow to the atom at the head of the arrow.* We'll use this curved-arrow notation frequently in the remainder of this text to indicate electron flow during reactions.

PRACTICE PROBLEM 1.8

Using curved arrows, show how acetaldehyde, CH_3CHO, can act as a Lewis base.

STRATEGY A Lewis base donates an electron pair to a Lewis acid. We therefore need to locate the electron lone pairs on acetaldehyde and use a curved arrow to show the movement of one pair from the oxygen toward a Lewis acid such as H^+.

SOLUTION

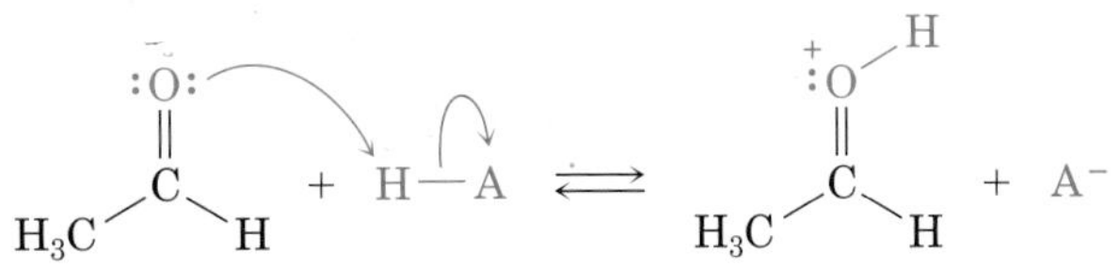

$$H_3C-C(=\ddot{O}:)-H + H-A \rightleftharpoons H_3C-C(=\overset{+}{O}-H)-H + A^-$$

Acetaldehyde

PROBLEM 1.23 Which of the following are likely to act as Lewis acids, which as Lewis bases, and which as both?

(a) CH_3CH_2OH (b) $(CH_3)_2NH$ (c) $MgBr_2$
(d) $(CH_3)_3B$ (e) H_3C^+ (f) $(CH_3)_3P$

PROBLEM 1.24 Show how the species in part (a) can act as Lewis bases in their reactions with HCl, and show how the species in part (b) can act as Lewis acids in their reaction with OH^-.

(a) CH_3CH_2OH, $(CH_3)_2NH$, $(CH_3)_3P$ (b) H_3C^+, $(CH_3)_3B$, $MgBr_2$

PROBLEM 1.25 The organic compound imidazole can act as both an acid and a base. Look at the following electrostatic potential map, and identify the most acidic hydrogen atom and the most basic nitrogen atom in imidazole.

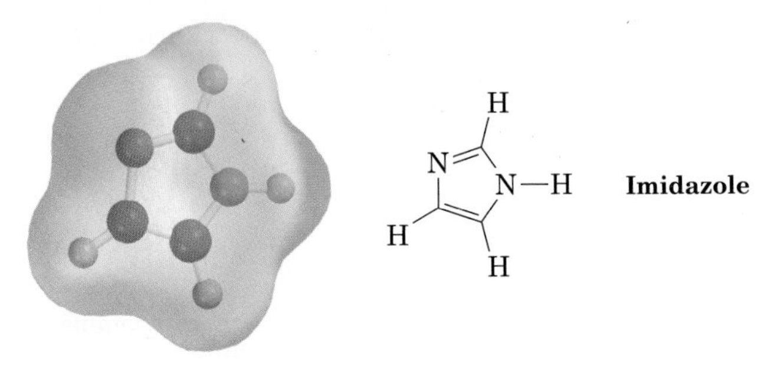

INTERLUDE

In the Field with Triazine Herbicides: Risk versus Benefits

Throughout the text, we'll look at the chemistry of atrazine and related herbicides.

Contrary to what a grocer advertises, all tomatoes are organic: tomatoes are complex assemblies of organic molecules. Grocers use the word *organic* to mean a freedom from synthetic chemicals, typically pesticides. How concerned should we be about traces of pesticides in our foods? Of toxins in our water? Of pollutants in the air? Are we being put at risk?

Life is not risk-free—we all take many risks each day without even thinking about it. We decide to ride a bike rather than drive, even though there is a ten times greater likelihood per mile of dying in a bicycling accident than in a car accident. We decide to walk down stairs rather than take an elevator, even though 7000 people die from falls each year in the United States. We decide to smoke cigarettes, even though it increases our chance of getting cancer by 50%. So what about risks from chemicals like pesticides?

Without pesticides, whether they be herbicides, which target weeds; insecticides, which kill insects; or fungicides, which target molds and fungi, crop production would drop significantly. Consider atrazine, one of the most commonly used herbicides. In the United States alone, approximately 100 million lb of atrazine is used each year to kill weeds in corn, sorghum, and sugarcane fields. Atrazine use continues to be an area of concern because it persists in the environment. Although atrazine exposure can pose health risks to humans and other animal species, the U.S. Environmental Protection Agency (EPA) is unwilling to ban its use because doing so would result in lower crop yields and increased costs, and currently no

Continued

suitable alternative for atrazine is available. Atrazine can be removed from drinking water by using charcoal filters, but what about its presence in lakes and ponds?

In such cases, we need a way to evaluate the risks presented by chemicals. One way to quantify these risks is to expose test animals to specific substances in controlled quantities and then monitor them for signs of harm, including death. The data obtained are then reduced to an *LD_{50} value* (LD stands for lethal dose), the amount of material per kilogram body weight that was lethal to 50% of the test animals. For the pesticide atrazine, the LD_{50} value is between 1 and 4 g atrazine/kg, depending on the animal species. Aspirin, for sake of comparison, has an LD_{50} of 1.1 g/kg. Table 1.5 lists the LD_{50} values of some familiar substances. However, be aware that LD_{50} values do *not* tell us about the risks of long-term exposure, including a molecule's ability to cause cancer or interfere with the development of an unborn child.

TABLE 1.5 Some LD_{50} Values

Substance	LD_{50} (g/kg)	Substance	LD_{50} (g/kg)
Strychnine	0.005	Iron(II) sulfate	1.5
Arsenic trioxide	0.015	Chloroform	3.2
DDT	0.115	Ethyl alcohol	10.6
Aspirin	1.1	Sodium cyclamate	17

So, should we still use atrazine? All decisions involve tradeoffs and the answer is rarely obvious. Does the benefit of increased food production outweigh possible health risks of a pesticide? Currently, atrazine is approved for continued use in the United States, but it is being phased out in Europe. The EPA believes that the benefits of increased food production outweigh possible health risks.

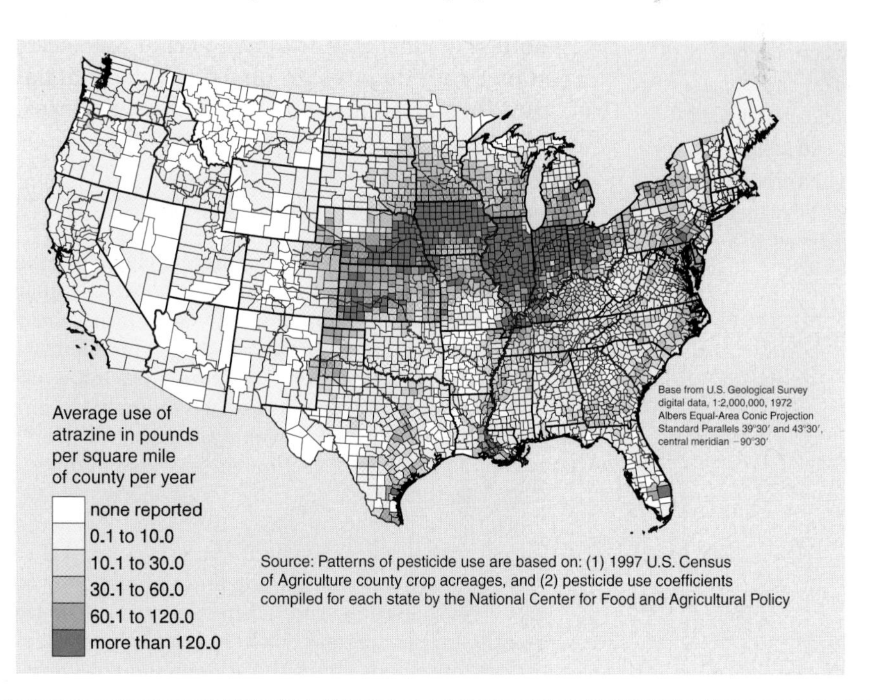

Estimated annual agricultural use of atrazine in the United States.

Summary and Key Words

Organic chemistry is the study of carbon compounds. Although a division into inorganic and organic chemistry occurred historically, there is no scientific reason for the division.

Atoms are composed of a positively charged nucleus surrounded by negatively charged electrons that occupy specific regions of space called **orbitals**. Different orbitals have different energy levels and shapes. For example, *s* orbitals are spherical and *p* orbitals are dumbbell-shaped.

There are two fundamental kinds of chemical bonds: **ionic bonds** and **covalent bonds**. The ionic bonds commonly found in inorganic salts result from the electrical attraction of unlike charges. The covalent bonds found in organic molecules result from the sharing of one or more electron pairs between atoms. Electron sharing occurs when two atoms approach and their atomic orbitals overlap. Bonds formed by head-on overlap of atomic orbitals are called **sigma (σ) bonds**; bonds formed by sideways overlap of *p* orbitals are called **pi (π) bonds**.

Carbon uses hybrid orbitals to form bonds in organic compounds. When forming only single bonds, carbon has four equivalent **sp^3 hybrid orbitals** with tetrahedral geometry. When forming double bonds, carbon has three equivalent **sp^2 orbitals** with planar geometry and one unhybridized *p* orbital. When forming triple bonds, carbon has two equivalent ***sp* orbitals** with linear geometry and two unhybridized *p* orbitals.

Organic molecules often have **polar covalent bonds** because of unequal electron sharing caused by the **electronegativity** of atoms. For example, a carbon–oxygen bond is polar because oxygen attracts the bonding electrons more strongly than carbon does. Carbon–metal bonds, by contrast, are polarized in the opposite sense because carbon attracts electrons more strongly than metals do.

A **Brønsted–Lowry acid** is a substance that can donate a proton (hydrogen ion, H^+); a **Brønsted–Lowry base** is a substance that can accept a proton. The strength of an acid is given by its acidity constant, K_a. A **Lewis acid** is a substance that can accept an electron pair. A **Lewis base** is a substance that donates an unshared electron pair. Most organic molecules that contain oxygen and nitrogen are Lewis bases.

Working Problems

There's no surer way to learn organic chemistry than by working problems. Although careful reading and rereading of the text is important, reading alone isn't enough. You must also be able to apply the information you read and be able to use your knowledge in new situations. Working problems gives you practice at doing this.

Each chapter in this book provides many problems of different sorts. The in-chapter problems are placed for immediate reinforcement of ideas just learned. The end-of-chapter problems provide additional practice and are of several types. They begin with a section called "Visualizing Chemistry," which contains problems that help you see the microscopic world of molecules. These visual problems are then followed by many Additional Problems. Early ones tend to be the drill type so that you can practice your command of the fundamentals; later problems tend to be more challenging and thought provoking.

As you study organic chemistry, take the time to work the problems. Do the ones you can and ask for help on the ones you can't. If you're stumped by a particular problem, check the accompanying *Study Guide and Solutions Manual* for an explanation that will help clarify the difficulty. Working problems takes effort, but the payoff in knowledge and understanding is immense.

EXERCISES

Visualizing Chemistry

1.26 Convert each of the following molecular models into a line-bond structure, and give the formula of each (gray = C, red = O, blue = N, ivory = H).

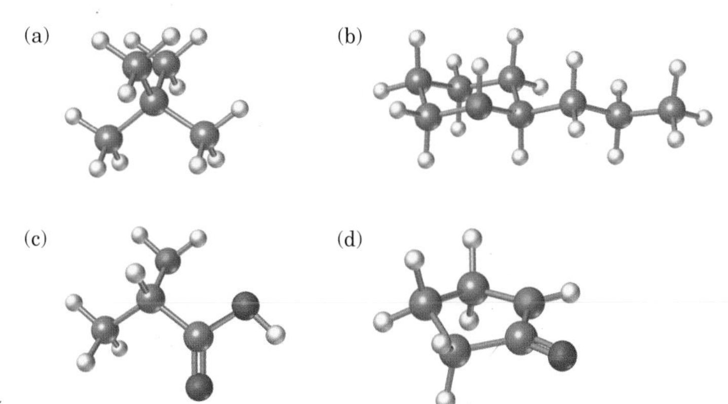

1.27 Following is a model of acetaminophen, a pain reliever sold in drugstores under various trade names, such as Tylenol. Identify the hybridization of each carbon atom in acetaminophen, and tell which atoms have lone pairs of electrons (gray = C, red = O, blue = N, ivory = H).

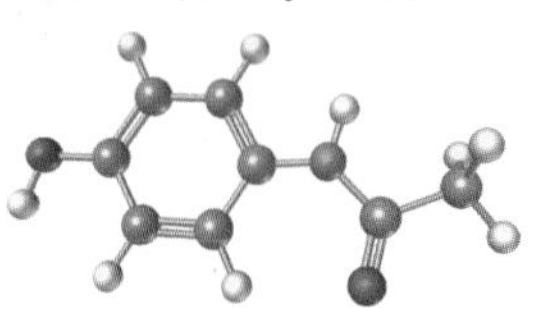

1.28 Following is a model of aspartame, $C_{14}H_{18}N_2O_5$, known commercially under a variety of trade names, including NutraSweet. Only the connections between atoms are shown; multiple bonds are not indicated. Complete the structure by indicating the positions of multiple bonds (gray = C, red = O, blue = N, ivory = H).

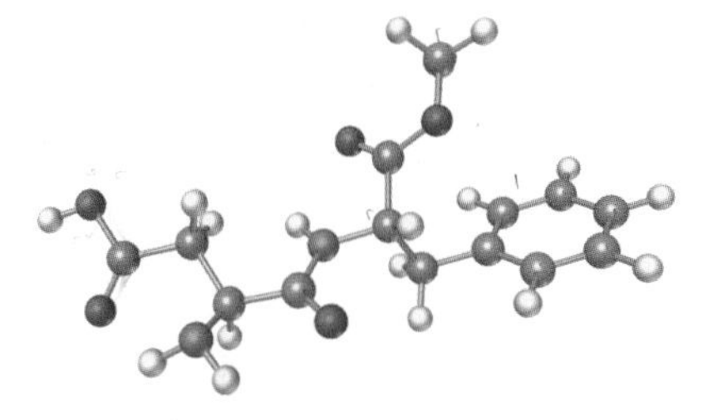

1.29 Electrostatic potential maps of (a) acetamide and (b) methylamine are shown. Which of the two has the more basic nitrogen atom? Which of the two has the more acidic hydrogen atoms?

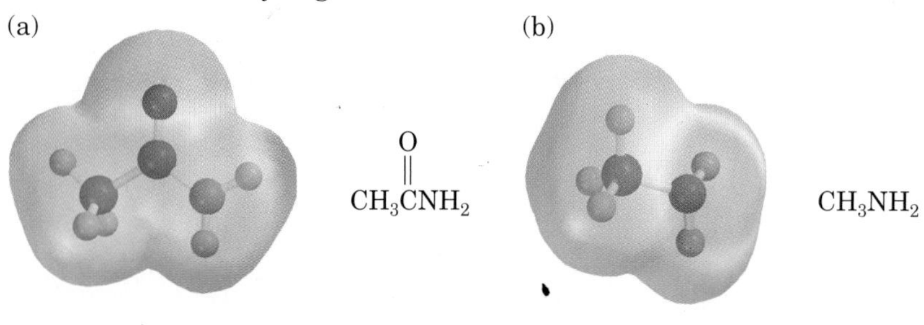

Additional Problems

Electron Configuration

1.30 How many valence (outer-shell) electrons does each of the following atoms have?
(a) Oxygen (b) Magnesium (c) Fluorine

1.31 Give the ground-state electron configuration of the following elements. For example, carbon is $1s^2\ 2s^2\ 2p^2$.
(a) Lithium (b) Sodium (c) Aluminum (d) Sulfur

Lewis Structures

1.32 What are the likely formulas of the following molecules?
(a) $AlCl_?$ (b) $CF_2Cl_?$ (c) $NI_?$

1.33 Write a Lewis (electron-dot) structure for acetonitrile, $CH_3C{\equiv}N$. How many electrons does the nitrogen atom have in its valence shell? How many are used for bonding, and how many are not used for bonding?

1.34 Fill in any unshared electrons that are missing from the following line-bond structures:

(a) $CH_3{-}O{-}CH_3$ (b) CH_3NH_2 (c) CH_2Cl_2 (d) $H_3C{-}C({=}O){-}CH_3$

1.35 Which atoms in the following structures have unshared valence electrons? Draw in these unshared electrons.

(a) CH_3SH (b) $CH_3{-}N(CH_3){-}CH_3$ (c) CH_3CH_2Br

(d) $CH_3C({=}O){-}OH$ (e) $CH_3C({=}O){-}Cl$

Structural Formulas

1.36 Draw both a Lewis structure and a line-bond structure for vinyl chloride, C_2H_3Cl, the starting material from which PVC [poly(vinyl chloride)] plastic is made.

1.37 There are two structures with the formula C_4H_{10}. Draw them and tell how they differ.

1.38 Convert the following molecular formulas into line-bond structures:
(a) C_3H_8 (b) C_3H_7Br (two possibilities)
(c) C_3H_6 (two possibilities) (d) C_2H_6O (two possibilities)

1.39 Draw a three-dimensional representation of chloroform, $CHCl_3$, using the standard convention of solid, wedged, and dashed lines. Do the same for ethanol, CH_3CH_2OH.

1.40 Draw line-bond structures for the following molecules:
(a) Ethyl methyl ether, C_3H_8O, which contains an oxygen atom bonded to two carbons
(b) Butane, C_4H_{10}, which contains a chain of four carbon atoms
(c) Cyclohexene, C_6H_{10}, which contains a ring of six carbon atoms and one carbon–carbon double bond

Electronegativity

1.41 Identify the bonds in the following molecules as covalent, polar covalent, or ionic:
(a) BeF_2 (b) SiH_4 (c) CBr_4

1.42 Indicate which of the bonds in the following molecules are polar covalent, using the symbols $\delta+$ and $\delta-$.
(a) Br_2 (b) CH_3Cl (c) HF (d) CH_3CH_2OH

1.43 Sodium methoxide, $NaOCH_3$, contains both ionic and covalent bonds. Indicate which is which.

1.44 Identify the most electronegative element in each of the following molecules:
(a) CH_2FCl (b) $FCH_2CH_2CH_2Br$ (c) $HOCH_2CH_2NH_2$ (d) CH_3OCH_2Li

1.45 Use the electronegativity values in Figure 1.12 to predict which of the indicated bonds in each of the following sets is more polar. Tell the direction of the polarity in each.
(a) Cl—CH_3 or Cl—Cl (b) H—CH_3 or H—Cl
(c) HO—CH_3 or $(CH_3)_3Si$—CH_3

1.46 Use Figure 1.12 to order the following molecules according to increasing positive character of the carbon atom:

CH_3F, CH_3OH, CH_3Li, CH_3I, CH_3CH_3, CH_3NH_2

1.47 Organic molecules can be classified according to the *functional groups* they contain, where a functional group is a collection of atoms with a characteristic chemical reactivity. Use the electronegativity values given in Figure 1.12 to predict the polarity of the following functional groups.

(a) —C≡N **Nitrile**
(b) 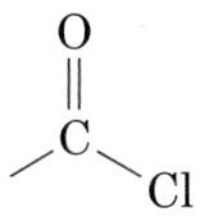 C(=O)Cl **Acid chloride**
(c) 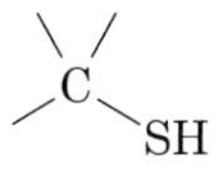 C—SH **Thiol**

HYBRIDIZATION

1.48 What is the hybridization of each carbon atom in acetonitrile, $CH_3C{\equiv}N$?

1.49 Indicate the kind of hybridization you expect for each carbon atom in the following molecules:
(a) Butane, $CH_3CH_2CH_2CH_3$ (b) But-1-ene, $CH_3CH_2CH{=}CH_2$
(c) Cyclobutene, (d) But-1-en-3-yne, $H_2C{=}CH{-}C{\equiv}CH$

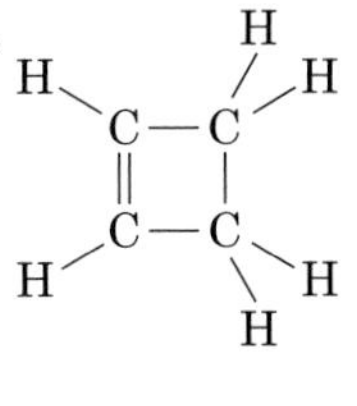

[Cyclobutene structure: H—C=C—H ring with CH₂—CH₂]

1.50 What is the hybridization of each carbon atom in benzene? What shape do you expect benzene to have?

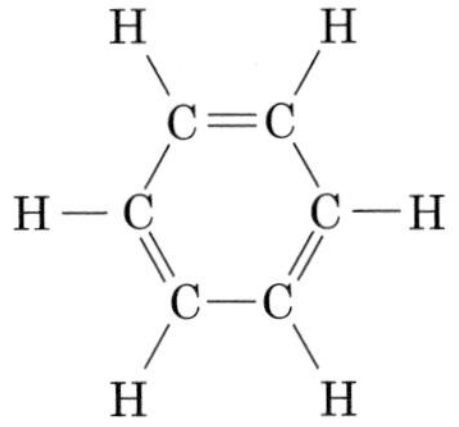

Benzene

1.51 Propose structures for molecules that meet the following descriptions:
(a) Contains two sp^2-hybridized carbons and two sp^3-hybridized carbons
(b) Contains only four carbons, all of which are sp^2-hybridized
(c) Contains two sp-hybridized carbons and two sp^2-hybridized carbons

ACID–BASE CHEMISTRY

1.52 Ammonia, $H_2N{-}H$, has $pK_a \approx 36$ and acetone has $pK_a \approx 19$. Will the following reaction take place? Explain.

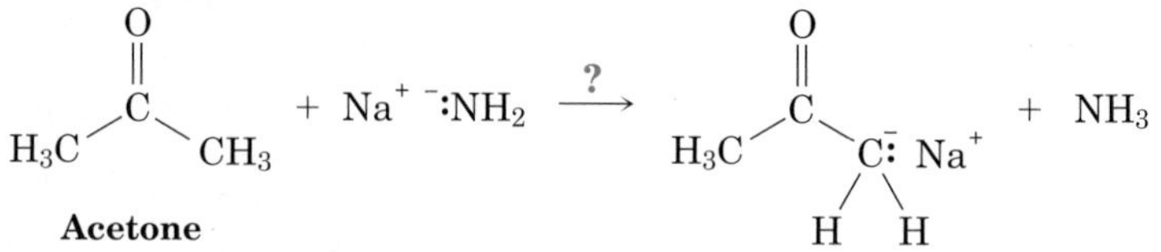

$$H_3C{-}C(=O){-}CH_3 + Na^+\ {}^-{:}NH_2 \xrightarrow{?} H_3C{-}C(=O){-}\ddot{C}H_2^-\ Na^+ + NH_3$$

Acetone

1.53 Which of the following substances are likely to behave as Lewis acids and which as Lewis bases?
(a) $AlBr_3$ (b) $CH_3CH_2NH_2$ (c) HF (d) CH_3SCH_3

1.54 Is the bicarbonate anion (HCO_3^-) a strong enough base to react with methanol (CH_3OH)? In other words, does the following reaction take place as written? (The pK_a of methanol is 15.5; the pK_a of H_2CO_3 is 6.4.)

$$CH_3OH + HCO_3^- \xrightarrow{?} CH_3O^- + H_2CO_3$$

1.55 Identify the acids and bases in the following reactions:
(a) $CH_3OH + H^+ \longrightarrow CH_3\overset{+}{O}H_2$
(b) $CH_3OH + {}^-NH_2 \longrightarrow CH_3O^- + NH_3$
(c) 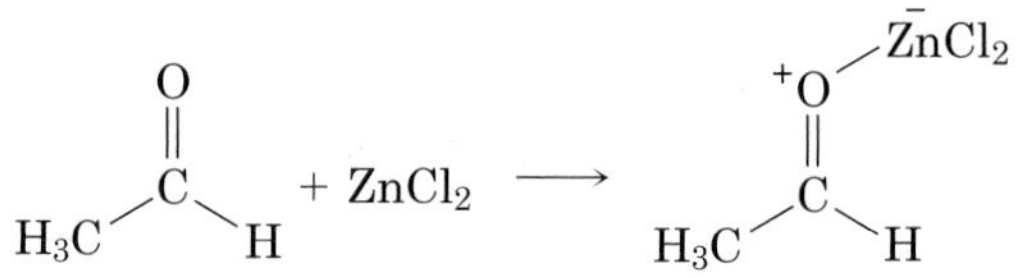 $H_3C{-}C(=O){-}H + ZnCl_2 \longrightarrow H_3C{-}C(={}^+O{-}\bar{Z}nCl_2){-}H$

1.56 Rank the following substances in order of increasing acidity:

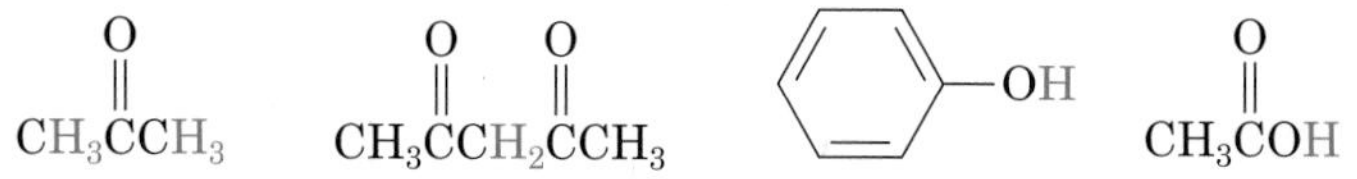

$CH_3\overset{O}{\overset{\|}{C}}CH_3$	$CH_3\overset{O}{\overset{\|}{C}}CH_2\overset{O}{\overset{\|}{C}}CH_3$	$C_6H_5{-}OH$	$CH_3\overset{O}{\overset{\|}{C}}OH$
Acetone (pK_a = 19)	**Pentane-2,4-dione** (pK_a = 9)	**Phenol** (pK_a = 9.9)	**Acetic acid** (pK_a = 4.76)

1.57 Which, if any, of the four substances in Problem 1.56 are strong enough acids to react almost completely with NaOH? (The pK_a of H_2O is 15.7.)

1.58 The ammonium ion (NH_4^+, pK_a = 9.25) has a lower pK_a than the methylammonium ion ($CH_3NH_3^+$, pK_a=10.66). Which is the stronger base, ammonia (NH_3) or methylamine (CH_3NH_2)? Explain.

INTEGRATED PROBLEMS

1.59 Why can't molecules with the following formulas exist?
(a) CH_5 (b) C_2H_6N (c) $C_3H_5Br_2$

1.60 Complete the Lewis (electron-dot) structure of caffeine, showing all lone-pair electrons, and identify the hybridization of the indicated atoms.

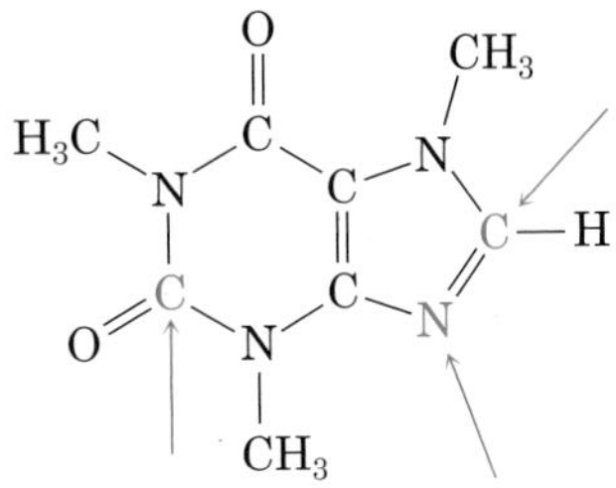

Caffeine

1.61 The ammonium ion, NH_4^+, has a geometry identical to that of methane, CH_4. What kind of hybridization do you think the nitrogen atom has? Explain.

1.62 Indicate the kind of hybridization you would expect for each carbon atom in the following molecules:

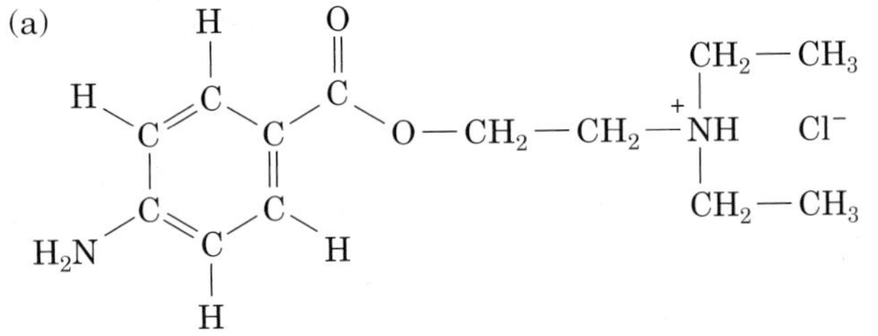

(a) Procaine

(b) Vitamin C

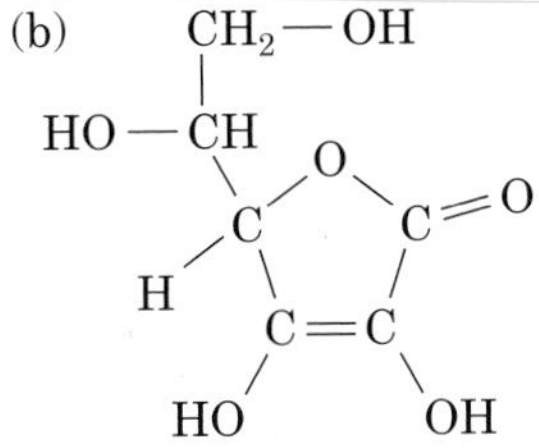

1.63 Why do you suppose no one has ever been able to make cyclopentyne as a stable molecule?

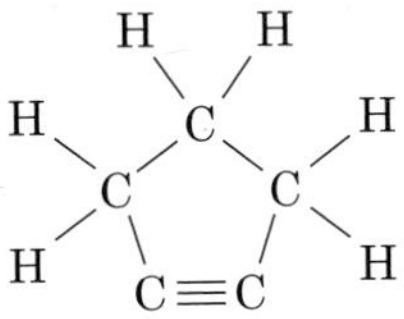

Cyclopentyne

1.64 Draw an orbital picture of allene, $H_2C{=}C{=}CH_2$. What hybridization must the central carbon atom have to form two double bonds? What shape does allene have?

1.65 Draw a Lewis structure and an orbital picture for carbon dioxide, CO_2. What kind of hybridization does the carbon atom have? What is the relationship between CO_2 and allene (Problem 1.64)?

1.66 Although most stable organic compounds have tetravalent carbon atoms, high-energy species with trivalent carbon atoms also exist. *Carbocations* are one such class of compounds. If the positively charged carbon atom has planar geometry, what hybridization do you think it has? How many valence electrons does the carbon have?

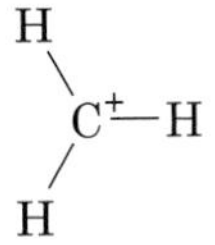

A carbocation

IN THE MEDICINE CABINET

1.67 Nonsteroidal anti-inflammatory drugs, referred to as NSAIDs, are commonly used to treat minor aches and pains. Four of the most common NSAIDs are:

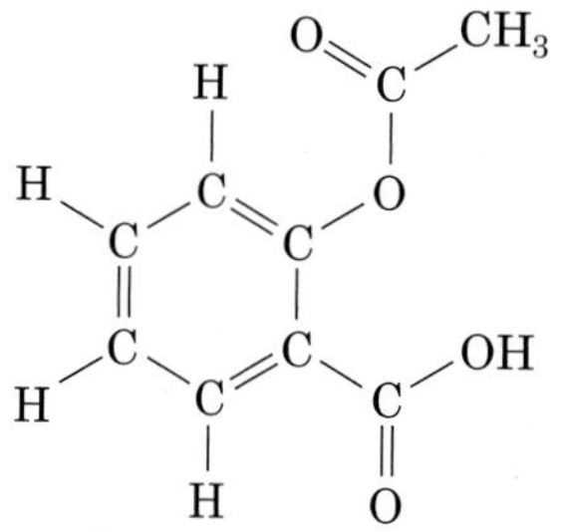

Acetylsalicylic acid
(aspirin)

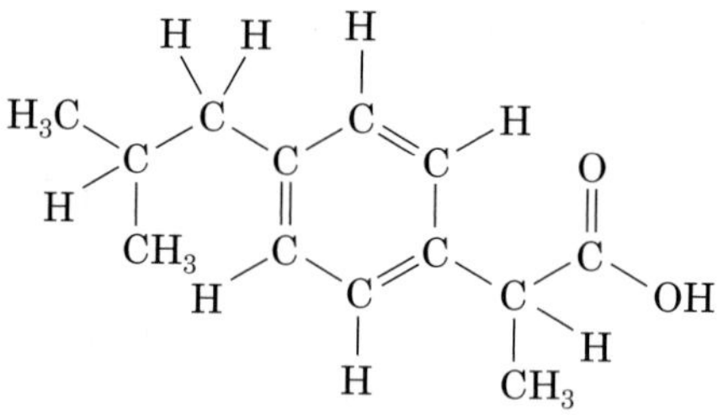

Ibuprofen
(Advil, Nuprin, Motrin)

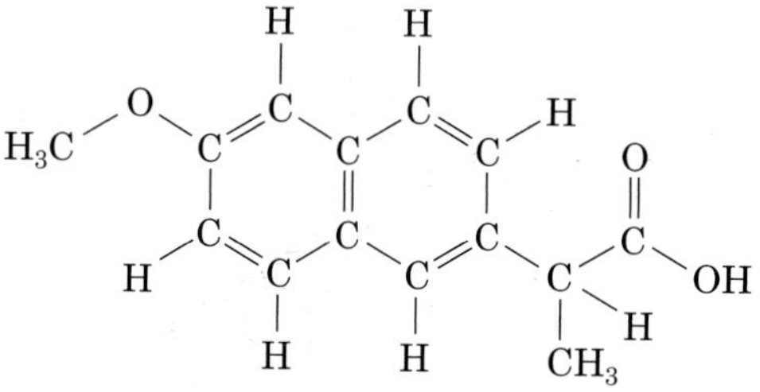

Naproxen
(Naprosyn, Aleve)

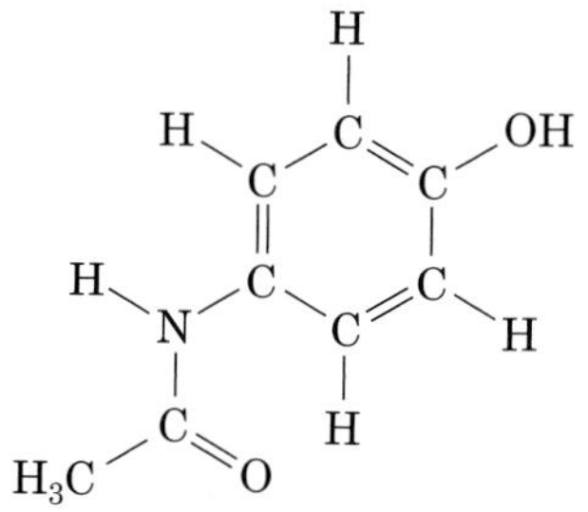

Acetaminophen
(Tylenol)

(a) How many *sp*^{3}-hybridized carbons are present in aspirin?
(b) How many *sp*^{2}-hybridized carbons are present in naproxen?
(c) What is the molecular formula of acetaminophen?
(d) Aspirin, ibuprofen, and naproxen are all believed to target the same protein, cyclooxygenase, which produces prostaglandins that mediate swelling. Discuss any similarities in the structures of these drugs.

In the Field with Agrochemicals

1.68 Herbicides that differ wildly in chemical formula and structure often differ in mode of action.

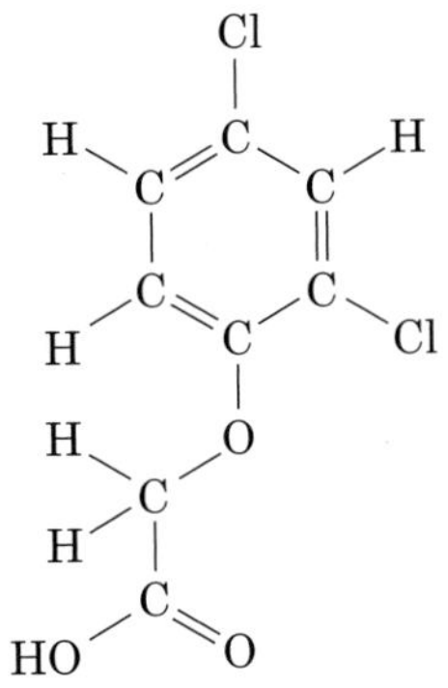

2,4-D (Hi-Dep or Weedar 64)
Disrupts growth regulation signals

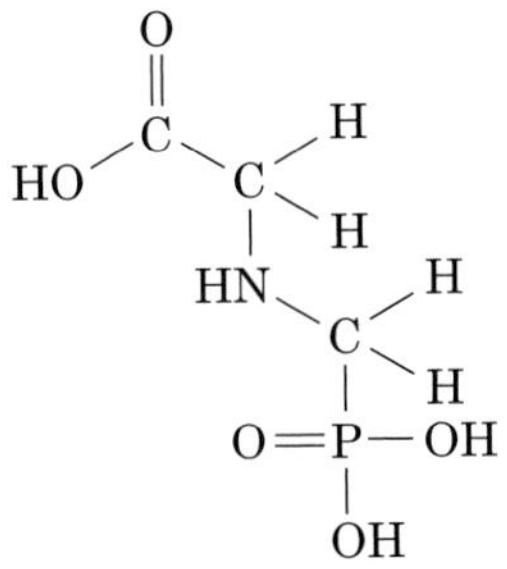

Glyphosate (Roundup)
Inhibits amino acid synthesis

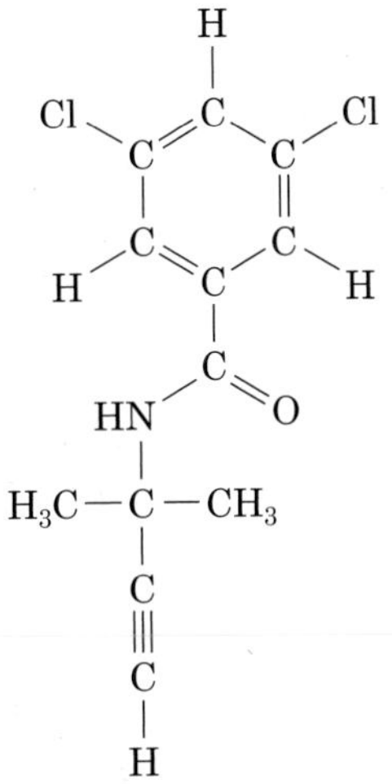

Pronamide (Propyzamide)
Interferes with cell division

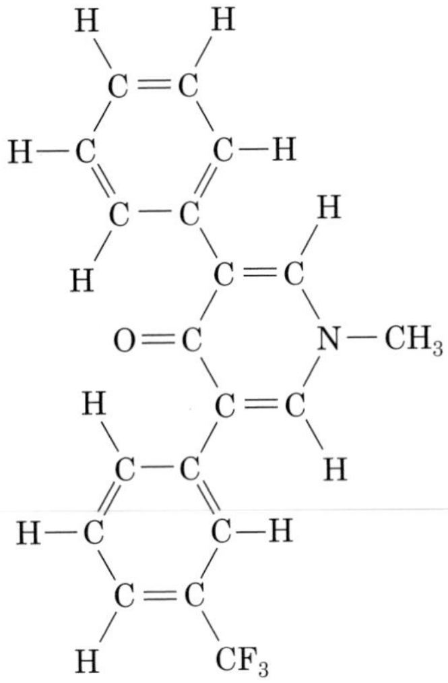

Fluridone (Sonar)
Inhibits pigment biosynthesis

(a) How many sp^3-hybridized carbons are present in 2,4-D?
(b) How many sp^2-hybridized carbons are present in Roundup?
(c) How many *sp*-hybridized carbons are present in Pronamide?
(d) What is the molecular formula of Fluridone?
(e) Triclopyr has a very similar mechanism of action to that of one of the four herbicides pictured. Which plant process do you think it will interrupt? Why?

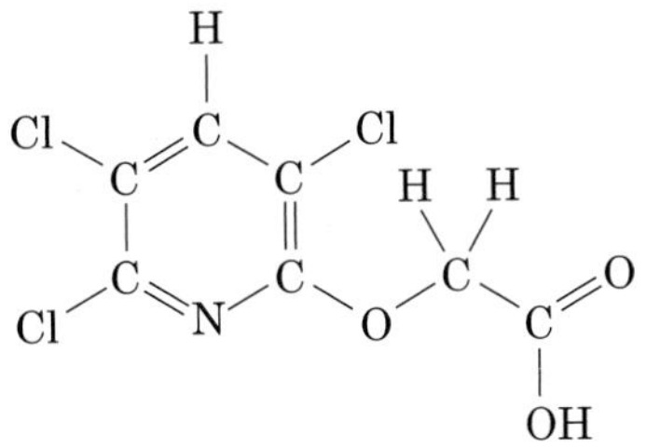

Triclopyr (Garlon, Grazon)

The cancer drug Taxol, initially obtained from the bark of the Pacific yew tree, is a complex molecule built from simple functional groups. (Getty Images/Inga Spence)

CHAPTER 2

The Nature of Organic Molecules

There are more than *27 million* known organic molecules, each of which has its own unique physical and chemical properties. Some of these compounds occur in nature. Others are the creation of chemists. The "rules" that nature and people use to make organic molecules rest largely on understanding the chemistry of small combinations of atoms. So, instead of 27 million compounds with random reactivity, there are a few dozen families of compounds whose chemistry is reasonably predictable. We'll study the chemistry of the most common families throughout this book.

2.1 Functional Groups

The structural features that make it possible to classify compounds by reactivity are called *functional groups.* A **functional group** is a group of atoms within a larger molecule that has a characteristic chemical behavior. Chemically, a given functional group behaves almost the same way in every molecule it's in. For example, one of the simplest functional groups is the carbon–carbon double bond. Because the electronic structure of the carbon–carbon double bond remains essentially the same in all molecules where it occurs, its chemical reactivity also remains the same. For instance, ethylene, the simplest compound with a carbon–carbon double bond, undergoes reactions that are identical to those of menthene, a substantially larger molecule found in peppermint oil. Both, for example, react with Br_2 to give products in which a bromine atom has added to each of the double-bond carbons (Figure 2.1). The example shown in Figure 2.1 is typical. *The chemistry of every organic molecule, regardless of size and complexity, is determined by the functional groups it contains.*

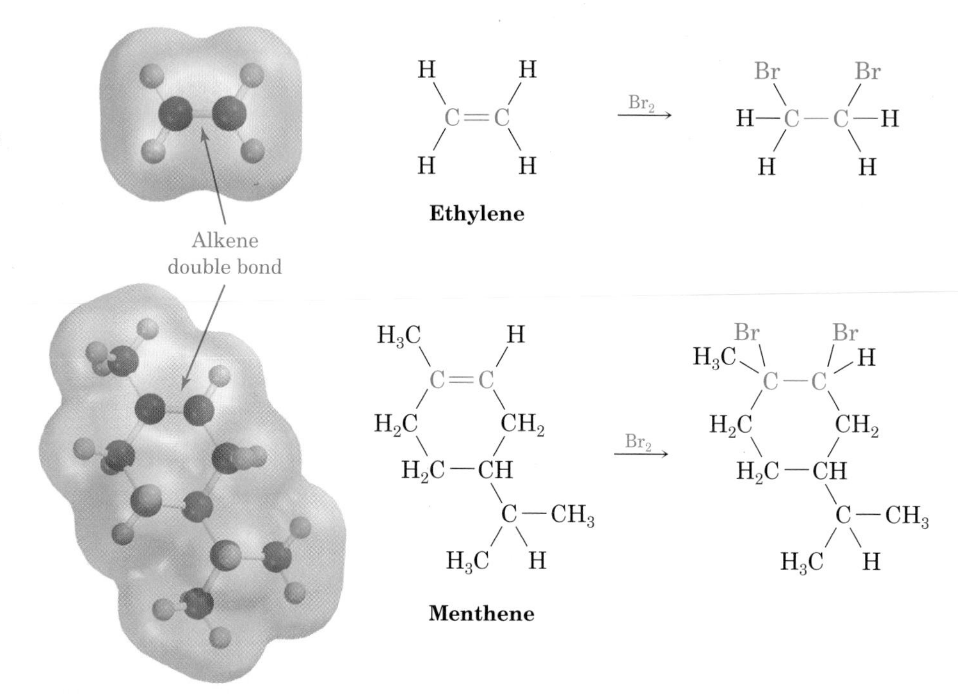

FIGURE 2.1 The reactions of ethylene and menthene with Br_2. In both cases, Br_2 reacts with the C=C functional group in exactly the same way. The size and nature of the remainder of the molecule are not important.

The structures of many common functional groups are shown in Table 2.1. Some functional groups have only carbon–carbon multiple bonds; others have a carbon atom singly bonded to an electronegative halogen, oxygen, nitrogen, or sulfur atom; and still others have carbon–oxygen double bonds. It's a good idea at this point to familiarize yourself with the structures of these functional groups so that you'll recognize them when you see them again. One way to do this is by making flash cards. Draw the functional group on one side, and write its name on the other side. Later, you can expand this card set to include chemical reactions.

Functional Groups with Carbon–Carbon Multiple Bonds

Alkenes, alkynes, and arenes (aromatic compounds) all contain carbon–carbon multiple bonds. *Alkenes* have a double bond, *alkynes* have a triple bond, and *arenes* have three alternating double and single bonds in a six-membered ring of carbon atoms. Alkenes and alkynes undergo similar chemical reactions. The cycle of π bonds in arenes leads to different reactivity: arenes are a distinct functional group because of their unique reactivity.

Table 2.1 Structures of Some Common Functional Groups

Name	Structure[a]	Name ending	Example
Alkene (double bond)	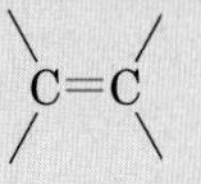 C=C	*-ene*	$H_2C{=}CH_2$ Ethene
Alkyne (triple bond)	—C≡C—	*-yne*	HC≡CH Ethyne
Arene (aromatic ring)	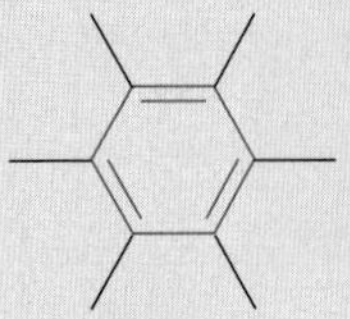	None	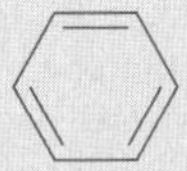 Benzene
Halide	 C–X (X = F, Cl, Br, I)	None	CH_3Cl Chloromethane
Alcohol	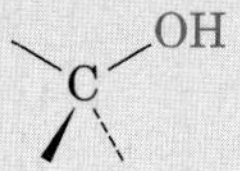 C–OH	*-ol*	CH_3OH Methanol
Ether	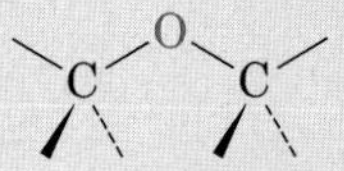 C–O–C	*ether*	CH_3OCH_3 Dimethyl ether
Monophosphate	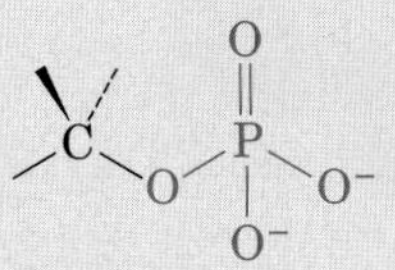 C–O–P(=O)(O⁻)–O⁻	*phosphate*	$CH_3OPO_3^{2-}$ Methyl phosphate
Diphosphate	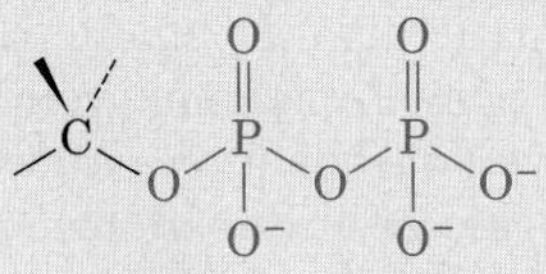 C–O–P(=O)(O⁻)–O–P(=O)(O⁻)–O⁻	*diphosphate*	$CH_3OP_2O_6^{3-}$ Methyl diphosphate
Amine	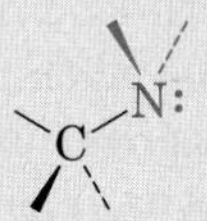 C–N:	*-amine*	CH_3NH_2 Methylamine
Imine (Schiff base)	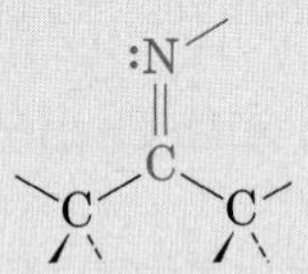 :N= C(C)C	None	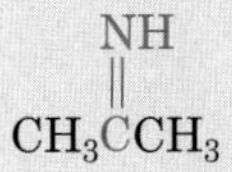 NH ‖ CH_3CCH_3 Acetone imine
Nitrile	—C≡N	*-nitrile*	$CH_3C{\equiv}N$ Ethanenitrile
Thiol	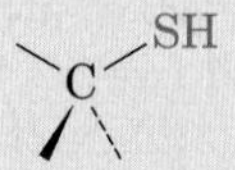 C–SH	*-thiol*	CH_3SH Methanethiol

[a]The bonds whose connections aren't specified are assumed to be attached to carbon or hydrogen atoms in the rest of the molecule.

Table 2.1 Structures of Some Common Functional Groups, cont'd

Name	Structure[a]	Name ending	Example
Sulfide	$C{-}S{-}C$	*sulfide*	CH_3SCH_3 Dimethyl sulfide
Disulfide	$C{-}S{-}S{-}C$	*disulfide*	CH_3SSCH_3 Dimethyl disulfide
Sulfoxide	$C{-}\overset{O^{-}}{\overset{\vert}{S^{+}}}{-}C$	*sulfoxide*	$CH_3\overset{O^{-}}{\overset{\vert}{S^{+}}}CH_3$ Dimethyl sulfoxide
Aldehyde	$C{-}\overset{O}{\overset{\Vert}{C}}{-}H$	*-al*	$CH_3\overset{O}{\overset{\Vert}{C}}H$ Ethanal
Ketone	$C{-}\overset{O}{\overset{\Vert}{C}}{-}C$	*-one*	$CH_3\overset{O}{\overset{\Vert}{C}}CH_3$ Propanone
Carboxylic acid	$C{-}\overset{O}{\overset{\Vert}{C}}{-}OH$	*-oic acid*	$CH_3\overset{O}{\overset{\Vert}{C}}OH$ Ethanoic acid
Ester	$C{-}\overset{O}{\overset{\Vert}{C}}{-}O{-}C$	*-oate*	$CH_3\overset{O}{\overset{\Vert}{C}}OCH_3$ Methyl ethanoate
Thioester	$C{-}\overset{O}{\overset{\Vert}{C}}{-}S{-}C$	*-thioate*	$CH_3\overset{O}{\overset{\Vert}{C}}SCH_3$ Methyl ethanethioate
Amide	$C{-}\overset{O}{\overset{\Vert}{C}}{-}\ddot{N}$	*-amide*	$CH_3\overset{O}{\overset{\Vert}{C}}NH_2$ Ethanamide
Acid chloride	$C{-}\overset{O}{\overset{\Vert}{C}}{-}Cl$	*-oyl chloride*	$CH_3\overset{O}{\overset{\Vert}{C}}Cl$ Ethanoyl chloride

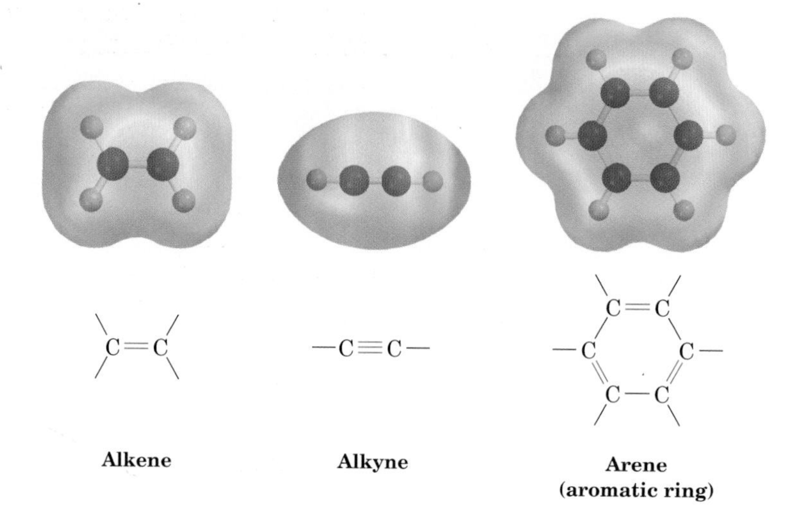

Functional Groups with Carbon Singly Bonded to an Electronegative Atom

Alkyl halides, alcohols, ethers, amines, thiols, and sulfides all have a carbon atom singly bonded to an electronegative atom—halogen, oxygen, nitrogen, or sulfur. Alkyl halides have a carbon atom bonded to a halogen (–X), alcohols have a carbon atom bonded to a hydroxyl (–OH) group, ethers have two carbon atoms bonded to the same oxygen, amines have a carbon atom bonded to a nitrogen, thiols have a carbon atom bonded to an –SH group, and sulfides have two carbon atoms bonded to the same sulfur. In all cases, the bonds are polar, with the carbon atom bearing a partial positive charge ($\delta+$) and the electronegative atom bearing a partial negative charge ($\delta-$).

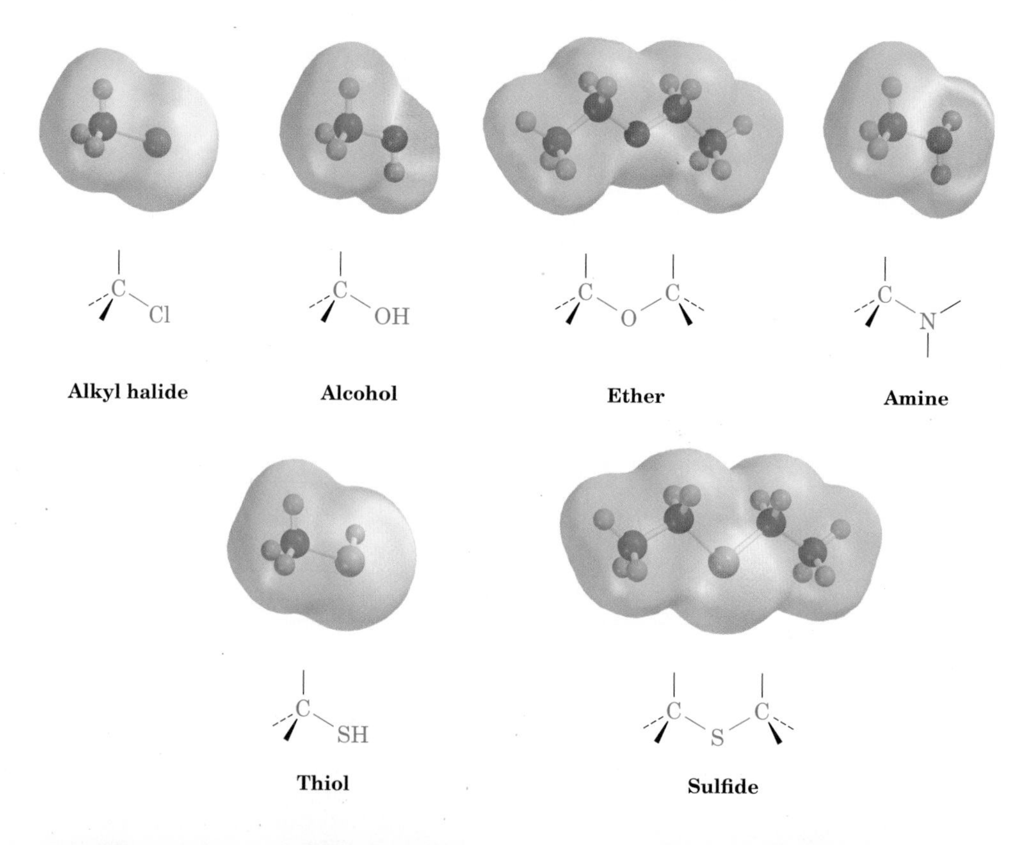

Functional Groups with a Carbon–Oxygen Double Bond (Carbonyl Groups)

The *carbonyl group*, C=O (pronounced car-bo-**neel**), is common to many of the families of compounds shown in Table 2.1. Carbon–oxygen double bonds are present in some of the most important compounds in organic chemistry. The chemistry of these compounds depends greatly on the identity of the atoms bonded to the carbonyl-group carbon. Accordingly, we subdivide these compounds into many different classes of functional groups. Aldehydes have at least one hydrogen bonded to the C=O, ketones have two carbons bonded to the C=O, carboxylic acids have one carbon and one –OH group bonded to the C=O, esters have one carbon and one ether-like oxygen bonded to the C=O, amides have one carbon and one nitrogen bonded to the C=O, acid chlorides have one carbon and one chlorine bonded to the C=O, and so on. The carbonyl carbon atom bears a partial positive charge ($\delta+$), and the oxygen bears a partial negative charge ($\delta-$).

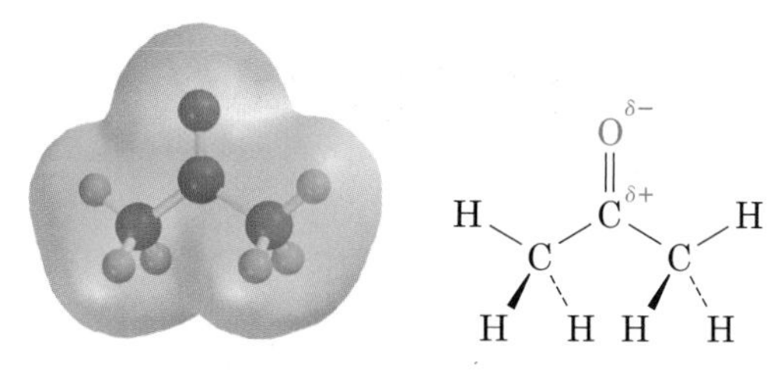

Acetone—a typical carbonyl compound

Aldehyde **Ketone** **Carboxylic acid** **Ester** **Amide** **Acid chloride**

PROBLEM 2.1 Circle and identify the functional groups in the following molecules:

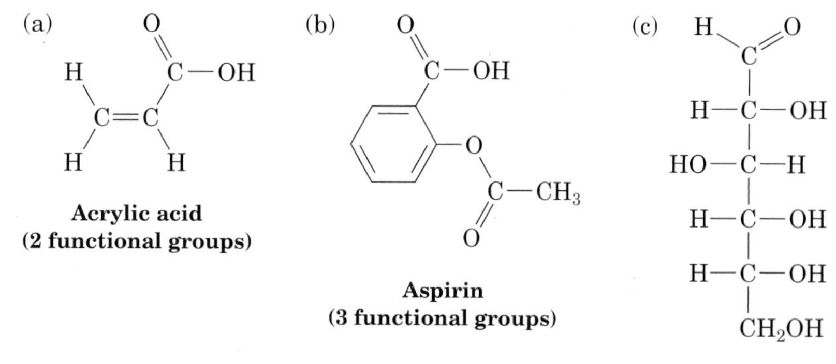

(a) **Acrylic acid (2 functional groups)**

(b) **Aspirin (3 functional groups)**

(c) **Glucose (6 functional groups)**

PROBLEM 2.2 Propose structures for simple molecules that contain the following functional groups:

(a) Alcohol (b) Aromatic ring (c) Carboxylic acid
(d) Amine (e) Both ketone and amine (f) Two double bonds

PROBLEM 2.3 Identify the functional groups in the following model of arecoline, a veterinary drug used to control worms in animals. Convert the drawing into a line-bond structure (gray = C, red = O, blue = N, ivory = H).

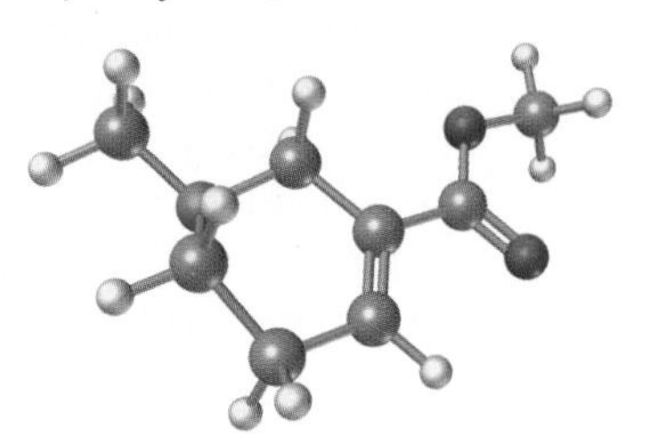

2.2 Alkanes and Alkyl Groups: Isomers

We saw in Section 1.7 that the C–C single bond in ethane results from σ (head-on) overlap of carbon sp^3 hybrid orbitals. If we imagine joining three, four, five, or even more carbon atoms by C–C single bonds, we generate the large family of molecules called **alkanes**.

```
   H         H  H          H  H  H           H  H  H  H
   |         |  |          |  |  |           |  |  |  |
H—C—H     H—C—C—H      H—C—C—C—H       H—C—C—C—C—H  . . . and so on
   |         |  |          |  |  |           |  |  |  |
   H         H  H          H  H  H           H  H  H  H

Methane     Ethane         Propane             Butane
```

Alkanes are often described as *saturated hydrocarbons*: **hydrocarbons** because they contain only carbon and hydrogen atoms; **saturated** because they have only C–C and C–H single bonds and thus contain the maximum possible number of hydrogens per carbon. They have the general formula C_nH_{2n+2}, where n is any integer. Alkanes are also occasionally called **aliphatic** compounds, a word derived from the Greek *aleiphas*, meaning "fat." We'll see in Chapter 16 that animal fats contain long carbon chains similar to alkanes.

Think about the ways that carbon and hydrogen can combine to make alkanes. With one carbon and four hydrogens, only one structure is possible: methane, CH_4. Similarly, there is only one possible combination of two carbons with six hydrogens (ethane, CH_3CH_3) and only one possible combination of three carbons with eight hydrogens (propane, $CH_3CH_2CH_3$). If larger numbers of carbons and hydrogens combine, however, more than one kind of molecule can form. For example, there are *two* ways that molecules with the formula C_4H_{10} can form: the four carbons can be in a row (butane), or they can branch (isobutane). Similarly, there are three ways in which C_5H_{12} molecules can form, and so on for larger alkanes:

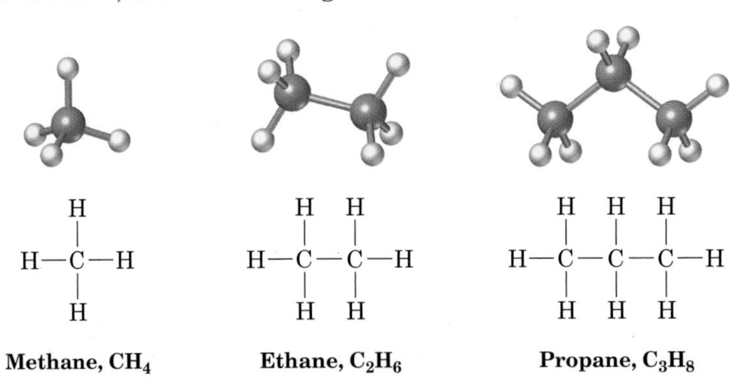

Methane, CH_4 **Ethane, C_2H_6** **Propane, C_3H_8**

C_4H_{10}

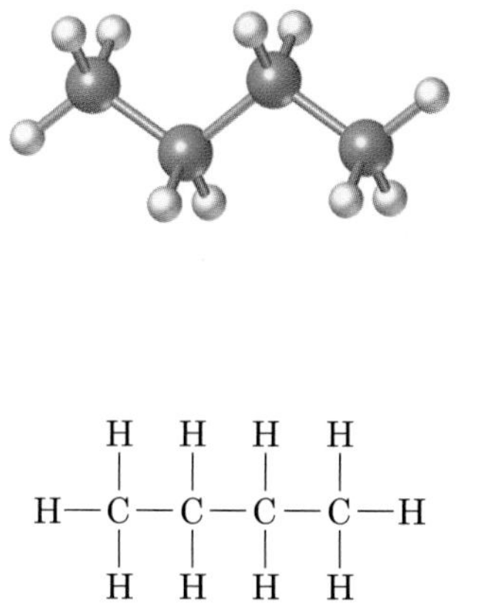

Butane

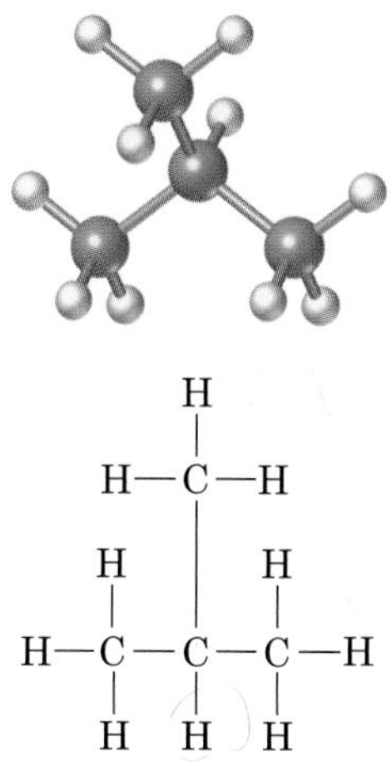

Isobutane
(2-methylpropane)

C_5H_{12}

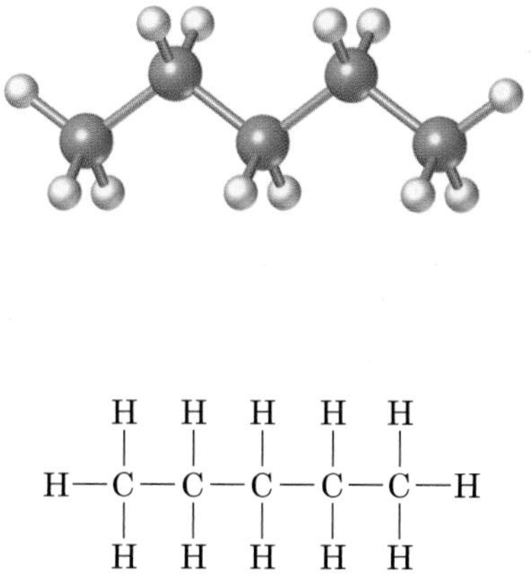

Pentane

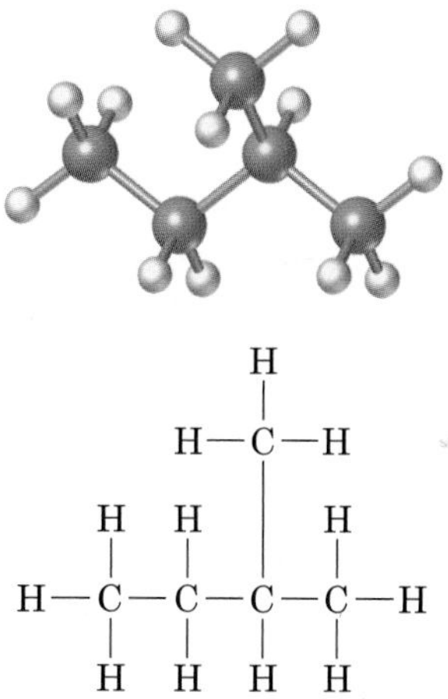

2-Methylbutane

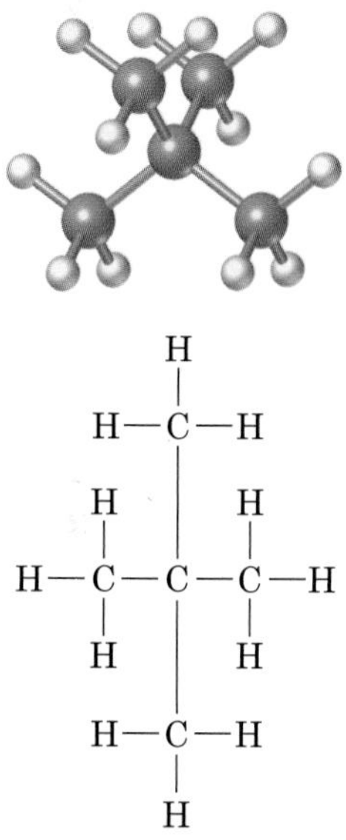

2,2-Dimethylpropane

Compounds like butane, whose carbons are connected in a row, are called **straight-chain alkanes**, or **normal (*n*) alkanes**, whereas compounds with branched carbon chains, such as isobutane (2-methylpropane), are called **branched-chain alkanes**.

Compounds like the two C_4H_{10} molecules, which have the same formula but different structures, are called *isomers*, from the Greek *isos* + *meros*, meaning "made of the same parts." **Isomers** have the same numbers and kinds of atoms but differ in the way the atoms are arranged. Compounds like butane and isobutane, whose atoms are connected differently, are called **constitutional isomers**. We'll see shortly that other kinds of isomerism are also possible, even among compounds whose atoms are connected in the same order.

A given alkane can be arbitrarily shown in many ways. For example, the straight-chain, four-carbon alkane called butane can be represented by any of the structures shown in Figure 2.2. These structures aren't intended to imply any particular three-dimensional geometry for butane; they only indicate the connections among atoms. In practice, chemists rarely draw all the bonds in a molecule and usually refer to butane by the *condensed structure*, $CH_3CH_2CH_2CH_3$ or $CH_3(CH_2)_2CH_3$. In such representations, the C–C and C–H bonds are "understood" rather than shown. If a carbon has three hydrogens bonded to it, we write CH_3; if a carbon has two hydrogens bonded to it, we write CH_2, and so on. Still more simply, butane can even be represented as n-C_4H_{10}, where n signifies *normal*, straight-chain butane.

FIGURE 2.2 Some representations of butane (n-C_4H_{10}). The molecule is the same regardless of how it's drawn. These structures imply only that butane has a continuous chain of four carbon atoms.

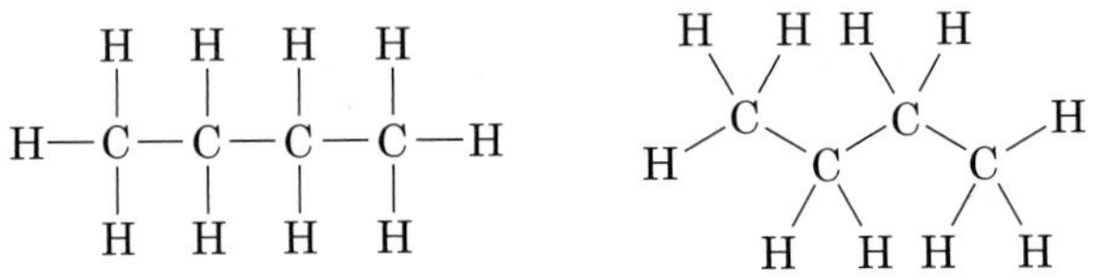

$CH_3—CH_2—CH_2—CH_3$ $CH_3CH_2CH_2CH_3$ $CH_3(CH_2)_2CH_3$

Straight-chain alkanes are named according to the number of carbon atoms they contain, as shown in Table 2.2. With the exception of the first four compounds—methane, ethane, propane, and butane—whose names have historical origins, the alkanes are named based on Greek numbers, according to the number of carbons. The suffix *-ane* is added to the end of each name to identify the molecule as an alkane.

If a hydrogen atom is removed from an alkane, the partial structure that remains is called an **alkyl group**. Alkyl groups are named by replacing the *-ane* ending of the parent alkane with an *-yl* ending. For example, removal of a hydrogen atom from methane, CH_4, generates a *methyl group*, $–CH_3$, and removal of a hydrogen atom from ethane, CH_3CH_3, generates an *ethyl group*, $–CH_2CH_3$. Similarly, removal of a hydrogen atom from the end carbon of any n-alkane gives the series of n-alkyl groups shown in Table 2.3.

TABLE 2.2 Names of Straight-Chain Alkanes

Number of carbons (n)	Name	Formula (C_nH_{2n+2})	Number of carbons (n)	Name	Formula (C_nH_{2n+2})
1	Methane	CH_4	9	Nonane	C_9H_{20}
2	Ethane	C_2H_6	10	Decane	$C_{10}H_{22}$
3	Propane	C_3H_8	11	Undecane	$C_{11}H_{24}$
4	Butane	C_4H_{10}	12	Dodecane	$C_{12}H_{26}$
5	Pentane	C_5H_{12}	13	Tridecane	$C_{13}H_{28}$
6	Hexane	C_6H_{14}	20	Icosane	$C_{20}H_{42}$
7	Heptane	C_7H_{16}	21	Henicosane	$C_{21}H_{44}$
8	Octane	C_8H_{18}	30	Triacontane	$C_{30}H_{62}$

TABLE 2.3 Some Straight-Chain Alkyl Groups

Alkane	Name	Alkyl group	Name (abbreviation)
CH_4	Methane	$-CH_3$	Methyl (Me)
CH_3CH_3	Ethane	$-CH_2CH_3$	Ethyl (Et)
$CH_3CH_2CH_3$	Propane	$-CH_2CH_2CH_3$	Propyl (Pr)
$CH_3CH_2CH_2CH_3$	Butane	$-CH_2CH_2CH_2CH_3$	Butyl (Bu)
$CH_3CH_2CH_2CH_2CH_3$	Pentane	$-CH_2CH_2CH_2CH_2CH_3$	Pentyl

Just as *n*-alkyl groups are generated by removing a hydrogen from an *end* carbon, branched alkyl groups are generated by removing a hydrogen atom from an *internal* carbon. Two 3-carbon alkyl groups and four 4-carbon alkyl groups are possible (Figure 2.3).

FIGURE 2.3 Generation of straight-chain and branched-chain alkyl groups from alkanes.

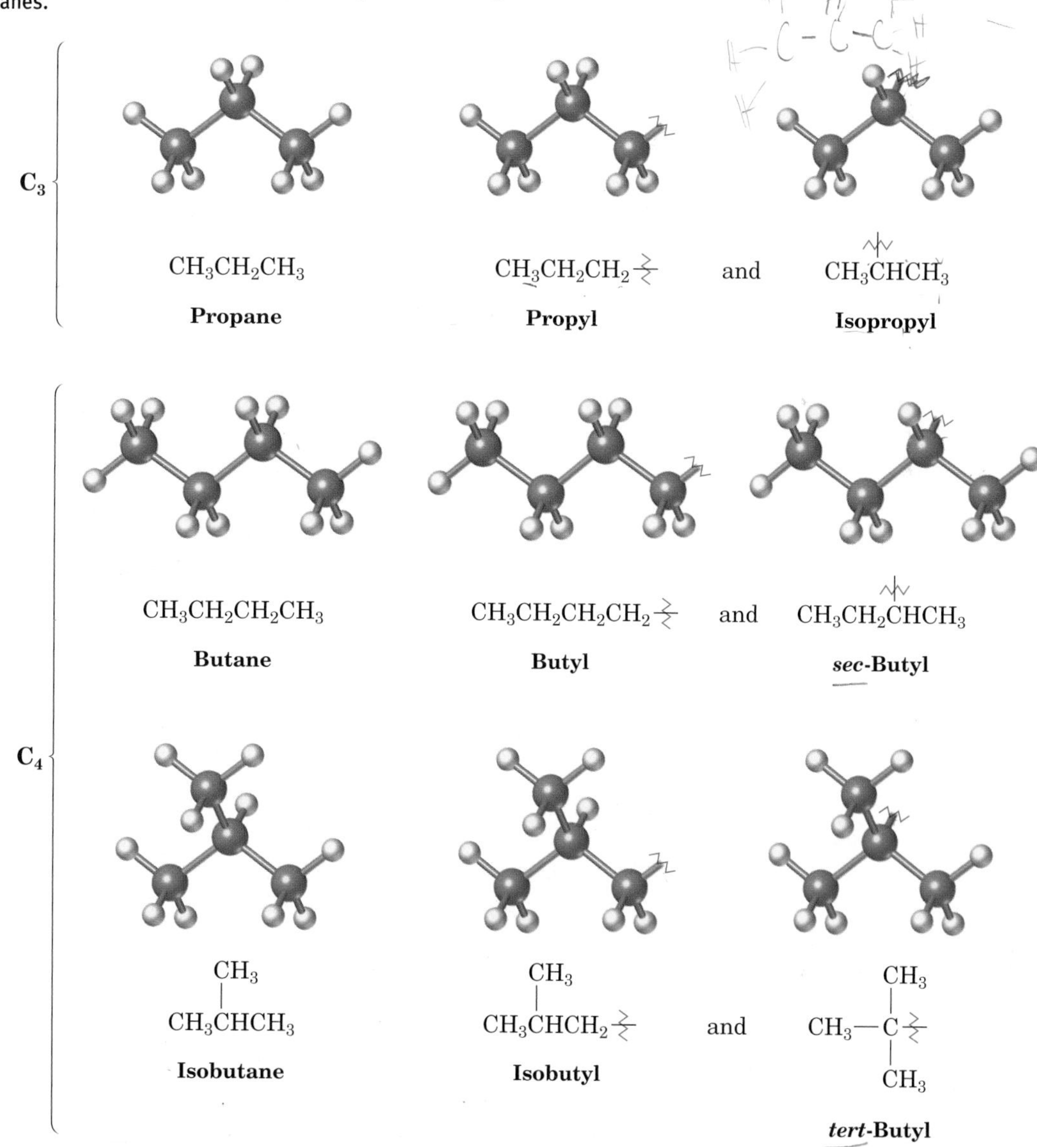

One further word about naming alkyl groups: the prefixes used for the C_4 alkyl groups in Figure 2.3—*sec* (for secondary) and *tert* (for tertiary)—refer to the number of other carbon atoms attached to the branched carbon atom. There are four possibilities: primary (1°), secondary (2°), tertiary (3°), and quaternary (4°).

$$\begin{array}{c} \mathrm{H} \\ | \\ \mathrm{R{-}C{-}H} \\ | \\ \mathrm{H} \end{array} \qquad \begin{array}{c} \mathrm{H} \\ | \\ \mathrm{R{-}C{-}H} \\ | \\ \mathrm{R} \end{array} \qquad \begin{array}{c} \mathrm{R} \\ | \\ \mathrm{R{-}C{-}H} \\ | \\ \mathrm{R} \end{array} \qquad \begin{array}{c} \mathrm{R} \\ | \\ \mathrm{R{-}C{-}R} \\ | \\ \mathrm{R} \end{array}$$

Primary carbon (1°) is bonded to one other carbon.

Secondary carbon (2°) is bonded to two other carbons.

Tertiary carbon (3°) is bonded to three other carbons.

Quaternary carbon (4°) is bonded to four other carbons.

The symbol **R** is used here and throughout this text to represent a *generalized* alkyl group. The R group can be methyl, ethyl, or any of a multitude of others. You might think of **R** as representing the **R**est of the molecule, which we aren't bothering to specify.

$$\begin{array}{c} \mathrm{H} \\ | \\ \mathrm{R{-}C{-}OH} \\ | \\ \mathrm{H} \end{array} \qquad \begin{array}{l} \mathrm{\ \ \ \ \ \ \ CH_3} \\ \mathrm{\ \ \ \ \ \ \ \ |} \\ \mathrm{CH_3CHCH_2CH_2OH} \end{array}$$

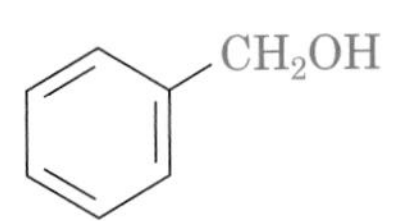

General class of primary alcohols, RCH_2OH

Specific examples of primary alcohols, RCH_2OH

PRACTICE PROBLEM 2.1

Propose structures for two isomers with the formula C_2H_6O.

STRATEGY We know that carbon forms four bonds, oxygen forms two, and hydrogen forms one. Put the pieces together by trial and error, along with intuition.

SOLUTION There are two possibilities:

$$\begin{array}{c} \mathrm{H\ \ \ H} \\ |\ \ \ \ | \\ \mathrm{H{-}C{-}C{-}O{-}H} \\ |\ \ \ \ | \\ \mathrm{H\ \ \ H} \end{array} \quad \text{and} \quad \begin{array}{c} \mathrm{H\ \ \ \ \ \ \ H} \\ |\ \ \ \ \ \ \ \ | \\ \mathrm{H{-}C{-}O{-}C{-}H} \\ |\ \ \ \ \ \ \ \ | \\ \mathrm{H\ \ \ \ \ \ \ H} \end{array}$$

PROBLEM 2.4 Draw structures for the five isomers of C_6H_{14}.

PROBLEM 2.5 Draw structures that meet the following descriptions:
(a) Three isomers with the formula C_8H_{18}
(b) Two isomers with the formula $C_4H_8O_2$

PROBLEM 2.6 Draw the eight possible five-carbon alkyl groups (pentyl isomers).

PROBLEM 2.7 Draw alkanes that meet the following descriptions:
(a) An alkane with two tertiary carbons
(b) An alkane that contains an isopropyl group
(c) An alkane that has one quaternary and one secondary carbon

PROBLEM 2.8 Identify the carbon atoms in the following molecules as primary, secondary, tertiary, or quaternary:

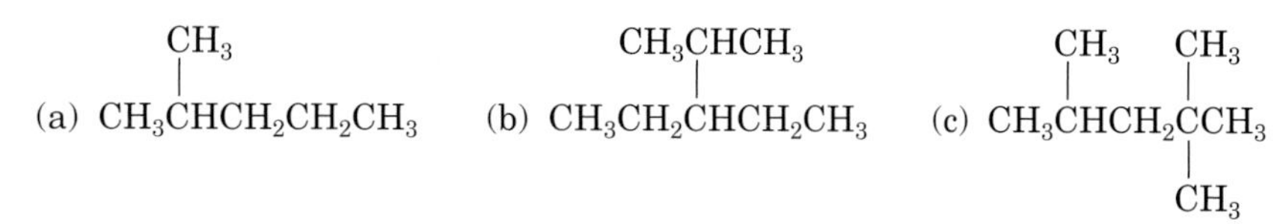

2.3 Naming Branched-Chain Alkanes

In earlier times, when few pure organic chemicals were known, new compounds were named at the whim of their discoverer. Thus, urea (CH_4N_2O) is a crystalline substance isolated from urine, and barbituric acid is a tranquilizing agent said to be named by its discoverer in honor of his friend Barbara. As the science of organic chemistry grew in the 19th century, so too did the number of known compounds and the need for a systematic method of naming them. The system of naming used in this book is that devised by the International Union of Pure and Applied Chemistry (IUPAC, usually spoken as **eye**-you-pac).

A chemical name has three parts in the **IUPAC system of nomenclature**: prefix, parent, and suffix. The parent name selects a main part of the molecule and tells how many carbon atoms are in that part, the suffix identifies the functional-group family that the molecule belongs to, and the prefix specifies the location(s) of various substituents on the main part:

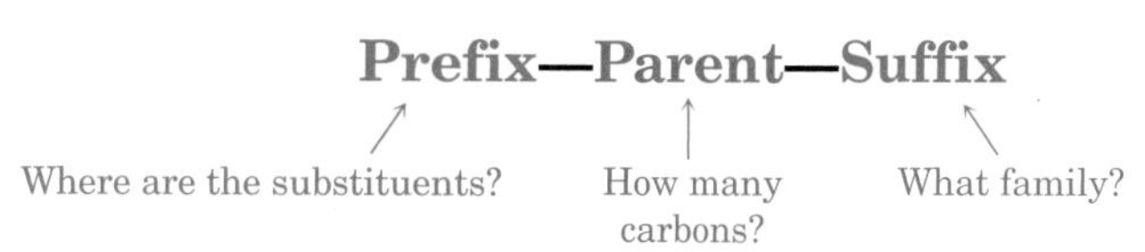

As we cover new functional groups in later chapters, the applicable IUPAC rules of nomenclature will be given. In addition, Appendix A gives an overall view of organic nomenclature and shows how compounds that contain more than one functional group are named. For the present, let's see how to name branched-chain alkanes. All but the most complex branched-chain alkanes can be named by following four steps:

STEP 1 **Find the parent hydrocarbon.**

(a) Find the *longest continuous carbon chain* in the molecule and use the name of that chain as the parent name. The longest chain may not always be obvious; you may have to "turn corners":

$$\begin{array}{c} \quad\quad\quad\quad\quad CH_2CH_3 \\ \quad\quad\quad\quad\quad | \\ CH_3CH_2CH_2CH{-}CH_3 \end{array}$$ Named as a substituted hexane

(b) If two chains of equal length are present, choose the one with the larger number of branch points as the parent:

Named as a hexane with *two* substituents NOT as a hexane with *one* substituent

STEP 2 **Number the atoms in the main chain.**

Beginning at the end *nearer the first branch point*, number each carbon atom in the parent chain:

```
       1CH3                                   7CH3
        |                                      |
       2CH2                                   6CH2
        |                                      |
CH3—CHCH—CH2CH3          NOT          CH3—CHCH—CH2CH3
     3  |4                                 5  |4
        CH2CH2CH3                             CH2CH2CH3
        5   6   7                             3   2   1
```

The first branch occurs at C3 in the proper system of numbering but at C4 in the improper system.

STEP 3 **Identify and number the substituents.**

Assign a number to each substituent according to its point of attachment on the parent chain. If there are two substituents on the same carbon, assign them both the same number. There must always be as many numbers in the name as there are substituents.

```
   9  8
   CH3CH2           CH3  CH2CH3
     |               |    |
CH3—CHCH2CH2CH—CHCH2CH3
     7  6  5   4    3  2   1
```

Substituents:
On C3, CH_2CH_3 (3-ethyl)
On C4, CH_3 (4-methyl)
On C7, CH_3 (7-methyl)

```
              CH3
               |
  CH3CH2—C—CH2CHCH3
   6   5   4|   3  |2  1
             CH2      CH3
              |
             CH3
```

Substituents:
On C2, CH_3 (2-methyl)
On C4, CH_3 (4-methyl)
On C4, CH_2CH_3 (4-ethyl)

STEP 4 **Write the name as a single word.**

Use hyphens to separate the various prefixes and commas to separate numbers. If two or more different side chains are present, cite them in alphabetical order. If two or more identical side chains are present, use one of the prefixes *di-*, *tri-*, *tetra-*, and so forth. Don't use these prefixes for alphabetizing, though.

```
              2  1
             CH2CH3
 6   5   4   3|
CH3CH2CH2CH—CH3
```

3-Methylhexane

```
    CH3
     |
CH3CHCHCH2CH2CH3
 1   2  3|  4   5   6
        CH2CH3
```

3-Ethyl-2-methylhexane

```
       1CH3
        |
       2CH2
        |
CH3—CHCH—CH2CH3
     3  |4
        CH2CH2CH3
        5   6   7
```

4-Ethyl-3-methylheptane

```
   9  8
   CH3CH2           CH3  CH2CH3
     |               |    |
CH3—CHCH2CH2CH—CHCH2CH3
     7  6  5   4    3  2   1
```

3-Ethyl-4,7-dimethylnonane

PRACTICE PROBLEM 2.2

What is the IUPAC name of the following alkane?

$$\begin{array}{l} \quad\; CH_2CH_3 \qquad\quad CH_3 \\ \quad\;\; | \qquad\qquad\qquad\quad | \\ CH_3CHCH_2CH_2CH_2CHCH_3 \end{array}$$

STRATEGY The molecule has a chain of eight carbons (octane) with two methyl substituents. Numbering from the end nearer the first methyl substituent indicates that the methyls are at C2 and C6.

SOLUTION The alkane is named 2,6-dimethyloctane.

$$\begin{array}{l} \quad\;\; {}^{7}\;\;\; {}^{8} \\ \quad\; CH_2CH_3 \qquad\quad CH_3 \\ \quad\;\; | \qquad\qquad\qquad\quad | \\ CH_3CHCH_2CH_2CH_2CHCH_3 \\ \quad\;\; {}_{6}\;\; {}_{5}\;\;\; {}_{4}\;\;\; {}_{3}\;\;\; {}_{2}\;\; {}_{1} \end{array}$$

PRACTICE PROBLEM 2.3

Draw the structure of 3-isopropyl-2-methylhexane.

STRATEGY First, look at the parent name (hexane) and draw its carbon structure:

C—C—C—C—C—C **Hexane**

Next, find the substituents (3-isopropyl and 2-methyl), and place them on the proper carbons:

$$\begin{array}{l} \qquad\;\; CH_3CHCH_3 \longleftarrow \text{An isopropyl group at C3} \\ \qquad\qquad\; | \\ C—C—C—C—C—C \\ {}_{1}\quad {}_{2}\quad {}_{3}\quad {}_{4}\quad {}_{5}\quad {}_{6} \\ \qquad | \\ \quad\;\; CH_3 \longleftarrow \text{A methyl group at C2} \end{array}$$

Finally, add hydrogens to complete the structure.

SOLUTION

$$\begin{array}{l} \quad\; CH_3CHCH_3 \\ \qquad\quad | \\ CH_3CHCHCH_2CH_2CH_3 \qquad \textbf{3-Isopropyl-2-methylhexane} \\ \qquad | \\ \quad\;\; CH_3 \end{array}$$

PROBLEM 2.9 Give IUPAC names for the following alkanes:

(a) The three isomers of C_5H_{12}

(b)
$$\begin{array}{l} \qquad\qquad\quad CH_3 \\ \qquad\qquad\quad | \\ CH_3CH_2CHCHCH_3 \\ \qquad\qquad\qquad | \\ \qquad\qquad\quad CH_2CH_3 \end{array}$$

(c)
$$\begin{array}{l} \quad\;\; CH_3 \quad\; CH_3 \\ \quad\;\; | \qquad\quad | \\ CH_3CHCH_2CHCH_3 \end{array}$$

(d)
$$\begin{array}{l} \qquad\quad CH_3 \qquad\qquad CH_2CH_3 \\ \qquad\quad | \qquad\qquad\quad | \\ CH_3—C—CH_2CH_2CHCH_3 \\ \qquad\quad | \\ \qquad\quad CH_3 \end{array}$$

PROBLEM 2.10 Draw structures corresponding to the following IUPAC names:

(a) 3,4-Dimethylnonane (b) 3-Ethyl-4,4-dimethylheptane
(c) 2,2-Dimethyl-4-propyloctane (d) 2,2,4-Trimethylpentane

PROBLEM 2.11 Name the following alkane:

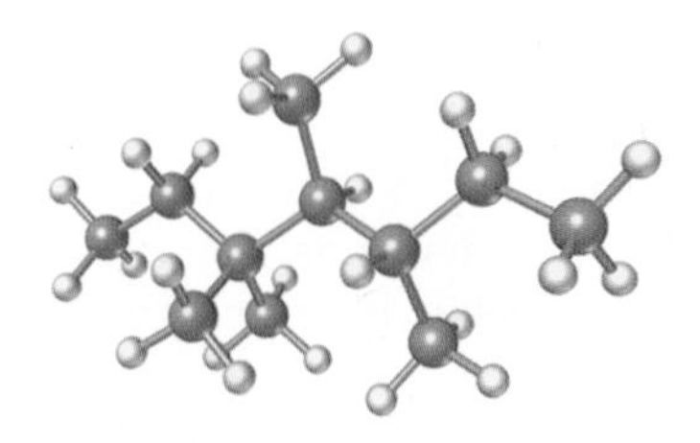

2.4 Properties of Alkanes

Many alkanes occur naturally in the plant and animal world. For example, the waxy coating on cabbage leaves contains nonacosane ($C_{29}H_{60}$), and the wood oil of the Jeffrey pine common to the Sierra Nevada Mountains of California contains heptane (C_7H_{16}). Regardless of the molecule, however, the average C–C bond parameters are nearly the same in all alkanes, with bond lengths of 154 ± 1 pm and bond strengths of 355 ± 20 kJ/mol (85 ± 5 kcal/mol). C–H bond parameters are also nearly constant at 109 ± 1 pm and 400 ± 20 kJ/mol (95 ± 5 kcal/mol).

By far, the major sources of alkanes are the world's natural gas and petroleum deposits. Laid down eons ago, these natural deposits are derived from the decomposition of plant and animal matter, primarily of marine origin. *Natural gas* consists chiefly of methane but also contains ethane, propane, and butane. *Petroleum* is a complex mixture of hydrocarbons that must be separated into fractions and then further refined before it can be used.

This separation is easy because alkanes show regular increases in both boiling point and melting point as molecular weight increases (Figure 2.4). Separating molecules based on differences in boiling points is called **distillation**. When multiple fractions are separated during a single distillation, the process is called **fractional distillation**.

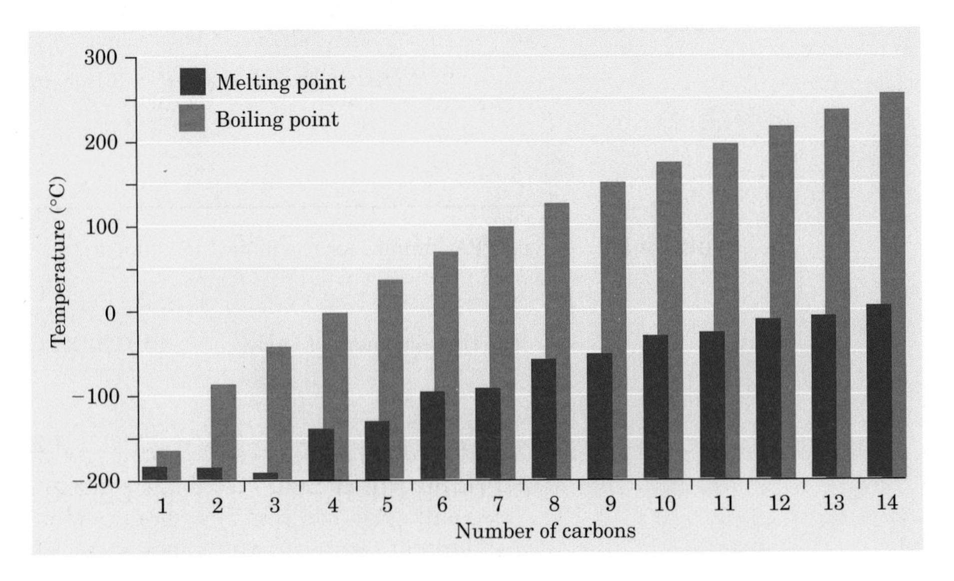

FIGURE 2.4 A plot of melting and boiling points versus number of carbons for the C_1–C_{14} alkanes. There is a regular increase with molecular size.

Modern petroleum refining begins by fractional distillation of crude oil into three principal cuts according to their boiling points (bp): gasoline (bp 20–200 °C), kerosene (bp 175–275 °C), and heating oil or diesel fuel (bp 250–400 °C). Finally, distillation under reduced pressure yields lubricating oils and waxes and leaves an undistillable tarry residue called asphalt. This process is illustrated in Figure 2.5.

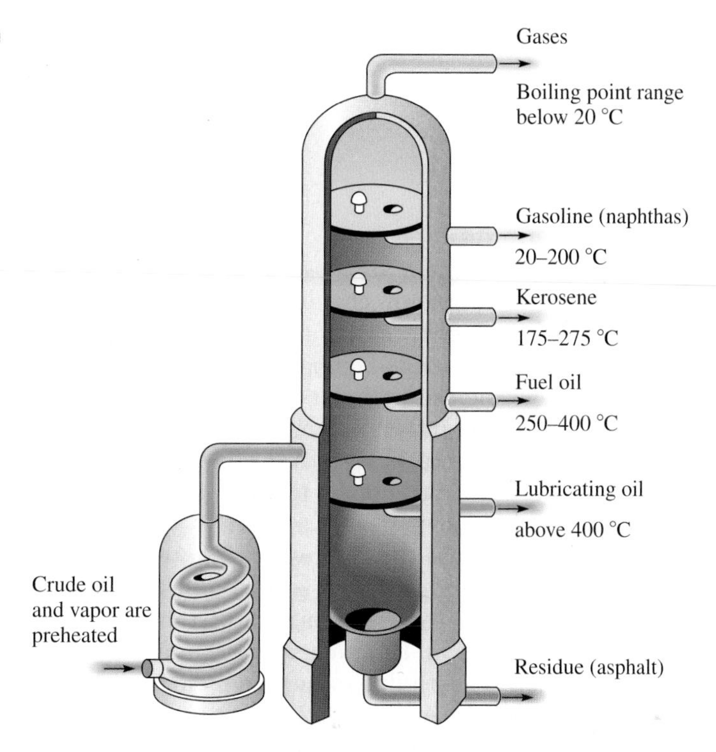

FIGURE 2.5 Fractional distillation separates crude mixtures of alkanes based on differences in boiling point. Temperature decreases in the tower with increasing height, allowing condensation of the gases and fractional collection of components.

These alkane fractions are the raw materials that become the building blocks of polymers, pharmaceuticals, industrial solvents, and sources of energy. Given our reliance on these alkanes, you might be surprised that alkanes are sometimes referred to as *paraffins*, a word derived from the Latin *parum affinis*, meaning "slight affinity." This term aptly describes their behavior, for alkanes show little chemical affinity for other substances and are inert to most laboratory reagents. They do, however, react under appropriate conditions with oxygen, chlorine, and a few other substances.

Combustion

The reaction of an alkane with O_2 occurs during combustion in an engine or furnace when the alkane is used as a fuel. Carbon dioxide and water are formed as products, and a large amount of heat is released. For example, methane (natural gas) reacts with oxygen according to the equation

$$CH_4 + 2\,O_2 \longrightarrow CO_2 + 2\,H_2O + 890\text{ kJ/mol (213 kcal/mol)}$$

Halogenation

The reaction of an alkane with Cl_2 occurs when a mixture of the two is irradiated with ultraviolet light (denoted $h\nu$, where ν is the lowercase Greek letter nu). Depending on the relative amounts of the two reactants and the time allowed for reaction, sequential replacement of the alkane hydrogen atoms by chlorine occurs, leading to a mixture of chlorinated products. Methane, for instance, reacts with chlorine to yield a mixture of chloromethane (CH_3Cl), dichloromethane (CH_2Cl_2), trichloromethane ($CHCl_3$), and tetrachloromethane (CCl_4).

$$CH_4 + Cl_2 \xrightarrow{h\nu} CH_3Cl + HCl$$
$$CH_3Cl \xrightarrow{h\nu,\ Cl_2} CH_2Cl_2 + HCl$$
$$CH_2Cl_2 \xrightarrow{h\nu,\ Cl_2} CHCl_3 + HCl$$
$$CHCl_3 \xrightarrow{h\nu,\ Cl_2} CCl_4 + HCl$$

Refining

Because linear alkanes obtained from the distillation of crude oil are such poor fuels, petroleum chemists have devised numerous refining methods for producing higher-quality fuels. One of these methods, *catalytic cracking*, involves taking the high-boiling kerosene cut (C_{11}–C_{14}) and "cracking" it into smaller molecules suitable for use in gasoline. The process takes place on a silica–alumina catalyst at temperatures of 400 to 500 °C, and the major products are light hydrocarbons in the C_3–C_5 range. These small hydrocarbons are then catalytically recombined to yield useful C_7–C_{10} alkanes. Reforming is a process used to convert C_6–C_8 alkanes to arenes, including benzene and toluene.

Because the end result of these refining processes are complex mixtures, the fuel industry has settled on a rating system, the **octane number**, that describes the performance of the mixture instead of the exact chemical composition. Heptane, a particularly poor fuel, has an octane number of 0; 2,2,4-trimethylpentane (commonly known as isooctane) has an octane number of 100. Adding a 92 octane fuel to our cars corresponds to a mixture that performs similarly to a mixture of 92% 2,2,4-trimethylpentane and 8% heptane. Octane numbers higher than 100 are measured by adding a performance enhancer, tetraethyllead, to isooctane.

$CH_3CH_2CH_2CH_2CH_2CH_2CH_3$

Heptane

$(CH_3)_3CCH_2CH(CH_3)CH_3$

Isooctane (2,2,4-trimethylpentane)

2.5 Conformations of Ethane

We saw earlier that C–C bonds in alkanes result from σ overlap of two tetrahedrally oriented sp^3 orbitals. Let's now look into the three-dimensional consequences of such bonding. What are the spatial relationships between the hydrogens on one carbon and the hydrogens on the other?

We know that a σ bond results from the head-on overlap of two atomic orbitals. Because this motion does not change the amount of orbital overlap, *rotation* is possible around C–C single bonds: a slice cut through the bond looks like a circle (Figure 2.6). The different arrangements of atoms that result from rotation around a single bond are called **conformations**, and they interconvert too rapidly for them to be isolated.

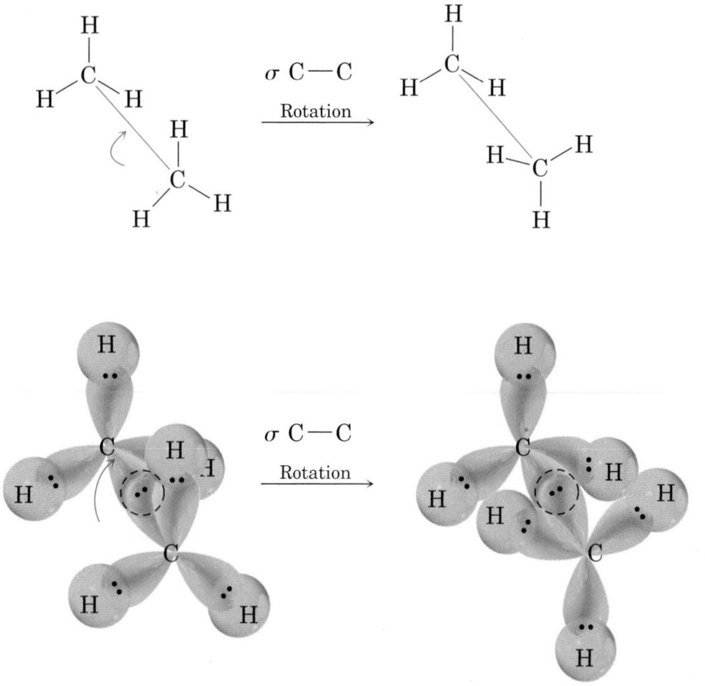

FIGURE 2.6 Two conformations of ethane. Rotation around the C–C single bond (green) interconverts the different conformations. A slice through the bond is symmetrical; a circle (red).

Chemists represent different conformations in two ways, as shown in Figure 2.7. **Sawhorse representations** view the C–C bond from an oblique angle and indicate spatial relationships by showing all the C–H bonds. **Newman projections** view the C–C bond end-on. In Newman projections, the carbon in front is represented by a dot from which three bonds radiate. The carbon in back is represented by a circle. Bonds from the rear carbon are shown radiating from the edge of the circle.

FIGURE 2.7 A sawhorse representation and a Newman projection of ethane. The sawhorse projection views the molecule from an oblique angle, while the Newman projection views the molecule end-on.

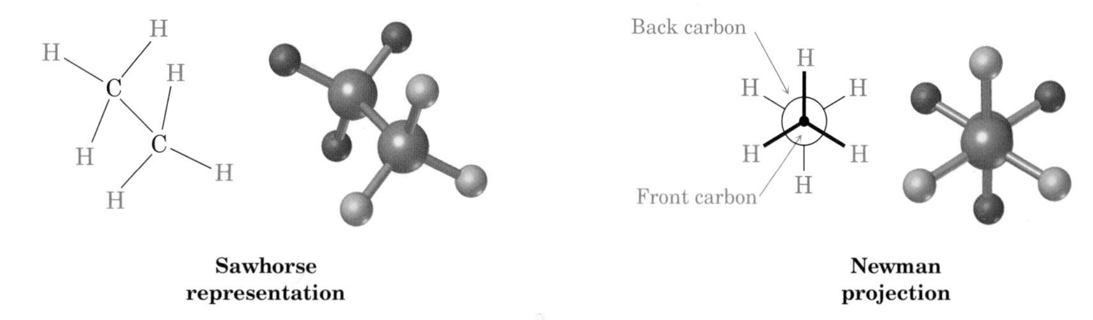

Let's look at the C–C bond rotation in ethane using Newman projections. Despite what we've just said, we actually don't observe *perfectly* free rotation in ethane. Experiments show that there is a slight (12 kJ/mol; 2.9 kcal/mol) barrier to rotation and that some conformations are more stable than others. The lowest-energy, most stable conformation is the one in which all six C–H bonds are as far away from one another as possible (**staggered** when viewed end-on in a Newman projection). The highest-energy, least stable conformation is the one in which the six C–H bonds are as close as possible (**eclipsed** in a Newman projection). At any given instant, about 99% of ethane molecules have an approximately staggered conformation, and only about 1% are close to the eclipsed conformation (Figure 2.8).

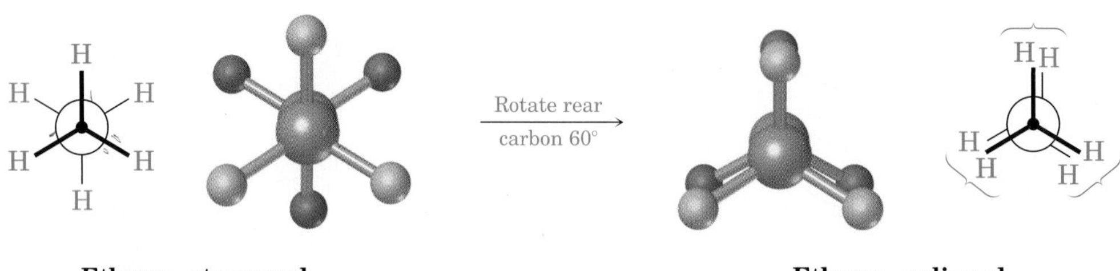

FIGURE 2.8 Staggered and eclipsed conformations of ethane. The staggered conformation is lower in energy and more stable by 12.0 kJ/mol.

What is true for ethane is also true for propane, butane, and all higher alkanes. The most favored conformation for any alkane is the one in which all bonds have staggered arrangements (Figure 2.9).

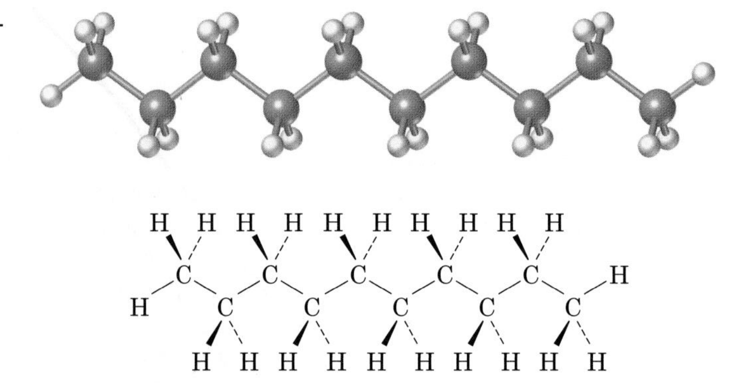

FIGURE 2.9 The most stable conformation of any alkane is the one in which the bonds on adjacent carbons are staggered and the carbon chain is fully extended, as in this structure of decane.

PROBLEM 2.12 Sight along a C–C bond of propane and draw a Newman projection of the most stable conformation. Draw a Newman projection of the least stable conformation.

PROBLEM 2.13 Looking along the C2–C3 bond of butane, there are two different staggered conformations and two different eclipsed conformations. Draw them.

PROBLEM 2.14 Which of the butane conformations you drew in Problem 2.13 do you think is the most stable? Explain.

2.6 Drawing Chemical Structures

In the structures we've been using, a line between atoms has represented the two electrons in a covalent bond. Drawing every bond and identifying every atom are tedious tasks, however, so chemists have devised a shorthand way of drawing **skeletal structures** that greatly simplifies matters. Drawing skeletal structures is easy:

- Carbon atoms usually aren't shown. Instead, a carbon atom is assumed to be at the intersection of two lines (bonds) and at the end of each line. Occasionally, a carbon atom might be indicated for emphasis or clarity.
- Hydrogen atoms bonded to carbon aren't shown. Because carbon always has a valence of four, we mentally supply the correct number of hydrogen atoms for each carbon.
- All atoms other than carbon and hydrogen *are* shown. (Hydrogen atoms attached to all noncarbon atoms must also be shown.)

The following structures give some examples.

Isoprene, C_5H_8

Methylcyclohexane, C_7H_{14}

PRACTICE PROBLEM 2.4

Carvone, a substance responsible for the odor of spearmint, has the following structure. Tell how many hydrogens are bonded to each carbon, and give the molecular formula of carvone.

Carvone

STRATEGY Remember that the end of a line represents a carbon atom with three hydrogens, CH_3; a two-way intersection is a carbon atom with two hydrogens, CH_2; a three-way intersection is a carbon atom with one hydrogen, CH; and a four-way intersection is a carbon atom with no attached hydrogens.

SOLUTION

Carvone, $C_{10}H_{14}O$

PROBLEM 2.15 Convert the following skeletal structures into molecular formulas, and tell how many hydrogens are bonded to each carbon:

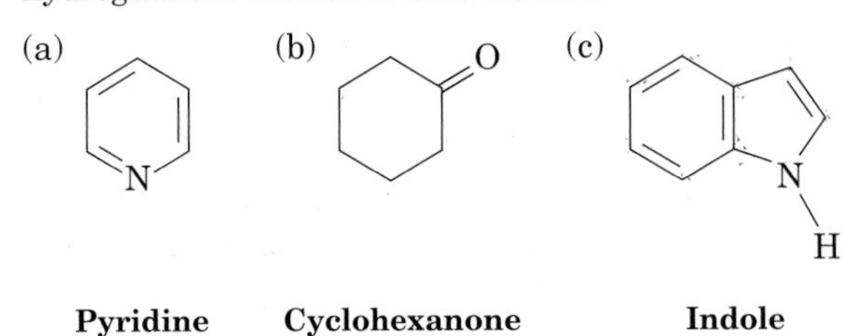

PROBLEM 2.16 Propose skeletal structures for the following molecular formulas:
(a) C_4H_8 (b) C_3H_6O (c) C_4H_9Cl

PROBLEM 2.17 The following molecular model is a representation of *para*-aminobenzoic acid (PABA), the active ingredient in many sunscreens. Indicate the positions of the multiple bonds, and draw a skeletal structure (gray = C, red = O, blue = N, ivory = H).

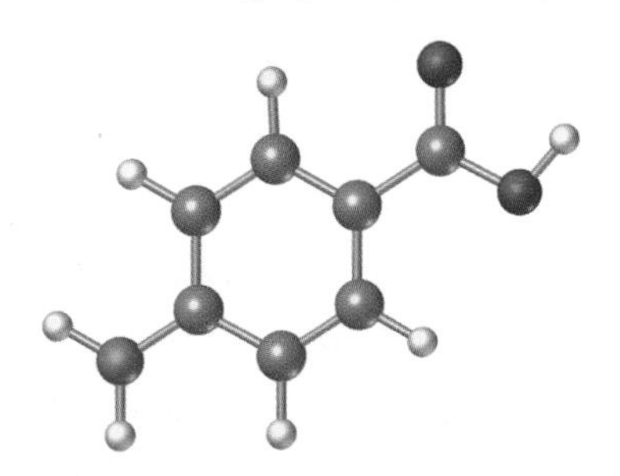

2.7 Cycloalkanes

Thus far, we've discussed only open-chain alkanes, but chemists have known for more than a century that compounds with *rings* of carbon atoms also exist. Such compounds are called **cycloalkanes**, or *alicyclic* (*ali*phatic *cyclic*) compounds. Since cycloalkanes consist of rings of $-CH_2-$ units, they have the general formula $(CH_2)_n$, or C_nH_{2n}, and are represented by polygons in skeletal drawings:

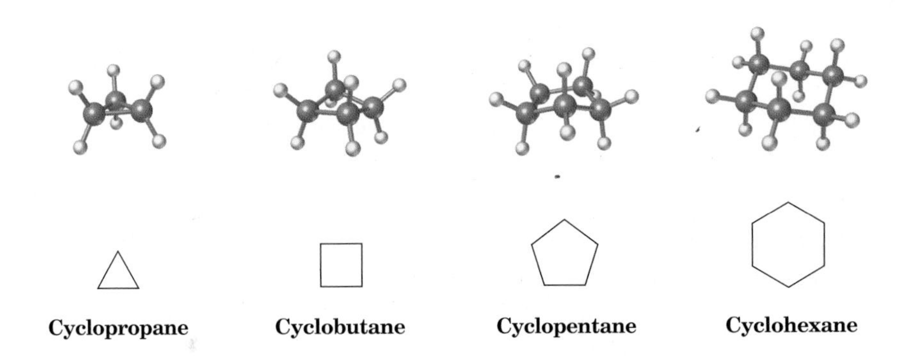

Substituted cycloalkanes are named by rules similar to those for open-chain alkanes. For most compounds, there are only two steps:

STEP 1 Count the number of carbon atoms in the ring, and add the prefix *cyclo-* to the name of the corresponding alkane. If a substituent is present on the ring, the compound is named as an alkyl-substituted cycloalkane rather than as a cycloalkyl-substituted alkane.

$-CH_3$ **Methylcyclopentane**

STEP 2 For substituted cycloalkanes, start at a point of attachment and number around the ring. If two substituents are present, begin numbering at the group that has alphabetical priority and proceed around the ring so as to give the second substituent the lowest number.

1,3-Dimethylcyclohexane *NOT* **1,5-Dimethylcyclohexane**

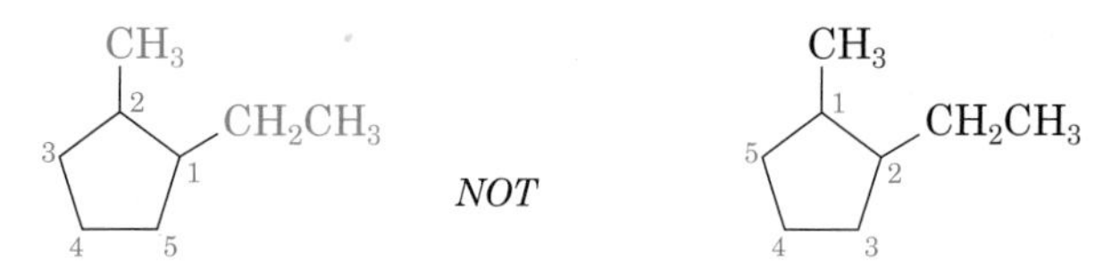

1-Ethyl-2-methylcyclopentane **2-Ethyl-1-methylcyclopentane**

PROBLEM 2.18 Give IUPAC names for the following cycloalkanes:

(a) CH_3, H_3C (b) CH_2CH_3, CH_3 (c) CH_3, CH_3

PROBLEM 2.19 Draw structures corresponding to the following IUPAC names:

(a) 1-*tert*-Butyl-2-methylcyclopentane (b) 1,1-Dimethylcyclobutane
(c) 1-Ethyl-4-isopropylcyclohexane

2.8 Cis–Trans Isomerism in Cycloalkanes

In many respects, the behavior of cycloalkanes is similar to that of open-chain, acyclic alkanes. Both are nonpolar and chemically inert to most reagents. There are, however, some important differences. One difference is that cycloalkanes are less flexible than their open-chain counterparts. Although open-chain alkanes have nearly free rotation around their C–C single bonds, cycloalkanes are more constrained. Cyclopropane, for example, is geometrically constrained to be a rigid, planar molecule. No rotation around a C–C bond is possible in cyclopropane without breaking open the ring (Figure 2.10).

FIGURE 2.10 The structure of cyclopropane. No rotation is possible around the C–C bonds without breaking open the ring.

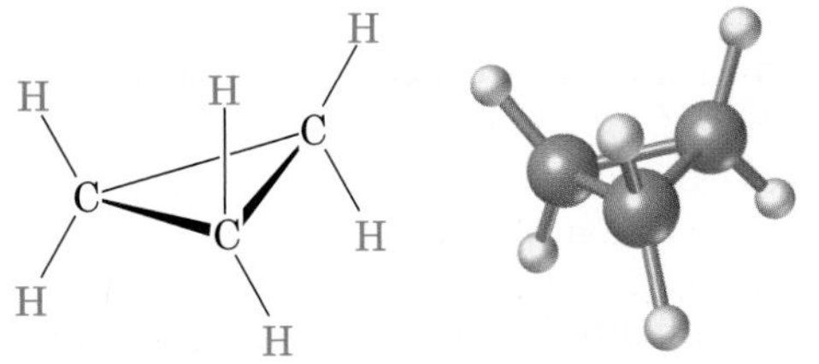

FIGURE 2.11 There are two different 1,2-dimethylcyclopropane isomers: one with the methyl groups on the same side of the ring (cis) and the other with the methyl groups on opposite sides of the ring (trans). The two isomers do not interconvert.

Because of their cyclic structures, cycloalkanes have two sides: a "top" side and a "bottom" side. As a result, isomerism is possible in substituted cycloalkanes. For example, there are two 1,2-dimethylcyclopropane isomers: one with the two methyl groups on the same side of the ring and one with the methyls on opposite sides. Both isomers are stable compounds, and neither can be converted into the other without breaking bonds (Figure 2.11).

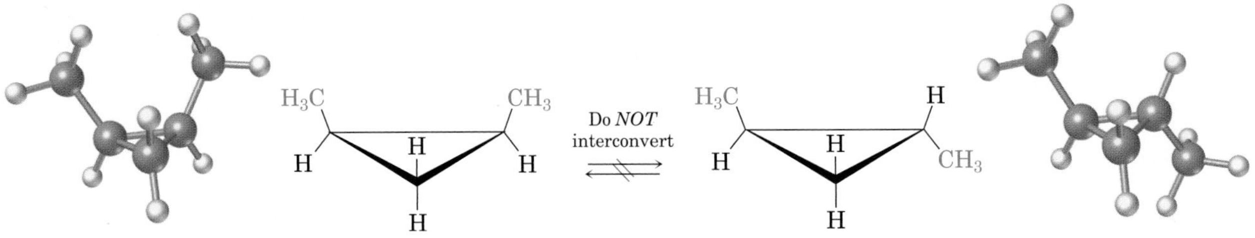

Do *NOT* interconvert

cis-1,2-Dimethylcyclopropane

trans-1,2-Dimethylcyclopropane

Unlike the constitutional isomers butane and isobutane (see Section 2.2), which have different connections among atoms, the two 1,2-dimethylcyclopropanes have the *same* connections but differ in the spatial orientation of their atoms. Such compounds, which have their atoms connected in the same way but differ in three-dimensional orientation, are called **stereoisomers**. The 1,2-dimethylcyclopropanes are special kinds of stereoisomers called **cis–trans isomers**. The prefixes *cis-* (Latin, "on the same side") and *trans-* (Latin, "across") are used to distinguish between them.

PRACTICE PROBLEM 2.5

Draw *cis*-1,4-dimethylcyclohexane.

STRATEGY *cis*-1,4-Dimethylcyclohexane contains a ring of six carbon atoms (cyclohexane) with methyl substituents on the same (cis) side of the ring at carbons 1 and 4.

SOLUTION

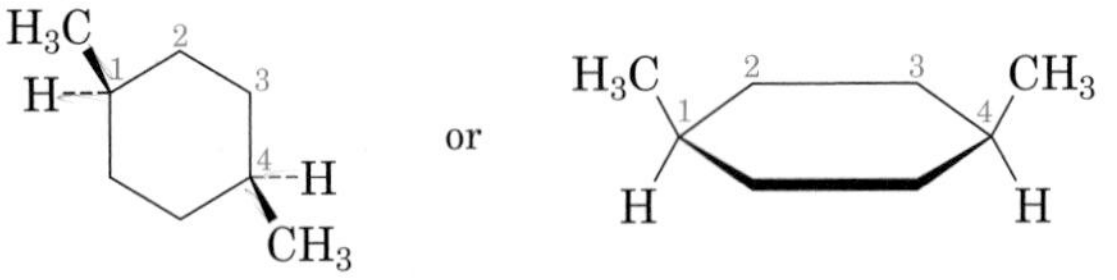

or

cis-1,4-Dimethylcyclohexane

PROBLEM 2.20 Draw *cis*-1-chloro-3-methylcyclopentane.

PROBLEM 2.21 Draw both cis and trans isomers of 1,2-dibromocyclobutane.

2.9 Conformations of Some Cycloalkanes

In the early days of organic chemistry, cycloalkanes provoked a good deal of consternation among chemists. The problem was that if carbon prefers to have bond angles of 109.5°, how is it possible for cyclopropane and cyclobutane to exist? After all, cyclopropane must have a triangular shape with bond angles near 60°, and cyclobutane must have a square or rectangular shape with bond angles near 90°. Nonetheless, these compounds *do* exist and are stable. Let's look at the most common cycloalkanes.

Cyclopropane, Cyclobutane, and Cyclopentane

Cyclopropane is a symmetrical molecule with C–C–C bond angles of 60°, as indicated in Figure 2.12a. This deviation from the normal 109.5° tetrahedral angle causes an **angle strain** in the molecule that raises its energy and makes it more reactive than unstrained alkanes. All six of the C–H bonds have an eclipsed, rather than staggered, arrangement with their neighbors.

Cyclobutane and cyclopentane are slightly puckered rather than flat, as indicated in Figure 2.12b–c. This puckering makes the C–C–C bond angles a bit smaller than they would otherwise be and increases the angle strain. At the same time, though, the puckering relieves the eclipsing interactions of adjacent C–H bonds that would occur if the rings were flat (see Section 2.5).

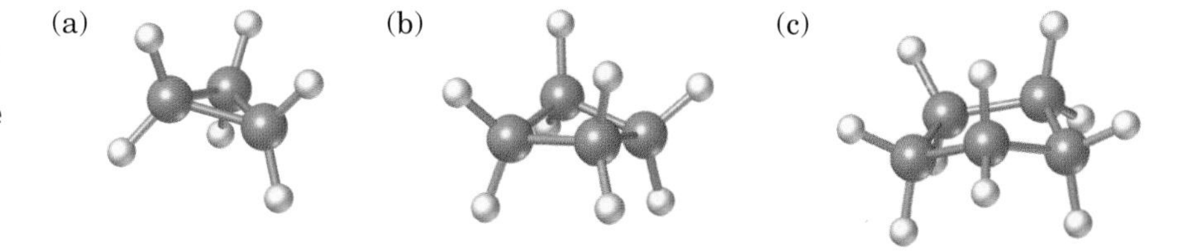

FIGURE 2.12 The structures of (a) cyclopropane, (b) cyclobutane, and (c) cyclopentane. Cyclopropane is planar, but cyclobutane and cyclopentane are slightly puckered.

Cyclohexane

Substituted cyclohexanes are the most common cycloalkanes because of their wide occurrence in nature. A large number of compounds, including steroids and numerous pharmaceutical agents, have cyclohexane rings. Cholesterol, for instance, has 3 six-membered rings and 1 five-membered ring.

HO

Cholesterol

Cyclohexane is not flat. Rather, it is puckered into a strain-free, three-dimensional shape called a **chair conformation**, in which the C–C–C bond angles are close to the ideal 109.5° tetrahedral value (Figure 2.13). In addition to being free of angle strain, chair cyclohexane is also free of all C–H eclipsing interactions because neighboring C–H bonds are staggered.

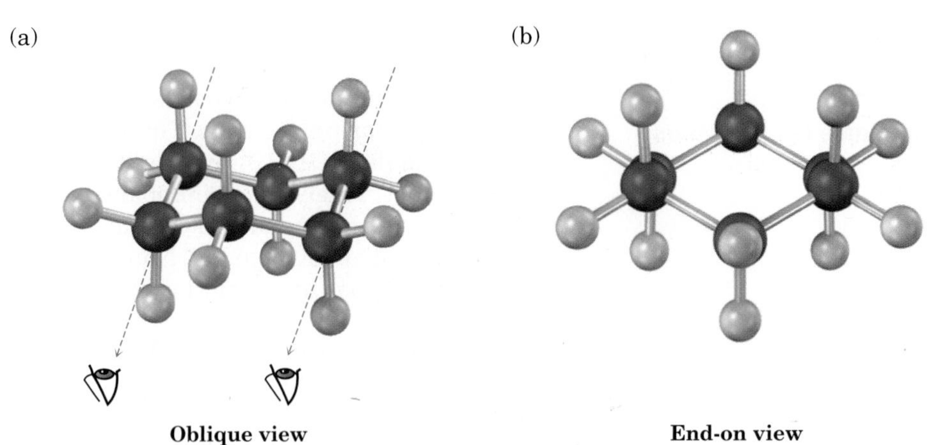

FIGURE 2.13 The strain-free, chair conformation of cyclohexane. All C–C–C bond angles are close to 109°, and all neighboring C–H bonds are staggered, as evident in the end-on view in **(b)**.

A chair conformation is drawn in three steps:

STEP 1 Draw two parallel lines, slanted downward and slightly offset from each other. This means that four of the cyclohexane carbons lie in a plane.

STEP 2 Place the topmost carbon atom above and to the right of the plane of the other four, and connect the bonds.

STEP 3 Place the bottommost carbon atom below and to the left of the plane of the middle four, and connect the bonds. Note that the bonds to the bottommost carbon atom are parallel to the bonds to the topmost carbon.

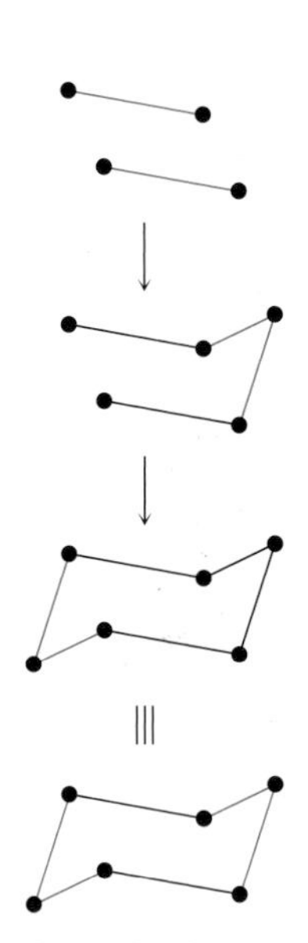

When viewing cyclohexane, it's important to remember that the lower bond is in front and the upper bond is in back. If this convention is not defined, an optical illusion can make it appear that the reverse is true. For clarity, all the cyclohexane rings drawn in this book have the front (lower) bond heavily shaded to indicate its nearness to the viewer.

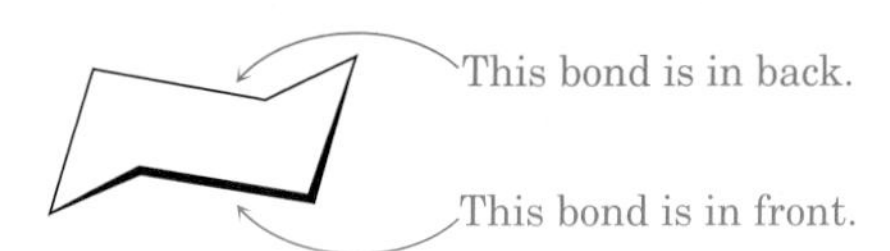

2.10 Axial and Equatorial Bonds in Cyclohexane

The chair conformation of cyclohexane has many chemical consequences. One such consequence is that there are two kinds of positions for hydrogens on the ring—**axial positions** and **equatorial positions** (Figure 2.14). Each carbon of a chair-shaped cyclohexane has one axial and one equatorial hydrogen. The equatorial hydrogen on each carbon appears in the rough plane of the ring (around the ring *equator*). Axial hydrogens alternately appear either above or below the ring (parallel to the ring axis).

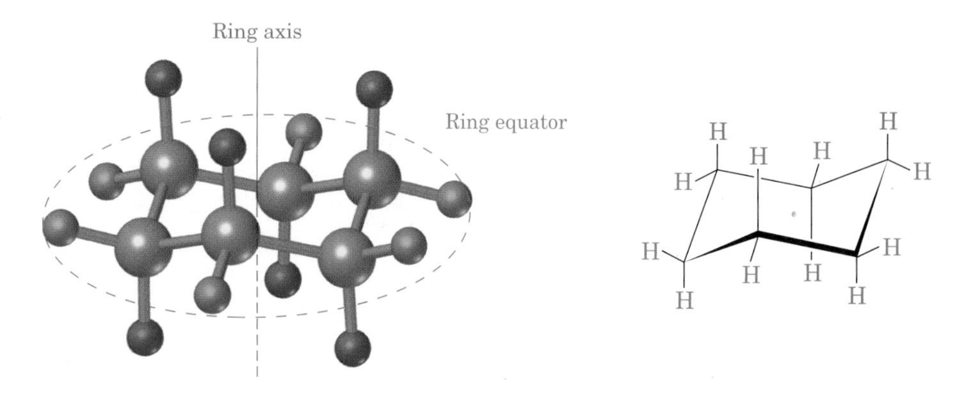

FIGURE 2.14 Axial (red) and equatorial (blue) hydrogen atoms in cyclohexane. The six axial C–H bonds are parallel to the ring axis, and the six equatorial C–H bonds are in a band around the ring equator.

Note that we haven't used the words *cis* and *trans* in this discussion of cyclohexane geometry. Two hydrogens on the same side of a ring are always cis, regardless of whether they're axial or equatorial and regardless of whether they're adjacent. Similarly, two hydrogens on opposite sides of the ring are always trans.

Axial and equatorial bonds can be drawn following the procedure shown in Figure 2.15.

FIGURE 2.15 A procedure for drawing axial and equatorial bonds in cyclohexane.

Axial bonds: The six axial bonds, one on each carbon, are parallel and alternate up–down.

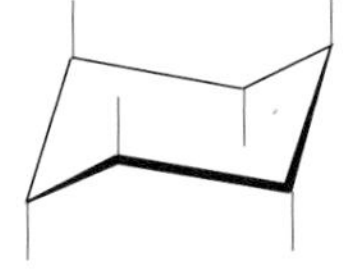

Equatorial bonds: The six equatorial bonds, one on each carbon, come in three sets of two parallel lines. Each set is also parallel to two ring bonds. Equatorial bonds alternate between sides around the ring.

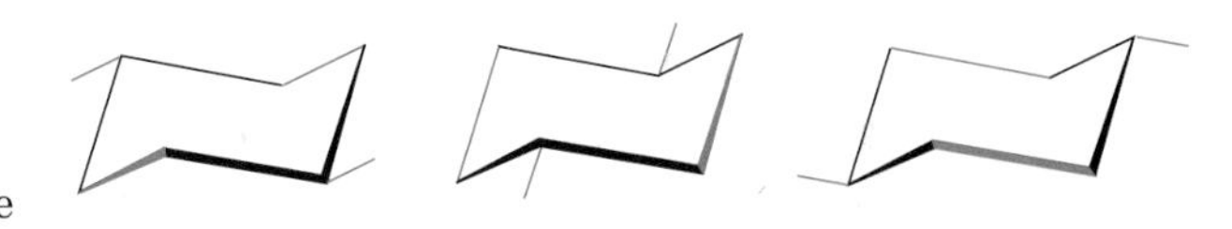

Completed cyclohexane

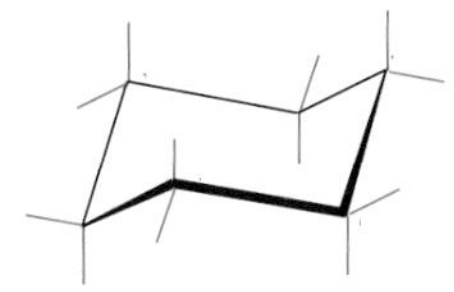

PROBLEM 2.22 Draw two chair structures for methylcyclohexane, one with the methyl group axial and one with the methyl group equatorial.

2.11 Conformational Mobility of Cyclohexane

Because chair cyclohexane has two kinds of positions, axial and equatorial, we might expect to find two isomeric forms of a monosubstituted cyclohexane. In fact, though, there is only *one* methylcyclohexane, *one* bromocyclohexane, and so forth, because cyclohexane rings are *conformationally mobile* at room temperature. Different chair cyclohexane conformations readily interconvert, resulting in the exchange of axial and equatorial positions. This interconversion of chair conformations, usually referred to as a **ring-flip**, is shown in Figure 2.16.

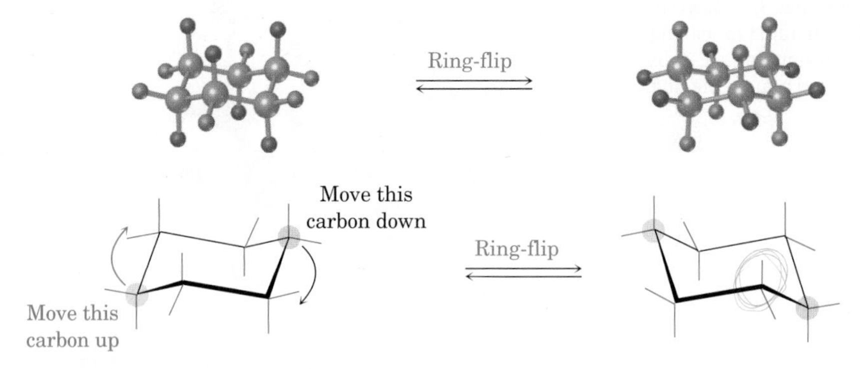

FIGURE 2.16 A ring-flip in chair cyclohexane interconverts axial and equatorial positions.

Conceptually, a chair cyclohexane can be ring-flipped by keeping the middle four carbon atoms in place while folding the two ends in opposite directions. Axial substituents in one chair form become equatorial substituents in the ring-flipped chair form and vice versa. For example, axial methylcyclohexane becomes equatorial methylcyclohexane after ring-flip. Because this interconversion occurs rapidly at room temperature, we can isolate only an interconverting mixture rather than distinct axial and equatorial isomers.

Although axial and equatorial methylcyclohexanes interconvert rapidly, they aren't equally stable. The equatorial conformation is more stable than the axial conformation by 7.6 kJ/mol (1.8 kcal/mol), meaning that about 95% of methylcyclohexane molecules have their methyl group equatorial at any given instant. The energy difference is due to an unfavorable *steric* (spatial) interaction that occurs in the axial conformation between the methyl group on carbon 1 and the axial hydrogen atoms on carbons 3 and 5. This so-called *1,3-diaxial interaction* introduces 7.6 kJ/mol (1.8 kcal/mol) of **steric strain** into the molecule because the axial methyl group and the nearby axial hydrogen are too close together (Figure 2.17).

FIGURE 2.17 Axial versus equatorial methylcyclohexane. The 1,3-diaxial steric interactions in axial methylcyclohexane (easier to see in space-filling models) make the equatorial conformation more stable by 7.6 kJ/mol.

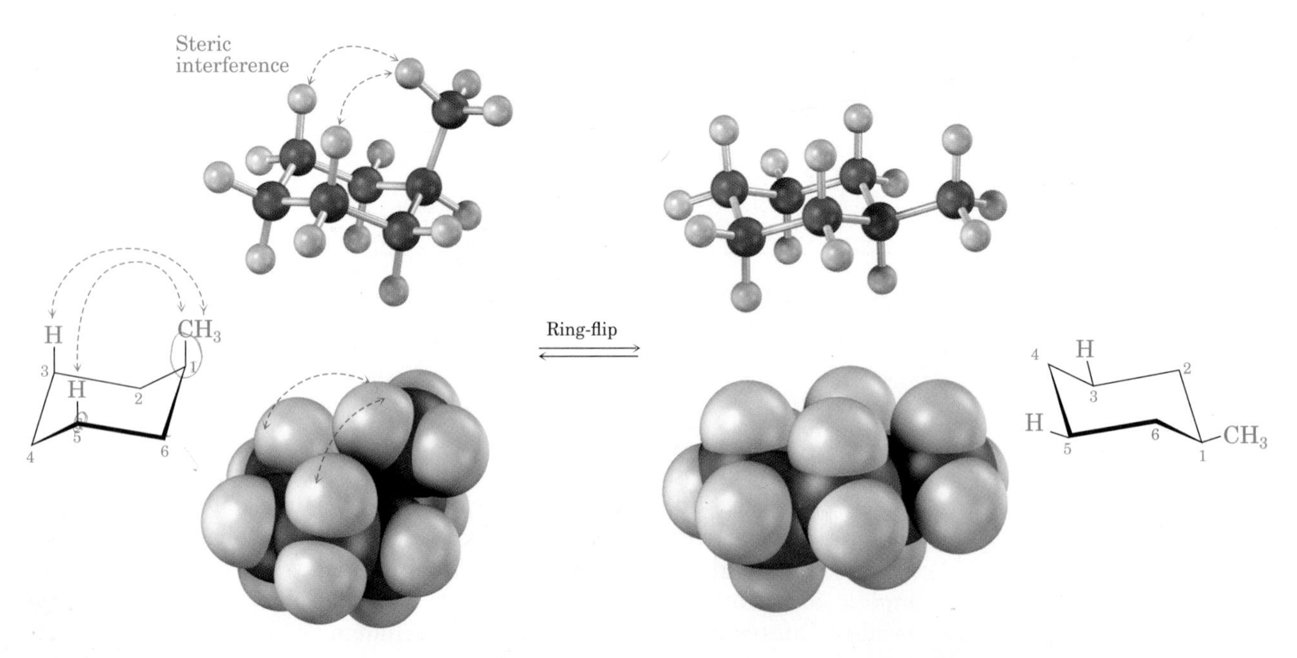

What is true for methylcyclohexane is also true for other monosubstituted cyclohexanes: a substituent is always more stable in an equatorial position than in an axial position. As you might expect, the amount of steric strain increases as the size of the axial substituent group increases.

Practice Problem 2.6

Draw 1,1-dimethylcyclohexane, indicating the axial and equatorial methyl groups.

Strategy Draw a chair cyclohexane ring, and then put two methyl groups on the same carbon. The methyl group in the rough plane of the ring is equatorial, and the other (above or below the ring) is axial.

Solution

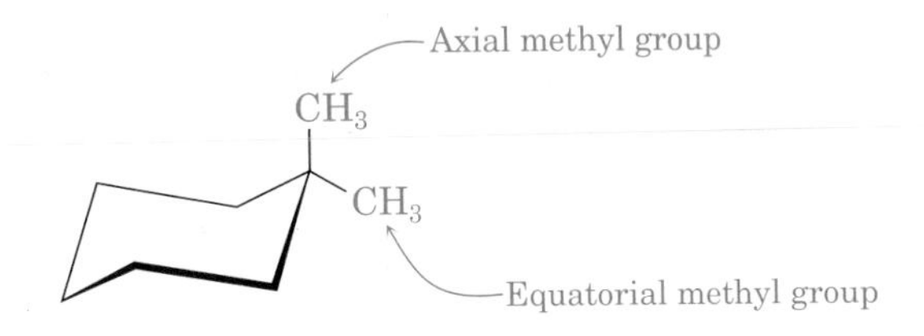

Problem 2.23 Draw two different chair conformations of bromocyclohexane showing all hydrogen atoms. Label all positions as axial or equatorial. Which of the two conformations do you think is more stable?

Problem 2.24 Draw *cis*-1,2-dichlorocyclohexane in chair conformation, and explain why one group must be axial and one equatorial.

Problem 2.25 Draw *trans*-1,2-dichlorocyclohexane in chair conformation, and explain why both groups must be axial or both equatorial.

Problem 2.26 Name the following compound, identify each substituent as axial or equatorial, and tell whether the conformation shown is the more stable or less stable chair form (gray = C, yellow-green = Cl, ivory = H).

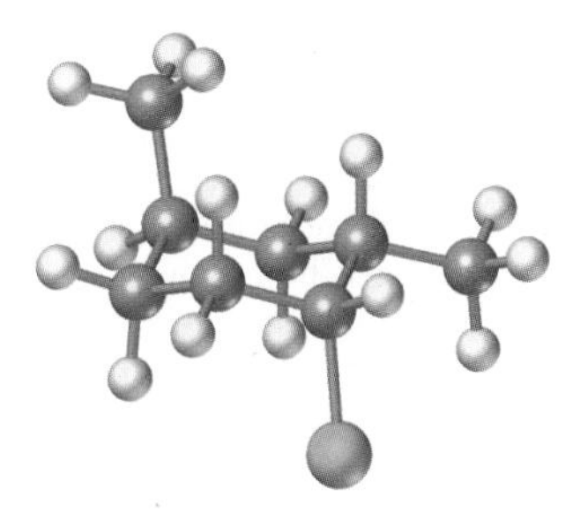

INTERLUDE

The Sources and Development of Drugs

Where do drugs come from? The answer to this question might be "land," "sea," "hard work," or "luck," depending on whom we ask. Historically, the origins of drugs are medicinal plants. Hippocrates recommended chewing the bark of a willow tree to relieve toothaches. More than 2000 years later, the German chemist Bayer made, in the laboratory, aspirin, a derivative of the same active ingredient in the bark of willow tree—salicylic acid. Indeed, whether it is a drug for malaria or the most exciting new cancer therapy, nature often provides the lead. In fact, 67% of the 1031 new drug leads reported to the U.S. Food and Drug Administration (FDA) between 1981 and 2002 were derived from nature, according to a study conducted by the U.S. National Cancer Institute. (However, as we'll see shortly, only a small fraction of these molecules become medicines.)

We can depict the origins of these drugs on a pie chart. The blue region represents a relatively new and growing source of drugs: these are the vaccines and protein and peptide drugs produced in fermentations (15%) using genetically engineered bacteria and other organisms. The green region corresponds to molecules used exactly as they are found in nature (28%). For economic reasons, these molecules might be produced either by nature and subsequently isolated or prepared synthetically in the laboratory. The yellow portion refers to molecules that do not occur in nature, but where the active part of the molecule is taken from nature and incorporated into an unnatural molecule (24%). The final region shown in red (33%) refers to molecules that do not have natural origins and are prepared by chemists in the laboratory.

Sources of the 1031 new potential drugs reported between 1981 and 2002.

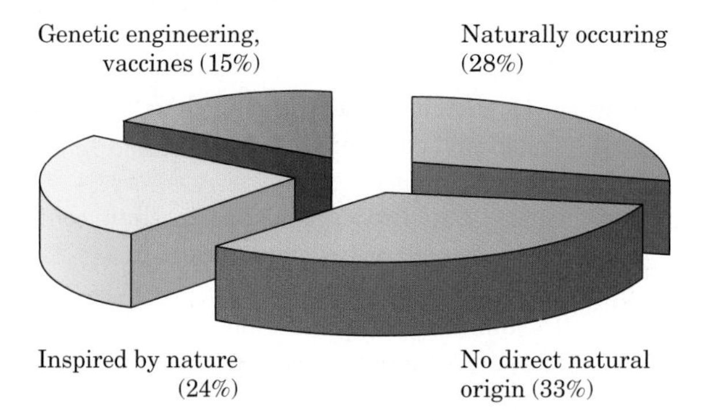

With all these sources of compounds, why are medicines so slow to come to market and so costly when they arrive? The answer is *time*. Once scientists identify a potential drug, it takes, on average, 11 years for the drug to come to market. These 11 years can be divided into different phases.

Preclinical development (red box) focuses primarily on scientific and economic issues. The drug must be proved safe in animals and cheap enough to produce at the large scales required for human (clinical) trials. If the drug looks feasible, it is put on record with the FDA like any one of the 1031 molecules shown in the previous illustration. At this point, the FDA evaluates the scientific data and makes a decision as to whether it can be tested in people.

This begins the first phase of clinical trials, called phase I. In phase I the drug is administered to healthy people (orange box) who are then monitored for side effects. Based on the side effects observed, the drug might be withdrawn, administered in a different way in an additional phase I study, or considered for phase II. The latter involves decisions by both the FDA and the company testing the drug. If, based on the phase I data, the FDA determines the drug is suitable for phase II trials, the company must decide whether to move forward and incur the expense of phase II trials.

Continued

In phase II trials, the drug is administered to a relatively small number of patients who have the disease or condition the drug is expected to improve or cure (yellow box). Typically, these patients have been treated with traditional therapies, but the treatments failed. Based on the results of phase II studies, the company and FDA both decide whether phase III studies are warranted (blue box). These studies allow a broader range of patients to take the drug. If these studies are successful, the company will request approval for widespread clinical use from the FDA. A newly approved drug can be worth *billions of dollars* to a company.

Timeline for bringing drugs to the market.

	Year
Choose candidate drug	0
Animal safety and economic analysis	1
Go/no go decision	2
Preclinical development and manufacturing	3
File with FDA for phase I trials	4
Phase I clinical trials—healthy people	4–6
Go/no go decision after phase I	
Phase II trials—small target population	5–9
Go/no go decision after phase II	
Phase III trials—general people	8–11
Go/no go decision after phase III	
Seek FDA approval for clinical use	~11

Passing phase III still does not ensure a successful product. At times, small populations of patients develop unfavorable and serious side effects after the drug has been in widespread circulation for a few years, leading the company to withdraw the drug from the market. In addition, drugs do not always produce a significant profit. One reason is that competitors constantly introduce new drugs. Another reason is that the patents protecting these drugs are good for only 20 years. Since patents are filed during the preclinical trials, the company has exclusive rights to the drug for only about 6 years. Afterward, generic drug manufacturers can enter the market and sell the drug for significantly less money because these companies have no expenses associated with preclinical or clinical trials.

Summary and Key Words

A **functional group** is an atom or group of atoms within a larger molecule that has a characteristic chemical reactivity. Because functional groups behave approximately the same way in all molecules where they occur, the reactions of an organic molecule are largely determined by its functional groups.

Alkanes are a class of **saturated hydrocarbons** having the general formula C_nH_{2n+2}. They contain no functional groups, are chemically rather inert, and can be either **straight-chain** or **branched**. Alkanes are named by a series of **IUPAC** rules of nomenclature. **Isomers**—compounds that have the same chemical formula but different structures—exist for all but the simplest alkanes. Compounds such as butane and isobutane, which have the same formula but differ in the way their atoms are connected, are called **constitutional isomers**.

Because they are formed by head-on orbital overlap, rotation is possible about C—C single bonds. Alkanes can therefore adopt any of a large number of rapidly interconverting **conformations**. A **staggered conformation** is more stable than an **eclipsed conformation**.

Cycloalkanes contain rings of carbon atoms and have the general formula C_nH_{2n}. Because complete rotation around C–C bonds is restricted in cycloalkanes, conformational mobility is reduced and disubstituted cycloalkanes can exist as **cis–trans stereoisomers**. In a cis isomer, both substituents are on the same side of the ring, whereas in a trans isomer, the substituents are on opposite sides of the ring.

Cyclohexanes are the most common of all rings because of their wide occurrence in nature. Cyclohexane exists in a puckered, strain-free **chair conformation** in which all bond angles are near 109° and all neighboring C–H bonds are staggered. Chair cyclohexane has two kinds of bonds: axial and equatorial. **Axial bonds** are directed up and down, parallel to the ring axis; **equatorial bonds** lie in a belt around the ring equator. Chair cyclohexanes can undergo a **ring-flip** that interconverts axial and equatorial positions. Substituents on the ring are more stable in the equatorial than in the axial position.

EXERCISES

Visualizing Chemistry

2.27 Give IUPAC names for the following substances, and convert each drawing into a skeletal structure.

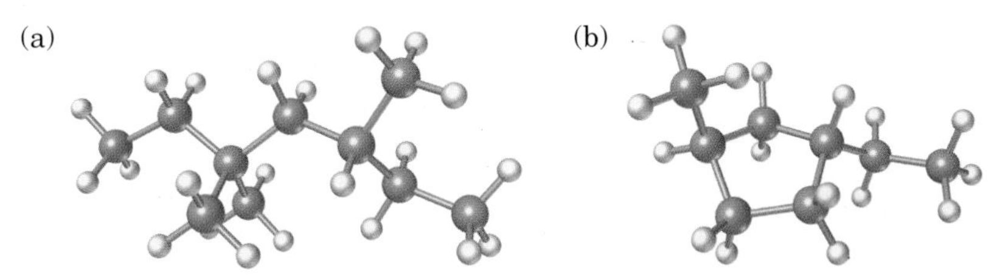

2.28 Identify the functional groups in the following substances, and convert each drawing into a molecular formula (gray = C, red = O, blue = N, ivory = H).

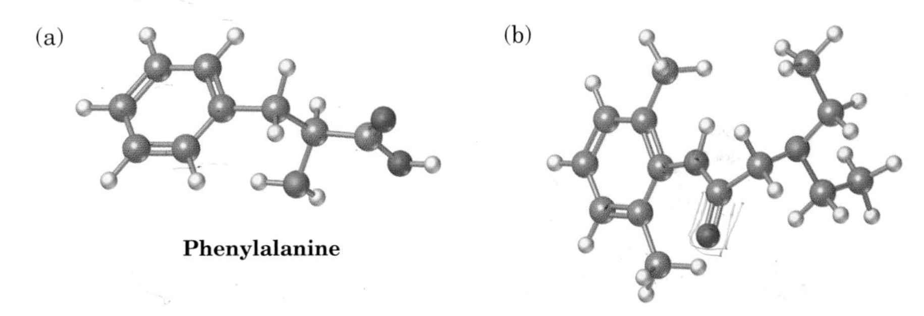

Phenylalanine

Lidocaine

2.29 The following cyclohexane derivative has three substituents—red, green, and blue. Identify each substituent as axial or equatorial, and identify each pair of relationships (red-blue, red-green, and blue-green) as cis or trans.

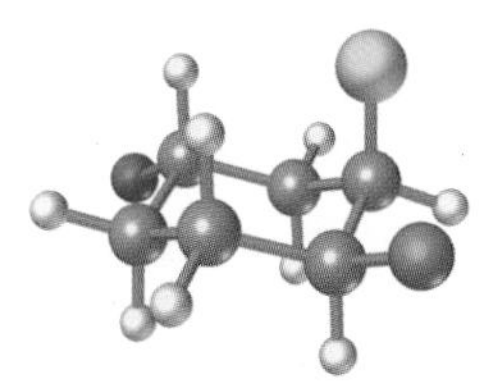

2.30 A trisubstituted cyclohexane with three substituents—red, yellow, and blue—undergoes a ring-flip to its alternative chair conformation. Identify each substituent as axial or equatorial, and show the positions occupied by the three substituents in the ring-flipped form.

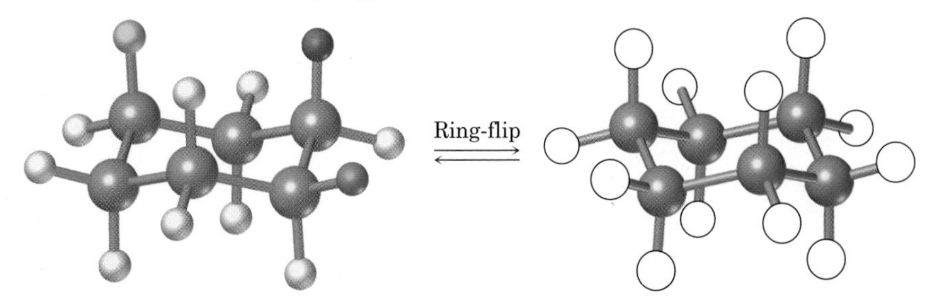

Additional Problems

FUNCTIONAL GROUPS AND ISOMERISM

2.31 Locate and identify the functional groups in the following molecules:

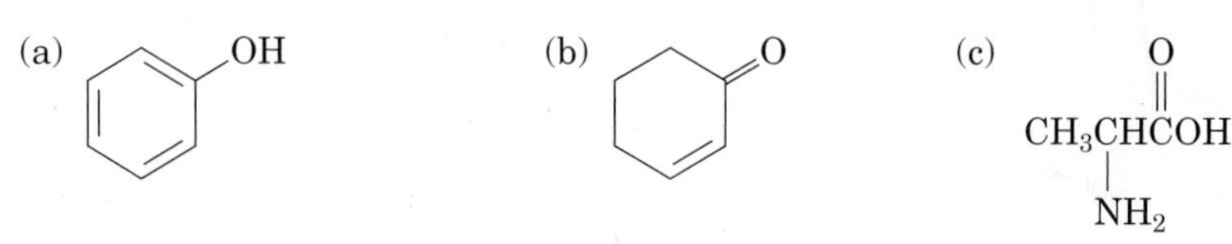

Phenol   Cyclohex-2-enone   Alanine

(d)   (e)

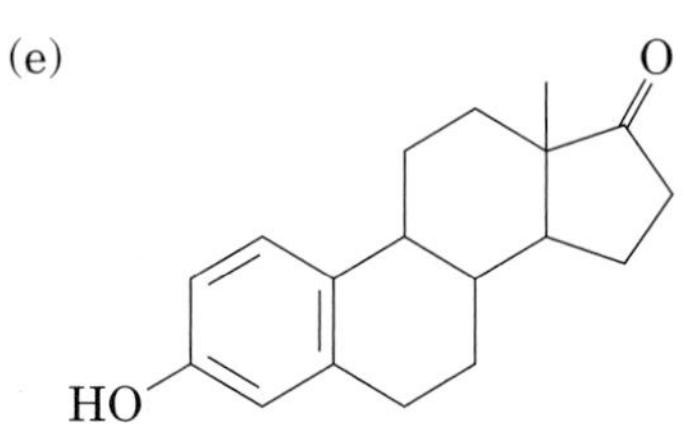

Nootkatone (from grapefruit)   Estrone

2.32 Propose structures for molecules that fit the following descriptions:
(a) An alkene with six carbons (b) A cycloalkene with five carbons
(c) A ketone with five carbons (d) An amide with four carbons
(e) A five-carbon ester (f) An aromatic alcohol

2.33 Propose suitable structures for the following:
(a) An alkene, C_7H_{14} (b) A cycloalkene, C_3H_4
(c) A ketone, C_4H_8O (d) A nitrile, C_5H_9N
(e) A dialkene, C_5H_8 (f) A dialdehyde, $C_4H_6O_2$

2.34 Write as many structures as you can that fit the following descriptions:
(a) Alcohols with formula $C_4H_{10}O$ (b) Amines with formula $C_5H_{13}N$
(c) Ketones with formula $C_5H_{10}O$ (d) Aldehydes with formula $C_5H_{10}O$
(e) Ethers with formula $C_4H_{10}O$ (f) Esters with formula $C_4H_8O_2$

2.35 Draw all monobromo derivatives of pentane, $C_5H_{11}Br$.

2.36 Draw all monochloro derivatives of 2,5-dimethylhexane.

2.37 How many constitutional isomers are there with the formula C_3H_8O? Draw them.

2.38 Propose structures for compounds that contain the following:
(a) A quaternary carbon (b) Four methyl groups
(c) An isopropyl group (d) Two tertiary carbons
(e) An amino group ($-NH_2$) bonded to a secondary carbon

2.39 What hybridization do you expect for the carbon atom in the following functional groups?
(a) Ketone (b) Nitrile (c) Ether (d) Alcohol

NAMING AND REPRESENTING CHEMICAL STRUCTURES

2.40 Which of the structures in each of the following sets represent the same compound and which represent different compounds?

(a)

```
                                            H
                                            |
                                         H—C—H
                                            |
  H   H   H   H                   H   H     |                 H   H   H
  |   |   |   |                   |   |     |                 |   |   |
H—C—C—C—C—H                 H—C—C—C—H                 H—C—C—C—H
  |   |   |   |                   |   |   |                   |   |   |
  H   H   H   H                   H   H   H                   H   |   H
                                                                   |
                                                                H—C—H
                                                                   |
                                                                   H
```

(b)

```
                                                                H
                                                                |
                                                             H—C—H
                                                                |
  H   H   Br  H                 H   Br  H   H                   |   H   H
  |   |   |   |                 |   |   |   |                   |   |   |
H—C—C—C—C—H               H—C—C—C—C—H                 H—C—C—C—H
  |   |   |   |                 |   |   |   |                   |   |   |
  H   H   H   H                 H   H   H   H                   H   Br  H
```

(c)

```
            CH3              CH3
            |                |
CH3CH(Br)CHCH3        CH3CHCH(Br)CH3        (CH3)2CHCH(Br)CH2CH3
```

(d) OH, OH (1,2-positions on benzene ring) OH, OH (1,3-positions on benzene ring) HO, HO (1,2-positions on benzene ring)

2.41 What is the molecular formula of each of the following condensed structures?

(a) (b) (c)

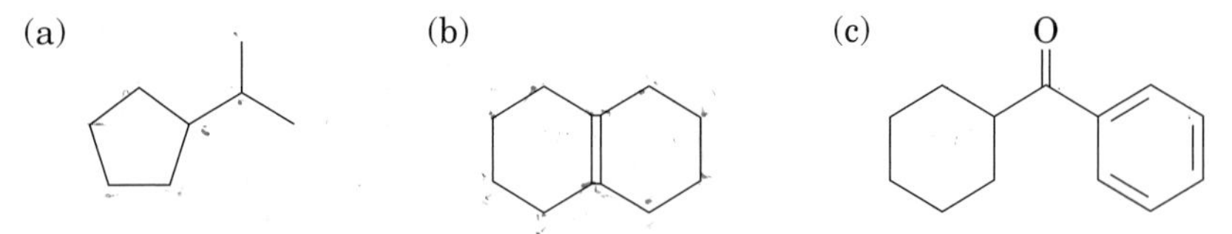

2.42 Draw the structure of a constitutional isomer for each of the molecules shown in Problem 2.41.

2.43 Draw structural formulas for the following substances:

(a) 2-Methylheptane
(b) 4-Ethyl-2-methylhexane
(c) 4-Ethyl-3,4-dimethyloctane
(d) 2,4,4-Trimethylheptane
(e) 1,1-Dimethylcyclopentane
(f) 4-Isopropyl-3-methylheptane

2.44 Give IUPAC names for the following alkanes:

(a) 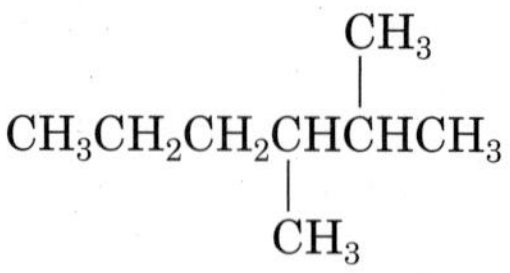 $CH_3CH_2CH_2CHCHCH_3$ (with CH_3 above and CH_3 below)

(b) 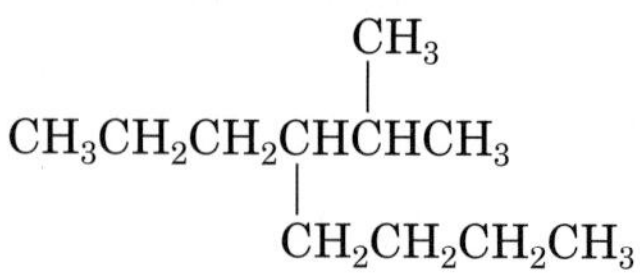 $CH_3CH_2CH_2CHCHCH_3$ (with CH_3 above and $CH_2CH_2CH_2CH_3$ below)

(c) 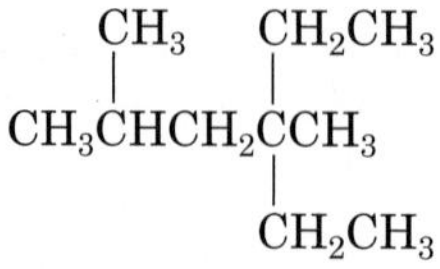 $CH_3CHCH_2CCH_3$ (with CH_3 and CH_2CH_3 above, CH_2CH_3 below)

(d) 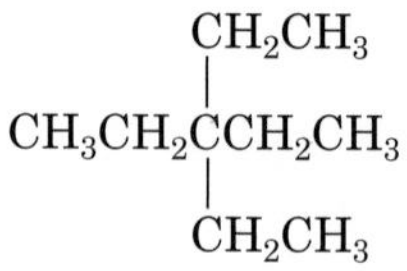 $CH_3CH_2CCH_2CH_3$ (with CH_2CH_3 above and CH_2CH_3 below)

2.45 For each of the following compounds, draw a constitutional isomer with the same functional groups:

(a) 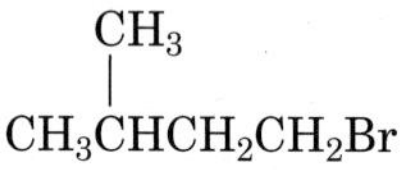 $CH_3CHCH_2CH_2Br$ (with CH_3 above)

(b) 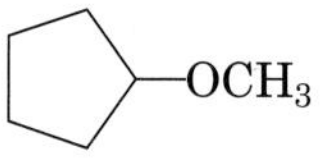 cyclopentyl–OCH_3

(c) $CH_3CH_2CH_2C{\equiv}N$

(d) 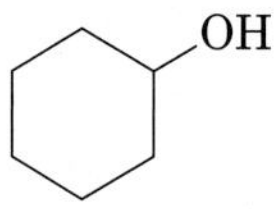 cyclohexyl–OH

(e) 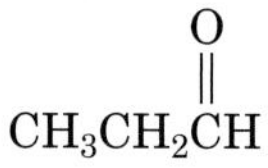 CH_3CH_2CH (with =O, i.e. CH_3CH_2CHO)

(f) 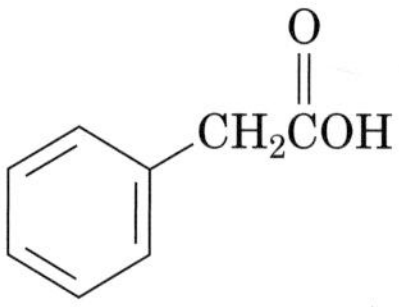 phenyl–CH_2COH (with =O)

2.46 Give IUPAC names for the following compounds:

(a) 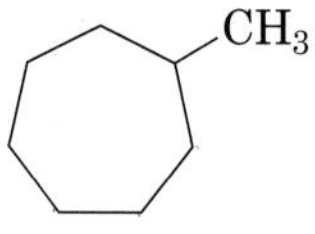 cycloheptane with CH_3

(b) 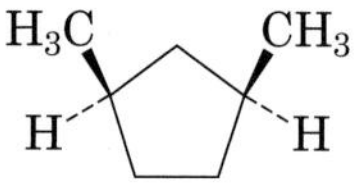 cyclopentane with H_3C, CH_3, H, H

(c) 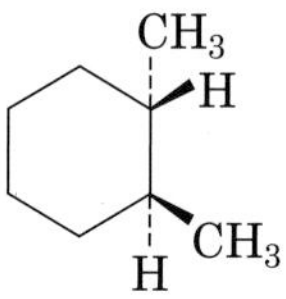 cyclohexane with CH_3, H, CH_3, H

(d) 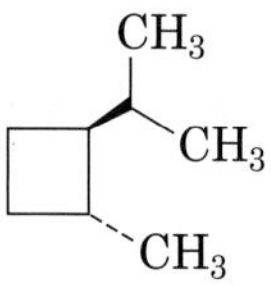 cyclobutane with CH_3, CH_3, CH_3

(e) 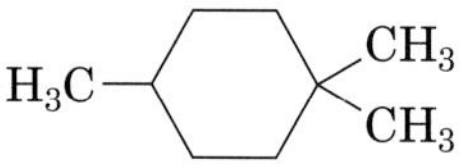 H_3C–cyclohexane with CH_3, CH_3

2.47 Give IUPAC names for the five isomers of C_6H_{14}.

2.48 Draw structures for the nine isomers of C_7H_{16}.

2.49 Propose structures and give correct IUPAC names for the following:
(a) A dimethyloctane (b) A diethyldimethylhexane
(c) A cycloalkane with three methyl groups

2.50 The following names are *incorrect*. Give the proper IUPAC names.
(a) 2,2-Dimethyl-6-ethylheptane (b) 4-Ethyl-5,5-dimethylpentane
(c) 3-Ethyl-4,4-dimethylhexane (d) 5,5,6-Trimethyloctane

2.51 Draw the structures of the following molecules:
(a) *Biacetyl*, $C_4H_6O_2$, a substance with the aroma of butter; it contains no rings or carbon–carbon multiple bonds.
(b) *Ethylenimine*, C_2H_5N, a substance used in the synthesis of melamine polymers; it contains no multiple bonds.
(c) *Glycerol*, $C_3H_8O_3$, a substance used in cosmetics; it has an –OH group on each carbon.

Conformations and Cis–Trans Isomerism

2.52 Sighting along the C2–C3 bond of 2-methylbutane, there are two different staggered conformations. Draw them both in Newman projections, tell which is more stable, and explain your choice.

2.53 Sighting along the C2–C3 bond of 2-methylbutane (see Problem 2.52), there are also two possible eclipsed conformations. Draw them both in Newman projections, tell which you think is lower in energy, and explain.

2.54 *cis*-1-*tert*-Butyl-4-methylcyclohexane exists almost exclusively in the conformation shown. What does this tell you about the relative sizes of a *tert*-butyl substituent and a methyl substituent?

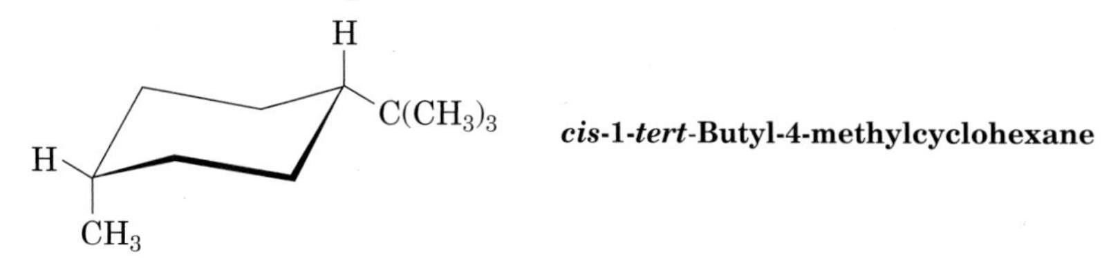

2.55 The barrier to rotation about the C–C bond in bromoethane is 15.0 kJ/mol (3.6 kcal/mol). If each hydrogen–hydrogen interaction in the eclipsed conformation is responsible for 3.8 kJ/mol (0.9 kcal/mol), how much is the hydrogen–bromine eclipsing interaction responsible for?

2.56 Tell whether the following pairs of compounds are identical, constitutional isomers, or stereoisomers.

(a) *cis*-1,3-Dibromocyclohexane and *trans*-1,4-dibromocyclohexane

(b) 2,3-Dimethylhexane and 2,5,5-trimethylpentane (named incorrectly)

(c)

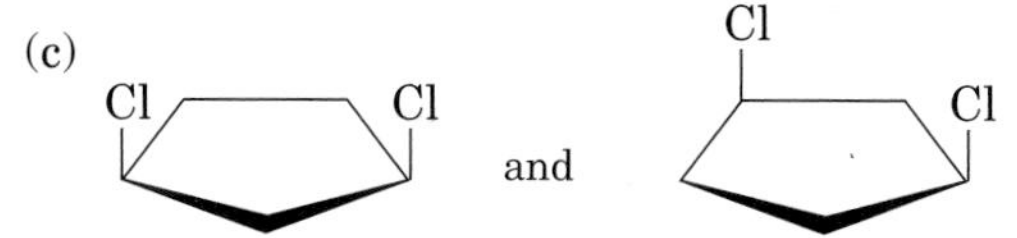

2.57 Draw two constitutional isomers of *cis*-1,2-dibromocyclopentane.

2.58 Draw a stereoisomer of *trans*-1,3-dimethylcyclobutane.

2.59 Draw *trans*-1,2-dimethylcyclohexane in its more stable chair conformation. Are the methyl groups axial or equatorial?

2.60 Draw *cis*-1,2-dimethylcyclohexane in its more stable chair conformation. Are the methyl groups axial or equatorial? Which is more stable, *cis*-1,2-dimethylcyclohexane or *trans*-1,2-dimethylcyclohexane (Problem 2.59)? Explain.

2.61 Which is more stable, *cis*-1,3-dimethylcyclohexane or *trans*-1,3-dimethylcyclohexane? Draw chair conformations of both, and explain your answer.

Octane Ratings

2.62 A fuel is described as having an octane rating of 87. How would you describe the performance of this fuel in terms of isooctane and heptane?

2.63 If a 35:65 ratio of isooctane and heptane best describes the performance of a fuel sample, what octane rating would you give this fuel?

2.64 Octane ratings at the pump are described as (R + M)/2 ratings. Measuring performance can be done using a variety of methods. Test conditions that mimic normal engine operation give *r*esearch octane *n*umbers (RON, R). Conditions that mimic harsh conditions like high speed give *m*otor octane *n*umbers (MON, M). Averaging the two numbers gives the octane rating that appears at the pump. Fill in the missing data for these fuels:

Fuel	R	M	R + M/2
Pentane	62	62	?
Hexane	25	?	25.5
Heptane	?	0	0
2,3-Dimethylpentane	93	96	?

INTEGRATED PROBLEMS

2.65 Malic acid, $C_4H_6O_5$, has been isolated from apples. Because malic acid reacts with 2 equivalents of base, it can be formulated as a dicarboxylic acid (that is, it has two —CO_2H groups).

(a) Draw at least five possible structures for malic acid.

(b) If malic acid is also a secondary alcohol (has an —OH group attached to a secondary carbon), what is its structure?

2.66 Cyclopropane was first prepared by reaction of 1,3-dibromopropane with sodium.

(a) Formulate the reaction.

(b) What product might the following reaction give? What geometry would you expect for the product?

$$BrCH_2-\underset{CH_2Br}{\overset{CH_2Br}{C}}-CH_2Br \xrightarrow{4\ Na} ?$$

2.67 *N*-Methylpiperidine has the conformation shown. What does this tell you about the relative steric requirements of a methyl group versus an electron lone pair?

N̈ CH3

***N*-Methylpiperidine**

2.68 Glucose contains a six-membered ring in which all the substituents are equatorial. Draw glucose in its more stable chair conformation.

HO O CH_2OH
HO OH
OH

Glucose

2.69 Draw 1,3,5-trimethylcyclohexane using a hexagon to represent the ring. How many cis–trans stereoisomers are possible?

2.70 One of the two chair structures of *cis*-1-chloro-3-methylcyclohexane is more stable than the other by 15.5 kJ/mol (3.7 kcal/mol). Which is it?

2.71 Draw the three cis–trans isomers of menthol.

CH_3
OH
$(CH_3)_2CH$

Menthol

2.72 "Reflect" on this tough one. There are two different substances named *trans*-1,2-dimethylcyclopentane. What is the relationship between them? (We'll explore this kind of isomerism in Chapter 6.)

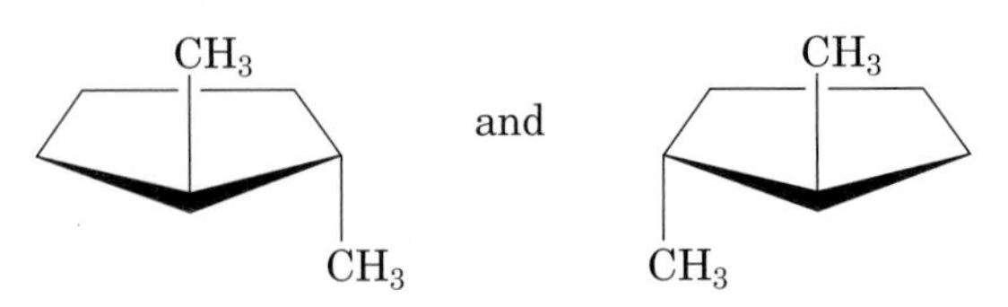

In the Medicine Cabinet

2.73 Amantadine is an antiviral agent that is active against influenza A infection. Draw a three-dimensional representation of amantadine showing the chair cyclohexane rings.

$-NH_2$ **Amantadine**

2.74 Statin drugs are used to treat high cholesterol.

Zocor (Merck) **Pravachol (Bristol-Myers Squibb)** **Lipitor (Pfizer)**

(a) Identify the functional (or alkyl) groups labeled A–I.
(b) On Pravachol, are the groups C and E cis or trans?
(c) Why can groups G, H, and I not be identified as cis or trans? (*Hint*: It has nothing to do with the fact that there are only five atoms or a nitrogen in the ring.)
(d) What are the molecular formulas of Zocor, Pravachol, and Lipitor?

2.75 The anticancer drug Taxol is a complex molecule. Identify the functional groups indicated.

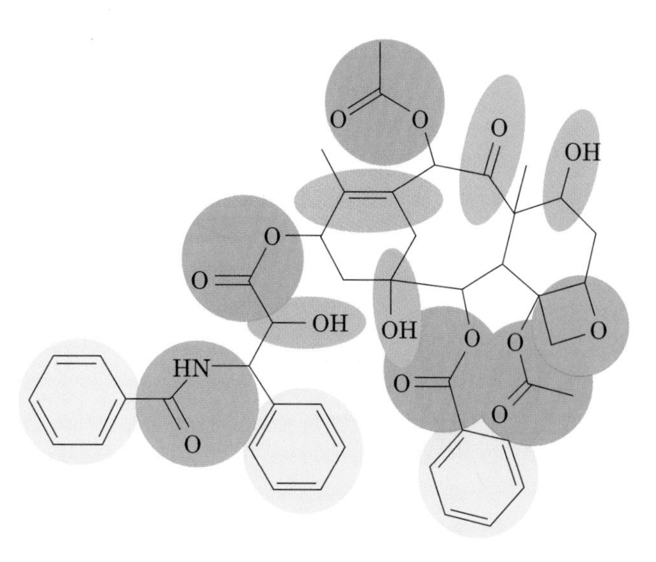

Taxol

In the Field with Agrochemicals

2.76 Identify functional groups A–G on the popular herbicides shown and give the molecular formula of each molecule.

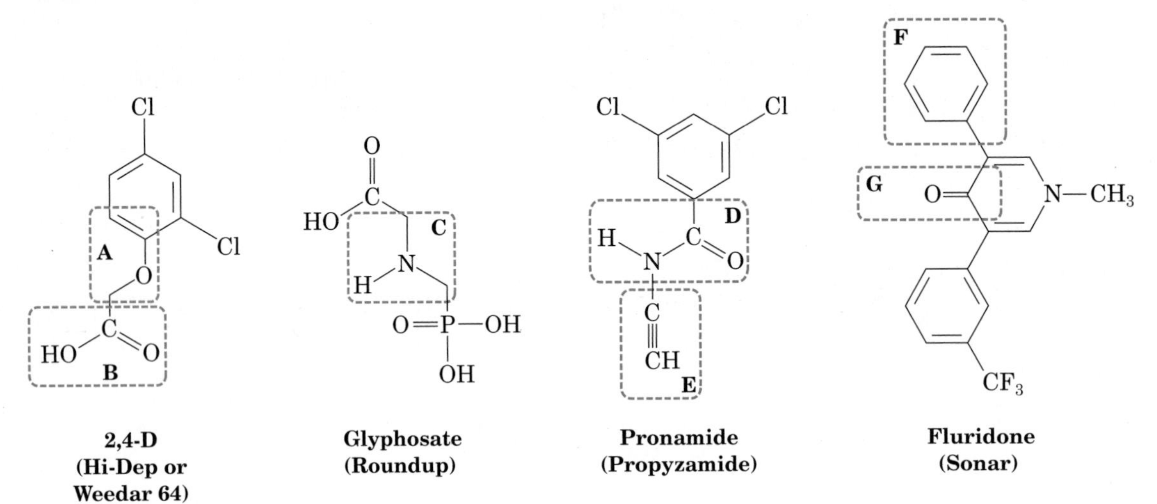

2,4-D (Hi-Dep or Weedar 64) **Glyphosate (Roundup)** **Pronamide (Propyzamide)** **Fluridone (Sonar)**

2.77 Metolachlor is marketed under the names of Bicep, CGA-24705, Dual, Pennant, and Pimagram. It is used to control weeds and grasses in fields of a variety of plants, including corn, soybeans, cotton, and peanuts. Metolachlor is degraded through oxidation in the environment to produce contaminants that are more problematic because of increased water solubility. Identify the three functional groups in Metolachlor and the new functional group in the derivative.

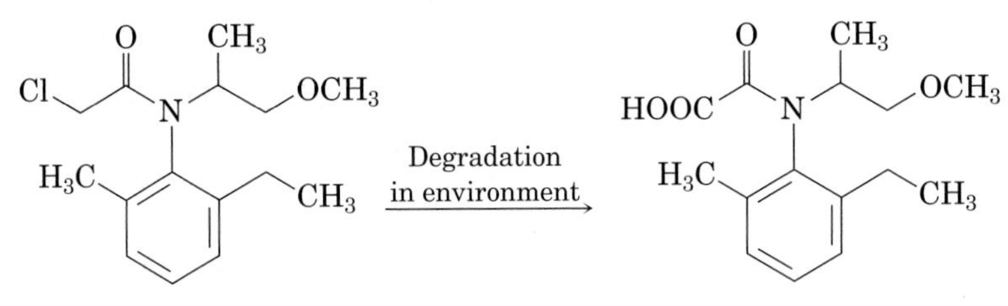

Metolachlor **Metolachlor oxanilic acid**

The double bonds of β-carotene not only make this molecule orange but allow it to serve a critical role in photosynthesis. (Getty Images/Todd Gipstein/National Geographic)

CHAPTER 3

The Nature of Organic Reactions: Alkenes

Historically called *olefins* (ō-la-fins), **alkenes** are hydrocarbons that contain a carbon–carbon double bond, C=C. Alkenes occur abundantly in nature, and many have important biological roles. Ethylene, for example, is a plant hormone that induces ripening in fruit, and α-pinene is the major constituent of turpentine. Life itself would be impossible without such alkenes as β-carotene, which contains 11 double bonds. An orange pigment responsible for the color of carrots, β-carotene is a valuable dietary source of vitamin A.

Ethylene

α-Pinene

β-Carotene
(orange pigment and vitamin A precursor)

We'll see in this chapter how and why alkenes behave the way they do, and we'll develop some general ideas about organic chemical reactivity that can be applied to all molecules.

3.1 Naming Alkenes

Because of their double bond, alkenes have fewer hydrogens per carbon than related alkanes and are therefore referred to as **unsaturated**. Ethylene, for example, has the formula C_2H_4, whereas ethane has the formula C_2H_6.

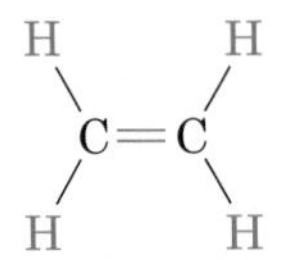

Ethylene: C_2H_4
(fewer hydrogens—*unsaturated*)

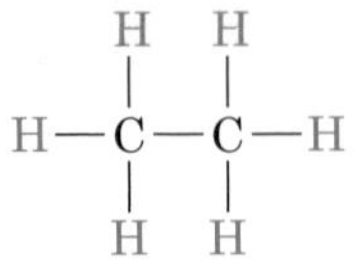

Ethane: C_2H_6
(more hydrogens—*saturated*)

Alkenes are named according to a series of rules similar to those used for alkanes, with the suffix *-ene* used in place of *-ane* to identify the family. There are three steps:

STEP 1 **Name the parent hydrocarbon.**

Find the longest carbon chain that contains the double bond, and name the compound using the suffix *-ene*.

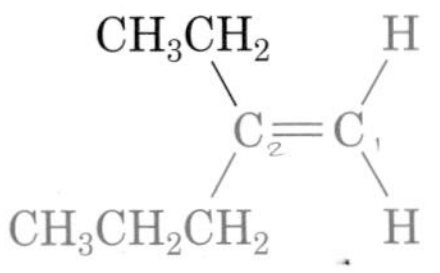

Named as a *pentene* *NOT* as a hexene, since the double bond is not contained in the six-carbon chain

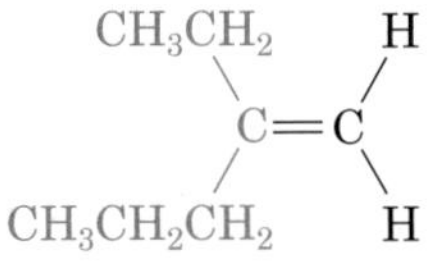

STEP 2 **Number the carbon atoms in the chain.**

Begin numbering at the end nearer the double bond. If the double bond is equidistant from the two ends, begin numbering at the end nearer the first branch point. This rule ensures that the double-bond carbons receive the lowest possible numbers:

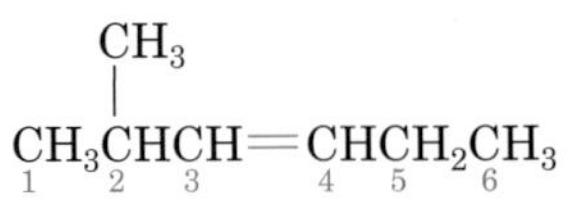

$CH_3CH_2CH_2CH{=}CHCH_3$ (numbered 6 5 4 3 2 1)

$CH_3CH(CH_3)CH{=}CHCH_2CH_3$ (numbered 1 2 3 4 5 6)

STEP 3 **Write the full name.**

Number the substituents according to their positions in the chain, and list them alphabetically. Indicate the position of the double bond by giving the number of the first alkene carbon and placing that number directly before the *-ene* suffix. If more than one double bond is present, indicate the position of each and use one of the suffixes *-diene*, *-triene*, and so on. We might also note that prior to 1993 the number locating the position of the double bond was placed at the beginning of the name rather than before the *-ene* suffix: 2-butene rather than but-2-ene, for instance. This book places the number immediately preceding the suffix, but you may encounter the other style in other sources.

$CH_3CH_2CH_2CH{=}CHCH_3$ (carbons numbered 6 5 4 3 2 1)

Hex-2-ene
2-Hexene (pre-1993)

$CH_3CH(CH_3)CH{=}CHCH_2CH_3$ (carbons numbered 1 2 3 4 5 6)

2-Methylhex-3-ene
2-Methyl-3-hexene (pre-1993)

$(CH_3CH_2)(CH_3CH_2CH_2)C{=}CH_2$ (C2=C1; propyl carbons numbered 5 4 3)

2-Ethylpent-1-ene
2-Ethyl-1-pentene (pre-1993)

$H_2C{=}C(CH_3){-}CH{=}CH_2$ (carbons numbered 1 2 3 4)

2-Methylbuta-1,3-diene
2-Methyl-1,3-butadiene (pre-1993)

Cycloalkenes are named in a similar way, but because there is no chain end to begin from, we number the cycloalkene so that the double bond is between C1 and C2 and so that the first substituent has as low a number as possible. Note that it's not necessary to specify the position of the double bond in the name because it's always between C1 and C2:

1-Methylcyclohexene **Cyclohexa-1,4-diene** **1,5-Dimethylcyclopentene**

For historical reasons, there are a few alkenes whose names don't conform to the rules. For example, the alkene corresponding to ethane should be called *ethene*, but the name *ethylene* has been used for so long that it is accepted by IUPAC. Table 3.1 lists some other common names.

TABLE 3.1 Common Names of Some Alkenes[a]

Compound	Systematic name	Common name
$H_2C{=}CH_2$	Ethene	Ethylene
$CH_3CH{=}CH_2$	Propene	Propylene
$CH_3C(CH_3){=}CH_2$	2-Methylpropene	Isobutylene
$H_2C{=}C(CH_3){-}CH{=}CH_2$	2-Methylbuta-1,3-diene	Isoprene

[a]Both common and systematic names are recognized by IUPAC.

PRACTICE PROBLEM 3.1

What is the IUPAC name of the following alkene?

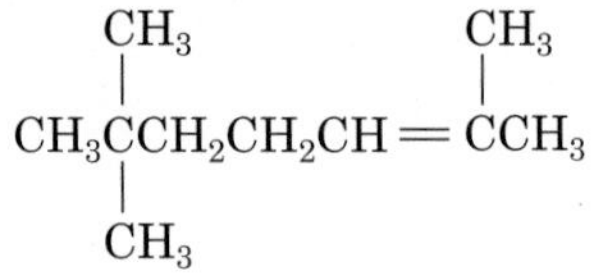

$$\begin{array}{c} \quad CH_3 \qquad\qquad\quad CH_3 \\ CH_3\overset{|}{\underset{|}{C}}CH_2CH_2CH{=}\overset{|}{C}CH_3 \\ \quad CH_3 \end{array}$$

STRATEGY First, find the longest chain containing the double bond—in this case, a heptene. Next, number the chain beginning at the end nearer the double bond, and identify the substituents at each position. In this case, there are methyl groups at C2 and C6 (two):

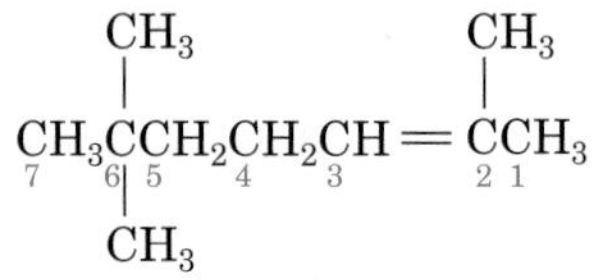

$$\begin{array}{c} \quad CH_3 \qquad\qquad\quad CH_3 \\ \underset{7}{CH_3}\overset{|}{\underset{6|}{C}}\underset{5}{C}H_2\underset{4}{C}H_2\underset{3}{C}H{=}\overset{|}{\underset{2}{C}}\underset{1}{C}H_3 \\ \quad CH_3 \end{array}$$

SOLUTION The full name is 2,6,6-trimethylhept-2-ene.

PROBLEM 3.1 Give IUPAC names for the following compounds:

(a)  $H_2C{=}CHCH_2\overset{CH_3}{\overset{|}{C}}HCH_3$ (b) $CH_3CH_2CH{=}CHCH_2CH_2CH_3$

(c) $H_2C{=}CHCH_2CH_2CH{=}CHCH_3$ (d) $CH_3CH_2CH{=}CHCH(CH_3)_2$

PROBLEM 3.2 Name the following cycloalkenes:

(a) 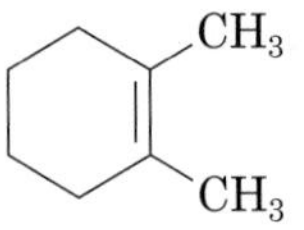 cyclohexene with CH_3 and CH_3 on the double-bond carbons (b) 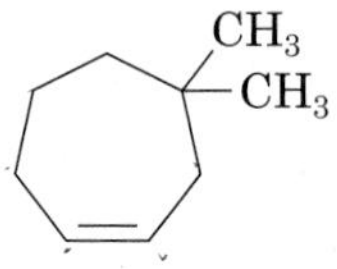 cycloheptene with CH_3, CH_3 (c) 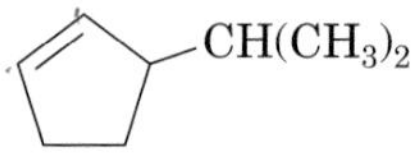 cyclopentene with $CH(CH_3)_2$

PROBLEM 3.3 Draw structures corresponding to the following IUPAC names:

(a) 2-Methylhex-1-ene (b) 4,4-Dimethylpent-2-ene

(c) 2-Methylhexa-1,5-diene (d) 3-Ethyl-2,2-dimethylhept-3-ene

3.2 Electronic Structure of Alkenes

We know from Section 1.8 that the carbon atoms in a double bond have three equivalent sp^2 hybrid orbitals, which lie in a plane at angles of 120° to one another. The fourth carbon orbital is an unhybridized p orbital perpendicular to the sp^2 plane. When two such carbon atoms approach each other, they form a σ bond by head-on overlap of sp^2 orbitals and a π bond by sideways overlap of p orbitals. The doubly bonded carbons and the four attached atoms lie in a plane, with bond angles of approximately 120° (Figure 3.1).

We also know from Section 2.5 that rotation can occur around single bonds and that open-chain alkanes like ethane and propane therefore have many rapidly interconverting conformations. The same is not true for double bonds, however. For rotation to take place around a double bond, the π part of the bond would have to break temporarily (Figure 3.1). Thus, the energy barrier to rotation around a double bond must be at least as great as the strength of the π bond itself, an estimated 268 kJ/mol (64 kcal/mol). Recall that the rotation barrier for a single bond is only about 12 kJ/mol.

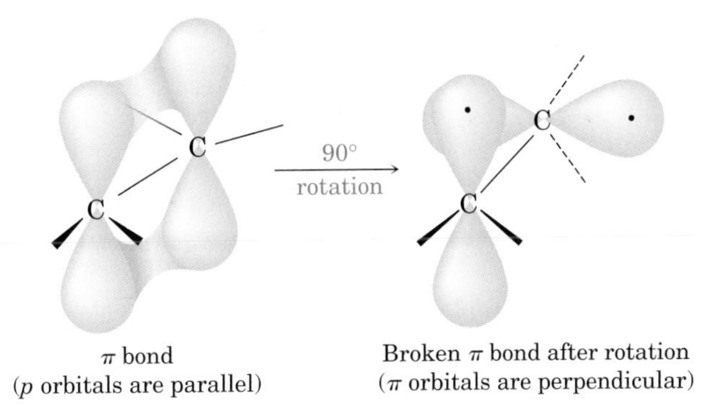

FIGURE 3.1 The π bond must break for rotation around a carbon–carbon double bond to take place.

3.3 Cis–Trans Isomers of Alkenes

The lack of rotation around carbon–carbon double bonds is of more than just theoretical interest; it also has chemical consequences. Imagine the situation for a disubstituted alkene such as but-2-ene. (*Disubstituted* means that two substituents other than hydrogen are bonded to the double-bond carbons.) The two methyl groups in but-2-ene can be either on the same side of the double bond or on opposite sides, a situation reminiscent of substituted cycloalkanes (see Section 2.8). Figure 3.2 shows the two but-2-ene isomers.

Because bond rotation can't occur at room temperature, the two but-2-ene isomers can't spontaneously interconvert and are different chemical compounds. As with disubstituted cycloalkanes (see Section 2.8), we call such compounds *cis–trans isomers.* The isomer with both substituents on the same side is *cis*-but-2-ene, and the isomer with substituents on opposite sides is *trans*-but-2-ene.

FIGURE 3.2 Cis and trans isomers of but-2-ene. The cis isomer has the two methyl groups on the same side of the double bond, and the trans isomer has the methyl groups on opposite sides.

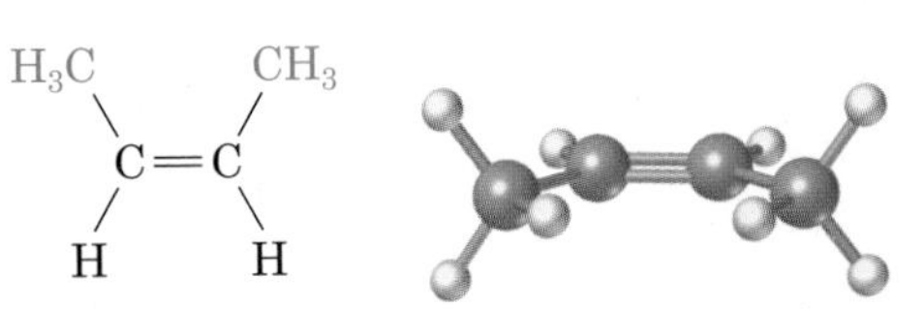

H_3C / CH_3 — C=C — H / H

***cis*-But-2-ene**

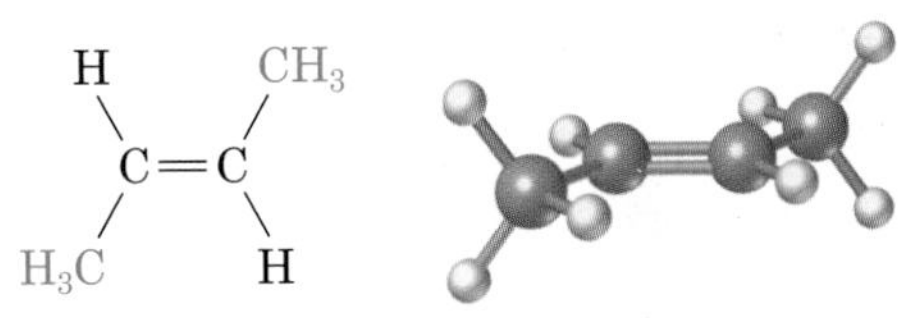

H / CH_3 — C=C — H_3C / H

***trans*-But-2-ene**

Cis–trans isomerism is not limited to disubstituted alkenes. It occurs whenever both double-bond carbons are attached to two different groups. If one of the double-bond carbons is attached to two identical groups, however, then cis–trans isomerism is not possible (Figure 3.3).

FIGURE 3.3 The requirement for cis–trans isomerism in alkenes. Both double-bond carbons must be bonded to two different groups.

$$(A)(B)C{=}C(D)(D) \;=\; (B)(A)C{=}C(D)(D)$$

These two compounds are identical; they are not cis–trans isomers.

$$(A)(B)C{=}C(D)(E) \;\neq\; (B)(A)C{=}C(D)(E)$$

These two compounds are not identical; they are cis–trans isomers.

Although the interconversion of cis and trans alkene isomers doesn't occur spontaneously, it can be brought about by treating the alkene with a strong acid catalyst. Alternatively, if a molecule contains multiple double bonds, energy from light can lead to cis–trans interconversion. If we interconvert *cis*-but-2-ene with *trans*-but-2-ene and allow them to reach equilibrium, we find that they aren't of equal stability. The trans isomer is more favored than the cis isomer by a ratio of 76 : 24.

$$(H)(H_3C)C{=}C(CH_3)(H) \underset{\text{catalyst}}{\overset{\text{Acid}}{\rightleftharpoons}} (H_3C)(H)C{=}C(CH_3)(H)$$

trans (76%) **cis (24%)**

Cis alkenes are less stable than their trans isomers because of steric (spatial) interference between the large substituents on the same side of the double bond. This is the same kind of interference, or *steric strain*, that we saw in axial methylcyclohexane (see Section 2.11).

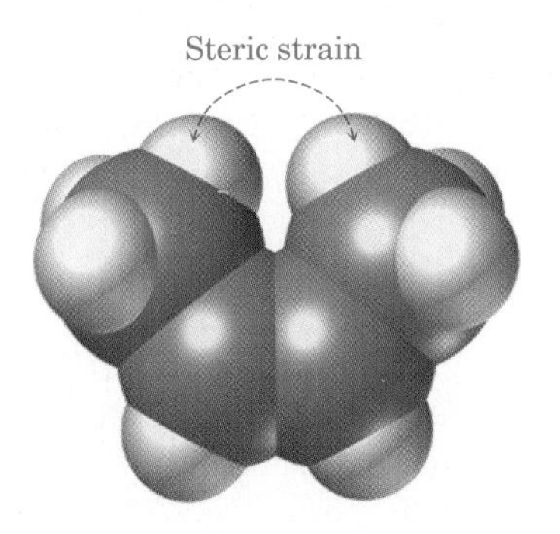

***cis*-But-2-ene**

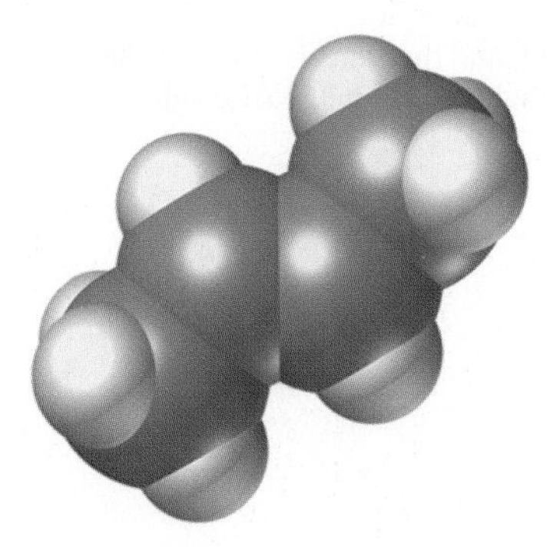

***trans*-But-2-ene**

PRACTICE PROBLEM 3.2

Draw the cis and trans isomers of 5-chloropent-2-ene.

STRATEGY First, draw the molecule without indicating isomers to see the overall structure: $ClCH_2CH_2CH{=}CHCH_3$. Then locate the two substituent groups on the same side of the double bond for the cis isomer and on opposite sides for the trans isomer.

SOLUTION

$ClCH_2CH_2$ / H on C=C, CH_3 / H on C	H / $ClCH_2CH_2$ on C=C, CH_3 / H on C
***cis*-5-Chloropent-2-ene**	***trans*-5-Chloropent-2-ene**

PROBLEM 3.4 Which of the following compounds can exist as cis–trans isomers? Draw each cis–trans pair.

(a) $CH_3CH{=}CH_2$ (b) $(CH_3)_2C{=}CHCH_3$

(c) $ClCH{=}CHCl$ (d) $CH_3CH_2CH{=}CHCH_3$

(e) $CH_3CH_2CH{=}C(Br)CH_3$ (f) 3-Methylhept-3-ene

PROBLEM 3.5 Name the following alkenes, including the cis or trans designation:

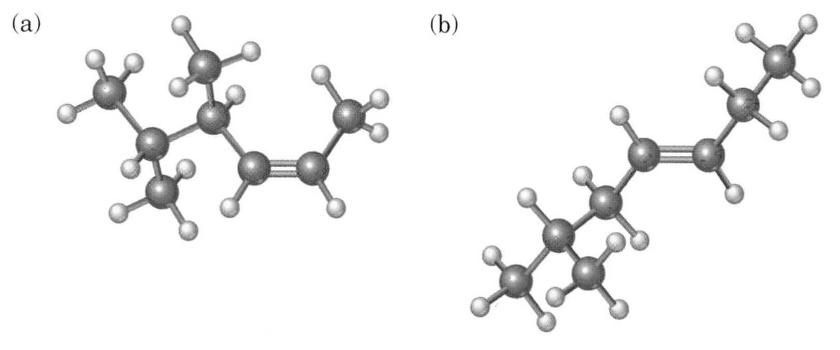

3.4 Sequence Rules: The *E,Z* Designation

The cis–trans naming system used in the previous section works only with disubstituted alkenes. With trisubstituted and tetrasubstituted double bonds, however, a more general method is needed for describing double-bond geometry. (*Trisubstituted* means three substituents other than hydrogen on the double bond; *tetrasubstituted* means four substituents other than hydrogen.)

According to the ***E,Z* system** of nomenclature, a set of **sequence rules** is used to assign priorities to the substituent groups on the double-bond carbons. Considering the double-bond carbons separately, we decide which of the two attached groups on each is higher in priority. If the higher-priority groups on each carbon are on opposite sides of the double bond, the alkene is designated ***E***, for the German *entgegen*, meaning "opposite." If the higher-priority groups are on the same side, the alkene is designated **Z**, for the German *zusammen*, meaning "together." (You can remember which is which by noting that the groups are on "ze zame zide" in the *Z* isomer.)

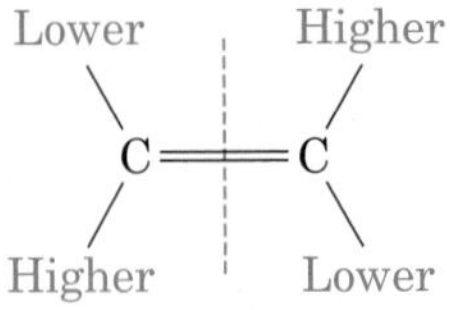

E double bond (Higher-priority groups are on opposite sides.)

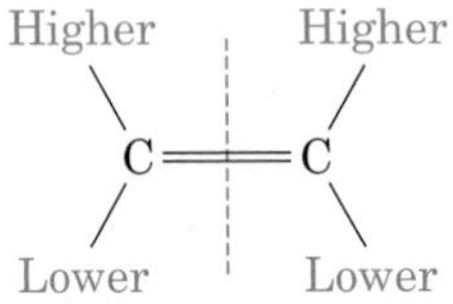

Z double bond (Higher-priority groups are on the same side.)

Called the *Cahn–Ingold–Prelog rules* after the chemists who proposed them, the sequence rules are as follows:

RULE 1 **Taking the double-bond carbons separately, look at the atoms directly attached to each carbon and rank them according to atomic number.** An atom with a higher atomic number is higher in priority than an atom with a lower atomic number. Thus, the atoms that we commonly find attached to a double-bond carbon are assigned the following priorities:

$$\overset{35}{Br} > \overset{17}{Cl} > \overset{8}{O} > \overset{7}{N} > \overset{6}{C} > \overset{1}{H}$$

For example:

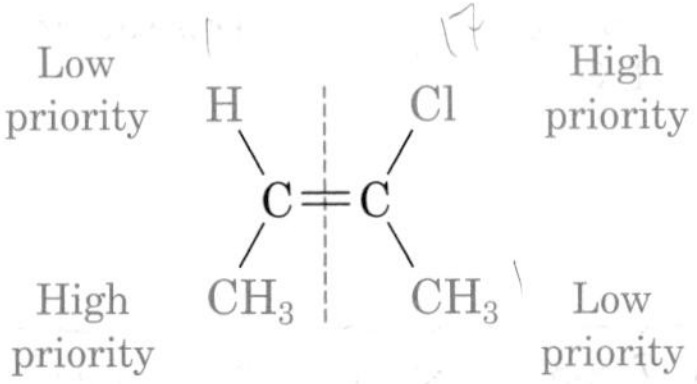

(a) (*E*)-2-Chlorobut-2-ene

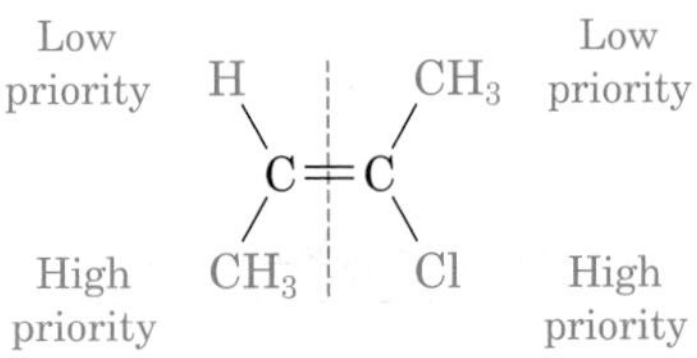

(b) (*Z*)-2-Chlorobut-2-ene

Because chlorine has a higher atomic number than carbon, it receives higher priority than a methyl (CH_3) group. Methyl receives higher priority than hydrogen, however, and isomer (a) is therefore assigned *E* geometry (high-priority groups on opposite sides of the double bond). Isomer (b) has *Z* geometry (high-priority groups on "ze zame zide" of the double bond).

RULE 2 **If a decision can't be reached by ranking the first atoms in the substituents, look at the second, third, or fourth atoms away from the double-bond carbons until the first difference is found.** Thus, an ethyl substituent, $-CH_2CH_3$, and a methyl substituent, $-CH_3$, are equivalent by rule 1 because both have carbon as the first atom. By rule 2, however, ethyl receives higher priority than methyl because ethyl has a *carbon* as its highest *second* atom, while methyl has only hydrogen as its second atom. Look at the following examples to see how the rule works:

$-CH_3$ (Lower) vs. $-CH_2CH_3$ (Higher)

$-CH_2OH$ (Lower) vs. $-CH_2OCH_3$ (Higher)

$-C(CH_3)_2H$ (Higher) vs. $-CH_2CH_3$ (Lower)

$-CH_2Cl$ (Higher) vs. $-CH(NH_2)CH_3$ (Lower)

Rule 3 **Multiple-bonded atoms are equivalent to the same number of single-bonded atoms.** For example, an aldehyde substituent (–CH=O), which has a carbon atom *doubly* bonded to *one* oxygen, is equivalent to a substituent with a carbon atom *singly* bonded to *two* oxygens:

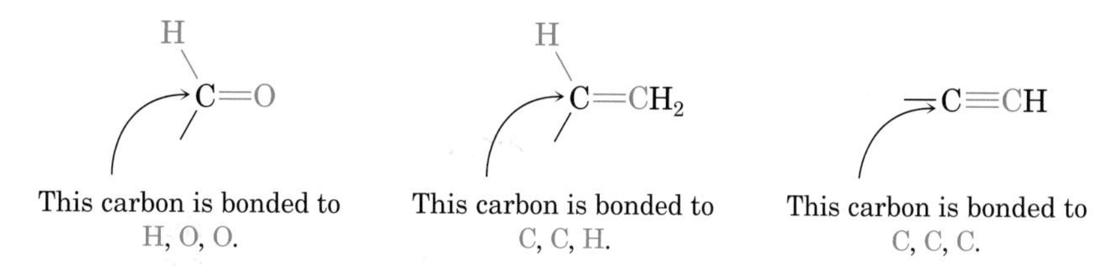

Practice Problem 3.3

Assign *E* or *Z* configuration to the double bond in the following compound:

$(H)(H_3C)C{=}C(CH(CH_3)_2)(CH_2OH)$

Strategy Look at each double-bond carbon individually, and assign priorities. Then see whether the two high-priority groups are on the same or opposite sides of the double bond.

Solution The left-hand carbon has two substituents, –H and $-CH_3$, of which $-CH_3$ receives higher priority by rule 1. The right-hand carbon also has two substituents, $-CH(CH_3)_2$ and $-CH_2OH$, which are equivalent by rule 1. By rule 2, however, $-CH_2OH$ receives higher priority than $-CH(CH_3)_2$ because $-CH_2OH$ has an *oxygen* as its highest second atom, whereas $-CH(CH_3)_2$ has *carbon* as its highest second atom. The two high-priority groups are on the same side of the double bond, so the compound has *Z* configuration.

Low — H and $CH(CH_3)_2$ — Low (C, C, H bonded to this carbon)

$C=C$

High — H_3C and CH_2OH — High (O, H, H bonded to this carbon)

***Z* configuration**

PROBLEM 3.6 Which member in each of the following sets is higher in priority?

(a) $-H$ or $-Br$ (b) $-Cl$ or $-Br$ (c) $-CH_3$ or $-CH_2CH_3$
(d) $-NH_2$ or $-OH$ (e) $-CH_2OH$ or $-CH_3$ (f) $-CH_2OH$ or $-CH{=}O$

PROBLEM 3.7 Which is higher in priority,

$$-C(=O)-OH \quad \text{or} \quad -C(=O)-OCH_3 \; ?$$

PROBLEM 3.8 Assign *E* or *Z* configuration to the following compounds:

(a) $(CH_3O)(H)C=C(Cl)(CH_3)$ (b) $(H_3C)(H)C=C(C(=O)OCH_3)(OCH_3)$

PROBLEM 3.9 Assign *E* or *Z* configuration to the following compound (blue = N):

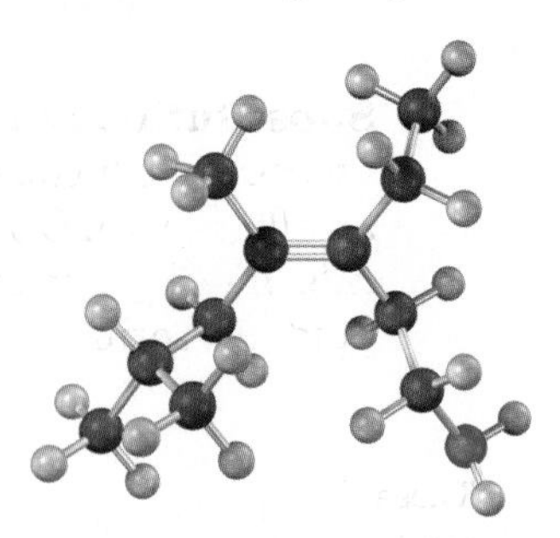

3.5 Kinds of Organic Reactions

Now that we know something about alkenes, let's learn about their chemical reactivity. As an introduction, we'll first look at some of the basic principles that underlie all organic reactions. In particular, we'll develop some general notions about why compounds react the way they do, and we'll see some methods that have been developed to help understand how reactions take place.

Organic chemical reactions can be organized either by *what kinds* of reactions occur or by *how* reactions occur. Let's look first at the kinds of reactions that take place. There are four particularly broad types of organic reactions: *additions*, *eliminations*, *substitutions*, and *rearrangements*.

- **Addition reactions** occur when two reactants add together to form a single new product with no atoms "left over." An example that we'll be studying soon is the reaction of an alkene with HCl to yield an alkyl chloride:

These two reactants . . . H—Cl + $H_2C=CH_2$ ⟶ H—C(H)(H)—C(Cl)(H)—H . . . add to give this product.

Ethylene (an alkene) **Chloroethane (an alkyl halide)**

- **Elimination reactions** are, in a sense, the opposite of addition reactions and occur when a single reactant splits into two products. An example is the reaction of an alkyl halide with base to yield an acid and an alkene:

This one reactant . . . H—C(H)(H)—C(Cl)(H)—H $\xrightarrow{\text{NaOH}}$ $H_2C=CH_2$ + H—Cl . . . gives these two products.

Chloroethane (an alkyl halide) **Ethylene (an alkene)**

- **Substitution reactions** occur when two reactants exchange parts to give two new products. An example that we saw in Section 2.4 is the reaction of an alkane with Cl_2 in the presence of ultraviolet light to yield an alkyl chloride. A –Cl group substitutes for the –H group of the alkane, and two new products result:

These two reactants . . . H—C(H)(H)—H + Cl—Cl $\xrightarrow{\text{Light}}$ H—C(H)(H)—Cl + H—Cl . . . give these two products.

Methane (an alkane) **Chloromethane (an alkyl halide)**

- **Rearrangement reactions** occur when a single reactant undergoes a reorganization of bonds and atoms to yield a single isomeric product. An example that we saw in Section 3.3 is the conversion of *cis*-but-2-ene into its isomer *trans*-but-2-ene by treatment with an acid catalyst:

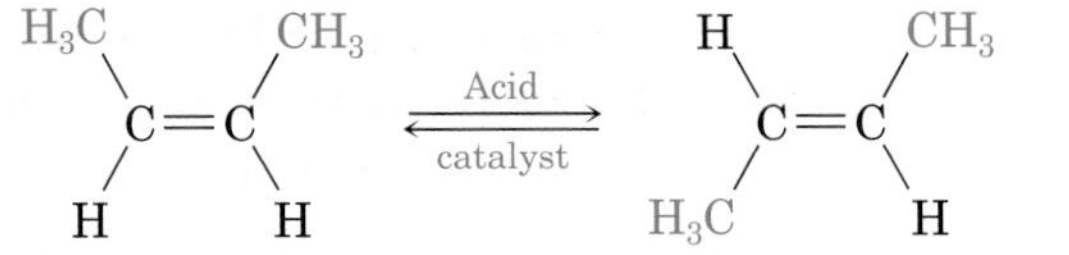

$$\underset{\textit{cis}\text{-But-2-ene (24\%)}}{(H_3C)(H)C{=}C(CH_3)(H)} \xrightleftharpoons[\text{catalyst}]{\text{Acid}} \underset{\textit{trans}\text{-But-2-ene (76\%)}}{(H)(H_3C)C{=}C(CH_3)(H)}$$

Problem 3.10 Classify the following reactions as additions, eliminations, substitutions, or rearrangements:

(a) $CH_3Br + KOH \longrightarrow CH_3OH + KBr$

(b) $CH_3CH_2OH \longrightarrow H_2C{=}CH_2 + H_2O$

(c) $H_2C{=}CH_2 + H_2 \longrightarrow CH_3CH_3$

3.6 How Reactions Occur: Mechanisms

How do reactions occur? How do atoms rearrange? The answer lies in the valence electrons. Most reactions occur with a flurry of electron activity. A **reaction mechanism** is a detailed account of these electron motions. Understanding reaction mechanisms takes time and patience, but the rewards are great. Mechanisms form the basis of a real chemical "intuition" and allow chemists to "dream" new reactions, reagents, and processes. Mechanisms, however, are models. The mechanism that we write for a chemical reaction represents our best guess as to what is occurring: No mechanism is known with 100% certainty.

In practice, a reaction mechanism is written as a series of chronologically ordered steps, with each step describing the motion of one or more electrons. When the steps are added together, the result is the net reaction. From a mechanism, we can tell when specific bonds are broken and when others are formed.

The electrons that initiate chemistry are usually *not* hidden between nuclei in a strong C–C σ bond. More often, these electrons are in accessible positions such as π bonds or lone pairs. The number of electrons that move in any mechanistic step depends greatly on the type of reaction and conditions. Most of this book deals with **polar reactions** in which two electrons move together when bonds are formed or broken. **Radical reactions** typically involve the movement of a lone electron.

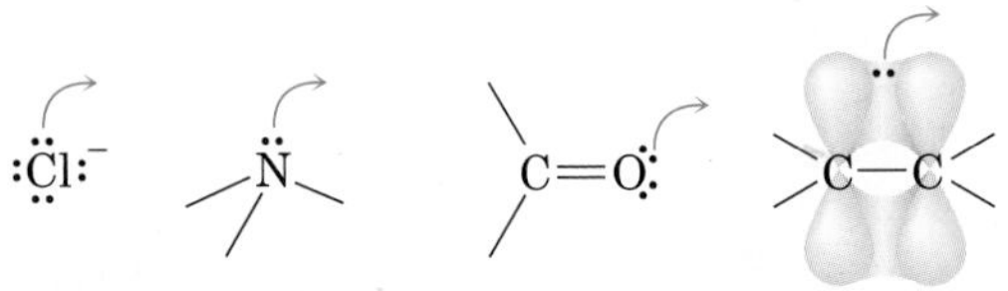

$:\ddot{C}l:^-$ $\quad$ $\ddot{N}$ $\quad$ $C{=}\ddot{O}:$ $\quad$ $C{-}C$

The electrons in lone pairs of ions or on oxygen or nitrogen as well as the π electrons of double bonds can initiate chemical reactions.

When electrons move in pairs, bonds break in a **heterolytic** manner and form in a **heterogenic** manner. That is, the electrons are distributed unequally: *both* electrons depart with a single atom or *both* are received from a single atom. In radical processes, in which electrons distribute evenly between atoms, bond breaking and forming reactions are called **homolytic** and **homogenic**, respectively. Figure 3.4 shows three one-step mechanisms, including a reaction where electrons move in pairs (polar reaction), a reaction where electrons move alone (radical reaction), and a reaction where three pairs of electrons move simultaneously.

FIGURE 3.4 Some one-step reaction mechanisms. The motion and number of electrons are depicted with single- and double-headed arrows that start with the electron(s) and end at an atom or bond.

A polar substitution reaction showing heterolysis of the C–Br bond and heterogenesis of the C–I bond.

$$:\ddot{\underset{..}{I}}:^{-} + H_3C-Br \longrightarrow I-CH_3 + :\ddot{\underset{..}{Br}}:^{-}$$

A radical elimination reaction featuring homolysis of the O–O bond.

$$-\overset{|}{\underset{|}{C}}-\ddot{\underset{..}{O}}-\ddot{\underset{..}{O}}-\overset{|}{\underset{|}{C}}- \longrightarrow 2\cdot\ddot{\underset{..}{O}}-\overset{|}{\underset{|}{C}}-$$

A rearrangement reaction showing three electron pairs moving simultaneously.

Notice how the electron motion is shown using curved arrows. Each arrow starts with the electrons and points to their destination—either an atom or a bond. The number of electrons moving is indicated with the type of arrow: a double-headed arrow denotes two electrons; a single-headed (fish-hook) arrow represents a single electron.

3.7 Mechanisms of Polar Reactions

Let's refocus our attention on polar reactions. To fully understand polar reactions, we first need to look more deeply into the effects of bond polarity on organic molecules. We saw in Section 1.9 that certain bonds in a molecule, particularly the bonds in functional groups, are often polar. When carbon bonds to a more electronegative atom such as chlorine or oxygen, the bond is polarized so that the carbon bears a partial positive charge ($\delta+$) and the electronegative atom bears a partial negative charge ($\delta-$). When carbon bonds to a less electronegative atom such as a metal, the opposite polarity results. Remember that electrostatic potential maps show electron-rich regions of a molecule in red and electron-poor regions in blue.

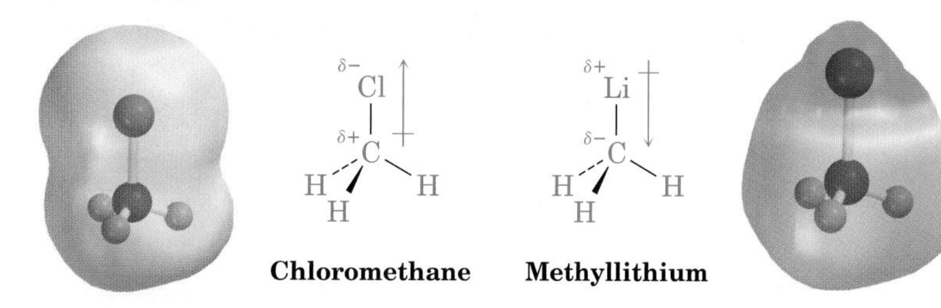

What effect does bond polarity have on chemical reactions? *Because opposite charges attract each other, the fundamental characteristic of all polar reactions is that electron-rich sites in one molecule react with electron-poor*

sites in another molecule. Bonds are made when an electron-rich atom donates a pair of electrons to an electron-poor atom. Bonds are broken when one atom leaves with both electrons.

A generalized polar reaction

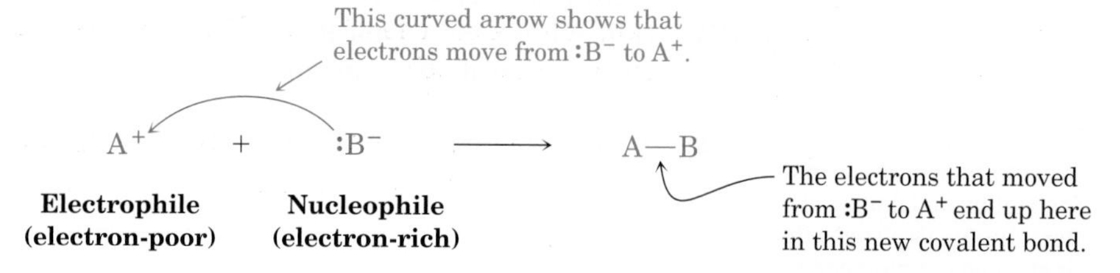

In referring to polar reactions, chemists use the words *nucleophile* and *electrophile*. Electron-rich species are called **nucleophiles** because they are "nucleus-loving" and attracted to a positive charge. Accordingly, nucleophiles frequently have one or more lone pairs of electrons and are often negatively charged. Nucleophiles form bonds with electron-poor atoms by donating an electron pair. Examples of nucleophiles are ammonia, water, hydroxide ion, and chloride ion.

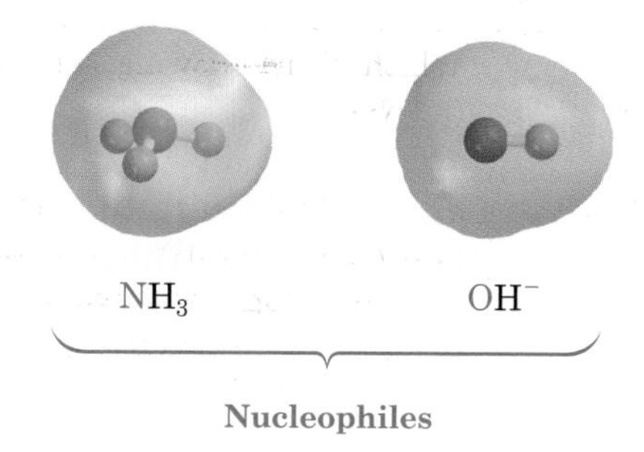

An **electrophile**, by contrast, is "electron-loving" and attracted to a negative charge. An electrophile has an electron-poor atom and can form a bond by accepting an electron pair from a nucleophile. Electrophiles are often, although not always, positively charged. Examples of electrophiles are acids (H^+ donors), alkyl halides, and carbonyl compounds.

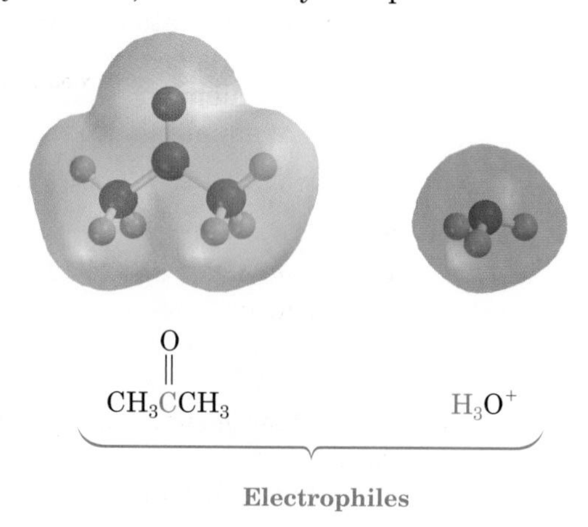

PRACTICE PROBLEM 3.4

What is the direction of bond polarity in the amine functional group C—NH_2?

STRATEGY Look at the electronegativity values in Figure 1.12 to see which atoms withdraw electrons more strongly.

SOLUTION Nitrogen is more electronegative than carbon according to Figure 1.12, so an amine is polarized with carbon $\delta+$ and nitrogen $\delta-$.

$\delta-$ NH_2 / $\delta+$ C

An amine

PRACTICE PROBLEM 3.5

Which of the following species is likely to be an electrophile, and which a nucleophile?
(a) NO_2^+ (b) CH_3OH

STRATEGY Electrophiles have an electron-poor site, either because they are positively charged or because they have a functional group containing an atom that is positively polarized. Nucleophiles have an electron-rich site, either because they are negatively charged or because they have a functional group containing an atom that has a lone pair of electrons.

SOLUTION (a) NO_2^+ (nitronium ion) is likely to be an electrophile because it is positively charged.
(b) CH_3OH (methyl alcohol) can be either a nucleophile, because it has two lone pairs of electrons on oxygen, or an electrophile, because it has polar C–O and O–H bonds.

$\overset{\delta+}{CH_3}—\overset{\delta-}{\ddot{O}}—\overset{\delta+}{H}$

Electrophilic Nucleophilic Electrophilic

PROBLEM 3.11 What is the direction of bond polarity in the following functional groups? (See Figure 1.12 for electronegativity values.)
(a) Aldehyde (b) Ether
(c) Ester (d) Alkylmagnesium bromide, R—MgBr

PROBLEM 3.12 Which of the following are most likely to behave as electrophiles, and which as nucleophiles? Explain.
(a) NH_4^+ (b) CN^- (c) Br^+ (d) CH_3NH_2 (e) H—C≡C—H
e n e n

Problem 3.13 An electrostatic potential map of boron trifluoride is shown. Is BF_3 likely to be an electrophile or a nucleophile? Draw a Lewis structure for BF_3, and explain the result.

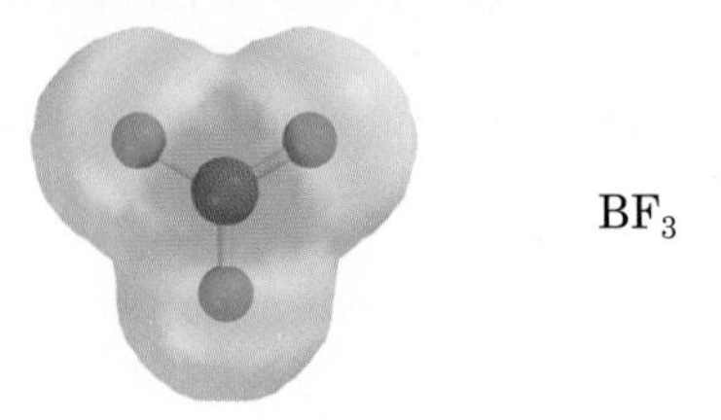

3.8 The Mechanism of an Organic Reaction: Addition of HCl to Ethylene

Let's look in detail at a typical polar reaction, the reaction of ethylene with HCl. When ethylene is treated with hydrogen chloride at room temperature, chloroethane is produced. Overall, the reaction can be written as:

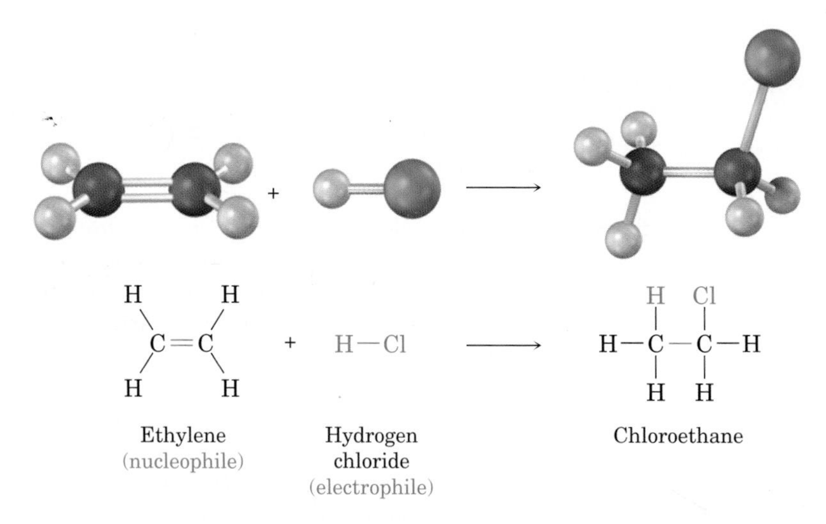

This reaction, an example of a general polar reaction type known as an *electrophilic addition to an alkene,* can be understood using the general concepts discussed in the previous section. Let's begin by looking at the nature of the two reactants.

What do we know about ethylene? We know that a carbon–carbon double bond results from orbital overlap of two sp^2-hybridized carbon atoms. The σ part of the double bond results from sp^2–sp^2 overlap, and the π part results from p–p overlap. What kind of chemical reactivity might we expect of a C=C bond? Unlike the electrons in *alkanes*, which are inaccessible because they are tied up in strong, nonpolar, σ C–C and C–H bonds, the π electrons of alkenes are accessible and reactive (Figure 3.5). These weakly held electrons, which make the C=C bond behave as a nucleophile, can be donated to form a new bond to an electron-poor atom, an electrophile. In fact, reaction with electrophiles is the most common reaction of alkenes.

What about HCl? As a strong acid, HCl is a powerful proton (H^+) donor. Because a proton is positively charged and electron-poor, it is a good electrophile. Thus, the reaction of H^+ with ethylene is a typical electrophile–nucleophile combination, characteristic of all polar reactions.

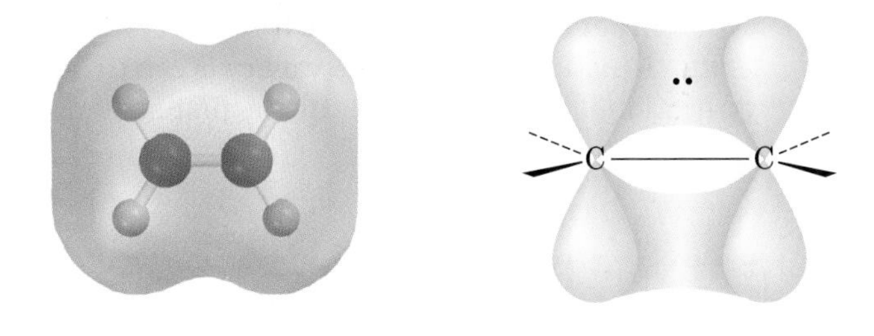

Carbon–carbon π bond

FIGURE 3.5 A carbon–carbon double bond is electron-rich (nucleophilic), and its π electrons are relatively accessible to reaction with external reagents.

We can imagine that the electrophilic addition reaction of ethylene with HCl proceeds by the mechanism shown:

Step 1 $\quad >C{=}C< \; + \; H^+ \longrightarrow \; >C^+{-}C(H)<$

Step 2 $\quad :\ddot{Cl}:^- \; + \; >C^+{-}C(H)< \longrightarrow \; >C(Cl){-}C(H)<$

Sum of steps $\quad >C{=}C< \; + \; H^+ \; + \; Cl^- \longrightarrow \; >C(Cl){-}C(H)<$

As always, the reaction begins with electrons. In this case, two electrons from the ethylene π bond move to form a new σ bond between the H^+ and one of the ethylene carbon atoms. This electron motion is indicated with a double-headed, curved arrow that starts with the electrons and points to the H^+. The other ethylene carbon atom, having lost its share of the π electrons, is now left with a vacant p orbital and only six valence electrons. This carbon atom carries a positive charge and is called a carbon cation, or a **carbocation**. Carbocations are electrophiles that readily accept electron pairs.

In the second step of the mechanism, the nucleophilic chlorine ion donates an electron pair to form a new σ bond with the carbocation. Once again, a curved arrow shows the path of the electron-pair movement from Cl^- ion to carbon. In this case, the result is a stable and neutral addition product.

Notice that the sum of the mechanistic steps is the overall reaction, and throughout each step of the mechanism, both atoms and charge are conserved.

Throughout this book, we'll show mechanisms vertically as in Figure 3.6. This depiction allows the entire process to be seen more efficiently than the traditional stepwise mechanisms just shown.

FIGURE 3.6 MECHANISM: The mechanism of the electrophilic addition of HCl to ethylene. The reaction takes place in two steps and involves an intermediate carbocation.

The electrophile H^+ is attacked by the π electrons of the double bond, and a new C–H σ bond is formed. This leaves the other carbon atom with a + charge and a vacant p orbital.

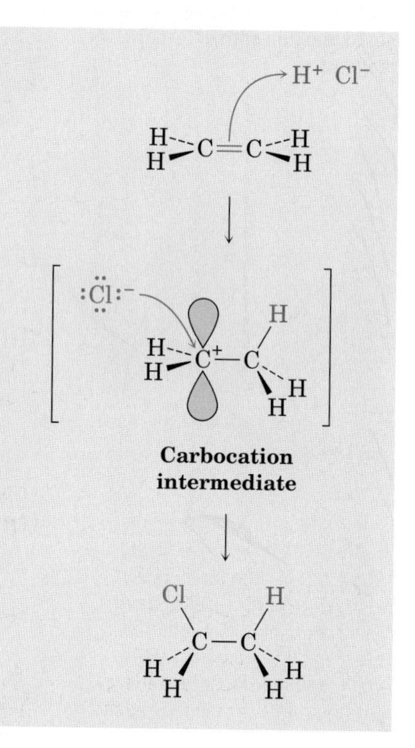

Cl^- donates an electron pair to the positively charged carbon atom, forming a C–Cl σ bond and yielding the neutral addition product.

PRACTICE PROBLEM 3.6

What product would you expect from reaction of HCl with cyclohexene?

STRATEGY HCl adds to the double-bond functional group in cyclohexene in exactly the same way it adds to ethylene, yielding an addition product.

SOLUTION

+ HCl ⟶

Cyclohexene

Chlorocyclohexane

PROBLEM 3.14 Reaction of HCl with 2-methylpropene yields 2-chloro-2-methylpropane. What is the structure of the carbocation formed during the reaction? Show the mechanism of the reaction.

$$(CH_3)_2C{=}CH_2 + HCl \longrightarrow (CH_3)_3C{-}Cl$$

PROBLEM 3.15 Reaction of HCl with pent-2-ene yields a mixture of two addition products. Write the reaction, and show the two products.

3.9

Describing a Reaction: Reaction Energy Diagrams and Transition States

For a reaction to take place, reactant molecules must collide and a reorganization of atoms and bonds must occur. Over the years, chemists have developed a method for depicting the energy changes that occur during a reaction by using **reaction energy diagrams** of the sort shown in Figure 3.7. The vertical axis of the diagram represents the total energy of all reactants, and the horizontal axis represents the progress of the reaction from beginning (left) to end (right). Notice how we can break this diagram into regions corresponding to each step of our mechanism. Let's look again at the addition of HCl to ethylene as an example.

At the beginning of the reaction, ethylene and HCl have the total amount of energy indicated by the reactant level on the left side of the diagram (Figure 3.7a). As the two molecules crowd together, their electron clouds repel each other, causing the energy level to rise (Figure 3.7b). If the collision has occurred with sufficient force and proper orientation, the reactants continue to approach each other despite the repulsion until the new C–H bond starts to form and the H–Cl bond starts to break (Figure 3.7c). At some point, a structure of maximum energy is reached, a point we call the **transition state** (Figure 3.7d).

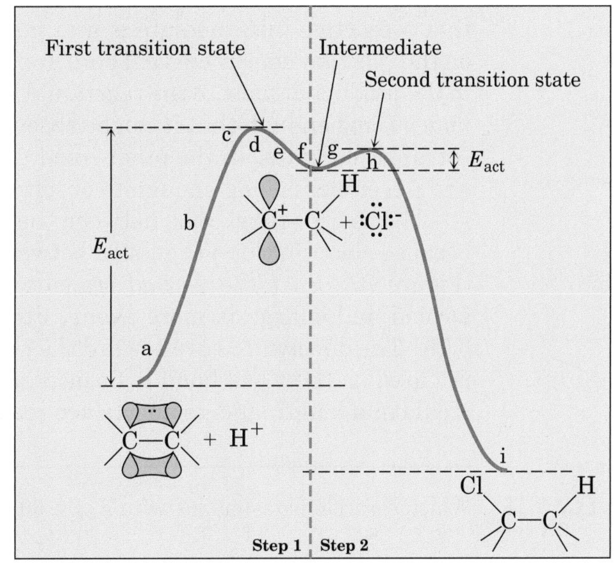

FIGURE 3.7 Energy increases from the ground state (a) as the reactants approach each other (b and c); this increase continues until a point of maximum energy, the transition state (d), is reached. Successful reaction releases energy (e) until a high-energy, reactive intermediate (f) is formed. As the C–Cl bond forms, energy increases again (g) until a second transition state is reached (h). The reaction is favored because the energy of the products (i) is lower than that of the reagents.

The transition state represents the highest-energy structure involved in this step of the reaction and can't be isolated or directly observed. Nevertheless, we can imagine it to be a kind of activated complex of the two reactants in which the C=C π bond is partially broken and the new C–H bond is partially formed.

The energy difference between reactants and transition state, called the **activation energy, E_{act}**, helps determine how rapidly a reaction occurs. A large activation energy results in a slow reaction because few of the reacting molecules collide with enough energy to reach the transition state. A small activation energy results in a rapid reaction because almost all reacting molecules are energetic enough to climb to the transition state. As an analogy, think about hikers climbing over a mountain pass. If the pass is a high one, the hikers need a lot of energy and surmount the barrier slowly. If the pass is low, however, the hikers need less energy and reach the top quickly.

Most organic reactions have activation energies in the range 40 to 125 kJ/mol (10–30 kcal/mol). Reactions with activation energies less than 80 kJ/mol take place spontaneously at or below room temperature, whereas reactions with higher activation energies normally require heating. Heat gives the colliding molecules enough energy to climb the activation barrier.

Once the transition state has been reached, the reaction proceeds to yield carbocation product (Figure 3.7e). Energy is released as the new C–H bond forms fully, and the curve in the reaction energy diagram in Figure 3.7 therefore turns downward until it reaches a minimum. The structure represented by this minimum point is called an **intermediate** (Figure 3.7f). The carbocation intermediate in our reaction is formed in the first step of our mechanism and consumed in the second step.

The overall energy change for this first step of the mechanism is the difference between the energy levels of the reactants and the carbocation as shown in Figure 3.7. Since the carbocation is less stable than the alkene reactant, its energy level is higher and energy is absorbed in this step of the mechanism. Because the energy level of this intermediate is higher than that of either the initial reactants (ethylene + HCl) or the final product (chloroethane), the intermediate is reactive and cannot be isolated. Note, however, that a reactive intermediate is at a minimum, although a high energy one, on the reaction energy curve. There are small energy barriers for it to react on to the product or back to the reactant. Under some circumstances, its lifetime may be long enough that it can be observed by modern analytical equipment. A transition state, on the other hand, is at a maximum on the energy curve and never exists long enough to observe directly.

The barrier to reaction between the carbocation and chlorine anion is low because electron cloud repulsion between these molecules is greatly reduced (Figure 3.7g). At the second transition state, the C–Cl bond is partially formed and charge is more evenly distributed between the atoms (Figure 3.7h). The downward turn toward the products indicates that energy is being released as the C–Cl bond is formed and continues to be released until the ideal bond length and geometry are reached in the product (Figure 3.7i).

PROBLEM 3.16 Which reaction is faster, one with $E_{act} = 60$ kJ/mol or one with $E_{act} = 80$ kJ/mol?

3.10 Describing a Reaction: Energetics and Catalysis

How do we know when reactions will occur? How fast will they occur? First, let's talk about *when*. Reactions occur when the overall change in energy—the difference between the energy of the initial reactants (far left) and energy of the final products (far right)—is favorable, or downhill. Said another way, reactions where the product energy is less than the reactant energy proceed spontaneously. This observation is always true, regardless of the shape of the reaction energy curve. *If the energy level of the final products is lower than that of the reactants, energy is released and the reaction is favorable. If the energy level of the final products is higher than that of the reactants, energy is absorbed and the reaction is not favorable.* Note, for example, that the energy diagram for the reaction of HCl with ethylene in Figure 3.7 shows that the energy level of the final product is lower than that of the reactants. Thus, the reaction is favorable and occurs spontaneously with a release of energy.

How fast a reaction occurs cannot be easily predicted from the difference in energy between reactants and products. The rate of a reaction depends on its activation energy. The higher the activation energy, the higher the barrier to reaction and the slower the reaction will proceed.

How can we speed up a reaction? There are two ways. We could give the reactants more energy so that they are better able to pass over the activation energy barrier. In some cases, this is done easily: we simply increase the temperature. However, even at high temperatures, some reactions are still too slow. Alternatively, we can look for another mechanism to accomplish the reaction by changing the species present. Adding new species can sometimes provide a different mechanism with a different activation barrier and, accordingly, a different rate. Chemists describe this activity as trying to *catalyze* a reaction. A **catalyst** is a compound that changes the rate of a chemical reaction by providing a new mechanism. In the process of changing the rate, the catalyst is not consumed. Common catalysts include acid (H^+), metals, and enzymes.

The best way to illustrate the role of a catalyst is with an energy diagram, as shown in Figure 3.8. As an example, let's look at the reaction between alkenes and H_2 as expressed by:

$$H_2 + H_2C{=}CH_2 \longrightarrow H_3C{-}CH_3$$

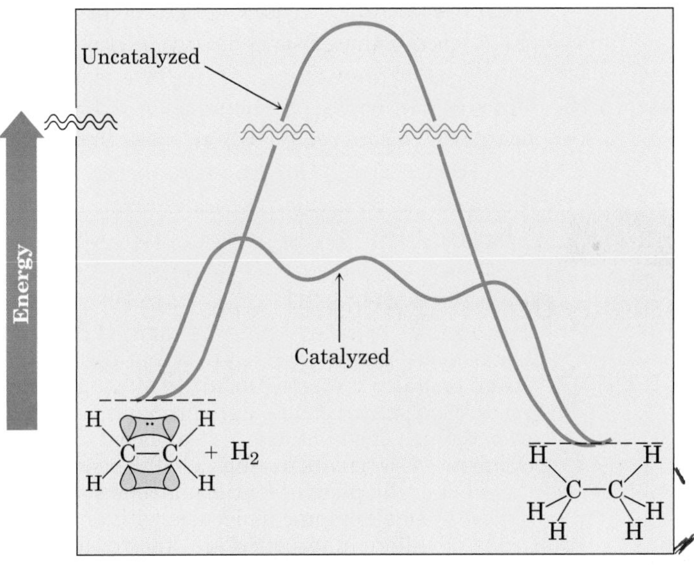

FIGURE 3.8 A reaction energy diagram showing the catalyzed and uncatalyzed reaction of ethylene and H_2. The barrier in the uncatalyzed reaction is sufficiently high that the reaction does not proceed readily. A catalyst increases the rate of reaction by introducing a different reaction mechanism. The actual diagram for the catalyzed reaction is hypothetical: this simple reaction still remains the focus of chemical research.

Even under even extreme conditions, the reaction of ethylene with hydrogen will not proceed because the activation energy barrier is too high. However, by introducing a catalyst in the form of a metal such as platinum or palladium, the reaction proceeds quickly under mild conditions. The metal catalyst provides a new reaction path for the reagents to follow. This new reaction path comes with new intermediates and new activation energy barriers for reactions. Catalysts increase reaction rates by lowering the activation barrier to reaction. It is important to remember that adding more steps to a reaction mechanism as denoted by more transition states and more intermediates has no impact on the rate of the reaction. Reaction rates are affected only by the height of the activation energy barrier.

PRACTICE PROBLEM 3.7

Sketch a reaction energy diagram for a one-step reaction that is fast and releases a large amount of energy.

STRATEGY A fast reaction has a low E_{act}, and a reaction that releases a large amount of energy forms products that are much more stable than reactants.

SOLUTION

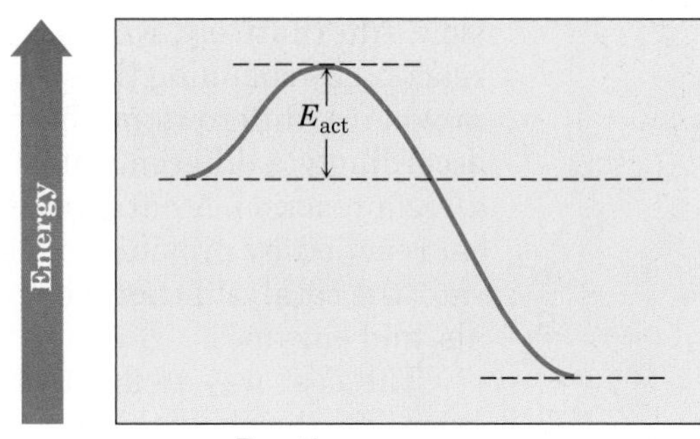

PROBLEM 3.17 Sketch reaction energy diagrams to represent each of the following situations.

(a) A reaction that releases energy and takes place in one step

(b) A reaction that absorbs energy and takes place in one step

PROBLEM 3.18 Draw a reaction energy diagram for a two-step reaction whose first step absorbs energy and whose second step releases energy. Label the intermediate.

INTERLUDE

Terpenes: Naturally Occurring Alkenes

It has been known for centuries that distillation of many plant materials with steam produces a fragrant mixture of liquids called *essential oils.* For hundreds of years, such plant extracts have been used as medicines, spices, and perfumes. The investigation of essential oils also played a major role in the emergence of organic chemistry as a science during the 19th century.

Chemically, plant essential oils consist largely of mixtures of compounds called *terpenes*—small organic molecules with an immense diversity of structure. Thousands of different terpenes are known, and many have carbon–carbon double bonds. Some are hydrocarbons, and others contain oxygen; some are open-chain molecules, and others contain rings. For example:

H_3C CH_3 CH_3 O

Myrcene (oil of bay) **α-Pinene (turpentine)** **Carvone (spearmint oil)**

Continued

© CORBIS

The fragrance of these oranges is due primarily to limonene, a simple terpene.

All terpenes are related, regardless of their apparent structural differences. According to a formalism called the *isoprene rule*, terpenes can be thought of as arising from head-to-tail joining of five-carbon isoprene (2-methylbuta-1,3-diene) units. Carbon 1 is the head of the isoprene unit, and carbon 4 is the tail. For example, myrcene contains two isoprene units joined head to tail, forming an eight-carbon chain with 2 one-carbon branches. α-Pinene similarly contains two isoprene units assembled into a more complex cyclic structure. (See if you can identify the isoprene units in α-pinene.)

Head Tail 1 2 3 4 →

Isoprene **Myrcene**

Terpenes are classified according to the number of isoprene units they contain. Thus, *monoterpenes* are 10-carbon substances biosynthesized from two isoprene units, *sesquiterpenes* are 15-carbon molecules from three isoprene units, *diterpenes* are 20-carbon substances from four isoprene units, and so on. Monoterpenes and sesquiterpenes are found primarily in plants, but the higher terpenes occur in both plants and animals and many have important biological roles. The triterpene lanosterol, for example, is the precursor from which all steroid hormones are made.

CH_3 H CH_3 CH_3 HO H H H_3C CH_3

Lanosterol, a triterpene (C_{30})

Research has shown that isoprene itself is not the true biological precursor of terpenes. Nature instead uses two "isoprene equivalents"—isopentenyl diphosphate and dimethylallyl diphosphate—five-carbon molecules that are themselves made from acetic acid. Every step in the biological conversion from acetic acid through lanosterol to human steroids has been worked out—an immense achievement for which several Nobel Prizes have been awarded.

3 $H_3C-C(=O)-OH$ →

Acetic acid

Isopentenyl diphosphate

Dimethylallyl diphosphate

Summary and Key Words

Activation energy, E_{act}, p. 91
Addition reaction, p. 83
Alkene, p. 73
Carbocation, p. 89
Catalyst, p. 93
Electrophile, p. 86
Elimination reaction, p. 83
E,Z system, p. 79
Heterogenic, p. 84
Heterolytic, p. 84
Homogenic, p. 85
Homolytic, p. 85
Intermediate, p. 92
Nucleophile, p. 86
Polar reaction, p. 84
Radical reaction, p. 84
Reaction energy diagram, p. 91
Reaction mechanism, p. 84
Rearrangement reaction, p. 83
Sequence rules, p. 79
Substitution reaction, p. 83
Transition state, p. 91
Unsaturated, p. 74

Alkenes are hydrocarbons that contain one or more carbon–carbon double bonds. Because they contain fewer hydrogens than related alkanes, alkenes are often referred to as **unsaturated**.

A double bond consists of two parts: a σ bond formed by head-on overlap of two sp^2 orbitals and a π bond formed by sideways overlap of two *p* orbitals. Because rotation around the double bond is not possible, substituted alkenes can exist as **cis–trans stereoisomers**. The geometry of a double bond can be described as either ***Z*** (*zusammen*) or ***E*** (*entgegen*) by application of a series of sequence rules.

A full description of how a reaction occurs is called its **mechanism**. There are two kinds of organic mechanisms: polar and radical. **Polar reactions**, the most common kind, occur when an electron-rich reagent, or **nucleophile**, donates an electron pair to an electron-poor reagent, or **electrophile**, in forming a new bond. **Radical reactions** involve odd-electron species and occur when each reactant donates one electron in forming a new bond.

A reaction can be described pictorially by using a **reaction energy diagram**, which follows the course of the reaction from reactant to product. Every reaction proceeds through a **transition state**, which represents the highest-energy point reached and is a kind of activated complex between reactants. The amount of energy needed by reactants to reach the transition state is the **activation energy**, E_{act}. The larger the activation energy, the slower the reaction.

Many reactions take place in more than one step and involve the formation of an **intermediate**. An intermediate is a species that is formed during the course of a multistep reaction and that lies in an energy minimum between two transition states. Intermediates are more stable than transition states but are often too reactive to be isolated. **Catalysts** increase the rates of reaction by providing an alternative mechanism.

EXERCISES

Visualizing Chemistry

3.19 Give IUPAC names for the following alkenes, and convert each drawing into a skeletal structure:

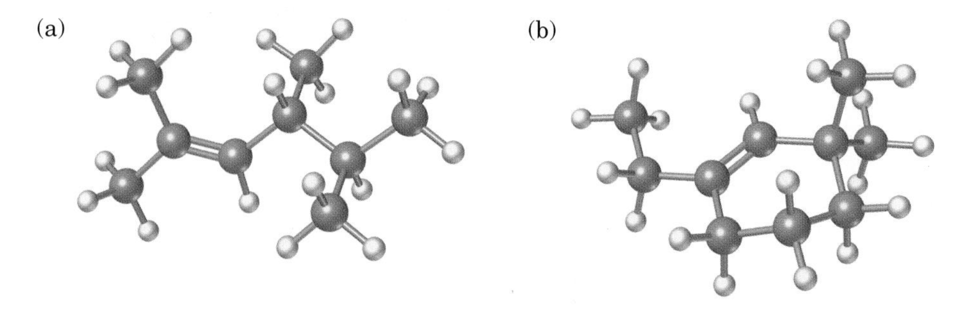

3.20 Assign stereochemistry (E or Z) to each of the following alkenes, and convert each drawing into a skeletal structure (red = O, yellow-green = Cl):

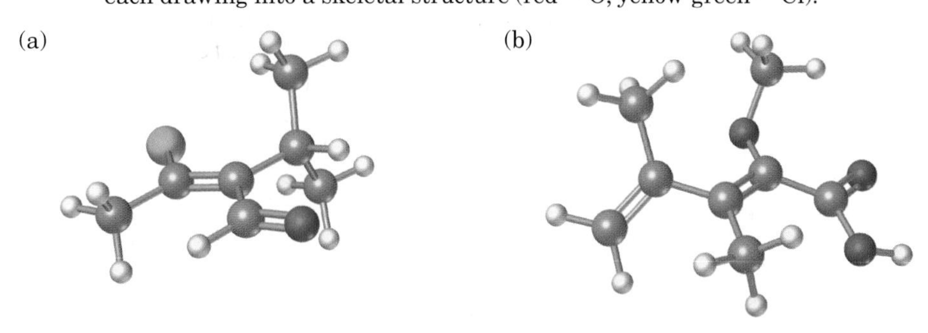

3.21 The following alkyl chloride can be prepared by addition of HCl to two different alkenes. Name and draw the structures of both (yellow-green = Cl).

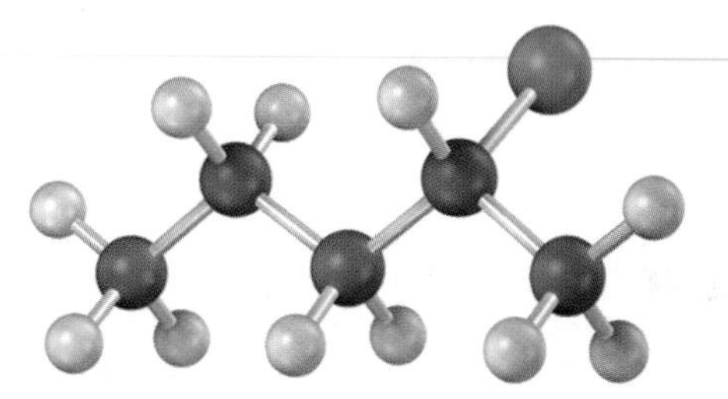

3.22 The following carbocation is a possible intermediate in the electrophilic addition of HCl with two different alkenes. Write structures for both.

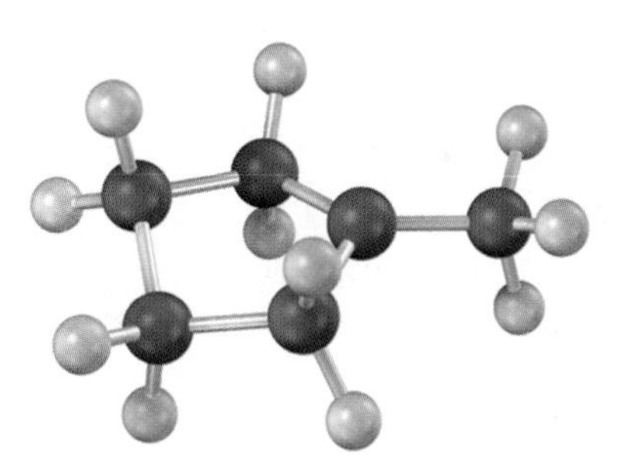

3.23 An electrostatic potential map of formaldehyde is shown. Is the formaldehyde carbon atom likely to be electrophilic or nucleophilic? Explain.

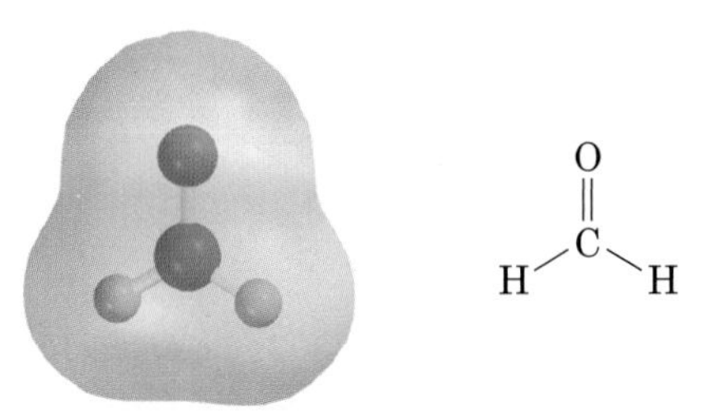

Formaldehyde

Additional Problems

Functional Groups

3.24 Identify the functional groups in each of the following molecules:

(a) $CH_3CH_2C{\equiv}N$ (b) [cyclopentyl–OCH_3] (c) $CH_3\overset{O}{\overset{\|}{C}}CH_2\overset{O}{\overset{\|}{C}}OCH_3$

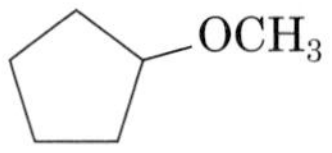

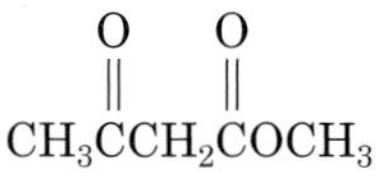

(d) [cyclohexa-2,5-diene-1,4-dione] (e) [$CH_3CH{=}CHC(=O)NH_2$] (f) [benzene–C(=O)H]

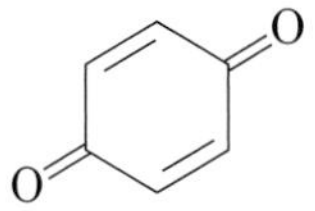

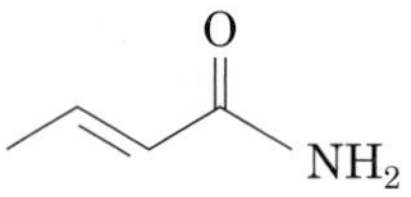

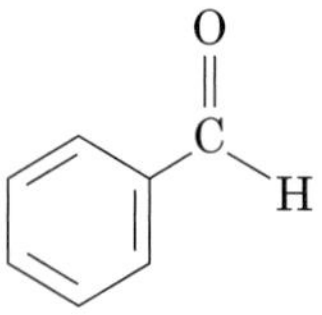

3.25 Predict the direction of polarization of the functional groups you identified in Problem 3.24.

3.26 Identify the functional groups in each of the following molecules:

(a) [benzene–$CH_2CH(NH_2)CH_3$]

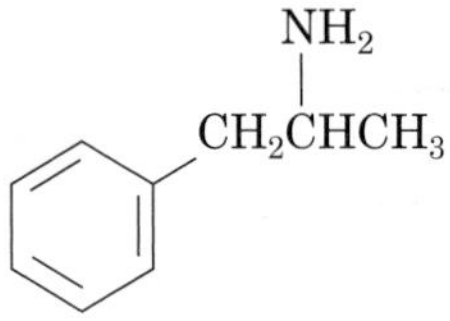

Amphetamine

(b) [H_3C-pyrimidine with NH_2, CH_2–thiazolium (N^+, S) with CH_3 and CH_2CH_2OH; Cl^-]

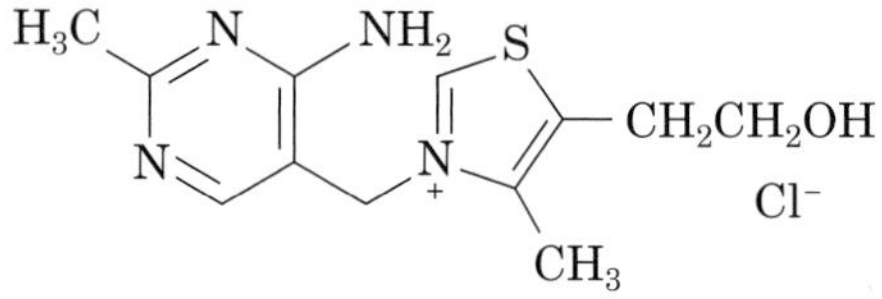

Thiamine

Reactions, Nucleophiles, and Electrophiles

3.27 Explain the differences between addition, elimination, substitution, and rearrangement reactions.

3.28 Which of the following are most likely to behave as electrophiles and which as nucleophiles?

(a) Cl^- (b) $N(CH_3)_3$ (c) Hg^{2+} (d) CN^- (e) CH_3^+

Nomenclature

3.29 Give IUPAC names for the following alkenes:

(a) $CH_3CH{=}CH\underset{}{C}H(CH_3)CH_2CH_3$ (b) $CH_3CH{=}CHCH(CH_2CH_2CH_3)CH_2CH_2CH_3$

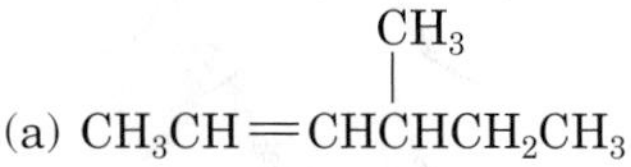

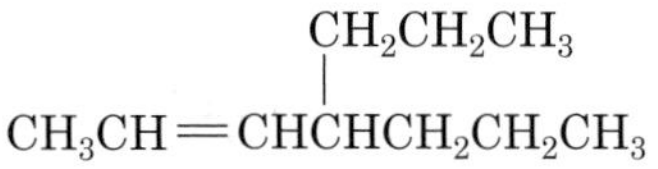

(c) $H_2C{=}C(CH_2CH_3)CH_2CH_3$ (d) $H_2C{=}C{=}CHCH_3$

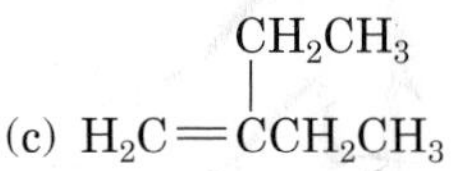

3.30 Name the following cycloalkenes by IUPAC rules:

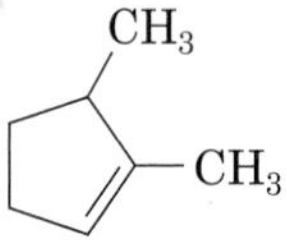

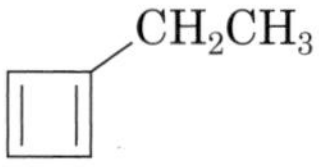

(a) [cyclohexene with CH_3] (b) [cyclopentene with two CH_3] (c) [cyclobutene with CH_2CH_3] (d) [cyclohexene with two CH_3]

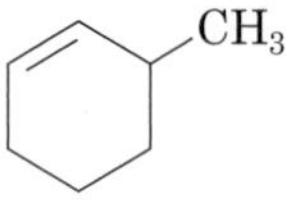

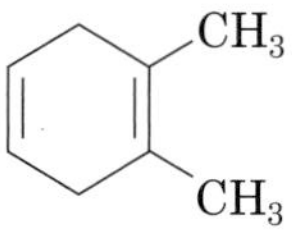

3.31 Draw structures corresponding to the following IUPAC names:

(a) 3-Propylhept-2-ene (b) 2,4-Dimethylhex-2-ene
(c) Octa-1,5-diene (d) 4-Methylpenta-1,3-diene
(e) *cis*-4,4-Dimethylhex-2-ene (f) (*E*)-3-Methylhept-3-ene

3.32 Draw the structures of the following cycloalkenes:

(a) *cis*-4,5-Dimethylcyclohexene (b) 3,3,4,4-Tetramethylcyclobutene

3.33 The following names are incorrect. Draw each molecule, and give its correct name.

(a) 1-Methylcyclopent-2-ene (b) 1-Methylpent-1-ene
(c) 6-Ethylcycloheptene (d) 2-Ethyl-3-methylcyclohexene

3.34 A compound of formula C_9H_{14} contains no rings or triple bonds. How many double bonds does it have?

CIS–TRANS ISOMERS

3.35 Which of the following molecules show cis–trans isomerism?

(a)
```
   CH3
   |
CH3C=CHCH2CH3
```

(b)
```
        H3C  CH3
          |  |
ClCH2CH2C=CCH2CH2Cl
```

(c) HO (cyclohexane ring bearing HO and an ethylidene double bond)

3.36 Draw and name molecules that meet the following descriptions:
(a) An alkene, C_6H_{12}, that does not show cis–trans isomerism
(b) The E isomer of a trisubstituted alkene, C_6H_{12}
(c) A cycloalkene, C_7H_{12}, with a tetrasubstituted double bond

3.37 Neglecting cis–trans isomers, there are five substances with the formula C_4H_8. Draw and name them.

3.38 Which of the molecules you drew in Problem 3.37 show cis–trans isomerism? Draw and name their cis–trans isomers.

3.39 Draw four possible structures for each of the following formulas:
(a) C_6H_{10} (b) C_8H_8O (c) $C_7H_{10}Cl_2$

3.40 How can you explain the fact that cyclohexene does not show cis–trans isomerism but cyclodecene does?

3.41 Rank the following pairs of substituents in order of priority according to the sequence rules:

(a) $-CH_2CH_3$, $-CH_2CH_2CH_3$

(b)
```
  O
  ||
—C—CH3        —CH2OH
```

(c) —(cyclopentyl) —$CH(CH_3)_2$

(d)
```
     CH3
     |
—CH2CHCH2Cl       —CH2CH2CH2Br
```

3.42 Rank the following sets of substituents in order of priority according to the sequence rules:
(a) $-CH_3$, $-Br$, $-H$, $-I$
(b) $-OH$, $-OCH_3$, $-H$, $-CO_2H$
(c) $-CH_3$, $-CO_2H$, $-CH_2OH$, $-CHO$
(d) $-CH_3$, $-CH{=}CH_2$, $-CH_2CH_3$, $-CH(CH_3)_2$

3.43 Assign E or Z configuration to the double bonds in the following compounds:

(a)
```
HOCH2     CH3
     \   /
      C=C
     /   \
  CH3     H
```

(b)
```
       O
       ||
   HO—C      H
       \    /
        C=C
       /   \
     Cl     OCH3
```

3.44 Draw and name the five C_5H_{10} alkene isomers. Ignore cis–trans isomers.

3.45 Draw and name all possible stereoisomers of hepta-2,4-diene.

3.46 Menthene, a hydrocarbon found in mint plants, has the IUPAC name 1-isopropyl-4-methylcyclohexene. What is the structure of menthene?

3.47 α-Farnesene is a constituent of the natural waxy coating found on apples. What is its IUPAC name?

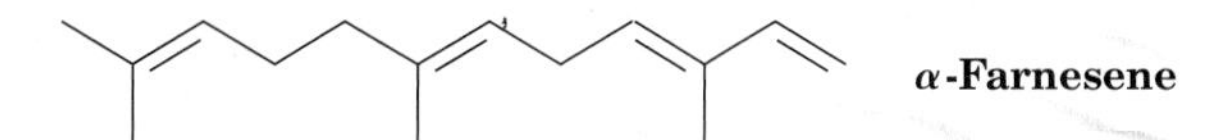

α-Farnesene

3.48 Indicate E or Z configuration for each of the double bonds in α-farnesene (see Problem 3.47).

REACTION ENERGY DIAGRAMS

3.49 If a reaction has $E_{\text{act}} = 15$ kJ/mol, is it likely to be fast or slow at room temperature? Explain.

3.50 Draw a reaction energy diagram for a two-step reaction that releases energy and whose first step is faster than its second step. Label the parts of the diagram corresponding to reactants, products, transition states, intermediate, activation energies, and overall energy change.

3.51 Draw a reaction energy diagram for a two-step reaction whose second step is faster than its first step.

3.52 Draw a reaction energy diagram for a reaction whose products and reactants are of equal stability.

3.53 Describe the difference between a transition state and a reaction intermediate.

3.54 Consider the reaction energy diagram shown, and answer the following questions:

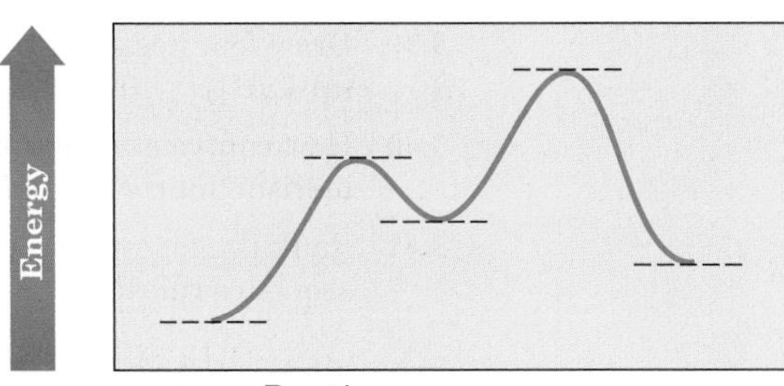

(a) Indicate the overall energy change for the reaction. Is it positive or negative?
(b) How many steps are involved in the reaction?
(c) Which step is faster?
(d) How many transition states are there? Label them.

INTEGRATED PROBLEMS

3.55 Name the following cycloalkenes:

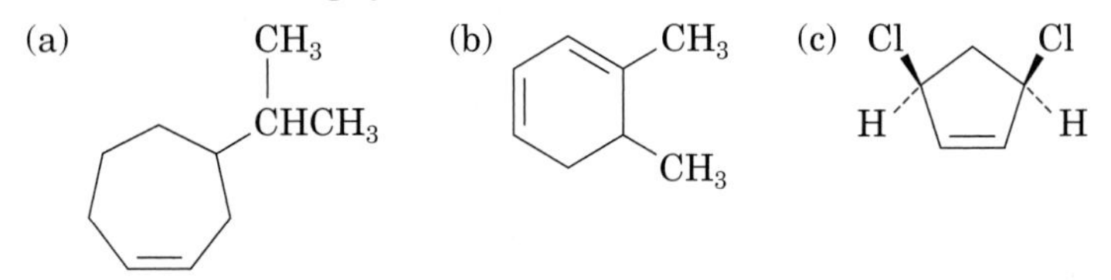

3.56 Reaction of 2-methylpropene with HCl might, in principle, lead to a mixture of two products. Draw them.

3.57 Give an example of each of the following:
(a) An electrophile
(b) A nucleophile
(c) An oxygen-containing functional group

3.58 Hydroxide ion reacts with chloromethane in a single step according to the following equation:

$$\text{H}\ddot{\underset{\cdot\cdot}{\text{O}}}\text{:}^- + \text{H}-\overset{\overset{\text{H}}{|}}{\underset{\underset{\text{H}}{|}}{\text{C}}}-\text{Cl} \longrightarrow \text{HO}-\overset{\overset{\text{H}}{|}}{\underset{\underset{\text{H}}{|}}{\text{C}}}-\text{H} + :\ddot{\underset{\cdot\cdot}{\text{Cl}}}:^-$$

Identify the bonds broken and formed, and draw curved arrows to represent the flow of electrons during the reaction.

3.59 Methoxide ion (CH_3O^-) reacts with bromoethane in a single step according to the following equation:

$$CH_3\ddot{O}:^- + H_3C{-}CH_2Br \longrightarrow H_2C{=}CH_2 + CH_3OH + :\ddot{Br}:^-$$

Identify the bonds broken and formed, and draw curved arrows to represent the flow of electrons during the reaction.

3.60 Follow the flow of electrons indicated by the curved arrows in the following reaction, and predict the products that result:

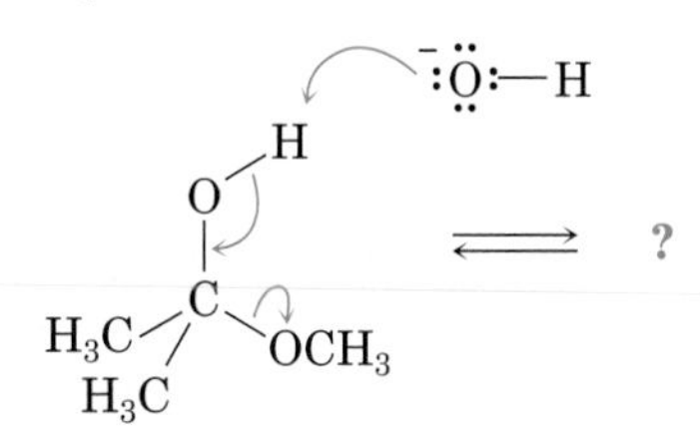

3.61 When isopropylidenecyclohexane is treated with strong acid at room temperature, isomerization occurs by the mechanism shown below to yield 1-isopropylcyclohexene:

Acid catalyst

Isopropylidenecyclohexane **1-Isopropylcyclohexene**

At equilibrium, the product mixture contains about 30% isopropylidenecyclohexane and about 70% 1-isopropylcyclohexene.

(a) What kind of reaction is occurring? Is the mechanism polar or radical?

(b) Draw curved arrows to indicate electron flow in each step.

3.62 We'll see in the next chapter that the stability of carbocations depends on the number of alkyl groups attached to the positively charged carbon—the more alkyl groups, the more stable the cation. Draw the two possible carbocation intermediates that might be formed in the reaction of HCl with 2-methylpropene (Problem 3.56), tell which is more stable, and predict which product will form.

IN THE MEDICINE CABINET

3.63 Tamoxifen and clomiphene have very similar structures (because they mimic similar steroids) but drastically different uses.

(a) Identify four different functional groups in tamoxifen and clomiphene.

(b) Classify the alkene double bond as *E* or *Z* in each molecule.

Tamoxifen, to treat cancer **Clomiphene, for fertility**

3.64 Retin A is commonly used to reduce wrinkles and treat severe acne.
(a) What two functional groups are present in Retin A?
(b) What is the molecular formula of Retin A?
(c) Given that the double bond in the cyclohexene ring cannot isomerize, how many different molecules arising from double-bond isomerization are possible?

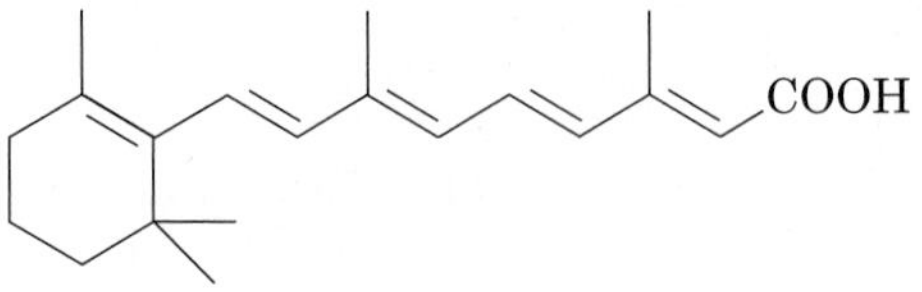

COOH

Retin A (retinoic acid), "wrinkle remover"

IN THE FIELD WITH AGROCHEMICALS

3.65 Plants use isoprene units to make larger molecules such as lycopene, which gives tomatoes their red color, and β-carotene.
(a) Each carbon in lycopene or β-carotene comes from an isoprene unit. Starting at one end of the molecule, circle all the contiguous isoprene groups in both pigments. (The *Interlude* might help.)
(b) Calculate the molecular formula of lycopene and β-carotene.

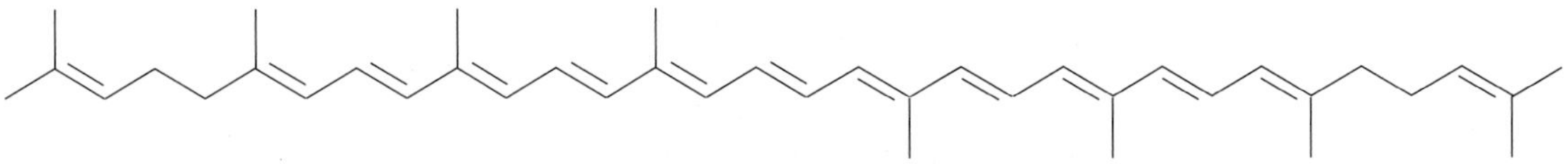

Lycopene

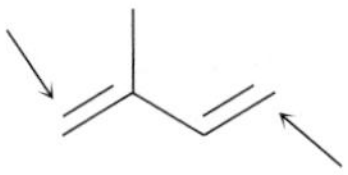

Isoprene

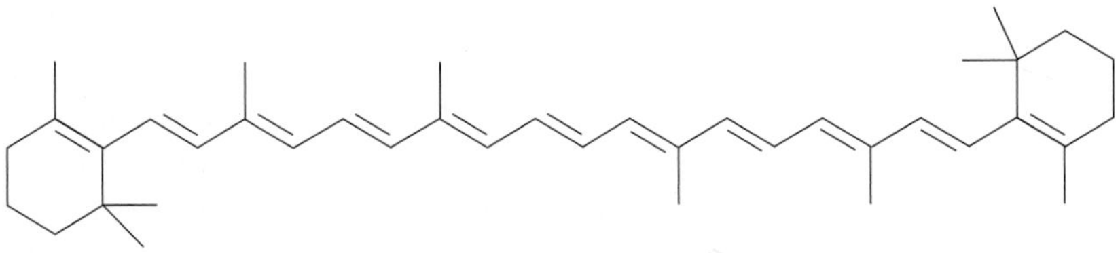

β-Carotene

3.66 Based on your answer to Problem 3.65, what is the relationship between lycopene and β-carotene?

3.67 Take a close look at lycopene and β-carotene (Problem 3.65). Plants make β-carotene from lycopene. Identify the two σ bonds in carotene that are created when lycopene undergoes two cyclization reactions.

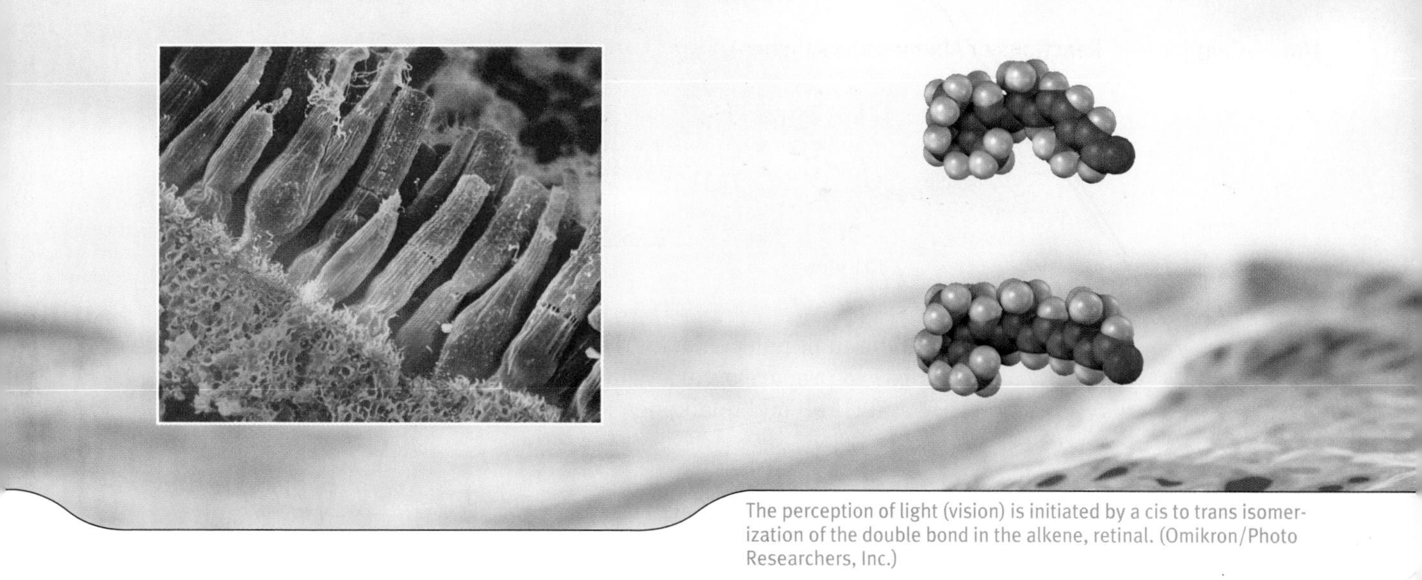

The perception of light (vision) is initiated by a cis to trans isomerization of the double bond in the alkene, retinal. (Omikron/Photo Researchers, Inc.)

CHAPTER 4

Reactions of Alkenes and Alkynes

We saw in the previous chapter how organic reactions can be classified, and we developed some general ideas about how reactions can be described. In this chapter, we'll apply those ideas to a systematic study of the alkene and alkyne families of compounds. In particular, we'll see that the most important reaction of these two functional groups is the addition to the C=C and C≡C multiple bonds of various reagents X–Y to yield saturated products:

$$\text{>C=C<} + \text{X—Y} \longrightarrow \text{—}\overset{\text{X}}{\overset{|}{\text{C}}}\text{—}\overset{\text{Y}}{\overset{|}{\text{C}}}\text{—}$$

An alkene — An addition product

By the time we are finished, we'll see that all the reactions that we discuss in this chapter follow this pattern. This material is easy to master by identifying the X and Y of X–Y and understanding the reactivity preferences.

4.1 Addition of HX to Alkenes: Hydrohalogenation

We saw in Section 3.8 that alkenes react with HCl to yield alkyl chloride addition products. For example, ethylene reacts with HCl to give chloroethane. The reaction takes place in two steps and involves a carbocation intermediate.

$$H_2C{=}CH_2 + H^+ \longrightarrow \left[H{-}\underset{H}{\overset{H}{C}}{-}\underset{H}{\overset{+}{C}}{-}H \right] \xrightarrow{Cl^-} H{-}\underset{H}{\overset{H}{C}}{-}\underset{H}{\overset{Cl}{C}}{-}H$$

Ethylene — Carbocation intermediate — **Chloroethane**

The addition of halogen acids, HX, to alkenes is a general reaction that allows chemists to prepare a variety of halo-substituted alkane products. Thus, HCl, HBr, and HI all add to alkenes:

$$(CH_3)_2C{=}CH_2 + HCl \xrightarrow{\text{Ether}} CH_3{-}\underset{CH_3}{\overset{Cl}{C}}{-}CH_3$$

2-Methylpropene — **2-Chloro-2-methylpropane**

$$\text{1-Methylcyclohexene} + HBr \xrightarrow{\text{Ether}} \text{1-Bromo-1-methylcyclohexane}$$

1-Methylcyclohexene — **1-Bromo-1-methylcyclohexane**

$$CH_3CH_2CH_2CH{=}CH_2 + HI \xrightarrow{\text{Ether}} CH_3CH_2CH_2\overset{I}{C}HCH_3$$

Pent-1-ene — **2-Iodopentane**

4.2 Orientation of Alkene Addition Reactions: Markovnikov's Rule

Look carefully at the three reactions shown at the end of the previous section. In each case, an unsymmetrically substituted alkene has given a single addition product rather than the mixture that might have been expected. For example, 2-methylpropene *might* have reacted with HCl to give 1-chloro-2-methylpropane in addition to 2-chloro-2-methylpropane, but it didn't. We say that reactions are *regiospecific* (**ree**-jee-oh-specific) when only one of the two possible directions of addition occurs.

A regiospecific reaction:

$$(CH_3)_2C{=}CH_2 + HCl \longrightarrow CH_3{-}\underset{CH_3}{\overset{Cl}{C}}{-}CH_3 \qquad \left[CH_3\overset{CH_3}{C}HCH_2Cl \right]$$

2-Methylpropene — **2-Chloro-2-methylpropane** (*sole product*) — **1-Chloro-2-methylpropane** (*NOT* formed)

After looking at many such reactions, the Russian chemist Vladimir Markovnikov proposed in 1869 what has become known as **Markovnikov's rule**:

Markovnikov's rule In the addition of HX to an alkene, the H attaches to the carbon with fewer alkyl substituents, and the X attaches to the carbon with more alkyl substituents.

No alkyl groups on this carbon

2 alkyl groups on this carbon

$$(CH_3)_2C{=}CH_2 + HCl \xrightarrow{\text{Ether}} CH_3{-}C(Cl)(CH_3){-}CH_3$$

2-Methylpropene **2-Chloro-2-methylpropane**

2 alkyl groups on this carbon

1 alkyl group on this carbon

$$\text{1-Methylcyclohexene} + HBr \xrightarrow{\text{Ether}} \text{1-Bromo-1-methylcyclohexane}$$

1-Methylcyclohexene **1-Bromo-1-methylcyclohexane**

When both double-bond carbon atoms have the same degree of substitution, a mixture of addition products results:

1 alkyl group on this carbon

1 alkyl group on this carbon

$$CH_3CH_2CH{=}CHCH_3 + HBr \xrightarrow{\text{Ether}} CH_3CH_2CH_2CH(Br)CH_3 + CH_3CH_2CH(Br)CH_2CH_3$$

Pent-2-ene **2-Bromopentane** **3-Bromopentane**

Since carbocations are involved as intermediates in these reactions (see Section 3.8), another way to express Markovnikov's rule is to say that, in the addition of HX to alkenes, the more highly substituted carbocation intermediate is formed rather than the less highly substituted one. For example, addition of H^+ to 2-methylpropene yields the intermediate *tertiary* carbocation rather than the alternative primary carbocation. Why should this be?

$$(CH_3)_2C{=}CH_2 + HCl \longrightarrow \left[CH_3{-}\overset{+}{C}(CH_3){-}CH_2H\right] \xrightarrow{Cl^-} CH_3{-}C(Cl)(CH_3){-}CH_3$$

2-Methylpropene *tert*-**Butyl carbocation (tertiary; 3°)** **2-Chloro-2-methylpropane**

$$(CH_3)_2C{=}CH_2 + HCl \not\longrightarrow \left[CH_3{-}C(H)(CH_3){-}\overset{+}{C}H_2\right] \xrightarrow{Cl^-} CH_3{-}C(H)(CH_3){-}CH_2Cl$$

Isobutyl carbocation (primary; 1°) **1-Chloro-2-methylpropane** ***(NOT formed)***

PRACTICE PROBLEM 4.1

What product would you expect from the reaction of HCl with 1-ethylcyclopentene?

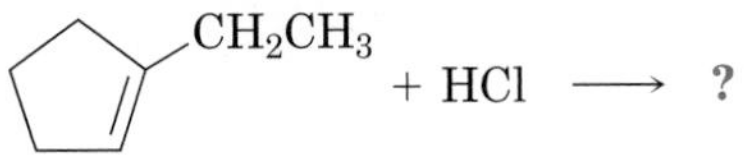

$$\text{(1-ethylcyclopentene)}\ CH_2CH_3 + HCl \longrightarrow ?$$

STRATEGY When solving a problem that asks you to predict a reaction product, begin by looking at the functional group(s) in the reactants and deciding what kind of reaction is likely to occur. In the present instance, the reactant is an alkene that will probably undergo an electrophilic addition reaction with HCl. Next, recall what you know about electrophilic addition reactions, and use your knowledge to predict the product. You know that electrophilic addition reactions follow Markovnikov's rule, so H^+ will add to the double-bond carbon that has one alkyl group (C2 on the ring), and the Cl will add to the double-bond carbon that has two alkyl groups (C1 on the ring).

SOLUTION The expected product is 1-chloro-1-ethylcyclopentane.

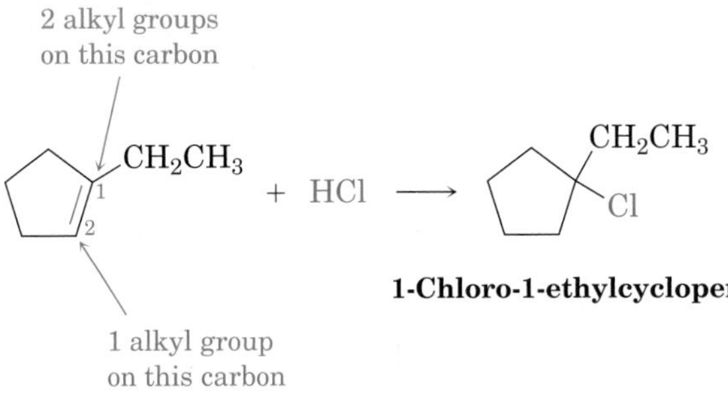

2 alkyl groups on this carbon

$$CH_2CH_3 + HCl \longrightarrow CH_2CH_3,\ Cl$$

1 alkyl group on this carbon

1-Chloro-1-ethylcyclopentane

PROBLEM 4.1 Predict the products of the following reactions:

(a) $CH_3CH_2CH{=}CH_2 + HCl \longrightarrow ?$ (b) $CH_3\overset{CH_3}{\overset{|}{C}}{=}CHCH_2CH_3 + HI \longrightarrow ?$ (c) 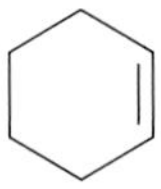 (cyclohexene) $+ HCl \longrightarrow ?$

PROBLEM 4.2 What alkenes would you start with to prepare the following alkyl halides?

(a) Bromocyclopentane

(b) $CH_3CH_2\overset{Br}{\overset{|}{C}}HCH_2CH_2CH_3$

(c) 1-Iodo-1-isopropylcyclohexane

(d) 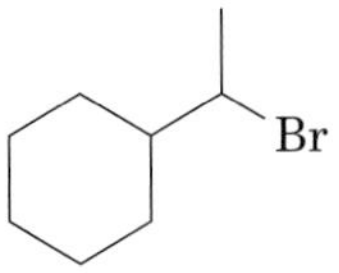 (cyclohexyl–CH(CH$_3$)Br) Br

4.3 Carbocation Structure and Stability

To understand why Markovnikov's rule works, we need to learn more about the structure and stability of substituted carbocations. Regarding structure, evidence shows that carbocations are *planar*. The positively charged carbon atom is sp^2-hybridized, and the three substituents bonded to it are oriented to the corners of an equilateral triangle (Figure 4.1). Because there are only six electrons in the carbon valence shell and all six are used in the three σ bonds, the p orbital extending above and below the plane is vacant.

FIGURE 4.1 The electronic structure of a carbocation. The carbon is sp^2-hybridized and has a vacant *p* orbital.

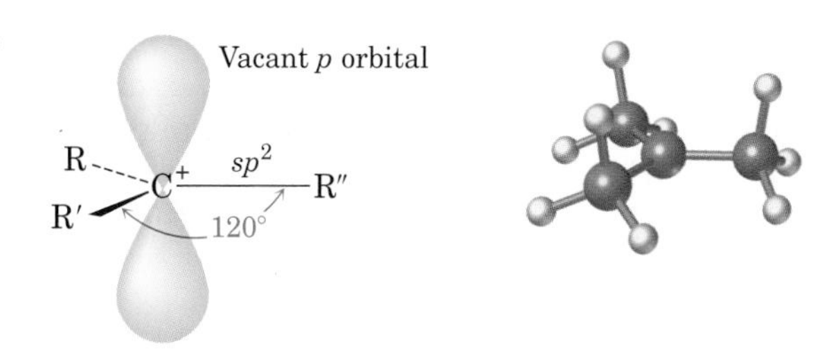

Regarding stability, measurements show that carbocation stability increases with increasing alkyl substitution: more highly substituted carbocations are more stable than less highly substituted ones because alkyl groups tend to donate electrons to the positively charged carbon atom. The more alkyl groups there are, the more electron donation there is and the more stable the carbocation.

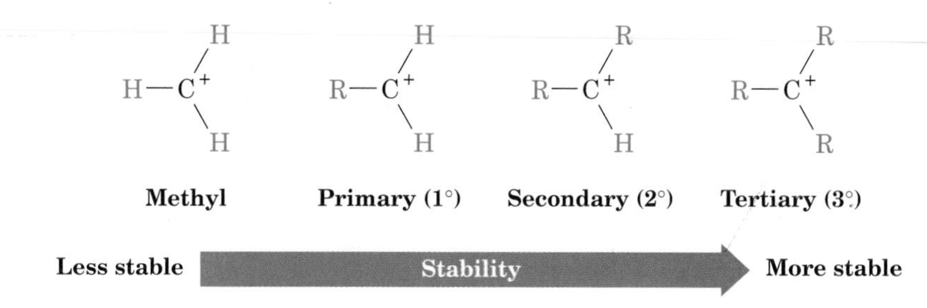

With this knowledge, we can now explain Markovnikov's rule. In the reaction of 1-methylcyclohexene with HBr, for example, the intermediate carbocation might have either *three* alkyl substituents (a tertiary cation, 3°) or *two* alkyl substituents (a secondary cation, 2°). Because the tertiary cation is more stable than the secondary one, it's the tertiary cation that forms as the reaction intermediate, thus leading to the observed tertiary alkyl bromide product.

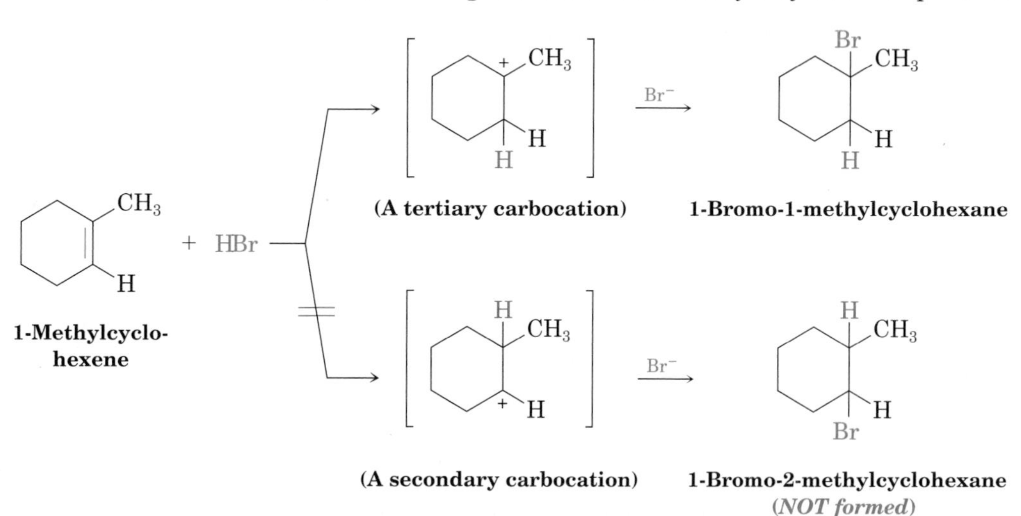

PROBLEM 4.3 Show the structures of the two carbocation intermediates you would expect in each of the following reactions, and identify which is more stable:

(a) $CH_3CH_2C(CH_3){=}CHCH(CH_3)CH_3 + HBr \longrightarrow ?$ (b) (cyclopentylidene)$=CHCH_3 + HI \longrightarrow ?$

4.4

Addition of H_2O to Alkenes: Hydration

Just as HX adds to alkenes to yield alkyl halides, water adds to yield alcohols, ROH, a process called **hydration**. Industrially, more than 300,000 tons of ethanol are produced each year in the United States by this method:

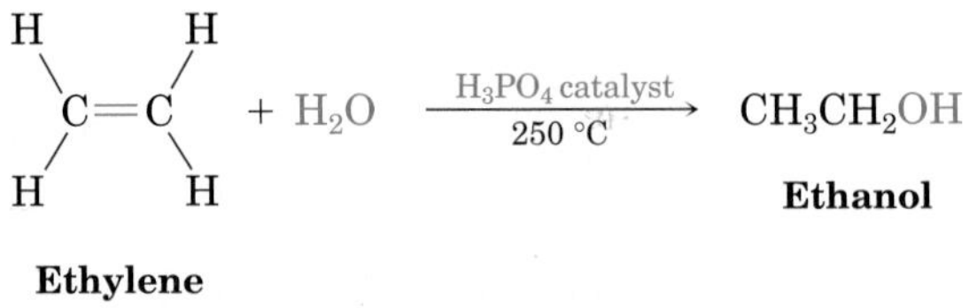

$$H_2C{=}CH_2 + H_2O \xrightarrow[250\ °C]{H_3PO_4\ \text{catalyst}} CH_3CH_2OH$$

Ethylene **Ethanol**

Hydration takes place on reaction of an alkene with aqueous acid by a mechanism similar to that of HX addition. Reaction of the alkene double bond with H^+ yields a carbocation intermediate, which then reacts with water as nucleophile to yield a protonated alcohol (ROH_2^+) product. Loss of H^+ from the protonated alcohol gives the neutral alcohol and regenerates the acid catalyst (Figure 4.2). The addition of water to an unsymmetrical alkene follows Markovnikov's rule, just as addition of HX does, giving the more highly substituted alcohol as product.

FIGURE 4.2 MECHANISM: Mechanism of the acid-catalyzed hydration of an alkene. Protonation of the alkene gives a carbocation intermediate, which reacts with water.

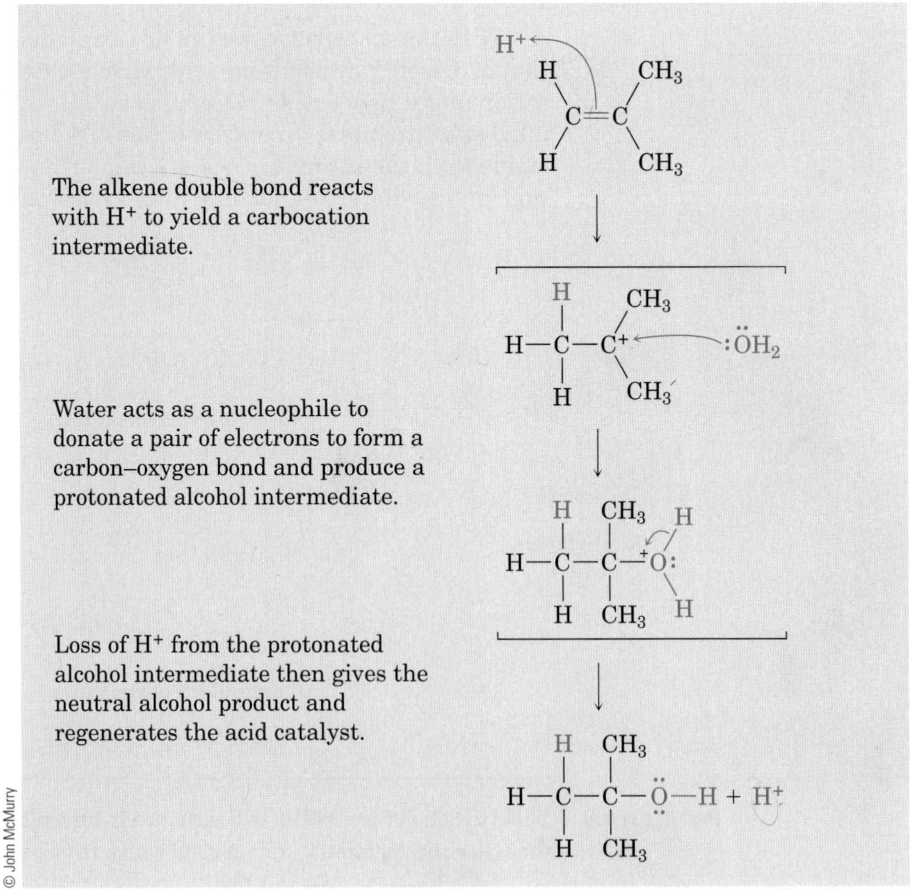

The alkene double bond reacts with H^+ to yield a carbocation intermediate.

$$H_2C{=}C(CH_3)_2 + H^+ \rightarrow CH_3\overset{+}{C}(CH_3)_2$$

Water acts as a nucleophile to donate a pair of electrons to form a carbon–oxygen bond and produce a protonated alcohol intermediate.

$$CH_3\overset{+}{C}(CH_3)_2 + :\ddot{O}H_2 \rightarrow CH_3C(CH_3)_2\overset{+}{O}H_2$$

Loss of H^+ from the protonated alcohol intermediate then gives the neutral alcohol product and regenerates the acid catalyst.

$$CH_3C(CH_3)_2\overset{+}{O}H_2 \rightarrow CH_3C(CH_3)_2\ddot{O}H + H^+$$

Unfortunately, the reaction conditions required for hydration are so severe that molecules are sometimes destroyed by the high temperatures and strongly acidic conditions. For example, the hydration of ethylene to produce ethanol requires a phosphoric acid catalyst and reaction temperatures of up to 250 °C.

As a result, chemists are constantly searching for better procedures that use new reactions and catalysts. Figure 4.3 shows three examples. The first two examples are new reactions. With an understanding of mechanisms, chemists have even developed strategies to overcome Markovnikov's rule, a so-called non-Markovnikov reaction. The last example shows a remarkable catalyst, a protein isolated from cells. While fumarase catalyzes hydration of this alkene called fumaric acid, it does not catalyze the same reaction with the isomeric *cis*-alkene called maleic acid. Indeed, the reactant selectivity that nature displays in its catalysts is largely unmatched in the catalysts chemists create.

FIGURE 4.3 Newer methods for hydration include the use of mercury or boron reagents (that lead to non-Markovnikov products) and catalysts like the protein fumarase.

(a) Oxymercuration–demercuration (Markovnikov addition)

CH_3, H — 1. $Hg(OAc)_2$, H_2O 2. $NaBH_4$ → H_3C, OH, H, H

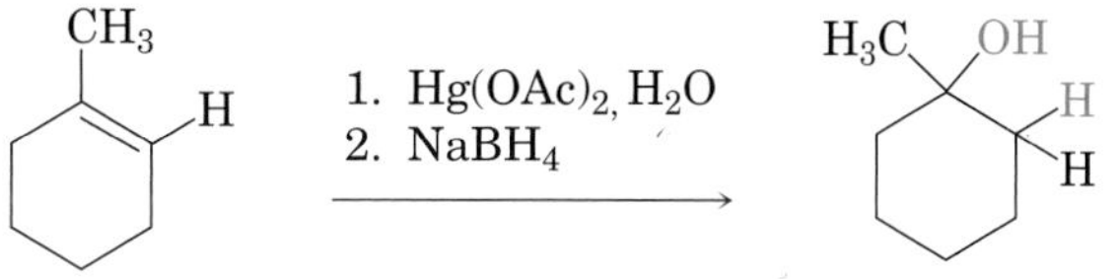

(b) Hydroboration (non-Markovnikov addition) Anti

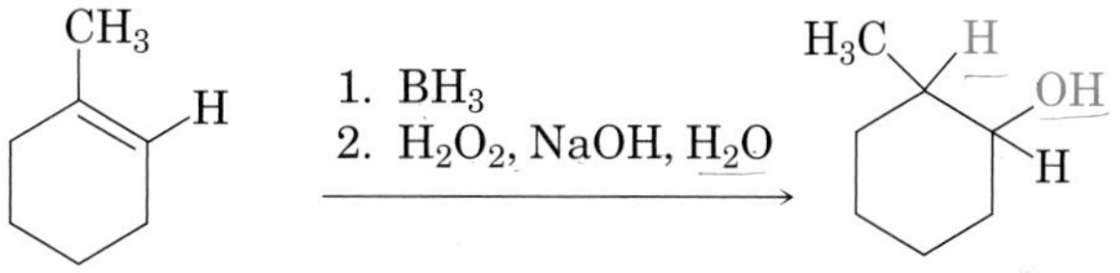

CH_3, H — 1. BH_3 2. H_2O_2, NaOH, H_2O → H_3C, H, OH, H

(c) Enzymatic hydration

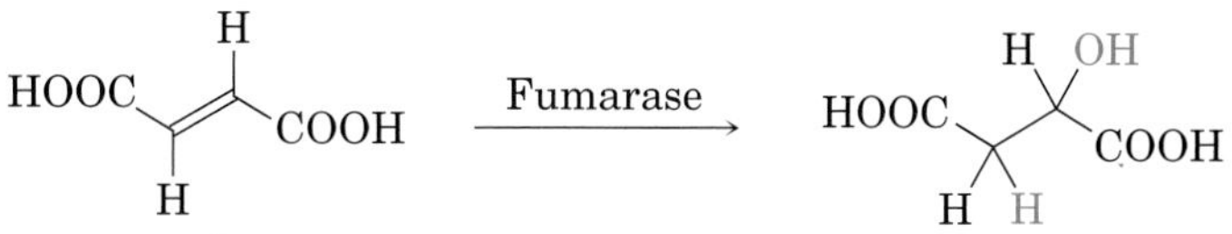

H, HOOC, COOH, H — Fumarase → H OH, HOOC, COOH, H H

PRACTICE PROBLEM 4.2

What product would you expect from the acid-catalyzed addition of water to methylenecyclopentane?

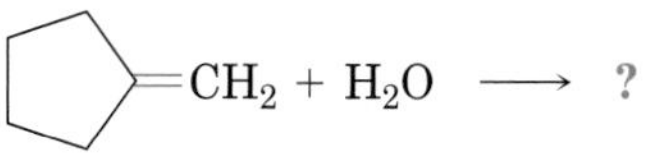

$$\text{C}_5\text{H}_8{=}CH_2 + H_2O \longrightarrow ?$$

Methylenecyclopentane

STRATEGY According to Markovnikov's rule, H^+ adds to the carbon that already has more hydrogens (the $=CH_2$ carbon) and OH adds to the carbon that has fewer hydrogens (the ring carbon). Thus, the product will be a tertiary alcohol.

SOLUTION

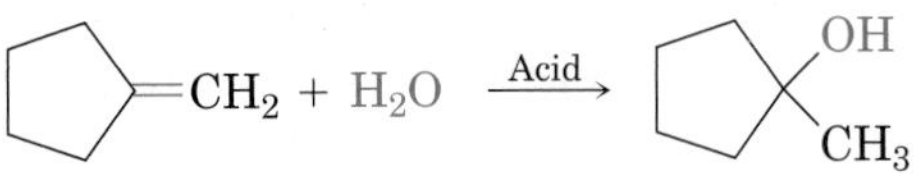

$$\text{(methylenecyclopentane)}{=}CH_2 + H_2O \xrightarrow{\text{Acid}} \text{(1-methylcyclopentanol: OH, } CH_3\text{)}$$

PROBLEM 4.4 What product would you expect to obtain from the acid-catalyzed addition of water to the following alkenes?

(a) $CH_3CH_2C(CH_3){=}CHCH_2CH_3$ (b) 1-Methylcyclopentene

(c) 2,5-Dimethylhept-2-ene

PROBLEM 4.5 What alkenes might the following alcohols be made from?

(a) $CH_3CH_2CH(OH)CH_3$ (b) $CH_3CH_2{-}C(OH)(CH_3){-}CH_2CH_3$ (c) cyclohexane ring bearing OH, CH_3, CH_3

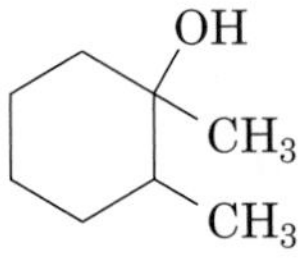

4.5 Addition of X_2 to Alkenes: Halogenation

Many other substances besides HX and H_2O add to alkenes. Bromine and chlorine, for instance, add readily to yield 1,2-dihaloalkanes. More than 6 million tons of 1,2-dichloroethane (also called ethylene dichloride) are synthesized each year in the United States by addition of Cl_2 to ethylene. The product is used both as a solvent and as a starting material for the synthesis of poly(vinyl chloride), PVC.

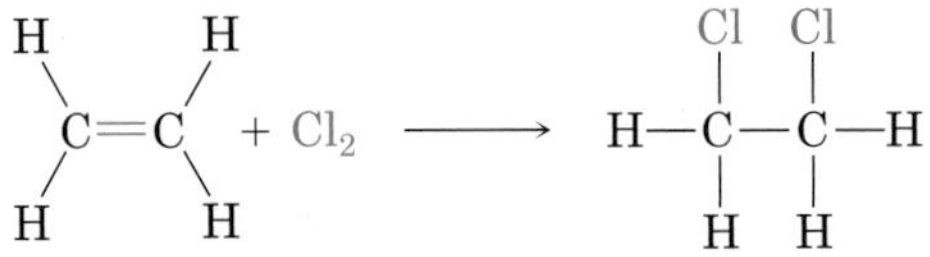

$$H_2C{=}CH_2 + Cl_2 \longrightarrow H{-}CHCl{-}CHCl{-}H$$

Ethylene **1,2-Dichloroethane (ethylene dichloride)**

Addition of Br_2 also acts as a simple and rapid laboratory test for unsaturation. A sample of unknown structure is dissolved in dichloromethane, CH_2Cl_2, and several drops of Br_2 are added. Immediate disappearance of the reddish Br_2 color signals a positive test and indicates that the sample is an alkene.

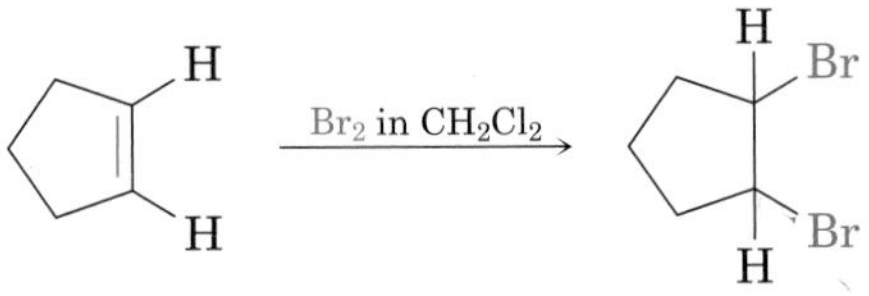

$$\text{Cyclopentene} \xrightarrow{Br_2 \text{ in } CH_2Cl_2} \text{1,2-Dibromocyclopentane (95\%)}$$

Cyclopentene **1,2-Dibromocyclopentane (95%)**

Based on what we've seen thus far, a possible mechanism for the reaction of Br_2 or Cl_2 with an alkene might involve formation of a bond between carbon and a Br^+ ion, displacing the other atom as Br^- anion. The net result would be that electrophilic Br^+ adds to the alkene in much the same way that H^+ does, giving a carbocation intermediate that reacts further with Br^- to yield the dibromo addition product.

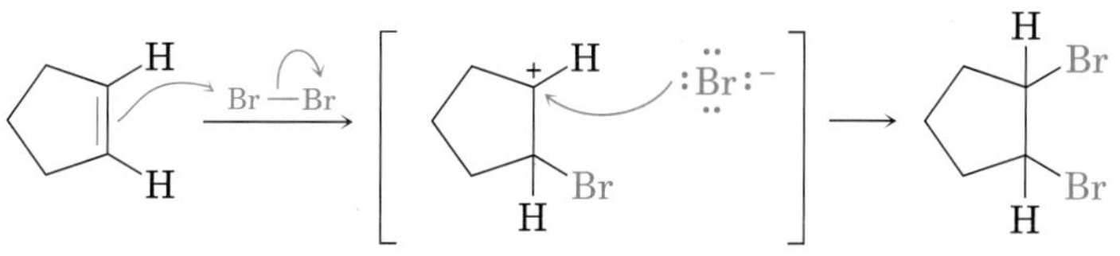

Although this mechanism looks reasonable, it's not completely consistent with known facts. In particular, the mechanism doesn't explain the *stereochemistry* of halogen addition. That is, the mechanism doesn't explain what product stereoisomers (see Section 2.8) are formed in the reaction.

Let's look again at the reaction of Br_2 with cyclopentene and assume that Br^+ adds from the bottom side of the molecule to form the carbocation intermediate shown in Figure 4.4. (The addition could just as well occur from the top side, but we'll consider only one possibility for simplicity.) Because the positively charged carbon is planar and sp^2-hybridized, it could react with Br^- anion in the second step of the reaction from either the top or the bottom side to give a *mixture* of products. One product has the two Br atoms on the same side of the ring (cis), and the other has them on opposite sides (trans). We find, however, that only *trans*-1,2-dibromocyclopentane is produced; none of the cis product is formed.

FIGURE 4.4 Stereochemistry of the addition of bromine to cyclopentene. Only the trans product is formed.

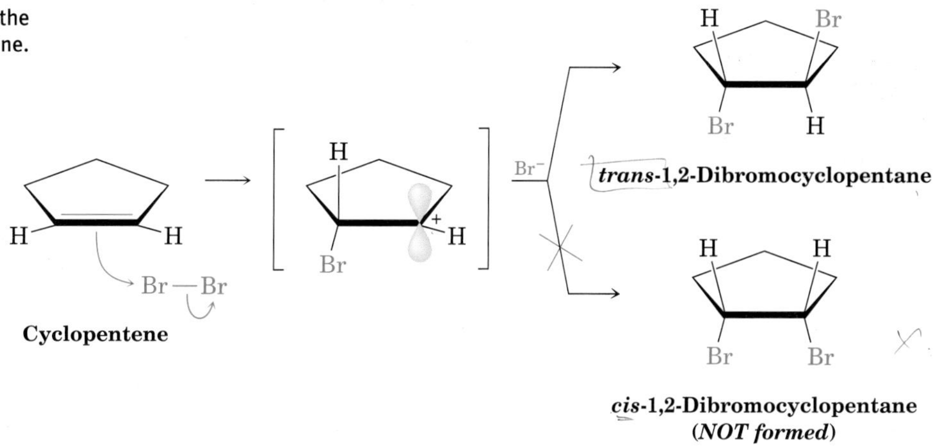

Cyclopentene

***trans*-1,2-Dibromocyclopentane**

***cis*-1,2-Dibromocyclopentane (*NOT formed*)**

Because the two Br atoms add to opposite faces of the cyclopentene double bond, we say that the reaction occurs with **anti stereochemistry**, meaning that the two bromines come from directions approximately 180° apart. This result is best explained by imagining that the reaction intermediate is not a true carbocation but is instead a *bromonium ion*, R_2Br^+, formed by overlap of bromine lone-pair electrons with the vacant *p* orbital of the neighboring carbon (Figure 4.5). Since the bromine atom effectively "shields" one side of the molecule, reaction with Br^- ion in the second step occurs from the opposite, more accessible side to give the anti product.

FIGURE 4.5 MECHANISM: Mechanism of the addition of Br_2 to an alkene. A bromonium ion intermediate is formed, shielding one face of the double bond and resulting in formation of anti stereochemistry for the addition product.

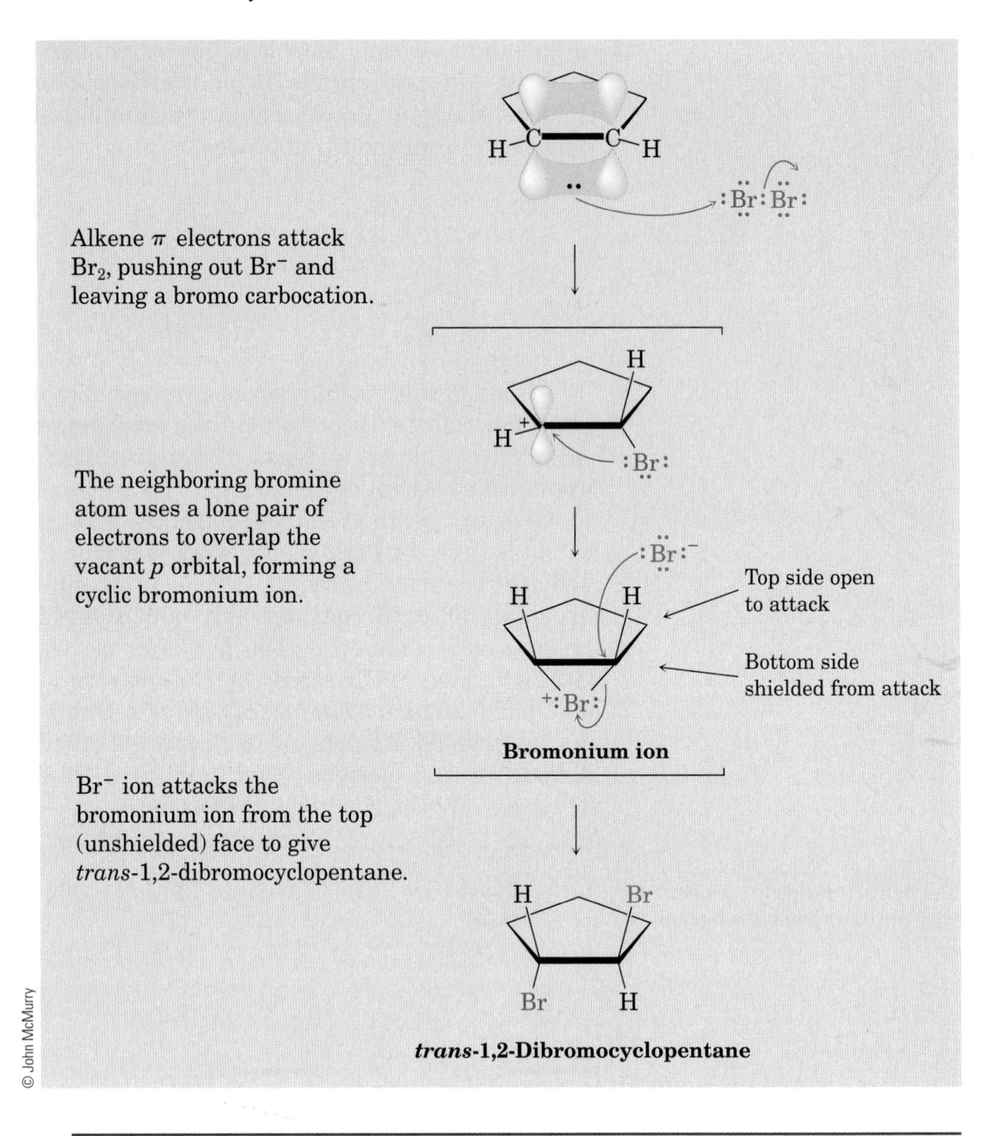

Alkene π electrons attack Br_2, pushing out Br^- and leaving a bromo carbocation.
The neighboring bromine atom uses a lone pair of electrons to overlap the vacant p orbital, forming a cyclic bromonium ion.
Top side open to attack
Bottom side shielded from attack
Bromonium ion
Br^- ion attacks the bromonium ion from the top (unshielded) face to give trans-1,2-dibromocyclopentane.
trans-1,2-Dibromocyclopentane

PROBLEM 4.6 What product would you expect to obtain from addition of Br_2 to 1,2-dimethylcyclohexene? Show the stereochemistry of the product.

PROBLEM 4.7 Show the structure of the intermediate bromonium ion formed in Problem 4.6.

4.6 Addition of H_2 to Alkenes: Hydrogenation

Addition of H_2 to the C=C bond occurs when alkenes are exposed to an atmosphere of hydrogen gas in the presence of a catalyst. We describe the result by saying that the double bond is **hydrogenated**, or *reduced.* (The word **reduction** in organic chemistry usually refers to the addition of hydrogen or removal of oxygen from a molecule.) For most alkene hydrogenations, either palladium metal or platinum (as PtO_2) is used as the catalyst.

$$>C{=}C< + H{-}H \xrightarrow{\text{Catalyst}} H{-}C{-}C{-}H$$

An alkene **An alkane**

syn

Catalytic hydrogenation of alkenes, unlike most other organic reactions, is a *heterogeneous* process, rather than a homogeneous one. That is, the hydrogenation reaction occurs on the surface of solid catalyst particles rather than in solution. The reaction occurs with **syn stereochemistry** (the opposite of *anti*), meaning that both hydrogens add to the double bond from the same side.

CH_3 CH_3 $\xrightarrow[CH_3CO_2H]{H_2,\ PtO_2}$ CH_3 H H CH_3

1,2-Dimethylcyclohexene ***cis*-1,2-Dimethylcyclohexane (82%)**

In addition to its usefulness in the laboratory, alkene hydrogenation is also of great commercial value. In the food industry, unsaturated vegetable oils are catalytically hydrogenated on a vast scale to produce the saturated fats used in margarine.

PROBLEM 4.8 What product would you expect to obtain from catalytic hydrogenation of the following alkenes?
(a) $(CH_3)_2C{=}CHCH_2CH_3$ (b) 3,3-Dimethylcyclopentene

4.7 Oxidation of Alkenes: Hydroxylation and Cleavage

Hydroxylation of an alkene—the addition of an –OH group to each of the alkene carbons—can be carried out by reaction of the alkene with potassium permanganate, $KMnO_4$, in basic solution. Since oxygen adds to the alkene during the reaction, we call this an **oxidation**. The reaction occurs with syn stereochemistry and yields a 1,2-dialcohol, or **diol**, product (also called a *glycol*). For example, cyclohexene gives *cis*-1,2-cyclohexanediol in 37% yield.

$+ KMnO_4 \xrightarrow[NaOH]{H_2O}$ H OH OH H

Cyclohexene ***cis*-Cyclohexane-1,2-diol (37%)**

When oxidation of the alkene is carried out with $KMnO_4$ in acidic rather than basic solution, *cleavage* of the double bond occurs and carbonyl-containing products are obtained. If the double bond is tetrasubstituted, the two carbonyl-containing products are ketones; if a hydrogen is present on the double bond, one of the carbonyl-containing products is a carboxylic acid; and if two hydrogens are present on one carbon, CO_2 is formed:

$$\text{C}_6\text{H}_{10}{=}\text{C}(\text{CH}_3)_2 + \text{KMnO}_4 \xrightarrow{\text{H}_3\text{O}^+} \text{C}_6\text{H}_{10}{=}\text{O} + \text{O}{=}\text{C}(\text{CH}_3)_2$$

Isopropylidenecyclohexane → **Cyclohexanone** **Acetone** (two ketones)

$$\text{CH}_3\text{CH}_2\text{CH}(\text{CH}_3)\text{CH}{=}\text{CH}_2 + \text{KMnO}_4 \xrightarrow{\text{H}_3\text{O}^+} \text{CH}_3\text{CH}_2\text{CH}(\text{CH}_3)\text{COOH} + \text{CO}_2$$

3-Methylpent-1-ene → **2-Methylbutanoic acid (45%)**

Practice Problem 4.3

Predict the product of reaction of pent-2-ene with aqueous acidic $KMnO_4$.

Strategy When trying to predict the products of a reaction, identify the functional groups and reagents involved and ask yourself what you know about these topics. In this instance, reaction of an alkene with acidic $KMnO_4$ yields carbonyl-containing products in which the double bond is broken and the two fragments have a C=O group in place of the original alkene C=C. If a hydrogen is present on the double bond, a carboxylic acid is produced.

Solution Pent-2-ene gives the following reaction:

$$\text{CH}_3\text{CH}_2\text{CH}{=}\text{CHCH}_3 + \text{KMnO}_4 \xrightarrow{\text{H}_3\text{O}^+} \text{CH}_3\text{CH}_2\text{COOH} + \text{HOOCCH}_3$$

Pent-2-ene → **Propanoic acid** **Acetic acid**

Practice Problem 4.4

What alkene gives a mixture of acetone and propanoic acid on reaction with acidic $KMnO_4$?

$$? \xrightarrow[\text{H}_3\text{O}^+]{\text{KMnO}_4} \text{CH}_3\text{COCH}_3 + \text{CH}_3\text{CH}_2\text{COOH}$$

Acetone **Propanoic acid**

Strategy When solving a problem that asks how to prepare a given product, *always work backward.* Look at the product, identify the functional group(s) it contains, and ask yourself, "How can I prepare that functional group?" In the present instance, the products are a ketone and a carboxylic acid, which can be prepared by reaction of an alkene with acidic $KMnO_4$. To find the starting alkene that gives the cleavage products shown, remove the oxygen atoms from the two products, join the fragments with a double bond, and replace the –OH by –H.

SOLUTION

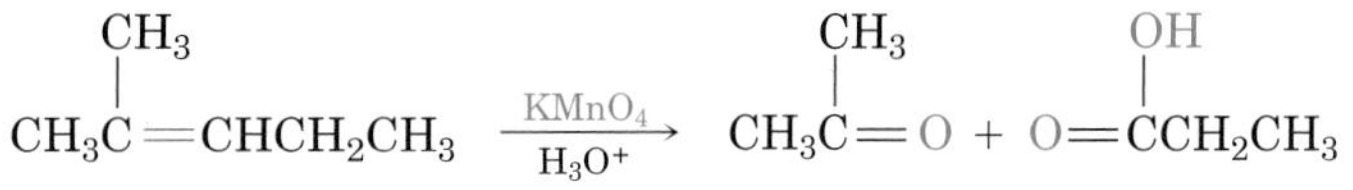

$$\underset{\textbf{2-Methylpent-2-ene}}{CH_3\overset{\overset{CH_3}{|}}{C}{=}CHCH_2CH_3} \xrightarrow[H_3O^+]{KMnO_4} \underset{\textbf{Acetone}}{CH_3\overset{\overset{CH_3}{|}}{C}{=}O} + \underset{\textbf{Propanoic acid}}{O{=}\overset{\overset{OH}{|}}{C}CH_2CH_3}$$

PROBLEM 4.9 Predict the product of the reaction of 1,2-dimethylcyclohexene with the following:
(a) $KMnO_4$, H_3O^+ (b) $KMnO_4$, OH^-, H_2O

PROBLEM 4.10 Propose structures for alkenes that yield the following products on treatment with acidic $KMnO_4$:
(a) $(CH_3)_2C{=}O + CO_2$ (b) 2 equiv $CH_3CH_2CO_2H$

4.8 Addition of Radicals to Alkenes: Polymers

No other group of synthetic chemicals has had as great an impact on our day-to-day lives as *polymers.* From carpeting to clothing to foam coffee cups, we are surrounded by polymers.

A **polymer** is a large (sometimes *very* large) molecule built up by repetitive bonding together of many smaller molecules, called **monomers**. As we'll see in later chapters, nature makes wide use of biological polymers. Cellulose, for example, is a polymer built of repeating sugar units, proteins are polymers built of repeating amino acid units, and nucleic acids are polymers built of repeating nucleotide units. Although synthetic polymers are chemically much simpler than biopolymers, there is an immense diversity to the structures and properties of synthetic polymers, depending on the nature of the monomers and on the reaction conditions used for polymerization.

Many simple alkenes undergo rapid polymerization when treated with a small amount of a radical as catalyst. Ethylene, for example, yields polyethylene, an enormous alkane that may have up to several thousand monomer units incorporated into its long hydrocarbon chain. Ethylene polymerization is usually carried out at high pressure (1000–3000 atm) and high temperature (100–250 °C) with a radical catalyst such as benzoyl peroxide.

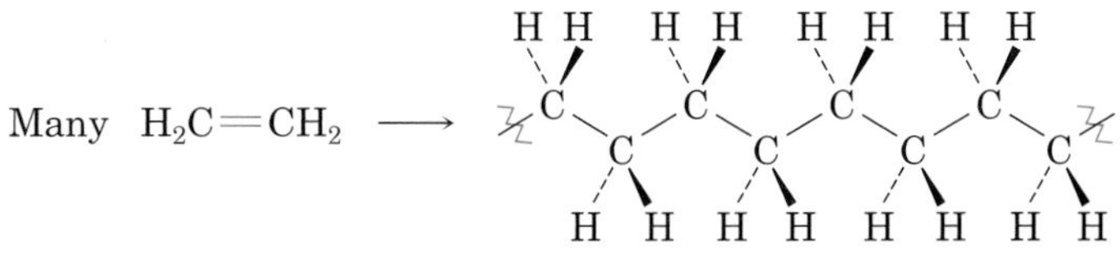

$$\text{Many } H_2C{=}CH_2 \longrightarrow \text{—}(CH_2CH_2CH_2CH_2CH_2CH_2CH_2CH_2)\text{—}$$

Ethylene **A section of polyethylene**

Radical polymerization of an alkene involves three kinds of steps: *initiation*, *propagation*, and *termination.* The key step is the addition of a *radical* to the ethylene double bond in a process similar to what takes place in the addition of an *electrophile* to an alkene (see Section 3.8). In writing the mechanism, a curved half-arrow, or "fishhook," is used to show the movement of a single electron, as opposed to the full curved arrow used to show the movement of an electron pair in a polar reaction.

STEP 1 **Initiation**: Reaction begins when a few radicals are generated by the catalyst. For example, when benzoyl peroxide is used as initiator, the O–O bond is broken on heating to yield benzoyloxy radicals. The benzoyloxy radical then adds to the C=C bond of ethylene to generate a carbon radical. One electron from the carbon–carbon double bond pairs up with the odd electron on the benzoyloxy radical to form a C–O bond, and the other electron remains on carbon.

$$C_6H_5C(=O)O{-}OC(=O)C_6H_5 \xrightarrow{\text{Heat}} 2\ C_6H_5C(=O)O\cdot = BzO\cdot$$

Benzoyl peroxide **Benzoyloxy radical**

$$BzO\cdot \ \ H_2C{=}CH_2 \longrightarrow BzO{-}CH_2CH_2\cdot$$

STEP 2 **Propagation**: Polymerization occurs when the carbon radical formed in step 1 adds to another ethylene molecule. Repetition of this step for hundreds or thousands of times builds the polymer chain.

$$BzOCH_2CH_2\cdot \ \ H_2C{=}CH_2 \longrightarrow BzOCH_2CH_2CH_2CH_2\cdot \xrightarrow[\text{many times}]{\text{Repeat}} BzO(CH_2CH_2)_nCH_2CH_2\cdot$$

STEP 3 **Termination**: Polymerization eventually stops when a reaction that consumes the radical occurs. For example, combination of two chains by chance meeting is a possible chain-terminating reaction:

$$2\ R{-}CH_2CH_2\cdot \longrightarrow R{-}CH_2CH_2CH_2CH_2{-}R$$

Many substituted ethylenes, called *vinyl monomers*, undergo radical-initiated polymerization, yielding polymers with substituent groups regularly spaced along the polymer chain. Propylene, for example, yields polypropylene when polymerized (although a nonradical method of polymerization is used industrially).

$$H_2C{=}CHCH_3 \longrightarrow {-}\!\left(CH_2CH(CH_3)CH_2CH(CH_3)CH_2CH(CH_3)CH_2CH(CH_3)\right)\!{-}$$

Propylene **Polypropylene**

Table 4.1 shows some commercially important vinyl monomers and lists some industrial uses of the different polymers that result.

PRACTICE PROBLEM 4.5

Show the structure of poly(vinyl chloride), a polymer made from $H_2C{=}CHCl$, by drawing several repeating units.

STRATEGY Mentally break the carbon–carbon double bond in the monomer unit, and form single bonds by connecting numerous units together.

TABLE 4.1 Some Alkene Polymers and Their Uses

Monomer name	Formula	Trade or common name of polymer	Uses
Ethylene	$H_2C{=}CH_2$	Polyethylene	Packaging, bottles, cable insulation, films and sheets
Propene (propylene)	$H_2C{=}CHCH_3$	Polypropylene	Automotive moldings, rope, carpet fibers
Chloroethylene (vinyl chloride)	$H_2C{=}CHCl$	Poly(vinyl chloride), Tedlar	Insulation, films, pipes
Styrene	$H_2C{=}CHC_6H_5$	Polystyrene, Styron	Foam and molded articles
Tetrafluoroethylene	$F_2C{=}CF_2$	Teflon	Valves and gaskets, coatings
Acrylonitrile	$H_2C{=}CHCN$	Orlon, Acrilan	Fibers
Methyl methacrylate	$H_2C{=}C(CH_3)CO_2CH_3$	Plexiglas, Lucite	Molded articles, paints
Vinyl acetate	$H_2C{=}CHOCOCH_3$	Poly(vinyl acetate)	Paints, adhesives

SOLUTION The general structure of poly(vinyl chloride) is

$$\underset{\textbf{Vinyl chloride}}{H_2C{=}CH(Cl)} \longrightarrow \underset{\textbf{Poly(vinyl chloride)}}{\left(CH_2CH(Cl)CH_2CH(Cl)CH_2CH(Cl) \right)}$$

PROBLEM 4.11 Show the structure of Teflon by drawing several repeating units. The monomer unit is tetrafluoroethylene, $F_2C{=}CF_2$.

4.9 Conjugated Dienes

Multiple bonds that alternate with single bonds are said to be **conjugated**. Thus, buta-1,3-diene is a conjugated diene, whereas penta-1,4-diene is a nonconjugated diene with isolated double bonds.

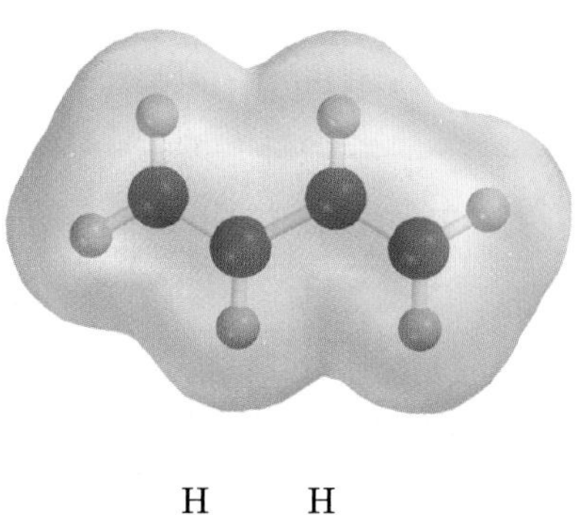

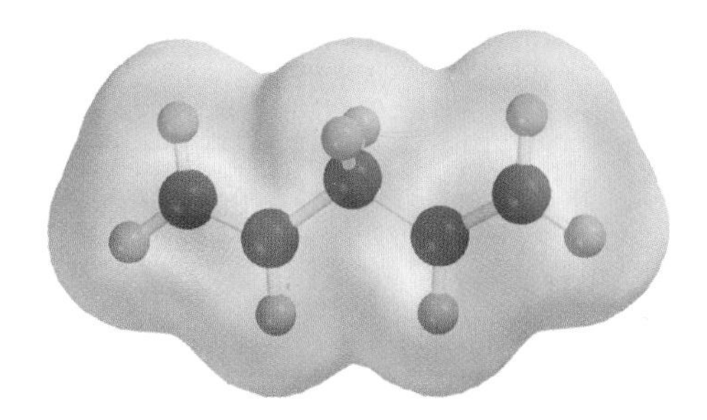

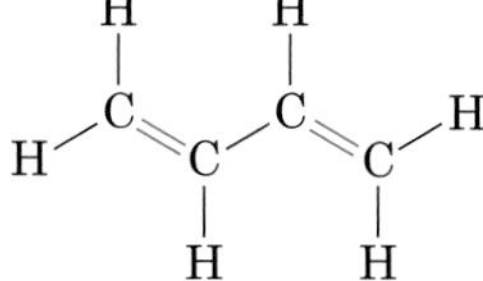

Buta-1,3-diene (conjugated)

Penta-1,4-diene (nonconjugated)

What's so special about conjugated dienes that we need to look at them separately? The orbital view of buta-1,3-diene shown in Figure 4.6 provides a clue to the answer: *there is an electronic interaction between the two double bonds of a conjugated diene* because of *p* orbital overlap across the central single bond. This interaction of *p* orbitals across a single bond gives conjugated dienes some unusual properties.

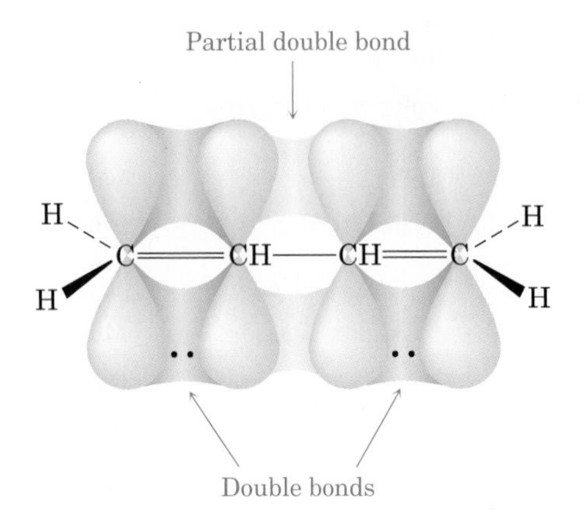

FIGURE 4.6 An orbital view of buta-1,3-diene. Each of the four carbon atoms has a *p* orbital, allowing for an electronic interaction across the C2–C3 single bond.

Although much of the chemistry of conjugated dienes and other alkenes is similar, there is a striking difference in their addition reactions with electrophiles like HX and X_2. When HX adds to an isolated alkene, Markovnikov's rule usually predicts the formation of a single product. When HX adds to a conjugated diene, though, mixtures of products are usually obtained. For example, reaction of HBr with buta-1,3-diene yields two products:

Buta-1,3-diene (a conjugated diene) + HBr →

3-Bromobut-1-ene (71%; 1,2-addition)

1-Bromobut-2-ene (29%; 1,4-addition)

3-Bromobut-1-ene is the typical product of Markovnikov addition, but 1-bromobut-2-ene appears unusual. The double bond in this product has moved to a position between C2 and C3, while HBr has added to C1 and C4, a result described as **1,4-addition**.

How can we account for the formation of the 1,4-addition product? The answer is that an *allylic carbocation* is involved as an intermediate in the reaction, where the word **allylic** means "next to a double bond." When H^+ adds to an electron-rich π bond of buta-1,3-diene, two carbocation intermediates are possible—a primary nonallylic carbocation and a secondary allylic carbocation. Allylic carbocations are more stable and therefore form faster than less stable, nonallylic carbocations.

Buta-1,3-diene + HBr

Secondary, allylic carbocation + Br^-

Primary carbocation (*NOT formed*) + Br^-

4.10

Stability of Allylic Carbocations: Resonance

Why are allylic carbocations particularly stable? To see the answer, look at the orbital picture of an allylic carbocation in Figure 4.7. The positively charged carbon atom has a vacant *p* orbital that can overlap the *p* orbitals of the neighboring double bond.

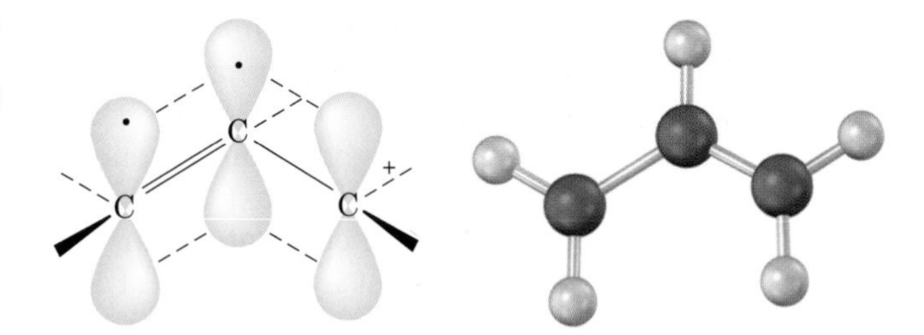

FIGURE 4.7 An orbital picture of an allylic carbocation. The vacant *p* orbital on the positively charged carbon can overlap the double-bond *p* orbitals.

From an electronic viewpoint, an allylic carbocation is symmetrical. All three carbon atoms are sp^2-hybridized, and each has a *p* orbital. Thus, the *p* orbital on the central carbon can overlap equally well with *p* orbitals on *either* of the two neighboring carbons, and the two electrons are free to move about over the entire three-orbital array.

One consequence of this orbital picture is that there are two ways to draw an allylic carbocation. We can draw it with the vacant *p* orbital on the right and the double bond on the left, or we can draw it with the vacant *p* orbital on the left and the double bond on the right. *Neither structure is correct by itself; the true structure of the allylic carbocation is somewhere between the two.*

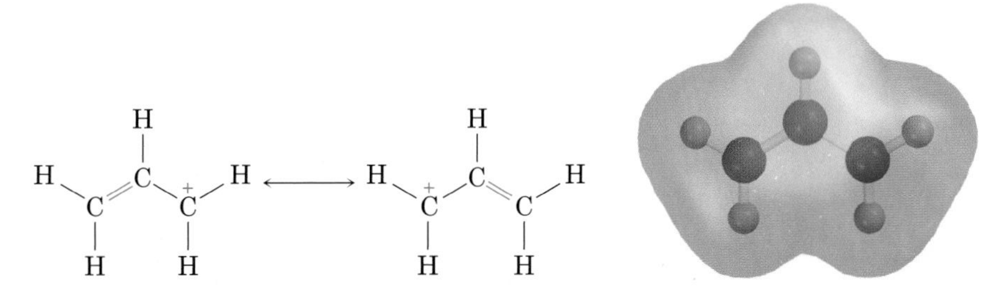

An allylic carbocation

The two individual structures of an allylic carbocation are called **resonance forms**, and their special relationship is indicated by the double-headed arrow between them. The only difference between the resonance forms is the position of the bonding electrons. The atoms themselves remain in exactly the same place in both resonance forms.

The best way to think about resonance is to realize that a species like an allylic carbocation is no different from any other organic substance. An allylic carbocation doesn't jump back and forth between two resonance forms, spending part of its time looking like one and the rest of its time looking like the other. Rather, an allylic carbocation has a single, unchanging structure called a **resonance hybrid**. A useful analogy is to think of a resonance hybrid as being like a mixed-breed dog. Just as a dog that's a mixture of dachshund and German shepherd doesn't change back and forth from one to the other, a resonance hybrid doesn't change back and forth between forms.

The difficulty in understanding resonance hybrids is visual rather than chemical, because we can't draw an accurate single picture of a resonance hybrid by using familiar kinds of structures. The line-bond structures that work so well to represent most organic molecules just don't work well for resonance hybrids like allylic carbocations because the two C–C bonds are equivalent and each is midway between single and double.

One of the most important consequences of resonance is that *the greater the number of possible resonance forms, the greater the stability.* Because an allylic carbocation is a resonance hybrid of two forms, it is more stable than a typical nonallylic carbocation, which has only one form. This stability is due to the fact that the π electrons can be spread out or shared over an extended p orbital network rather than being centered in only one bond.

In addition to its effect on stability, the resonance picture of an allylic carbocation also has chemical consequences. When the allylic carbocation produced by protonation of buta-1,3-diene reacts with Br^- ion to complete the addition, reaction can occur at either C1 or C3, because both share the positive charge. The result is a mixture of 1,2- and 1,4-addition products.

Br^-

1,4-addition (29%)

1,2-addition (71%)

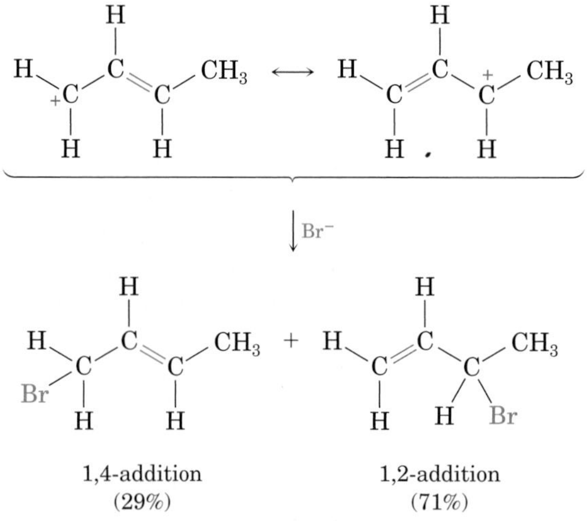

PROBLEM 4.12 Buta-1,3-diene, $H_2C{=}CH{-}CH{=}CH_2$, reacts with Br_2 to yield a mixture of 1,2- and 1,4-addition products. Show the structure of each.

4.11 Drawing and Interpreting Resonance Forms

Resonance is an extremely useful concept for explaining a variety of chemical phenomena. In the acetate ion, for instance, the lengths of the two C–O bonds are identical. Although there is no single line-bond structure that can account for this equivalence of C–O bonds, resonance theory accounts for it nicely. The acetate ion is simply a resonance hybrid of two resonance forms, with both oxygens sharing the π electrons and the negative charge equally:

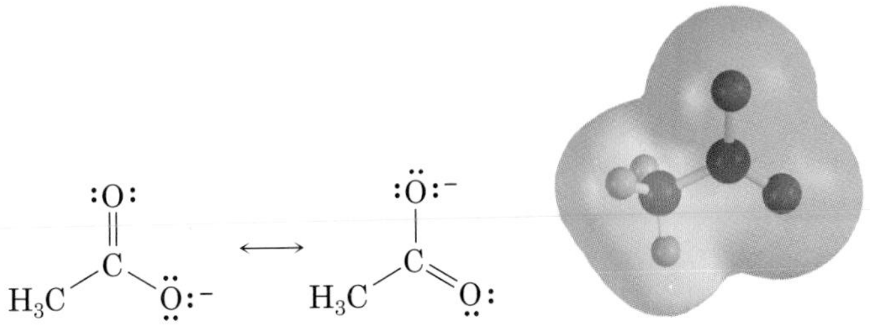

Acetate ion

As another example, we'll see in the next chapter that the six carbon–carbon bonds in aromatic compounds like benzene are equivalent because benzene is a resonance hybrid of two forms. Each form has alternating single and double bonds, and neither form is correct by itself. The true benzene structure is a hybrid of the two forms.

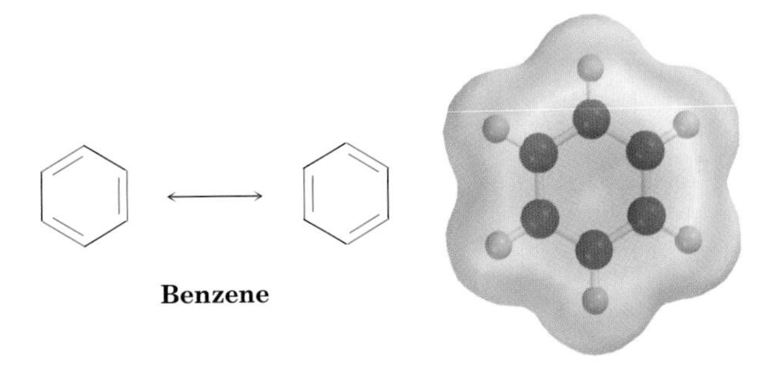

Benzene

When first dealing with resonance theory, it's often useful to have a set of guidelines that describe how to draw and interpret resonance forms. The following points should prove helpful:

- **Resonance forms are imaginary.** The real structure is a composite, or hybrid, of the different forms. Substances like the allylic carbocation, the acetate ion, and benzene are no different from any other substance: they have single, unchanging structures. The only difference between these and other substances is in the way they must be represented on paper.
- **Resonance forms differ only in the placement of their π or nonbonding electrons.** Neither the position nor the hybridization of atoms changes from one resonance form to another. In benzene, for example, the π electrons in the double bonds move, but the six carbon and six hydrogen atoms remain in place. By contrast, two structures such as cyclohexa-1,3-diene and cyclohexa-1,4-diene are *not* resonance structures because their

hydrogen atoms don't occupy the same positions. Instead, the two dienes are constitutional isomers.

Constitutional isomers, *NOT* resonance forms

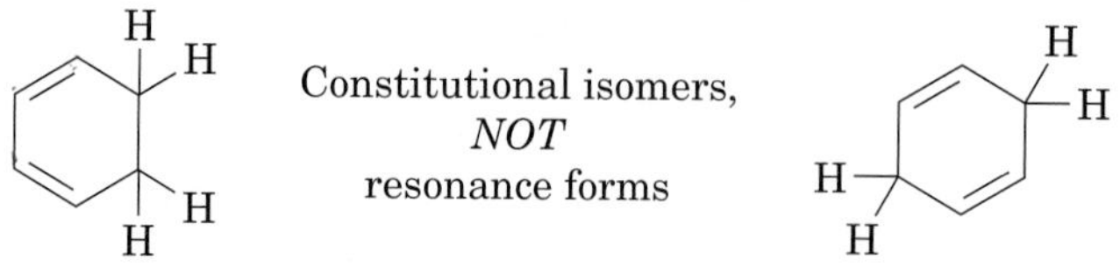

Cyclohexa-1,3-diene **Cyclohexa-1,4-diene**

- **Different resonance forms of a substance don't have to be equivalent.** For example, the allylic carbocation obtained by reaction of buta-1,3-diene with H^+ is unsymmetrical. One end of the π electron system has a methyl substituent, and the other end is unsubstituted. Even though the two resonance forms aren't equivalent, both contribute to the overall resonance hybrid.

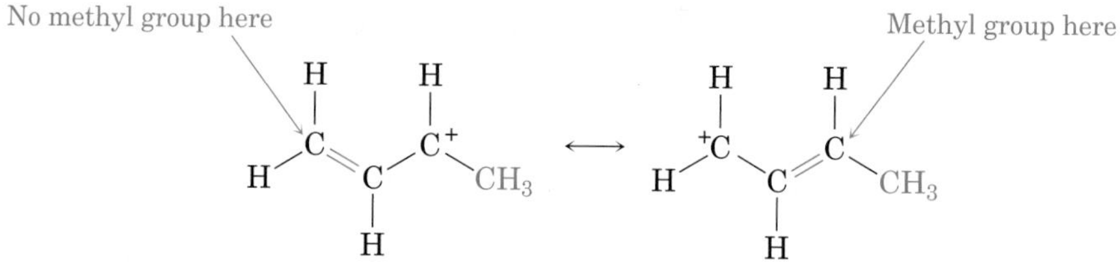

No methyl group here

Methyl group here

When two resonance forms are not equivalent, the actual structure of the resonance hybrid is closer to the more stable form than to the less stable form. Thus, we might expect the butenyl carbocation to look a bit more like a secondary carbocation than like a primary one.

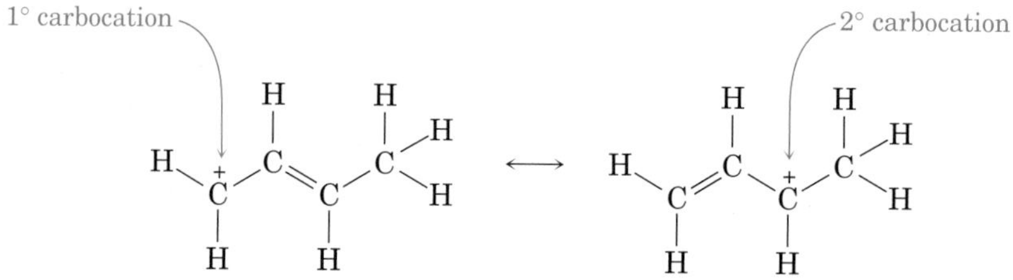

1° carbocation

2° carbocation

Less important resonance form

More important resonance form

- **Resonance forms must be valid Lewis structures and obey normal rules of valency.** A resonance form is like any other structure: the octet rule still applies. For example, one of the following structures for the acetate ion is not a valid resonance form because the carbon atom has five bonds and ten electrons:

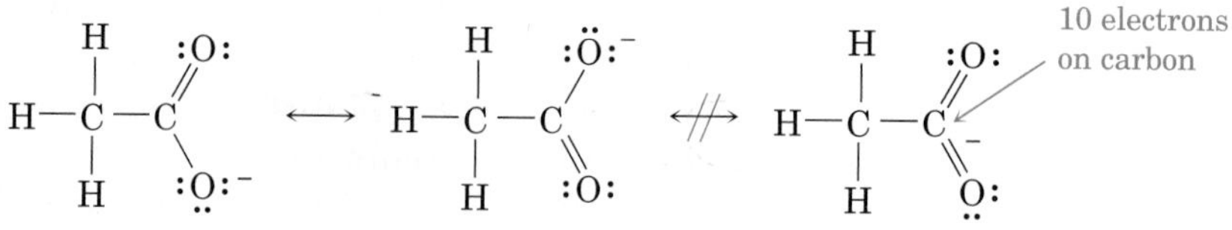

10 electrons on carbon

Acetate ion

NOT a valid resonance form

- **Resonance leads to stability.** The greater the number of resonance forms, the more stable the substance. We've already seen, for example, that an allylic carbocation is more stable than a nonallylic one. In a similar manner, we'll see in the next chapter that a benzene ring is more stable than a cyclic alkene.

PRACTICE PROBLEM 4.6

Use resonance structures to explain why the two N–O bonds of nitromethane are equivalent.

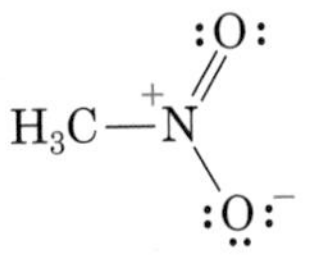

$H_3C-\overset{+}{N}(=O)-O^-$ **Nitromethane**

STRATEGY Resonance forms differ only in the placement of π (that is, multiple-bond) and nonbonding electrons. Nitromethane has two equivalent resonance forms, which can be drawn by showing the double bond either to the top oxygen or to the bottom oxygen. Only the positions of the electrons are different in the two forms.

SOLUTION

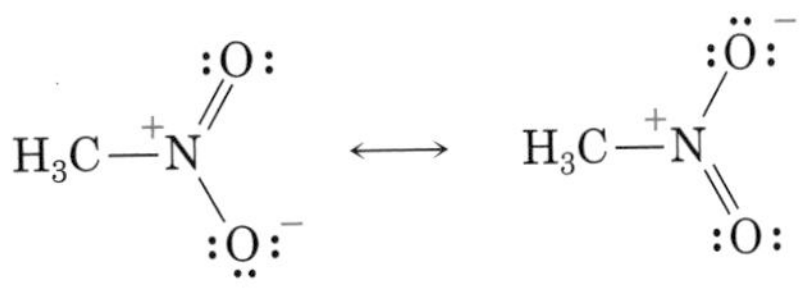

$H_3C-\overset{+}{N}(=O)-O^- \longleftrightarrow H_3C-\overset{+}{N}(-O^-)=O$ **Nitromethane**

PROBLEM 4.13 Give the structure of all possible monoadducts of HCl and penta-1,3-diene, $CH_3CH{=}CH{-}CH{=}CH_2$.

PROBLEM 4.14 Look at the possible carbocation intermediates produced during addition of HCl to penta-1,3-diene (Problem 4.13), and predict which is the most stable.

PROBLEM 4.15 Draw resonance structures for the following species:

(a) 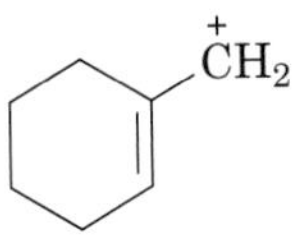 cyclohexene ring bearing $\overset{+}{C}H_2$ (b) $CH_3-C(=O)-\ddot{C}H_2^-$ (c) cyclohexadienyl ring with $^+$ and H

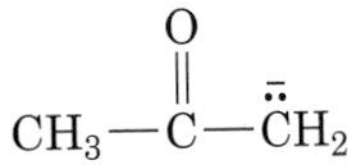

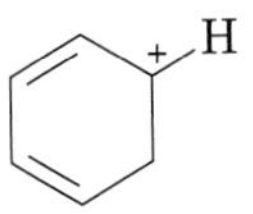

4.12 Alkynes and Their Reactions

Alkynes are hydrocarbons that contain a C≡C triple bond. As we saw in Section 1.8, a C≡C bond results from the overlap of two *sp*-hybridized carbon atoms and consists of one *sp*–*sp* σ bond and two *p*–*p* π bonds. Since four hydrogens must be removed from an alkane, C_nH_{2n+2}, to produce a triple bond, the general formula for an alkyne is C_nH_{2n-2}. Because alkynes occur much less commonly than alkenes, we'll look at them only briefly.

Alkynes are named by general rules similar to those used for alkanes (see Section 2.3) and alkenes (see Section 3.1). The suffix *-yne* is used in the parent hydrocarbon name to denote an alkyne, and the position of the triple bond is indicated by its number in the chain. Numbering begins at the chain end nearer the triple bond so that the triple bond receives as low a number as possible.

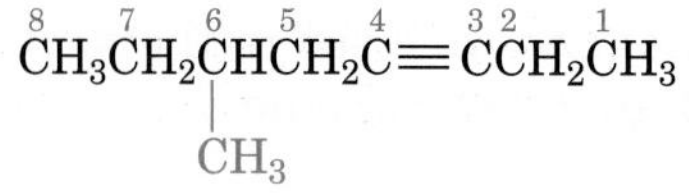

$$\overset{8}{C}H_3\overset{7}{C}H_2\overset{6}{C}H(CH_3)\overset{5}{C}H_2\overset{4}{C}\equiv\overset{3}{C}\overset{2}{C}H_2\overset{1}{C}H_3$$

Begin numbering at the end nearer the triple bond.

6-Methyloct-3-yne

Compounds containing both double and triple bonds are called *enynes* (not ynenes). Numbering of the hydrocarbon chain starts from the end nearer the first multiple bond, whether double or triple. If there is a choice in numbering, double bonds receive lower numbers than triple bonds. For example,

$$\underset{7}{HC}\equiv\underset{6}{C}\underset{5}{C}H_2\underset{4}{C}H_2\underset{3}{C}H_2\underset{2}{C}H=\underset{1}{C}H_2$$

Hept-1-en-6-yne

$$\underset{1}{HC}\equiv\underset{2}{C}\underset{3}{C}H_2\underset{4}{C}H(CH_3)\underset{5}{C}H_2\underset{6}{C}H_2\underset{7}{C}H=\underset{8}{C}H\underset{9}{C}H_3$$

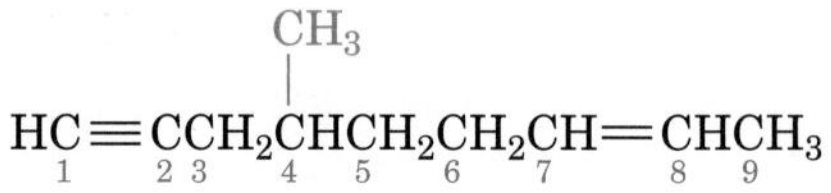

4-Methylnon-7-en-1-yne

PROBLEM 4.16 Give IUPAC names for the following compounds:

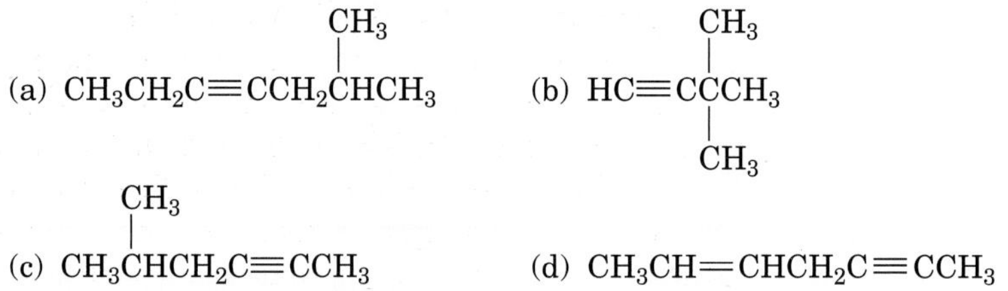

(a) $CH_3CH_2C\equiv CCH_2CH(CH_3)CH_3$

(b) $HC\equiv CC(CH_3)_2CH_3$

(c) $CH_3CH(CH_3)CH_2C\equiv CCH_3$

(d) $CH_3CH=CHCH_2C\equiv CCH_3$

Alkyne Reactions: Addition of H_2

Alkynes are easily converted into alkanes by reduction with 2 equivalents of H_2 over a palladium catalyst. The reaction proceeds through an alkene intermediate, and the reaction can be stopped at the alkene stage if the right catalyst is used. The catalyst most often used for this purpose is the Lindlar catalyst, a specially prepared form of palladium metal. Because hydrogenation occurs with syn stereochemistry, alkynes give cis alkenes when reduced. For example,

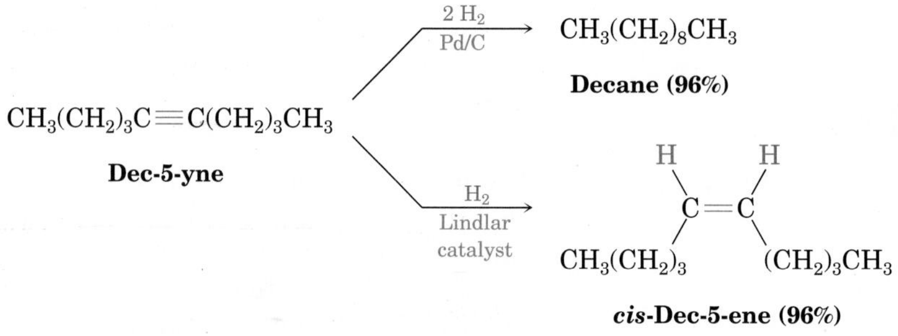

$$CH_3(CH_2)_3C\equiv C(CH_2)_3CH_3 \xrightarrow[Pd/C]{2\ H_2} CH_3(CH_2)_8CH_3$$

Dec-5-yne → **Decane (96%)**

$$CH_3(CH_2)_3C\equiv C(CH_2)_3CH_3 \xrightarrow[\text{Lindlar catalyst}]{H_2} \text{cis-}CH_3(CH_2)_3CH=CH(CH_2)_3CH_3$$

Dec-5-yne → ***cis*-Dec-5-ene (96%)**

Alkyne Reactions: Addition of HX

Alkynes give addition products on reaction with HCl, HBr, and HI just as alkenes do. Although the reaction can usually be stopped after addition of 1 equivalent of HX to yield a *vinylic* halide (**vinylic** means "on the C=C bond"), an excess of HX leads to formation of a dihalide product. As the following example indicates, the regioselectivity of addition to monosubstituted alkynes usually follows Markovnikov's rule. The H atom adds to the terminal carbon of the triple bond, and the X atom adds to the internal, more highly substituted carbon.

$$CH_3CH_2CH_2CH_2C{\equiv}CH + HBr \longrightarrow CH_3CH_2CH_2CH_2C(Br){=}CH_2$$

Hex-1-yne → **2-Bromohex-1-ene**

Alkyne Reactions: Addition of X_2

Bromine and chlorine add to alkynes to give dihalide addition products with anti stereochemistry:

$$CH_3CH_2CH_2CH_2C{\equiv}CH + Br_2 \xrightarrow{CCl_4} (CH_3CH_2CH_2CH_2)(Br)C{=}C(Br)(H)$$

Hex-1-yne → **(*E*)-1,2-Dibromohex-1-ene**

Alkyne Reactions: Addition of H_2O

Addition of water takes place when an alkyne is treated with aqueous sulfuric acid in the presence of mercuric sulfate catalyst. Markovnikov regioselectivity is found for the hydration reaction, with the H attaching to the less substituted carbon and the OH attaching to the more substituted carbon. Interestingly, though, the expected vinylic alcohol, or *enol* (*ene* = alkene; *ol* = alcohol), is not isolated. Instead, the enol rearranges to a more stable isomer, a ketone ($R_2C{=}O$). It turns out that enols and ketones rapidly interconvert—a process we'll discuss in more detail in Section 11.1. With few exceptions, the keto–enol equilibrium heavily favors the ketone. Enols are almost never isolated.

$$CH_3CH_2CH_2C{\equiv}CH + H_2O \xrightarrow[HgSO_4]{H_2SO_4} \left[CH_3CH_2CH_2C(OH){=}CH_2\right] \longrightarrow CH_3CH_2CH_2C({=}O)CH_3$$

Pent-1-yne → An enol → **Pentan-2-one (78%)**

catalyst

A mixture of both possible ketones results when an internal alkyne (R–C≡C–R′) is hydrated, but only a single product is formed from reaction of a terminal alkyne (R–C≡C–H).

$$CH_3CH_2C{\equiv}CCH_3 + H_2O \xrightarrow[HgSO_4]{H_2SO_4} CH_3CH_2C({=}O)CH_2CH_3 + CH_3CH_2CH_2C({=}O)CH_3$$

Pent-2-yne (an internal alkyne) → **Pentan-3-one** + **Pentan-2-one**

$$CH_3CH_2CH_2C{\equiv}CH + H_2O \xrightarrow[HgSO_4]{H_2SO_4} CH_3CH_2CH_2C({=}O)CH_3$$

Pent-1-yne (a terminal alkyne) → **Pentan-2-one**

Alkyne Reactions: Formation of Acetylide Anions

The most striking difference between the chemistry of alkenes and alkynes is that terminal alkynes (R–C≡C–H) are weakly acidic, with $pK_a \approx 25$ (see Section 1.10). Alkenes, by contrast, have $pK_a \approx 44$. When a terminal alkyne is treated with a strong base such as sodium amide, $NaNH_2$, the terminal hydrogen is removed and an **acetylide anion** is formed:

$$R{-}C{\equiv}C{-}H + :\ddot{N}H_2\ Na^+ \longrightarrow R{-}C{\equiv}C:^-\ Na^+ + :NH_3$$

Acetylide anion

The presence of an unshared electron pair on the negatively charged alkyne carbon makes acetylide anions both basic and nucleophilic. As a result, acetylide anions react with alkyl halides such as bromomethane to substitute for the halogen and yield a new alkyne product. We won't study the mechanism of this substitution reaction until Chapter 7, but will note for the present that it is a very useful method for preparing larger alkynes from simpler precursors. Terminal alkynes can be prepared by reaction of acetylene itself, and internal alkynes can be prepared by further reaction of a terminal alkyne:

$$HC{\equiv}CH \xrightarrow{NaNH_2} HC{\equiv}C^-\ Na^+ \xrightarrow{RCH_2Br} HC{\equiv}CCH_2R$$

Acetylene → **A terminal alkyne**

$$RC{\equiv}CH \xrightarrow{NaNH_2} RC{\equiv}C^-\ Na^+ \xrightarrow{R'CH_2Br} RC{\equiv}CCH_2R'$$

A terminal alkyne → **An internal alkyne**

The one limitation to the reaction of an acetylide anion with an alkyl halide is that only primary alkyl halides, RCH_2X, can be used, for reasons discussed in Chapter 7.

PRACTICE PROBLEM 4.7

What product would you obtain by hydration of 4-methylhex-1-yne?

STRATEGY Ask yourself what you know about alkyne addition reactions. Addition of water to 4-methylhex-1-yne according to Markovnikov's rule will yield a product with the OH group attached to C2 rather than C1. This enol then isomerizes to yield a ketone.

SOLUTION

$$CH_3CH_2CH(CH_3)CH_2C{\equiv}CH + H_2O \xrightarrow[HgSO_4]{H_2SO_4} \left[CH_3CH_2CH(CH_3)CH_2C(OH){=}CH_2\right] \longrightarrow CH_3CH_2CH(CH_3)CH_2C({=}O)CH_3$$

4-Methylhex-1-yne → **4-Methylhexan-2-one**

PRACTICE PROBLEM 4.8

What alkyne and what alkyl halide would you use to prepare pent-1-yne?

STRATEGY As always when synthesizing a compound, work the problem backward. Draw the structure of the target molecule, and identify the alkyl group(s) attached to the triple-bonded carbons. In the present case, one of the alkyne carbons has a propyl group attached to it and the other has a hydrogen attached to it. Thus, pent-1-yne could be prepared by treatment of acetylene with $NaNH_2$ to yield sodium acetylide, followed by reaction with 1-bromopropane

SOLUTION

$$\underset{\textbf{Acetylene}}{H{-}C{\equiv}C{-}H} + :\ddot{N}H_2^-\ Na^+ \longrightarrow \underset{\textbf{Sodium acetylide}}{H{-}C{\equiv}C{:}^-\ Na^+} + :NH_3$$

$$H{-}C{\equiv}C{:}^-\ Na^+ + \underset{\textbf{1-Bromopropane}}{CH_3CH_2CH_2Br} \longrightarrow \underset{\textbf{Pent-1-yne}}{H{-}C{\equiv}C{-}CH_2CH_2CH_3}$$

This propyl group comes from 1-bromopropane.

PROBLEM 4.17 What products would you expect from the following reactions?

(a) $CH_3CH_2CH_2C{\equiv}CH + 1\ \text{equiv}\ Cl_2 \longrightarrow\ ?$

(b) $CH_3CH_2CH_2C{\equiv}CCH_2CH_3 + 1\ \text{equiv}\ HBr \longrightarrow\ ?$

(c) $CH_3\overset{\overset{\large CH_3}{|}}{C}HCH_2C{\equiv}CCH_2CH_3 + H_2 \xrightarrow[\text{catalyst}]{\text{Lindlar}}\ ?$

PROBLEM 4.18 What product would you obtain by hydration of oct-4-yne?

PROBLEM 4.19 What alkynes would you start with to prepare the following ketones by a hydration reaction?

(a) $CH_3CH_2CH_2\overset{\overset{\large O}{\|}}{C}CH_3$ (b) $CH_3CH_2CH_2\overset{\overset{\large O}{\|}}{C}CH_2CH_3$

PROBLEM 4.20 Show the alkyne and alkyl halide from which the following products can be obtained. Where two routes look feasible, list both.

(a) $CH_3\overset{\overset{\large CH_3}{|}}{C}HCH_2CH_2C{\equiv}CH$ (b) $CH_3CH_2CH_2C{\equiv}CCH_3$ (c) $CH_3\overset{\overset{\large CH_3}{|}}{C}HC{\equiv}CCH_3$

INTERLUDE

Natural Rubber

Rubber—an unusual name for an unusual substance—is a naturally occurring alkene polymer produced by more than 400 different plants. The major source, however, is the so-called rubber tree, *Hevea brasiliensis*, from which the crude material is harvested as it drips from a slice made through the bark. The name *rubber* was coined by Joseph Priestley, the discoverer of oxygen and early researcher of rubber chemistry, for the simple reason that one of rubber's early uses was to rub out pencil marks on paper.

Unlike polyethylene and other simple alkene polymers, natural rubber is a polymer of a conjugated diene, *isoprene*, or 2-methylbuta-1,3-diene. The polymerization takes place by 1,4-addition (see Section 4.9) of isoprene monomers to the growing chain, leading to formation of a polymer that still contains double bonds spaced regularly at four-carbon intervals. As the following structure shows, these double bonds have *Z* stereochemistry:

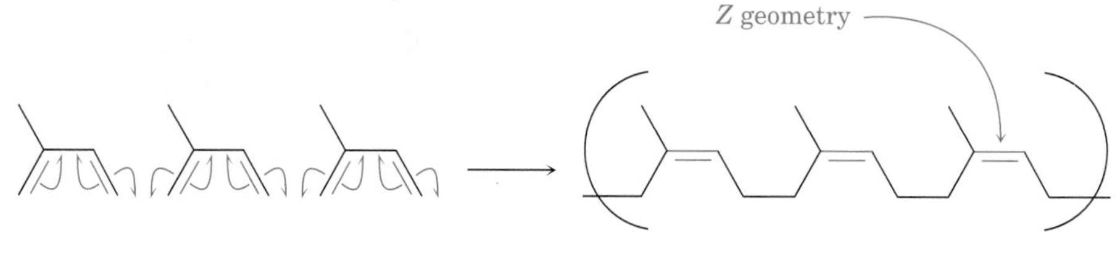

Crude rubber is harvested from the rubber tree, *Hevea brasiliensis.*

Crude rubber, called *latex*, is collected from the tree as an aqueous dispersion that is washed, dried, and coagulated by warming in air to give a polymer with chains that average about 5000 monomer units in length and have molecular weights of 200,000 to 500,000 amu. This crude coagulate is too soft and tacky to be useful until it is hardened by heating with elemental sulfur, a process called *vulcanization.* By mechanisms that are still not fully understood, vulcanization cross-links the rubber chains together by forming carbon–sulfur bonds between them, thereby hardening and stiffening the polymer. The exact degree of hardening can be varied, yielding material soft enough for automobile tires or hard enough for bowling balls (*ebonite*).

The remarkable ability of rubber to stretch and then contract to its original shape is due to the irregular shapes of the polymer chains caused by the double bonds. These double bonds introduce bends and kinks into the polymer chains, thereby preventing neighboring chains from nestling together into tightly packed, semicrystalline regions. When stretched, the randomly coiled chains straighten out and orient along the direction of the pull but are kept from sliding over each other by the cross-links. When the stretch is released, the polymer reverts to its original random state.

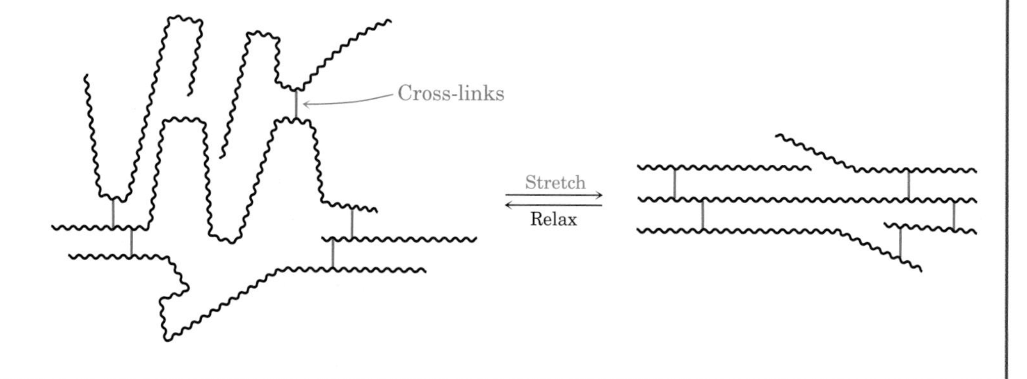

Summary and Key Words

Acetylide anion, p. 126
1,4-Addition, p. 118
Alkyne, p. 123
Allylic, p. 118
Anti stereochemistry, p. 111
Conjugation, p. 117
Diol, p. 113
Hydration, p. 108
Hydrogenation, p. 112
Hydroxylation, p. 113
Markovnikov's rule, p. 104
Monomer, p. 115
Oxidation, p. 113
Polymer, p. 115
Reduction, p. 112
Resonance form, p. 120
Resonance hybrid, p. 120
Syn stereochemistry, p. 113
Vinylic, p. 124

The chemistry of alkenes is dominated by addition reactions of electrophiles. When HX reacts with an alkene, **Markovnikov's rule** predicts that the hydrogen will add to the carbon that has fewer alkyl substituents and the X group will add to the carbon that has more alkyl substituents. Many electrophiles besides HX add to alkenes. Thus, Br_2 and Cl_2 add to give 1,2-dihalide addition products having **anti stereochemistry**. Addition of H_2O (**hydration**) takes place on reaction of the alkene with aqueous acid, and addition of H_2 (**hydrogenation**) occurs in the presence of a metal catalyst such as platinum or palladium. **Oxidation** of alkenes is carried out using potassium permanganate, $KMnO_4$. Under basic conditions, $KMnO_4$ reacts with alkenes to yield cis **1,2-diols**. Under neutral or acidic conditions, $KMnO_4$ cleaves double bonds to yield carbonyl-containing products.

Conjugated dienes, such as buta-1,3-diene, contain alternating single and double bonds. Conjugated dienes undergo **1,4-addition** of electrophiles through the formation of a resonance-stabilized **allylic** carbocation intermediate. No single line-bond representation can depict the true structure of an allylic carbocation. Rather, the true structure is a **resonance hybrid** somewhere intermediate between two contributing resonance forms. The only difference between two **resonance forms** is in the location of double-bond and lone-pair electrons. The atoms remain in the same places in both structures.

Many simple alkenes undergo **polymerization** when treated with a radical catalyst. **Polymers** are large molecules built up by the repetitive bonding together of many small **monomer** units.

Alkynes are hydrocarbons that contain carbon–carbon triple bonds. Much of the chemistry of alkynes is similar to that of alkenes. For example, alkynes react with 1 equivalent of HBr and HCl to yield **vinylic** halides, and with 1 equivalent of Br_2 and Cl_2 to yield 1,2-dihalides. Alkynes can also be hydrated by reaction with aqueous sulfuric acid in the presence of mercuric sulfate catalyst. The reaction leads to an intermediate enol that immediately isomerizes to a ketone. Alkynes can be hydrogenated with the Lindlar catalyst to yield a cis alkene. Terminal alkynes are weakly acidic and can be converted into **acetylide anions** by treatment with a strong base. Reaction of the acetylide anion with a primary alkyl halide then gives an internal alkyne.

Summary of Reactions

Note: No stereochemistry is implied unless specifically stated or indicated with wedged, solid, and dashed lines.

1. Reactions of alkenes
   (a) Addition of HX, where X = Cl, Br, or I (Sections 4.1–4.2)

$$\text{>C=C<} \xrightarrow[\text{Ether}]{\text{HX}} \text{>C(H)–C(X)<}$$

Markovnikov's rule: H adds to the less highly substituted carbon, and X adds to the more highly substituted one.

(b) Addition of H_2O (Section 4.4)

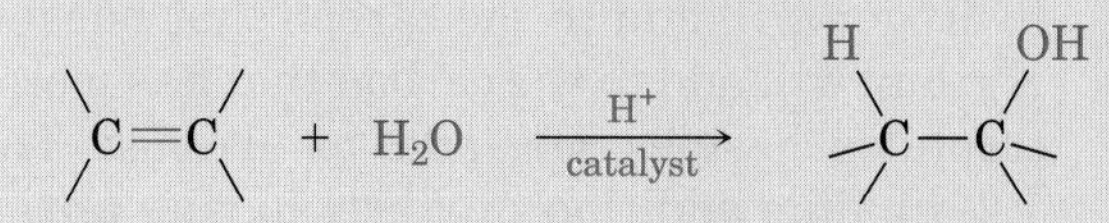

$$\mathrm{>C{=}C<} + H_2O \xrightarrow[\text{catalyst}]{H^+} \mathrm{-\overset{H}{C}-\overset{OH}{C}-}$$

Markovnikov's rule: H adds to the less highly substituted carbon, and OH adds to the more highly substituted one.

(c) Addition of X_2, where X = Cl, Br (Section 4.5)

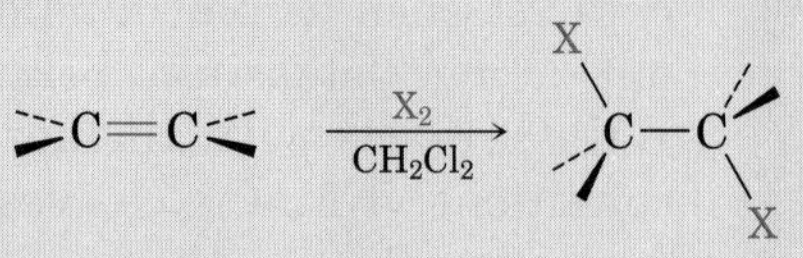

$$\mathrm{>C{=}C<} \xrightarrow[CH_2Cl_2]{X_2} \mathrm{\overset{X}{C}-\underset{X}{C}}$$

Anti addition

(d) Addition of H_2 (Section 4.6)

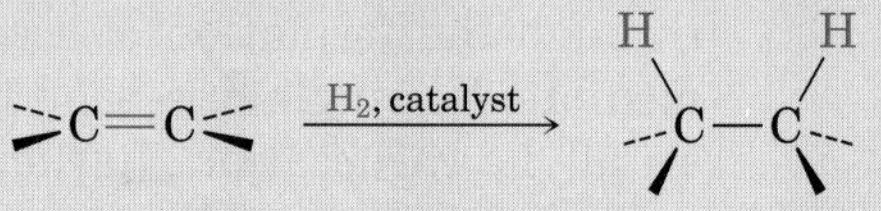

$$\mathrm{>C{=}C<} \xrightarrow{H_2,\ \text{catalyst}} \mathrm{\overset{H}{C}-\overset{H}{C}}$$

Syn addition

(e) Hydroxylation with $KMnO_4$ (Section 4.7)

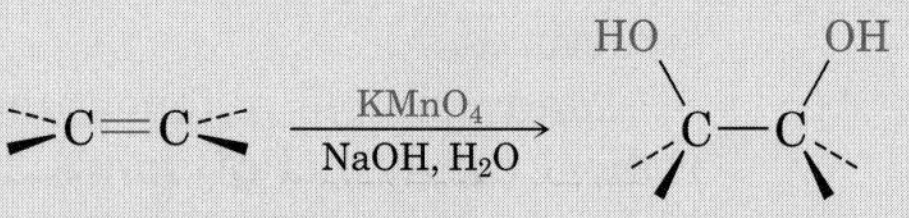

$$\mathrm{>C{=}C<} \xrightarrow[NaOH,\ H_2O]{KMnO_4} \mathrm{\overset{HO}{C}-\overset{OH}{C}}$$

Syn addition

(f) Oxidative cleavage of alkenes with acidic $KMnO_4$ (Section 4.7)

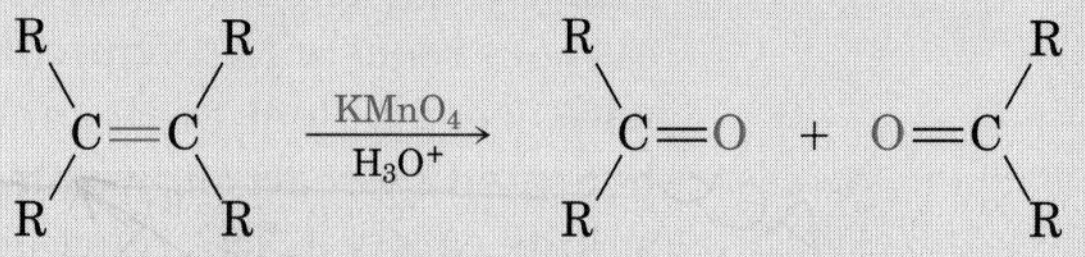

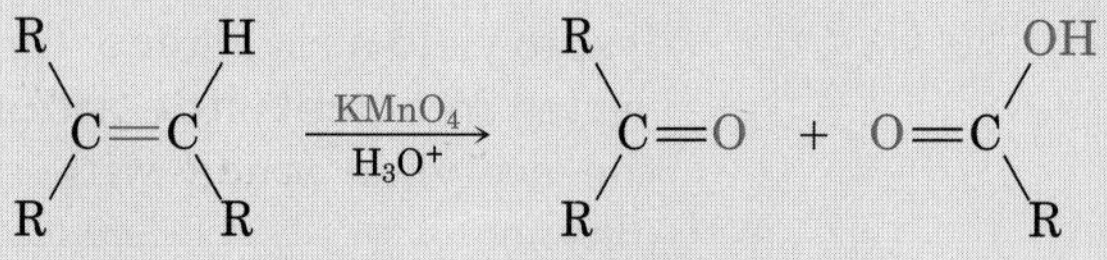

$$R_2C{=}CR_2 \xrightarrow[H_3O^+]{KMnO_4} R_2C{=}O + O{=}CR_2$$

$$R_2C{=}CHR \xrightarrow[H_3O^+]{KMnO_4} R_2C{=}O + O{=}C(OH)R$$

(g) Polymerization of alkenes (Section 4.8)

$$n\ H_2C{=}CH_2 \xrightarrow[\text{initiator}]{\text{Radical}} \mathrm{+\!(CH_2CH_2)\!+_{\mathit{n}}}$$

2. Reactions of alkynes (Section 4.12)
(a) Addition of H_2

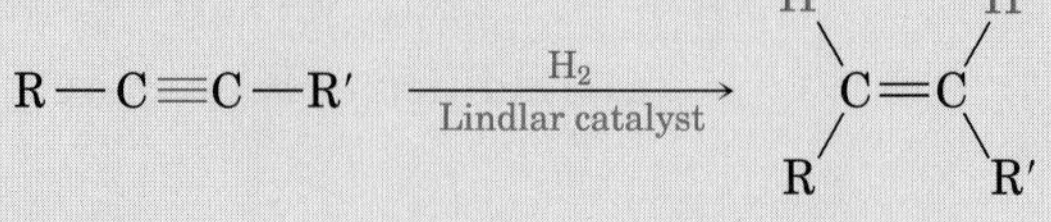

$$R{-}C{\equiv}C{-}R' \xrightarrow[\text{Lindlar catalyst}]{H_2} \mathrm{\underset{R}{\overset{H}{C}}{=}\underset{R'}{\overset{H}{C}}}$$

A cis alkene

Syn addition

(b) Addition of HX, where X = Cl, Br, or I

$$-C{\equiv}C- + HX \longrightarrow \overset{H\quad\ X}{C{=}C}$$

Markovnikov's rule:
H adds to the less highly substituted carbon, and X adds to the more highly substituted one.

(c) Addition of X_2, where X = Cl, Br

$$-C{\equiv}C- + X_2 \longrightarrow \overset{X}{C}{=}\underset{X}{C}$$

Anti addition

(d) Addition of H_2O

$$-C{\equiv}C- + H_2O \xrightarrow[HgSO_4]{H_2SO_4} \left[\overset{OH\quad H}{C{=}C}\right] \longrightarrow \overset{O\quad\ H}{C{-}C{-}H}$$

(e) Acetylide anion formation

$$R{-}C{\equiv}C{-}H \xrightarrow{NaNH_2} R{-}C{\equiv}C{:}^{-}\ Na^{+} + NH_3$$

(f) Reaction of acetylide anions with alkyl halides

$$R{-}C{\equiv}C{:}^{-}\ Na^{+} + R'CH_2X \longrightarrow R{-}C{\equiv}C{-}CH_2R' + NaX$$

EXERCISES

Visualizing Chemistry

4.21 Name the following alkenes, and predict the products of their reaction with (i) $KMnO_4$ in aqueous acid and (ii) $KMnO_4$ in aqueous NaOH:

(a)

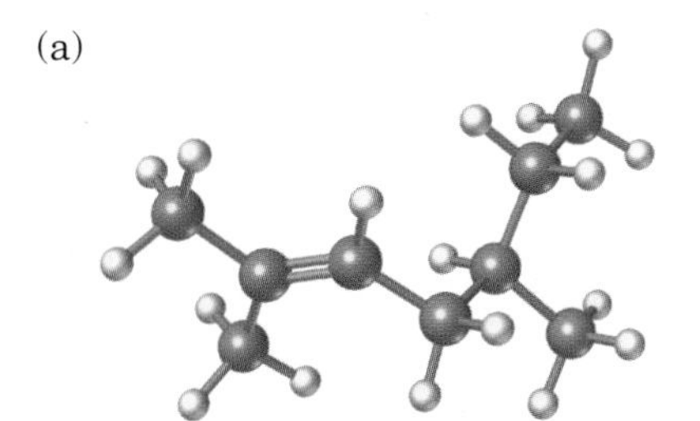

(b)

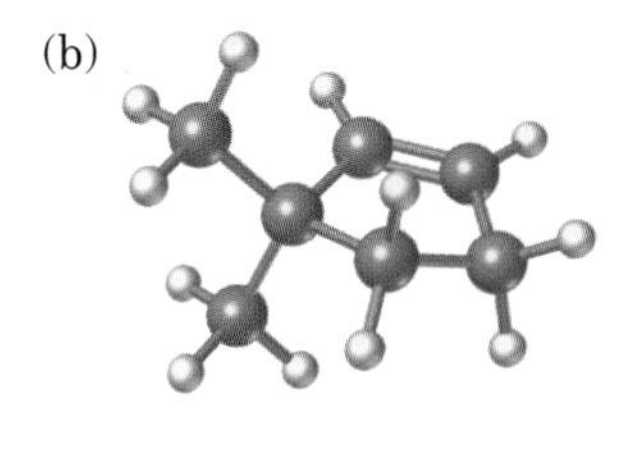

4.22 Name the following alkynes, and predict the products of their reaction with (i) H_2 in the presence of a Lindlar catalyst and (ii) H_3O^+ in the presence of $HgSO_4$:

(a) (b)

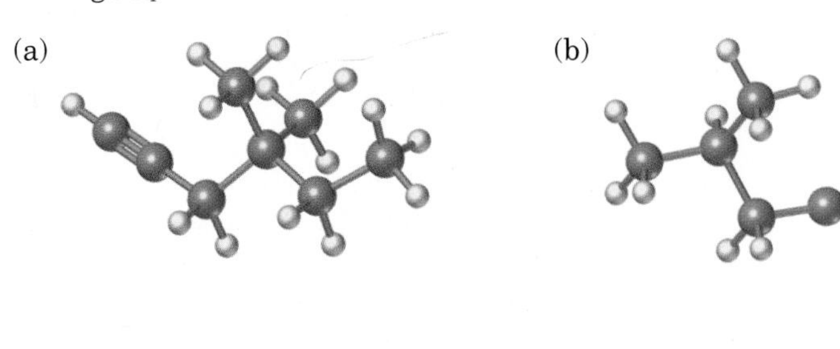

4.23 What alkenes would give the following alcohols on hydration? (Red = O.)

(a) (b)

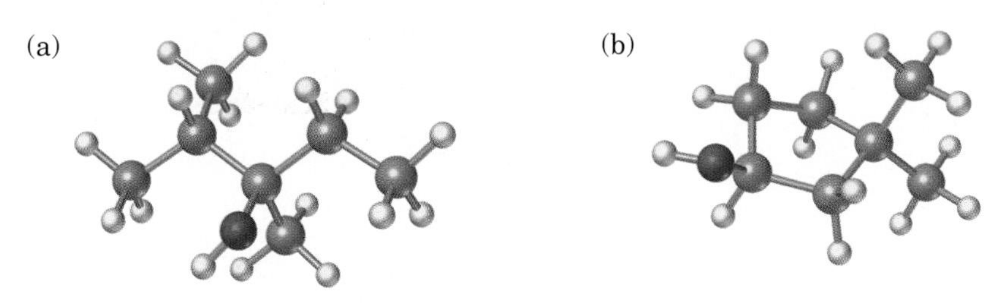

4.24 From what alkyne might each of the following substances have been made? (Red = O, yellow-green = Cl.)

(a) (b)

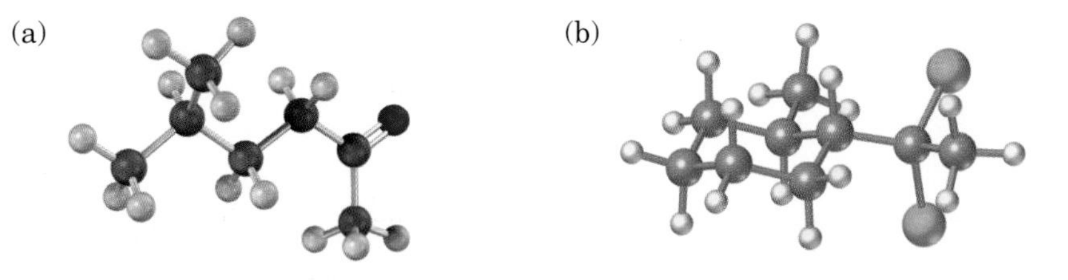

Additional Problems

NOMENCLATURE

4.25 Give IUPAC names for the following compounds:

(a) $CH_3CH{=}CHC(CH_3){=}CHCH_3$ (the CH_3 branch is on the third carbon)

(b) $CH_3CH{=}CHCH(CH_2CH_2CH_3)CH_2C{\equiv}CH$ (the $CH_2CH_2CH_3$ branch is on the fourth carbon)

(c) $H_2C{=}C{=}C(CH_3)CH_3$ (the CH_3 branch is on the third carbon)

(d) $HC{\equiv}CCH_2C{\equiv}CCH(CH_3)CH_3$ (the CH_3 branch is on the sixth carbon)

4.26 Draw structures corresponding to the following IUPAC names:

(a) 3-Ethylhept-1-yne (b) 3,5-Dimethylhex-4-en-1-yne

(c) Hepta-1,5-diyne (d) 1-Methylcyclopenta-1,3-diene

4.27 Draw structures corresponding to the following IUPAC names:

(a) Hept-3-yne (b) 3,3-Dimethyloct-4-yne

(c) 3,4-Dimethylcyclodecyne (d) 2,2,5,5-Tetramethylhex-3-yne

4.28 The following two hydrocarbons have been isolated from plants in the sunflower family. Name them according to IUPAC rules.
(a) $CH_3CH{=}CHC{\equiv}CC{\equiv}CCH{=}CHCH{=}CHCH{=}CH_2$ (all trans)
(b) $CH_3C{\equiv}CC{\equiv}CC{\equiv}CC{\equiv}CC{\equiv}CCH{=}CH_2$

ISOMERISM

4.29 Draw and name all the possible pentyne isomers, C_5H_8.

4.30 Draw and name the six possible diene isomers of formula C_5H_8. Which of the six are conjugated dienes?

4.31 Draw three possible structures for both of the following formulas:
(a) C_6H_8 (b) C_6H_8O

PREDICT THE PRODUCT

4.32 Predict the products of the following reactions. Indicate regioselectivity where relevant. (The aromatic ring is inert to all the indicated reagents.)

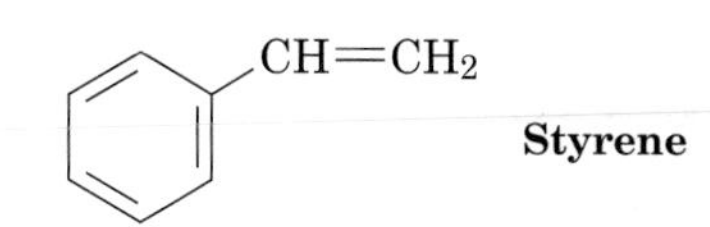

(a) Styrene + H_2 $\xrightarrow{\text{Pd}}$? (b) Styrene + Br_2 $\longrightarrow$?

(c) Styrene + HBr $\longrightarrow$? (d) Styrene + $KMnO_4$ $\xrightarrow{\text{NaOH, } H_2O}$?

4.33 Using an oxidative cleavage reaction, explain how you would distinguish between the following two isomeric cyclohexadienes:

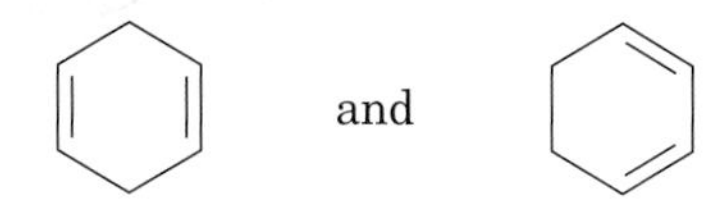

4.34 Formulate the reaction of cyclohexene with Br_2, showing the reaction intermediate and the final product with correct stereochemistry.

4.35 What products would you expect to obtain from reaction of cyclohexa-1,3-diene with each of the following?
(a) 1 mol Br_2 in CH_2Cl_2 (b) 1 mol HCl
(c) 1 mol DCl (D = deuterium, ^2H) (d) H_2 over a Pd catalyst

4.36 Predict the products of the following reactions on hex-1-yne:

(a) $\xrightarrow{\text{1 equiv HBr}}$? (b) $\xrightarrow{\text{1 equiv } Cl_2}$? (c) $\xrightarrow{H_2\text{, Lindlar catalyst}}$?

4.37 Predict the products of the following reactions on dec-5-yne:

(a) $\xrightarrow{H_2\text{, Lindlar catalyst}}$? (b) $\xrightarrow{\text{2 equiv } Br_2}$? (c) $\xrightarrow{H_2O,\ H_2SO_4,\ HgSO_4}$?

4.38 Suggest structures for alkenes that give the following reaction products. There may be more than one answer for some cases.

(a) ? $\xrightarrow{H_2\text{/Pd catalyst}}$ 2-Methylhexane

(b) ? $\xrightarrow{Br_2 \text{ in } CH_2Cl_2}$ 2,3-Dibromo-5-methylhexane

(c) ? $\xrightarrow{\text{HBr}}$ 2-Bromo-3-methylheptane

(d) ? $\xrightarrow[H_2O]{KMnO_4,OH^-}$ $CH_3CH(CH_3)CH_2CH(OH)CH(OH)CH_2CH_3$

PREDICT THE REACTANT

4.39 Draw the structure of a hydrocarbon that reacts with only 1 equivalent of H_2 on catalytic hydrogenation and gives only pentanoic acid, $CH_3CH_2CH_2CH_2CO_2H$, on treatment with acidic $KMnO_4$. Write the reactions involved.

4.40 Give the structure of an alkene that yields the following keto acid on reaction with $KMnO_4$ in aqueous acid:

$$? \xrightarrow[H_3O^+]{KMnO_4} HO\overset{O}{\overset{\|}{C}}CH_2CH_2CH_2CH_2\overset{O}{\overset{\|}{C}}CH_3$$

4.41 What alkenes would you hydrate to obtain the following alcohols?

(a) $CH_3CH_2\overset{OH}{\overset{|}{C}}HCH_3$ (b) 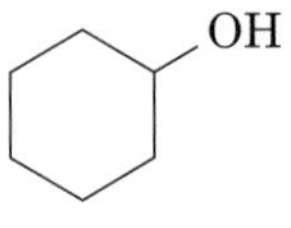 cyclohexyl–OH (c) 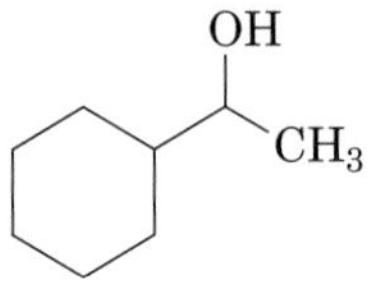 cyclohexyl–CH(OH)–CH_3

4.42 What alkynes would you hydrate to obtain the following ketones?

(a) $CH_3\overset{CH_3}{\overset{|}{C}}HCH_2\overset{O}{\overset{\|}{C}}CH_3$ (b) 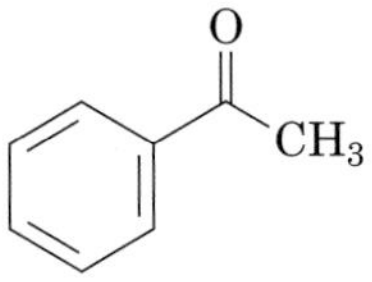 phenyl–C(=O)–CH_3

4.43 Draw the structure of a hydrocarbon that reacts with 2 equivalents of H_2 on catalytic hydrogenation and gives only succinic acid on reaction with acidic $KMnO_4$.

$$HO\overset{O}{\overset{\|}{C}}CH_2CH_2\overset{O}{\overset{\|}{C}}OH \quad \textbf{Succinic acid}$$

SYNTHESIS

4.44 In planning the synthesis of a compound, it's as important to know what *not* to do as to know what to do. What is wrong with each of the following reactions?

(a) $CH_3\overset{CH_3}{\overset{|}{C}}{=}CHCH_3 \xrightarrow{HBr} CH_3\overset{CH_3}{\overset{|}{C}}H\underset{Br}{\underset{|}{C}}HCH_3$

(b) 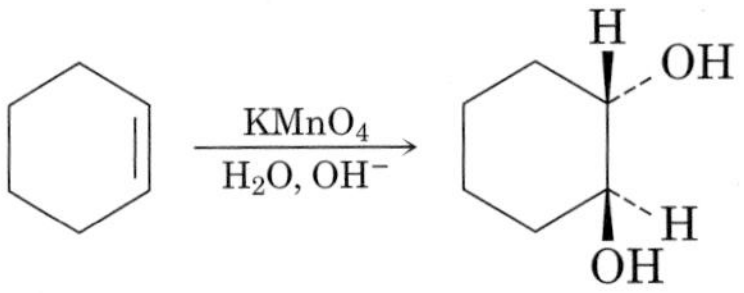 cyclohexene $\xrightarrow[H_2O,\ OH^-]{KMnO_4}$ trans-cyclohexane-1,2-diol (H, OH; H, OH)

(c) $CH_3CH_2\overset{CH_3}{\overset{|}{C}}HCH_2C{\equiv}CH \xrightarrow[HgSO_4]{H_2O,\ H_2SO_4} CH_3CH_2\overset{CH_3}{\overset{|}{C}}HCH_2CH_2\overset{O}{\overset{\|}{C}}H$

4.45 How would you prepare *cis*-but-2-ene starting from propyne, an alkyl halide, and any other reagents needed? (This problem can't be worked in a single step. You'll have to carry out more than one reaction.)

4.46 Using but-1-yne as the only organic starting material, along with any inorganic reagents needed, how would you synthesize the following compounds? (More than one step may be needed.)

(a) Butane (b) 1,1,2,2-Tetrachlorobutane
(c) 2-Bromobutane (d) Butan-2-one ($CH_3CH_2COCH_3$)

RESONANCE

4.47 Draw the indicated number of additional resonance forms for each of the following substances:

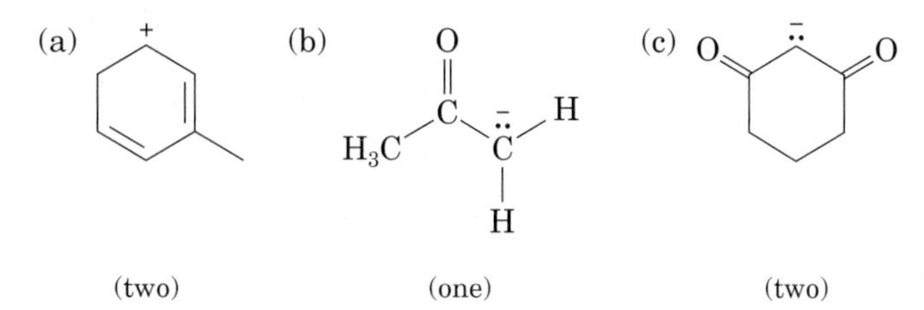

4.48 One of the following pairs of structures represents resonance forms, and one does not. Explain which is which.

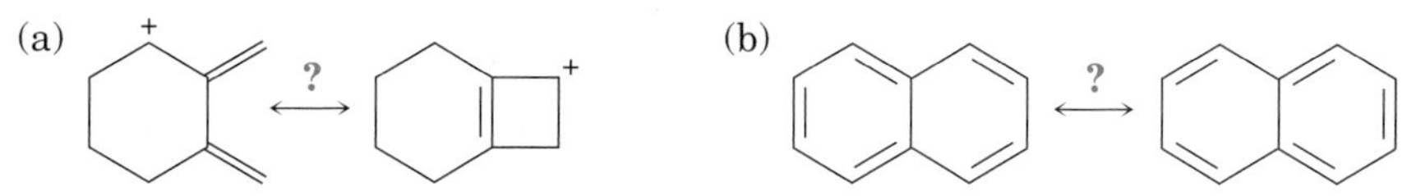

4.49 Draw three additional resonance forms for the benzyl cation.

$\overset{+}{C}H_2$

Benzyl cation

4.50 In light of your answer to Problem 4.49, what product would you expect from the following reaction? Explain.

+ HCl ⟶ ?

POLYMERS

4.51 Plexiglas, a clear plastic used to make many molded articles, is made by polymerization of methyl methacrylate. Draw a representative segment of Plexiglas.

$$(H)_2C{=}C(CO_2CH_3)(CH_3)$$

Methyl methacrylate

4.52 What monomer unit might be used to prepare the following polymer?

$$\text{-}CH_2\underset{Cl}{\overset{CH_3}{C}}CH_2\underset{Cl}{\overset{CH_3}{C}}CH_2\underset{Cl}{\overset{CH_3}{C}}CH_2\underset{Cl}{\overset{CH_3}{C}}CH_2\underset{Cl}{\overset{CH_3}{C}}\text{-}$$

4.53 Poly(vinyl pyrrolidone), prepared from *N*-vinyl pyrrolidone, is used both in cosmetics and as a synthetic blood substitute. Draw a representative segment of the polymer.

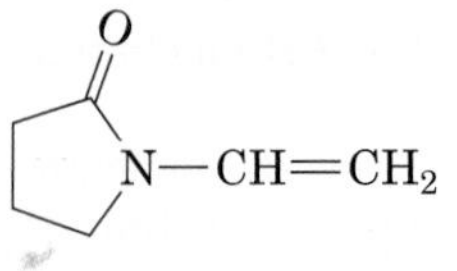

$N{-}CH{=}CH_2$ **N-Vinylpyrrolidone**

INTEGRATED PROBLEMS

4.54 Reaction of 2-methylpropene with CH_3OH in the presence of H_2SO_4 catalyst yields methyl *tert*-butyl ether, $CH_3OC(CH_3)_3$, by a mechanism analogous to that of acid-catalyzed alkene hydration. Write the mechanism.

4.55 Compound A has the formula C_8H_8. It reacts rapidly with acidic $KMnO_4$ but reacts with only 1 equivalent of H_2 over a palladium catalyst. On hydrogenation under conditions that reduce aromatic rings, A reacts with 4 equivalents of H_2 and hydrocarbon B, C_8H_{16}, is produced. The reaction of A with $KMnO_4$ gives CO_2 and a carboxylic acid C, $C_7H_6O_2$. What are the structures of A, B, and C? Write all the reactions.

4.56 Compound A, C_9H_{12}, absorbs 3 equivalents of H_2 on catalytic reduction over a palladium catalyst to give B, C_9H_{18}. On reaction with $KMnO_4$, compound A gives, among other things, a ketone that was identified as cyclohexanone. On treatment with $NaNH_2$ in NH_3, followed by addition of iodomethane, compound A gives a new hydrocarbon C, $C_{10}H_{14}$. What are the structures of A, B, and C?

4.57 The sex attractant of the common housefly is a hydrocarbon named *muscalure,* $C_{23}H_{46}$. On treatment of muscalure with aqueous acidic $KMnO_4$, two products are obtained: $CH_3(CH_2)_{12}CO_2H$ and $CH_3(CH_2)_7CO_2H$. Propose a structure for muscalure.

4.58 How would you synthesize muscalure (Problem 4.57) starting from acetylene and any alkyl halides needed? (The double bond in muscalure is cis.)

4.59 Give the structure of an alkene that provides only acetone, $(CH_3)_2C{=}O$, on reaction with acidic $KMnO_4$.

4.60 Draw a reaction energy diagram for the addition of HBr to pent-1-ene. Let one curve on your diagram show the formation of 1-bromopentane product and another curve on the same diagram show the formation of 2-bromopentane product. Label the positions for all reactants, intermediates, and products.

4.61 Make sketches of what you imagine the transition-state structures to look like in the reaction of HBr with pent-1-ene (Problem 4.60).

4.62 Methylenecyclohexane, on treatment with strong acid, isomerizes to yield 1-methylcyclohexene. Propose a mechanism by which the reaction might occur.

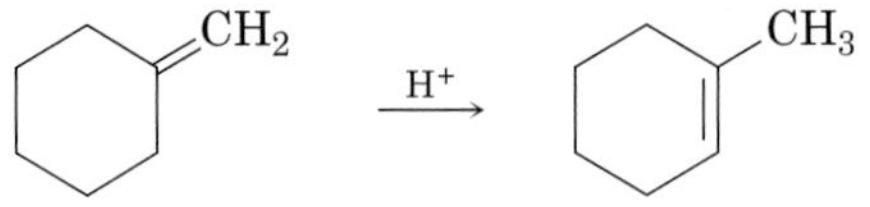

CH_2 $\xrightarrow{H^+}$ CH_3

Methylenecyclohexane **1-Methylcyclohexene**

4.63 α-Terpinene, $C_{10}H_{16}$, is a pleasant-smelling hydrocarbon that has been isolated from oil of marjoram. On hydrogenation over a palladium catalyst, α-terpinene reacts with 2 mol equiv of hydrogen to yield a new hydrocarbon, $C_{10}H_{20}$. On reaction with acidic $KMnO_4$, α-terpinene yields oxalic acid and 6-methylheptane-2,5-dione. Propose a structure for α-terpinene.

$$HOOC{-}COOH \qquad CH_3\overset{O}{\overset{\|}{C}}CH_2CH_2\overset{O}{\overset{\|}{C}}\underset{CH_3}{\underset{|}{C}H}CH_3$$

Oxalic acid **6-Methylheptane-2,5-dione**

4.64 Explain the observation that hydroxylation of *cis*-but-2-ene with basic $KMnO_4$ yields a different product than hydroxylation of *trans*-but-2-ene. First draw the structure and show the stereochemistry of each product, and then make molecular models. We'll explore the stereochemistry of the products in more detail in Chapter 6.

IN THE MEDICINE CABINET

4.65 Steroid analogs of progesterone and estrogen are the basis for birth control. Norethindrone, one of the most common components of the "pill," is prepared using alkyne chemistry.

CH3 OH C≡CH

Norethindrone

(a) What four functional groups are present in norethindrone?
(b) What is the molecular formula of norethindrone?
(c) To incorporate the alkyne group, acetylene is treated with a strong base to make reagent A. What is the structure of reagent A?

$$HC{\equiv}CH \;+\; \text{Base} \longrightarrow \mathbf{A}$$

Acetylene

(d) Reagent A is added to a carbonyl group. To understand this chemistry, we can draw a resonance structure of the C=O bond based on electronegativity. In the circles shown on the abbreviated structure, indicate where the "+" and "−" charges belong.

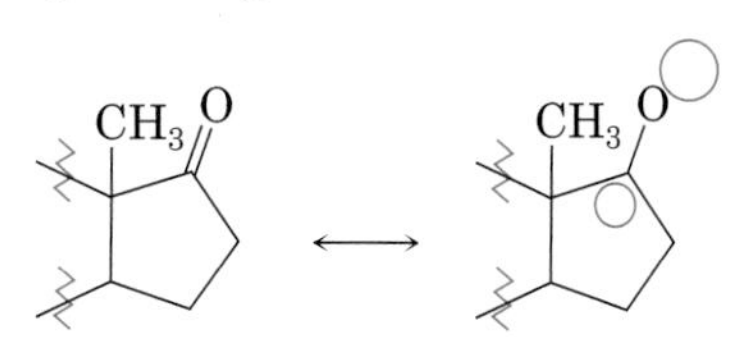

(e) Draw a mechanism for reagent A reacting with the resonance structure shown in part (d). For the second step, add a source of "H^+" to obtain the final product, norethindrone.

In the Field with Agrochemicals

4.66 Oct-1-en-3-ol is a potent mosquito attractant commonly used in mosquito traps. A number of reactions, including hydrogenation, will transform oct-1-en-3-ol into a less effective molecule. Draw the product of this reaction. (*Hint:* The 3-ol corresponds to an —OH group on carbon 3.)

4.67 Natural oils derived from seeds or beans of specific plants are complex mixtures of molecules. The main component of castor oil is a triester.

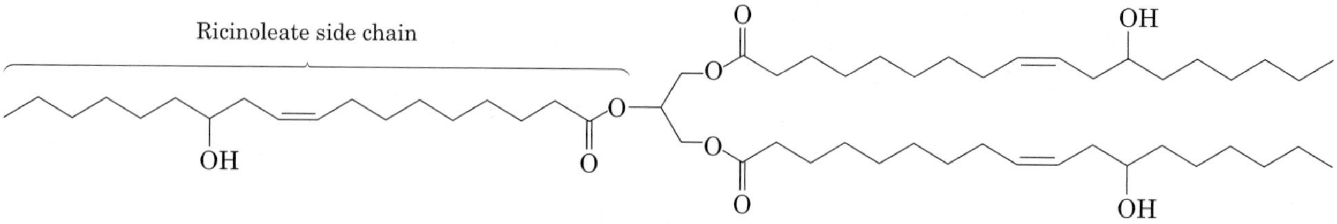

Ricinoleate side chain

OH O O OH O O OH O

Castor oil

(a) Identify the three esters as well as the alkene and alcohol functional groups in this component of castor oil.

(b) What is the molecular formula of this molecule?

(c) Chemical modification to this triester can include complete hydrogenation and complete halogenation with Cl_2 or Br_2. Draw the products of these three separate reactions on the abbreviated structure that follows.

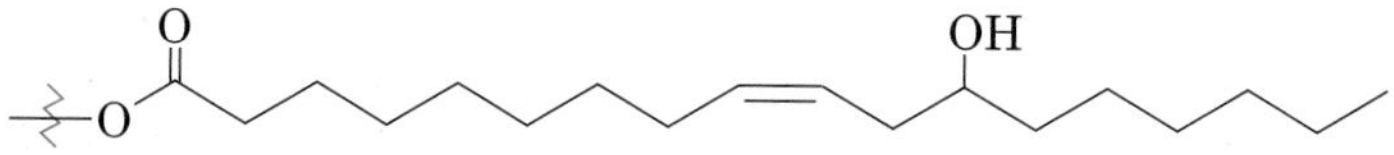

O OH O

Ricinoleate side chain

(d) Dehydration can yield cis and trans isomers of two different alkenes. Draw the two conjugated and two nonconjugated products of this reaction (assuming isomerization of the original *cis*-alkene does not occur).

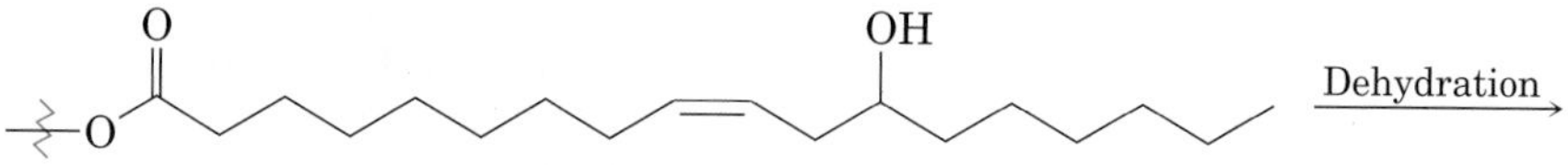

O OH O Dehydration

4.68 How important is double-bond geometry? Natural rubber contains *cis*-alkenes. The material that results is elastic: it bounces. Synthetic rubber is an isomer containing *trans*-alkenes. This material has been used as sealants on root canals and as the covering of golf balls because it is much less elastic. Draw the structure of one repeat unit of synthetic rubber.

One repeat unit, or monomer, of natural rubber

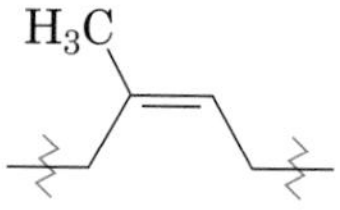

H_3C

Capsaicin, an aromatic compound, is responsible for the burning flavor of hot peppers. It is also the active ingredient in Zostrix and other creams used to treat arthritis-related joint pain. (© Lois Ellen Frank/CORBIS)

CHAPTER 5

Aromatic Compounds

In the early days of organic chemistry, the word *aromatic* was used to describe fragrant substances such as benzaldehyde (from cherries, peaches, and almonds), toluene (from tolu balsam), and benzene (from coal distillate). It was soon realized, however, that the substances grouped as aromatic differed from other compounds in their chemical behavior. A quick inspection of these molecules reveals that they each contain a benzene-like ring—a six-membered ring with alternating double and single bonds.

This **aromatic** benzene-like group is present in many important molecules, including the steroid hormone estrone and the analgesic ibuprofen. We'll see in this chapter how aromatic substances behave and why they're different from the alkanes, alkenes, and alkynes we've studied up to this point.

H_3C O H H H HO CH_3 O CH_3CHCH_2 CHCOH CH_3

Benzene **Estrone** **Ibuprofen**

5.1

Structure of Benzene: The Kekulé Proposal

By the mid-1800s, benzene was known to have the molecular formula C_6H_6 and its chemistry was being actively explored. It was known that although benzene is relatively unreactive toward most reagents that attack alkenes, it reacts with Br_2 in the presence of iron to give the *substitution* product C_6H_5Br rather than the *addition* product $C_6H_6Br_2$. Furthermore, only one monobromo substitution product was known; no isomers had been prepared.

On the basis of these and other results, August Kekulé proposed in 1865 that benzene contains a ring of carbon atoms and can be formulated as cyclohexa-1,3,5-triene. Kekulé reasoned that this structure would readily account for the isolation of only a single monobromo substitution product because all six carbon atoms and all six hydrogens in cyclohexa-1,3,5-triene are equivalent.

Br_2, Iron

Benzene
(all six hydrogens are equivalent)

Bromobenzene

Kekulé's proposal was widely criticized at the time. Although it satisfactorily accounts for the correct number of monosubstituted benzene isomers, it fails to answer two critical questions: why is benzene unreactive compared with other alkenes, and why does benzene give a substitution product rather than an addition product on reaction with Br_2?

The unusual stability of benzene was a great puzzle to early chemists. Although its formula, C_6H_6, indicates that unsaturation must be present, benzene shows none of the behavior characteristic of alkenes. For example, alkenes readily react with $KMnO_4$ to give 1,2-diols, they react with aqueous acid to give alcohols, and they react with HCl to give chloroalkanes. Benzene does none of these things. *Benzene does not undergo electrophilic addition reactions.*

Cyclohexene: $KMnO_4$, H_2O → (OH, OH); H_3O^+ → (OH); HCl, Ether → (Cl)

Benzene: $KMnO_4$ → No reaction; H_3O^+ → No reaction; HCl, Ether → No reaction

Further evidence for the unusual nature of benzene is that all carbon–carbon bonds in benzene have the same length, intermediate between a typical single bond and a typical double bond. Most C–C single bonds have lengths near 154 pm, and most C–C double bonds are about 134 pm long, but all the C–C bonds in benzene are 139 pm long.

PROBLEM 5.1 How many dibromobenzene derivatives, $C_6H_4Br_2$, are possible if benzene were cyclohexa-1,3,5-triene? Draw them.

5.2 Structure of Benzene: The Resonance Proposal

To account for benzene's properties, we need to look again at resonance theory. We saw in Sections 4.10 and 4.11 that an allylic carbocation can be described as a resonance hybrid of two contributing forms. Neither resonance form is correct by itself; the true structure is intermediate between the two forms.

In the same way, resonance theory says that benzene can't be described satisfactorily by a single line-bond structure but is instead a resonance hybrid of two forms. Benzene doesn't oscillate back and forth between two forms; its true structure is somewhere between the two. Each carbon–carbon connection is an average of 1.5 bonds, midway between a single bond and a double bond.

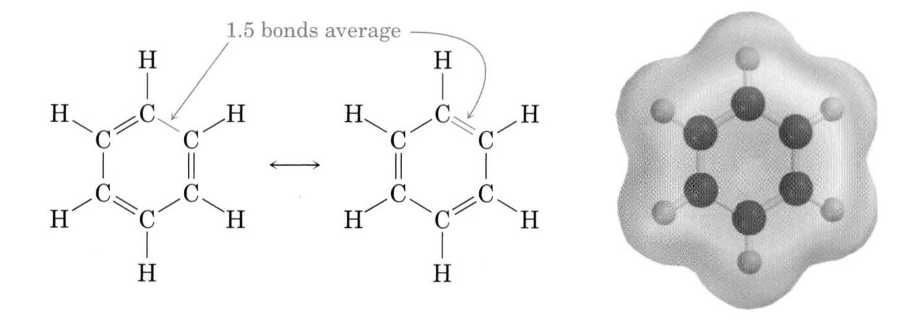

An orbital view of benzene shows the situation more clearly, emphasizing the cyclic conjugation of the benzene molecule and the equivalence of the six carbon–carbon bonds. Benzene is a flat, symmetrical molecule with the shape of a regular hexagon. All C–C–C bond angles are 120°, each carbon atom is sp^2-hybridized, and each carbon has a *p* orbital perpendicular to the plane of the six-membered ring. Because all six *p* orbitals are equivalent, it's impossible to define three localized alkene π bonds in which a given *p* orbital overlaps only *one* neighboring *p* orbital. Rather, each *p* orbital overlaps equally well with *both* neighboring *p* orbitals, leading to a structure for benzene in which the π electrons are shared around the ring in two doughnut-shaped clouds (Figure 5.1).

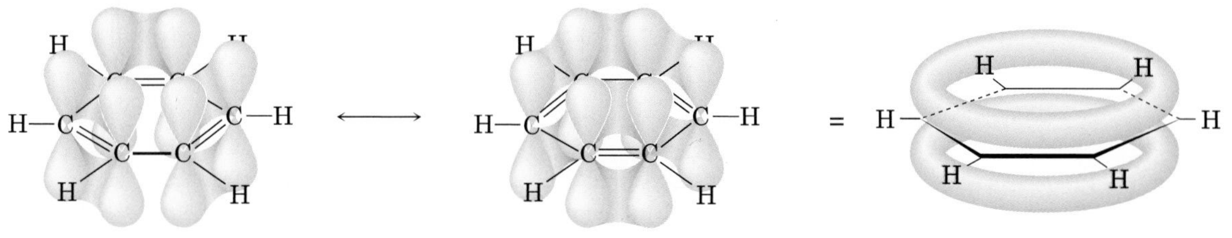

FIGURE 5.1 An orbital picture of benzene. Each of the six carbon atoms has a *p* orbital that can overlap equally well with neighboring *p* orbitals on both sides. The π electrons are thus shared around the ring in two doughnut-shaped clouds.

We can now see why benzene is stable: according to resonance theory, the more resonance forms a substance has, the more stable it is. Benzene, with two resonance forms of equal energy, is therefore more stable and less reactive than a typical alkene.

PROBLEM 5.2 How does resonance theory account for the fact that there is only one 1,2-dibromobenzene rather than the two isomers that Kekulé's theory would suggest?

5.3 Naming Aromatic Compounds

Aromatic substances, more than any other class of organic compounds, have acquired a large number of common names. Although the use of such names is discouraged, IUPAC rules allow for those shown in Table 5.1 to be retained. Thus, methylbenzene is commonly known as toluene; hydroxybenzene, as phenol; aminobenzene, as aniline; and so on.

Monosubstituted benzenes are systematically named in the same manner as other hydrocarbons, with *-benzene* used as the parent name. Thus, C_6H_5Br is bromobenzene, and $C_6H_5CH_2CH_3$ is ethylbenzene. The name **phenyl** (**fen**-nil) is used for the $-C_6H_5$ unit when the benzene ring is considered as a substituent, and the name **benzyl** is used for the $C_6H_5CH_2-$ group.

Br | CH_2CH_3 | | CH_2

Bromobenzene | Ethylbenzene | A phenyl group | A benzyl group

TABLE 5.1 Common Names of Some Aromatic Compounds

Formula	Name	Formula	Name
$C_6H_5-CH_3$	Toluene (bp 111 °C)	C_6H_5-CHO	Benzaldehyde (bp 178 °C)
C_6H_5-OH	Phenol (mp 43 °C)	C_6H_5-COOH	Benzoic acid (mp 122 °C)
$C_6H_5-NH_2$	Aniline (bp 184 °C)	C_6H_5-CN	Benzonitrile (bp 191 °C)
$C_6H_5-C(=O)-CH_3$	Acetophenone (mp 21 °C)	$C_6H_4(CH_3)_2$ (CH₃, CH₃ adjacent)	*ortho*-Xylene (bp 144 °C)

Disubstituted benzenes are named using one of the prefixes *ortho-* (*o*), *meta-* (*m*), or *para-* (*p*). An ortho-disubstituted benzene has its two substituents in a 1,2 relationship on the ring; a meta-disubstituted benzene has its two substituents in a 1,3 relationship; and a para-disubstituted benzene has its substituents in a 1,4 relationship:

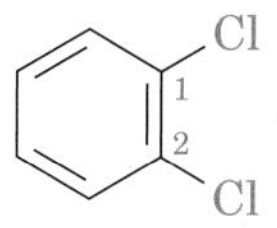

ortho-Dichlorobenzene **1,2 disubstituted**   **meta-Xylene** **1,3 disubstituted**   **para-Chlorobenzaldehyde** **1,4 disubstituted**

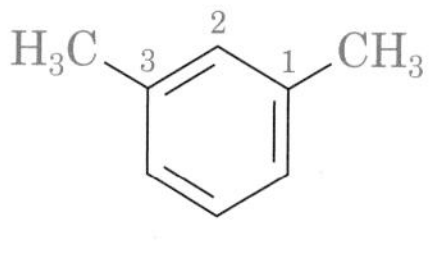

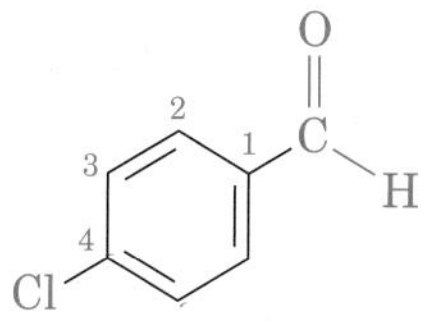

Benzenes with more than two substituents are named by numbering the position of each substituent on the ring so that the lowest possible numbers are used. The substituents are listed alphabetically when writing the name.

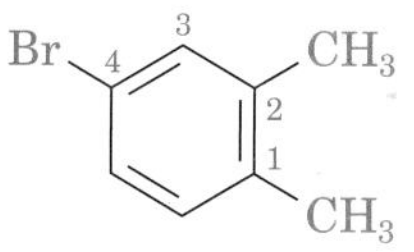

4-Bromo-1,2-dimethylbenzene 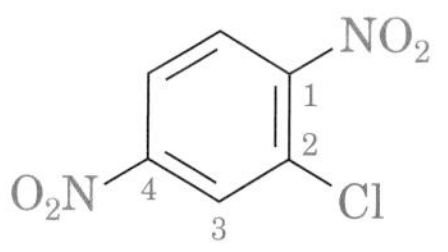   **2-Chloro-1,4-dinitrobenzene** 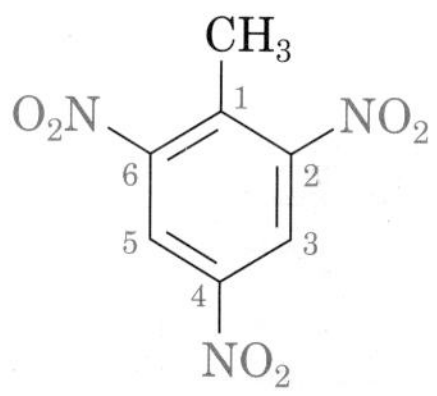   **2,4,6-Trinitrotoluene (TNT)**

In the third example shown, note that *-toluene* is used as the parent name rather than *-benzene*. Any of the monosubstituted aromatic compounds shown in Table 5.1 can serve as a parent name, with the principal substituent (–CH_3 in toluene, for example) assumed to be on carbon 1. The following two examples further illustrate this practice:

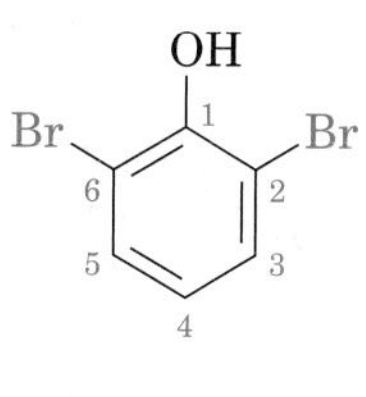

2,6-Dibromophenol 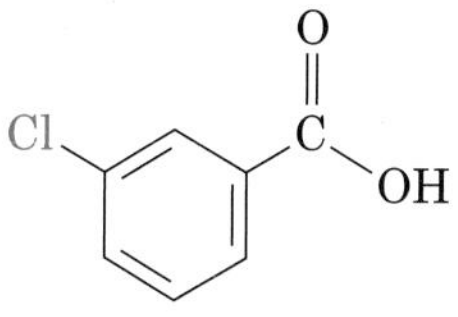   ***m*-Chlorobenzoic acid**

PRACTICE PROBLEM 5.1

What is the IUPAC name of the following compound?

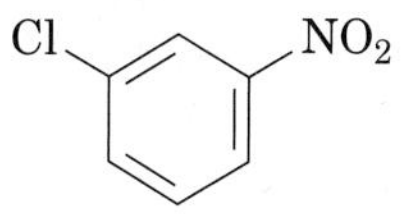

SOLUTION Because the nitro group (–NO_2) and chloro group are on carbons 1 and 3, they have a meta relationship. Citing the two substituents in alphabetical order gives the IUPAC name *m*-chloronitrobenzene.

PROBLEM 5.3 Tell whether the following compounds are ortho, meta, or para disubstituted:

(a) 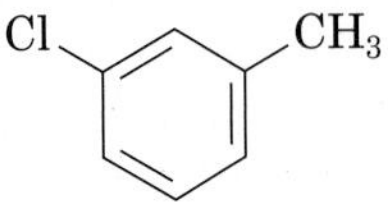 Cl, CH_3 (b) 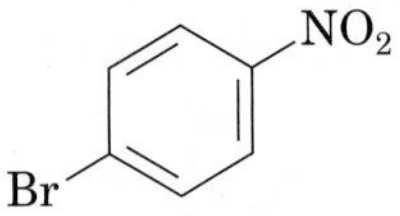 NO_2, Br (c) 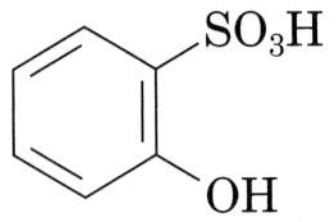 SO_3H, OH

PROBLEM 5.4 Give IUPAC names for the following compounds:

(a) 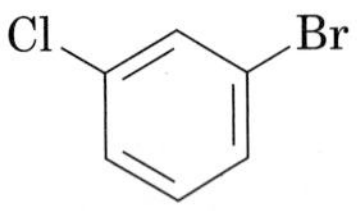 Cl, Br (b) 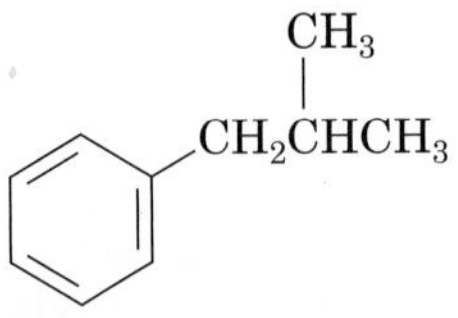 CH_2CHCH_3 with CH_3 (c) 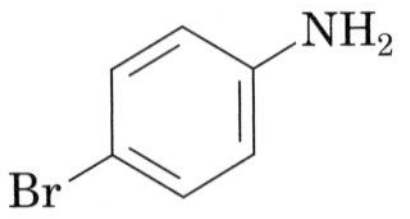 NH_2, Br

PROBLEM 5.5 Draw structures corresponding to the following IUPAC names:
(a) *p*-Bromochlorobenzene (b) *p*-Bromotoluene
(c) *m*-Chloroaniline (d) 1-Chloro-3,5-dimethylbenzene

5.4 Electrophilic Aromatic Substitution Reactions: Bromination

The most common reaction of aromatic compounds is **electrophilic aromatic substitution**. That is, an electron-poor reagent (an electrophile, E^+) reacts with the electron-rich aromatic ring (a nucleophile) and substitutes for one of the ring hydrogens.

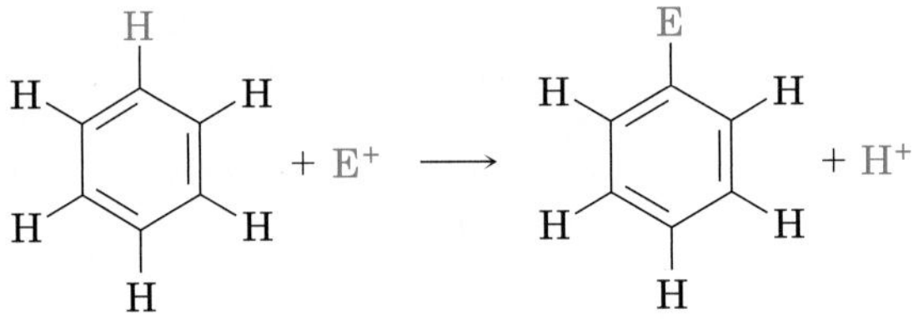

$$C_6H_6 + E^+ \longrightarrow C_6H_5E + H^+$$

Many different substituents can be introduced onto the aromatic ring by electrophilic substitution. By carrying out the right reactions, it is possible to *halogenate* the aromatic ring (substitute a halogen: –F, –Cl, –Br, or –I), *nitrate* it (substitute a nitro group: $-NO_2$), *sulfonate* it (substitute a sulfonic acid group: $-SO_3H$), *alkylate* it (substitute an alkyl group: –R), or *acylate* it (substitute an acyl group: –COR). Starting with only a few simple materials, we can prepare many thousands of substituted aromatic compounds (Figure 5.2).

All these reactions—and many more as well—take place by a similar mechanism. Let's begin a study of this fundamental reaction type by looking at one reaction in detail, the bromination of benzene. Benzene reacts with Br_2 in the presence of $FeBr_3$ as catalyst to yield the substitution product bromobenzene:

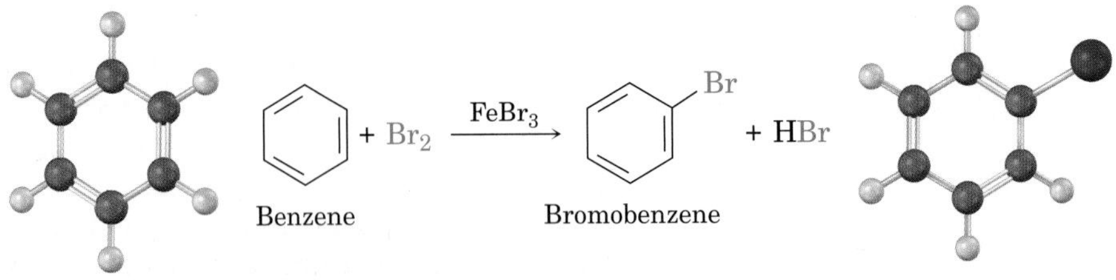

$$\text{Benzene} + Br_2 \xrightarrow{FeBr_3} \text{Bromobenzene} + HBr$$

Benzene Bromobenzene

FIGURE 5.2 Some electrophilic aromatic substitution reactions.

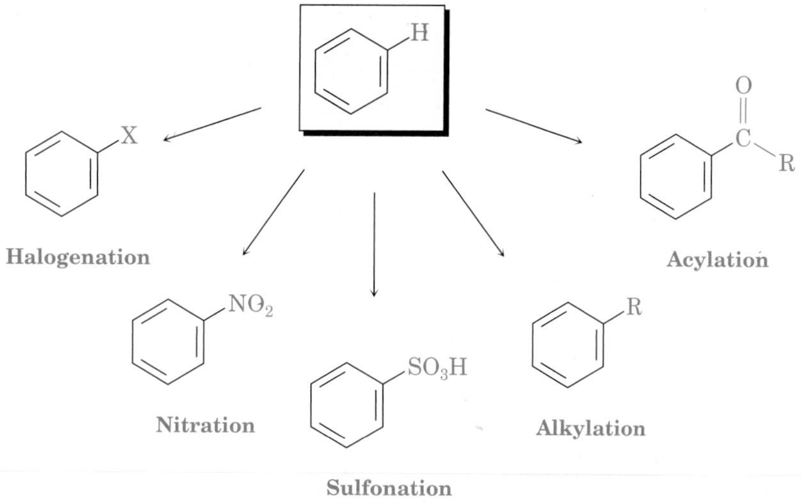

Before seeing how this electrophilic *substitution* reaction occurs, let's briefly recall what was said in Sections 3.8 through 3.10 about electrophilic *addition* reactions of alkenes. When a reagent such as HCl adds to an alkene, the electrophilic H^+ approaches the *p* orbitals of the double bond and forms a bond to one carbon, leaving a positive charge on the other carbon. The carbocation intermediate is then attacked by the nucleophile Cl^- ion to yield the addition product (Figure 5.3).

FIGURE 5.3 The mechanism of an electrophilic addition reaction of an alkene.

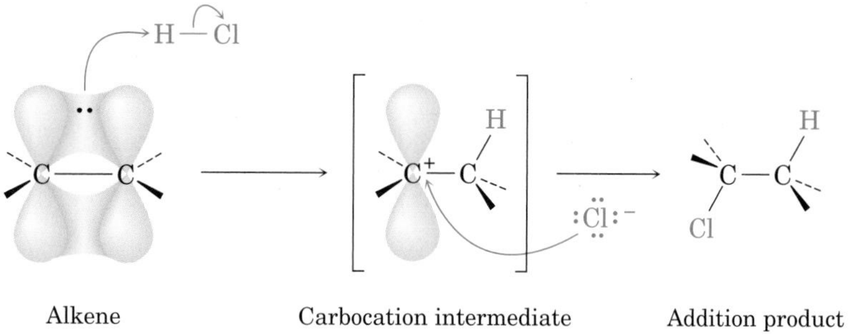

An electrophilic aromatic substitution reaction begins in a similar way, but there are a number of differences. One difference is that aromatic rings are less reactive toward electrophiles than alkenes are. For example, Br_2 in CH_2Cl_2 solution reacts instantly with most alkenes but does not react with benzene. For bromination of benzene to take place, a catalyst such as $FeBr_3$ is needed. The catalyst makes the Br_2 molecule more electrophilic by reacting with it to give $FeBr_4^-$ and Br^+.

$$FeBr_3 + Br_2 \rightarrow FeBr_4^- + Br^+$$

The electrophilic Br^+ then reacts with the electron-rich (nucleophilic) benzene ring to yield a nonaromatic carbocation intermediate. This carbocation is allylic (see Section 4.10) and is a hybrid of three resonance forms:

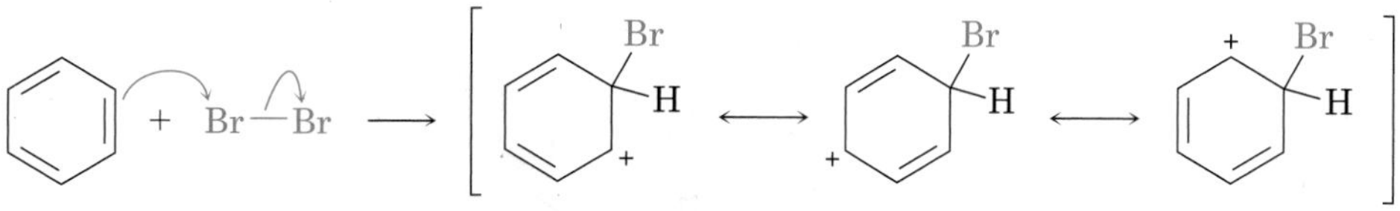

Although more stable than a typical nonallylic carbocation, the intermediate in electrophilic aromatic substitution is nevertheless much less stable than the starting aromatic reactant. Thus, reaction of an electrophile with a benzene ring has a relatively high activation energy and is rather slow. Figure 5.4 gives reaction energy diagrams that compare the reaction of an electrophile E^+ with an alkene and with benzene. The benzene reaction is slower (has a higher E_{act}) because the starting material is so stable.

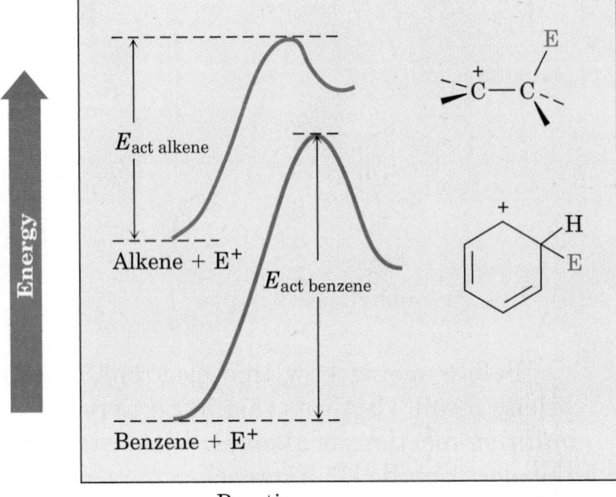

FIGURE 5.4 A comparison of the reactions of an electrophile (E^+) with an alkene and with benzene: E_{act} (alkene) < E_{act} (benzene).

A second difference between alkene addition reactions and aromatic substitution reactions occurs after the electrophile has added to the benzene ring to give the carbocation intermediate. Instead of Br^- adding to the carbocation intermediate to yield an addition product, a base removes H^+ from the bromine-bearing carbon to yield the neutral aromatic substitution product. The net effect is the substitution of Br^+ for H^+ by the overall mechanism shown in Figure 5.5.

FIGURE 5.5 MECHANISM: The mechanism of the electrophilic bromination of benzene. The reaction occurs in two steps and involves a carbocation intermediate.

An electron pair from the benzene ring attacks Br_2, forming a new C–Br bond and leaving a nonaromatic, carbocation intermediate.

Br—Br

Slow

Br
H
$FeBr_4^-$

Nonaromatic carbocation

The carbocation intermediate loses H^+, and the neutral substitution product forms as two electrons from the C–H bond move to regenerate the aromatic ring.

Fast

Br
+ HBr + $FeBr_3$

Why does the reaction of Br_2 with benzene take a different course than its reaction with an alkene? The answer is straightforward: if *addition* occurred, the stability of the aromatic ring would be lost, energy would be absorbed, and the overall reaction would be unfavorable. When *substitution* occurs, though, the stability of the aromatic ring is retained, energy is released, and the reaction is favorable. A reaction energy diagram for the overall process is shown in Figure 5.6.

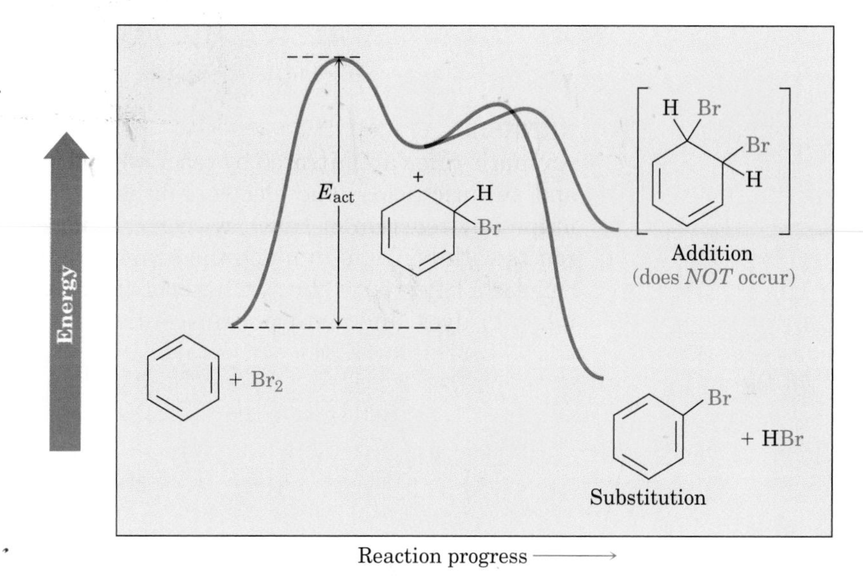

FIGURE 5.6 A reaction energy diagram for the electrophilic bromination of benzene. The reaction occurs in two steps and releases energy.

PROBLEM 5.6 There are three products that might form on reaction of toluene (methylbenzene) with Br_2. Draw and name them.

5.5 Other Electrophilic Aromatic Substitution Reactions

Other electrophilic aromatic substitutions occur by the same general mechanism as bromination.

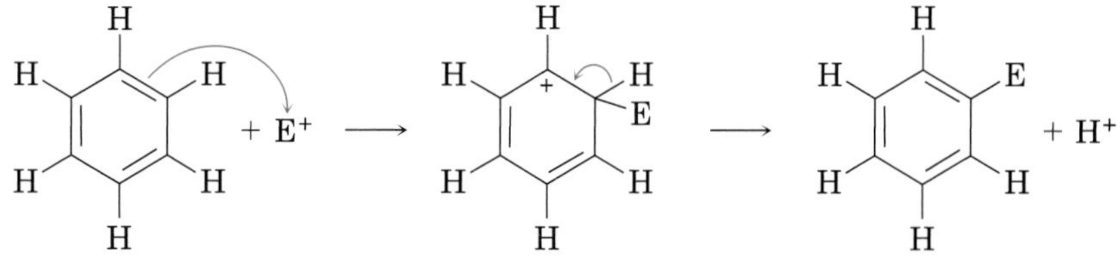

The only real variation in all these reactions is in how the electrophile, E^+, is generated. Let's look briefly at some of these other reactions.

Chlorination

Aromatic rings are chlorinated by reaction with Cl_2 in the presence of $FeCl_3$ catalyst. This kind of reaction is used in the synthesis of numerous pharmaceutical agents, including the tranquilizer diazepam (Valium).

$$\text{Benzene} + Cl_2 \xrightarrow[\text{catalyst}]{FeCl_3} \text{Chlorobenzene (86\%)} + HCl$$

Diazepam

Nitration

Aromatic rings are nitrated by reaction with a mixture of concentrated nitric and sulfuric acids. The electrophile is the nitronium ion, NO_2^+, which is formed by reaction of HNO_3 with H^+, followed by loss of water, and which reacts with benzene in much the same way Br^+ does. Nitration of aromatic rings is a key step in the synthesis of explosives such as TNT (2,4,6-trinitrotoluene), dyes, and many pharmaceutical agents.

$$HNO_3 \underset{}{\overset{H^+}{\rightleftharpoons}} H_2NO_3^+ \rightleftharpoons NO_2^+ + H_2O$$

$$\text{Benzene} + HNO_3 \xrightarrow[\text{catalyst}]{H_2SO_4} \text{Nitrobenzene (85\%)}$$

Trinitrotoluene (TNT)

Sulfonation

Aromatic rings are sulfonated by reaction with so-called fuming sulfuric acid, a mixture of SO_3 and H_2SO_4. The reactive electrophile is HSO_3^+, and substitution occurs by the usual two-step mechanism seen for bromination. Aromatic sulfonation is a key step in the synthesis of such compounds as the sulfa drug family of antibiotics.

$$\text{Benzene} + SO_3 \xrightarrow[\text{catalyst}]{H_2SO_4} \text{Benzenesulfonic acid (95\%)}$$

Sulfanilimide (a sulfa drug)

$$SO_3 + H_2SO_4 \longrightarrow SO_3H^+ + HSO_4^-$$

PRACTICE PROBLEM 5.2

Show the mechanism of the reaction of benzene with fuming sulfuric acid to yield benzenesulfonic acid.

STRATEGY The reaction of benzene with fuming sulfuric acid to yield benzenesulfonic acid is a typical electrophilic aromatic substitution reaction, all of which occur by the same two-step mechanism. An electrophile first adds to the aromatic ring, and H^+ is then lost. In sulfonation reactions, the electrophile is HSO_3^+.

SOLUTION

H O O S+ OH → [SO₃H H +] → SO₃H + H⁺

Carbocation intermediate

PROBLEM 5.7 Show the mechanism of the reaction of benzene with nitric acid and sulfuric acid to yield nitrobenzene.

PROBLEM 5.8 Chlorination of *o*-xylene (*o*-dimethylbenzene) yields a mixture of two products, but chlorination of *p*-xylene yields a single product. Explain.

PROBLEM 5.9 How many products might be formed on chlorination of *m*-xylene?

5.6 The Friedel–Crafts Alkylation and Acylation Reactions

An alkyl group is attached to an aromatic ring on reaction with an alkyl chloride, RCl, in the presence of $AlCl_3$ catalyst, a process called the **Friedel–Crafts alkylation reaction**. For example, benzene reacts with 2-chloropropane in the presence of $AlCl_3$ to yield isopropylbenzene (also called cumene):

$$\text{C}_6\text{H}_6 + CH_3CH(Cl)CH_3 \xrightarrow{AlCl_3} C_6H_5CH(CH_3)CH_3 + HCl$$

Benzene **2-Chloropropane** **Cumene (85%) (Isopropylbenzene)**

The Friedel–Crafts alkylation reaction is an aromatic substitution in which the electrophile is a carbocation, R^+. Aluminum chloride catalyzes the reaction by helping the alkyl chloride ionize, in much the same way that $FeBr_3$ helps Br_2 ionize (see Section 5.4). The overall Friedel–Crafts mechanism for the synthesis of isopropylbenzene is shown in Figure 5.7.

Although useful, the Friedel–Crafts alkylation reaction has several limitations. For one thing, only *alkyl* halides can be used; aryl halides such as chlorobenzene don't react. In addition, Friedel–Crafts reactions don't succeed on aromatic rings that are already substituted by the groups $—NO_2$, $—C{\equiv}N$, $—SO_3H$, or $—COR$. Such aromatic rings are much less reactive than benzene for reasons we'll discuss in Sections 5.7 and 5.8.

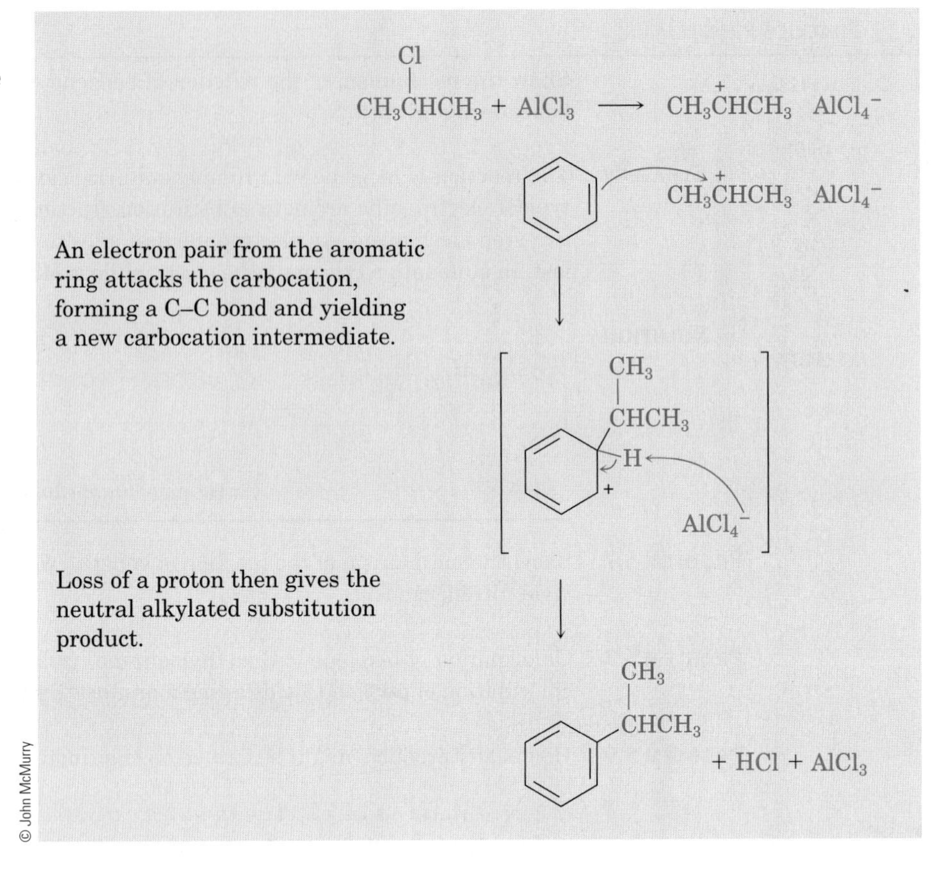

Cl
CH3CHCH3 + AlCl3 ⟶ CH3CHCH3 AlCl4−
CH3CHCH3 AlCl4−
An electron pair from the aromatic ring attacks the carbocation, forming a C–C bond and yielding a new carbocation intermediate.
CH3
CHCH3
H
AlCl4−
Loss of a proton then gives the neutral alkylated substitution product.
CH3
CHCH3
+ HCl + AlCl3
© John McMurry

FIGURE 5.7 MECHANISM: Mechanism of the Friedel–Crafts alkylation reaction. The electrophile is a carbocation, generated by $AlCl_3$-assisted ionization of an alkyl chloride.

Closely related to the Friedel–Crafts alkylation reaction is the **Friedel–Crafts acylation reaction**. When an aromatic compound is treated with a carboxylic acid chloride, RCOCl, in the presence of $AlCl_3$, an **acyl** (**a**-sil) **group**, –COR, is introduced onto the ring. For example, reaction of benzene with acetyl chloride yields acetophenone, a ketone:

$$C_6H_6 + CH_3COCl \xrightarrow[80\ °C]{AlCl_3} C_6H_5COCH_3 + HCl$$

Benzene **Acetyl chloride** **Acetophenone (95%)**

PROBLEM 5.10 What products would you expect to obtain from the reaction of the following compounds with chloroethane and $AlCl_3$?
(a) Benzene (b) *p*-Xylene

PROBLEM 5.11 What products would you expect to obtain from the reaction of benzene with the following reagents?
(a) $(CH_3)_3CCl$, $AlCl_3$ (b) CH_3CH_2COCl, $AlCl_3$

5.7 Substituent Effects in Electrophilic Aromatic Substitution

Only one product can form when an electrophilic substitution occurs on benzene, but what would happen if we were to carry out an electrophilic substitution reaction on a ring that already has a substituent? Substituents already present on an aromatic ring have two effects:

- **Substituents affect the *reactivity* of the aromatic ring.** Some groups activate the ring for further electrophilic substitution, and some deactivate it. An –OH group activates the ring, for instance, making phenol (hydroxybenzene) 1000 times more reactive than benzene toward nitration. An $-NO_2$ group deactivates the ring, however, making nitrobenzene 20 million times less reactive than benzene.

	NO_2	Cl	H	OH
Relative rate of nitration	6×10^{-8}	0.033	1	1000

Reactivity →

- **Substituents affect the *orientation* of the reaction.** The three possible disubstituted products—ortho, meta, and para—are usually not formed in equal amounts. Instead, the nature of the substituent already present on the ring determines the position of the second substitution. An –OH group directs further substitution toward the ortho and para positions, for instance, while a –CN group directs further substitution primarily toward the meta position.

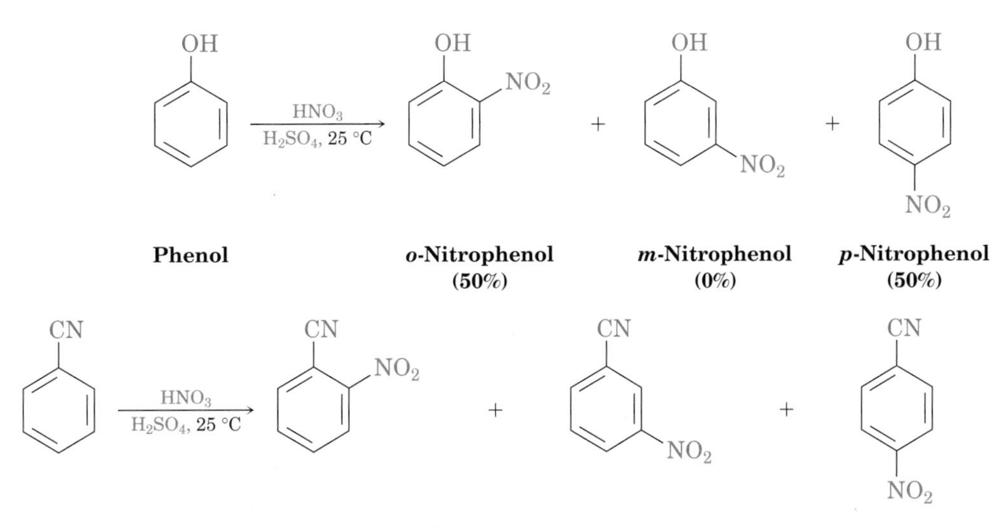

Substituents can be classified into three groups based on *site of reaction* and *reactivity*:

- **Ortho- and para-directing activators:** Groups like –OH and $-NH_2$ present on a ring direct an electrophile, E^+, to ortho or para positions, *and* they react faster than benzene.
- **Ortho- and para-directing deactivators:** Halogens present on a ring direct an electrophile, E^+, to ortho or para positions, *and* they react slower than benzene.
- **Meta-directing deactivators:** Groups containing a carbonyl (C=O) or a –CN group direct an electrophile, E^+, to the meta positions, *but* they react slower than benzene.

No meta-directing activators are known. Figure 5.8 shows how the directing effects of the groups correlate with their reactivities. All meta-directing groups are deactivating, and most ortho- and para-directing groups are activating. The halogens are unique in being ortho- and para-directing *and* deactivating.

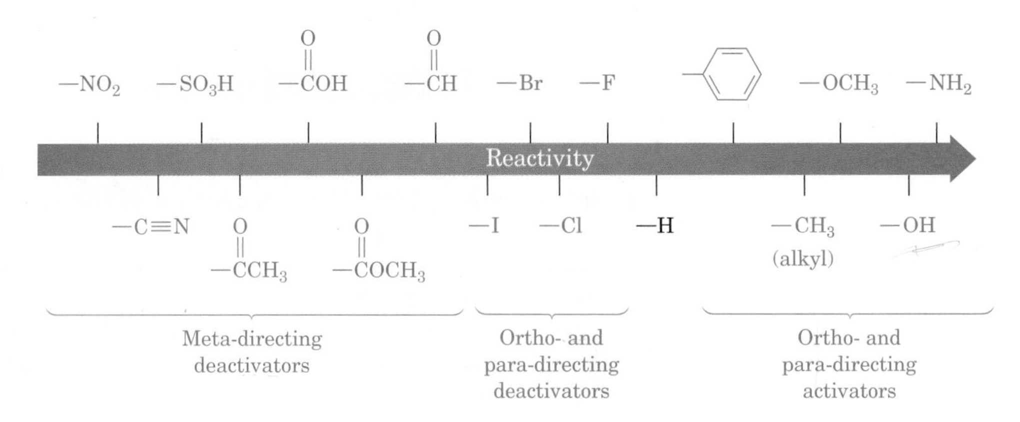

FIGURE 5.8 Substituent effects in electrophilic aromatic substitutions. All activating groups are ortho- and para-directing, and all deactivating groups other than halogen are meta-directing. The halogens are ortho- and para-directing deactivators.

PRACTICE PROBLEM 5.3

Which would you expect to react faster in an electrophilic aromatic substitution reaction, chlorobenzene or ethylbenzene? Explain.

STRATEGY Look at Figure 5.8, and compare the relative reactivities of chloro and alkyl groups.

SOLUTION A chloro substituent is deactivating, whereas an alkyl group is activating. Thus, ethylbenzene is more reactive than chlorobenzene.

PROBLEM 5.12 Use Figure 5.8 to rank the compounds in each of the following groups in order of their reactivity to electrophilic aromatic substitution:

(a) Nitrobenzene, phenol (hydroxybenzene), toluene
(b) Phenol, benzene, chlorobenzene, benzoic acid
(c) Benzene, bromobenzene, benzaldehyde, aniline (aminobenzene)

PROBLEM 5.13 Draw and name the products you would expect to obtain by reaction of the following substances with Cl_2 and $FeCl_3$ (blue = N, reddish brown = Br):

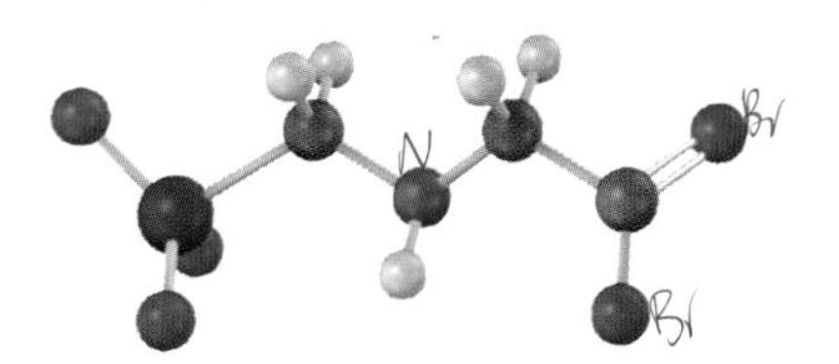

5.8 An Explanation of Substituent Effects

We have seen that substituents affect both the reactivity and the site of reaction of an aromatic ring. Let's look at these effects separately.

Reactivity: Activating and Deactivating Effects in Aromatic Rings

What makes a group either activating or deactivating? *The common characteristic of all activating groups is that they donate electrons to the ring.* Recall that these reactions rely on electrons from the aromatic ring attacking an electrophile, E^+. The result is a positive charge on the aromatic ring. If the aromatic ring is electron-rich, the carbocation intermediate will be more stable and the activation energy barrier for its formation will be lower.

The common characteristic of all deactivating groups is that they withdraw electrons from the ring, thereby making the ring more electron-poor, destabilizing the carbocation intermediate, and raising the activation energy for its formation. Compare the electrostatic potential maps of phenol (activated), chlorobenzene (weakly deactivated), and nitrobenzene (strongly deactivated), for instance. The –OH substituent makes the ring more negative (red), while –Cl makes the ring less negative (yellow) and $-NO_2$ makes the ring much less negative (green).

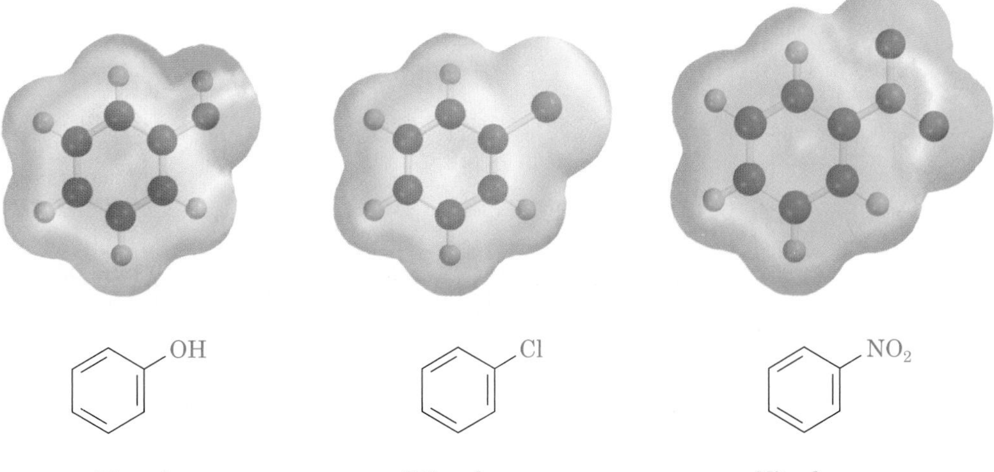

Phenol (activated, electron-rich ring)

Chlorobenzene (slightly deactivated, electron-poor ring)

Nitrobenzene (strongly deactivated, very electron-poor ring)

The electron donation or withdrawal may occur by either an inductive effect (see Section 1.9) or a resonance effect. Recall that an inductive effect is due to an electronegativity difference between the ring and the attached substituent, while a resonance effect is due to orbital overlap between a *p* orbital on the ring and a *p* orbital on the substituent.

Site of Reaction: Ortho and Para Directors

Let's look at the nitration of phenol as an example of how ortho- and para-directing substituents work. In the first step, reaction with the electrophilic nitronium ion (NO_2^+) can occur either ortho, meta, or para to the –OH group, giving the carbocation intermediates shown in Figure 5.9. The ortho and para intermediates are more stable than the meta intermediate because they have more resonance forms, including one that allows the positive charge to be stabilized by electron donation from the substituent oxygen atom. Since the ortho and para intermediates are more stable than the meta intermediate, they are formed faster.

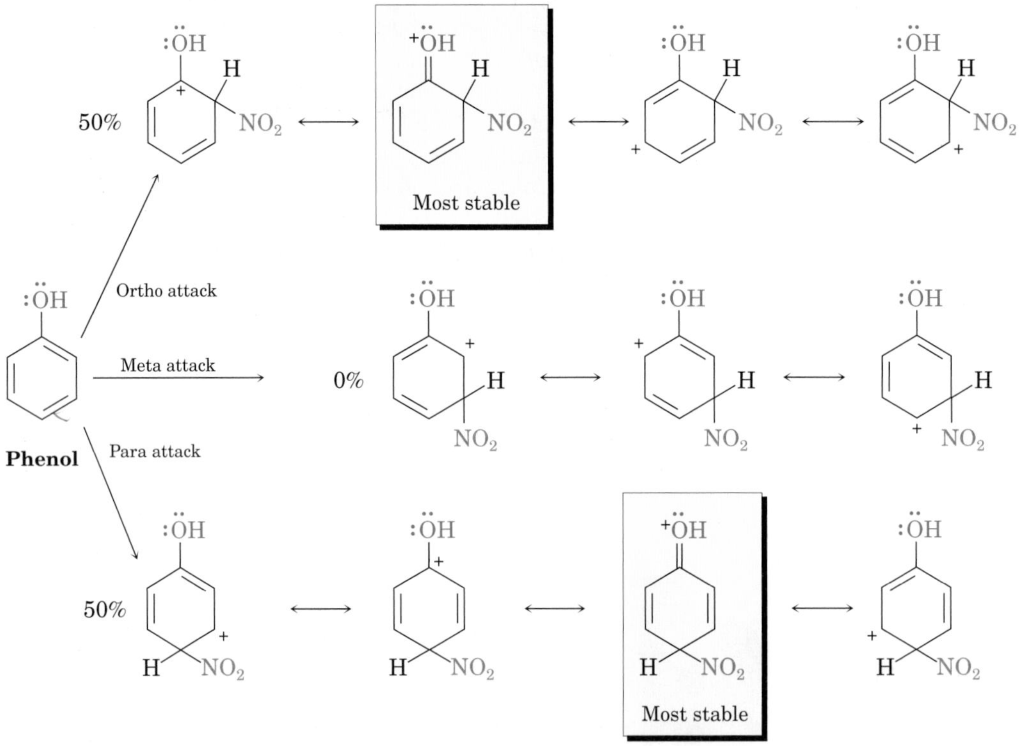

FIGURE 5.9 Intermediates in the nitration of phenol. The ortho and para intermediates are more stable than the meta intermediate because they have more resonance forms, including one that involves electron donation from the oxygen atom.

In general, any substituent that has a lone pair of electrons on the atom directly bonded to the aromatic ring allows an electron-donating resonance interaction to occur and thus acts as an **ortho** and **para director**:

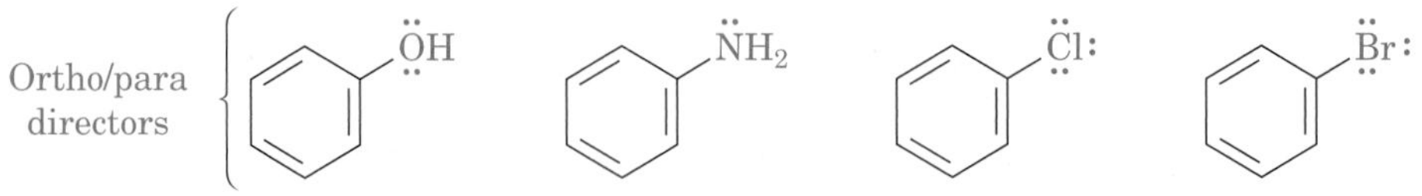

Site of Reaction: Meta Directors

The influence of meta-directing substituents can be explained using the same kinds of arguments used for ortho and para directors. Look at the chlorination of benzaldehyde, for instance (Figure 5.10). Of the three possible carbocation intermediates, the meta intermediate has three favorable resonance forms, while the ortho and para intermediates have only two favorable forms. In both ortho and para intermediates, the third possible resonance form is unfavorable because it places the positive charge directly on the carbon that bears the aldehyde group, where it is disfavored by a repulsive interaction with the positively polarized carbon atom of the C=O group. Hence, the meta intermediate is more favored and is formed faster than the ortho and para intermediates.

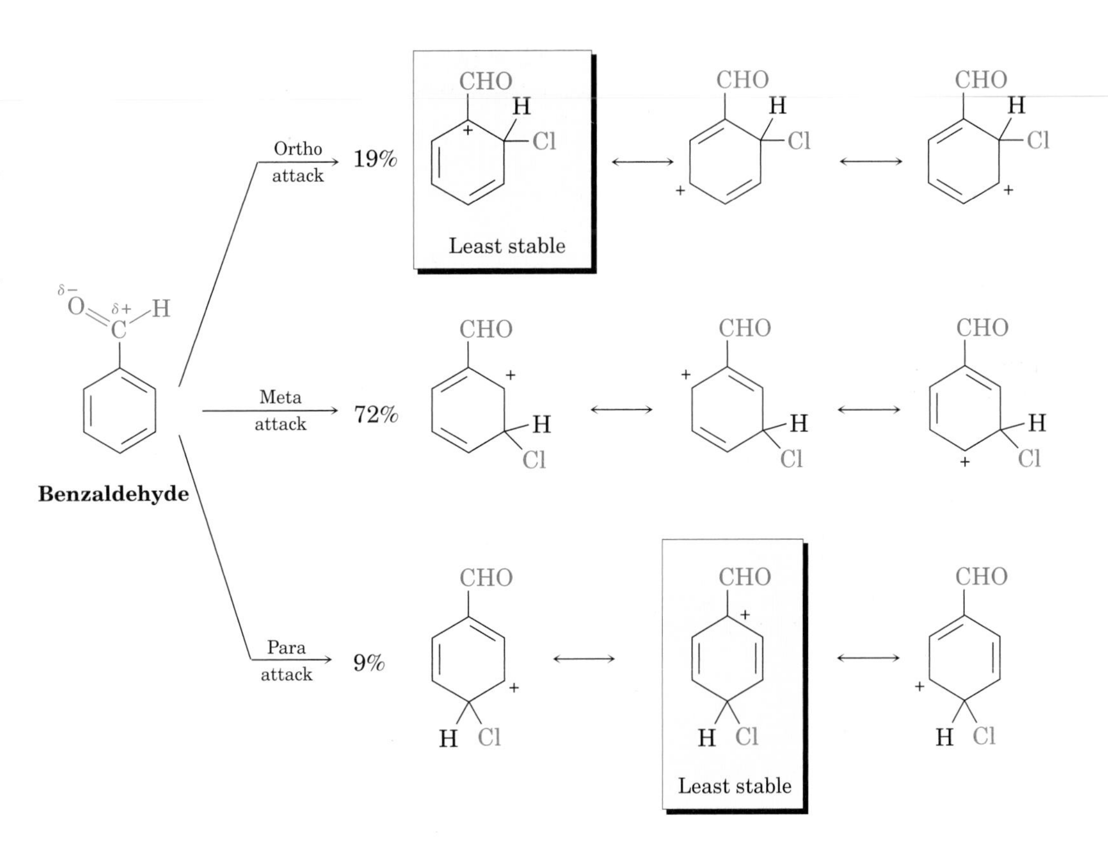

FIGURE 5.10 Intermediates in the chlorination of benzaldehyde. The meta intermediate is more favorable than ortho and para intermediates because it has three favorable resonance forms rather than two.

In general, any substituent that has a positively polarized atom (δ+) directly attached to the ring makes one of the resonance forms of the ortho and para intermediates unfavorable and thus acts as a **meta director**:

Meta directors

PRACTICE PROBLEM 5.4

What product(s) would you expect from bromination of aniline, $C_6H_5NH_2$?

STRATEGY Look at Figure 5.8 to see whether the $-NH_2$ substituent is ortho- and para-directing or meta-directing. Since an amino group is ortho- and para-directing, we expect to obtain a mixture of *o*-bromoaniline and *p*-bromoaniline.

SOLUTION

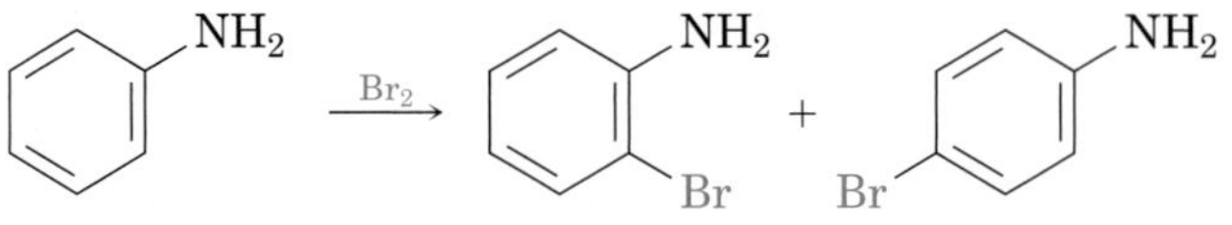

NH_2 $\xrightarrow{Br_2}$ NH_2, Br + Br, NH_2

Aniline **o-Bromoaniline** **p-Bromoaniline**

PROBLEM 5.14 What product(s) would you expect from sulfonation of the following compounds?
(a) Nitrobenzene (b) Bromobenzene (c) Toluene
(d) Benzoic acid (e) Benzonitrile

PROBLEM 5.15 Draw resonance forms of the three possible carbocation intermediates to show how a methoxyl group ($-OCH_3$) directs bromination toward ortho and para positions.

PROBLEM 5.16 Draw resonance forms of the three possible carbocation intermediates to show how an acetyl group, $CH_3C{=}O$, directs bromination toward the meta position.

5.9 Oxidation and Reduction of Aromatic Compounds

Despite its unsaturation, a benzene ring does not usually react with strong oxidizing agents such as $KMnO_4$. (Recall from Section 4.7 that $KMnO_4$ cleaves alkene C=C bonds.) Alkyl groups attached to the aromatic ring are readily attacked by oxidizing agents, however, and are converted into carboxyl groups ($-CO_2H$). For example, butylbenzene is oxidized by $KMnO_4$ to give benzoic acid. The mechanism of this reaction is complex and involves attack on the side-chain C–H bonds at the position next to the aromatic ring (the **benzylic position**) to give radical intermediates.

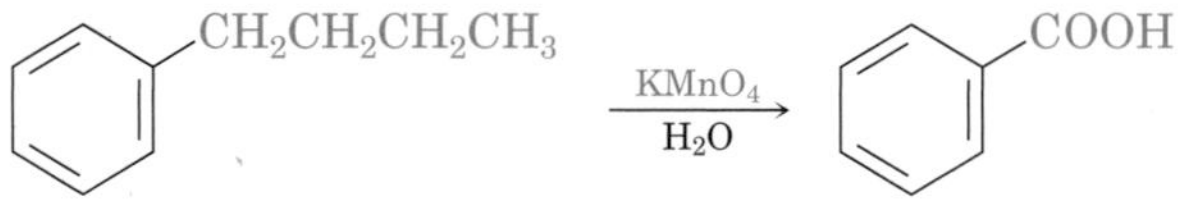

$CH_2CH_2CH_2CH_3$ $\xrightarrow[H_2O]{KMnO_4}$ COOH

Butylbenzene **Benzoic acid (85%)**

Just as aromatic rings are usually inert to oxidation, they are also inert to reduction under typical alkene hydrogenation conditions. Only if high temperatures and pressures are used does reduction of an aromatic ring occur. For example, *o*-dimethylbenzene (*o*-xylene) gives 1,2-dimethylcyclohexane if reduced at high pressure:

CH_3, CH_3 — $\xrightarrow[\text{2000 psi, 25 °C}]{H_2\text{, Pt; ethanol}}$ — CH_3, CH_3

***o*-Xylene** **1,2-Dimethylcyclohexane (100%)**

PROBLEM 5.17 What aromatic products do you expect to obtain from oxidation of the following substances with $KMnO_4$?

(a) *m*-Chloroethylbenzene (b) **Tetralin**

5.10 Aromaticity in Nonbenzene Rings and Polycyclic Compounds

As chemists began to understand the special reactivity of benzene, other aromatic compounds—cyclic conjugated molecules that have chemical stability—were identified. Extending the concept of aromaticity beyond simple monocyclic compounds to include **polycyclic aromatic compounds** is relatively straightforward. Naphthalene, familiar for its use in mothballs, has two benzene-like rings fused together and is the simplest and best-known polycyclic aromatic compound. Benz[*a*]pyrene is found in chimney soot, cigarette smoke, and charcoal-broiled meat. Once in the body, it is oxidized and becomes carcinogenic.

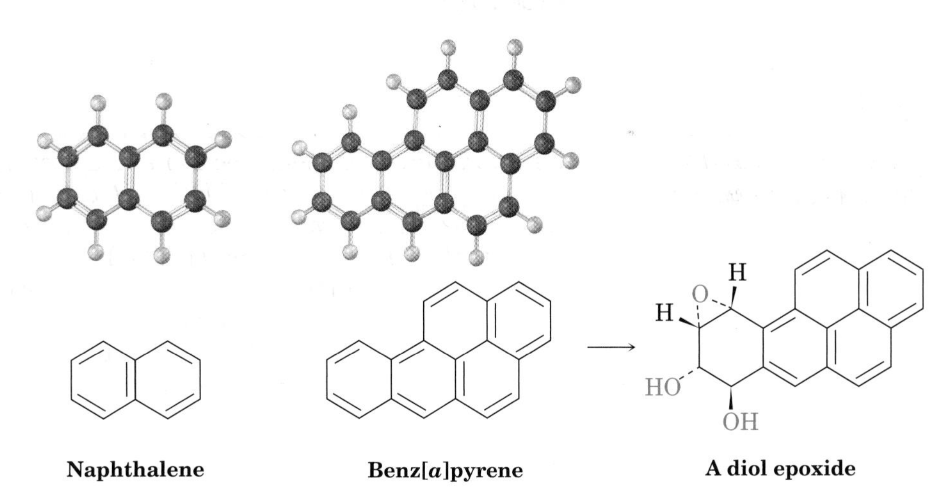

Naphthalene **Benz[*a*]pyrene** **A diol epoxide**

Pyridine and pyrrole are examples of nitrogen-containing aromatic compounds. Pyridine has the same shape as benzene and displays a similar series of alternating π electrons. But what about pyrrole? Chemists can rapidly determine whether a molecule is aromatic by applying the $4n + 2$ π electron rule. *If the number of π electrons in a flat, cyclic, conjugated molecule is an equal to $4n + 2$ (where n is an integer), the molecule is aromatic.* Let's look back at benzene. Benzene has 6 π electrons and fits the $4n + 2$ rule, where $n = 1$. Why is pyrrole aromatic? The lone pair of electrons on nitrogen leads to a total of 6 π electrons and completes the ring.

PRACTICE PROBLEM 5.5

Which of the following molecules is aromatic?

Cyclobuta-1,3-diene **Cycloocta-1,3,5,7-tetraene** **Hexahydrocoronene**

STRATEGY Count the number of π electrons and determine whether it matches an integer solution to the $4n + 2$ rule. Remember, the system must be cyclic.

SOLUTION Cyclobuta-1,3-diene (4 electrons) and cycloocta-1,3,5,7-tetraene (8 electrons) are not aromatic. Hexahydrocoronene (18 electrons) is aromatic.

PROBLEM 5.18 Which of these molecules is aromatic?

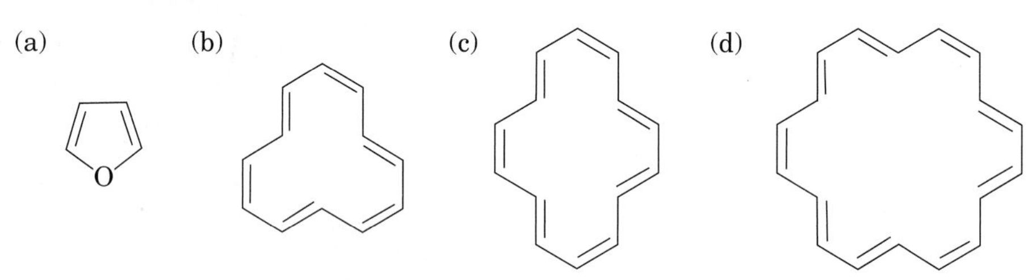

PROBLEM 5.19 There are three resonance structures of naphthalene, of which only one is shown. Draw the other two.

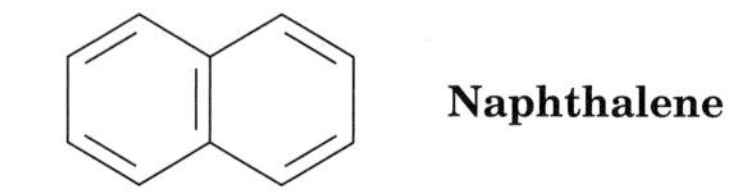

5.11 Organic Synthesis

The laboratory synthesis of organic molecules from simple precursors might be carried out for many reasons. In the pharmaceutical industry, new organic molecules are often designed and synthesized for evaluation as medicines. In the chemical industry, syntheses are often undertaken to devise more economical routes to known compounds. In this book, too, we'll sometimes devise syntheses of complex molecules from simpler precursors, but the purpose here is simply to help you learn organic chemistry. Devising a route for the synthesis of an organic molecule requires that you approach chemical problems in a logical way, draw on all your knowledge of organic reactivity, and organize that knowledge into a workable plan.

The only trick to devising an organic synthesis is to *work backward*, a process called *retrosynthesis*. Look at the product and ask yourself, "What is the immediate precursor of that product?" Having found an immediate precursor, work backward again, one step at a time, until a suitable starting material is found. Let's try some examples.

PRACTICE PROBLEM 5.6

Synthesize *m*-chloronitrobenzene starting from benzene.

STRATEGY Work backward by first asking, "What is an immediate precursor of *m*-chloronitrobenzene?"

? ⟶ [structure: NO_2 and Cl on benzene ring, meta]

***m*-Chloronitrobenzene**

There are two substituents on the ring, a –Cl group, which is ortho- and para-directing, and an $-NO_2$ group, which is meta-directing. We can't nitrate chlorobenzene because the wrong isomers (*o*- and *p*-chloronitrobenzenes) would result, but chlorination of nitrobenzene should give the desired product.

Chlorobenzene —(HNO_3, H_2SO_4; crossed out)→ ***m*-Chloronitrobenzene**

Nitrobenzene —(Cl_2, $FeCl_3$)→ ***m*-Chloronitrobenzene**

"What is an immediate precursor of nitrobenzene?" Benzene, which can be nitrated.

SOLUTION We've solved the problem in two steps:

Benzene —(HNO_3, H_2SO_4)→ **Nitrobenzene** —(Cl_2, $FeCl_3$)→ ***m*-Chloronitrobenzene**

PRACTICE PROBLEM 5.7

Synthesize *p*-bromobenzoic acid starting from benzene.

STRATEGY Work backward by first asking, "What is an immediate precursor of *p*-bromobenzoic acid?"

? ⟶ Br–C_6H_4–COOH

***p*-Bromobenzoic acid**

There are two substituents on the ring, a $-CO_2H$ group, which is meta-directing, and a –Br atom, which is ortho- and para-directing. We can't brominate benzoic acid because the wrong isomer (*m*-bromobenzoic acid) would be formed. We've seen, however, that oxidation of alkylbenzene side chains yields benzoic acids. An immediate precursor of our target molecule might therefore be *p*-bromotoluene.

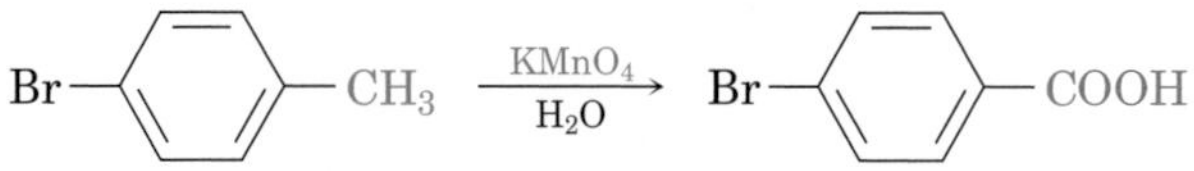

$$\text{Br–C}_6\text{H}_4\text{–CH}_3 \xrightarrow[\text{H}_2\text{O}]{\text{KMnO}_4} \text{Br–C}_6\text{H}_4\text{–COOH}$$

***p*-Bromotoluene** ***p*-Bromobenzoic acid**

"What is an immediate precursor of *p*-bromotoluene?" Perhaps toluene, because the methyl group would direct bromination to the ortho and para positions, and we could then separate isomers. Alternatively, bromobenzene might be an immediate precursor because we could carry out a Friedel–Crafts alkylation and obtain the para product. Both methods are satisfactory.

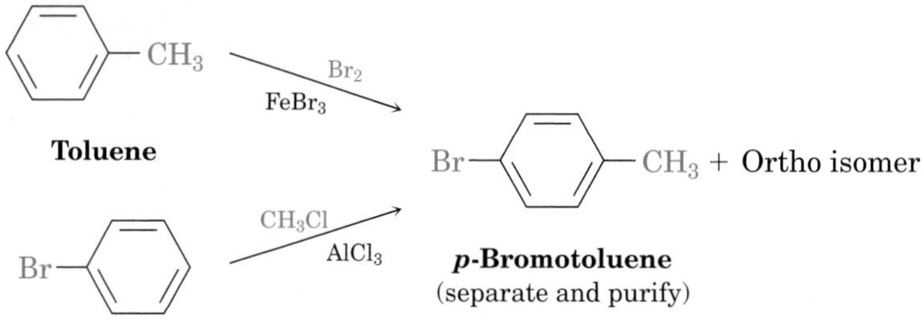

$$\text{C}_6\text{H}_5\text{–CH}_3 \xrightarrow[\text{FeBr}_3]{\text{Br}_2} \text{Br–C}_6\text{H}_4\text{–CH}_3 + \text{Ortho isomer}$$

$$\text{Br–C}_6\text{H}_5 \xrightarrow[\text{AlCl}_3]{\text{CH}_3\text{Cl}} \text{Br–C}_6\text{H}_4\text{–CH}_3$$

Toluene

Bromobenzene

***p*-Bromotoluene**
(separate and purify)

"What is an immediate precursor of toluene?" Benzene, which can be methylated in a Friedel–Crafts reaction:

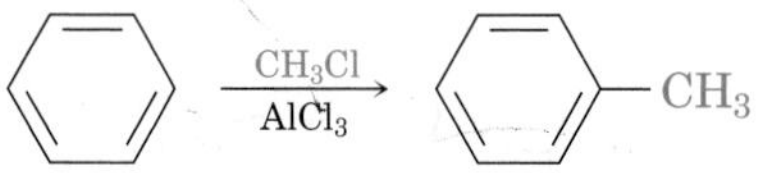

$$\text{C}_6\text{H}_6 \xrightarrow[\text{AlCl}_3]{\text{CH}_3\text{Cl}} \text{C}_6\text{H}_5\text{–CH}_3$$

Benzene **Toluene**

"Alternatively, what is an immediate precursor of bromobenzene?" Benzene, which can be brominated:

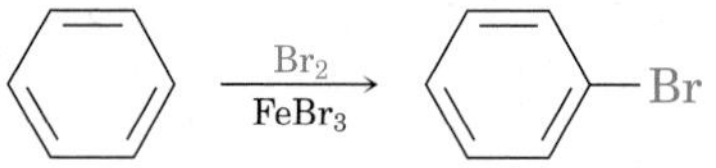

$$\text{C}_6\text{H}_6 \xrightarrow[\text{FeBr}_3]{\text{Br}_2} \text{C}_6\text{H}_5\text{–Br}$$

Benzene **Bromobenzene**

SOLUTION Our backward synthetic (*retrosynthetic*) analysis has provided two workable routes from benzene to *p*-bromobenzoic acid.

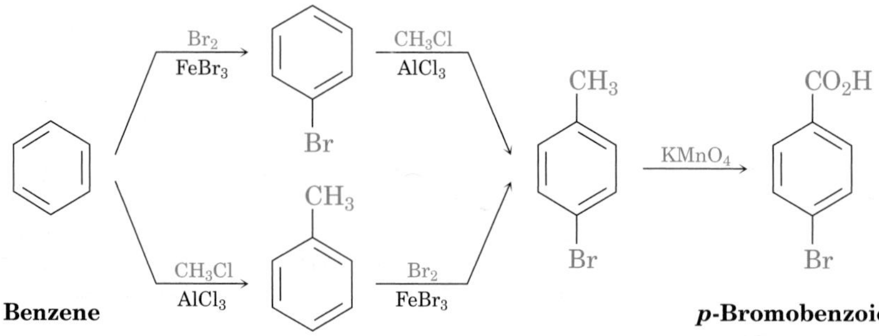

$$\text{C}_6\text{H}_6 \xrightarrow[\text{FeBr}_3]{\text{Br}_2} \text{C}_6\text{H}_5\text{Br} \xrightarrow[\text{AlCl}_3]{\text{CH}_3\text{Cl}} p\text{-BrC}_6\text{H}_4\text{CH}_3$$

$$\text{C}_6\text{H}_6 \xrightarrow[\text{AlCl}_3]{\text{CH}_3\text{Cl}} \text{C}_6\text{H}_5\text{CH}_3 \xrightarrow[\text{FeBr}_3]{\text{Br}_2} p\text{-BrC}_6\text{H}_4\text{CH}_3$$

$$p\text{-BrC}_6\text{H}_4\text{CH}_3 \xrightarrow{\text{KMnO}_4} p\text{-BrC}_6\text{H}_4\text{CO}_2\text{H}$$

Benzene ***p*-Bromobenzoic acid**

PROBLEM 5.20 Propose syntheses of the following substances starting from benzene:

(a) H_3C–C_6H_4–C(=O)–CH_3 (b) Cl–C_6H_4–NO_2

PROBLEM 5.21 Synthesize the following substances from benzene:
(a) *o*-Bromotoluene (b) 2-Bromo-1,4-dimethylbenzene

PROBLEM 5.22 How would you prepare the following substance from benzene? (Yellow-green = Cl.)

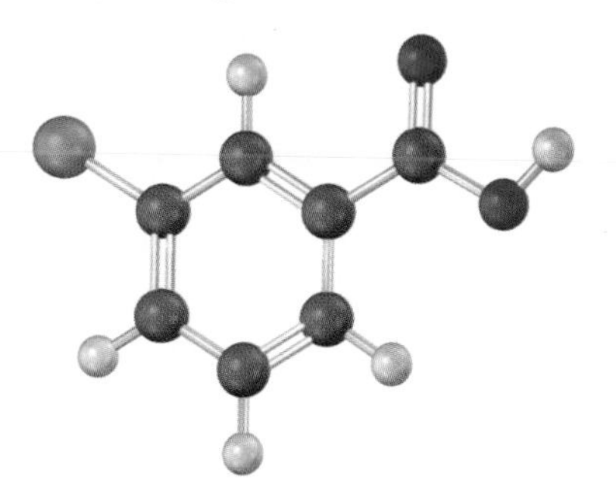

INTERLUDE

In the Field with Triazine Herbicides

Holt Studios/Photo Researchers, Inc.

Organic chemistry of aromatic rings.

A field of growing corn is chemistry in action. In the ordered rows of stalks, the photosynthetic machinery is turning inorganic CO_2 and water into complex organic molecules. The narrow paths of bare earth that separate row from row are often the result of aromatic herbicides.

We've seen that aromatic rings can react with *electrophiles* through a substitution reaction. But what about reactions between aromatic rings and *nucleophiles*? What factors would favor such a reaction?

$$+ \; E^+ \longrightarrow \quad (\text{E}) \quad + \; H^+$$

$$+ \; Nu^- \longrightarrow \; ?$$

In an electrophilic substitution, the *aromatic ring donates electrons to an electrophile* during the initial bond-forming reaction. In the case of a nucleophilic substitution, the opposite happens and the *nucleophile donates the electrons to the aromatic ring*. Accordingly, electron-deficient aromatic rings are more likely to undergo nucleophilic aromatic substitution reactions than are electron-rich rings.

Continued

Triazine herbicides are produced by nucleophilic aromatic substitution reactions. In fact, almost 100,000,000 lb of these materials are produced each year in the form of two major weed killers, atrazine and simazine. Let's look more carefully at this reaction sequence.

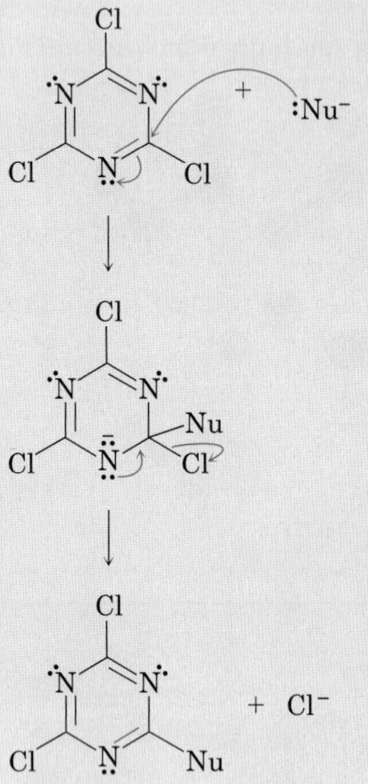

The three electronegative chlorine atoms make the aromatic ring electron-deficient. When a nucleophile attacks one of the carbon atoms, aromaticity is lost and the π electrons move onto nitrogen.

The aromaticity of the ring is reestablished when chloride anion is expelled by a lone pair of electrons on nitrogen.

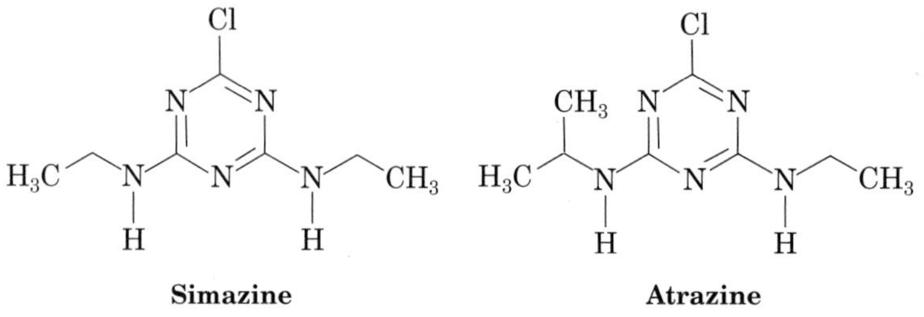

Simazine **Atrazine**

The herbicides simazine and atrazine are produced by a nucleophillic aromatic substitution reaction.

In the case of atrazine and simazine, two nucleophilic aromatic substitution reactions are carried out in the presence of a base, sodium hydroxide. For simazine, ethylamine ($H_2NCH_2CH_3$) is the nucleophile. Consider these two questions: What two nucleophiles are used for the production of atrazine? Why do you think these two nucleophiles are added to triazine trichloride in a stepwise fashion instead of at the same time as is done for simazine?

Summary and Key Words

Acyl group, p. 150
Aromatic, p. 139
Benzyl group, p. 142
Benzylic position, p. 156
Electrophilic aromatic substitution reaction, p. 144
Friedel–Crafts acylation reaction, p. 150
Friedel–Crafts alkylation reaction, p. 149
Meta director, p. 155
Ortho director, p. 154
Para director, p. 154
Phenyl group, p. 142
Polycyclic aromatic compound, p. 157

The word **aromatic** refers to the class of compounds structurally related to benzene. Aromatic compounds are named according to IUPAC rules, with disubstituted benzenes referred to as either **ortho** (1,2 disubstituted), **meta** (1,3 disubstituted), or **para** (1,4 disubstituted). Benzene is a resonance hybrid of two equivalent forms, neither of which is correct by itself. The true structure of benzene is intermediate between the two.

The most common reaction of aromatic compounds is **electrophilic aromatic substitution**. In this two-step polar reaction, the π electrons of the aromatic ring first attack the electrophile to yield a resonance-stabilized carbocation intermediate, which then loses H^+ to give a substituted aromatic product. Bromination, chlorination, iodination, nitration, sulfonation, **Friedel–Crafts alkylation**, and **Friedel–Crafts acylation** can all be carried out. Friedel–Crafts alkylation is particularly useful for preparing a variety of alkylbenzenes but is limited because only alkyl halides can be used and strongly deactivated rings do not react.

Substituents on the benzene ring affect both the reactivity of the ring toward further substitution and the orientation of that further substitution. Substituents can be classified either as **activators** or **deactivators**, and either as **ortho and para directors** or as **meta directors**.

The side chains of alkylbenzenes have unique reactivity because of the neighboring aromatic ring. Thus, an alkyl group attached to the aromatic ring can be degraded to a carboxyl group ($-CO_2H$) by oxidation with aqueous $KMnO_4$. In addition, aromatic rings can be reduced to yield cyclohexanes on catalytic hydrogenation at high pressure.

Summary of Reactions

1. Electrophilic aromatic substitution
   (a) Bromination (Section 5.4)

$$C_6H_6 \xrightarrow[FeBr_3]{Br_2} C_6H_5Br + HBr$$

   (b) Chlorination (Section 5.5)

$$C_6H_6 \xrightarrow[FeCl_3]{Cl_2} C_6H_5Cl + HCl$$

   (c) Nitration (Section 5.5)

$$C_6H_6 \xrightarrow[H_2SO_4]{HNO_3} C_6H_5NO_2 + H_2O$$

   (d) Sulfonation (Section 5.5)

$$C_6H_6 \xrightarrow[H_2SO_4]{SO_3} C_6H_5SO_3H$$

(e) Friedel–Crafts alkylation (Section 5.6)

$$\text{Benzene} + CH_3Cl \xrightarrow{AlCl_3} \text{C}_6\text{H}_5\text{CH}_3 + HCl$$

(f) Friedel–Crafts acylation (Section 5.6)

$$\text{Benzene} + CH_3\overset{O}{\overset{\|}{C}}Cl \xrightarrow{AlCl_3} \text{C}_6\text{H}_5\overset{O}{\overset{\|}{C}}CH_3 + HCl$$

2. Oxidation of aromatic side chains (Section 5.9)

$$\text{C}_6\text{H}_5\text{R} \xrightarrow[H_2O]{KMnO_4} \text{C}_6\text{H}_5\overset{O}{\overset{\|}{C}}OH$$

3. Hydrogenation of aromatic rings (Section 5.9)

$$\text{Benzene} \xrightarrow[PtO_2]{H_2} \text{Cyclohexane}$$

EXERCISES

Visualizing Chemistry

5.23 Give IUPAC names for the following substances (red = O, blue = N):

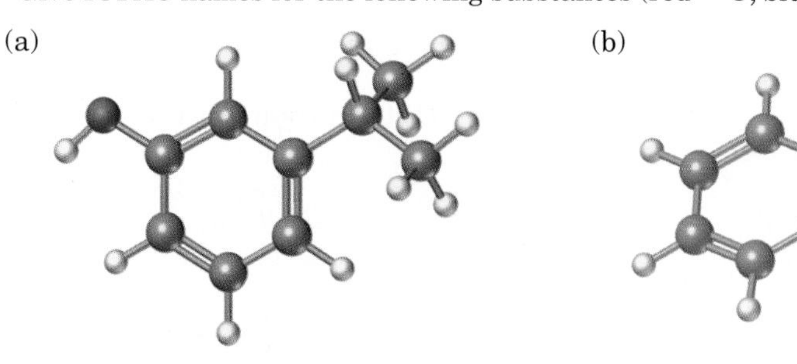

5.24 Draw and name the product from reaction of each of the following substances with (i) Br_2, $FeBr_3$ and (ii) CH_3COCl, $AlCl_3$ (red = O):

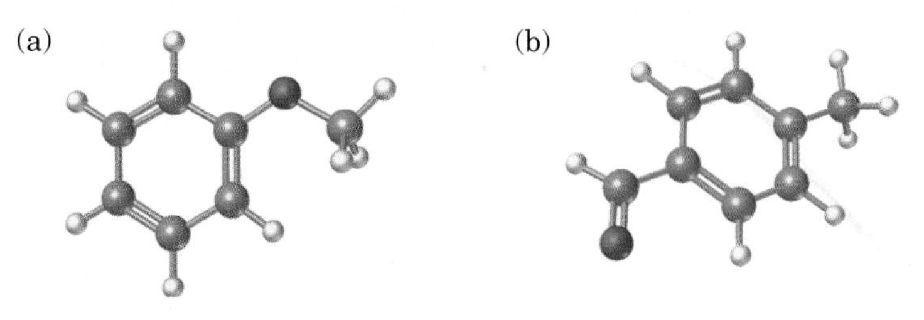

5.25 The following structure represents a carbocation. Draw two resonance forms, indicating the positions of the double bonds.

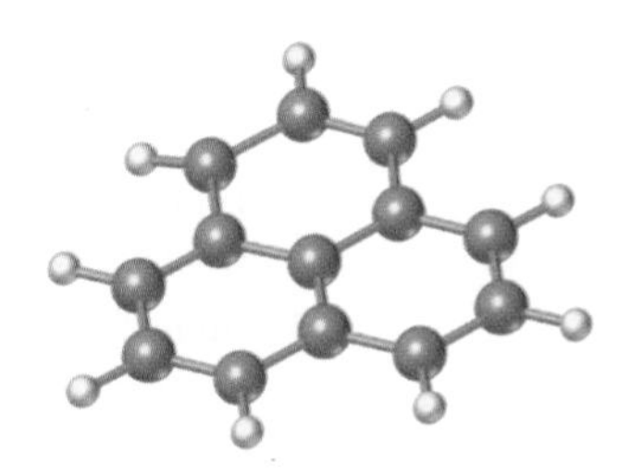

5.26 How would you synthesize the following compound starting from benzene? More than one step is needed.

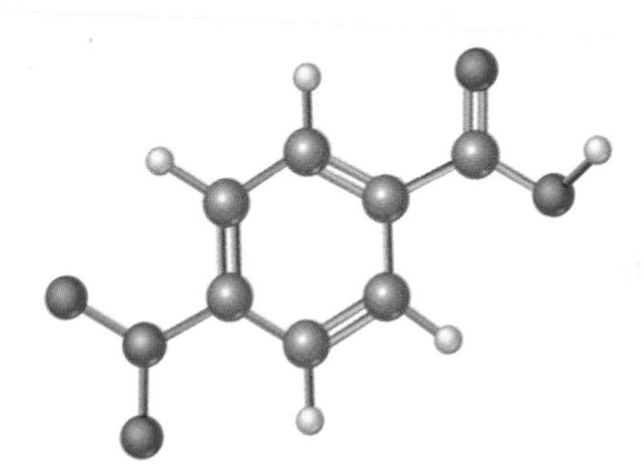

Additional Problems

NOMENCLATURE

5.27 Give IUPAC names for the following compounds:

(a) CH_3 / $CH_2CH_2CH_2CHCH_3$ (benzene ring attached)

(b) CO_2H / Br (benzene ring)

(c) Br / H_3C / CH_3 (benzene ring)

(d) Br / $CH_2CH_2CH_3$ (benzene ring)

5.28 Draw structures corresponding to the following names:

(a) *m*-Bromophenol
(b) 1,3,5-Benzenetriol
(c) *p*-Iodonitrobenzene
(d) 2,4,6-Trinitrotoluene (TNT)
(e) *o*-Aminobenzoic acid
(f) 3-Methyl-2-phenylhexane

ISOMERS

5.29 Draw and name all aromatic compounds with the formula C_7H_7Cl.

5.30 Draw and name all isomeric (a) dinitrobenzenes and (b) bromodimethylbenzenes.

5.31 Propose structures for aromatic hydrocarbons meeting the following descriptions:
(a) C_9H_{12}; can give only one product on aromatic bromination
(b) C_8H_{10}; can give three products on aromatic chlorination
(c) $C_{10}H_{14}$; can give two products on aromatic nitration

REACTIONS AND SUBSTITUENT EFFECTS

5.32 Formulate the reaction of benzene with 2-chloro-2-methylpropane in the presence of $AlCl_3$ catalyst to give *tert*-butylbenzene.

5.33 Identify each of the following groups as an activator or deactivator and as an *o,p*-director or *m*-director:

(a) $-N(CH_3)_2$ (b) 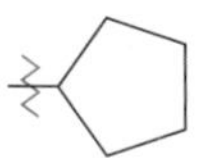 [cyclopentyl] (c) $-OCH_2CH_3$ (d) 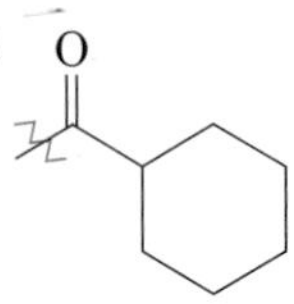 O [cyclohexyl ketone]

5.34 Predict the major product(s) of the following reactions:

(a) 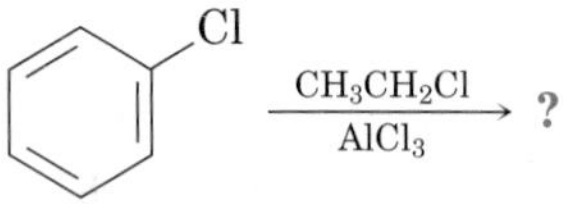 Chlorobenzene $\xrightarrow[AlCl_3]{CH_3CH_2Cl}$?

(b) 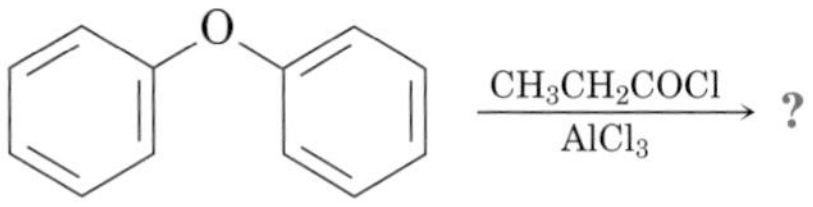 Diphenyl ether $\xrightarrow[AlCl_3]{CH_3CH_2COCl}$?

(c) 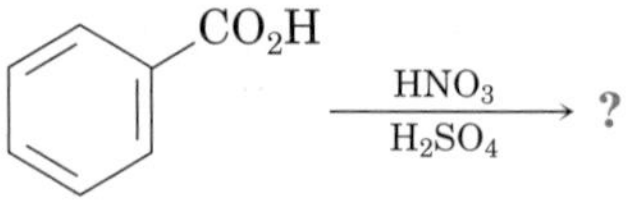 Benzoic acid (CO_2H) $\xrightarrow[H_2SO_4]{HNO_3}$?

(d) 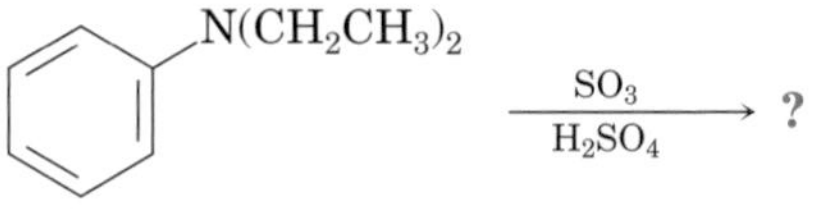 $N(CH_2CH_3)_2$-benzene $\xrightarrow[H_2SO_4]{SO_3}$?

5.35 Predict the major product(s) of mononitration of the following substances:
(a) Bromobenzene (b) Benzonitrile (cyanobenzene) (c) Benzoic acid
(d) Nitrobenzene (e) Phenol (f) Benzaldehyde

5.36 Which of the substances listed in Problem 5.35 react faster than benzene and which react slower?

5.37 Rank the compounds in each group according to their reactivity toward electrophilic substitution:
(a) Chlorobenzene, *o*-dichlorobenzene, benzene
(b) *p*-Bromonitrobenzene, nitrobenzene, phenol
(c) Fluorobenzene, benzaldehyde, *o*-dimethylbenzene

5.38 The orientation of electrophilic aromatic substitution on a disubstituted benzene ring is usually controlled by whichever of the two groups already on the ring is the more powerful activator. Name and draw the structure(s) of the major product(s) of electrophilic chlorination of these substances:
(a) *m*-Nitrophenol (b) *o*-Methylphenol (c) *p*-Chloronitrobenzene

5.39 Predict the major product(s) you would expect to obtain from sulfonation of the following substances (see Problem 5.38):
(a) *o*-Chlorotoluene (b) *m*-Bromophenol (c) *p*-Nitrotoluene

5.40 Rank the following aromatic compounds in the expected order of their reactivity toward Friedel–Crafts acylation. Which compounds are unreactive?
(a) Bromobenzene (b) Toluene (c) Anisole ($C_6H_5OCH_3$)
(d) Nitrobenzene (e) *p*-Bromotoluene

5.41 In some cases, the Friedel–Crafts acylation reaction can occur *intramolecularly,* that is, within the same molecule. Predict the product of the following reaction:

$$\xrightarrow{AlCl_3} ?$$

5.42 What is the structure of the compound with formula C_8H_9Br that gives *p*-bromobenzoic acid on oxidation with $KMnO_4$?

5.43 Draw the three additional resonance forms of anthracene.

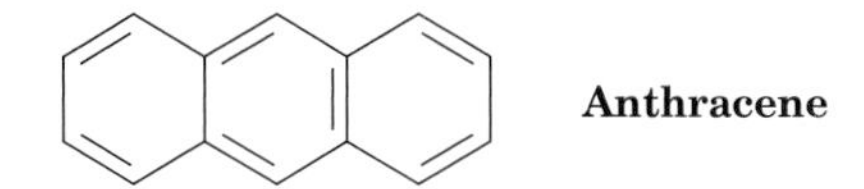

MECHANISMS

5.44 Show the steps involved in the Friedel–Crafts reaction of benzene with CH_3Cl.

5.45 Propose a mechanism to explain the fact that deuterium (D, ^2H) slowly replaces hydrogen (^1H) in the aromatic ring when benzene is treated with D_2SO_4.

5.46 Explain why bromination of biphenyl occurs at the ortho and para positions rather than at the meta positions. Use resonance forms of the carbocation intermediates in your explanation.

Biphenyl

5.47 In light of your answer to Problem 5.46, at what position and on which ring would you expect nitration of 4-bromobiphenyl to occur?

Br **4-Bromobiphenyl**

SYNTHESIS

5.48 Starting with benzene, how would you synthesize the following substances? Assume that you can separate ortho and para isomers if necessary.
(a) *m*-Bromobenzenesulfonic acid (b) *o*-Chlorobenzenesulfonic acid
(c) *p*-Chlorotoluene

5.49 Starting from any aromatic hydrocarbon of your choice, how would you synthesize the following substances? Ortho and para isomers can be separated if necessary.
(a) *o*-Nitrobenzoic acid (b) *p-tert*-Butylbenzoic acid

5.50 Explain by drawing resonance forms of the intermediate carbocations why naphthalene undergoes electrophilic aromatic substitution at C1 rather than at C2.

$$+ Br_2 \xrightarrow{FeBr_3} + HBr$$

INTEGRATED PROBLEMS

5.51 We said in Section 4.10 that allylic carbocations are stabilized by resonance. Draw resonance forms to account for a similar stabilization of benzylic carbocations.

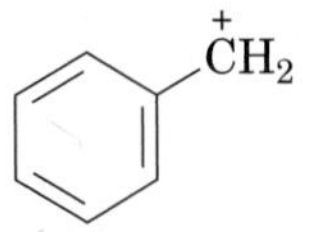

$\overset{+}{C}H_2$

A benzylic carbocation

5.52 Addition of HBr to 1-phenylpropene yields (1-bromopropyl)benzene as the exclusive product. Propose a mechanism for the reaction, and explain why none of the other regioisomer is produced (see Problem 5.51).

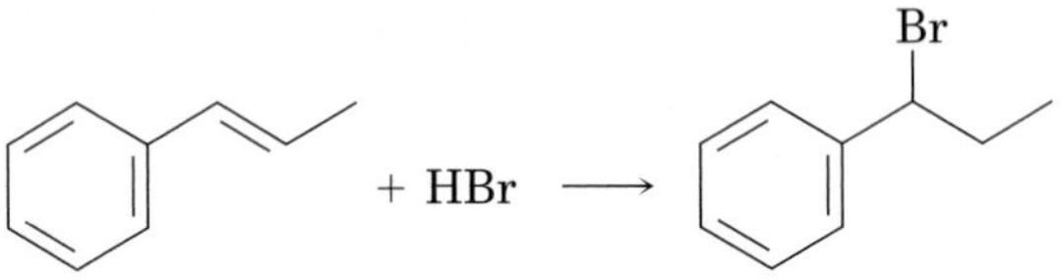

Br

+ HBr ⟶

5.53 The following syntheses have flaws in them. What is wrong with each?

(a) CH_3 — 1. Cl_2, $FeCl_3$; 2. $KMnO_4$ ⟶ COOH, Cl

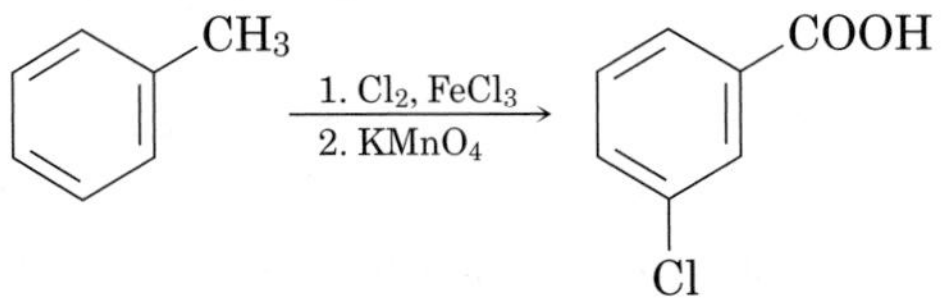

(b) Cl — 1. $(CH_3)_3CCl$, $AlCl_3$; 2. $KMnO_4$, H_2O ⟶ Cl, COOH

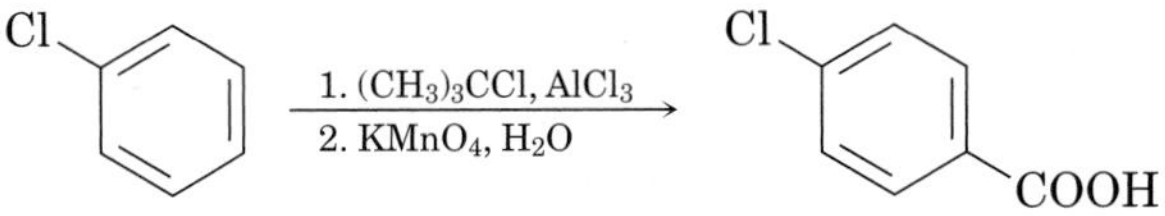

5.54 Pyridine is a cyclic nitrogen-containing compound that shows many of the properties associated with aromaticity. For example, pyridine undergoes electrophilic substitution reactions. Draw an orbital picture of pyridine, and account for its aromatic properties.

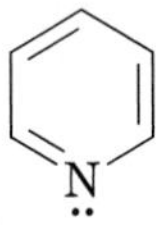

N

Pyridine

5.55 Would you expect the trimethylammonium group to be an activating or deactivating substituent? Explain.

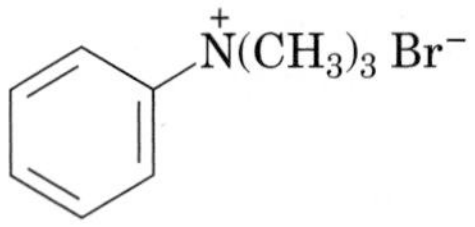

$\overset{+}{N}(CH_3)_3\ Br^-$

Phenyltrimethylammonium bromide

5.56 Starting with toluene, how would you synthesize the three nitrobenzoic acids?

5.57 Carbocations generated by reaction of an alkene with a strong acid catalyst can react with aromatic rings in a Friedel–Crafts reaction. Propose a mechanism to account for the industrial synthesis of the food preservative BHT from *p*-cresol and 2-methylpropene:

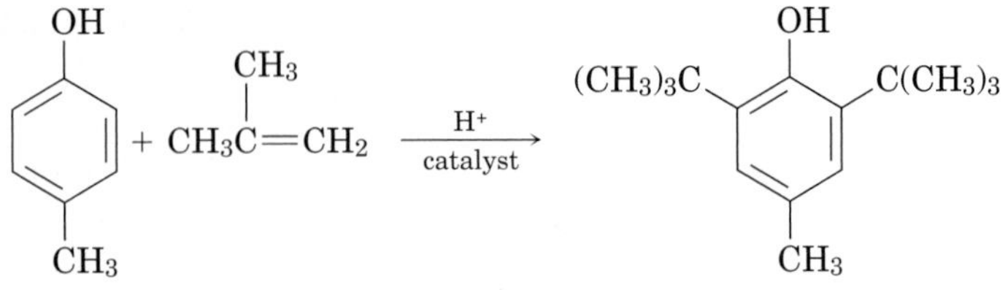

OH, CH_3 + $CH_3\overset{CH_3}{C}{=}CH_2$ $\xrightarrow[\text{catalyst}]{H^+}$ OH, $(CH_3)_3C$, $C(CH_3)_3$, CH_3

***p*-Cresol** **BHT**

5.58 You know the mechanism of HBr addition to alkenes, and you know the effects of various substituent groups on aromatic substitution. Use this knowledge to predict which of the following two alkenes reacts faster with HBr. Explain your answer by drawing resonance forms of the carbocation intermediates.

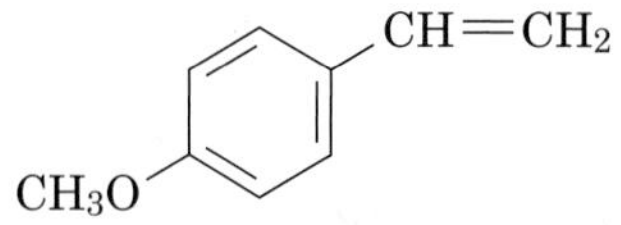

$CH{=}CH_2$ CH_3O and $CH{=}CH_2$ O_2N

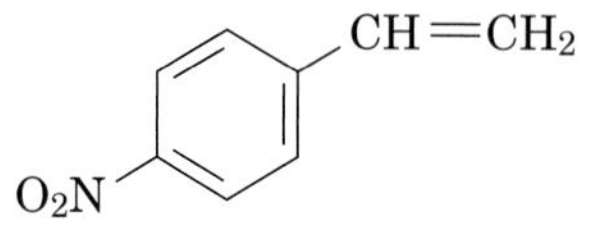

5.59 Identify the reagents represented by the letters a through d in the following scheme:

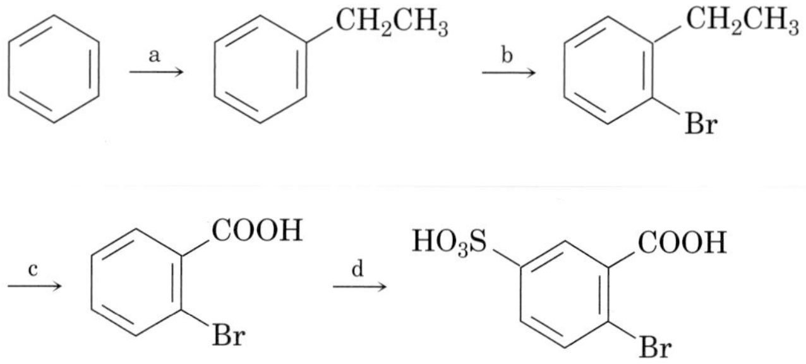

a → CH_2CH_3 b → CH_2CH_3 Br

c → COOH Br d → HO_3S COOH Br

In the Medicine Cabinet

5.60 Acetaminophen is marketed under the name of Tylenol (among others) and is prepared in only three steps from phenol.

(a) What three functional groups are present in acetaminophen?
(b) What is the molecular formula of acetaminophen?
(c) Nitration of phenol can produce two products. What is the desired product **A**? Using resonance structures, explain why nitration occurs at the desired position.
(d) Of the two potential choices, ortho or meta, formation of which other isomer is supported by the resonance structures you drew?
(e) After reduction of the nitro group, the last step to install the acetyl (–$COCH_3$) group is accomplished with a reagent containing what functional group?

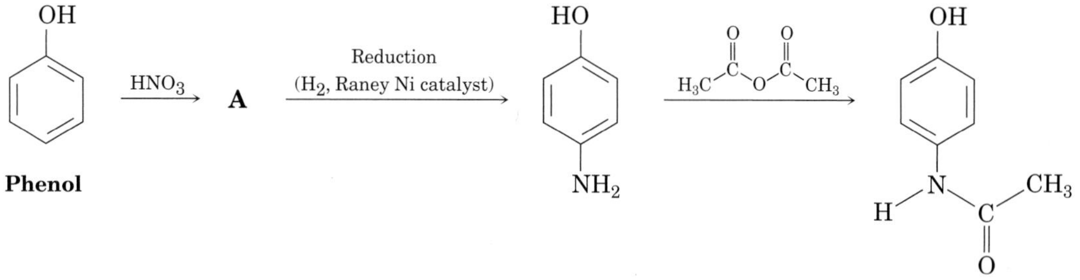

OH **Phenol** $\xrightarrow{HNO_3}$ **A** $\xrightarrow{\text{Reduction } (H_2\text{, Raney Ni catalyst})}$ HO NH_2 $\xrightarrow{H_3C-C(=O)-O-C(=O)-CH_3}$ OH H N C O CH_3

Acetaminophen (Tylenol)

5.61 Like Tylenol, ibuprofen is categorized as a nonsteroidal anti-inflammatory drug (NSAID). Draw the mechanism for the Friedel–Crafts acylation of isobutylbenzene to form intermediate A. Use resonance structures to show the formation of the desired product and the undesired isomer.

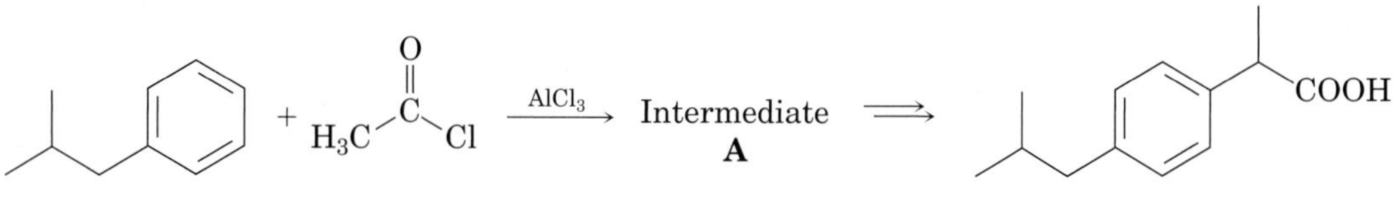

Isobutylbenzene + $H_3C-C(=O)-Cl$ $\xrightarrow{AlCl_3}$ Intermediate **A** ⟹ COOH

Ibuprofen

5.62 Recently, controversy has surrounded NSAIDs with more complex structures that target enzymes called cyclooxygenases. Several of these so-called COX-2 inhibitors have been withdrawn from the market. Identify five aromatic rings in the structures of Celebrex and Vioxx.

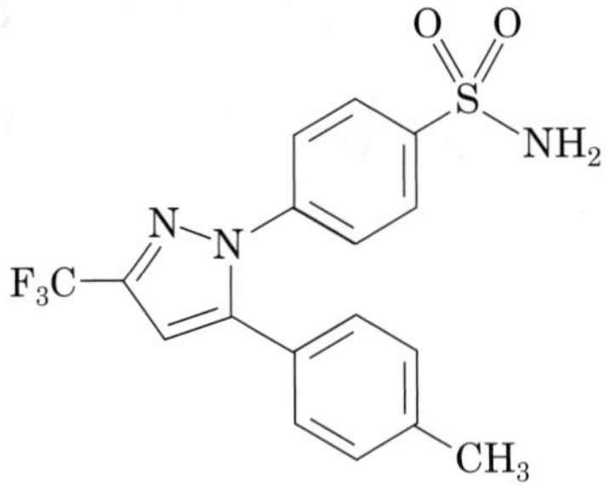

O O S NH_2 N N F_3C CH_3

Celecoxib (Celebrex)

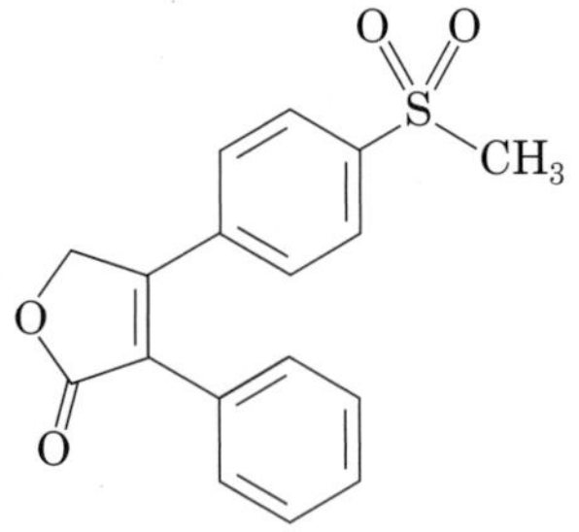

O O S CH_3 O O

Rofecoxib (Vioxx)

IN THE FIELD WITH AGROCHEMICALS

5.63 As we learned in Problem 2.77, Metolachlor is broadly used in the United States to control weeds but it is being phased out in Europe due to environmental risks. Usually marketed under the name Dual, approximately 50 million pounds of Metolachlor are applied on crops each year in the United States. The preparation of Metolachlor relies on using 2-ethyl-6-methylacetanilide as an intermediate. How might you prepare this material from acetanilide?

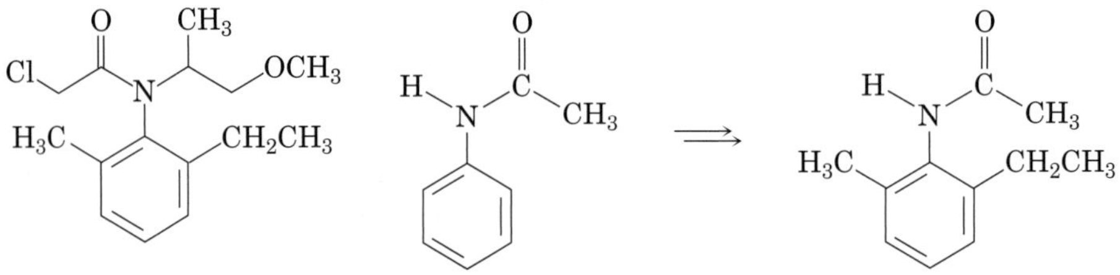

Cl O N CH_3 OCH_3 H_3C CH_2CH_3

H N C O CH_3 ⟹ H N C O CH_3 H_3C CH_2CH_3

Metolachlor **Acetanilide** **2-Ethyl-6-methylanilide**

5.64 To prepare the herbicide 2,4-D, phenol is chlorinated and then reacted with a base and chloroacetic acid. Draw a mechanism showing relevant resonance structures to explain the pattern of chlorination in the first step.

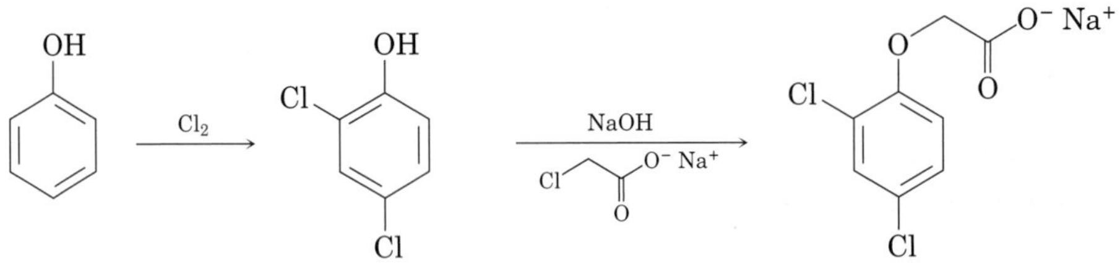

OH → Cl_2 → OH Cl Cl → NaOH, Cl–CH_2–C(=O)–O^- Na^+ → O–CH_2–C(=O)–O^- Na^+ Cl Cl

Phenol **2,4-Dichlorophenol** **2,4-D**

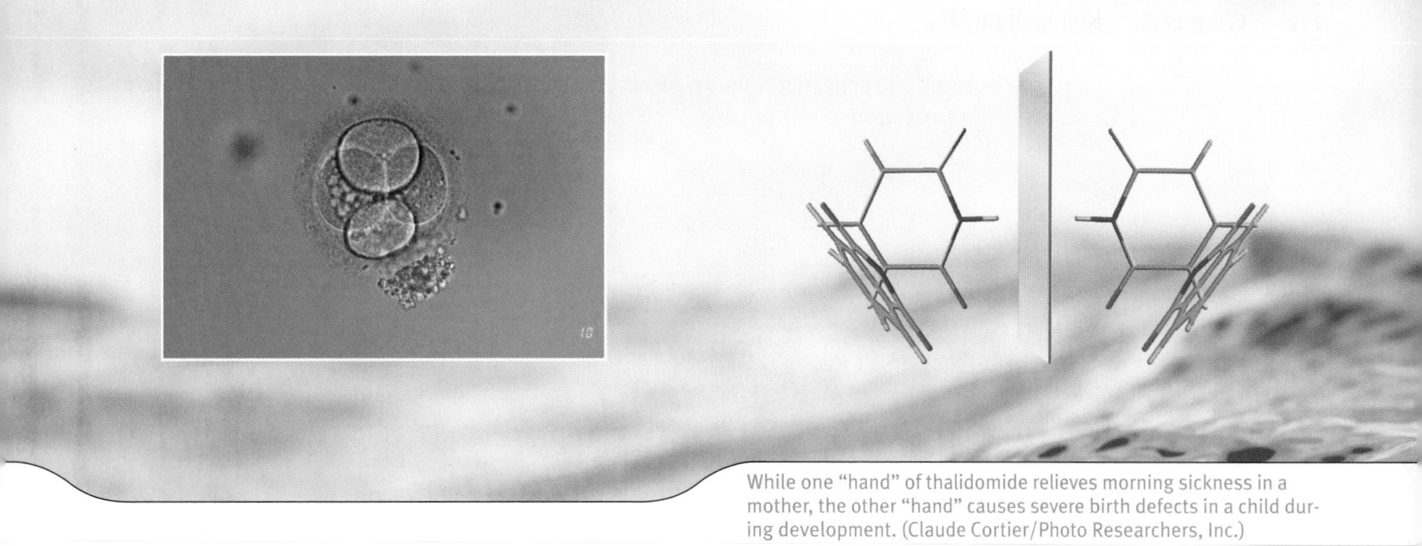

While one "hand" of thalidomide relieves morning sickness in a mother, the other "hand" causes severe birth defects in a child during development. (Claude Cortier/Photo Researchers, Inc.)

CHAPTER 6

Stereochemistry

Stereochemistry is the branch of chemistry concerned with the three-dimensional structures of molecules. We talked about cis–trans stereoisomers of cycloalkanes in Section 2.8 and cis–trans alkenes in Section 3.3. In both cases, the order of atom connectivity is not changed; only the spatial arrangement of the atoms differs. In this chapter we will focus on chirality (**ky**-ral-i-tee), or handedness (Greek *cheir*, meaning "hand"), in molecules. The difference in shape between "left-handed" and "right-handed" molecules can have significant biological consequences, particularly in medicine.

6.1 Stereochemistry and the Tetrahedral Carbon

Our right and left hands are mirror images of each other. The result is that our hands, while similar, are not identical. This handedness, or chirality, has profound implications. Consider what happens when a right-handed baseball player ends up with a left-handed glove. It is important to study and understand chirality because many of the molecules of life are **chiral**. Many biologically important molecules behave like a right- or left-handed baseball glove. In doing so, they provide a **chiral environment** that will selectively recognize a second molecule of one handedness, but not its mirror image. As shown in the opener to this chapter, the chiral environment of a pregnant mother recognizes the two hands of the drug thalidomide differently. One hand is recognized as an effective drug while the other hand is recognized as a mutagen.

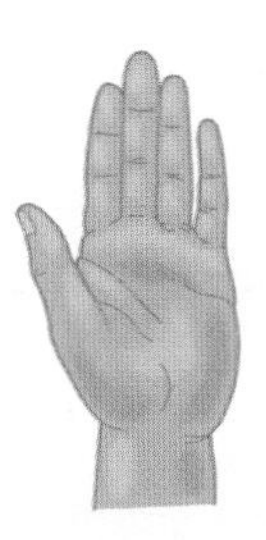

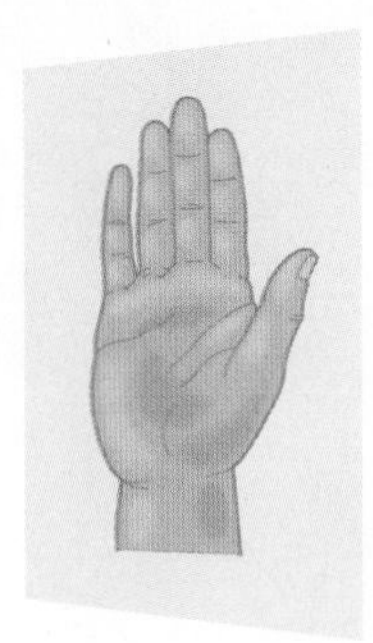

Left hand Right hand

What makes the molecules of life—carbohydrates, amino acids, nucleic acids, and many other naturally occurring molecules—chiral? The answer is the tetrahedral carbon. Look at the molecules shown in Figure 6.1. On the left of Figure 6.1 are three molecules, and on the right are their images reflected in a mirror. The CH_3X and CH_2XY molecules are identical to their mirror images and thus are not handed. If you make molecular models of each molecule and of its mirror image, you find that you can superimpose one on the other. By contrast, the CHXYZ molecule is *not* identical to its mirror image. You can't superimpose a model of the molecule on a model of its mirror image for the same reason that you can't superimpose a left hand on a right hand: they simply aren't the same.

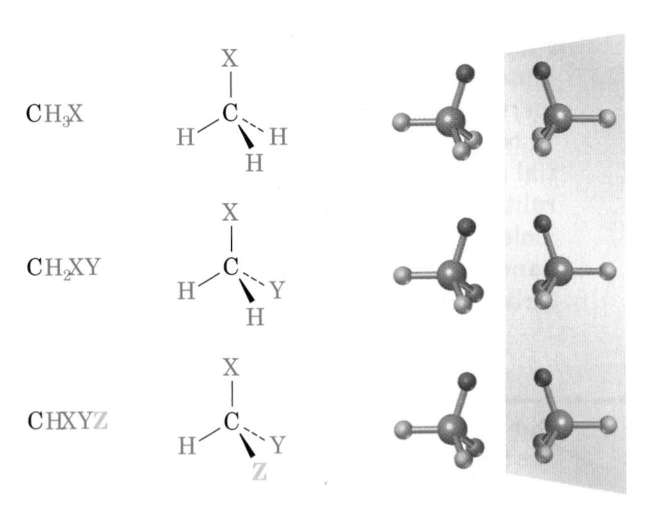

FIGURE 6.1 Three tetrahedral carbon atoms and their mirror images. Molecules of the type CH_3X and CH_2XY are identical to their mirror images, but a molecule of the type CHXYZ is not. A CHXYZ molecule is related to its mirror image in the same way that a right hand is related to a left hand.

A molecule that is not identical to its mirror image is a kind of stereoisomer called an **enantiomer** (e-**nan**-tee-o-mer; Greek *enantio*, meaning "opposite"). Enantiomers are related to each other as a right hand is related to a left hand and result whenever a tetrahedral carbon atom is bonded to four different substituents (one need not be H). For example, lactic acid (2-hydroxypropanoic acid) exists as a pair of enantiomers because there are four different groups (–H, –OH, $-CH_3$, $-CO_2H$) bonded to the central carbon atom:

Lactic acid: a molecule of general formula CHXYZ

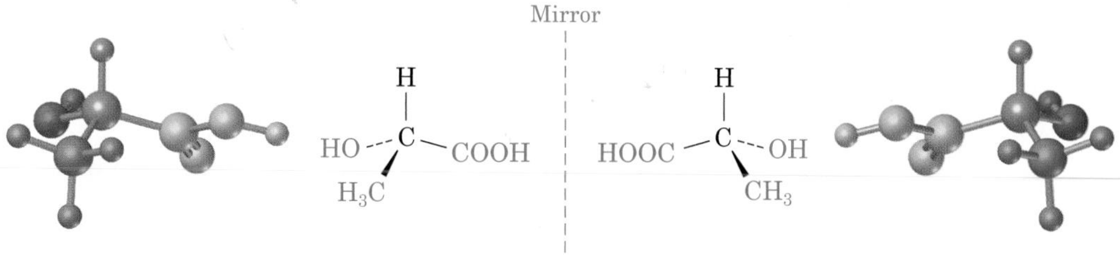

Mirror

(+)-Lactic acid (−)-Lactic acid

No matter how hard you try, you can't superimpose a molecule of "right-handed" lactic acid on top of a molecule of "left-handed" lactic acid. If any two groups match up, say –H and $-CO_2H$, the other two groups don't match (Figure 6.2).

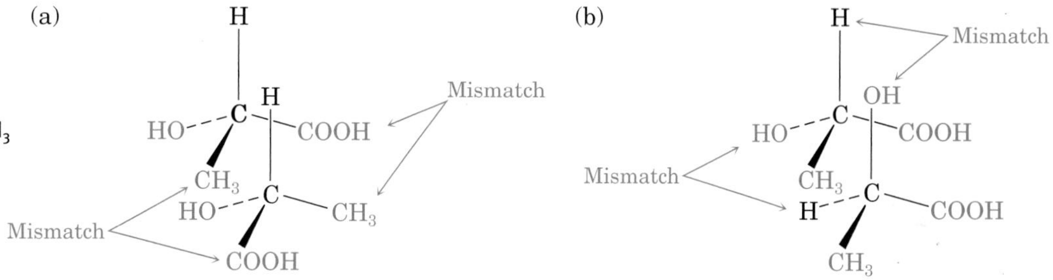

FIGURE 6.2 Attempts at superimposing the mirror-image forms of lactic acid: (a) When the –H and –OH substituents match up, the $-CO_2H$ and $-CH_3$ substituents don't. (b) When $-CO_2H$ and $-CH_3$ match up, –H and –OH don't. Regardless of how the molecules are oriented, they aren't identical.

(a) Mismatch Mismatch

(b) Mismatch Mismatch

6.2 Finding Handedness in Molecules

How can you predict whether a given molecule is or is not chiral? Throughout this chapter and the text, we will focus on the most common (although not the only) cause of chirality in an organic molecule: the presence of a carbon atom bonded to four different groups. Such a carbon is called a **stereocenter**, or *chirality center*. The middle carbon atom in lactic acid (Figure 6.2) is a chirality center because it has four different groups attached. Accordingly, lactic acid is a chiral molecule. Note that chirality is a *property* of the entire molecule, whereas a stereocenter is the *cause* of chirality.

Identifying stereocenters in a complex molecule takes practice, because it's not always apparent that four different groups are bonded to a given carbon. The differences don't necessarily appear right next to the stereocenter. For example, 5-bromodecane is a chiral molecule because four different groups are bonded to C5 (marked by an asterisk). A butyl substituent is very *similar* to a pentyl substituent, but it isn't identical. The difference isn't apparent until four carbons away from the stereocenter, but there's still a difference.

$$CH_3CH_2CH_2CH_2CH_2\overset{Br}{\underset{H}{C}}{}^{*}CH_2CH_2CH_2CH_3$$

5-Bromodecane (chiral)

Substituents on carbon 5

—H

Br

—$CH_2CH_2CH_2CH_3$ (butyl)

—$CH_2CH_2CH_2CH_2CH_3$ (pentyl)

Several examples of chiral molecules are shown in the following illustration; check for yourself that the labeled atoms are indeed stereocenters. (It's helpful to note that CH_2, CH_3, C=C, C≡C, and C=O carbons *can't* be stereocenters because they have at least two identical bonds.)

Carvone (spearmint oil) **Nootkatone (grapefruit oil)**

Another way to identify chiral molecules is to look for a plane of symmetry. A plane of symmetry is a plane that cuts through the middle of a molecule or other object so that one half of the object is a mirror image of the other half. *A molecule is not chiral if it contains a plane of symmetry.* A laboratory flask, for example, has a plane of symmetry (Figure 6.3). If you were to cut the flask into left and right halves, one half would be an exact mirror image of the other half. A spoon is not chiral for the same reason. A hand, however, does not have a plane of symmetry. One "half" of a hand is not a mirror image of the other "half." The barber's pole is also chiral because it *lacks* a plane of symmetry.

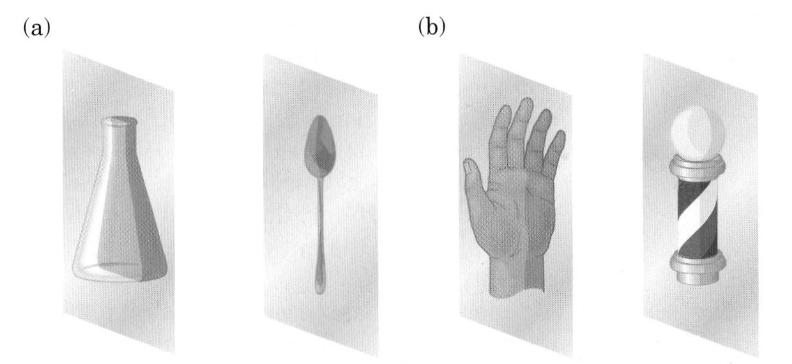

FIGURE 6.3 The meaning of a symmetry plane. (a) Objects like this flask or spoon have planes of symmetry passing through them, making the right and left halves mirror images. (b) Objects like this hand or barber's pole have no symmetry plane; the right "half" is not a mirror image of the "left" half.

A molecule that has a plane of symmetry in any of its possible conformations must be identical to its mirror image and hence must be nonchiral, or **achiral**. Thus, propanoic acid, $CH_3CH_2CO_2H$, contains a plane of symmetry when lined up as shown in Figure 6.4 and is therefore achiral. Lactic acid, $CH_3CH(OH)CO_2H$, however, has no plane of symmetry in any conformation and is chiral.

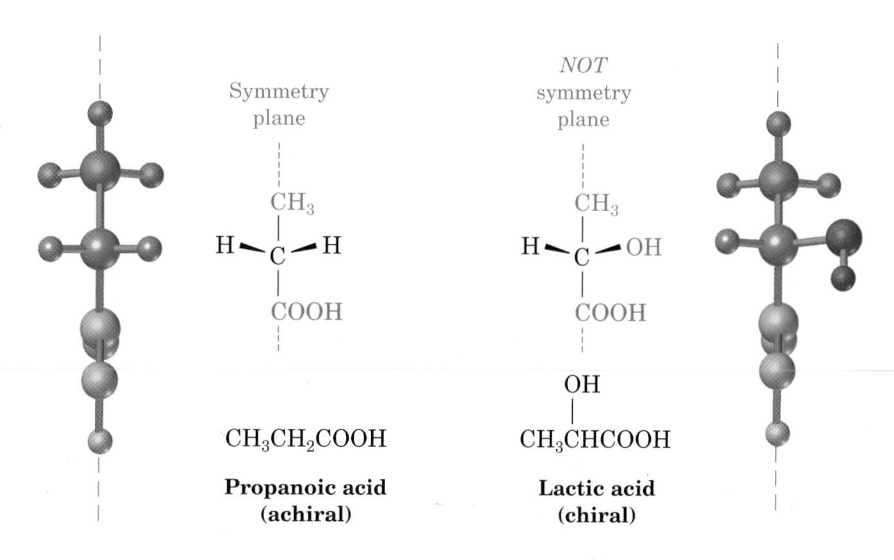

FIGURE 6.4 The achiral propanoic acid molecule versus the chiral lactic acid molecule. Propanoic acid has a plane of symmetry that makes one side of the molecule a mirror image of the other side. Lactic acid has no such symmetry plane.

PRACTICE PROBLEM 6.1

Draw the structure of a chiral alcohol.

STRATEGY An alcohol is a compound that contains the –OH functional group. To make an alcohol chiral, we need to have four different groups bonded to a single carbon atom, say –H, –OH, $–CH_3$, and $–CH_2CH_3$.

SOLUTION

$$CH_3CH_2-\overset{\displaystyle OH}{\underset{\displaystyle H}{C}}-CH_3 \quad \textbf{Butan-2-ol}$$

PRACTICE PROBLEM 6.2

Is 3-methylhexane chiral?

STRATEGY Draw the structure of 3-methylhexane, and cross out all the CH_2 and CH_3 carbons because they can't be stereocenters. Then look closely at any carbon that remains to see if it's bonded to four different groups.

SOLUTION Carbon 3 is bonded to –H, $–CH_3$, $–CH_2CH_3$, and $–CH_2CH_2CH_3$, so the molecule is chiral.

$$CH_3CH_2CH_2-\overset{\displaystyle CH_3}{\underset{\displaystyle H}{\overset{*}{C}}}-CH_2CH_3 \quad \textbf{3-Methylhexane (chiral)}$$

PRACTICE PROBLEM 6.3

Is 2-methylcyclohexanone chiral?

2-Methylcyclohexanone

STRATEGY Ignore the CH_3 carbon, the four CH_2 carbons in the ring, and the C=O carbon because they can't be stereocenters. Then look carefully at C2, the only carbon that remains.

SOLUTION Carbon 2 is bonded to four different groups: a $-CH_3$ group, an –H atom, a –C=O carbon in the ring, and a $-CH_2-$ ring carbon, so 2-methylcyclohexanone is chiral.

PROBLEM 6.1 Which of the following objects are chiral (handed)?
(a) Bean stalk (b) Screwdriver (c) Screw (d) Shoe

PROBLEM 6.2 Which of the following molecules are chiral?
(a) 3-Bromopentane (b) 1,3-Dibromopentane
(c) 3-Methylhex-1-ene (d) *cis*-1,4-Dimethylcyclohexane

PROBLEM 6.3 Which of the following molecules are chiral? Identify the stereocenter(s) in each.

(a)

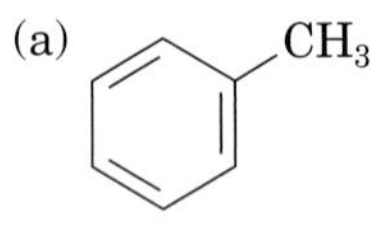

Toluene

(b)

Coniine (from poison hemlock)

(c)

Phenobarbital (tranquilizer)

PROBLEM 6.4 Place asterisks at the stereocenters in the following molecules:

(a) **Menthol**

(b) **Camphor**

(c) **Dextromethorphan (a cough suppressant)**

PROBLEM 6.5 Alanine, an amino acid found in proteins, is a chiral molecule. Use the standard convention of wedged, solid, and dashed lines to draw the two enantiomers of alanine.

Alanine

6.3 Optical Activity

The study of stereochemistry originated in the early 19th century during investigations by the French physicist Jean Baptiste Biot into the nature of *plane-polarized light.* A beam of ordinary light consists of electromagnetic waves that oscillate in an infinite number of planes at right angles to the direction of light travel. When a beam of ordinary light passes through a device called a *polarizer*, though, only the light waves oscillating in a *single* plane pass through and the light is said to be plane-polarized.

Biot made the remarkable observation that when a beam of plane-polarized light passes through a solution of certain organic molecules, such as sugar or camphor, the plane of polarization is *rotated.* Not all organic molecules exhibit this property, but those that do are said to be **optically active**.

The amount of rotation can be measured with an instrument called a *polarimeter*, represented in Figure 6.5. A solution of optically active organic molecules is placed in a sample tube, plane-polarized light is passed through the tube, and rotation of the plane occurs. The light then goes through a second polarizer called the *analyzer.* By rotating the analyzer until light passes through *it*, we can find the new plane of polarization and can tell to what extent rotation has occurred. The amount of rotation observed is denoted by α (Greek alpha) and is expressed in degrees.

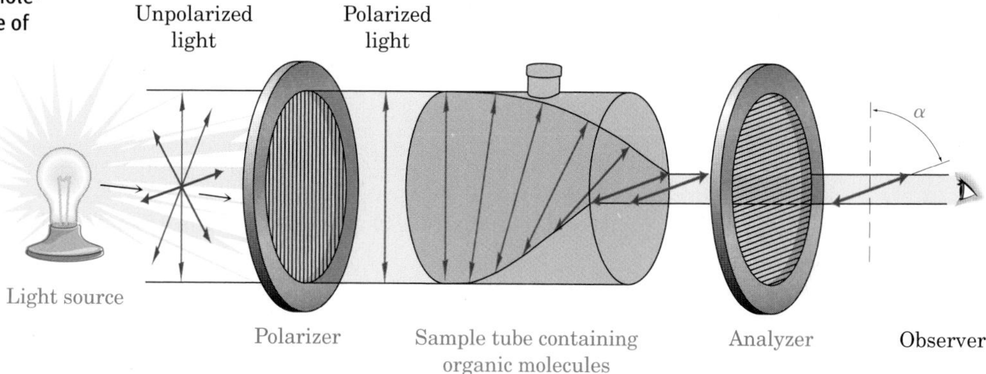

FIGURE 6.5 Schematic representation of a polarimeter. Plane-polarized light passes through a solution of optically active molecules, which rotate the plane of polarization.

In addition to determining the extent of rotation, we can also find the direction. From the vantage point of the observer looking at the analyzer, some optically active molecules rotate plane-polarized light to the left (counterclockwise) and are said to be **levorotatory**; other molecules rotate light to the right (clockwise) and are said to be **dextrorotatory**. By convention, rotation to the left is given a minus sign (−), and rotation to the right is given a plus sign (+). For example, (−)-morphine is levorotatory and (+)-sucrose is dextrorotatory.

6.4 Specific Rotation

The degree of rotation observed in a polarimetry experiment depends on the structure of the sample molecules and on the number of molecules encountered by the light beam. The number of molecules encountered depends, in turn, on sample concentration and sample path length. If the concentration of the sample in a tube is doubled, the observed rotation is doubled. If the concentration is kept constant but the length of the sample tube is doubled, the observed rotation is doubled.

TABLE 6.1 Specific Rotations of Some Organic Molecules

Compound	$[\alpha]_D$ (degrees)	Compound	$[\alpha]_D$ (degrees)
Penicillin V	+233	Cholesterol	−31.5
Sucrose	+66.47	Morphine	−132
Camphor	+44.26	Acetic acid	0
Monosodium glutamate	+25.5	Benzene	0

To express optical rotation data so that comparisons can be made, we have to choose standard conditions. The **specific rotation, $[\alpha]_D$,** of a compound is defined as the observed rotation when light of 589.6 nanometer (nm; 1 nm = 10^{-9} m) wavelength is used with a sample path length l of 1 decimeter (dm; 1 dm = 10 cm) and a sample concentration C of 1 g/mL. (Light of 589.6 nm, the so-called sodium D line, is the yellow light emitted from common sodium lamps.)

$$[\alpha]_D = \frac{\text{Observed rotation (degrees)}}{\text{Path length, } l \text{ (dm)} \times \text{Concentration, } C \text{ (g/mL)}} = \frac{\alpha}{l \times C}$$

When optical rotation data are expressed in this standard way, the specific rotation $[\alpha]_D$ is a physical constant characteristic of a given optically active compound. Some examples are listed in Table 6.1.

PRACTICE PROBLEM 6.4

A 1.20 g sample of cocaine, $[\alpha]_D = -16°$, was dissolved in 7.50 mL of chloroform and placed in a sample tube having a path length of 5.00 cm. What was the observed rotation?

STRATEGY Observed rotation, α, is equal to specific rotation $[\alpha]_D$ times sample concentration C times path length l: $\alpha = [\alpha]_D \times C \times l$, where $[\alpha]_D = -16°$, $l = 5.00$ cm $= 0.500$ dm, and $C = 1.20$ g/7.50 mL $= 0.160$ g/mL.

SOLUTION $\alpha = -16° \times 0.500 \times 0.160 = -1.3°$.

PROBLEM 6.6 Is cocaine (Practice Problem 6.4) dextrorotatory or levorotatory?

PROBLEM 6.7 A 1.50 g sample of coniine, the toxic extract of poison hemlock, was dissolved in 10.0 mL of ethanol and placed in a sample tube with a path length of 5.00 cm. The observed rotation at the sodium D line was +1.21°. Calculate the specific rotation $[\alpha]_D$ for coniine.

6.5 Pasteur's Discovery of Enantiomers

Little was done after Biot's discovery of optical activity until 1848, when Louis Pasteur began work on a study of crystalline tartaric acid salts derived from wine. On recrystallizing a concentrated solution of sodium ammonium tartrate below 28 °C, Pasteur made the surprising observation that two distinct kinds of crystals precipitated. Furthermore, the two kinds of crystals were mirror images and were related in the same way that a right hand is related to a left hand.

Working carefully with a pair of tweezers, Pasteur was able to separate the crystals into two piles, one of "right-handed" crystals and one of "left-handed" crystals, like those shown in Figure 6.6. Although the original sample (a 50:50 mixture of right and left) was optically inactive, *solutions of crystals from each of the sorted piles were optically active* and their specific rotations were equal in amount but opposite in sign.

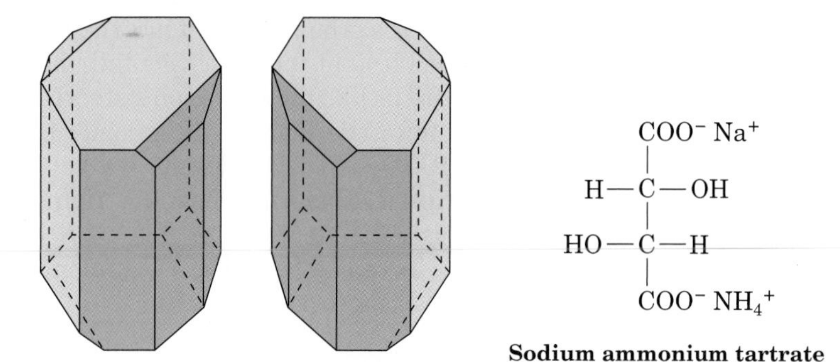

FIGURE 6.6 Crystals of sodium ammonium tartrate, taken from Pasteur's original sketches. One of the crystals is dextrorotatory in solution, and the other is levorotatory.

We call this original 50:50 *mixture* of enantiomers a **racemic** (rah-**see**-mic) **mixture**, or **racemate**. Racemates are often denoted by the symbol (±) to indicate that they contain equal amounts of dextrorotatory and levorotatory enantiomers. Such mixtures show no optical activity because the (+) rotation from one enantiomer exactly cancels the (–) rotation from the other. These enantiomers have identical chemical properties in the absence of a chiral environment.

Pasteur was far ahead of his time. Although the structural theory of Kekulé had not yet been proposed, Pasteur explained his results by speaking of the molecules themselves, saying, "There is no doubt that [in the *dextro* tartaric acid] there exists an asymmetric arrangement having a nonsuperimposable image. It is no less certain that the atoms of the *levo* acid possess precisely the inverse asymmetric arrangement." Pasteur's vision was extraordinary, for it was not until 25 years later that his theories regarding the asymmetry of chiral molecules were confirmed.

Today, we would describe Pasteur's work by saying that he had discovered enantiomers. Enantiomers (also called *optical isomers*) have identical physical properties, such as melting points and boiling points, but differ in the direction in which their solutions rotate plane-polarized light.

6.6 Sequence Rules for Specifying Configuration

Drawings provide visual representations of stereochemistry, but a verbal method for specifying the three-dimensional arrangement, or **configuration**, of substituents around a stereocenter is also necessary. The method used employs the same sequence rules given in Section 3.4 for specifying *E* and *Z* alkene stereochemistry. Let's briefly review these sequence rules and see how they're used to specify the configuration of a stereocenter. For a more thorough review, you should reread Section 3.4.

RULE 1 Look at the four atoms directly attached to the stereocenter, and assign priorities in order of decreasing atomic number. The atom with the highest atomic number is ranked first; the atom with the lowest atomic number is ranked fourth.

RULE 2 If a decision can't be reached by ranking the first atoms in the substituents, look at the second, third, or fourth atoms outward until the first difference is found.

RULE 3 Multiple-bonded atoms are equivalent to the same number of single-bonded atoms.

Having assigned priorities to the four groups attached to a stereocenter, we describe the stereochemical configuration around the carbon by orienting the molecule so that the group of lowest priority (4) is pointing directly back, away from us. We then look at the three remaining substituents, which now appear to radiate toward us like the spokes on a steering wheel (Figure 6.7). If a curved arrow drawn from the highest- to second-highest- to third-highest-priority substituent ($1 \rightarrow 2 \rightarrow 3$) is clockwise, we say that the stereocenter has an ***R* configuration** (Latin *rectus*, meaning "right"). If an arrow from $1 \rightarrow 2 \rightarrow 3$ is counterclockwise, the stereocenter has an ***S* configuration** (Latin *sinister*, meaning "left"). To remember these assignments, think of a car's steering wheel when making a *R*ight (clockwise) turn.

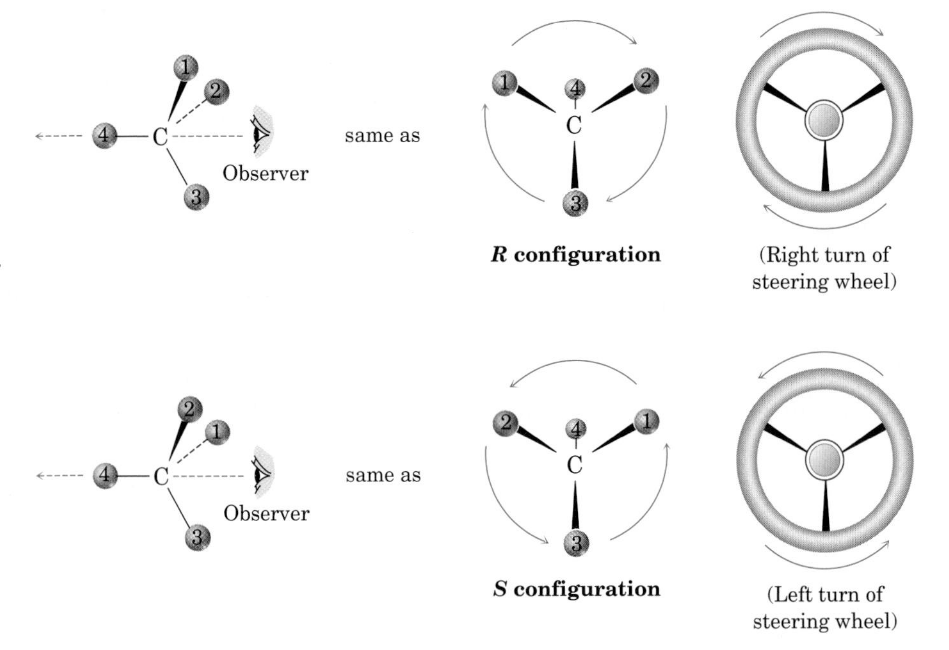

FIGURE 6.7 Assignment of configuration to a stereocenter. When the molecule is oriented so that the group of lowest priority (4) is toward the rear, the remaining three groups radiate toward the viewer like the spokes of a steering wheel. If the direction of travel $1 \rightarrow 2 \rightarrow 3$ is clockwise (right turn), the center has the *R* configuration. If the direction of travel $1 \rightarrow 2 \rightarrow 3$ is counterclockwise (left turn), the center is *S*.

Look at (−)-lactic acid in Figure 6.8 to see an example of how configuration is assigned. Sequence rule 1 says that –OH has priority 1 and –H has priority 4, but it doesn't allow us to distinguish between $–CH_3$ and $–CO_2H$ because both groups have carbon as their first atom. Sequence rule 2, however, says that $–CO_2H$ is higher priority than $–CH_3$ because O outranks H (the second atom in each group). Now, turn the molecule so that the fourth-priority group (–H) is oriented toward the rear, away from the observer. Since a curved arrow from 1 (–OH) to 2 ($–CO_2H$) to 3 ($–CH_3$) is clockwise (right turn of the steering wheel), (−)-lactic acid has the *R* configuration. Applying the same procedure to (+)-lactic acid leads to the opposite assignment.

Further examples are provided by naturally occurring (−)-glyceraldehyde and (+)-alanine, which have the *S* configurations shown in Figure 6.9. *Note that the sign of optical rotation, (+) or (−), is not related to the R,S designation.* (*S*)-Alanine happens to be dextrorotatory (+), and (*S*)-glyceraldehyde happens to be levorotatory (−), but there is no correlation between *R*,*S* configuration and direction of optical rotation.

FIGURE 6.8 Assignment of configuration to (a) (*R*)-(−)-lactic acid and (b) (*S*)-(+)-lactic acid.

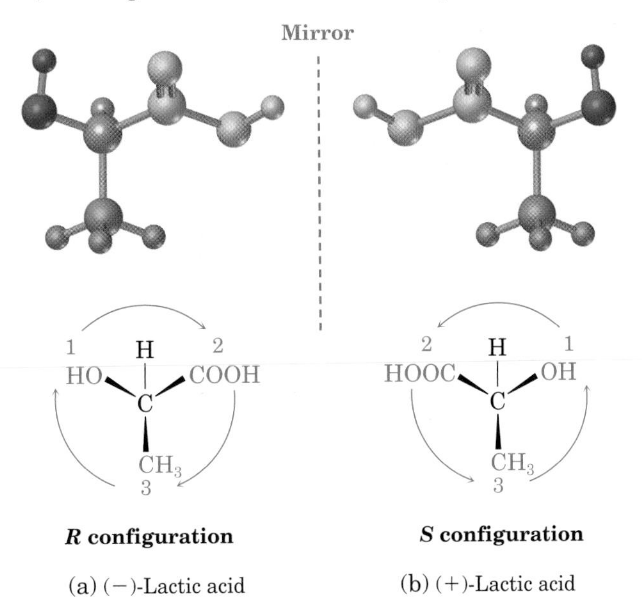

FIGURE 6.9 Assignment of configuration to (a) (−)-glyceraldehyde and (b) (+)-alanine. Both happen to have the *S* configuration, although one is levorotatory and the other is dextrorotatory.

(a)

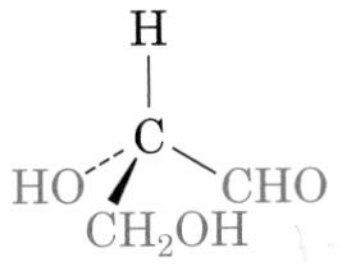

(*S*)-Glyceraldehyde
[(*S*)-(−)-2,3-Dihydroxypropanal]
$[\alpha]_D = -8.7°$

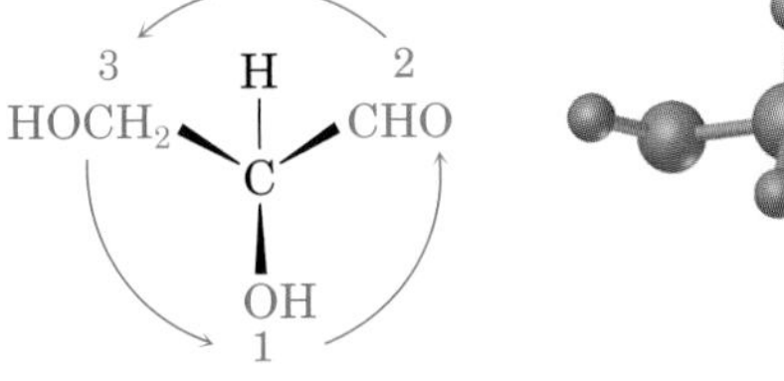

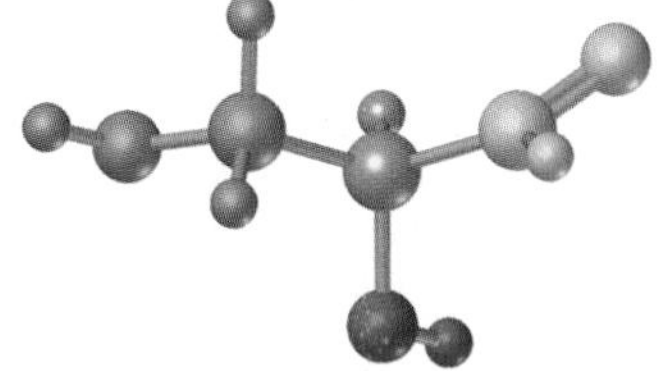

(b)

H
C
H_2N COOH
CH_3

(*S*)-Alanine
[(*S*)-(+)-2-Aminopropanoic acid]
$[\alpha]_D = +8.5°$

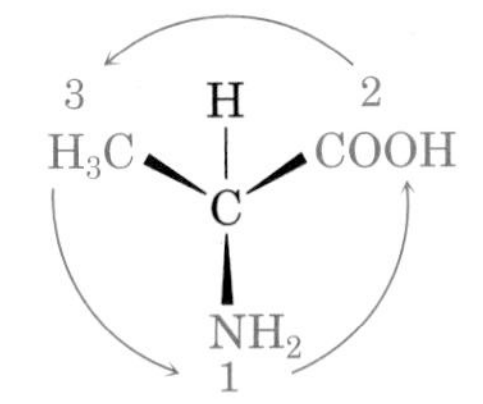

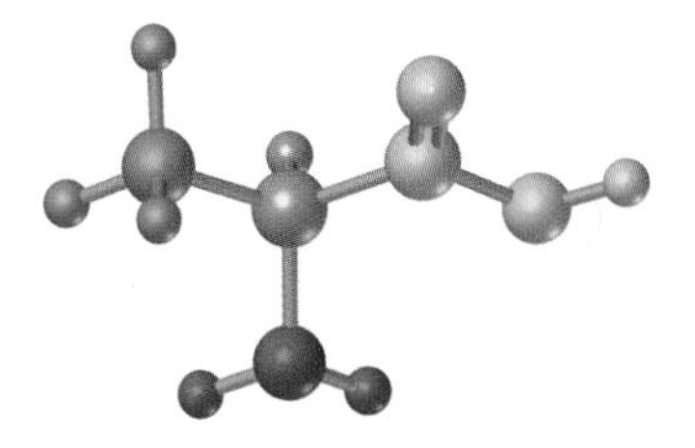

PRACTICE PROBLEM 6.5

Orient each of the following drawings so that the lowest-priority group is toward the rear, and then assign *R* or *S* configuration:

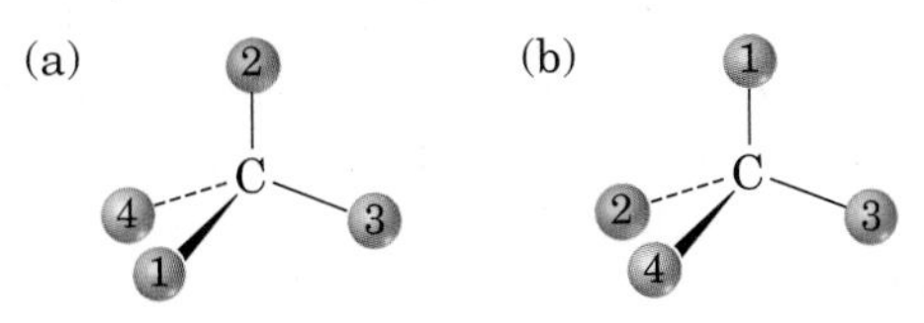

STRATEGY It takes practice to be able to visualize and orient a stereocenter in three dimensions. You might start by indicating where the observer must be located—180° opposite the lowest-priority group. Then imagine yourself in the position of the observer, and redraw what you would see.

SOLUTION In (a), you would be located in front of the page toward the top right of the molecule, and you would see group 2 to your left, group 3 to your right, and group 1 below you. This corresponds to an R configuration.

(a) Observer

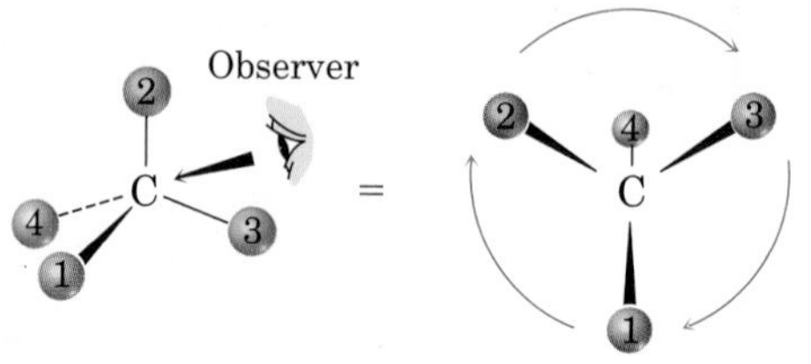

***R* configuration**

In (b), you would be located behind the page toward the top *left* of the molecule from your point of view, and you would see group 3 to your left, group 1 to your right, and group 2 below you. This also corresponds to an R configuration.

(b) Observer

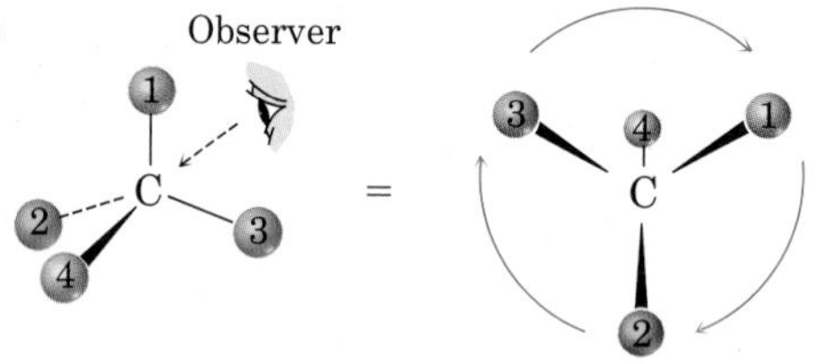

***R* configuration**

PRACTICE PROBLEM 6.6

Draw a tetrahedral representation of (R)-2-chlorobutane.

STRATEGY Identify the four substituents bonded to the stereocenter, and assign the priorities: (1) –Cl, (2) $-CH_2CH_3$, (3) $-CH_3$, (4) –H. To draw a tetrahedral representation of the molecule, orient the low-priority –H group away from you and imagine that the other three groups are coming out of the page toward you. Then place the remaining three substituents such that the direction of travel 1 → 2 → 3 is clockwise (right turn), and tilt the molecule toward you by 90° to bring the rear hydrogen into view. Using molecular models is a great help in working problems of this sort.

SOLUTION

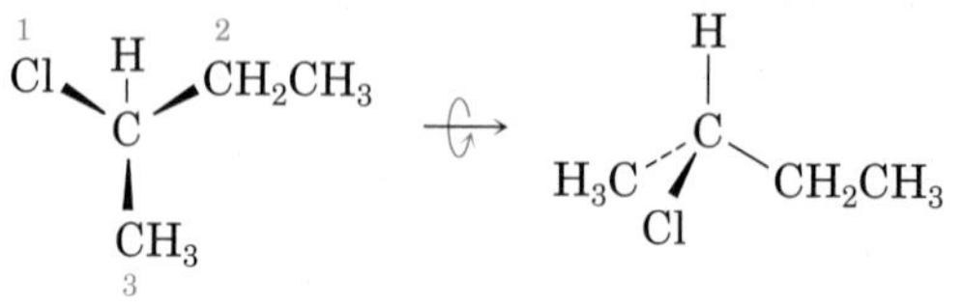

(*R*)-2-Chlorobutane

PROBLEM 6.8 Assign priorities to the substituents in each of the following sets:

(a) –H, –Br, $-CH_2CH_3$, $-CH_2CH_2OH$ (b) $-CO_2H$, $-CO_2CH_3$, $-CH_2OH$, –OH

(c) –Br, $-CH_2Br$, –Cl, $-CH_2Cl$

PROBLEM 6.9 Assign *R,S* configurations to the following molecules:

(a) Br, C, H_3C, COOH, H

(b) H, C, HO, COOH, CH_3

(c) NH_2, C, NC, CH_3, H

PROBLEM 6.10 Draw a tetrahedral representation of (*S*)-pentan-2-ol.

PROBLEM 6.11 Assign *R* or *S* configuration to the stereocenter in the following molecular model of the amino acid methionine (red = O, blue = N, yellow = S):

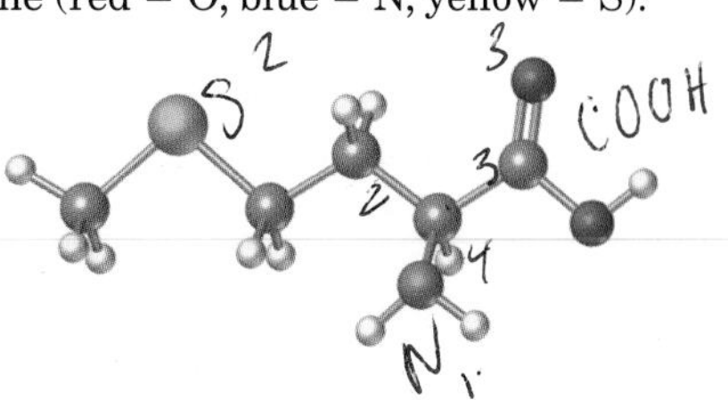

6.7 Enantiomers and Diastereomers

Molecules like lactic acid and glyceraldehyde are relatively simple to deal with because each has only one stereocenter and only two enantiomeric forms. The situation becomes more complex, however, for molecules that have more than one stereocenter. Take the amino acid threonine (2-amino-3-hydroxybutanoic acid), for example. Since threonine has two stereocenters (C2 and C3), there are four possible stereoisomers, as shown in Figure 6.10. Check for yourself that the *R,S* configurations are correct.

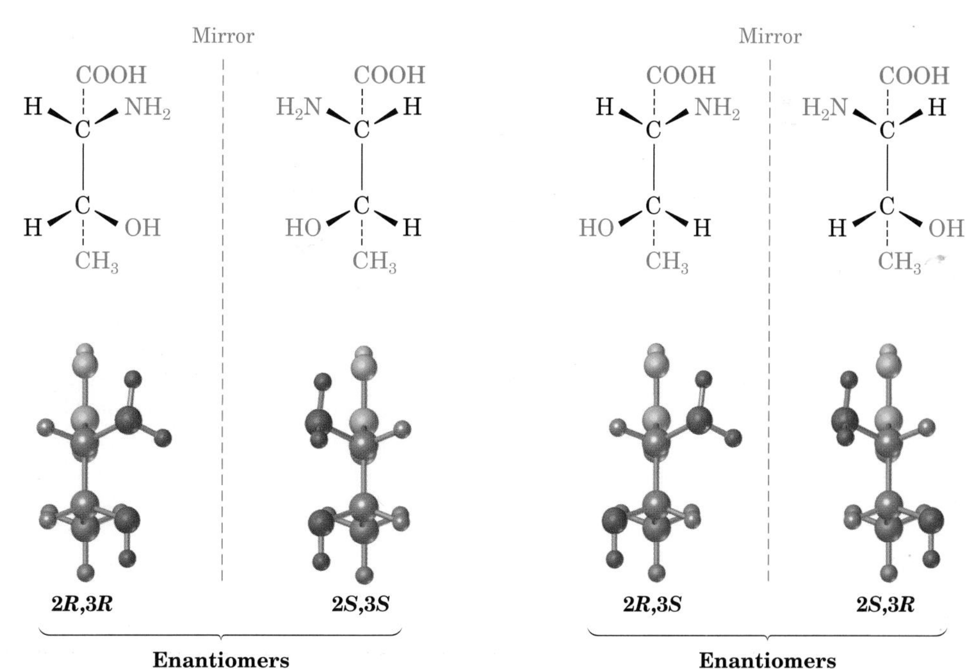

FIGURE 6.10 The four stereoisomers of 2-amino-3-hydroxybutanoic acid.

The four stereoisomers of 2-amino-3-hydroxybutanoic acid can be grouped into two pairs of mirror-image enantiomers. The 2*S*,3*S* stereoisomer

TABLE 6.2 Relationships among the Four Stereoisomers of Threonine

Stereoisomer	Enantiomeric with	Diastereomeric with
2*R*,3*R*	2*S*,3*S*	2*R*,3*S* and 2*S*,3*R*
2*S*,3*S*	2*R*,3*R*	2*R*,3*S* and 2*S*,3*R*
2*R*,3*S*	2*S*,3*R*	2*R*,3*R* and 2*S*,3*S*
2*S*,3*R*	2*R*,3*S*	2*R*,3*R* and 2*S*,3*S*

is the mirror image of 2*R*,3*R*, and the 2*S*,3*R* stereoisomer is the mirror image of 2*R*,3*S*. But what is the relationship between any two molecules that are not mirror images? What, for example, is the relationship between the 2*R*,3*R* isomer and the 2*R*,3*S* isomer? They are stereoisomers, yet they aren't enantiomers. To describe such a relationship, we need a new term—*diastereomer*. **Diastereomers** are stereoisomers that are not mirror images.

Note carefully the difference between enantiomers and diastereomers: enantiomers have opposite configurations at *all* stereocenters; diastereomers have opposite configurations at *some* (one or more) stereocenters but the same configuration at others. A full description of the four threonine stereoisomers is given in Table 6.2.

Of the four stereoisomers of threonine, only the 2*S*,3*R* isomer, $[\alpha]_D = -28.3°$, occurs naturally and is an essential human nutrient. Most biologically important organic molecules are chiral, and usually only one stereoisomer is found in nature.

PROBLEM 6.12 Assign *R* or *S* configuration to each stereocenter in the following molecules:

(a) Br, H, CH_3 on C; C with H, OH, CH_3

(b) CH_3, H, Br on C; C with H_3C, H, OH

(c) CH_3, Br, H on C; C with H, CH_3, OH

PROBLEM 6.13 Which of the compounds in Problem 6.12 are enantiomers, and which are diastereomers?

PROBLEM 6.14 Chloramphenicol is a powerful antibiotic isolated from the *Streptomyces venezuelae* bacterium. It is active against a broad spectrum of bacterial infections and is particularly valuable against typhoid fever. Assign *R* or *S* configuration to the stereocenters in chloramphenicol.

NO_2 (on benzene ring); HO, H on C; C with H, $NHCOCHCl_2$, CH_2OH

Chloramphenicol
$[\alpha]_D = +18.6°$

PROBLEM 6.15 Assign *R*,*S* configuration to each stereocenter in the following molecular model of the amino acid isoleucine (red = O, blue = N):

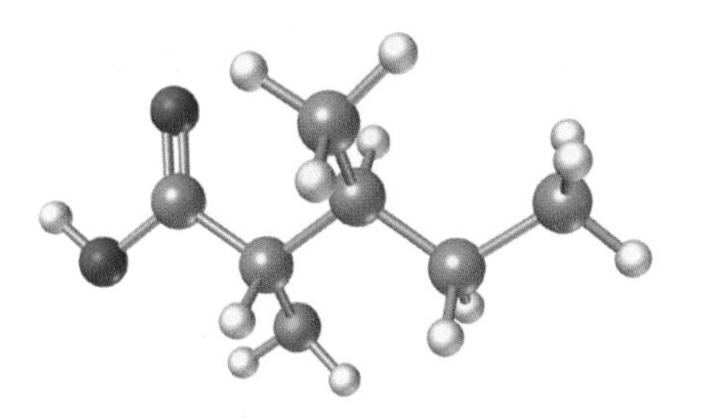

6.8 Meso Compounds

Let's look at one more example of a compound with two stereocenters: the tartaric acid used by Pasteur. The four stereoisomers can be drawn as follows:

Mirror | Mirror

2*R*,3*R* | **2*S*,3*S*** | **2*R*,3*S*** | **2*S*,3*R***

The mirror-image 2*R*,3*R* and 2*S*,3*S* structures are not identical and therefore represent an enantiomeric pair. A careful look, however, shows that the 2*R*,3*S* and 2*S*,3*R* structures *are* identical, as can be seen by rotating one structure 180°:

Rotate 180°

2*R*,3*S* | **2*S*,3*R***

Identical

The 2*R*,3*S* and 2*S*,3*R* structures are identical because the molecule has a plane of symmetry and is therefore achiral. The symmetry plane cuts through the C2–C3 bond, making one half of the molecule a mirror image of the other half (Figure 6.11).

Because of the plane of symmetry, the tartaric acid stereoisomer shown in Figure 6.11 is achiral, despite the fact that it has two stereocenters. Such compounds that are achiral, yet contain stereocenters, are called **meso** (**me**-zo) **compounds**. Thus, tartaric acid exists in three stereoisomeric forms: two enantiomers and one meso form.

Since tartaric acid exists in different stereoisomeric configurations, the question arises whether the different stereoisomers have different physical properties. The answer is yes, they do.

FIGURE 6.11 A symmetry plane cutting through the C2–C3 bond of *meso*-tartaric acid makes the molecule achiral.

Some physical properties of the three stereoisomers of tartaric acid and of the racemic mixture are shown in Table 6.3. The (+) and (−) enantiomers have identical melting points, solubilities, and densities. They differ only in the sign of their rotation of plane-polarized light. The meso isomer, in contrast, is diastereomeric with the (+) and (−) forms. It is therefore a different compound altogether and has different physical properties. The racemic mixture is different still. Although a mixture of enantiomers, racemates act as though they were pure compounds, different from either enantiomer or from the meso form.

PRACTICE PROBLEM 6.7

Does *cis*-1,2-dimethylcyclobutane have any stereocenters? Is it a chiral molecule?

STRATEGY To see whether a stereocenter is present, look for a carbon atom bonded to four different groups. To see whether the molecule is chiral, look for a symmetry plane. Not all molecules with stereocenters are chiral—meso compounds are an exception.

SOLUTION Looking at the structure of *cis*-1,2-dimethylcyclobutane, we see that both of the methyl-bearing ring carbons (C1 and C2) are stereocenters. Overall, though, the compound is achiral because there is a symmetry plane bisecting the ring between C1 and C2. Thus, *cis*-1,2-dimethylcyclobutane is a meso compound.

TABLE 6.3 Some Properties of the Stereoisomers of Tartaric Acid

Stereoisomer	Melting point (°C)	$[\alpha]_D$ (degrees)	Density (g/cm³)	Solubility at 20 °C (g/100 mL H_2O)
(+)	168–170	+12	1.7598	139.0
(−)	168–170	−12	1.7598	139.0
Meso	146–148	0	1.6660	125.0
(±)	206	0	1.7880	20.6

PROBLEM 6.16 Which of the following substances have meso forms?

(a) 2,3-Dibromobutane (b) 2,3-Dibromopentane (c) 2,4-Dibromopentane

PROBLEM 6.17 Which of the following structures represent meso compounds?

(a) 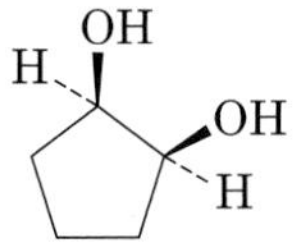

(b) 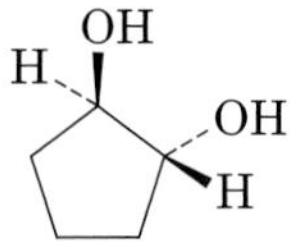

(c) 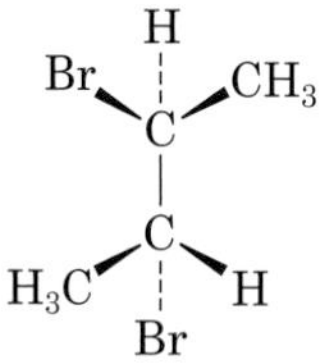

6.9

Molecules with More Than Two Stereocenters

One stereocenter gives rise to two stereoisomers (one pair of enantiomers), and two stereocenters give rise to a maximum of four stereoisomers (two pairs of enantiomers). In general, a molecule with n stereocenters has a maximum of 2^n stereoisomers (2^{n-1} pairs of enantiomers). For example, cholesterol has eight stereocenters. Thus, $2^8 = 256$ stereoisomers of cholesterol, or 128 pairs of enantiomers, are possible in principle, although many would be too strained to exist. Only one, however, is produced in nature.

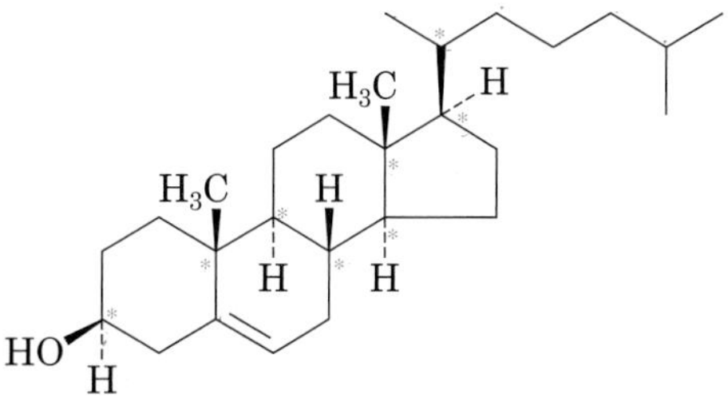

Cholesterol
(eight stereocenters)

PROBLEM 6.18 Nandrolone is an anabolic steroid used by some athletes to build muscle mass. How many stereocenters does nandrolone have? How many stereoisomers of nandrolone are possible in principle?

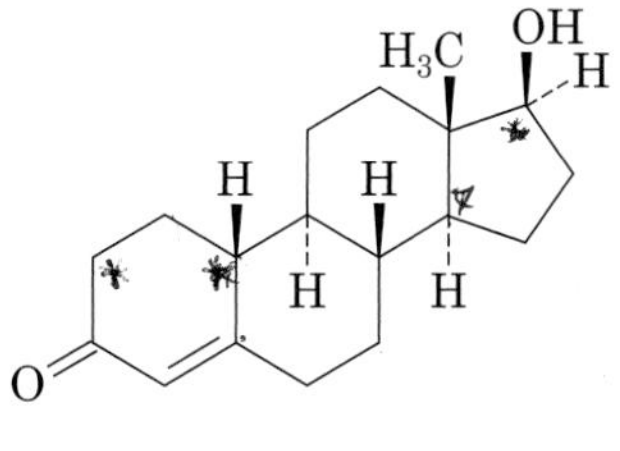

Nandrolone

6.10

The Chiral Environment

Most of our efforts in this chapter focus on skills including identifying chirality centers and assigning them as R or S. However, the single most important concept that should be distilled from these pages is the role of a chiral environment. We know by now that enantiomers have identical physical properties like melting and boiling points. Diastereomers, on the contrary, behave as completely different molecules. Similarly, in the presence of another source of chirality (a chiral environment), enantiomers exhibit very different properties. *The chiral environment acts like an additional stereocenter and turns a pair of enantiomers into diastereomers.* The different actions of the enantiomers of thalidomide are a result of the chiral environment inside a living cell.

When Pasteur took a racemate of tartaric acid and separated it into enantiomers by crystallization, crystals of (+)-tartaric acid provided a chiral environment that favored interaction of the (+)-crystal with soluble molecules of (+)-tartaric acid over (–)-tartaric acid. The relationship between an interaction (+)-tartaric acid crystal and soluble (+)-tartaric acid molecule is diastereomeric with that of a (+)-tartaric acid crystal and (–)-tartaric acid molecule. A similar situation arises for the (–)-tartaric acid crystals. It was only through good luck that Pasteur was able to achieve separation (or **resolution**) of these enantiomers. Ordinarily, a racemic solution gives racemic crystals.

The most common way to generate a chiral environment is illustrated with the resolution of racemic mixtures of carboxylic acids using a single enantiomer of an amine to yield diastereomeric ammonium salts. To understand how this method of resolution works, let's see what happens when a racemic mixture of chiral acids, such as (+)- and (−)-lactic acids, reacts with an achiral amine base, such as methylamine. Stereochemically, the situation is analogous to what happens when left and right hands (chiral) pick up a ball (achiral). Both left and right hands pick up the ball equally well, and the products—ball in right hand versus ball in left hand—are mirror images. In the same way, both (+)- and (−)-lactic acid react with methylamine equally well, and the product is a racemic mixture of methylammonium (+)-lactate and methylammonium (−)-lactate (Figure 6.12).

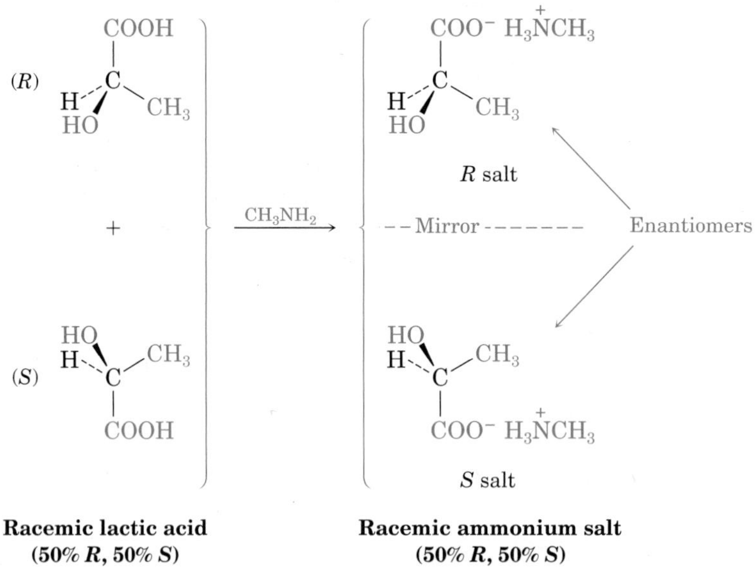

FIGURE 6.12 Reaction of racemic lactic acid with achiral methylamine leads to a racemic mixture of ammonium salts.

(*R*) COOH, H, HO, C, CH_3

+

(*S*) HO, H, C, CH_3, COOH

$\xrightarrow{CH_3NH_2}$

$COO^-\ H_3\overset{+}{N}CH_3$, H, HO, C, CH_3

R salt

Mirror — Enantiomers

HO, H, C, CH_3, $COO^-\ H_3\overset{+}{N}CH_3$

S salt

Racemic lactic acid (50% *R*, 50% *S*)

Racemic ammonium salt (50% *R*, 50% *S*)

Now let's see what happens when the racemic mixture of (+)- and (−)-lactic acids reacts with a *single* enantiomer of a *chiral* amine base, such as (*R*)-1-phenylethanamine. Stereochemically, this situation is analogous to what happens when a hand (a chiral reagent) puts on a glove (*also a chiral reagent*). *Left and right hands don't put on the same glove in the same way.*

The products—right hand in right glove versus left hand in right glove—are not mirror images; they're altogether different.

In the same way, (+)- and (−)-lactic acid react with (*R*)-1-phenylethanamine to give two different products (Figure 6.13). (*R*)-Lactic acid reacts with (*R*)-1-phenylethanamine to give the *R*,*R* salt, whereas (*S*)-lactic acid reacts with the same *R* amine to give the *S*,*R* salt. *The two salts are diastereomers* (see Section 6.7). They are different compounds and have different chemical and physical properties. It therefore may be possible to separate them by crystallization or some other means. Once separated, acidification of the two diastereomeric salts with HCl then allows us to isolate the two pure enantiomers of lactic acid and to recover the chiral amine for further use.

FIGURE 6.13 Reaction of racemic lactic acid with optically pure (*R*)-1-phenylethanamine leads to a mixture of diastereomeric salts, which have different properties and can, in principle, be separated.

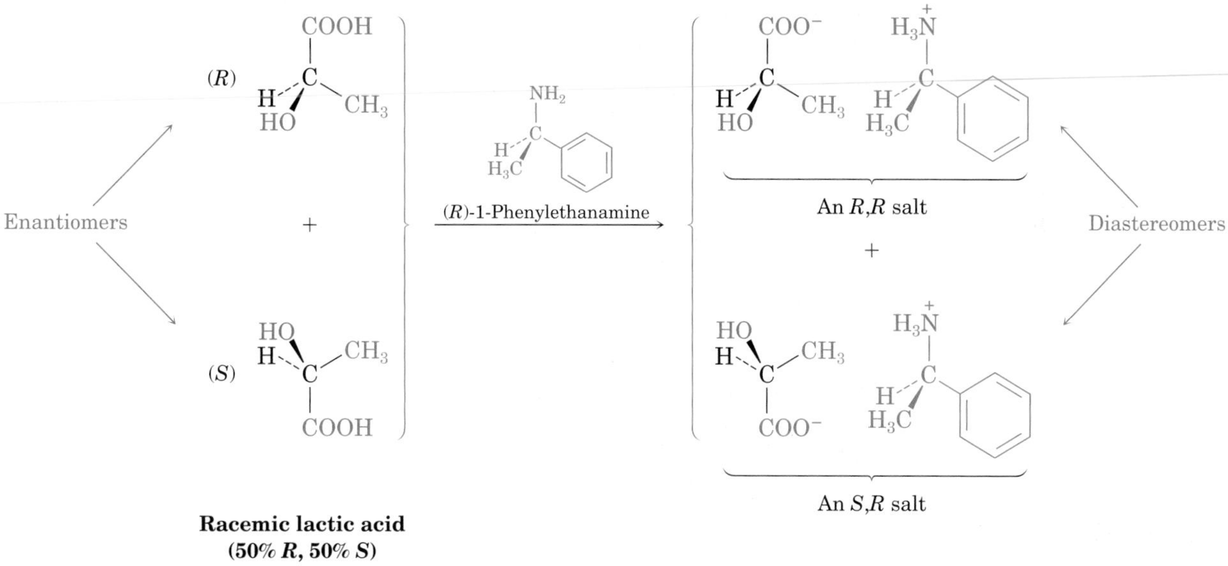

Enantiomers

(*R*) COOH, H, HO, C, CH_3

+

(*S*) HO, H, C, CH_3, COOH

Racemic lactic acid
(50% *R*, 50% *S*)

NH_2, H, H_3C, C

(*R*)-1-Phenylethanamine →

COO^-, H, HO, C, CH_3 $H_3\overset{+}{N}$, H, H_3C, C

An *R*,*R* salt

+

HO, H, C, CH_3, COO^- $H_3\overset{+}{N}$, H, H_3C, C

An *S*,*R* salt

Diastereomers

PRACTICE PROBLEM 6.8

We'll see in Section 10.8 that carboxylic acids (RCO_2H) react with alcohols (R′OH) to form esters (RCO_2R'). Suppose that racemic lactic acid reacts with CH_3OH to form the ester, methyl lactate. What stereochemistry would you expect the product(s) to have? What is the relationship of the products?

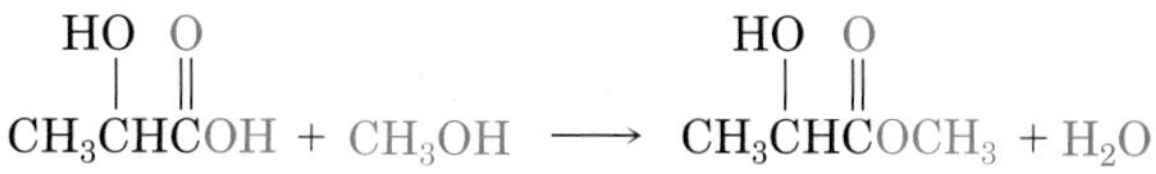

$$\underset{\textbf{Lactic acid}}{CH_3\overset{HO}{\overset{|}{C}}H\overset{O}{\overset{\|}{C}}OH} + \underset{\textbf{Methanol}}{CH_3OH} \longrightarrow \underset{\textbf{Methyl lactate}}{CH_3\overset{HO}{\overset{|}{C}}H\overset{O}{\overset{\|}{C}}OCH_3} + H_2O$$

SOLUTION Reaction of a racemic acid with an achiral alcohol such as methanol yields a racemic mixture of mirror-image (enantiomeric) products:

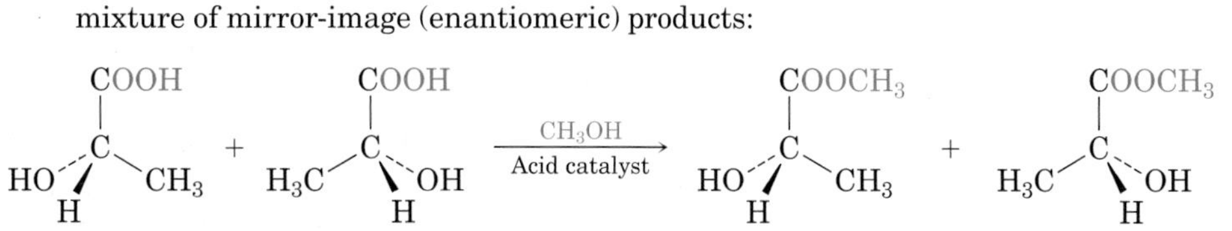

COOH, HO, C, CH_3, H   +   COOH, H_3C, C, OH, H $\xrightarrow[\text{Acid catalyst}]{CH_3OH}$ $COOCH_3$, HO, C, CH_3, H   + $COOCH_3$, H_3C, C, OH, H

(*S*)-Lactic acid   **(*R*)-Lactic acid**   **Methyl (*S*)-lactate**   **Methyl (*R*)-lactate**

PROBLEM 6.19 Suppose that acetic acid (CH_3CO_2H) reacts with (*S*)-butan-2-ol to form an ester (see Practice Problem 6.8). What stereochemistry would you expect the product(s) to have? What is the relationship of the products?

$$CH_3\overset{O}{\overset{\|}{C}}OH + HO\overset{CH_3}{\overset{|}{C}}HCH_2CH_3 \longrightarrow CH_3\overset{O}{\overset{\|}{C}}O\overset{CH_3}{\overset{|}{C}}HCH_2CH_3 + H_2O$$

Acetic acid **Butan-2-ol** ***sec*-Butyl acetate**

6.11 A Brief Review of Isomerism

As noted on several previous occasions, isomers are compounds that have the same chemical formula but different structures. We've seen several kinds of isomers in the past few chapters, and it's a good idea at this point to see how they relate to one another (Figure 6.14).

FIGURE 6.14 A summary of the different kinds of isomers.

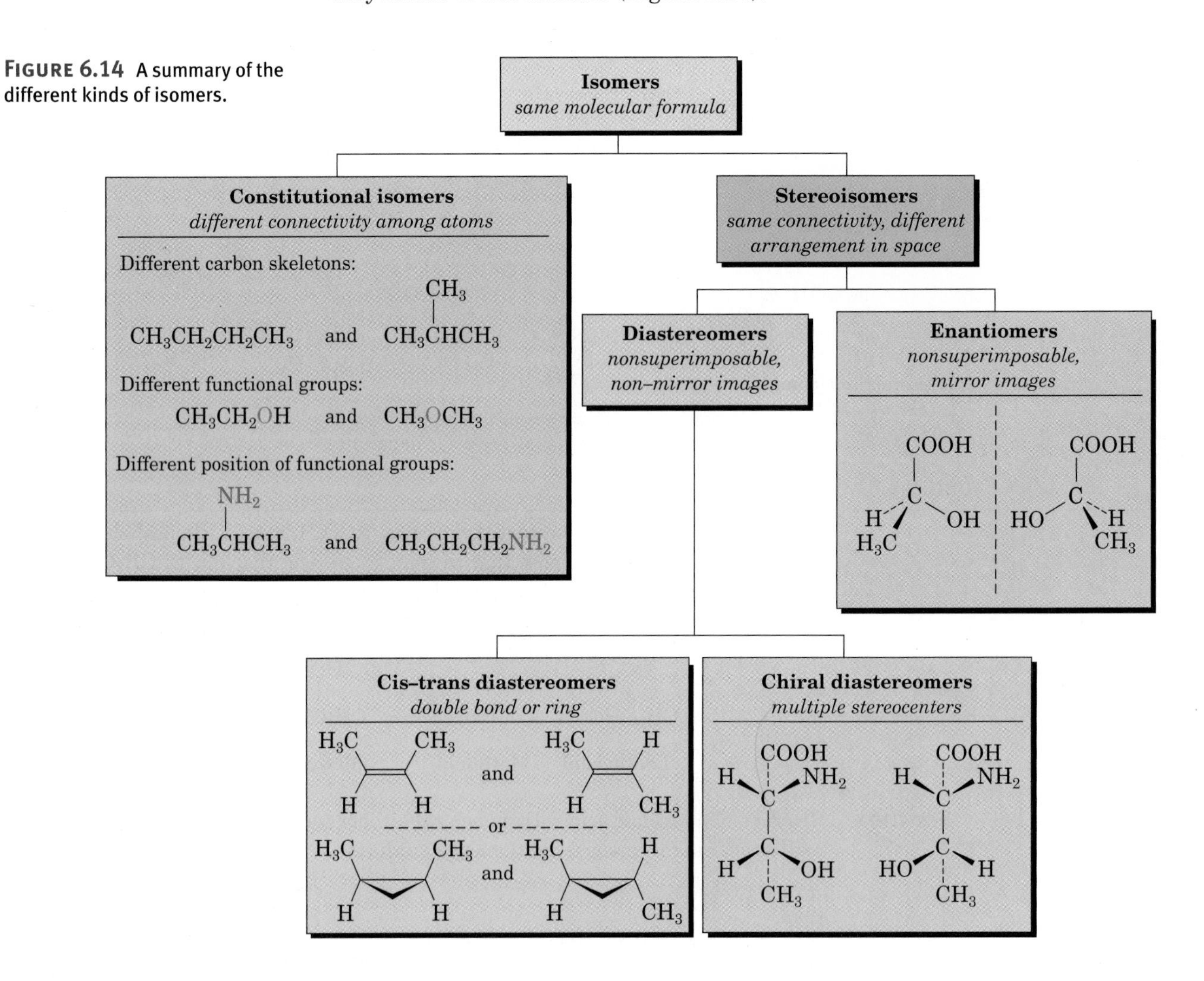

There are two fundamental types of isomerism, both of which we've now encountered: constitutional isomerism and stereoisomerism.

Constitutional isomers (see Section 2.2) are compounds whose atoms are connected differently. Among the kinds of constitutional isomers we've seen are skeletal, functional, and positional isomers.

Stereoisomers (see Section 2.8) are compounds whose atoms are connected in the same way but with a different geometry. Among the kinds of stereoisomers we've seen are enantiomers, diastereomers, and cis–trans isomers (both in alkenes and in cycloalkanes). Actually, though, cis–trans isomers are just a special kind of diastereomers because they are non–mirror-image stereoisomers.

PROBLEM 6.20 What kinds of isomers are the following pairs?

(a) (*S*)-5-Chlorohex-2-ene and chlorocyclohexane

(b) (2*R*,3*R*)-Dibromopentane and (2*S*,3*R*)-dibromopentane

6.12 Chirality in Nature

Just as different stereoisomeric forms of a chiral molecule have different physical properties, they usually have different biological properties as well. For example, the (+) enantiomer of limonene has the odor of oranges, but the (−) enantiomer has the odor of lemons.

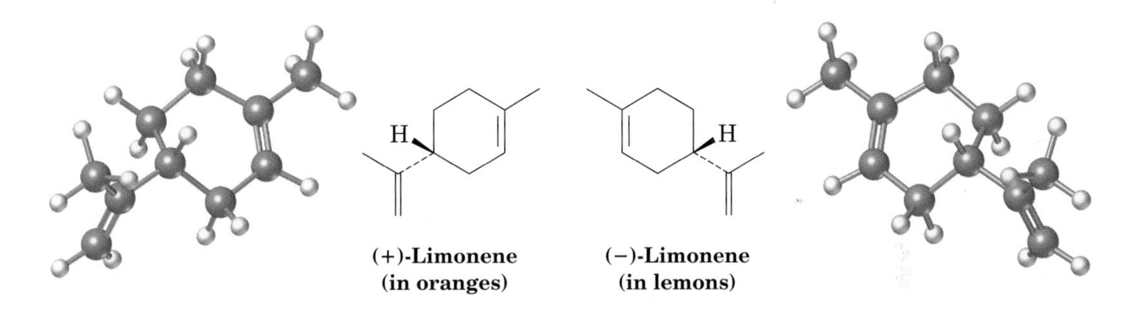

(+)-Limonene (in oranges) **(−)-Limonene (in lemons)**

More dramatic examples of how a change in chirality can affect the biological properties of a molecule are found in many drugs, such as fluoxetine, a commonly prescribed medication sold under the trade name Prozac. Racemic fluoxetine is an extraordinarily effective antidepressant, but it has no activity against migraine. The pure *S* enantiomer, however, works remarkably well in preventing migraine and is now undergoing clinical evaluation. The *Interlude* "Chiral Drugs" at the end of this chapter gives other examples.

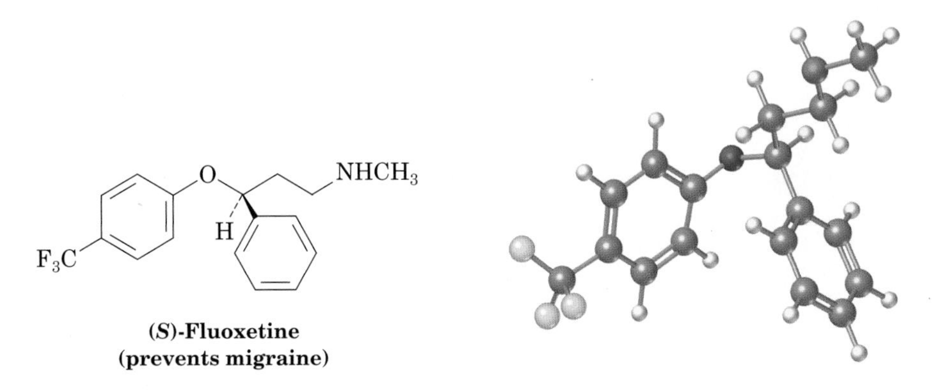

(*S*)-Fluoxetine (prevents migraine)

Why do different stereoisomers have different biological properties? To exert its biological action, a chiral molecule must fit into a chiral receptor at a target site, much as a hand fits into a glove. But just as a right hand can fit only into a right-hand glove, so a particular stereoisomer can fit only into a receptor having the proper complementary shape. Any other stereoisomer will be a misfit, like a right hand in a left-handed glove. A schematic representation of the interaction between a chiral molecule and a chiral biological receptor is shown in Figure 6.15. One enantiomer fits the receptor perfectly, but the other does not.

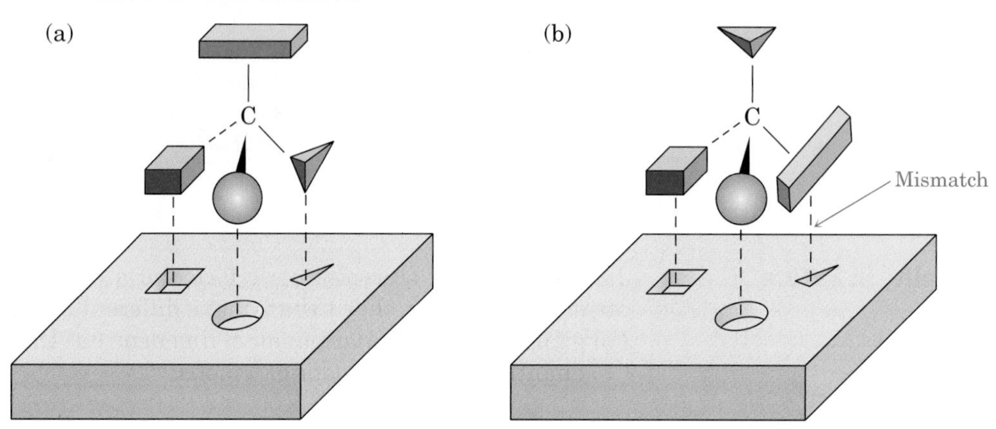

FIGURE 6.15 (a) One enantiomer fits easily into a chiral receptor site to exert its biological effect, but (b) the other enantiomer can't fit into the same receptor.

INTERLUDE

Chiral Drugs

The hundreds of different pharmaceutical agents approved for use by the U.S. Food and Drug Administration come from many sources. Some are isolated directly from plants or bacteria, others are made by chemical modification of naturally occurring compounds, and still others are made entirely in the laboratory and have no relatives in nature.

Those drugs that come from natural sources, either directly or after chemical modification, are usually chiral and are generally found only as a single enantiomer rather than as a racemic mixture. Penicillin V, for example, an antibiotic isolated from the *Penicillium* mold, has a 2*S*,5*R*,6*R* configuration. Its enantiomer, which does not occur naturally but can be made in the laboratory, has essentially no biological activity.

The *S* enantiomer of ibuprofen soothes the aches and pains of athletic injuries. The *R* enantiomer has no effect.

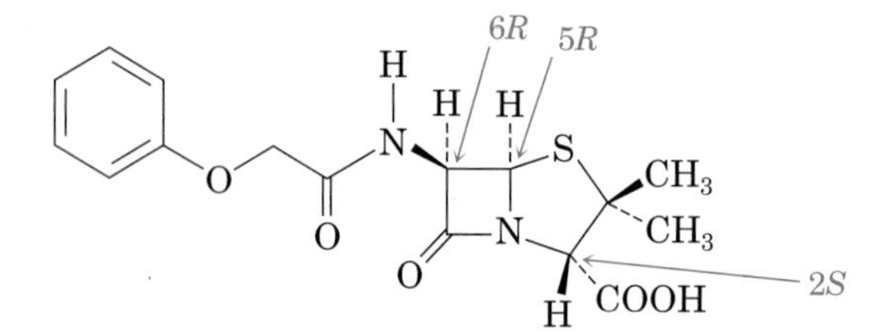

Penicillin V (2*S*,5*R*,6*R* configuration)

In contrast to drugs from natural sources, drugs that are made entirely in the laboratory are either achiral or, if chiral, are generally produced and sold as racemic mixtures. Ibuprofen, for example, contains one stereocenter, but only the

Continued

S enantiomer is an analgesic/anti-inflammatory agent useful in treating aches and pains. Even though the *R* enantiomer of ibuprofen is inactive, the substance marketed under such trade names as Advil, Nuprin, and Motrin is a racemic mixture of *R* and *S*.

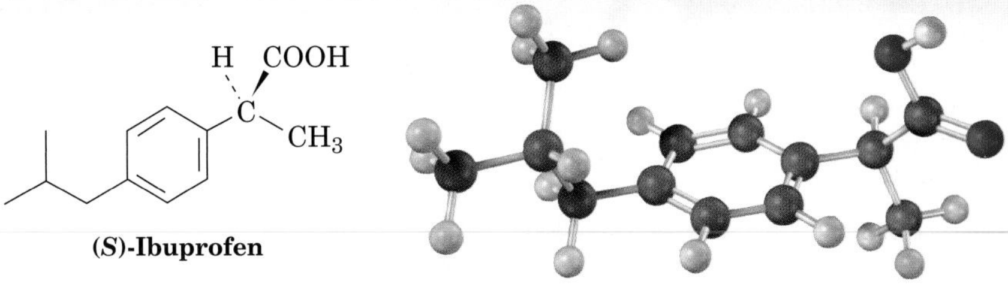

H COOH
C
CH_3

(*S*)-Ibuprofen

Not only is it wasteful to synthesize and administer a physiologically inactive enantiomer, many examples are now known where the presence of the "wrong" enantiomer in a racemic mixture either affects the body's ability to utilize the "right" enantiomer or has unintended effects of its own. The presence of (*R*)-ibuprofen in the racemic mixture, for instance, seems to slow substantially the rate at which the *S* enantiomer takes effect.

To get around this problem, pharmaceutical companies are rapidly devising methods of so-called enantioselective synthesis, which allows them to prepare only a single enantiomer rather than a racemic mixture. Viable methods have already been developed for the preparation of (*S*)-ibuprofen, and the time may not be far off when television ads show famous athletes talking about the advantages of chiral drugs.

Summary and Key Words

Achiral, p. 174
Chiral, p. 171
Chiral environment, p. 171
Configuration, p. 179
Dextrorotatory, p. 177
Diastereomers, p. 184
Enantiomers, p. 172
Levorotatory, p. 177
Meso compound, p. 185
Optical activity, p. 177
R configuration, p. 180
Racemic mixture (racemate), p. 179
Resolution, p. 188
S configuration, p. 180
Specific rotation, $[\alpha]_D$, p. 178
Stereocenter, p. 173
Stereochemistry, p. 171

A molecule that is not identical to its mirror image is said to be **chiral**, meaning "handed." A chiral molecule is one that does not contain a **plane of symmetry**. The usual cause of chirality is the presence of a tetrahedral carbon atom bonded to four different groups—a so-called **stereocenter** or chirality center. Chiral compounds can exist as a pair of mirror-image stereoisomers called **enantiomers**, which are related to each other as a right hand is related to a left hand. When a beam of plane-polarized light is passed through a solution of a pure enantiomer, the plane of polarization is rotated and the compound is said to be **optically active**.

The three-dimensional **configuration** of a stereocenter is specified as either ***R*** or ***S***. Sequence rules are used to assign priorities to the four substituents on the chiral carbon, and the molecule is then oriented so that the lowest-priority group points directly away from the viewer. If a curved arrow drawn in the direction of decreasing priority for the remaining three groups is clockwise, the stereocenter has the *R* configuration. If the direction is counterclockwise, the stereocenter has the *S* configuration.

Some molecules have more than one stereocenter. Enantiomers have opposite configurations at all stereocenters, whereas **diastereomers** have the same configuration in at least one center but opposite configurations at the others. **Meso compounds** contain stereocenters but are achiral overall because they contain a plane of symmetry. **Racemates** are 50:50 mixtures of (+) and (−) enantiomers. Racemic mixtures and individual diastereomers differ in both their physical properties and their biological properties. A **chiral environment** acts like an additional stereocenter and can be used to separate enantiomers.

EXERCISES

Visualizing Chemistry

6.21 Which of the following structures are identical? (Red = O, yellow-green = Cl.)

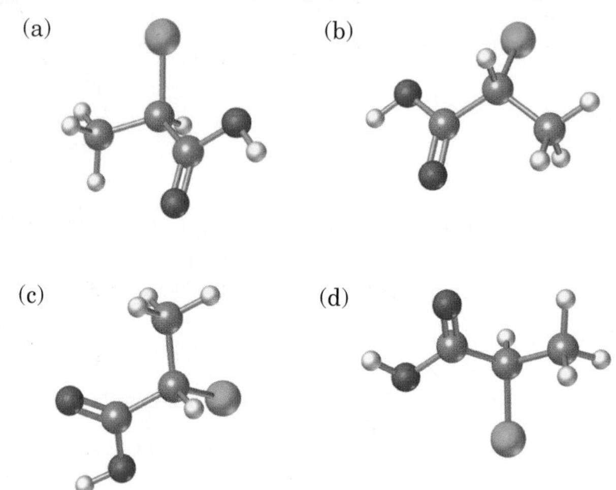

6.22 Assign *R* or *S* configuration to the following molecules (red = O, blue = N):

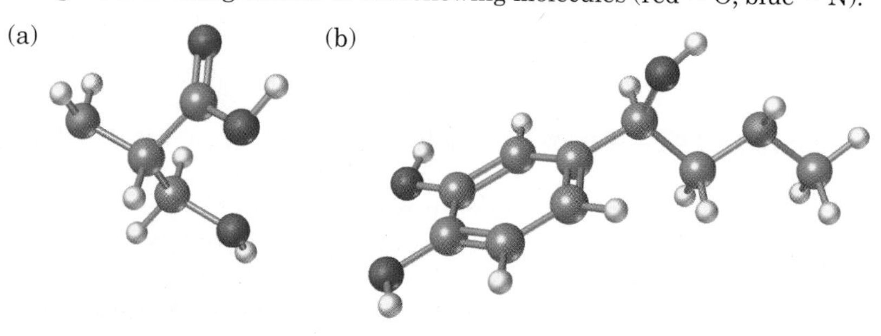

Serine **Adrenaline**

6.23 Which, if any, of the following structures represent meso compounds? (Red = O, blue = N, yellow-green = Cl.)

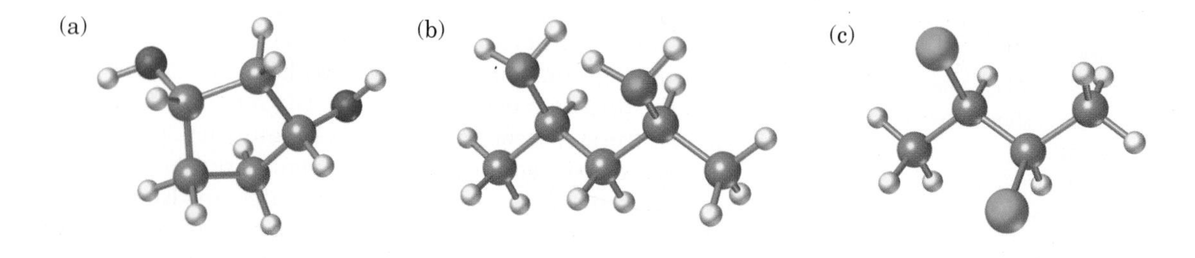

6.24 Assign *R* or *S* configuration to each stereocenter in pseudoephedrine, an over-the-counter decongestant found in cold remedies (red = O, blue = N).

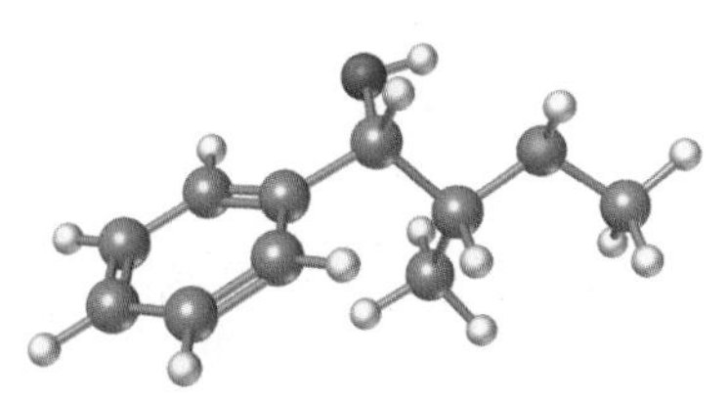

6.25 Orient each of the following drawings so that the lowest-priority group is toward the rear, and then assign *R* or *S* configuration:

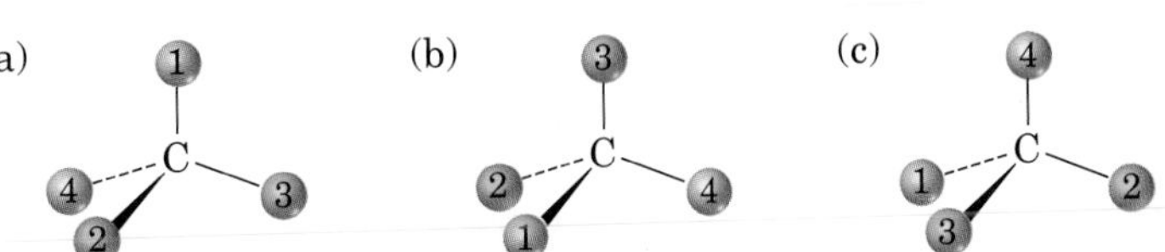

Additional Problems

Identifying Chiral Molecules and Chirality Centers

6.26 Which of the following objects are chiral?
(a) A basketball (b) A wine glass (c) An ear
(d) A snowflake (e) A coin (f) Scissors

6.27 Which of the following compounds are chiral?
(a) 2,4-Dimethylheptane (b) 5-Ethyl-3,3-dimethylheptane
(c) *cis*-1,3-Dimethylcyclohexane

6.28 Penicillin V is a broad-spectrum antibiotic that contains three stereocenters. Identify them with asterisks.

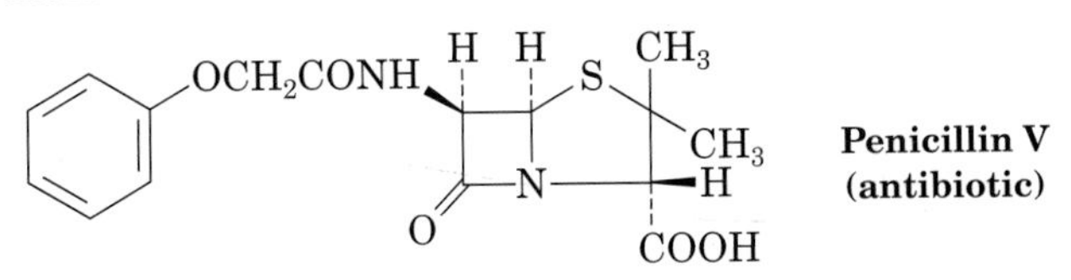

6.29 Draw chiral molecules that meet the following descriptions:
(a) A chloroalkane, $C_5H_{11}Cl$ (b) An alcohol, $C_6H_{14}O$
(c) An alkene, C_6H_{12} (d) An alkane, C_8H_{18}

6.30 Which of the following compounds are chiral? Label all stereocenters.

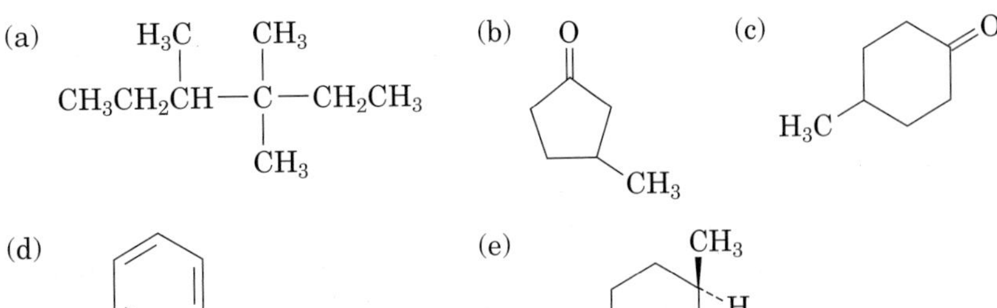

OPTICAL ROTATION

6.31 Cholic acid, the major steroid found in bile, was found to have a rotation of +2.22° when a 3.00 g sample was dissolved in 5.00 mL of alcohol in a sample tube with a 1.00 cm path length. Calculate $[\alpha]_D$ for cholic acid.

6.32 Polarimeters are so sensitive that they can measure rotations to the thousandth of a degree, an important advantage when only small amounts of a sample are available. For example, when 7.00 mg of ecdysone, an insect hormone that controls molting in the silkworm moth, was dissolved in 1.00 mL of chloroform in a cell with a 2.00 cm path length, an observed rotation of +0.087° was found. Calculate $[\alpha]_D$ for ecdysone.

6.33 Naturally occurring (*S*)-serine has $[\alpha]_D = -6.83°$. What specific rotation do you expect for (*R*)-serine?

DRAWING STEREOISOMERS

6.34 There are eight alcohols with the formula $C_5H_{12}O$. Draw them, and tell which are chiral.

6.35 Propose structures for compounds that meet the following descriptions:
(a) A chiral alcohol with four carbons
(b) A chiral carboxylic acid
(c) A compound with two stereocenters

PRIORITY RULES FOR THE *R,S* SYSTEM AND ASSIGNMENT OF STEREOCHEMISTRY

6.36 Assign priorities to the substituents in each of the following sets:
(a) $-H, -OH, -OCH_3, -CH_3$
(b) $-Br, -CH_3, -CH_2Br, -Cl$
(c) $-CH{=}CH_2, -CH(CH_3)_2, -C(CH_3)_3, -CH_2CH_3$
(d) $-CO_2CH_3, -COCH_3, -CH_2OCH_3, -OCH_3$

6.37 Assign priorities to the substituents in each of the following sets:

(a) 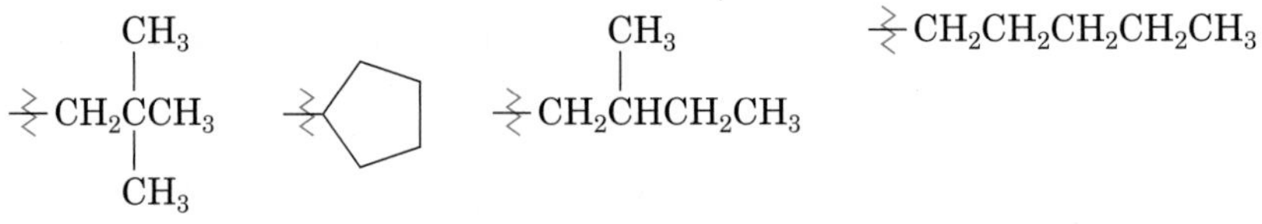 $-CH_2C(CH_3)_2CH_3$ cyclopentyl $-CH_2CH(CH_3)CH_2CH_3$ $-CH_2CH_2CH_2CH_2CH_3$

(b) $-SH$ $-NH_2$ $-SO_3H$ $-OCH_2CH_2OH$

6.38 One enantiomer of lactic acid is shown below. Is it *R* or *S*? Draw its mirror image in the standard tetrahedral representation.

COOH, H, OH, H_3C on C

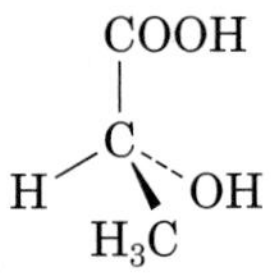

6.39 Draw tetrahedral representations of both enantiomers of the amino acid serine. Tell which of your structures is *S* and which is *R*.

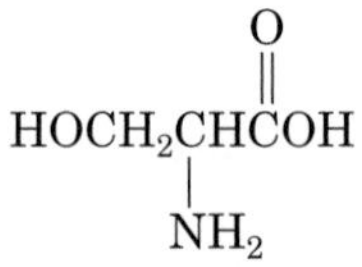

$HOCH_2CH(NH_2)COOH$ **Serine**

6.40 Assign *R* or *S* configuration to the stereocenters in the following molecules:

(a) 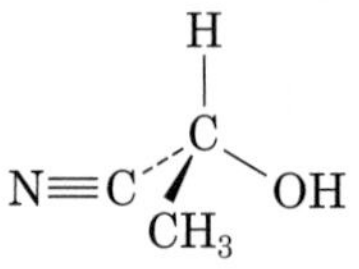 H, $N{\equiv}C$, OH, CH_3 on C   (b) OCH_2CH_3, H, CH_3, Cl on C   (c) OH, H_3C, CH_2OH, H on C

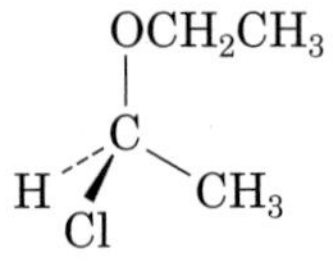

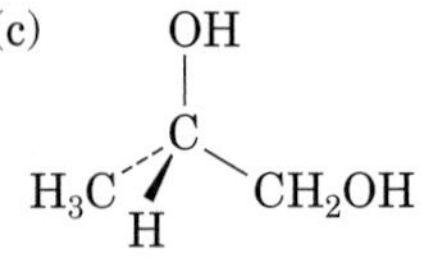

6.41 Assign *R* or *S* configuration to the stereocenters in the following molecules:

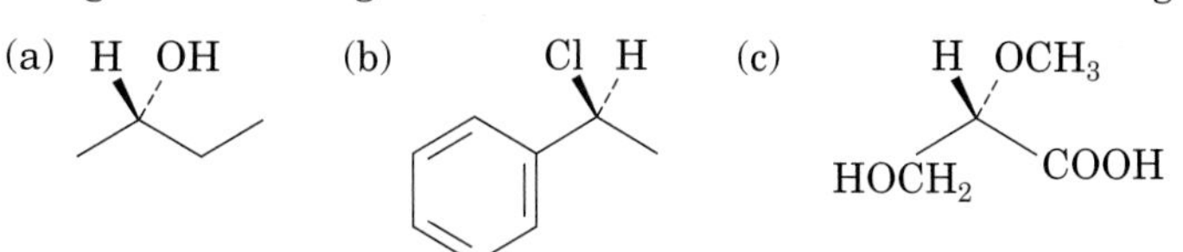

STEREOCHEMICAL RELATIONSHIPS

6.42 What is the relationship between the specific rotations of (2*R*,3*R*)-dihydroxypentane and (2*S*,3*S*)-dihydroxypentane? Between (2*R*,3*S*)-dihydroxypentane and (2*R*,3*R*)-dihydroxypentane?

6.43 What is the stereochemical configuration of the enantiomer of (2*S*,4*R*)-dibromooctane?

6.44 What are the stereochemical configurations of the two diastereomers of (2*S*,4*R*)-dibromooctane?

6.45 Draw examples of the following:
(a) A meso compound with the formula C_8H_{18}
(b) A compound with two stereocenters, one *R* and the other *S*

6.46 Tell whether the following Newman projection of 2-chlorobutane is *R* or *S*. (You might want to review Section 2.5.)

Cl
H　CH_3
H_3C　H
H

6.47 Draw a Newman projection that is enantiomeric with the one shown in Problem 6.46.

6.48 Draw a Newman projection of *meso*-tartaric acid.

6.49 Draw Newman projections of (2*R*,3*R*)- and (2*S*,3*S*)-tartaric acid, and compare them to the projection you drew in Problem 6.48 for the meso form.

INTEGRATED PROBLEMS

6.50 β-Glucose has the following structure. Identify the stereocenters in β-glucose, and tell how many stereoisomers of glucose are possible.

$HOCH_2$
HO　O
HO　OH
OH

β-Glucose

6.51 Draw a tetrahedral representation of (*R*)-3-chloropent-1-ene.

6.52 Draw all the stereoisomers of 1,2-dimethylcyclopentane. Assign *R*,*S* configurations to the stereocenters in all isomers, and indicate which stereoisomers are chiral and which, if any, are meso.

6.53 Draw the meso form of each of the following molecules, and indicate the plane of symmetry in each:

(a) $CH_3CH(OH)CH_2CH_2CH(OH)CH_3$

(b) CH_3 / CH_3 (dimethylcyclohexane)

(c) H_3C / H_3C / OH (cyclopentane)

6.54 Assign *R* or *S* configuration to each stereocenter in the following molecules:

(a) COOH, H, NH_2, CH_3 (b) HO, H, H_2N, CH_3, H

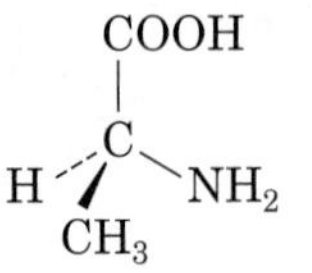

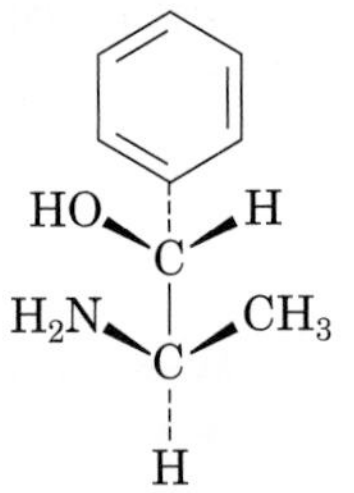

6.55 How many stereoisomers of 2,4-dibromo-3-chloropentane are there? Draw them, and indicate which are optically active.

6.56 Alkenes undergo reaction with peroxycarboxylic acids (RCO_3H) to give compounds called *epoxides*. For example, *cis*-but-2-ene gives 2,3-epoxybutane:

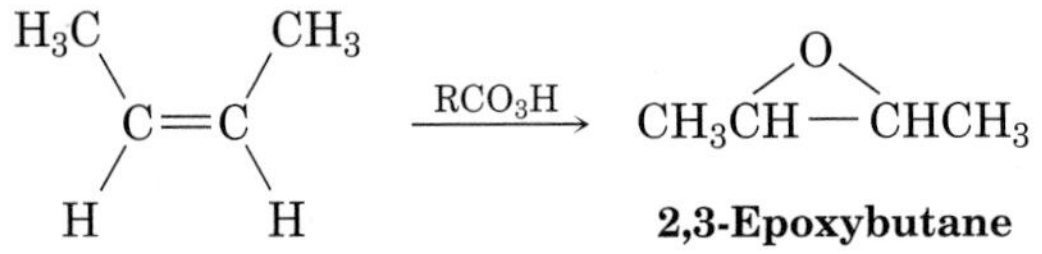

H_3C, CH_3, H, H (C=C) $\xrightarrow{RCO_3H}$ CH_3CH—$CHCH_3$ (O)

2,3-Epoxybutane

Assuming that both C–O bonds form from the same side of the molecule (syn stereochemistry; Section 4.6), show the stereochemistry of the product. Is the epoxide chiral? How many stereocenters does it have? How would you describe the product stereochemically?

6.57 Ribose, an essential part of ribonucleic acid (RNA), has the following structure:

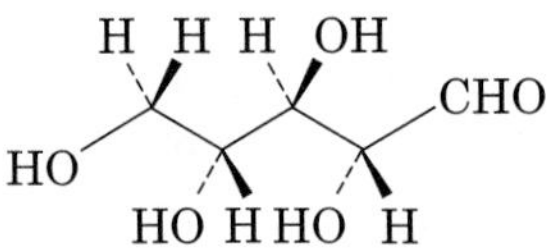

H H H OH, CHO, HO, HO H HO H

Ribose

How many stereocenters does ribose have? Identify them with asterisks. How many stereoisomers of ribose are there?

6.58 Draw the structure of the enantiomer of ribose (see Problem 6.57).

6.59 Draw the structure of a diastereomer of ribose (see Problem 6.57).

6.60 On catalytic hydrogenation over a platinum catalyst, ribose (see Problem 6.57) is converted into ribitol. Is ribitol optically active or inactive? Explain.

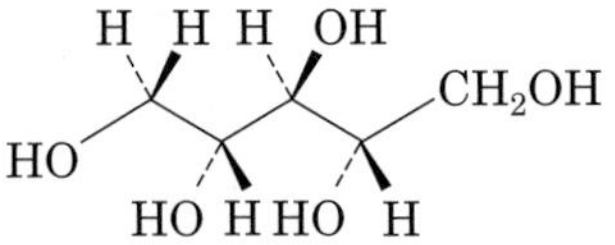

H H H OH, CH_2OH, HO, HO H HO H

Ribitol

6.61 Draw the two enantiomers of the amino acid cysteine, $HSCH_2CH(NH_2)CO_2H$, and identify each as *R* or *S*.

6.62 Draw the structure of (*R*)-2-methylcyclohexanone.

6.63 Compound A, C_7H_{14}, is optically active. On catalytic reduction over a palladium catalyst, 1 equivalent of H_2 is absorbed, yielding compound B, C_7H_{16}. On cleavage of A with acidic $KMnO_4$, two fragments are obtained. One fragment can be identified as acetic acid, CH_3CO_2H, and the other fragment, C, is an optically active carboxylic acid. Formulate the reactions, and propose structures for A, B, and C.

6.64 *Allenes* are compounds with adjacent C–C double bonds. Even though they don't contain stereocenters, many allenes are chiral. For example, mycomycin, an antibiotic isolated from the bacterium *Nocardia acidophilus*, is chiral and has $[\alpha]_D = -130°$. Can you explain why mycomycin is chiral? Making a molecular model should be helpful.

$$HC{\equiv}C{-}C{\equiv}C{-}CH{=}C{=}CH{-}CH{=}CH{-}CH{=}CH{-}CH_2COOH$$

Mycomycin (an allene)

IN THE MEDICINE CABINET

6.65 Let's look back at thalidomide from the chapter opener.
(a) Identify the stereocenters in these enantiomers of thalidomide.
(b) Assign the stereocenters in each molecule as *R* or *S*.

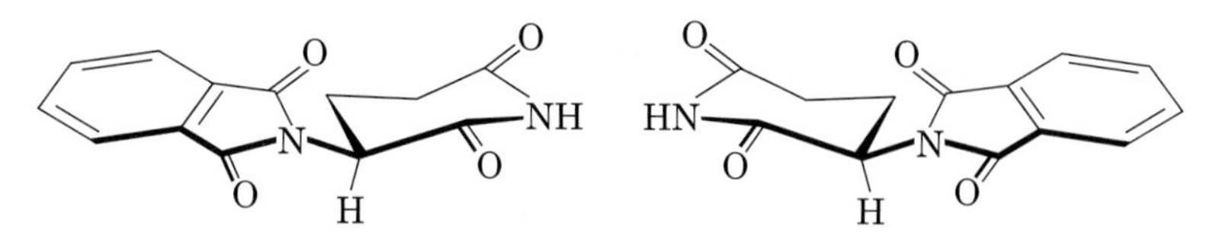

Thalidomide

6.66 The cancer drug Taxol is rich in stereochemistry.

Taxol

(a) Identify all the stereocenters in Taxol.
(b) Calculate the theoretical number of isomers that could exist for Taxol.
(c) Nature produces only one isomer. How?

6.67 In Chapter 3, we described that the hydrogenation of alkenes to produce alkanes can be catalyzed by certain metals. Hydrogenation of alkene A leads to a mixture of two enantiomers of dopamine, a molecule useful for treating Parkinson's disease.

A $\xrightarrow[\text{metal catalyst}]{H_2}$ **A racemate**

(a) Draw the two enantiomers and assign them as *R* and *S*.
(b) Do you expect the enantiomers to have similar physical properties? Why or why not?
(c) Do you expect the enantiomers to perform equally well as drugs? Why or why not?

(d) The Monsanto process, commercialized in 1974, uses a chiral metal catalyst to hydrogenate **A**, producing the following amide in nearly enantiomerically pure form. This amide is then converted to L-dopa. Show a simple reaction coordinate diagram for the formation of both enantiomers of the hydrogenation product in the presence of the chiral catalyst and use this to explain how one enantiomer can be favored.

L-Dopa

In the Field with Agrochemicals

6.68 Metolachlor, a clear, odorless, liquid herbicide, is used in corn, sorghum, and soybean fields. In 1996, it generated sales of $450 million, making it one of the most popular pesticides on the market.

Metolachlor

(a) Identify the stereocenter in Metolachlor.
(b) Draw the R and S isomers.

6.69 Metolachlor kills weeds by preventing the plants from making a waxy coating on the leaves. This wax is produced with the help of a enzyme called fatty acid elongase. When the enzyme activity is inhibited, the wax is not produced and the weed dies. The following graph shows the amount of R- and S-Metolachlor required to interfere with the enzyme fatty acid elongase.

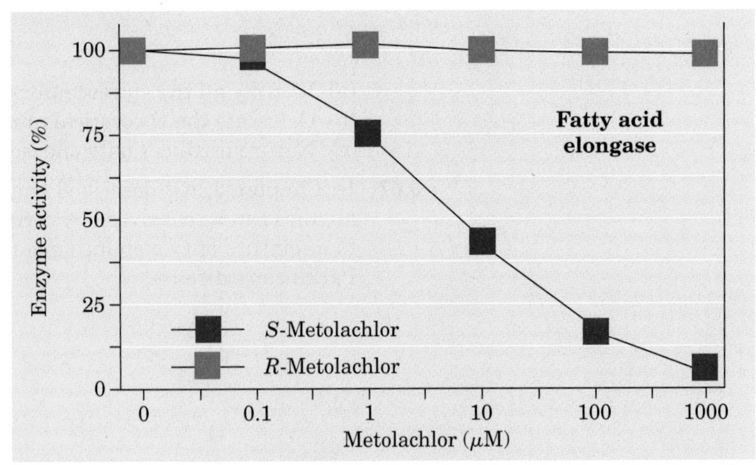

Reprinted from *Phytochemistry*, Vol. 64, pp. 1045–1554, © 2003, with permission from Elsevier.

(a) If you wanted to kill weeds, which enantiomer would you use?
(b) Why do these enantiomers have different activities?

6.70 Both the R- and S-enantiomers of Metolachlor are degraded in the environment, producing molecules that kill fish. Metolachlor is currently sold as a racemate. As an environmentalist, what legislation would you promote to reduce the levels of toxic metabolites by 50%?

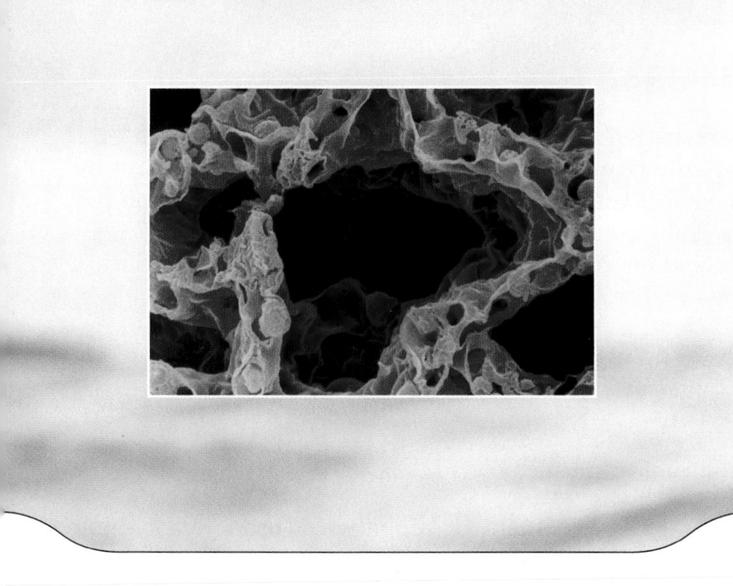

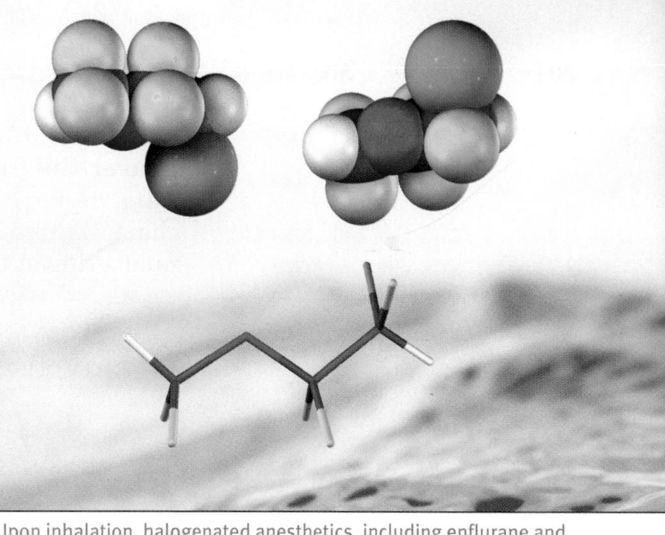

Upon inhalation, halogenated anesthetics, including enflurane and isoflurane, enter the bloodstream through the lungs. (David M. Phillips/Photo Researchers, Inc.)

CHAPTER 7 Alkyl Halides

Halogen-substituted organic compounds are widespread throughout nature and have a vast array of uses in modern industrial processes. Several thousand organohalides have been found in algae and various other marine organisms. Chloromethane, for instance, is released in large amounts by oceanic kelp, as well as by forest fires and volcanoes. Among their many uses, organohalides are valuable as industrial solvents, inhaled anesthetics in medicine, refrigerants, and pesticides. The modern electronics industry, for example, relies on halogenated solvents such as trichloroethylene for cleaning semiconductor chips and other components.

$Cl(H)C{=}CCl_2$	$CF_3{-}CHBrCl$	CCl_2F_2	CH_3Br
Trichloroethylene (a solvent)	**Halothane (an inhaled anesthetic)**	**Dichlorodifluoromethane (a refrigerant)**	**Bromomethane (a fumigant)**

7.1 Naming Alkyl Halides

Alkyl halides are named in the same way as alkanes (see Section 2.3), by considering the halogen as a substituent on the parent alkane chain. There are three steps:

STEP 1 **Find the longest chain, and name it as the parent.** If a multiple bond is present, the parent chain must contain it.

STEP 2 **Number the carbons of the parent chain beginning at the end nearer the first substituent, regardless of whether it is alkyl or halo.** Assign each substituent a number according to its position on the chain. If there are substituents the same distance from both ends, begin numbering at the end nearer the substituent with alphabetical priority.

CH_3 on C2, Br on C5, CH_3 on C4: $CH_3CHCH_2CHCHCH_2CH_3$ (numbered 1 2 3 4 5 6 7)

5-Bromo-2,4-dimethylheptane

Br on C2, CH_3 on C5, CH_3 on C4: $CH_3CHCH_2CHCHCH_2CH_3$ (numbered 1 2 3 4 5 6 7)

2-Bromo-4,5-dimethylheptane

STEP 3 **Write the name.** List all substituents in alphabetical order and use one of the prefixes *di-*, *tri-*, and so forth, if more than one of the same substituent is present.

Cl on C2, Cl on C3, CH_3 on C4: $CH_3CHCHCHCH_2CH_3$ (numbered 1 2 3 4 5 6)

2,3-Dichloro-4-methylhexane

In addition to their systematic names, many simple alkyl halides are also named by identifying first the alkyl group and then the halogen. For example, CH_3I can be called either iodomethane or methyl iodide.

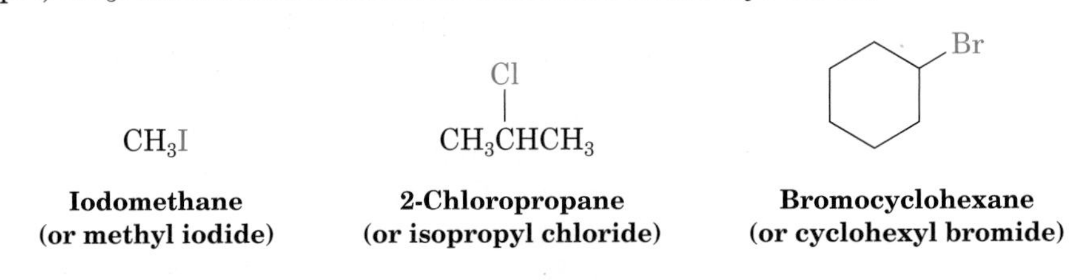

Iodomethane (or methyl iodide) **2-Chloropropane (or isopropyl chloride)** **Bromocyclohexane (or cyclohexyl bromide)**

PROBLEM 7.1 Give the IUPAC names of the following alkyl halides:

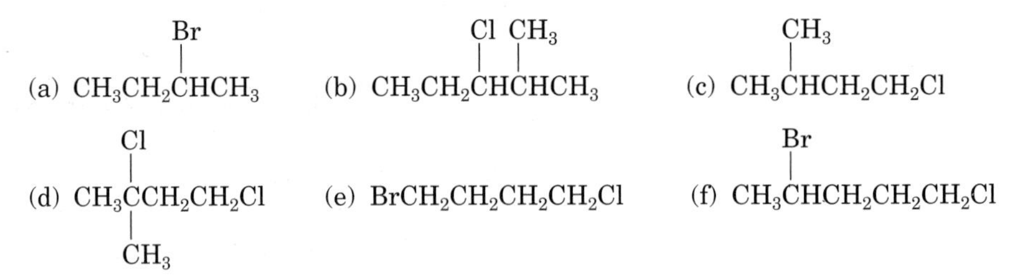

PROBLEM 7.2 Draw structures corresponding to the following names:

(a) 2-Chloro-3,3-dimethylhexane (b) 3,3-Dichloro-2-methylhexane
(c) 3-Bromo-3-ethylpentane (d) 2-Bromo-5-chloro-3-methylhexane

7.2 Preparing Alkyl Halides

We've already seen several methods for preparing alkyl halides, including the addition reactions of HX and X_2 with alkenes (see Sections 4.1 and 4.5) and the reaction of an alkane with Cl_2 (see Section 2.4).

$$\text{C=C} + HCl \longrightarrow \text{H–C–C–Cl}$$

$$\text{C=C} + Br_2 \longrightarrow \text{Br–C–C–Br}$$

$$CH_4 + Cl_2 \xrightarrow{h\nu} CH_3Cl + HCl$$

Methane **Chloromethane**

The most general method for preparing alkyl halides is to make them from alcohols, a reaction carried out simply by treating the alcohol with hydrogen halide, HX. 1-Methylcyclohexanol, for example, is converted into 1-chloro-1-methylcyclohexane by treating with HCl:

$$\text{1-Methylcyclohexanol} \xrightarrow[\text{Ether, 0 °C}]{\text{HCl (gas)}} \text{1-Chloro-1-methylcyclohexane (90\%)} + H_2O$$

1-Methylcyclohexanol **1-Chloro-1-methylcyclohexane (90%)**

For reasons that will be discussed in Section 7.6, the reaction works best with tertiary alcohols. Primary and secondary alcohols react much more slowly.

$$R{-}OH + HX \longrightarrow R{-}X + H_2O$$

$$H_3C{-}OH < RCH_2{-}OH < R_2CH{-}OH < R_3C{-}OH$$

Methyl 1° 2° 3°

Less reactive — Reactivity → More reactive

Primary and secondary alcohols are best converted into alkyl halides by treatment with either thionyl chloride ($SOCl_2$) or phosphorus tribromide (PBr_3). These reactions normally take place in high yield.

$$\text{Cyclopentanol} \xrightarrow{SOCl_2} \text{Chlorocyclopentane} + SO_2 + HCl$$

Cyclopentanol **Chlorocyclopentane**

$$3\ CH_3CH_2CH(OH)CH_3 \xrightarrow[\text{Ether, 35 °C}]{PBr_3} 3\ CH_3CH_2CH(Br)CH_3 + P(OH)_3$$

Butan-2-ol **2-Bromobutane (86%)**

PRACTICE PROBLEM 7.1

Predict the product of the following reaction:

$$C_6H_5\text{-}CH(OH)CH_3 \xrightarrow{SOCl_2}$$

STRATEGY A big part of learning organic chemistry is remembering reactions. Ask yourself what you know about alcohols, and then recall that alcohols yield alkyl chlorides on treatment with $SOCl_2$.

SOLUTION

$$C_6H_5\text{-}CH(OH)CH_3 \xrightarrow{SOCl_2} C_6H_5\text{-}CH(Cl)CH_3$$

PROBLEM 7.3 Alkane chlorination can occur at any position in the alkane chain. Draw and name all monochloro products you might obtain from radical chlorination of 3-methylpentane. Which, if any, are chiral?

PROBLEM 7.4 How would you prepare the following alkyl halides from the appropriate alcohols?

(a) 2-Chloro-2-methylpropane

(b) 2-Bromo-4-methylpentane

(c) $BrCH_2CH_2CH_2CH_2CH(CH_3)CH_3$

(d) $CH_3CH_2CH(CH_3)CH_2C(Cl)(CH_3)CH_3$

PROBLEM 7.5 Predict the products of the following reactions:

(a) $CH_3CH_2CH(OH)CH_2CH(CH_3)CH_3 + PBr_3 \longrightarrow ?$

(b) 1-methylcyclohexanol (cyclohexane with CH_3 and OH on the same carbon) $+ HCl \longrightarrow ?$

(c) 3,3-dimethylcyclopentanol (cyclopentane with two H_3C groups on one carbon and OH) $+ SOCl_2 \longrightarrow ?$

7.3 Reactions of Alkyl Halides: Grignard Reagents

Alkyl halides react with magnesium metal in ether solvent to yield organomagnesium halides, called **Grignard reagents** after their discoverer, Victor Grignard. Grignard reagents contain a carbon–metal bond and are thus *organometallic compounds*.

$$R\text{—}X + Mg \xrightarrow{\text{Ether}} R\text{—}Mg\text{—}X$$

where R = 1°, 2°, or 3° alkyl, aryl, or alkenyl
X = Cl, Br, or I

Chlorobenzene, for instance, reacts rapidly with magnesium metal in ether to give phenylmagnesium chloride.

$$\text{C}_6\text{H}_5\text{Cl} \xrightarrow[\text{Ether}]{\text{Mg}} \text{C}_6\text{H}_5^{\delta-}\text{MgCl}^{\delta+}$$

Chlorobenzene **Phenylmagnesium chloride**

As you might expect from a knowledge of electronegativities (see Section 1.9), a carbon–magnesium bond is strongly polarized, making the organic part both nucleophilic and basic. Note in the electrostatic potential map of phenylmagnesium chloride, for example, how the magnesium atom is electron-poor (positive; blue) while the benzene ring and the chlorine are electron-rich (negative; red).

Because of its electron-rich carbon, a Grignard reagent is both a base and a nucleophile; it therefore reacts with acids and electrophiles. For example, it reacts with HCl, H_2O, or an alcohol ROH by accepting H^+ and yielding a hydrocarbon. The overall sequence, $R{-}X \rightarrow R{-}MgX \rightarrow R{-}H$, is a useful method for converting an organic halide into a hydrocarbon:

$$CH_3CH_2CH_2CH_2CH_2CH_2Br \xrightarrow[\text{Ether}]{\text{Mg}} CH_3CH_2CH_2CH_2CH_2CH_2MgBr$$

1-Bromohexane **1-Hexylmagnesium bromide**

$$\downarrow H_2O$$

$$CH_3CH_2CH_2CH_2CH_2CH_3$$

Hexane (85%)

PRACTICE PROBLEM 7.2

By using several reactions in sequence, you can accomplish transformations that can't be done in a single step. How would you prepare the alkane methylcyclohexane from the alcohol 1-methylcyclohexanol?

$$\text{1-methylcyclohexanol (OH, CH}_3\text{)} \xrightarrow{?} \text{cyclohexyl–CH}_3$$

1-Methylcyclohexanol **Methylcyclohexane**

STRATEGY Working backward, we know that alkanes can be made from alkyl halides and that alkyl halides can be made from alcohols. Carrying out the two reactions sequentially thus converts 1-methylcyclohexanol into methylcyclohexane.

SOLUTION

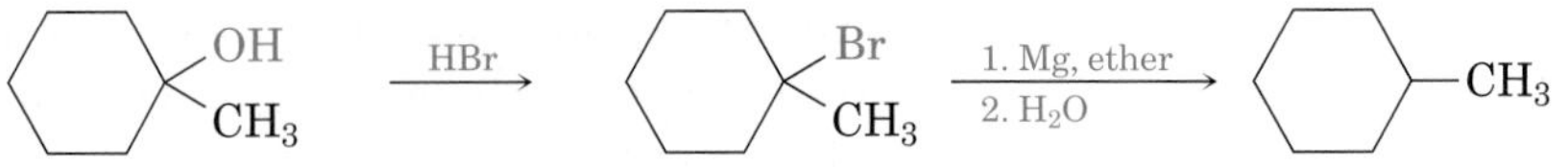

$$\text{1-Methylcyclohexanol (cyclohexane with OH and } CH_3 \text{ on C1)} \xrightarrow{HBr} \text{1-Bromo-1-methylcyclohexane (Br, } CH_3\text{)} \xrightarrow[2.\ H_2O]{1.\ Mg,\ ether} \text{Methylcyclohexane (}-CH_3\text{)}$$

1-Methylcyclohexanol **1-Bromo-1-methylcyclohexane** **Methylcyclohexane**

PROBLEM 7.6 An advantage to preparing an alkane from a Grignard reagent is that deuterium (D; the ^2H isotope of hydrogen) can be placed at a specific site in a molecule. How might you convert 2-bromobutane into 2-deuteriobutane?

$$CH_3\overset{Br}{\overset{|}{C}}HCH_2CH_3 \xrightarrow{?} CH_3\overset{D}{\overset{|}{C}}HCH_2CH_3$$

PROBLEM 7.7 How could you convert 4-methylpentan-1-ol into 2-methylpentane?

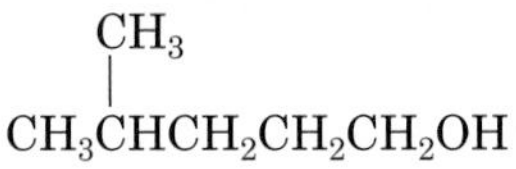

$$CH_3\overset{CH_3}{\overset{|}{C}}HCH_2CH_2CH_2OH$$ **4-Methylpentan-1-ol**

7.4 Nucleophilic Substitution Reactions

In 1896, the German chemist Paul Walden made the remarkable discovery that (+)- and (−)-malic acids could be interconverted by a series of simple substitution reactions. When Walden treated (−)-malic acid with PCl_5, he isolated (+)-chlorosuccinic acid. This, on reaction with wet Ag_2O, gave (+)-malic acid. Similarly, reaction of (+)-malic acid with PCl_5 gave (−)-chlorosuccinic acid, which was converted into (−)-malic acid when treated with wet Ag_2O. The full cycle of reactions reported by Walden is shown in Figure 7.1.

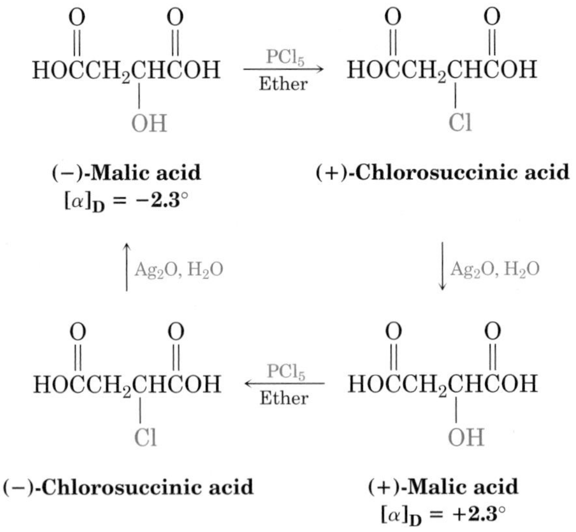

FIGURE 7.1 Walden's cycle of reactions interconverting (+)- and (−)-malic acids.

$$\underset{\textbf{(−)-Malic acid},\ [\alpha]_D = -2.3°}{HO\overset{O}{\overset{\|}{C}}CH_2\underset{OH}{\underset{|}{C}}H\overset{O}{\overset{\|}{C}}OH} \xrightarrow[Ether]{PCl_5} \underset{\textbf{(+)-Chlorosuccinic acid}}{HO\overset{O}{\overset{\|}{C}}CH_2\underset{Cl}{\underset{|}{C}}H\overset{O}{\overset{\|}{C}}OH}$$

$$\uparrow Ag_2O,\ H_2O \qquad\qquad \downarrow Ag_2O,\ H_2O$$

$$\underset{\textbf{(−)-Chlorosuccinic acid}}{HO\overset{O}{\overset{\|}{C}}CH_2\underset{Cl}{\underset{|}{C}}H\overset{O}{\overset{\|}{C}}OH} \xleftarrow[Ether]{PCl_5} \underset{\textbf{(+)-Malic acid},\ [\alpha]_D = +2.3°}{HO\overset{O}{\overset{\|}{C}}CH_2\underset{OH}{\underset{|}{C}}H\overset{O}{\overset{\|}{C}}OH}$$

At the time, the results were astonishing. Since (−)-malic acid was converted into (+)-malic acid, *some reactions in the cycle must have occurred with an inversion, or change, in the configuration of the stereocenter.* But which ones, and how? (Remember from Section 6.6 that you can't tell the configuration of a stereocenter by looking at the sign of optical rotation.)

Today we refer to the transformations taking place in Walden's cycle as **nucleophilic substitution reactions** because each step involves the substitution of one nucleophile (chloride ion, Cl^-, or hydroxide ion, OH^-) by another. Nucleophilic substitution reactions are one of the most common and versatile reaction types in organic chemistry.

Following the work of Walden, a series of investigations was undertaken during the 1920s and 1930s to clarify the mechanism of nucleophilic substitution reactions and to find out how inversions of configuration occur. We now know that nucleophilic substitutions occur by two major pathways, named the *S_N1 reaction* and the *S_N2 reaction*. In both cases, the "S_N" part of the name stands for *substitution, nucleophilic*. The meanings of the *1* and the *2* are discussed in the next two sections.

Regardless of mechanism, the overall change during all nucleophilic substitution reactions is the same: a *nucleophile* (symbolized **Nu:** or **Nu:$^-$**) reacts with a *substrate* R—X and substitutes for a *leaving group* **X:$^-$** to yield the product R—Nu. If the nucleophile is neutral (Nu:), then the product is positively charged to maintain charge conservation; if the nucleophile is negatively charged (Nu:$^-$), the product is neutral.

Neutral Nu: $\quad$ Nu: + R—X ⟶ R—Nu$^+$ + X:$^-$

Negatively charged Nu:$^-$ $\quad$ Nu:$^-$ + R—X ⟶ R—Nu + X:$^-$

A vast number of substances can be prepared using nucleophilic substitution reactions. In fact, we've already seen examples in previous chapters. The reaction of an acetylide anion with an alkyl halide (see Section 4.12), for instance, is an S_N2 reaction in which the acetylide nucleophile replaces halide. Table 7.1 lists other examples.

$$R—C{\equiv}C:^- + CH_3Br \xrightarrow[\text{reaction}]{S_N2} R—C{\equiv}C—CH_3 + Br:^-$$

An acetylide anion (nucleophile)

TABLE 7.1 Some Common Nucleophiles and the Products of Nucleophilic Substitution Reactions with Bromomethane

Common Nucleophiles			And Reaction Products with CH_3Br		
Formula	**Name**		**Formula**	**Name**	
H:$^-$	Hydride	+ CH_3Br ⟶	CH_4	Methane	+ Br^-
$CH_3S:^-$	Methanethiolate		CH_3SCH_3	Dimethyl sulfide	
HS:$^-$	Hydrosulfide		$HSCH_3$	Methanethiol	
N≡C:$^-$	Cyanide		N≡CCH_3	Acetonitrile	
:I:$^-$	Iodide		ICH_3	Iodomethane	
HO:$^-$	Hydroxide		$HOCH_3$	Methanol	
$CH_3O:^-$	Methoxide		CH_3OCH_3	Dimethyl ether	
$^-$:N=$\overset{+}{N}$=N:$^-$	Azide		N_3CH_3	Azidomethane	
:Cl:$^-$	Chloride		$ClCH_3$	Chloromethane	
$CH_3CO_2:^-$	Acetate		$CH_3CO_2CH_3$	Methyl acetate	
H_3N:	Ammonia		$H_3\overset{+}{N}CH_3Br^-$	Methylammonium bromide	
$(CH_3)_3N$:	Trimethylamine		$(CH_3)\overset{+}{N}CH_3Br^-$	Tetramethylammonium bromide	

PRACTICE PROBLEM 7.3

What is the substitution product from reaction of 1-chloropropane with NaOH?

STRATEGY Write the two reactants, and identify the nucleophile (in this instance, OH^-) and the leaving group (in this instance, Cl^-). Then replace the –Cl group by –OH and write the complete equation.

SOLUTION

$$\underset{\textbf{1-Chloropropane}}{CH_3CH_2CH_2Cl} + Na^+ \ ^-OH \longrightarrow \underset{\textbf{Propan-1-ol}}{CH_3CH_2CH_2OH} + Na^+ \ ^-Cl$$

PRACTICE PROBLEM 7.4

How would you prepare propane-1-thiol, $CH_3CH_2CH_2SH$, using a nucleophilic substitution reaction?

STRATEGY Identify the group in the product that is introduced by nucleophilic substitution. In this case, the product contains an –SH group, so it might be prepared by reaction of ^-SH (hydrosulfide ion) with an alkyl halide such as 1-bromopropane.

SOLUTION

$$\underset{\textbf{1-Bromopropane}}{CH_3CH_2CH_2Br} + Na^+ \ ^-SH \longrightarrow \underset{\textbf{Propane-1-thiol}}{CH_3CH_2CH_2SH} + Na^+ \ ^-Br$$

PROBLEM 7.8 What substitution products would you expect to obtain from the following reactions?

(a) $CH_3CH_2CH(Br)CH_3 + LiI \longrightarrow ?$

(b) $CH_3CH(CH_3)CH_2Cl + HS^- \longrightarrow ?$

(c) C_6H_5–$CH_2Br + NaCN \longrightarrow ?$

PROBLEM 7.9 How might you prepare the following substances by using nucleophilic substitution reactions?

(a) $CH_3CH_2CH_2CH_2OH$ (b) $(CH_3)_2CHCH_2CH_2N_3$

7.5 The S_N2 Reaction

An $\mathbf{S_N2}$ **reaction** takes place in a single step without intermediates when the entering nucleophile attacks the substrate from a direction 180° away from the leaving group. As the nucleophile comes in on one side of the molecule, an electron pair on the nucleophile $Nu:^-$ forces out the leaving group $X:^-$, which departs from the other side of the molecule and takes with it the electron pair from the C–X bond. In the transition state for the reaction, the new Nu–C bond is partially forming at the same time the old C–X bond is partially breaking, and the negative charge is shared by both the incoming nucleophile and the outgoing leaving group. The simultaneous bond-making and bond-breaking events prevent carbon from violating the octet rule. The mechanism is shown in Figure 7.2 for the reaction of OH^- with (*S*)-2-bromobutane.

FIGURE 7.2 MECHANISM: The mechanism of the S_N2 reaction. The reaction takes place in a single step when the incoming nucleophile (OH^-) approaches from a direction 180° away from the leaving group (Br^-), thereby inverting the configuration at carbon.

The nucleophile ^-OH uses its lone-pair electrons to attack the alkyl halide carbon 180° away from the departing halogen. This leads to a transition state with a partially formed C–OH bond and a partially broken C–Br bond.

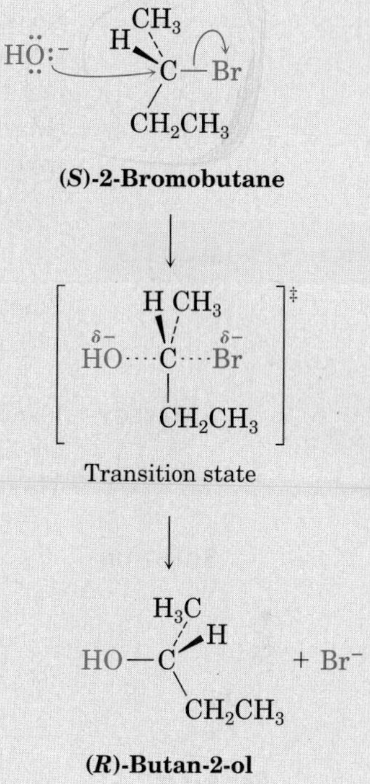

(S)-2-Bromobutane

Transition state

The stereochemistry at carbon is inverted as the C–OH bond forms fully and the bromide ion departs with the electron pair from the former C–Br bond.

(R)-Butan-2-ol

Let's see what evidence there is for this mechanism and what the chemical consequences are.

Rates of S_N2 Reactions

The exact speed at which a reaction occurs is called the *reaction rate* and is a quantity that can be measured. The determination of reaction rates and of how those rates depend on reactant concentrations is a powerful tool for probing mechanisms. As an example, let's look at the effect of reactant concentrations on the rate of the S_N2 reaction of OH^- with CH_3Br to yield CH_3OH:

$$HO:^- + CH_3{-}Br: \longrightarrow HO{-}CH_3 + :Br:^-$$

The S_N2 reaction of CH_3Br with OH^- takes place in a single step when substrate and nucleophile collide and react. At a given concentration of reactants, the reaction takes place at a certain rate. If we double the concentration of OH^-, the frequency of collision between the two reactants doubles and we therefore find that the reaction rate also doubles. Similarly, if we double the concentration of CH_3Br, the reaction rate doubles. Thus, the origin of the "2" in S_N2: S_N2 reactions are said to be **bimolecular** because the rate of the reaction depends on the concentrations of *two* substances—alkyl halide and nucleophile.

PROBLEM 7.10 What effects would the following changes have on the rate of the S_N2 reaction between CH_3I and sodium acetate?

(a) The CH_3I concentration is tripled.

(b) Both CH_3I and CH_3CO_2Na concentrations are doubled.

Stereochemistry of S_N2 Reactions

Look carefully at the mechanism of the S_N2 reaction shown in Figure 7.2. As the incoming nucleophile attacks the substrate and begins pushing out the leaving group on the opposite side, the configuration of the molecule *inverts* (Figure 7.3). (*S*)-2-Bromobutane gives (*R*)-butan-2-ol, for example, by an inversion of configuration that occurs through a planar transition state.

PRACTICE PROBLEM 7.5

What product would you expect to obtain from the S_N2 reaction of (*S*)-2-iodooctane with sodium cyanide, NaCN?

STRATEGY Identify the nucleophile (cyanide ion) and the leaving group (iodide ion). Then carry out the substitution, making sure to invert the configuration at the stereocenter. (*S*)-2-Iodooctane reacts with CN^- to yield (*R*)-2-methyloctanenitrile.

SOLUTION

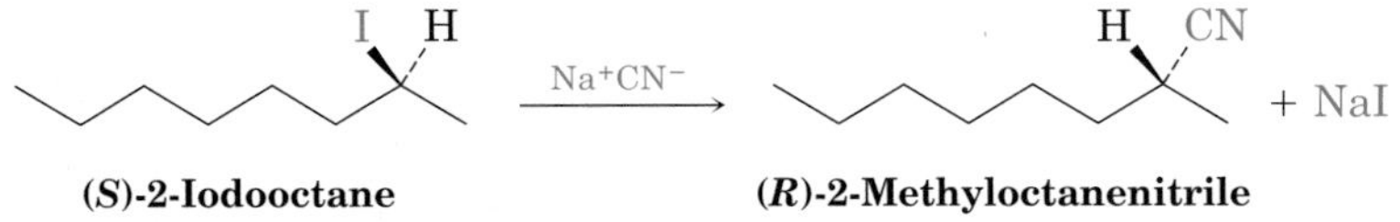

I H — $\xrightarrow{Na^+CN^-}$ — H CN + NaI

(*S*)-2-Iodooctane **(*R*)-2-Methyloctanenitrile**

PROBLEM 7.11 What product would you expect to obtain from the S_N2 reaction of (*S*)-2-bromohexane with sodium acetate, CH_3CO_2Na? Show the stereochemistry of both product and reactant.

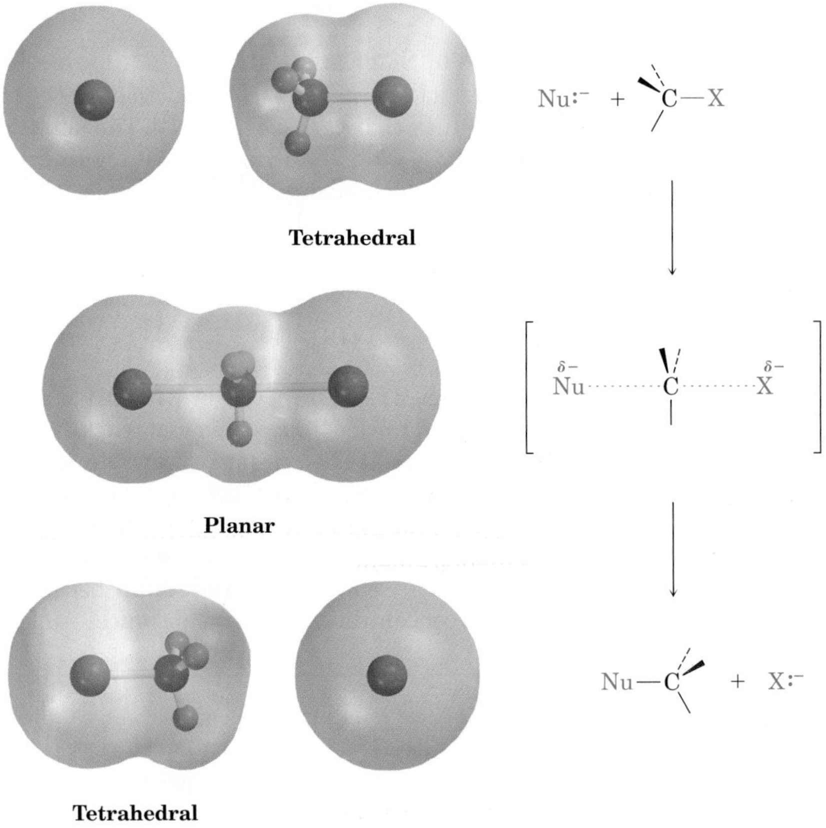

FIGURE 7.3 The transition state of the S_N2 reaction has a planar arrangement of the carbon atom and the three attached groups. Electrostatic potential maps show how the negative charge (red) is shared by the incoming nucleophile and the leaving group in the transition state. (The dotted red lines indicate partial bonding.)

Nu:⁻ + C—X

Tetrahedral

[δ− Nu·········C·········X δ−]

Planar

Nu—C + X:⁻

Tetrahedral

PROBLEM 7.12 Assign configuration to the following substance, and draw the structure of the product that would result on nucleophilic substitution reaction with HS^- (reddish brown = Br):

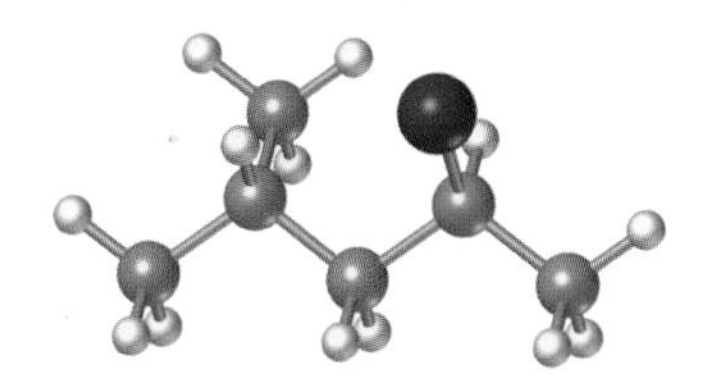

Steric Effects in S_N2 Reactions

The ease with which a nucleophile can approach a substrate to carry out an S_N2 reaction depends on spatial accessibility. Bulky substrates, in which the halide-bearing carbon atom is difficult to approach, react much more slowly than those in which the carbon is more accessible (Figure 7.4).

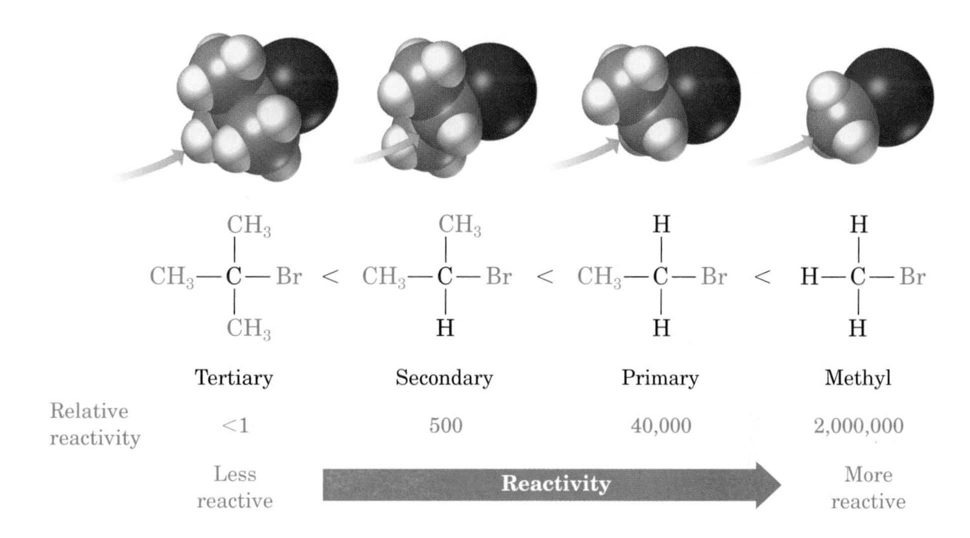

FIGURE 7.4 Steric hindrance to the S_N2 reaction. These models show that as the carbon atom becomes more accessible, the reaction rates of nucleophiles increase: 2-bromo-2-methylpropane < 2-bromopropane < bromoethane < methylbromide. Of these four substrates, methylbromide has the most accessible carbon and is therefore the most reactive.

Methyl halides (CH_3X) are the most reactive substrates, followed by primary alkyl halides (RCH_2—X) such as ethyl and propyl. Alkyl branching next to the leaving group slows the reaction greatly for secondary halides (R_2CH—X), and further branching effectively halts the reaction for tertiary halides (R_3C—X).

Vinylic ($R_2C{=}CRX$) and aryl (Ar—X) halides are not shown on this reactivity list because they are completely unreactive toward S_N2 displacements. This lack of reactivity is due to steric hindrance. The incoming nucleophile would have to burrow through part of the molecule to carry out a displacement.

R₂C=C(Cl)R + Nu:⁻ ⟶ No reaction

Vinylic halide

:Nu⁻ + Ar—Cl ⟶ No reaction

Aryl halide

PRACTICE PROBLEM 7.6

Which would you expect to be faster, the S_N2 reaction of OH^- ion with 1-bromopentane or with 2-bromopentane?

STRATEGY Decide which substrate is less hindered. Since 1-bromopentane is a 1° halide and 2-bromopentane is a 2° halide, reaction with the less hindered 1-bromopentane is faster.

SOLUTION

PROBLEM 7.13 Which of the following S_N2 reactions would you expect to be faster?
(a) Reaction of CN^- (cyanide ion) with $CH_3CH(Br)CH_3$ or with $CH_3CH_2CH_2Br$?
(b) Reaction of I^- with $(CH_3)_2CHCH_2Cl$ or with $H_2C{=}CHCl$?

The Leaving Group in S_N2 Reactions

Another variable that can affect the S_N2 reaction is the identity of the leaving group displaced by the attacking nucleophile. Because the leaving group is expelled with a negative charge in most S_N2 reactions, the best leaving groups are those that give the most stable anions (anions of strong acids). A halide ion (I^-, Br^-, or Cl^-) is the most common leaving group, although others are also possible. Anions such as F^-, OH^-, OR^-, and NH_2^- are rarely found as leaving groups.

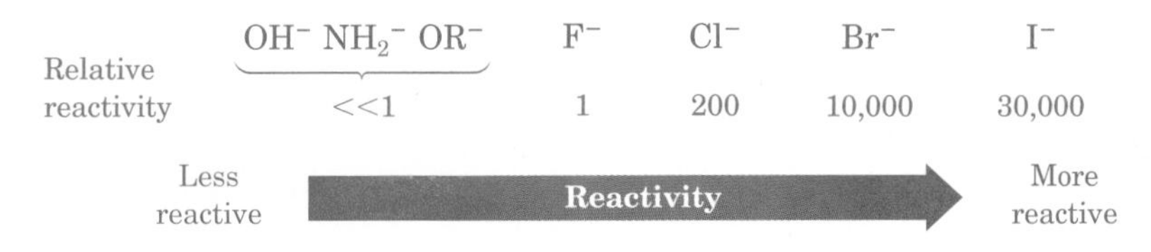

PROBLEM 7.14 Rank the following compounds in order of their expected reactivity toward S_N2 reaction: CH_3I, CH_3F, CH_3Br.

7.6 The S_N1 Reaction

Most nucleophilic substitutions take place by the S_N2 pathway just discussed, but an alternative called the **S_N1 reaction** can also occur. In general, S_N1 reactions take place only on *tertiary* substrates and only under neutral or acidic conditions in a hydroxylic solvent such as water or alcohol. We saw in Section 7.2, for example, that alkyl halides can be prepared from alcohols by treatment with HCl or HBr. Tertiary alcohols react rapidly, but primary and secondary alcohols react more slowly.

$$R_3COH >> R_2CHOH > R_2CH_2OH > CH_3OH$$

What's going on here? Clearly, a nucleophilic substitution reaction is taking place—a halogen is replacing a hydroxyl group—yet the reactivity order 3° > 2° > 1° is backward from the normal S_N2 order. Furthermore, an –OH group is being replaced, although we said in the previous section that OH^- is a poor leaving group. What's going on is that this is *not* an S_N2 reaction; it is an S_N1 reaction, whose mechanism is shown in Figure 7.5.

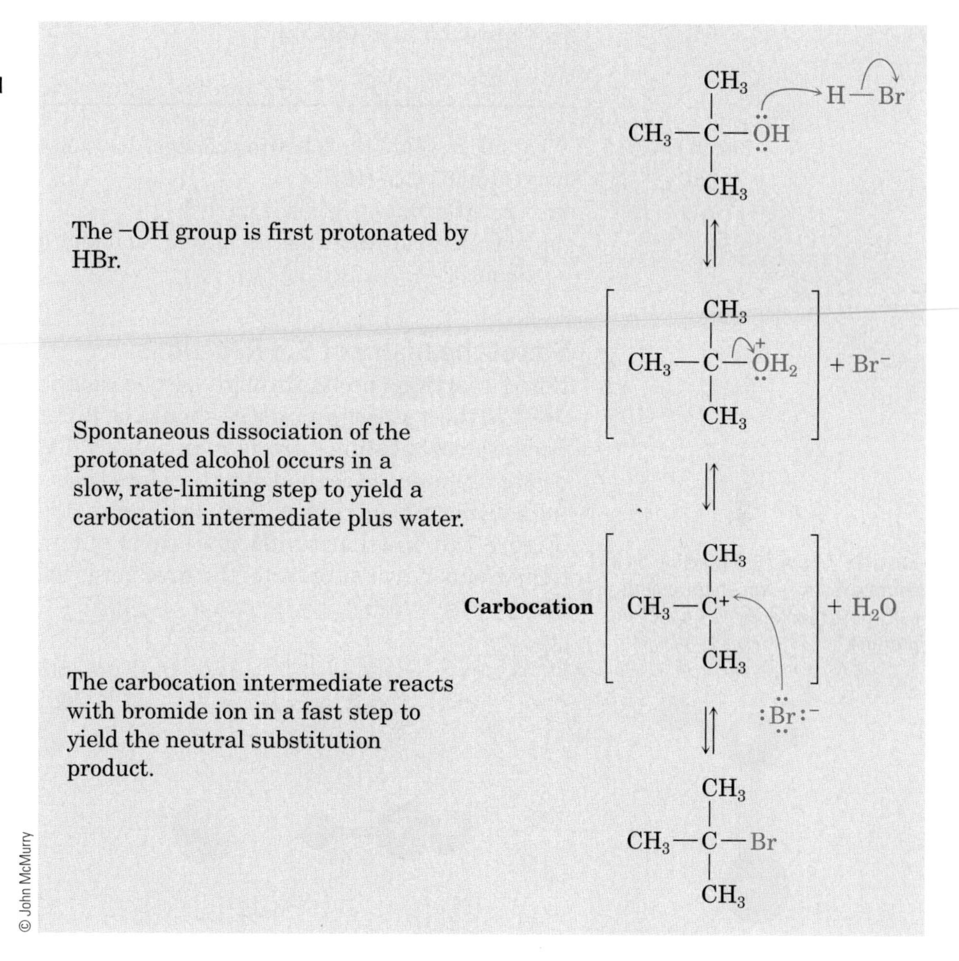

CH_3
H—Br
CH_3—C—OH
CH_3
The –OH group is first protonated by HBr.
CH_3
CH_3—C—$\overset{+}{O}H_2$
CH_3
+ Br^-
Spontaneous dissociation of the protonated alcohol occurs in a slow, rate-limiting step to yield a carbocation intermediate plus water.
Carbocation
CH_3
CH_3—C^+
CH_3
+ H_2O
The carbocation intermediate reacts with bromide ion in a fast step to yield the neutral substitution product.
:Br:$^-$
CH_3
CH_3—C—Br
CH_3
© John McMurry

FIGURE 7.5 MECHANISM: The mechanism of the S_N1 reaction of *tert*-butyl alcohol with HBr to yield an alkyl halide. Neutral H_2O is the leaving group.

Unlike what occurs in an S_N2 reaction, where the leaving group is displaced *at the same time* that the incoming nucleophile approaches, an S_N1 reaction occurs by spontaneous loss of the leaving group *before* the incoming nucleophile approaches. Loss of the leaving group gives a carbocation intermediate, which then reacts with nucleophile in a second step to yield the substitution product.

This two-step mechanism explains why tertiary alcohols react with HBr so much more rapidly than primary or secondary ones do: S_N1 reactions can occur only when stable carbocation intermediates are formed. The more stable the carbocation intermediate, the faster the S_N1 reaction. Thus, the reactivity order of alcohols with HBr is the same as the stability order of carbocations (see Section 4.3).

Rates of S_N1 Reactions

Unlike the rate of an S_N2 reaction, which depends on the concentrations of both substrate and nucleophile, *the rate of an S_N1 reaction depends only on the concentration of the substrate and is independent of the nucleophile concentration.* Thus, the origin of the "1" in S_N1: S_N1 reactions are **unimolecular** because the rate of the reaction depends on the concentration of only *one* substance—the substrate. The observation that S_N1 reactions are unimolecular means that the substrate must undergo a spontaneous reaction without involvement of the nucleophile, exactly what the mechanism shown in Figure 7.5 accounts for.

PROBLEM 7.15 What effect would the following changes have on the rate of the S_N1 reaction of *tert*-butyl alcohol with HBr?

(a) The HBr concentration is tripled.

(b) The HBr concentration is halved, and the *tert*-butyl alcohol concentration is doubled.

Stereochemistry of S_N1 Reactions

If S_N1 reactions occur through carbocation intermediates, as shown in Figure 7.5, their stereochemistry should be different from that of S_N2 reactions. Because carbocations are planar and sp^2-hybridized, they are achiral. The positively charged carbon can therefore react with a nucleophile equally well from either top or bottom face, leading to a racemic mixture of enantiomers (Figure 7.6). In other words, if we carry out an S_N1 reaction on a single enantiomer of a chiral substrate, the product is racemic.

FIGURE 7.6 An S_N1 reaction on a chiral substrate. An enantiomerically pure reactant gives a racemic product.

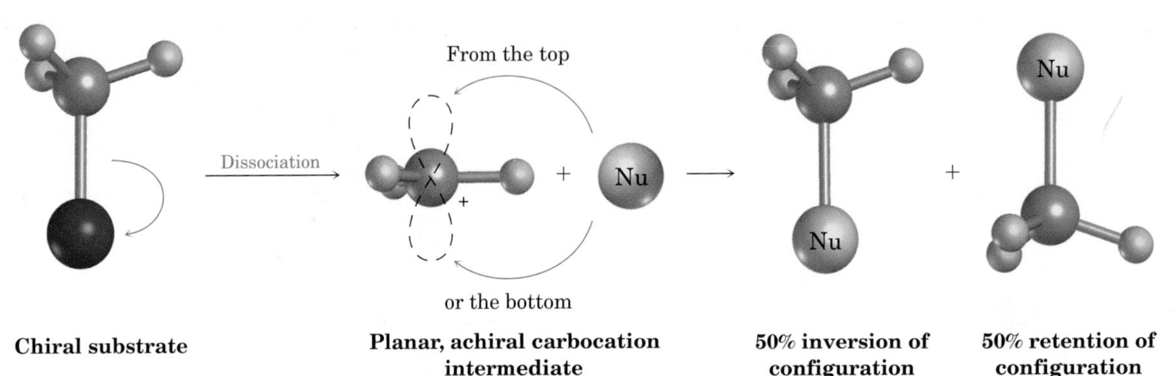

An example of what occurs when a chiral substrate undergoes S_N1 reaction is seen on treatment of (*R*)-1-phenylbutan-1-ol with HCl. A racemic alkyl chloride product is formed:

$$CH_3CH_2CH_2-C(H)(C_6H_5)-OH + HCl \longrightarrow CH_3CH_2CH_2-C(H)(C_6H_5)-Cl + Cl-C(H)(C_6H_5)-CH_2CH_2CH_3$$

(*R*)-Phenylbutan-1-ol — **(*R*)-1-Phenyl-1-chlorobutane** (50%, retention) — **(*S*)-1-Phenyl-1-chlorobutane** (50%, inversion)

PRACTICE PROBLEM 7.7

What stereochemistry would you expect for the S_N1 reaction of (*R*)-3-bromo-3-methylhexane with methanol to yield 3-methoxy-3-methylhexane?

STRATEGY First draw the starting alkyl halide, showing its correct stereochemistry. Then replace the –Br with a methoxy group (–OCH_3) to give the racemic product.

SOLUTION

H_3C Br

+ CH_3OH ⟶

(*R*)-3-Bromo-3-methylhexane

CH_3O CH_3

(*S*)-3-Methoxy-3-methylhexane (50%)

H_3C OCH_3

(*R*)-3-Methoxy-3-methylhexane (50%)

PROBLEM 7.16 What product would you expect to obtain from the S_N1 reaction of (*S*)-3-methyloctan-3-ol with HBr? Show the stereochemistry of both starting material and product.

PROBLEM 7.17 Assign configuration to the following substrate, and show the stereochemistry and identity of the product you would obtain by S_N1 reaction with H_2O (reddish brown = Br):

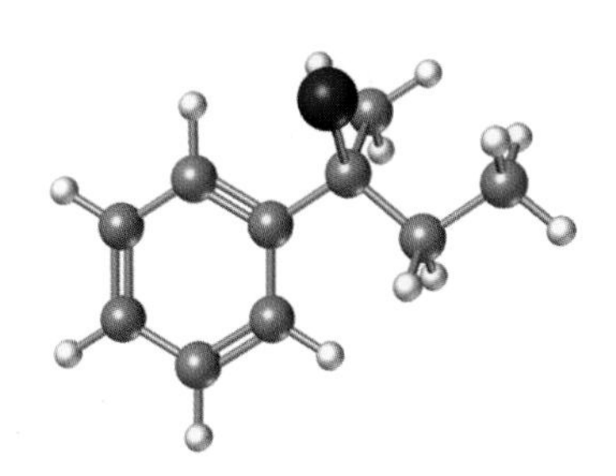

The Leaving Group in S_N1 Reactions

The best leaving groups in S_N1 reactions are those that give the most stable anions, just as in S_N2 reactions. Note that if an S_N1 reaction is carried out under acidic conditions, as occurs when a tertiary alcohol reacts with HX to yield an alkyl halide (Figure 7.5), neutral water can be the leaving group. The S_N1 reactivity order of leaving groups is:

$$F^- << Cl^- = H_2O < Br^- < I^-$$

Worse leaving group — **Reactivity** ⟶ Better leaving group

7.7 Eliminations: The E2 Reaction

Thus far, we've looked only at substitution reactions, but in fact *two* kinds of processes are possible when a nucleophile/base reacts with an alkyl halide. The nucleophile/base can *substitute* for the leaving group in an S_N1 or S_N2 reaction, or the nucleophile/base can cause *elimination* of HX, leading to formation of an alkene:

Substitution H–C–C–Br + OH^- ⟶ H–C–C–OH + Br^-

Elimination H–C–C–Br + OH^- ⟶ C=C + H_2O + Br^-

The elimination of HX from alkyl halides is an extremely useful method for preparing alkenes, but the topic is complex for several reasons. For one thing, there is the problem of regiochemistry: what product results from loss of HX from an unsymmetrical halide? In fact, elimination reactions almost always give *mixtures* of alkene products, and the best we can usually do is to predict which will be the major one.

According to **Zaitsev's rule**, a predictive guideline formulated by the Russian chemist Alexander Zaitsev in 1875, base-induced elimination reactions generally give the more highly substituted alkene product—that is, the alkene with the larger number of substituents on the double bond. For example, treatment of 2-bromobutane with KOH in ethanol gives primarily but-2-ene (disubstituted; two alkyl group substituents on the double-bond carbons) rather than but-1-ene (monosubstituted; one alkyl group substituent on the double-bond carbons).

$$CH_3CH_2CH(Br)CH_3 \xrightarrow[CH_3CH_2OH]{KOH} CH_3CH{=}CHCH_3 + CH_3CH_2CH{=}CH_2$$

2-Bromobutane **But-2-ene (81%)** **But-1-ene (19%)**

Zaitsev's rule: In the elimination of HX from an alkyl halide, the more highly substituted alkene product predominates.

A second complication is that eliminations can take place by several different mechanistic pathways, just as substitutions can. The most commonly occurring elimination mechanism is called the **E2 reaction** (for *elimination, bimolecular*). The process takes place when an alkyl halide is treated with a strong base, such as hydroxide ion or alkoxide ion (RO^-), and can be formulated as shown in Figure 7.7.

Like the S_N2 reaction, the E2 reaction takes place in one step without intermediates. As the attacking base begins to abstract H^+ from a carbon atom next to the leaving group, the C–H and C–X bonds begin to break and the C=C double bond begins to form. When the leaving group departs, it takes with it the two electrons from the former C–X bond.

FIGURE 7.7 MECHANISM: The mechanism of the E2 reaction. The reaction takes place in a single step through a transition state in which the double bond begins to form at the same time the H and X groups are leaving. (Dotted lines indicate partial bonding in the transition state.)

Base (B:) attacks a neighboring hydrogen and begins to remove the H at the same time as the alkene double bond starts to form and the X group starts to leave.

Neutral alkene is produced when the C–H bond is fully broken and the X group has departed with the C–X bond electron pair.

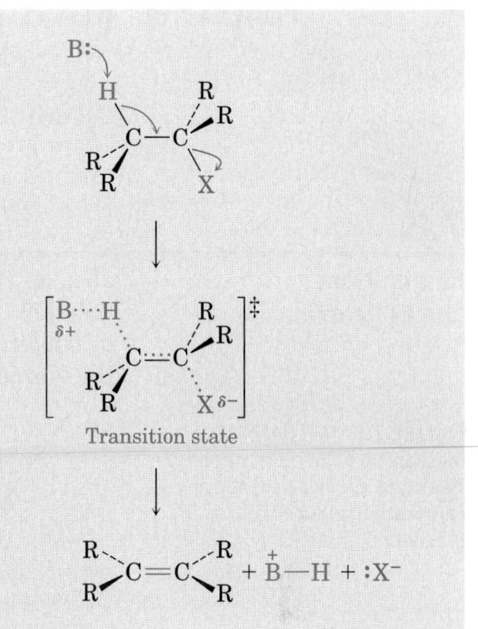

PRACTICE PROBLEM 7.8

What product would you expect from reaction of 1-chloro-1-methylcyclohexane with KOH in ethanol?

STRATEGY We know that treatment of an alkyl halide with a strong base such as KOH yields an alkene. To find the products in a specific case, draw the structure of the starting material and locate the hydrogen atoms on each neighboring carbon. Then generate the potential alkene products by removing HX in as many ways as possible. The major product will be the one that has the most highly substituted double bond—in this case, 1-methylcyclohexene.

SOLUTION

1-Chloro-1-methylcyclohexane + KOH $\xrightarrow{CH_3CH_2OH}$ 1-Methylcyclohexene (major) + Methylenecyclohexane (minor)

PROBLEM 7.18 Ignoring double-bond stereochemistry, what elimination products would you expect from the following reactions?

(a) $CH_3CH_2CH(Br)CH(CH_3)CH_3$

(b) $CH_3CH(CH_3)CH_2—C(Cl)(CH_3)—CH(CH_3)CH_3$

(c) Cyclohexyl—CH(Br)CH$_3$

PROBLEM 7.19 What alkyl halides might the following alkenes have been made from?

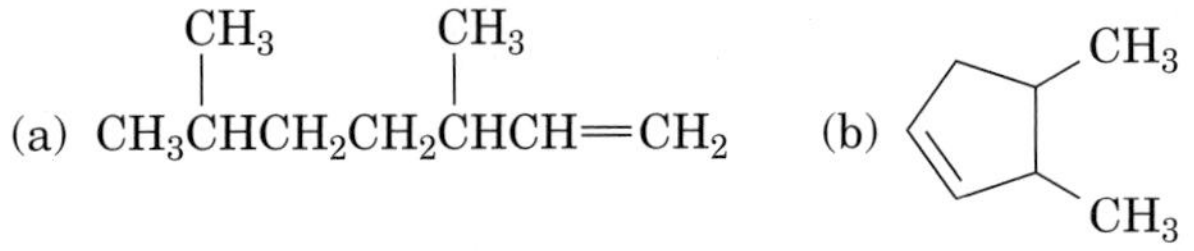

(a) $CH_3CH(CH_3)CH_2CH_2CH(CH_3)CH{=}CH_2$

(b) 3,4-dimethylcyclopentene (ring structure with two CH_3 groups)

7.8 Eliminations: The E1 Reaction

Just as the S_N2 reaction has a close analog in the E2 reaction, the S_N1 reaction has a close analog called the **E1 reaction** (for *elimination, unimolecular*). The E1 reaction can be formulated as shown in Figure 7.8 for the elimination of HCl from 2-chloro-2-methylpropane.

FIGURE 7.8 MECHANISM: Mechanism of the E1 reaction. Two steps are involved, and a carbocation intermediate is present.

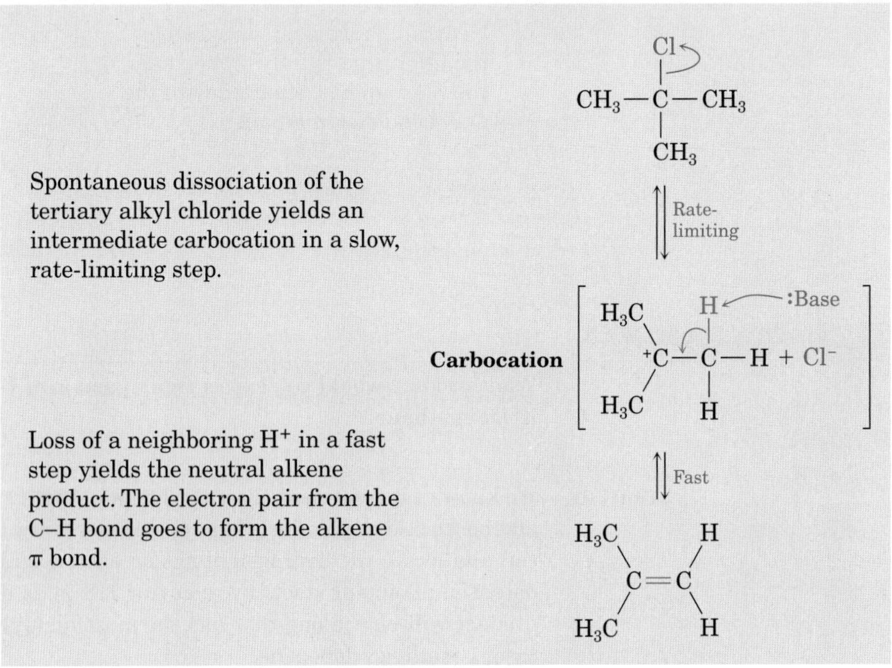

Spontaneous dissociation of the tertiary alkyl chloride yields an intermediate carbocation in a slow, rate-limiting step.

$$CH_3-C(Cl)(CH_3)-CH_3 \xrightleftharpoons{\text{Rate-limiting}} \left[(H_3C)_2\overset{+}{C}-CH_2-H \right] + Cl^-$$

Carbocation

Loss of a neighboring H^+ in a fast step yields the neutral alkene product. The electron pair from the C–H bond goes to form the alkene π bond.

$$\left[(H_3C)_2\overset{+}{C}-CH_3 \right] \xrightleftharpoons{\text{Fast}} (H_3C)_2C{=}CH_2$$

E1 eliminations begin with the same unimolecular dissociation we saw in the S_N1 reaction, but the dissociation is followed by loss of H^+ from the intermediate carbocation rather than by substitution. In fact, the E1 and S_N1 reactions normally occur in competition whenever an alkyl halide is treated in a hydroxylic solvent with a nonbasic nucleophile. Thus, the best E1 substrates are also the best S_N1 substrates, and mixtures of substitution and elimination products are usually obtained. For example, when 2-chloro-2-methylpropane is warmed to 65 °C in 80% aqueous ethanol, a 64:36 mixture of 2-methylpropan-2-ol (S_N1) and 2-methylpropene (E1) results:

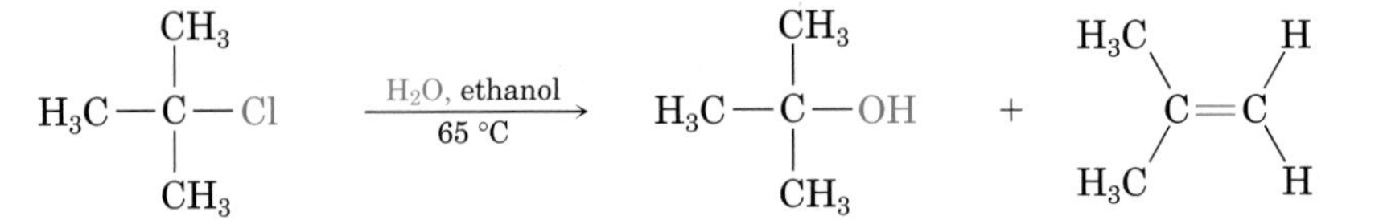

$$H_3C-C(CH_3)_2-Cl \xrightarrow[65\ °C]{H_2O,\ \text{ethanol}} H_3C-C(CH_3)_2-OH + (H_3C)_2C{=}CH_2$$

2-Chloro-2-methylpropane — **2-Methylpropan-2-ol (64%)** — **2-Methylpropene (36%)**

PROBLEM 7.20 What effect on the rate of an E1 reaction of 2-chloro-2-methylpropane would you expect if the concentration of the alkyl halide were tripled?

7.9 A Summary of Reactivity: S_N1, S_N2, E1, E2

Now that we've seen four different kinds of nucleophilic substitution/elimination reactions, you may be wondering how to predict what will take place in any given case. Will substitution or elimination occur? Will the reaction be unimolecular or bimolecular? There are no rigid answers to these questions, but it's possible to make some broad generalizations.

- **A primary substrate (RCH_2X)** reacts by an S_N2 pathway if a good nucleophile such as I^-, Br^-, RS^-, NH_3, or CN^-, is used. Some reaction by an E2 pathway may accompany the S_N2 reaction if a strong base such as hydroxide ion or an alkoxide ion (RO^-) is used.
- **A secondary substrate (R_2CHX)** reacts by both S_N2 and E2 pathways to give a mixture of substitution and elimination products.
- **A tertiary substrate (R_3CX)** reacts by an E2 pathway if a strong base is used or by a mixture of S_N1 and E1 pathways under neutral or acidic conditions.

PRACTICE PROBLEM 7.9

Tell whether the following reaction is S_N1, S_N2, E1, or E2:

Chlorocyclohexane (Cl, H on ring carbon) + KOH ⟶ Cyclohexene + KCl + H_2O

STRATEGY Look to see whether the substrate is primary, secondary, or tertiary, and determine whether substitution or elimination has occurred. Then apply the generalizations summarized earlier.

SOLUTION The substrate is a secondary alkyl halide, and an elimination has occurred to give an alkene. This is an E2 reaction.

PROBLEM 7.21 Tell whether the following reactions are S_N1, S_N2, E1, or E2:

(a) 1-Bromobutane + NaN_3 ⟶ 1-Azidobutane

(b) $CH_3CH_2CH(Cl)CH_2CH_3$ + KOH ⟶ $CH_3CH_2CH{=}CHCH_3$

(c) 1-Chloro-1-methylcyclohexane (Cl, CH_3 on ring carbon) + CH_3COOH ⟶ 1-methylcyclohexyl acetate ($OCCH_3$ with C=O, CH_3 on ring carbon) + HCl

7.10 Substitution Reactions in Living Organisms

All chemistry—whether carried out in flasks by chemists or in cells by living organisms—follows the same rules. Most biological reactions therefore occur by the same addition, substitution, elimination, and rearrangement mechanisms encountered in laboratory reactions.

Among the most common biological substitution reactions is *methylation*, the transfer of a $-CH_3$ group from an electrophilic donor to a nucleophile. A laboratory chemist might choose CH_3I for such a reaction, but living organisms use the complex molecule *S*-adenosylmethionine as the biological methyl-group donor. Since the sulfur atom in *S*-adenosylmethionine has a positive charge (a *sulfonium* ion, R_3S^+), it is an excellent leaving group for S_N2 displacements on the methyl carbon. An example of such a biological methylation takes place in the adrenal medulla during the biological synthesis of adrenaline from norepinephrine.

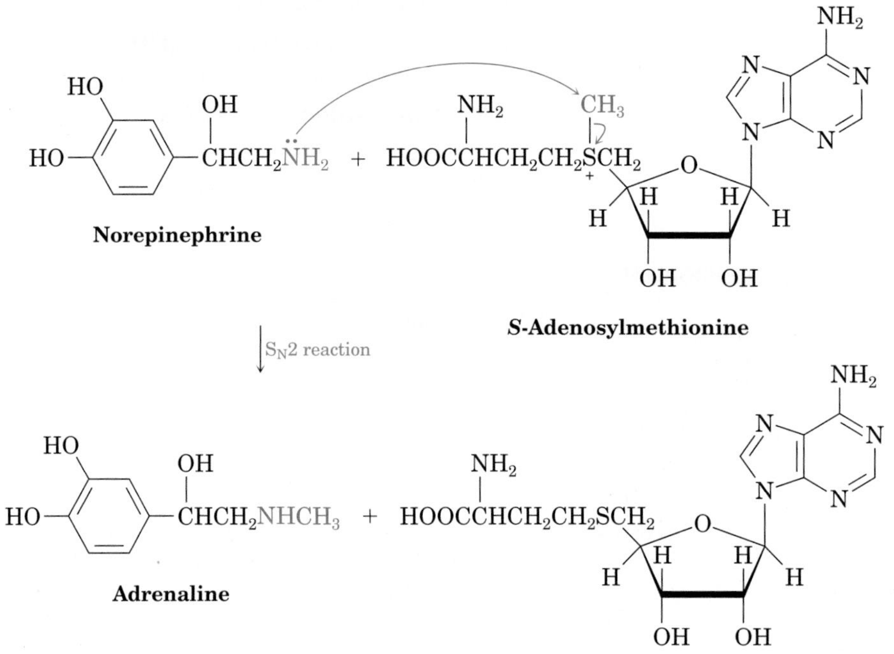

HO, HO, OH, CHCH₂NH₂ + HOOCCHCH₂CH₂SCH₂ (NH₂, CH₃, +), O, H, H, H, H, OH, OH, N, N, N, N, NH₂

Norepinephrine

***S*-Adenosylmethionine**

S_N2 reaction

HO, HO, OH, CHCH₂NHCH₃ + HOOCCHCH₂CH₂SCH₂ (NH₂), O, H, H, H, H, OH, OH, N, N, N, N, NH₂

Adrenaline

After dealing only with simple halides such as CH_3I up to this point, it's a shock to encounter a molecule as complex as *S*-adenosylmethionine. From a chemical standpoint, however, CH_3I and *S*-adenosylmethionine do exactly the same thing: both transfer a methyl group by an S_N2 reaction.

Another example of a biological S_N2 reaction is involved in the response of organisms to certain toxic chemicals. Many reactive S_N2 substrates with deceptively simple structures are quite toxic to living organisms. Methyl bromide, for example, has been widely used as a fumigant to kill termites. Its toxicity derives from its ability to transfer a methyl group to a nucleophilic amino group ($-NH_2$) or mercapto group ($-SH$) in enzymes, thus altering the enzyme's normal biological activity.

INTERLUDE

Naturally Occurring Organohalogen Compounds

As recently as 1970, only about 30 naturally occurring organohalogen compounds were known. It was simply assumed that chloroform, halogenated phenols, chlorinated aromatic compounds called PCBs, and other such substances found in the environment were industrial "pollutants." Now, only a third of a century later, the situation is quite different. More than 3000 organohalogen compounds have been found to occur naturally, and many thousands more surely exist. From simple compounds like chloromethane to extremely complex ones, a remarkably diverse range of organohalogen compounds exists in plants, bacteria, and animals. Many even have unusual physiological activity. For example, the bromine-containing substance called jasplakinolide, discovered by Phillip Crews at the University of California, Santa Cruz, disrupts formation of the actin microtubules that make up the skeleton of cellular organelles.

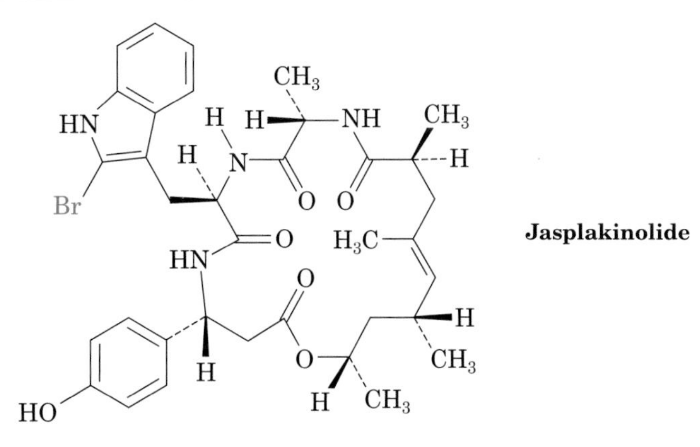

Jasplakinolide

Some naturally occurring organohalogen compounds are produced in massive quantities. Forest fires, volcanoes, and marine kelp release up to 5 *million* tons of CH_3Cl per year, for example, while annual industrial emissions total only about 26,000 tons. A detailed examination of one species of Okinawan acorn worm in a 1 km^2 study area showed that they released nearly 100 pounds per day of halogenated phenols, compounds previously thought to be nonnatural pollutants.

Why do organisms produce organohalogen compounds, many of which are undoubtedly toxic? The answer seems to be that many organisms use organohalogen compounds for self-defense, either as feeding deterrents, as irritants to predators, or as natural pesticides. Marine sponges, coral, and sea hares, for example, release foul-tasting organohalogen compounds that deter fish, starfish, and other predators from eating them. More remarkably, even humans appear to produce halogenated compounds as part of their defense against infection. The human immune system contains a peroxidase enzyme capable of carrying out halogenation reactions on fungi and bacteria, thereby killing the pathogen.

Much remains to be learned—only a few hundred of the more than 500,000 known species of marine organisms have been examined—but it's already clear that organohalogen compounds are an integral part of the world around us.

Marine corals secrete organohalogen compounds that act as a feeding deterrent to starfish.

Summary and Key Words

Alkyl halides are usually prepared from alcohols by treatment either with HX (for tertiary alcohols) or with $SOCl_2$ or PBr_3 (for primary and secondary alcohols). Alkyl halides react with magnesium metal to form organomagnesium halides, or **Grignard reagents**. These organometallic compounds react with acids to yield the corresponding alkanes.

Treatment of an alkyl halide with a nucleophile/base results either in substitution or in elimination. **Nucleophilic substitution reactions** occur by two mechanisms: S_N2 and S_N1. In the **S_N2 reaction**, the entering nucleophile attacks the substrate from a direction 180° away from the leaving group, resulting in an umbrella-like inversion of configuration at the carbon atom. S_N2 reactions are strongly inhibited by increasing steric bulk of the reagents and are favored only for primary substrates. In the **S_N1 reaction**, the substrate spontaneously dissociates to a carbocation, which then reacts with a nucleophile in a second step. As a consequence, S_N1 reactions take place with racemization of configuration at the carbon atom and are favored only for tertiary substrates.

Elimination reactions also occur by two mechanisms: E2 and E1. In the **E2 reaction**, a base abstracts a proton at the same time that the leaving group departs. In the **E1 reaction**, the substrate spontaneously dissociates to form a carbocation, which can subsequently lose H^+ from a neighboring carbon. The reaction occurs on tertiary substrates in neutral or acidic hydroxylic solvents. The major product formed by an elimination can be predicted by **Zaitsev's rule**.

Summary of Reactions

1. Synthesis of alkyl halides from alcohols (Section 7.2)
   (a) Reaction of tertiary alcohol with HX, where X = Cl, Br

$$R_3C{-}OH \xrightarrow[\text{Ether}]{HX} R_3C{-}X + H_2O$$

   (b) Reaction of primary and secondary alcohols with PBr_3 and $SOCl_2$

$$ROH + PBr_3 \longrightarrow RBr$$
$$ROH + SOCl_2 \longrightarrow RCl$$

2. Reactions of alkyl halides
   (a) Formation and protonation of Grignard reagents (Section 7.3)

$$RX + Mg \longrightarrow RMgX$$
$$RMgX \longrightarrow RH$$

   (b) S_N2 reaction: backside attack of nucleophile on alkyl halide (Section 7.5)

$$Nu{:}^- + C{-}X \longrightarrow Nu{-}C + X{:}^-$$

Substrate must be primary or secondary.

(c) S_N1 reaction: carbocation intermediate is involved (Section 7.6)

$$R_3C{-}X \longrightarrow \left[R_3C^+\right] \xrightarrow{:Nu^-} R_3C{-}Nu + :X^-$$

Substrate must be tertiary or (occasionally) secondary.

(d) E2 reaction (Section 7.7)

$$H{-}C{-}C{-}X \xrightarrow{OH^-} C{=}C + HX$$

(e) E1 reaction (Section 7.8)

$$H{-}C{-}C(X)(R){-}R \longrightarrow \left[H{-}C{-}\overset{+}{C}(R){-}R\right] \longrightarrow C{=}C(R)_2 + HX$$

Best for tertiary substrates in neutral or acidic solvents.
Carbocation intermediate is involved.

EXERCISES

Visualizing Chemistry

7.22 Write the product you would expect from reaction of each of the following molecules with (i) $Na^+\ {}^-SCH_3$ and (ii) NaOH (yellow-green = Cl):

(a)

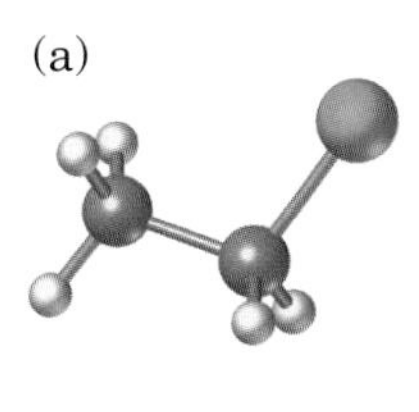

(b)

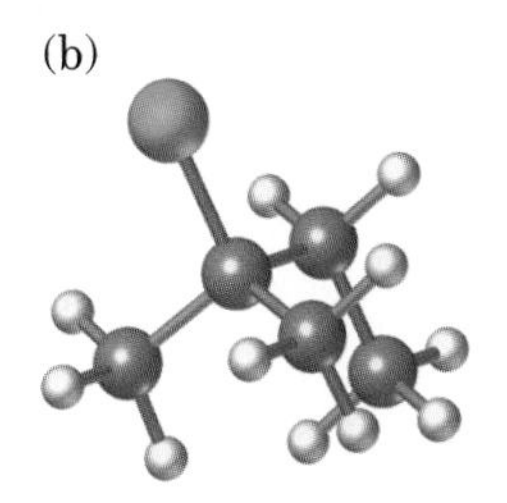

(c)

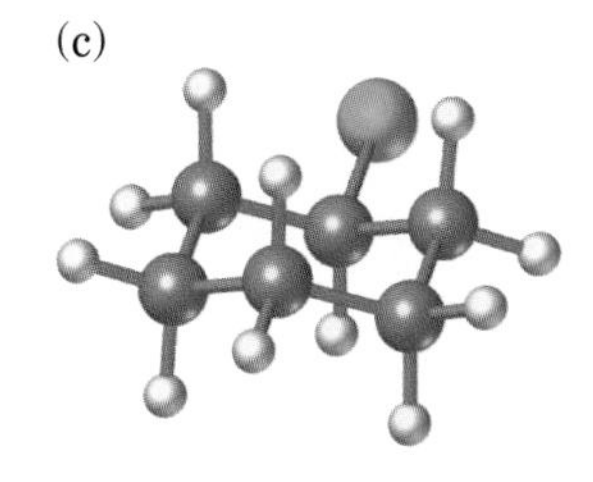

7.23 Assign *R* or *S* configuration to the following molecule, write the product you would expect from S_N2 reaction with NaCN, and assign *R* or *S* configuration to the product (red = O, yellow-green = Cl):

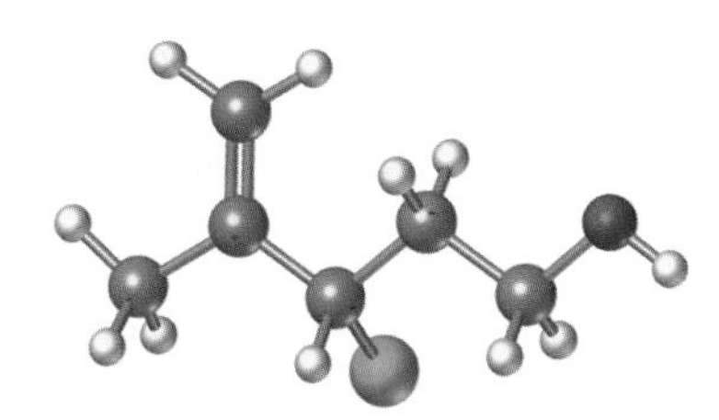

7.24 Draw the structure of the product you expect from E2 reaction of the following molecule with NaOH (yellow-green = Cl):

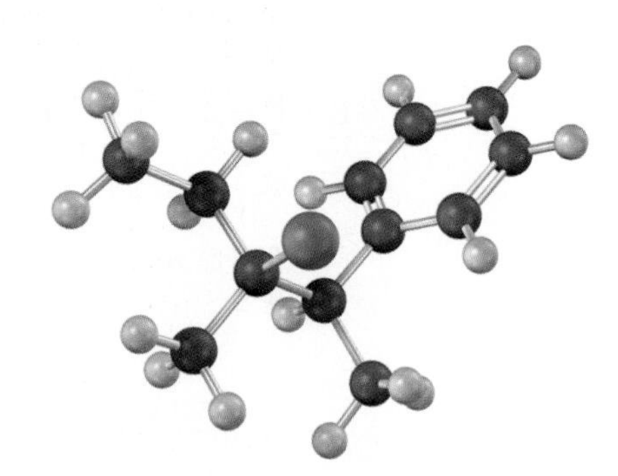

7.25 From what alkyl bromide was the following alkyl acetate made by S_N2 reaction? Write the reaction, showing all stereochemistry.

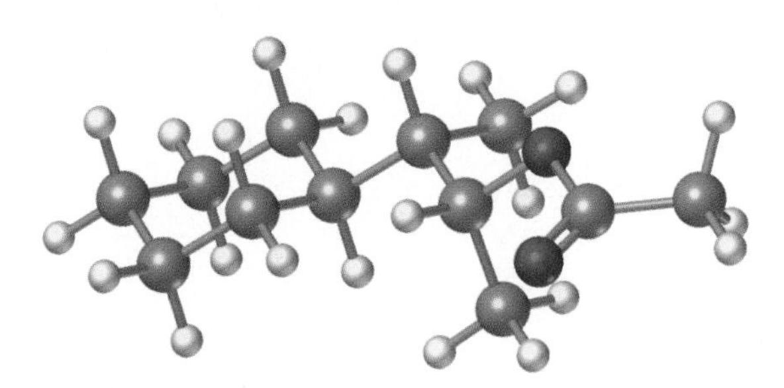

Additional Problems

NOMENCLATURE

7.26 Name the following alkyl halides according to IUPAC rules:

(a)
$$\begin{array}{c} \quad\ CH_3\ \ Br\quad\ CH_3 \\ \quad\ \ |\qquad |\qquad\ \ | \\ CH_3CHCHCHCH_2CHCH_3 \\ \qquad\quad | \\ \qquad\quad Br \end{array}$$

(b)
$$\begin{array}{c} \qquad\qquad\qquad\qquad\quad I \\ \qquad\qquad\qquad\qquad\quad | \\ CH_3CH{=}CHCH_2CHCH_3 \end{array}$$

(c)
$$\begin{array}{c} \quad Br\qquad\ Cl\ CH_3 \\ \quad |\qquad\quad\ |\ \ \ | \\ CH_3CCH_2CH_2CHCHCH_3 \\ \quad | \\ \quad CH_3 \end{array}$$

(d)
$$\begin{array}{c} \qquad\qquad CH_2Br \\ \qquad\qquad | \\ CH_3CH_2CHCH_2CH_2CH_3 \end{array}$$

7.27 Draw structures corresponding to the following IUPAC names:
(a) 2,3-Dichloro-4-methylhexane (b) 4-Bromo-4-ethyl-2-methylhexane
(c) 3-Iodo-2,2,4,4-tetramethylpentane

7.28 Draw and name the monochlorination products you might obtain by reaction of 2-methylpentane with Cl_2. Which of the products are chiral?

CHARACTERISTICS OF S_N1 AND S_N2 REACTIONS

7.29 Describe the effects of the following variables on both S_N2 and S_N1 reactions:
(a) Substrate structure (b) Leaving group

7.30 Which ion in each of the following pairs is a better leaving group?
(a) F^- or Br^- (b) Cl^- or NH_2^- (c) OH^- or I^-

7.31 Which alkyl halide in each of the following pairs will react faster in an S_N2 reaction with OH^-?
(a) Bromobenzene or benzyl bromide, $C_6H_5CH_2Br$
(b) CH_3Cl or $(CH_3)_3CCl$
(c) $CH_3CH{=}CHBr$ or $H_2C{=}CHCH_2Br$

7.32 What effect would you expect the following changes to have on the S_N2 reaction of CH_3Br and CN^- to give CH_3CN?
(a) The concentration of CH_3Br is tripled and that of CN^- is halved.
(b) The concentration of CH_3Br is halved and that of CN^- is tripled.
(c) The concentration of CH_3Br is tripled and that of CN^- is doubled.
(d) The reaction temperature is raised.
(e) The volume of the reacting solution is doubled by addition of more solvent.

7.33 What effect would you expect the following changes to have on the S_N1 reaction of $(CH_3)_3CBr$ with CH_3OH to give $(CH_3)_3COCH_3$?
(a) The concentration of $(CH_3)_3CBr$ is doubled and that of CH_3OH is halved.
(b) The concentration of $(CH_3)_3CBr$ is halved and that of CH_3OH is doubled.
(c) The concentrations of both $(CH_3)_3CBr$ and CH_3OH are tripled.
(d) The reaction temperature is lowered.

7.34 Order the following compounds with respect to both S_N1 and S_N2 reactivity:

$(CH_3)_3CCl$ (CH3–C(CH3)(Cl)–CH3) $C_6H_5CH_2Cl$ (benzene ring with CH_2Cl) C_6H_5Cl (benzene ring with Cl)

7.35 Order each set of compounds with respect to S_N2 reactivity:

(a) $CH_3C(CH_3)_2Cl$ $CH_3CH_2CH_2Cl$ $CH_3CH_2CH(Cl)CH_3$

(b) $CH_3CH(CH_3)CH(Br)CH_3$ $CH_3CH(CH_3)CH_2Br$ CH_3Br

SYNTHESIS

7.36 How would you prepare the following compounds, starting with cyclopentene and any other reagents needed?
(a) Chlorocyclopentane (b) Cyclopentanol
(c) Cyclopentylmagnesium chloride (d) Cyclopentane

7.37 Predict the product(s) of the following reactions:

(a) 1-Methylcyclohexanol (H_3C, OH on same ring carbon) $\xrightarrow[\text{Ether}]{\text{HBr}}$?

(b) $CH_3CH_2CH_2CH_2OH \xrightarrow{SOCl_2}$?

(c) Cyclohexanol (cyclohexane ring with OH) $\xrightarrow[\text{Ether}]{PBr_3}$?

(d) $CH_3CH_2CH(Br)CH_3 \xrightarrow[\text{Ether}]{\text{Mg}} A \xrightarrow{H_2O} B$

7.38 How might you prepare the following molecules using a nucleophilic substitution reaction at some step?

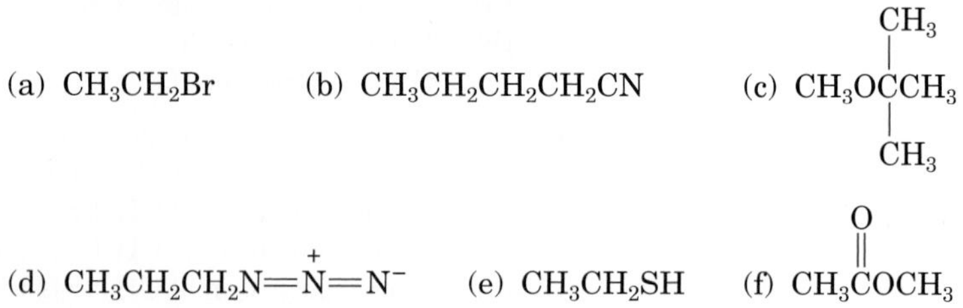

(a) CH_3CH_2Br (b) $CH_3CH_2CH_2CH_2CN$ (c) $CH_3OC(CH_3)_2CH_3$

(d) $CH_3CH_2CH_2N{=}\overset{+}{N}{=}N^-$ (e) CH_3CH_2SH (f) $CH_3C(=O)OCH_3$

7.39 What products do you expect from reaction of 1-bromopropane with the following reagents?
(a) NaI (b) NaCN (c) NaOH (d) Mg, then H_2O (e) $NaOCH_3$

7.40 What is wrong with each of the following reactions?

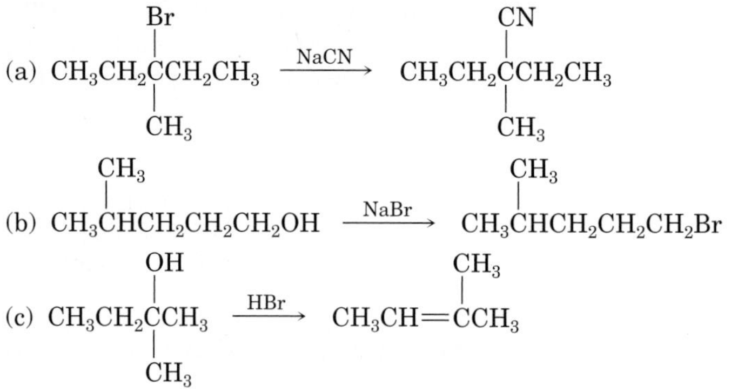

(a) $CH_3CH_2C(Br)(CH_3)CH_2CH_3 \xrightarrow{NaCN} CH_3CH_2C(CN)(CH_3)CH_2CH_3$

(b) $CH_3CH(CH_3)CH_2CH_2CH_2OH \xrightarrow{NaBr} CH_3CH(CH_3)CH_2CH_2CH_2Br$

(c) $CH_3CH_2C(OH)(CH_3)CH_3 \xrightarrow{HBr} CH_3CH{=}C(CH_3)CH_3$

ELIMINATION REACTIONS AND THE ELIMINATION/SUBSTITUTION CONTINUUM

7.41 Propose a structure for an alkyl halide that can give a mixture of three alkenes on E2 reaction.

7.42 Heating either *tert*-butyl chloride or *tert*-butyl bromide with ethanol yields the same reaction mixture: approximately 80% *tert*-butyl ethyl ether [$(CH_3)_3COCH_2CH_3$] and 20% 2-methylpropene. Explain why the identity of the leaving group has no effect on the product mixture.

7.43 What effect would you expect the following changes to have on the rate of the reaction of 1-iodo-2-methylbutane with CN^-?

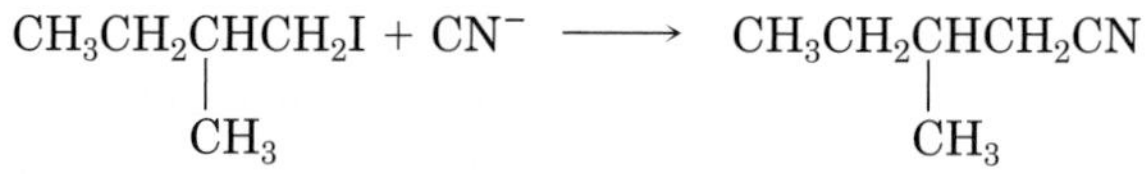

$$CH_3CH_2CH(CH_3)CH_2I + CN^- \longrightarrow CH_3CH_2CH(CH_3)CH_2CN$$

1-Iodo-2-methylbutane

(a) CN^- concentration is halved and 1-iodo-2-methylbutane concentration is doubled.
(b) Both CN^- and 1-iodo-2-methylbutane concentrations are tripled.

7.44 What effect would you expect on the rate of reaction of ethyl alcohol with 2-iodo-2-methylbutane if the concentration of the alkyl halide is tripled?

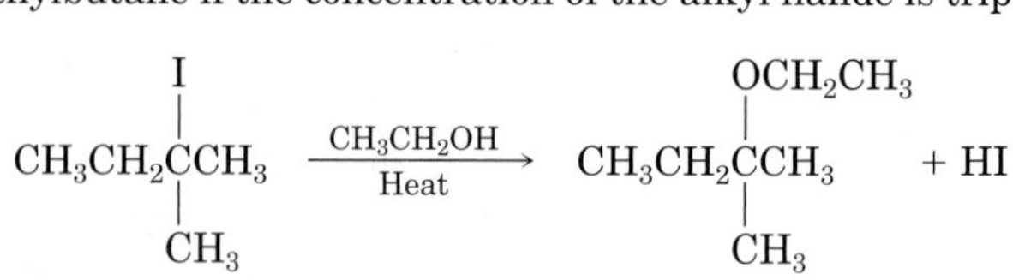

$$CH_3CH_2C(I)(CH_3)CH_3 \xrightarrow[\text{Heat}]{CH_3CH_2OH} CH_3CH_2C(OCH_2CH_3)(CH_3)CH_3 + HI$$

2-Iodo-2-methylbutane

7.45 Identify the following reactions as either S_N1, S_N2, E1, or E2:

(a) 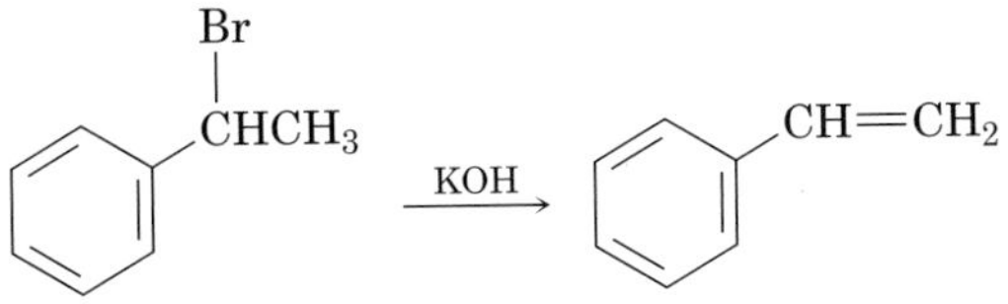 $C_6H_5CH(Br)CH_3 \xrightarrow{KOH} C_6H_5CH{=}CH_2$

(b) 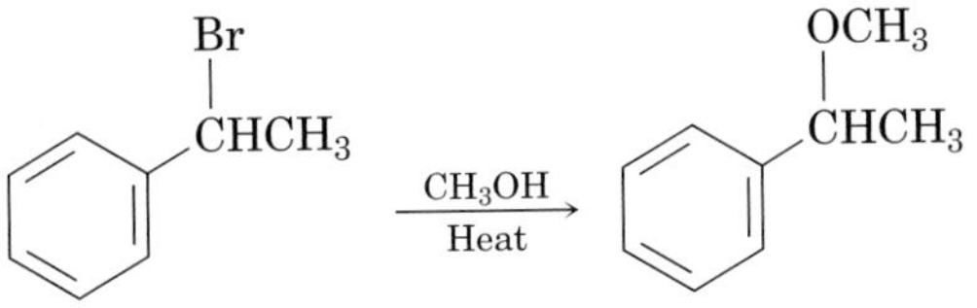 $C_6H_5CH(Br)CH_3 \xrightarrow[Heat]{CH_3OH} C_6H_5CH(OCH_3)CH_3$

7.46 Predict the major alkene product from the following eliminations:

(a) 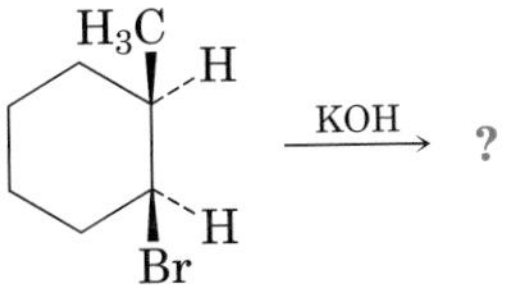 1-bromo-2-methylcyclohexane (H₃C and Br wedged, with H's) $\xrightarrow{KOH}$?

(b) 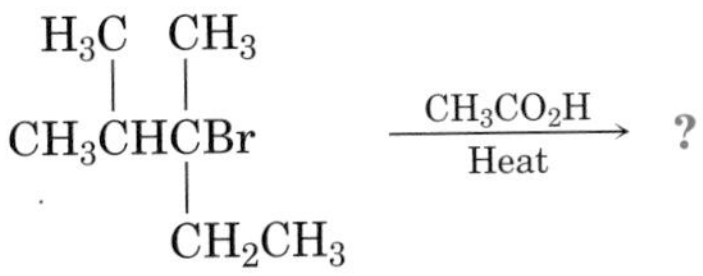 $CH_3CH(CH_3)C(CH_3)(Br)CH_2CH_3 \xrightarrow[Heat]{CH_3CO_2H}$?

7.47 Treatment of an alkyl chloride C_4H_9Cl with strong base gives a mixture of three isomeric alkene products. What is the structure of the alkyl chloride, and what are the structures of the three products?

Integrated Problems

7.48 Predict the product and give the stereochemistry of reactions of the following nucleophiles with (*R*)-2-bromooctane:
(a) CN^- (b) $CH_3CO_2^-$ (c) Br^-

7.49 Draw all isomers of C_4H_9Br, name them, and arrange them in order of decreasing reactivity in the S_N2 reaction.

7.50 Although the radical chlorination of alkanes with Cl_2 is usually unselective and gives a mixture of products, chlorination of propene, $CH_3CH{=}CH_2$, occurs almost exclusively on the methyl group rather than on a double-bond carbon. Draw resonance structures of the allyl radical $CH_2{=}CHCH_2\cdot$ to account for this result.

7.51 Draw resonance structures of the benzyl radical $C_6H_5CH_2\cdot$ to account for the fact that radical chlorination of toluene with Cl_2 occurs exclusively on the methyl group rather than on the aromatic ring.

7.52 Ethers can be prepared by S_N2 reaction of an alkoxide ion with an alkyl halide: $R{-}O^- + R'{-}Br \rightarrow R{-}O{-}R' + Br^-$. Suppose you wanted to prepare cyclohexyl methyl ether. Which route would be better, reaction of methoxide ion, CH_3O^-, with bromocyclohexane or reaction of cyclohexoxide ion with bromomethane? Explain.

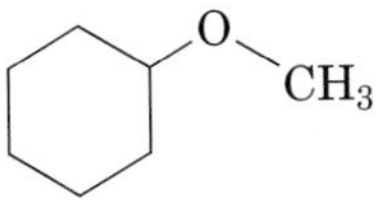

$C_6H_{11}{-}O{-}CH_3$ **Cyclohexyl methyl ether**

7.53 How could you prepare diethyl ether, $CH_3CH_2OCH_2CH_3$, starting from ethyl alcohol and any inorganic reagents needed? More than one step is needed. (See Problem 7.52.)

7.54 How could you prepare cyclohexane starting from 3-bromocyclohexene? More than one step is needed.

7.55 The S_N2 reaction can occur *intramolecularly*, meaning within the same molecule. What product would you expect from treatment of 4-bromobutan-1-ol with base?

$$BrCH_2CH_2CH_2CH_2OH \xrightarrow{\text{Base}} BrCH_2CH_2CH_2CH_2O^-\ Na^+ \longrightarrow ?$$

7.56 *trans*-1-Bromo-2-methylcyclohexane yields the non-Zaitsev elimination product 3-methylcyclohexene on treatment with KOH. What does this result tell you about the stereochemistry of E2 reactions?

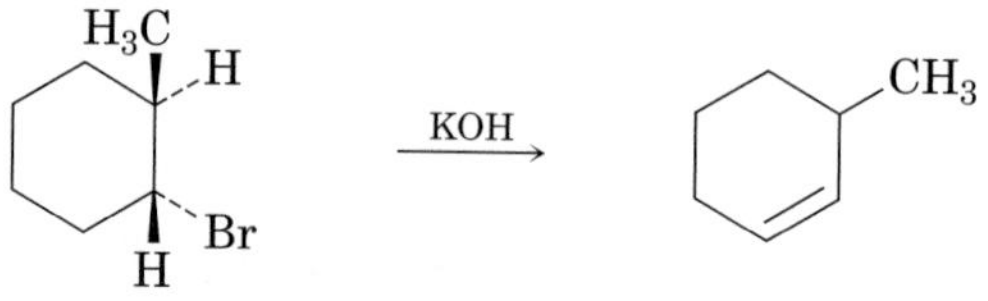

H_3C, H, Br, H $\xrightarrow{\text{KOH}}$ CH_3

***trans*-1-Bromo-2-methylcyclohexane** **3-Methylcyclohexene**

7.57 How can you explain the fact that treatment of (*R*)-2-bromohexane with NaBr yields *racemic* 2-bromohexane?

7.58 Reaction of HBr with (*R*)-3-methylhexan-3-ol yields (±)-3-bromo-3-methylhexane. Explain.

$$CH_3CH_2CH_2C(OH)(CH_3)CH_2CH_3$$

3-Methylhexan-3-ol

7.59 (*S*)-Butan-2-ol slowly racemizes to give (±)-butan-2-ol on standing in dilute sulfuric acid. Propose a mechanism to account for this observation.

$$CH_3CH_2CH(OH)CH_3$$

Butan-2-ol

7.60 Compound A is optically inactive and has the formula $C_{16}H_{16}Br_2$. On treatment with strong base, A gives hydrocarbon B, $C_{16}H_{14}$, which absorbs 2 equivalents of H_2 when reduced over a palladium catalyst. Hydrocarbon B also reacts with acidic $KMnO_4$ to give two carbonyl-containing products. One product, C, is a carboxylic acid with the formula $C_7H_6O_2$. The other product is oxalic acid, HO_2CCO_2H. Formulate the reactions involved, and suggest structures for A, B, and C.

7.61 Why do you suppose it's not possible to prepare a Grignard reagent from a bromoalcohol such as 4-bromopentan-1-ol?

$$CH_3CH(Br)CH_2CH_2CH_2OH \xrightarrow{\text{Mg}}\!\!\!\!\!\!\times\ CH_3CH(MgBr)CH_2CH_2CH_2OH$$

4-Bromopentan-1-ol

IN THE MEDICINE CABINET

7.62 The antidepressant fluoxetine, sold under the trade name Prozac, can be prepared by a substitution reaction of an alkyl chloride with a phenol using a weak base to generate the phenoxide ion.

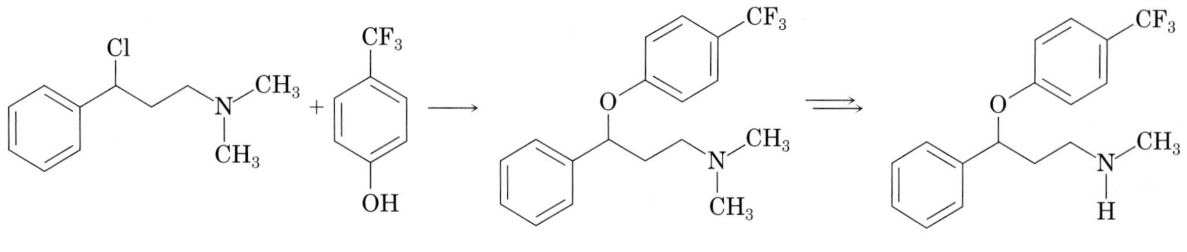

Prozac (fluoxetine)

(a) Identify the nucleophile in this reaction.
(b) Identify the electrophile in this reaction.
(c) If the rate of reaction depends on concentrations of both the alkyl chloride and phenol, is this an S_N1 or S_N2 reaction?
(d) Identify the chirality center in the alkyl chloride starting material.
(e) If you start with 100% *R*-alkyl chloride and the substitution proceeds with inversion, draw the chemical structure of Prozac ultimately produced and identify its chirality center as *R* or *S*.
(f) Given that amines such as Prozac form salts with carboxylic acids, draw a carboxylic acid that might allow you to resolve a racemate of Prozac.
(g) If a strong base is mixed with the alkyl chloride and phenol, less of the desired ether is produced and the solution gives a positive bromine test. What undesired product gives rise to the positive bromine test?

7.63 With only a few simple reactions, complex drugs can be prepared. The antipsychotic flupentixol is prepared as follows:

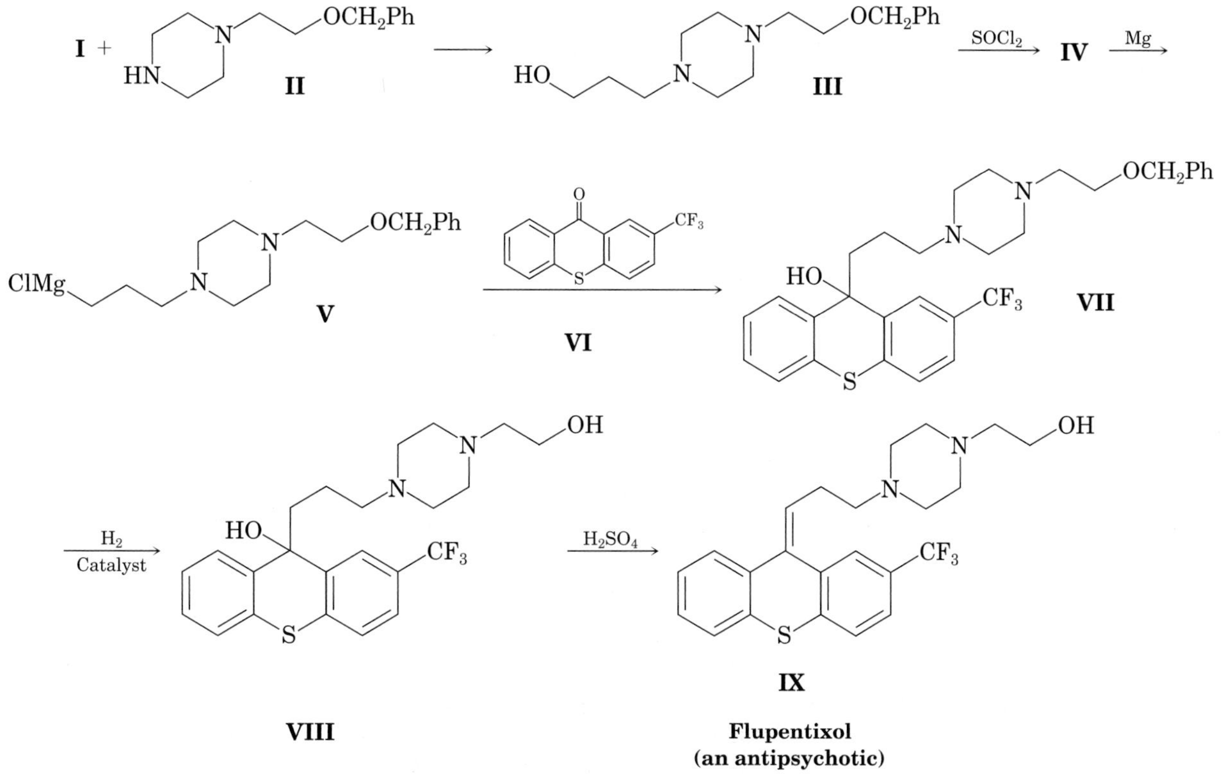

I + II → III $\xrightarrow{SOCl_2}$ IV $\xrightarrow{Mg}$

V + VI → VII

$\xrightarrow[\text{Catalyst}]{H_2}$ VIII $\xrightarrow{H_2SO_4}$ IX

Flupentixol (an antipsychotic)

(a) What alkyl chloride, **I**, is reacted with **II** to form **III**?
(b) The reaction of **I** and **II** to make **III** is an S_N2 reaction. Identify the nucleophile and electrophile, and describe the effect that doubling the concentration of each will have on the overall reaction rate.
(c) Molecule **III** is treated with $SOCl_2$. What is the structure of product **IV**? (*Hint*: You might look ahead to structure **V** for a clue.)
(d) What is the name of the reaction that converts **IV** to **V**?
(e) Later in the text, we'll introduce chemistry of the C=O (carbonyl) group. Here, the alkylmagnesium chloride reagent, **V**, reacts with ketone **VI** to produce **VII**. Because of the newly formed chirality center, compound **VII** exists as a pair of stereoisomers. Draw them both.
(f) After converting ether **VII** to alcohol **VIII**, an E1 reaction is catalyzed by HCl to form **IX**. Draw a mechanism for the conversion of **VIII** to **IX**.
(g) Two stereoisomers of **IX**, Flupentixol, are possible. Only one is shown. Draw the other isomer and describe this type of stereoisomerism.

IN THE FIELD WITH AGROCHEMICALS

7.64 All five of the herbicides shown contain trifluoromethyl groups attached to a benzene ring. This group is commonly included in many bioactive molecules because it slows down oxidation of the aromatic ring, a predominant decomposition pathway. Because the atmosphere is oxidizing, these CF_3-containing herbicides are more stable than their CH_3 analogs.

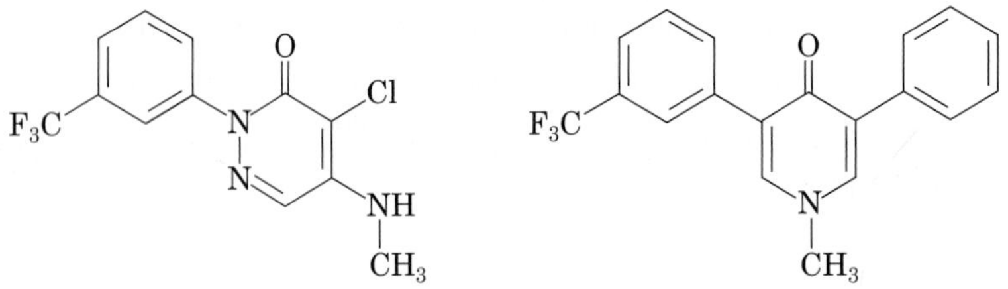

F_3C O Cl N N NH CH_3

Norflurazon

F_3C O N CH_3

Fluridone

Cl F_3C O O CH_3 NO_2

Oxyfluorfen

Cl F_3C O O O CH_3 NO_2 O O H_3C

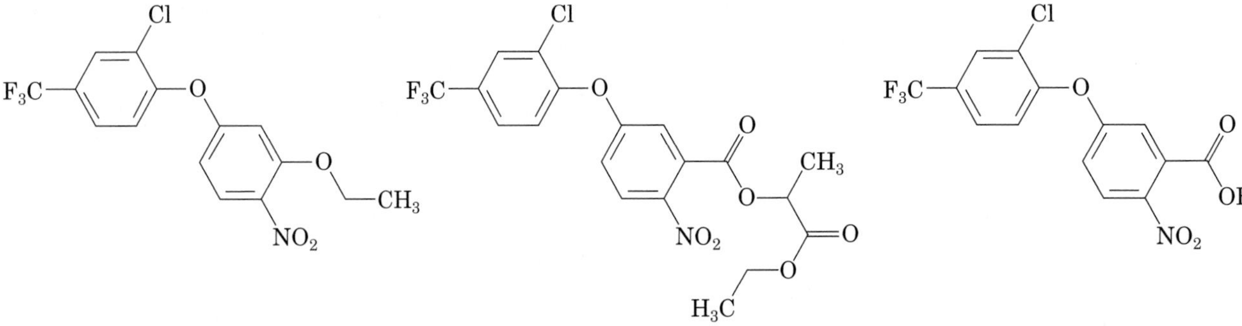

Lactofen

Cl F_3C O O OH NO_2

Acifluorfen

(a) Draw the bond dipoles for each of the C–F bonds in F_3C—Ar (where Ar stands for the *ar*omatic ring of any of these molecules). Would you describe the carbon of the CF_3 group as electron-rich or electron-deficient?
(b) Oxidation is the loss of electrons. Would it be easier to remove electrons from a benzene ring with a CF_3 or H group attached? Why?
(c) The molecules shown have two different mechanisms of action. Two of them inhibit pigment synthesis while three others inhibit lipid synthesis. Group these molecules by mechanism of action. What criteria did you use to make these groupings?

How do you get cations into organic solution? Simply *crown* them with cyclic polyethers whose oxygens mimic water molecules. (Charles D. Winters/Photo Researchers, Inc.)

CHAPTER 8

Alcohols, Phenols, and Ethers

In this and the next three chapters, we'll discuss the most common and important of all functional groups—those that contain *oxygen.* Let's begin in this chapter with a look at compounds that contain carbon–oxygen single bonds. An **alcohol** is a compound that has a hydroxyl group (–OH) bonded to a saturated, sp^3-hybridized carbon atom; a **phenol** has a hydroxyl group bonded to an aromatic ring; and an **ether** has an oxygen atom bonded to two organic groups.

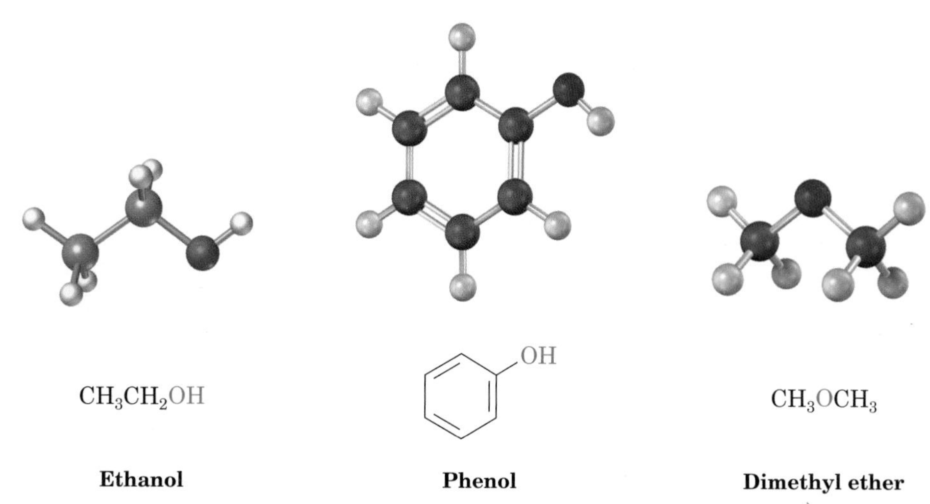

Alcohols, phenols, and ethers occur widely in nature and have many industrial, pharmaceutical, and biological applications. Ethanol, for instance, is a fuel additive, an industrial solvent, and a beverage; menthol is a flavoring agent; BHT (butylated hydroxytoluene) is an antioxidant food additive; and diethyl ether (the familiar "ether" of medical use) was once popular as an anesthetic agent.

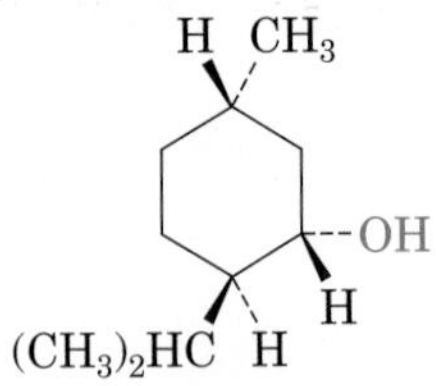

Menthol 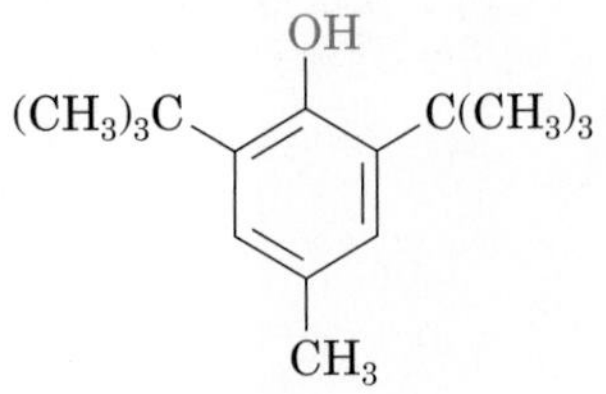 **BHT** **Diethyl ether** ($CH_3CH_2OCH_2CH_3$)

8.1 Naming Alcohols, Phenols, and Ethers

Alcohols

Alcohols are classified as primary (1°), secondary (2°), or tertiary (3°), depending on the number of carbon substituents bonded to the hydroxyl-bearing carbon:

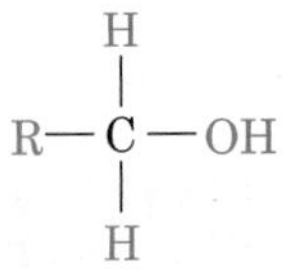

A primary alcohol (1°) — $R{-}CH_2{-}OH$ A secondary alcohol (2°) — $R_2CH{-}OH$ A tertiary alcohol (3°) — $R_3C{-}OH$

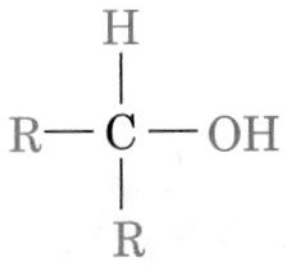

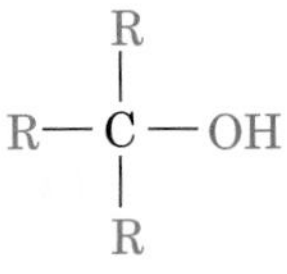

Simple alcohols are named in the IUPAC system as derivatives of the parent alkane, using the suffix *-ol*.

STEP 1 Select the longest carbon chain containing the hydroxyl group, and replace the *-e* ending of the corresponding alkane with *-ol*.

STEP 2 Number the carbons of the parent chain beginning at the end nearer the hydroxyl group.

STEP 3 Number all substituents according to their position on the chain, and write the name listing the substituents in alphabetical order.

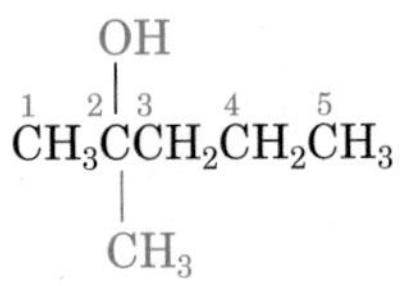

2-Methylpentan-2-ol ($CH_3C(OH)(CH_3)CH_2CH_2CH_3$) ***cis*-Cyclohexane-1,4-diol** 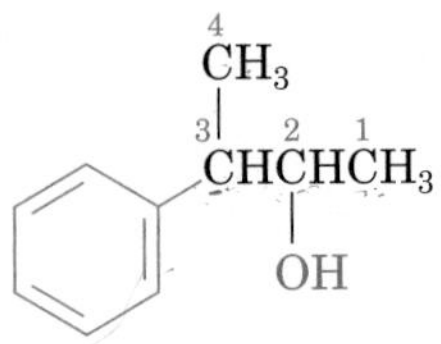 **3-Phenylbutan-2-ol**

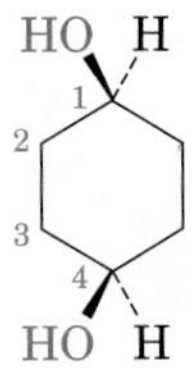

Some well-known alcohols also have common names. For example,

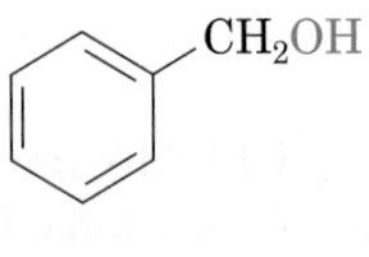

Benzyl alcohol (phenylmethanol) ($C_6H_5CH_2OH$) ***tert*-Butyl alcohol (2-methylpropan-2-ol)** ($(CH_3)_3COH$) **Ethylene glycol (ethane-1,2-diol)** ($HOCH_2CH_2OH$) **Glycerol (propane-1,2,3-triol)** ($HOCH_2CH(OH)CH_2OH$)

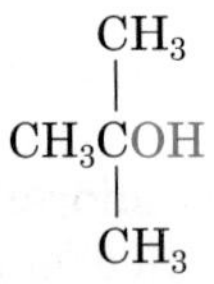

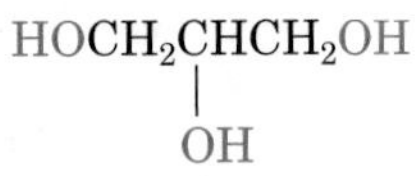

Phenols

The word *phenol* is used both as the name of a specific substance (hydroxybenzene) and as the family name for all hydroxy-substituted aromatic compounds. Phenols are named as substituted aromatic compounds according to the rules discussed in Section 5.3, with *-phenol* used as the parent name rather than *-benzene.*

m-Methylphenol (*m*-Cresol)

2,4-Dinitrophenol

Ethers

Simple ethers that contain no other functional groups are named by identifying the two organic groups and adding the word *ether.*

tert-Butyl **methyl ether**

Ethyl **phenyl ether**

If more than one ether linkage is present, or if other functional groups are present, the ether part is named as an *alkoxy* substituent on the parent compound.

p-Dimethoxy**benzene**

4-*tert*-Butoxy**cyclohex-1-ene**

PROBLEM 8.1 Give IUPAC names for the following compounds:

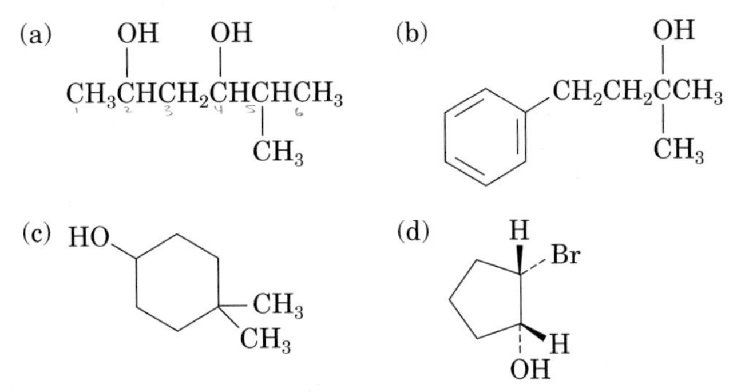

PROBLEM 8.2 Identify the alcohols in Problem 8.1 as primary, secondary, or tertiary.

PROBLEM 8.3 Draw structures corresponding to the following IUPAC names:

(a) 2-Methylhexan-2-ol (b) Hexane-1,5-diol
(c) 2-Ethylbut-2-en-1-ol (d) Cyclohex-3-en-1-ol
(e) *o*-Bromophenol (f) 2,4,6-Trinitrophenol

PROBLEM 8.4 Name the following ethers by IUPAC rules:

(a) $CH_3CHOCHCH_3$ (with CH_3 groups on the second and fourth carbons: CH_3 CH_3)

(b) Cyclopentyl–$OCH_2CH_2CH_3$

(c) Br–(benzene ring)–OCH_3

(d) $(CH_3)_2CHCH_2OCH_2CH_3$

8.2 Properties of Alcohols, Phenols, and Ethers: Hydrogen Bonding

Alcohols, phenols, and ethers can be thought of as organic derivatives of water in which one or both of the hydrogens have been replaced by organic parts: H–O–H becomes R–O–H or R–O–R′. Thus, all three classes of compounds have nearly the same geometry as water. The C–O–H or C–O–C bonds have an approximately tetrahedral angle—109° in methanol and 112° in dimethyl ether, for example—and the oxygen atom is sp^3-hybridized.

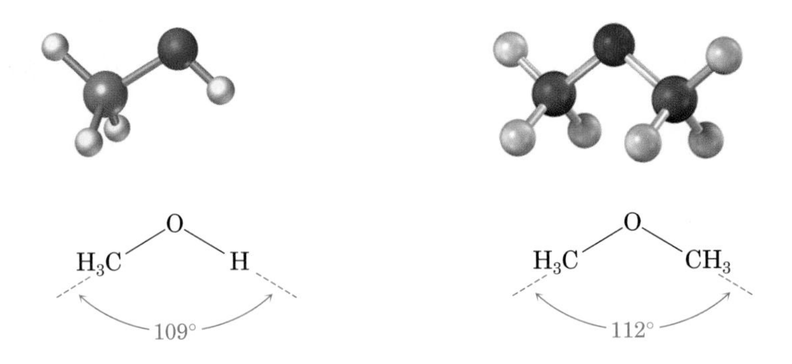

Alcohols and phenols differ significantly from the hydrocarbons and alkyl halides we've studied thus far. As shown in Figure 8.1, alcohols have higher boiling points than alkanes or haloalkanes of similar molecular weight. For example, the molecular weights of propan-1-ol (mol wt = 60 amu), butane (mol wt = 58 amu), and chloroethane (mol wt = 65 amu) are similar, but propan-1-ol boils at 97.4 °C, compared with −0.5 °C for the alkane and 12.3 °C for the chloroalkane. Similarly, phenols have higher boiling points than aromatic hydrocarbons. Phenol itself, for example, boils at 182 °C, whereas toluene boils at 110.6 °C.

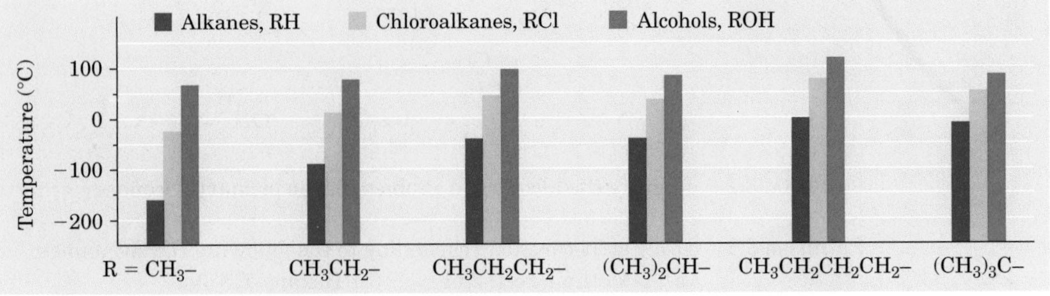

FIGURE 8.1 A comparison of boiling points for some alkanes, chloroalkanes, and alcohols. Alcohols have higher boiling points because of hydrogen bonding.

Alcohols and phenols have unusually high boiling points because, like water, they form hydrogen bonds in the liquid state. The positively polarized –OH hydrogen of one molecule is attracted to the negatively polarized oxygen of another molecule, resulting in a weak force that holds the molecules together (Figure 8.2). These forces must be overcome for a molecule to break free from the liquid and enter the vapor, so the boiling temperature is raised. Ethers, because they lack hydroxyl groups, can't form hydrogen bonds and therefore have lower boiling points.

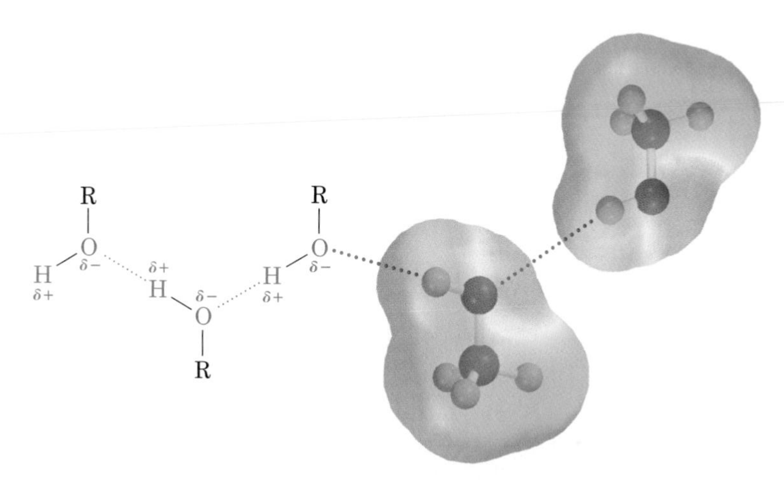

FIGURE 8.2 Hydrogen bonding in alcohols and phenols. The weak attraction between a positively polarized –OH hydrogen and a negatively polarized oxygen holds molecules together. The electrostatic potential map of methanol shows the polarization clearly.

8.3 Properties of Alcohols and Phenols: Acidity

Alcohols and phenols, like water, are both weakly basic and weakly acidic. As weak Lewis bases, alcohols and phenols are reversibly protonated by strong acids to yield oxonium ions, ROH_2^+:

$$R\ddot{O}H + H{-}X \rightleftharpoons R\overset{+}{O}H_2 \quad :X^-$$

An alcohol **An oxonium ion**

$$\left[\text{or } ArOH + HX \rightleftharpoons Ar\overset{+}{O}H_2\ X^-\right]$$

As weak acids, alcohols and phenols dissociate to a slight extent in dilute aqueous solution by donating a proton to water, generating H_3O^+ and an **alkoxide ion (RO^-)** or a **phenoxide ion (ArO^-)**.

$$R\ddot{O}{-}H + H_2\ddot{O}: \rightleftharpoons R\ddot{O}:^- + H_3O:^+$$

TABLE 8.1 Acidity Constants of Some Alcohols and Phenols

Alcohol or phenol	pK_a	
$(CH_3)_3COH$	18.00	Weaker acid
CH_3CH_2OH	16.00	
[HOH, water][a]	[15.74]	
CH_3OH	15.54	
p-Methylphenol	10.17	
Phenol	9.89	
p-Bromophenol	9.35	
p-Nitrophenol	7.15	Stronger acid

[a]Value for water is shown for reference.

Table 8.1 gives the pK_a values of some common alcohols and phenols. (You might want to review Sections 1.10 and 1.11 on the behavior of acids.)

The data in Table 8.1 show that alcohols are about as acidic as water. Thus, they react with alkali metals just as water does to yield metal alkoxides that are themselves strong bases.

$$2\ CH_3OH + 2\ Na \longrightarrow 2\ CH_3O^-\ Na^+ + H_2$$

Methanol → **Sodium methoxide**

$$2\ H_3C{-}C(CH_3)_2{-}OH + 2\ K \longrightarrow 2\ H_3C{-}C(CH_3)_2{-}O^-\ K^+ + H_2$$

***tert*-Butyl alcohol** → **Potassium *tert*-butoxide**

Phenols are about a million times more acidic than alcohols. In fact, some nitro-substituted phenols approach or surpass the acidity of carboxylic acids. One practical consequence of this acidity is that phenols are soluble in dilute aqueous NaOH, but alcohols are not.

$$C_6H_5{-}O{-}H + NaOH \longrightarrow C_6H_5{-}O^-\ Na^+ + H_2O$$

Phenol → **Sodium phenoxide**

Phenols are more acidic than alcohols because the phenoxide anion is resonance-stabilized by the aromatic ring. Sharing the negative charge over the ring increases the stability of the phenoxide anion and thus increases the tendency of the corresponding phenol to dissociate (Figure 8.3).

FIGURE 8.3 A phenoxide ion is more stable than an alkoxide ion because of resonance. Electrostatic potential maps show how the negative charge is less concentrated on the phenoxide oxygen than on the methoxide oxygen.

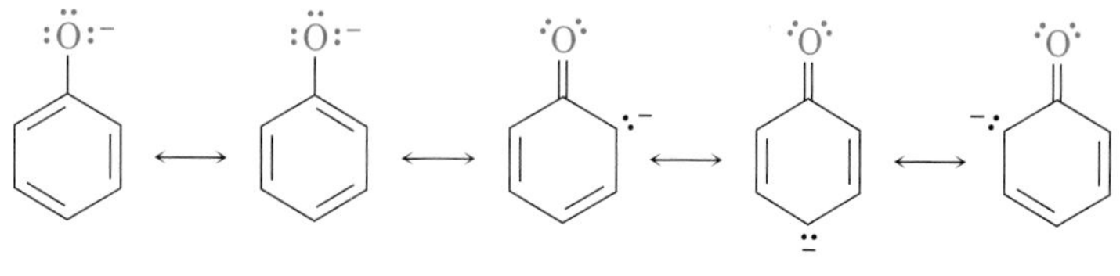

Phenoxide ion

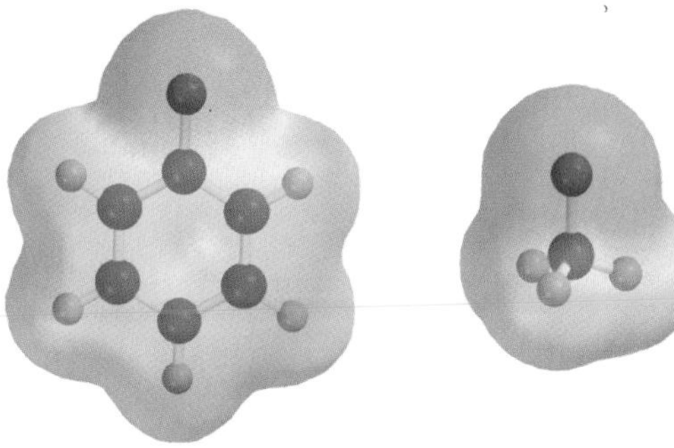

Methoxide ion

8.4 Synthesis of Alcohols

Alcohols occupy a central position in organic chemistry. They can be prepared from many other kinds of compounds (alkenes, alkyl halides, ketones, aldehydes, and esters, among others), and they can be transformed into an equally wide assortment of compounds. We've already seen, for instance, that alcohols can be prepared by hydration of alkenes (see Section 4.4). Treatment of the alkene with water and an acid catalyst leads to the Markovnikov product.

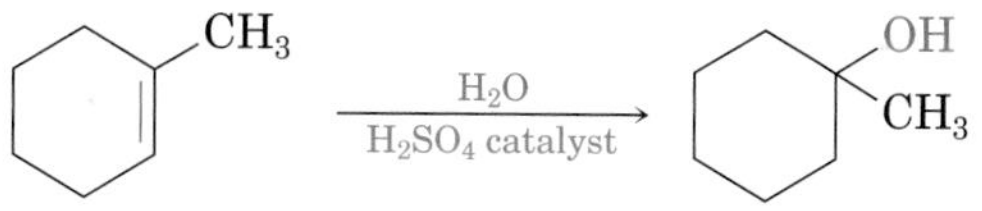

$$\text{1-Methylcyclohexene} \xrightarrow[H_2SO_4 \text{ catalyst}]{H_2O} \text{1-Methylcyclohexanol}$$

The most general method for preparing alcohols is by reduction of carbonyl compounds—the formal addition of H_2 to a C=O double bond. For the moment, we'll only list the *kinds* of carbonyl reductions that can be carried out, but we'll return for a look at the reduction mechanism in the next chapter (see Section 9.6).

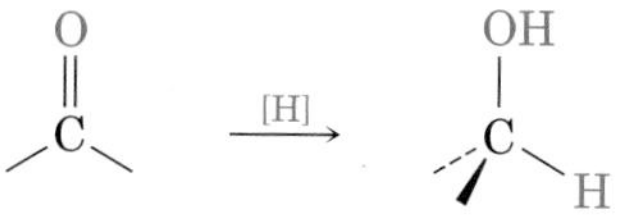

$$\text{A carbonyl compound} \xrightarrow{[H]} \text{An alcohol}$$

where [H] is a generalized reducing agent

Reduction of Aldehydes and Ketones

Aldehydes and ketones are easily reduced to yield alcohols. Aldehydes are converted into primary alcohols, and ketones are converted into secondary alcohols.

$$\text{An aldehyde (RCHO)} \xrightarrow{[H]} \text{A primary alcohol } (RCH_2OH)$$

$$\text{A ketone (RCOR')} \xrightarrow{[H]} \text{A secondary alcohol (RR'CHOH)}$$

Many reducing reagents are available, but sodium borohydride, $NaBH_4$, is usually chosen because of its safety and ease of handling.

Aldehyde reduction

$$CH_3CH_2CH_2\overset{O}{\overset{\|}{C}}H \xrightarrow[2.\ H_3O^+]{1.\ NaBH_4,\ ethanol} CH_3CH_2CH_2CH(OH)H$$

Butanal → **Butan-1-ol (85%) (a 1° alcohol)**

Ketone reduction

Dicyclohexyl ketone $\xrightarrow[2.\ H_3O^+]{1.\ NaBH_4,\ ethanol}$ **Dicyclohexylmethanol (88%) (a 2° alcohol)**

Reduction of Esters and Carboxylic Acids

Esters and carboxylic acids are reduced to give primary alcohols:

$$RCOOH \text{ or } RCOOR' \xrightarrow{[H]} RCH_2OH$$

A carboxylic acid, An ester → A primary alcohol

These reactions proceed more slowly than the corresponding reductions of aldehydes and ketones, so the more powerful reducing agent $LiAlH_4$ is used rather than $NaBH_4$. ($LiAlH_4$ will also reduce aldehydes and ketones.) Note that only one hydrogen is added to the carbonyl carbon atom during the reduction of an aldehyde or ketone, but two hydrogens are added to the carbonyl carbon during reduction of an ester or carboxylic acid.

Carboxylic acid reduction

$$CH_3(CH_2)_7CH{=}CH(CH_2)_7\overset{O}{\overset{\|}{C}}OH \xrightarrow[2.\ H_3O^+]{1.\ LiAlH_4,\ ether} CH_3(CH_2)_7CH{=}CH(CH_2)_7CH_2OH$$

Octadec-9-enoic acid (oleic acid) → **Octadec-9-en-1-ol (87%)**

Ester reduction

$$CH_3CH_2CH{=}CH\overset{O}{\overset{\|}{C}}OCH_3 \xrightarrow[2.\ H_3O^+]{1.\ LiAlH_4,\ ether} CH_3CH_2CH{=}CHCH_2OH + CH_3OH$$

Methyl pent-2-enoate → **Pent-2-en-1-ol (91%)**

Practice Problem 8.1

Predict the product of the following reaction:

$$CH_3CH_2CH_2\overset{\overset{\displaystyle O}{\|}}{C}CH_2CH_3 \xrightarrow[\text{2. } H_3O^+]{\text{1. } NaBH_4} ?$$

Strategy Ketones are reduced by treatment with $NaBH_4$ to yield secondary alcohols. Thus, reduction of hexan-3-one yields hexan-3-ol.

Solution

$$\underset{\textbf{Hexan-3-one}}{CH_3CH_2CH_2\overset{\overset{\displaystyle O}{\|}}{C}CH_2CH_3} \xrightarrow[\text{2. } H_3O^+]{\text{1. } NaBH_4} \underset{\textbf{Hexan-3-ol}}{CH_3CH_2CH_2\overset{\overset{\displaystyle OH}{|}}{C}HCH_2CH_3}$$

Practice Problem 8.2

What carbonyl compound(s) might you reduce to obtain the following alcohol?

$(CH_3)_2CHCH{=}CHCH_2OH$

Strategy Identify the alcohol as primary, secondary, or tertiary. A primary alcohol can be prepared by reduction of an aldehyde, an ester, or a carboxylic acid; a secondary alcohol can be prepared by reduction of a ketone; and a tertiary alcohol cannot be prepared by reduction. In this case the target molecule is a primary alcohol, which can be prepared by reduction of an aldehyde, an ester, or a carboxylic acid. $LiAlH_4$ is needed for the ester and carboxylic acid reductions.

Solution

$(CH_3)_2CHCH{=}CHCHO \xrightarrow[\text{2. } H_3O^+]{\text{1. } NaBH_4 \text{ or } LiAlH_4}$

$(CH_3)_2CHCH{=}CHCO_2CH_3 \xrightarrow[\text{2. } H_3O^+]{\text{1. } LiAlH_4} (CH_3)_2CHCH{=}CHCH_2OH$

$(CH_3)_2CHCH{=}CHCOOH \xrightarrow[\text{2. } H_3O^+]{\text{1. } LiAlH_4}$

Problem 8.5 How would you carry out the following reactions?

(a) $CH_3\overset{\overset{\displaystyle O}{\|}}{C}CH_2CH_2\overset{\overset{\displaystyle O}{\|}}{C}OCH_3 \xrightarrow{?} CH_3\overset{\overset{\displaystyle OH}{|}}{C}HCH_2CH_2\overset{\overset{\displaystyle O}{\|}}{C}OCH_3$

(b) $CH_3\overset{\overset{\displaystyle O}{\|}}{C}CH_2CH_2\overset{\overset{\displaystyle O}{\|}}{C}OCH_3 \xrightarrow{?} CH_3\overset{\overset{\displaystyle OH}{|}}{C}HCH_2CH_2CH_2OH$

PROBLEM 8.6 What carbonyl compounds give the following alcohols on reduction with $LiAlH_4$? Show all possibilities.

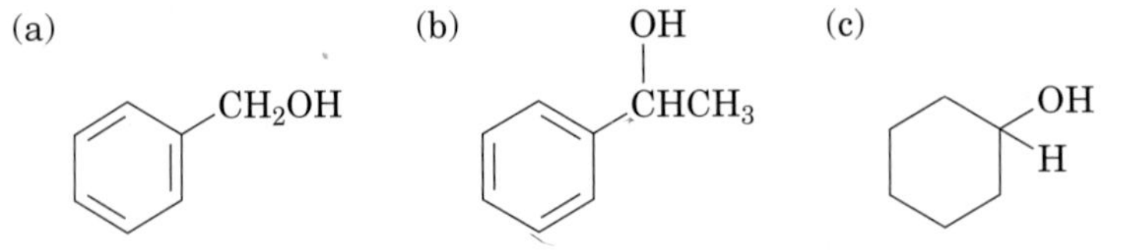

(a) $C_6H_5CH_2OH$ (b) $C_6H_5CH(OH)CH_3$ (c) cyclohexanol (OH, H)

8.5 Reactions of Alcohols

Dehydration of Alcohols

Alcohols undergo **dehydration**—the elimination of H_2O—to give alkenes. Although many ways of carrying out the reaction have been devised, a method that works particularly well for secondary and tertiary alcohols is treatment with a strong acid. For example, when 1-methylcyclohexanol is treated with aqueous sulfuric acid, dehydration occurs to yield 1-methylcyclohexene:

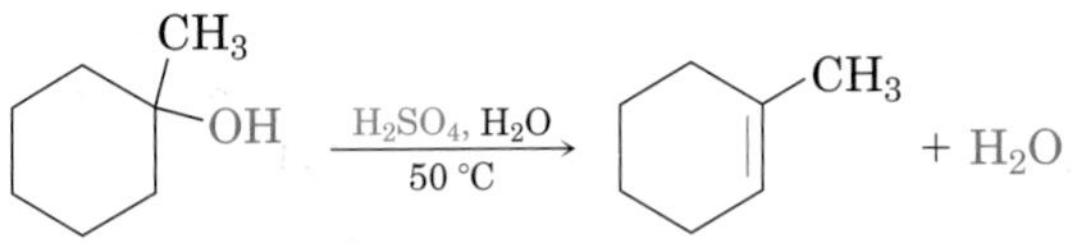

$$\text{1-Methylcyclohexanol} \xrightarrow[50\ °C]{H_2SO_4,\ H_2O} \text{1-Methylcyclohexene (91\%)} + H_2O$$

1-Methylcyclohexanol **1-Methylcyclohexene (91%)**

Acid-catalyzed dehydrations usually follow Zaitsev's rule (see Section 7.7) and yield the more highly substituted alkene as major product. Thus, 2-methylbutan-2-ol gives primarily 2-methylbut-2-ene (trisubstituted) rather than 2-methylbut-1-ene (disubstituted):

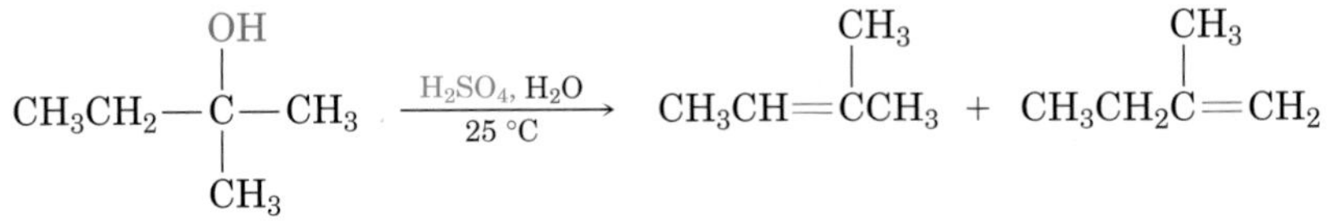

$$CH_3CH_2C(OH)(CH_3)CH_3 \xrightarrow[25\ °C]{H_2SO_4,\ H_2O} CH_3CH{=}C(CH_3)CH_3 + CH_3CH_2C(CH_3){=}CH_2$$

2-Methylbutan-2-ol **2-Methylbut-2-ene** (major) **2-Methylbut-1-ene** (minor)

The mechanism of this acid-catalyzed dehydration is simply an E1 process (see Section 7.8). Strong acid protonates the alcohol oxygen, the protonated intermediate spontaneously loses water to generate a carbocation, and loss of H^+ from a neighboring carbon atom then yields the alkene product (Figure 8.4).

PRACTICE PROBLEM 8.3

Predict the major product of the following reaction:

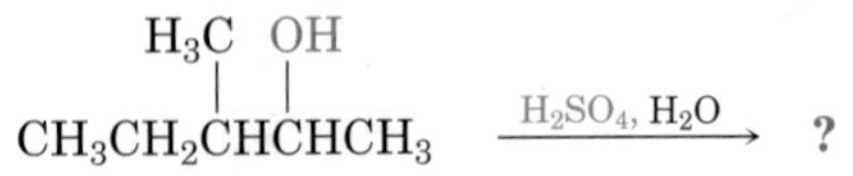

$$CH_3CH_2CH(CH_3)CH(OH)CH_3 \xrightarrow{H_2SO_4,\ H_2O} ?$$

STRATEGY Treatment of an alcohol with H_2SO_4 leads to dehydration and formation of the more highly substituted alkene product (Zaitsev's rule). Thus, dehydration of 3-methylpentan-2-ol yields 3-methylpent-2-ene as the major product rather than 3-methylpent-1-ene.

FIGURE 8.4 MECHANISM: Mechanism of the acid-catalyzed dehydration of a tertiary alcohol by an E1 reaction.

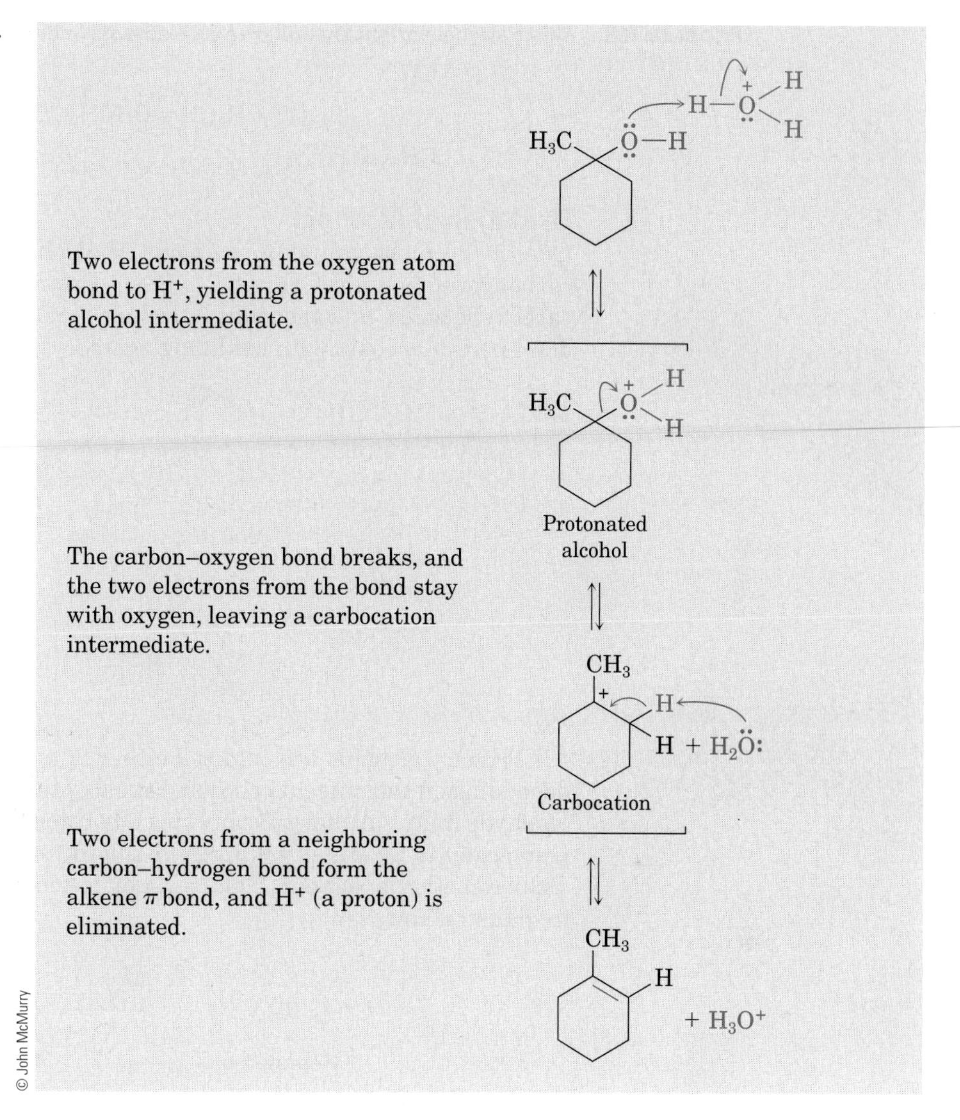
H₃C
Ö—H
H—O—H
H
H
Two electrons from the oxygen atom bond to H⁺, yielding a protonated alcohol intermediate.
H₃C
O
H
H
Protonated alcohol
The carbon–oxygen bond breaks, and the two electrons from the bond stay with oxygen, leaving a carbocation intermediate.
CH₃
H
H + H₂Ö:
Carbocation
Two electrons from a neighboring carbon–hydrogen bond form the alkene π bond, and H⁺ (a proton) is eliminated.
CH₃
H
+ H₃O⁺
© John McMurry

SOLUTION

$$\underset{\text{3-Methylpentan-2-ol}}{CH_3CH_2\overset{H_3C\ \ OH}{\overset{|\ \ \ |}{CHCH}}CH_3} \xrightarrow{H_2SO_4,\ H_2O} \underset{\substack{\text{3-Methylpent-2-ene}\\ \text{(major)}}}{CH_3CH_2\overset{CH_3}{\overset{|}{C}}{=}CHCH_3} + \underset{\substack{\text{3-Methylpent-1-ene}\\ \text{(minor)}}}{CH_3CH_2\overset{CH_3}{\overset{|}{C}}HCH{=}CH_2}$$

PROBLEM 8.7 Predict the products you would expect from the following reactions. Indicate the major product in each case.

(a) $CH_3\overset{H_3C}{\overset{|}{C}}H\underset{CH_3}{\underset{|}{\overset{OH}{\overset{|}{C}}}}CH_2CH_3 \xrightarrow{H_2SO_4}$

(b) $CH_3CH_2CH_2\underset{CH_3}{\underset{|}{\overset{OH}{\overset{|}{C}}}}CH_3 \xrightarrow{H_2SO_4}$

PROBLEM 8.8 What alcohols might the following alkenes have been made from?

(a) 1,2-dimethylcyclohexene ring structure with two CH_3 groups   (b) $CH_3CH_2CH{=}CHCH_2CH_2CH_3$

Oxidation of Alcohols

One of the most valuable reactions of alcohols is their oxidation to yield carbonyl compounds: CH—OH → C=O. Primary alcohols yield aldehydes or carboxylic acids, and secondary alcohols yield ketones, but tertiary alcohols don't normally react with oxidizing agents.

Primary alcohol

$$RCH_2OH \xrightarrow{[O]} RCHO \xrightarrow{[O]} RCOOH$$

An aldehyde   A carboxylic acid

Secondary alcohol

$$RR'CHOH \xrightarrow{[O]} RCOR'$$

A ketone

Primary alcohols are oxidized either to aldehydes or to carboxylic acids, depending on the reagents chosen. Probably the best method for preparing an aldehyde from a primary alcohol on a laboratory scale (as opposed to an industrial scale) is by use of pyridinium chlorochromate (PCC,$C_5H_6NCrO_3Cl$) in dichloromethane solvent. This reagent is too expensive for large-scale use in industry, however.

$$CH_3(CH_2)_5CH_2OH \xrightarrow[CH_2Cl_2]{PCC} CH_3(CH_2)_5CHO$$

Heptan-1-ol   **Heptanal (78%)**

Many oxidizing agents, such as chromium trioxide (CrO_3) and sodium dichromate ($Na_2Cr_2O_7$) in aqueous acid solution, oxidize primary alcohols to carboxylic acids. Although aldehydes are intermediates in these oxidations, they usually can't be isolated because further oxidation takes place too rapidly.

$$CH_3(CH_2)_8CH_2OH \xrightarrow[H_3O^+]{CrO_3} CH_3(CH_2)_8COOH$$

Decan-1-ol   **Decanoic acid (93%)**

Secondary alcohols are oxidized easily to produce ketones. Sodium dichromate in aqueous acetic acid is often used as the oxidant, although pyridinium chlorochromate also works.

$$(CH_3)_3C\text{–}C_6H_{10}\text{–}OH \xrightarrow[H_2O,\ CH_3CO_2H]{Na_2Cr_2O_7} (CH_3)_3C\text{–}C_6H_9{=}O$$

4-*tert*-Butylcyclohexanol   **4-*tert*-Butylcyclohexanone (91%)**

PRACTICE PROBLEM 8.4

What product would you expect from reaction of benzyl alcohol with CrO_3?

$C_6H_5-CH_2OH$ **Benzyl alcohol**

STRATEGY Treatment of a primary alcohol with CrO_3 yields a carboxylic acid. Thus, oxidation of benzyl alcohol yields benzoic acid.

SOLUTION

$$C_6H_5-CH_2OH \xrightarrow[H_3O^+]{CrO_3} C_6H_5-\overset{O}{\overset{\|}{C}}OH$$

Benzyl alcohol **Benzoic acid**

PROBLEM 8.9 What alcohols would give the following products on oxidation?

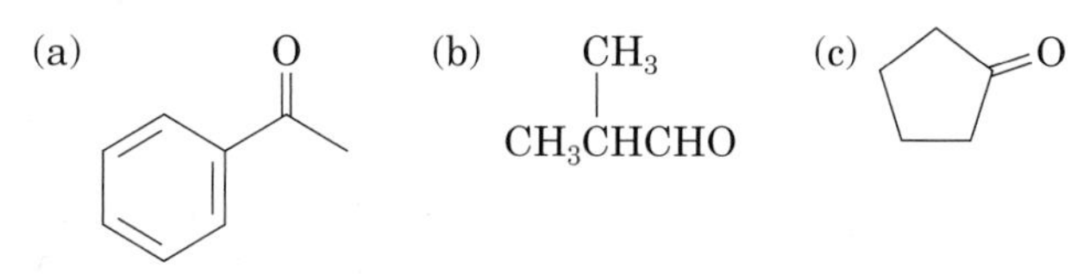

PROBLEM 8.10 What products would you expect to obtain from oxidation of the following alcohols with CrO_3?
(a) Cyclohexanol (b) Hexan-1-ol (c) Hexan-2-ol

PROBLEM 8.11 What products would you expect to obtain from oxidation of the alcohols in Problem 8.10 with pyridinium chlorochromate (PCC)?

Conversion into Ethers

Alcohols are converted into ethers by formation of the corresponding alkoxide ion followed by reaction with an alkyl halide, a reaction known as the *Williamson ether synthesis.* As noted in Section 8.3, the alkoxide ion needed in the reaction can be prepared by reaction of an alcohol with sodium or potassium metal.

$$2\ ROH + 2\ Na \longrightarrow 2\ RO^- Na^+ + H_2$$

$$C_5H_9-\ddot{O}:^- + CH_3-I \longrightarrow C_5H_9-O-CH_3 + I^-$$

Cyclopentoxide ion **Cyclopentyl methyl ether (74%)**

Mechanistically, the Williamson synthesis is an S_N2 reaction (see Section 7.5) and occurs by nucleophilic substitution of halide ion by the alkoxide ion. As with all S_N2 reactions, primary alkyl halides work best because competitive E2 elimination of HX can occur with more hindered substrates. Unsymmetrical ethers are therefore best prepared by reaction of the more hindered alkoxide partner with the less hindered alkyl halide partner, rather than vice versa. For example, *tert*-butyl methyl ether is best synthesized by reaction of *tert*-butoxide ion with iodomethane, rather than by reaction of methoxide ion with 2-chloro-2-methylpropane.

S_N2 reaction

$$(CH_3)_3C\text{—}\ddot{\underset{\cdot\cdot}{O}}{:}^- + CH_3\text{—}I \longrightarrow (CH_3)_3C\text{—}O\text{—}CH_3 + I^-$$

***tert*-Butoxide ion** **Iodomethane** ***tert*-Butyl methyl ether**

E2 reaction

$$CH_3\ddot{\underset{\cdot\cdot}{O}}{:}^- + H\text{—}CH_2\text{—}C(CH_3)_2\text{—}Cl \longrightarrow H_2C{=}C(CH_3)_2 + CH_3OH + Cl^-$$

Methoxide ion **2-Chloro-2-methylpropane** **2-Methylpropene**

PROBLEM 8.12 Treatment of cyclohexanol with sodium metal gives an alkoxide ion that undergoes reaction with iodoethane to yield cyclohexyl ethyl ether. Write the reaction, showing all the steps.

PROBLEM 8.13 How would you prepare the following ethers?

(a) $CH_3OCH_2CH_2CH_3$ (b) $C_6H_5OCH_3$ (c) $C_6H_5CH_2OCH(CH_3)CH_3$

PROBLEM 8.14 Rank the following alkyl halides in order of their expected reactivity toward an alkoxide ion in the Williamson ether synthesis: bromoethane, 2-bromopropane, chloroethane, 2-chloro-2-methylpropane.

PROBLEM 8.15 Draw the structure of the carbonyl compound(s) from which the following alcohol might have been prepared, and show the products you would obtain by treatment of the alcohol with (i) Na metal, followed by CH_3I, (ii) $SOCl_2$, and (iii) CrO_3.

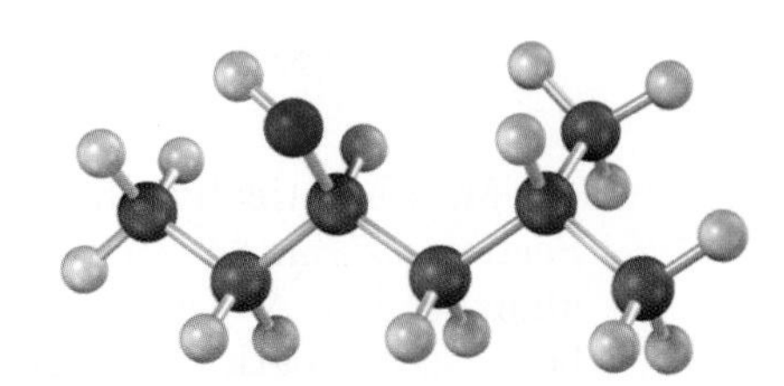

8.6

Synthesis and Reactions of Phenols

Synthesis of Phenols

Phenols can be synthesized from aromatic starting materials by a two-step sequence. The starting compound is first sulfonated by treatment with SO_3/H_2SO_4 (see Section 5.5), and the sulfonic acid product is then converted into a phenol by high-temperature reaction with NaOH.

SO_3H OH

$\xrightarrow[H_2SO_4]{SO_3}$ $\xrightarrow[300\ °C]{1.\ NaOH}$ 2. H_3O^+

CH_3 CH_3 CH_3

Toluene ***p*-Toluenesulfonic acid** ***p*-Methylphenol (72%)**

PROBLEM 8.16 *p*-Cresol (*p*-methylphenol) is used industrially both as an antiseptic and as a starting material to prepare the food additive BHT. Show how you could synthesize *p*-cresol from benzene.

Alcohol-Like Reactions of Phenols

Phenols and alcohols are very different despite the fact that both have –OH groups. Phenols can't be dehydrated by treatment with acid and can't be converted into halides by treatment with HX. Phenols can, however, be converted into ethers by S_N2 reaction with alkyl halides in the presence of base. Williamson ether synthesis with phenols occurs easily because phenols are more acidic than alcohols and are therefore more readily converted into their anions.

OH $OCH_2CH_2CH_2CH_3$

$+\ CH_3CH_2CH_2CH_2Br \xrightarrow[Acetone]{K_2CO_3}$

NO_2 NO_2

***o*-Nitrophenol** **1-Bromobutane** **Butyl *o*-nitrophenyl ether (80%)**

Electrophilic Aromatic Substitution Reactions of Phenols

The –OH group is an activating, ortho- and para-directing substituent in electrophilic aromatic substitution reactions (see Sections 5.7 and 5.8). As a result, phenols are reactive substrates for electrophilic halogenation, nitration, and sulfonation.

OH

Br_2, $FeBr_3$ HNO_3, H_2SO_4 SO_3, H_2SO_4

OH OH OH

Br NO_2 SO_3H

Oxidation of Phenols: Quinones

Treatment of a phenol with a strong oxidizing agent such as sodium dichromate yields a cyclohexadienedione, or **quinone**:

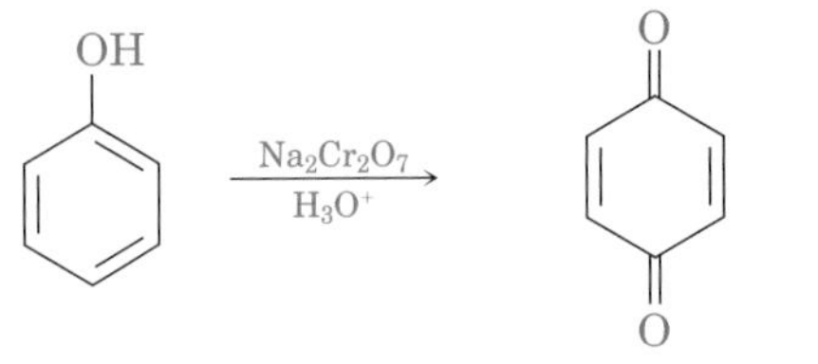

Phenol $\xrightarrow[H_3O^+]{Na_2Cr_2O_7}$ Benzoquinone

Phenol **Benzoquinone**

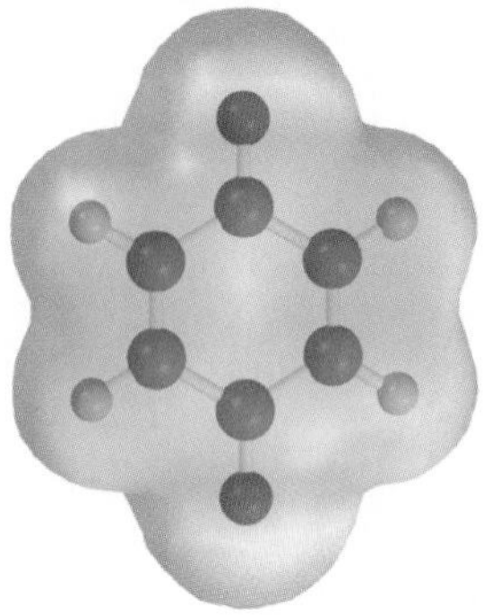

Quinones are an interesting and valuable class of compounds because of their oxidation–reduction properties. They can be easily reduced to **hydroquinones** (*p*-dihydroxybenzenes) by $NaBH_4$ or $SnCl_2$, and hydroquinones can be easily oxidized back to quinones by $Na_2Cr_2O_7$. Hydroquinone is used, among other things, as a photographic developer because it reduces Ag^+ on film to metallic silver.

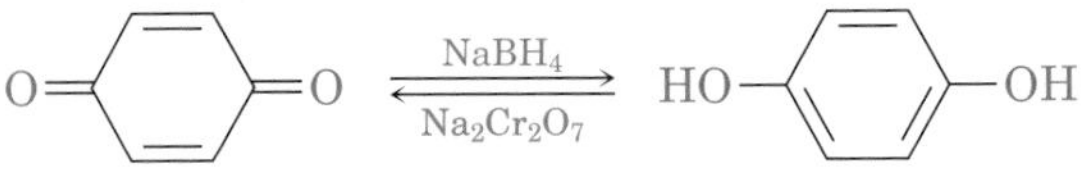

$$O=C_6H_4=O \underset{Na_2Cr_2O_7}{\overset{NaBH_4}{\rightleftharpoons}} HO-C_6H_4-OH$$

Benzoquinone **Hydroquinone**

The oxidation–reduction properties of quinones are crucial to the functioning of living cells, where compounds called *ubiquinones* act as biochemical oxidizing agents to mediate the electron-transfer processes involved in energy production. Ubiquinones, also called *coenzymes Q*, are components of the cells of all aerobic organisms, from the simplest bacterium to humans. They are so named because of their ubiquitous occurrence in nature.

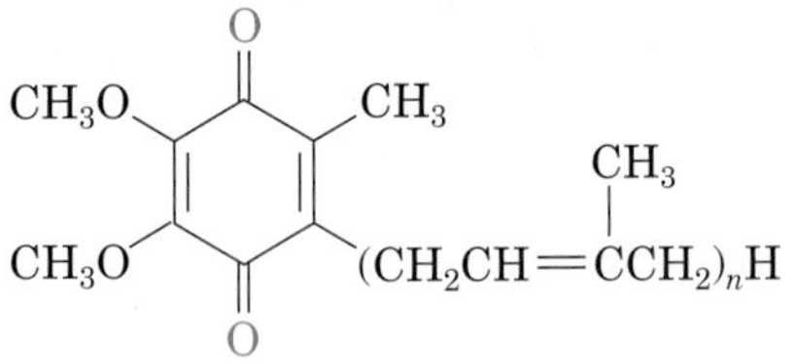

CH_3O, CH_3O, CH_3, $(CH_2CH{=}C(CH_3)CH_2)_nH$

Ubiquinones (n = 1–10)

Ubiquinones function within the mitochondria of cells to mediate the respiration process in which electrons are transported from the biological reducing agent NADH to molecular oxygen. Although a complex series of steps is involved in the overall process, the ultimate result is a cycle whereby NADH is oxidized to NAD^+, O_2 is reduced to water, and energy is produced. Ubiquinone acts only as an intermediary and is itself unchanged.

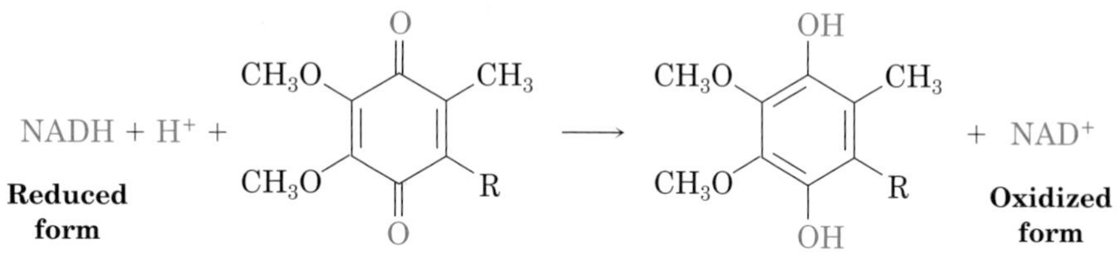

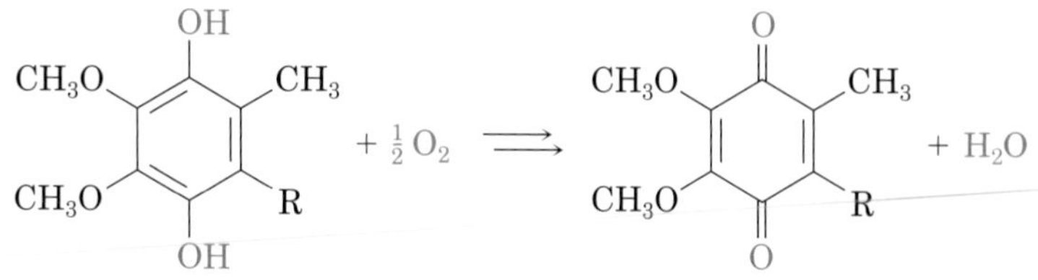

NADH + H^+ + [2,3-dimethoxy-5-methyl-6-R-1,4-benzoquinone] ⟶ [hydroquinone form] + NAD^+

Reduced form **Oxidized form**

[hydroquinone] + $\frac{1}{2}O_2$ ⇌ [quinone] + H_2O

8.7 Synthesis and Reactions of Ethers

Synthesis of Ethers

As noted in Section 8.5, ethers are easily prepared from alcohols by conversion to the alkoxide ion followed by S_N2 reaction with a primary alkyl halide.

$$\underset{\text{An alcohol}}{R{-}O{-}H} \longrightarrow \underset{\text{An alkoxide ion}}{R{-}O^-} \xrightarrow[S_N2]{R'X} \underset{\text{An ether}}{R{-}O{-}R'}$$

Reactions of Ethers

Ethers are unreactive to most common reagents, a property that accounts for their frequent use as reaction solvents. Halogens, mild acids, bases, and nucleophiles have no effect on most ethers. In fact, ethers undergo only one general reaction—they are cleaved by strong acids such as aqueous HI or HBr.

Acidic ether cleavages are typical nucleophilic substitution reactions. They take place by either an S_N1 or S_N2 pathway, depending on the structure of the ether. Ethers with only primary and secondary alkyl groups react by an S_N2 pathway, in which nucleophilic halide ion attacks the protonated ether at the less highly substituted site. The ether oxygen atom stays with the more hindered alkyl group, and the halide bonds to the less hindered group. For example, ethyl isopropyl ether yields isopropyl alcohol and iodoethane on cleavage by HI:

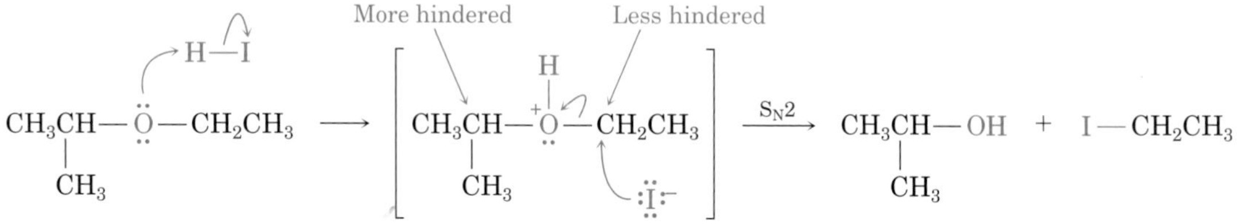

$$\underset{\text{Ethyl isopropyl ether}}{CH_3CH(CH_3){-}\ddot{O}{-}CH_2CH_3} \xrightarrow{H{-}I} \left[CH_3CH(CH_3){-}\overset{+}{O}(H){-}CH_2CH_3 \;\; :\ddot{I}:^-\right] \xrightarrow{S_N2} \underset{\text{Isopropyl alcohol}}{CH_3CH(CH_3){-}OH} + \underset{\text{Iodoethane}}{I{-}CH_2CH_3}$$

More hindered — Less hindered

Ethers with a tertiary alkyl group cleave by an S_N1 mechanism because they can produce stable intermediate carbocations. In such reactions, the ether oxygen atom stays with the *less* hindered alkyl group and the halide bonds to the tertiary group. Like most S_N1 reactions, the cleavage is fast and often takes place at room temperature or below.

Secondary — Tertiary

$$\text{(cyclohexyl)–O–C(CH}_3)_3 \xrightarrow[S_N1]{HBr,\ H_2O} \text{(cyclohexyl)–OH} + CH_3CBr(CH_3)_2$$

tert-Butyl cyclohexyl ether — Cyclohexanol — 2-Bromo-2-methylpropane

Practice Problem 8.5

Predict the products of the reaction of *tert*-butyl propyl ether with HBr.

Strategy Identify the substitution pattern of the two groups attached to oxygen—in this case a tertiary alkyl group and a primary alkyl group. Then recall the guidelines for ether cleavages. An ether with only primary and secondary alkyl groups usually undergoes cleavage by S_N2 attack of a nucleophile on the less hindered alkyl group, but an ether with a tertiary alkyl group usually undergoes cleavage by an S_N1 mechanism. In this case an S_N1 cleavage of the tertiary C–O bond will occur, giving propan-1-ol and a tertiary alkyl bromide.

Solution

$$CH_3C(CH_3)_2—O—CH_2CH_2CH_3 \xrightarrow{HBr} CH_3C(CH_3)_2—Br + HOCH_2CH_2CH_3$$

tert-Butyl propyl ether — 2-Bromo-2-methylpropane — Propan-1-ol

Problem 8.17 What products do you expect from the reaction of the following ethers with HI?

(a) $CH_3CH_2OCH_2CH_3$ (b) (cyclohexyl)OCH_2CH_3 (c) $CH_3C(CH_3)_2OCH_2CH_3$

8.8 Epoxides

For the most part, cyclic ethers behave like acyclic ethers. The chemistry of the ether functional group is the same whether it's in an open chain or in a ring. Thus, the cyclic ether tetrahydrofuran (THF) is often used as a solvent because of its inertness.

Tetrahydrofuran

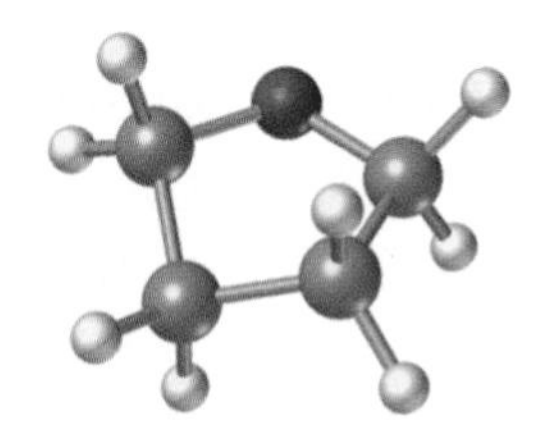

The one group of cyclic ethers that behave differently from open-chain ethers are the three-membered ring compounds called **epoxides**, or *oxiranes*. The strain of the three-membered ring gives epoxides unique chemical reactivity.

Epoxides are prepared by reaction of an alkene with a *peroxyacid*, RCO_3H. *m*-Chloroperoxybenzoic acid is often used because it is more stable and more easily handled than other peroxyacids.

CH_2Cl_2, 25 °C

Cycloheptene **1,2-Epoxycycloheptane (78%)**

Epoxide rings are cleaved by treatment with acid just like other ethers. The major difference is that epoxides react under much milder conditions because of the strain of the three-membered ring. Dilute aqueous acid at room temperature converts an epoxide to a 1,2-diol (also called a *glycol*). Two million tons of ethylene glycol, most of it used as automobile antifreeze, are produced every year in the United States by acid-catalyzed ring opening of ethylene oxide. Note that the name *ethylene glycol* refers to the glycol derived *from* ethylene. Similarly, *ethylene oxide* is the common name of the epoxide derived from ethylene.

H_3O^+

Ethylene oxide **Ethylene glycol (ethane-1,2-diol)**

Acid-catalyzed epoxide cleavage takes place by S_N2 attack of H_2O on the protonated epoxide.

H_3O^+ :$\ddot{O}H_2$:$\ddot{O}H_2$ + H_3O^+

1,2-Epoxycyclohexane ***trans*-Cyclohexane-1,2-diol (86%)**

Epoxide opening is also involved in the mechanism by which benzo[a]pyrene and other polycyclic aromatic hydrocarbons in chimney soot and cigarette smoke cause cancer (see Section 5.10). Benzo[a]pyrene is converted by metabolic oxidation into a diol epoxide, which then undergoes nucleophilic ring opening by reaction with an amino group in cellular DNA. With its DNA thus altered, the cell is unable to function normally.

OH, OH, OH, OH, O, $H_2\ddot{N}$—DNA, HN—DNA

Benzo[*a*]pyrene A diol epoxide

PROBLEM 8.18 Show the structure of the product you would obtain by treatment of the following epoxide with aqueous acid. What is the stereochemistry of the product if the ring opening takes place by normal backside S_N2 attack?

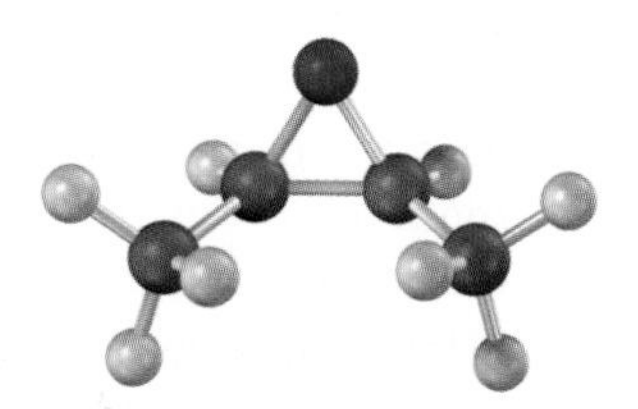

8.9 Thiols and Sulfides

Sulfur is the element just below oxygen in the periodic table, and many oxygen-containing organic compounds have sulfur analogs. **Thiols, R—SH,** are sulfur analogs of alcohols, and **sulfides, R—S—R′**, are sulfur analogs of ethers. Both classes of compounds are widespread in living organisms.

Thiols are named in the same way as alcohols, with the suffix *-thiol* used in place of *-ol*. The –SH group itself is referred to as a **mercapto group**.

CH_3CH_2SH — **Ethanethiol**

SH (on cyclohexane) — **Cyclohexanethiol**

CO_2H, SH (on benzene ring) — ***m*-Mercaptobenzoic acid**

Sulfides are named in the same way as ethers, with *sulfide* used in place of *ether* for simple compounds and with *alkylthio* used in place of *alkoxy* for more complex substances.

CH_3—S—CH_3 — **Dimethyl sulfide**

(phenyl)—S—CH_3 — **Methyl phenyl sulfide**

1, 2, 3 (cyclohexene)—S—CH_3 — **3-(Methylthio)cyclohexene**

Thiols can be prepared from the corresponding alkyl halide by S_N2 displacement with a sulfur nucleophile such as hydrosulfide anion, SH^-:

$$CH_3(CH_2)_6CH_2{-}Br + Na^+ \ {}^-{:}\ddot{\underset{\cdot\cdot}{S}}H \longrightarrow CH_3(CH_2)_6CH_2SH + NaBr$$

1-Bromooctane **Sodium hydrosulfide** **Octane-1-thiol**

Sulfides are prepared by treating a primary or secondary alkyl halide with a *thiolate ion,* RS^-, the sulfur analog of an alkoxide ion. Reaction occurs by an S_N2 mechanism analogous to the Williamson ether synthesis (see Section 8.5). Thiolate anions are among the best nucleophiles known, so these reactions usually work well.

$$C_6H_5S{:}^- \ Na^+ + CH_3{-}I \longrightarrow C_6H_5SCH_3 + NaI$$

Sodium benzenethiolate **Methyl phenyl sulfide (96%)**

Physically, the most unforgettable characteristic of thiols is their appalling odor. Skunk scent, in fact, is due primarily to the simple thiols 3-methylbutane-1-thiol and but-2-ene-1-thiol. Chemically, thiols can be oxidized by mild reagents such as bromine to yield *disulfides*, R—S—S—R, and disulfides can be reduced back to thiols by treatment with zinc metal and acetic acid:

$$2\,R{-}SH \underset{Zn,\ H^+}{\overset{Br_2}{\rightleftharpoons}} R{-}S{-}S{-}R + 2\,HBr$$

A thiol A disulfide

We'll see in Section 15.5 that this thiol–disulfide interconversion is extremely important in biochemistry because disulfide "bridges" form cross-links that help stabilize the three-dimensional structure of proteins.

Protein—SH + HS—Protein ⟶ Protein—S—S—Protein

A cross-linked protein

PROBLEM 8.19 Name the following thiols by IUPAC rules:

(a) $CH_3CH_2CH(SH)CH_3$

(b) $CH_3C(CH_3)_2CH_2CH(SH)CH_2CH(CH_3)CH_3$

(c) cyclopentene ring bearing —SH

PROBLEM 8.20 Name the following compounds by IUPAC rules:

(a) $CH_3CH_2SCH_3$ (b) 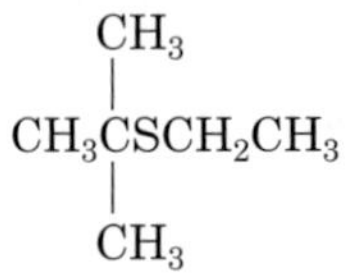 $CH_3\overset{CH_3}{\underset{CH_3}{C}}SCH_2CH_3$ (c) 1,2-bis(SCH_3)benzene (benzene ring with two adjacent SCH_3 groups)

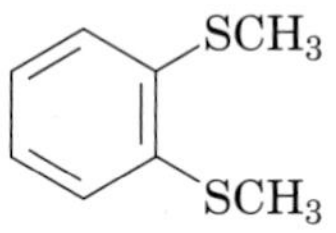

PROBLEM 8.21 But-2-en-1-thiol is a component of skunk spray. How would you synthesize this substance from but-2-en-1-ol? From methyl but-2-enoate, $CH_3CH{=}CHCO_2CH_3$? More than one step is required in both instances.

INTERLUDE

In the Field with Triazine Herbicides

The herbicide atrazine works by inhibiting photosynthesis in weeds like cocklebur, foxtails, ragweed, and wild cucumber. Photosynthesis is the process by which sunlight is converted to chemical energy. How and where does this happen? The "where" is in two clusters of proteins and pigment molecules in the membranes of plant cells, called photosystem I (PSI) and photosystem II (PSII). In PSII, these pigments are named P680, pheophytin, quinone A, and plastoquinone B.

We already know that the movement of electrons is responsible for chemical reactions, so let's start with PSII and follow the electron motion. When the energy from sunlight is transferred to P680, an electron migrates from P680 through the pigment molecules of PSII to end up at plastoquinone B. This process is repeated until plastoquinone B has been reduced to a plastohydroquinone (a bisphenol) by accepting two electrons in a reduction reaction (see Section 8.6).

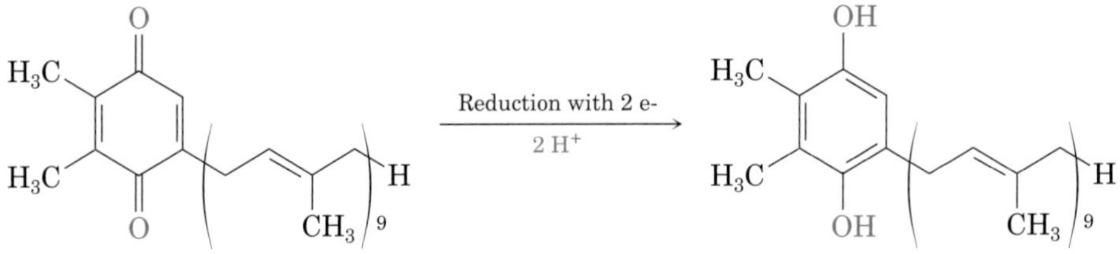

O, H_3C, H_3C, O, CH_3, H, 9 — Reduction with 2 e- / 2 H^+ → — OH, H_3C, H_3C, OH, CH_3, H, 9

A plastoquinone **A plastohydroquinone**

Plastoquinone B acts as an "electron shuttle" between PSII and PSI. After two electrons catch this shuttle, the newly formed plastohydroquinone departs PSII for the second protein, PSI. When the plastohydroquinone diffuses away from PSII, a new molecule of plastiquinone B can bind in the same site and repeat the process. However, if another molecule with a similar shape, such as atrazine (one of the triazine herbicides), is present, it may bind in place of the plastaquinone B. When atrazine binds, photosynthesis stops because plastoquinone B molecules are prevented from rebinding and carrying away more electrons. What happens to the sunlight-energized electrons when atrazine does not accept them? These electrons react with the surrounding lipids in the membrane, damage the cell membrane, and ultimately cause cell death.

Figure 8.5 shows the molecules involved in this process. PSII is so large that its atoms are replaced with a snaking ribbon in this figure. The pigment molecules, plastoquinone B, quinone A, pheophytin, and P680, are shown. Sunlight propels an electron along the red path, starting with a molecule called P680, to pheophytin, to quinone A, and finally to plastoquinone B, which is reduced twice before it diffuses away. The herbicide, atrazine (in the red circle with its green chlorine atom), binds to PSII in the plastoquinone B binding site, interfering with photosynthesis and causing cell damage.

Continued

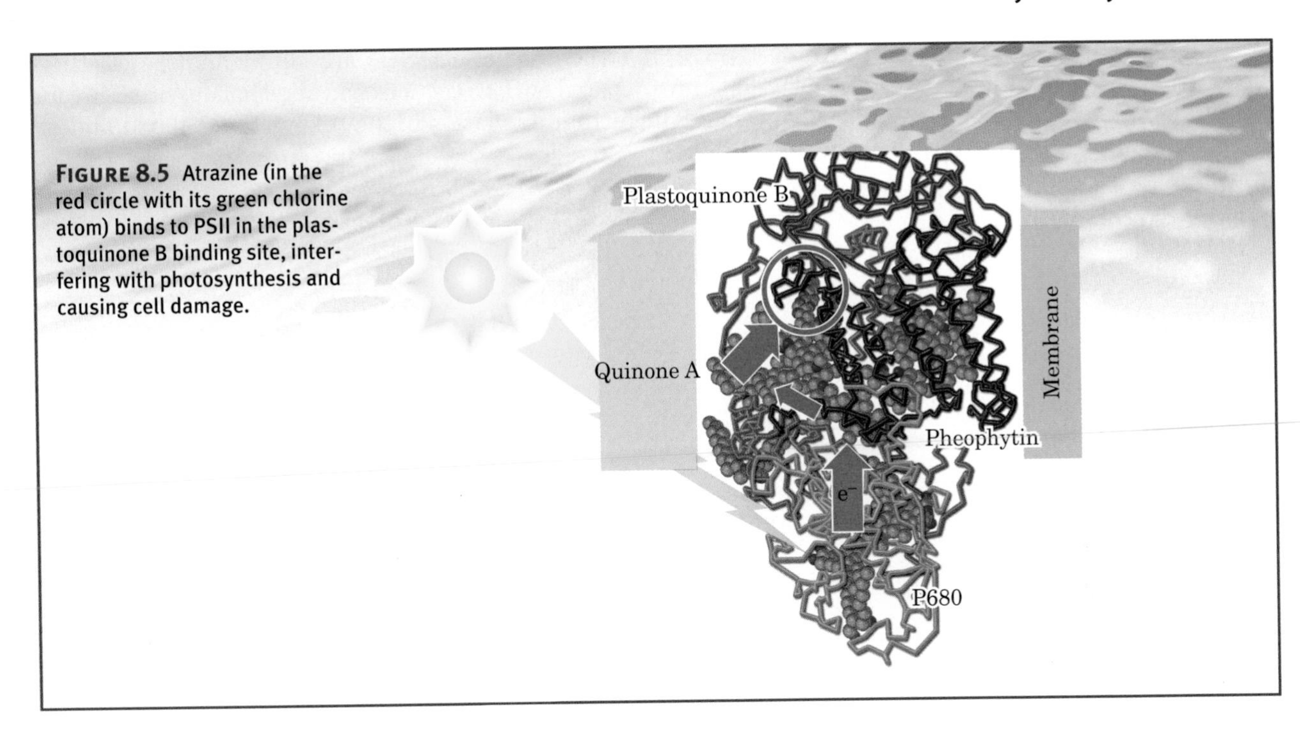

FIGURE 8.5 Atrazine (in the red circle with its green chlorine atom) binds to PSII in the plastoquinone B binding site, interfering with photosynthesis and causing cell damage.

Summary and Key Words

Alcohol, p. 231
Alkoxide ion (RO^-), p. 235
Dehydration, p. 240
Epoxide, p. 249
Ether, p. 231
Hydroquinone, p. 246
Mercapto group, p. 250
Phenol, p. 231
Phenoxide ion (ArO^-), p. 235
Quinone, p. 246
Sulfide, p. 250
Thiol, p. 250

Alcohols, phenols, and ethers are organic derivatives of water in which one or both of the water hydrogens have been replaced by organic groups.

Alcohols are compounds that have an –OH bonded to an alkyl group. They can be prepared in many ways, including hydration of alkenes. The most general method of alcohol synthesis involves reduction of a carbonyl compound. Aldehydes, esters, and carboxylic acids yield primary alcohols on reduction; ketones yield secondary alcohols.

Alcohols are weak acids and can be converted into their **alkoxide anions** on treatment with a strong base or with an alkali metal. Alcohols can also be dehydrated to yield alkenes, converted into ethers by reaction of their anions with alkyl halides, and oxidized to yield carbonyl compounds. Primary alcohols give either aldehydes or carboxylic acids when oxidized, secondary alcohols yield ketones, and tertiary alcohols are not oxidized.

Phenols are aromatic counterparts of alcohols. Although similar to alcohols in some respects, phenols are more acidic than alcohols because **phenoxide anions** are stabilized by resonance. Phenols undergo electrophilic aromatic substitution and can be oxidized to yield **quinones**.

Ethers have two organic groups bonded to the same oxygen atom. They are prepared by S_N2 reaction of an alkoxide ion with a primary alkyl halide—the Williamson synthesis. Ethers are inert to most reagents but are cleaved by the strong acids HBr and HI. **Epoxides**—cyclic ethers with an oxygen atom in a three-membered ring—differ from other ethers in their ease of cleavage. The high reactivity of the strained three-membered ether ring allows epoxides to react with aqueous acid, yielding diols (glycols).

Sulfides (R—S—R′) and **thiols (R—SH)** are sulfur analogs of ethers and alcohols. Thiols are prepared by S_N2 reaction of an alkyl halide with HS^-, and sulfides are prepared by further alkylation of the thiol with an alkyl halide.

Summary of Reactions

1. Synthesis of alcohols (Section 8.4)
   (a) Reduction of aldehydes to yield primary alcohols

$$\mathrm{R\overset{O}{\overset{\|}{C}}H} \xrightarrow[\text{2. } H_3O^+]{\text{1. NaBH}_4} \mathrm{RCH_2OH}$$

   (b) Reduction of ketones to yield secondary alcohols

$$\mathrm{R\overset{O}{\overset{\|}{C}}R'} \xrightarrow[\text{2. } H_3O^+]{\text{1. NaBH}_4} \mathrm{R\overset{OH}{\overset{|}{C}}HR'}$$

   (c) Reduction of esters to yield primary alcohols

$$\mathrm{R\overset{O}{\overset{\|}{C}}OR'} \xrightarrow[\text{2. } H_3O^+]{\text{1. LiAlH}_4} \mathrm{RCH_2OH + R'OH}$$

   (d) Reduction of carboxylic acids to yield primary alcohols

$$\mathrm{R\overset{O}{\overset{\|}{C}}OH} \xrightarrow[\text{2. } H_3O^+]{\text{1. LiAlH}_4} \mathrm{RCH_2OH}$$

2. Reactions of alcohols (Section 8.5)
   (a) Dehydration to yield alkenes

$$-\underset{|}{\overset{\overset{H}{|}}{C}}-\underset{|}{\overset{\overset{OH}{|}}{C}}- \xrightarrow{H_2SO_4} \mathrm{C{=}C} + H_2O$$

   (b) Oxidation to yield carbonyl compounds

$$\mathrm{RCH_2OH} \xrightarrow[\text{chlorochromate}]{\text{Pyridinium}} \mathrm{R\overset{O}{\overset{\|}{C}}H} \quad \text{An aldehyde}$$

$$\mathrm{RCH_2OH} \xrightarrow[H_3O^+]{CrO_3} \mathrm{R\overset{O}{\overset{\|}{C}}OH} \quad \text{A carboxylic acid}$$

$$\mathrm{R\overset{OH}{\overset{|}{C}}HR'} \xrightarrow[\text{chlorochromate}]{\text{Pyridinium}} \mathrm{R\overset{O}{\overset{\|}{C}}R'} \quad \text{A ketone}$$

(c) Conversion into ethers

$$2\,ROH + 2\,Na \longrightarrow 2\,RO^- \; Na^+ + H_2$$

$$RO^- \; Na^+ + R'Br \xrightarrow[\text{reaction}]{S_N2} ROR'$$

3. Synthesis of phenols (Section 8.6)

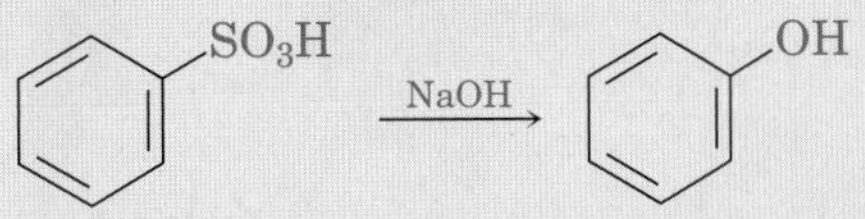

$$C_6H_5SO_3H \xrightarrow{NaOH} C_6H_5OH$$

4. Synthesis of ethers

(a) Williamson ether synthesis (Section 8.7)

$$RO^- \; Na^+ + R'Br \xrightarrow[\text{reaction}]{S_N2} ROR'$$

(b) Epoxides (Section 8.8)

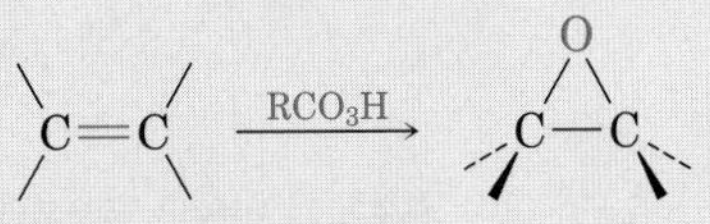

$$\text{C=C} \xrightarrow{RCO_3H} \text{C–C (with bridging O)}$$

5. Reactions of ethers

(a) Acidic cleavage with HBr or HI (Section 8.7)

$$ROR' + HI \longrightarrow ROH + R'I$$

(b) Epoxide cleavage with aqueous acid (Section 8.8)

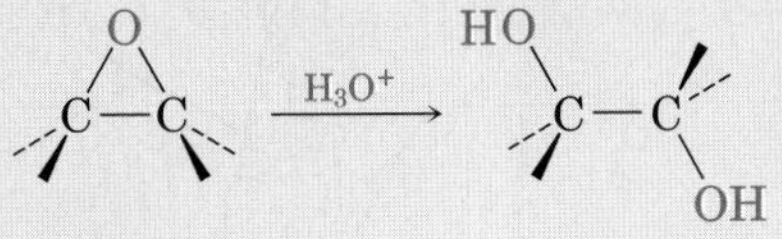

$$\text{C–C (with bridging O)} \xrightarrow{H_3O^+} \text{HO–C–C–OH}$$

6. Synthesis of thiols (Section 8.9)

$$Na^+ \; {}^-SH + RBr \xrightarrow[\text{reaction}]{S_N2} RSH$$

7. Synthesis of sulfides (Section 8.9)

$$RS^- \; Na^+ + R'Br \xrightarrow[\text{reaction}]{S_N2} RSR'$$

EXERCISES

Visualizing Chemistry

8.22 Give IUPAC names for the following compounds (red = O, blue = N):

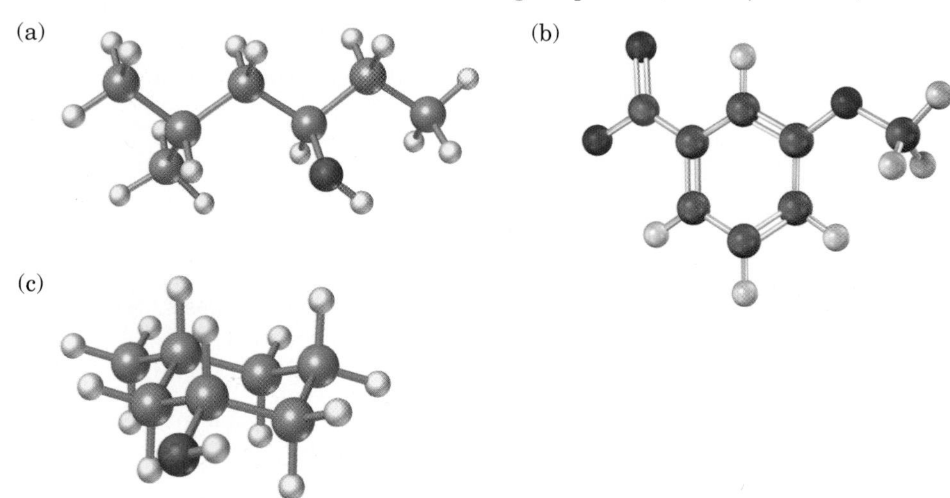

8.23 Predict the product of each of the following reactions (red = O):

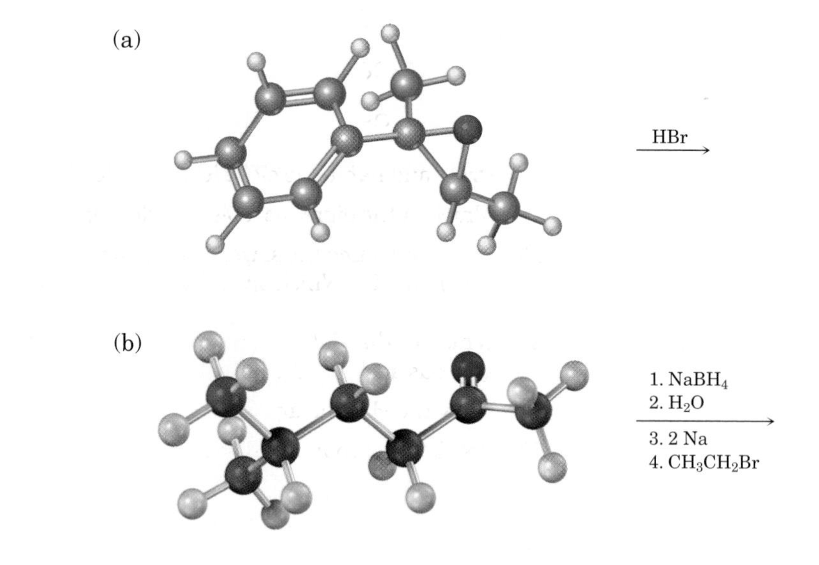

8.24 Show the product you would obtain by reaction of the following compound with $NaBH_4$ (red = O):

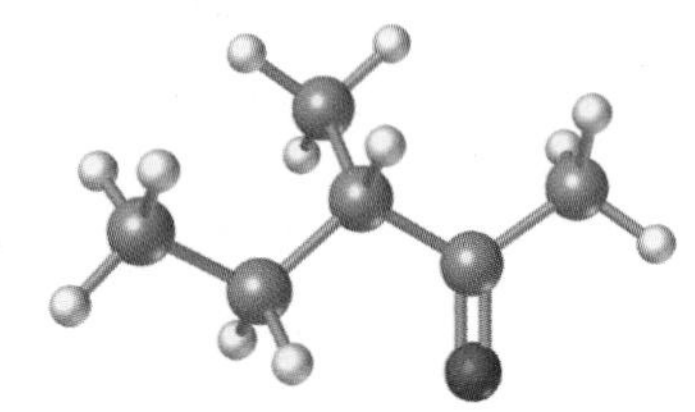

8.25 Show the structures of the carbonyl compounds that would give the following alcohol on reduction. Show also the structure of the product that would result by treating the alcohol with pyridinium chlorochromate (PCC) and with CrO_3.

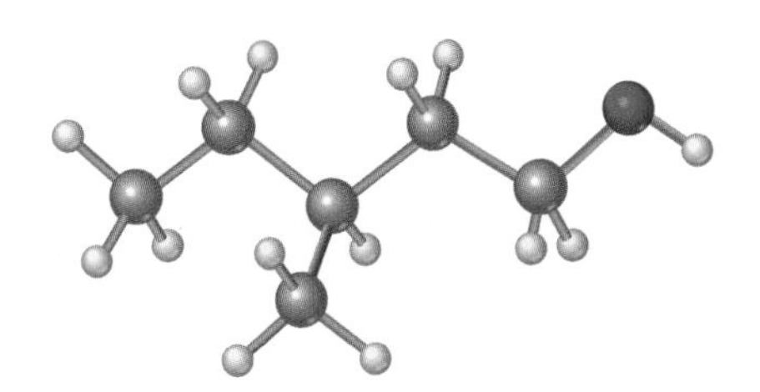

Additional Problems

Nomenclature and Isomerism

8.26 Draw structures corresponding to the following IUPAC names:
(a) Ethyl isopropyl ether (b) 3,4-Dimethoxybenzoic acid
(c) 2-Methylheptane-2,5-diol (d) *trans*-3-Ethylcyclohexanol
(e) 4-Allyl-2-methoxyphenol (eugenol, from oil of cloves)

8.27 Name the following compounds according to IUPAC rules:

(a) $HOCH_2CH_2CHCH_2OH$ (with CH_3 on the CH)

(b) $CH_3CHCHCH_2CH_3$ (with HO and $CH_2CH_2CH_3$ on the two CH carbons)

(c) Ph, H, OH, H (cyclopentane ring)

(d) $(CH_3)_2CHCCH_2CH_2CH_3$ (with SH and CH_3 on the C)

8.28 Draw and name the eight isomeric alcohols with the formula $C_5H_{12}O$.

8.29 Which of the eight alcohols you identified in Problem 8.28 are chiral?

8.30 Draw and name the six ethers that are isomeric with the alcohols you drew in Problem 8.28. Which are chiral?

Reactions and Synthesis

8.31 Which of the eight alcohols you identified in Problem 8.28 would react with aqueous acidic CrO_3? Show the products you would expect from each reaction.

8.32 Show the HI cleavage products of the ethers you drew in Problem 8.30.

8.33 Predict the product(s) of the following transformations:

(a) (cyclohexane)CH_2OH $\xrightarrow{\text{PCC}}$?

(b) (benzene)OCH_2CHCH_3 (with CH_3 on the CH) $\xrightarrow{\text{HBr}}$?

(c) $H_2C{=}CHCHCH_2COCH_3$ (with CH_3 on the CH and O on the C) $\xrightarrow[\text{2. } H_2O]{\text{1. LiAlH}_4}$?

(d) $H_2C{-}CHCH_2CH_3$ (epoxide, O bridging) $\xrightarrow{\text{HBr}}$?

8.34 Show how you could prepare the following substances from cyclohexanol:

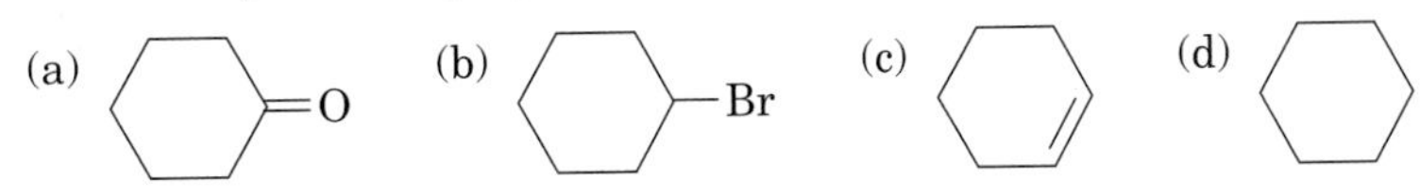

8.35 Show how you could prepare the following substances from propan-1-ol:

(a) $CH_3CH_2CH{=}O$ (b) $CH_3CH_2C(=O)OH$ (c) $CH_3CH_2CH_2O^-Na^+$

(d) $CH_3CH_2CH_2Cl$

8.36 Predict the likely products of reaction of the following ethers with HI:

(a) $CH_3CH_2OCH(CH_3)CH_3$ (b) $CH_3C(CH_3)_2CH_2OCH_3$

8.37 How would you prepare the following compounds from 2-phenylethanol?

(a) Benzoic acid (b) Ethylbenzene
(c) 2-Bromo-1-phenylethane (d) Phenylacetic acid ($C_6H_5CH_2CO_2H$)
(e) Phenylacetaldehyde ($C_6H_5CH_2CHO$)

8.38 Give the structures of the major products you would obtain from reaction of phenol with the following reagents:

(a) Br_2 (1 mol) (b) Br_2 (3 mol)
(c) NaOH, then CH_3I (d) $Na_2Cr_2O_7$, H_3O^+

8.39 What products would you obtain from reaction of butan-1-ol with the following reagents?

(a) PBr_3 (b) CrO_3, H_3O^+ (c) Na (d) Pyridinium chlorochromate (PCC)

8.40 What products would you obtain from reaction of 1-methylcyclohexanol with the following reagents?

(a) HBr (b) H_2SO_4 (c) CrO_3
(d) Na (e) Product of part (d), then CH_3I

8.41 What alcohols would you oxidize to obtain the following products?

(a) cyclopentanone (ring =O) (b) benzene ring —CHO (c) $CH_3CH(CH_3)COOH$

8.42 Show the alcohols you would obtain by reduction of the following carbonyl compounds:

(a) $CH_3CH(CH_3)CH_2CHO$ (b) benzene ring with two adjacent COOH groups (c) $CH_3CH_2C(=O)CH_2CH(CH_3)CH_3$

8.43 Predict the product(s) of the following reactions:

(a) HO–(benzene ring)–OH (para) $\xrightarrow{Na_2Cr_2O_7}$?

(b) $CH_3CH(CH_3)CH_2CH_2CH_2Br \xrightarrow{Na^+\ ^-SH}$?

(c) cyclopentane–SH $\xrightarrow{Br_2}$?

ACIDITY

8.44 Rank the following substances in order of increasing acidity:

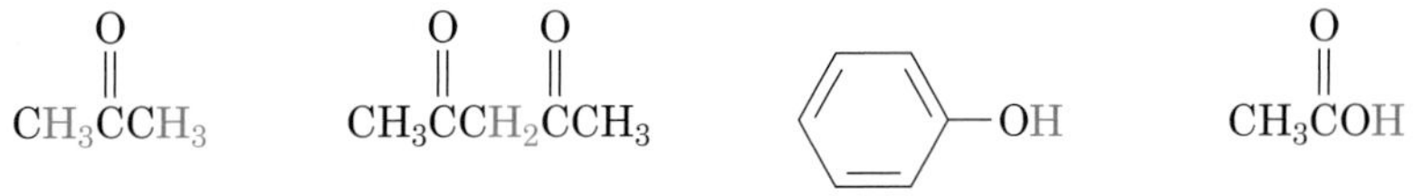

$CH_3\overset{O}{\overset{\|}{C}}CH_3$	$CH_3\overset{O}{\overset{\|}{C}}CH_2\overset{O}{\overset{\|}{C}}CH_3$	C_6H_5–OH	$CH_3\overset{O}{\overset{\|}{C}}OH$
Acetone	**Pentane-2,4-dione**	**Phenol**	**Acetic acid**
$pK_a = 19$	$pK_a = 9$	$pK_a = 9.9$	$pK_a = 4.7$

8.45 Which, if any, of the substances in Problem 8.44 are strong enough acids to react substantially with NaOH? (The pK_a of H_2O is 15.7.)

8.46 Is *tert*-butoxide anion a strong enough base to react with water? In other words, does the following reaction take place as written? (The pK_a of *tert*-butyl alcohol is 18.)

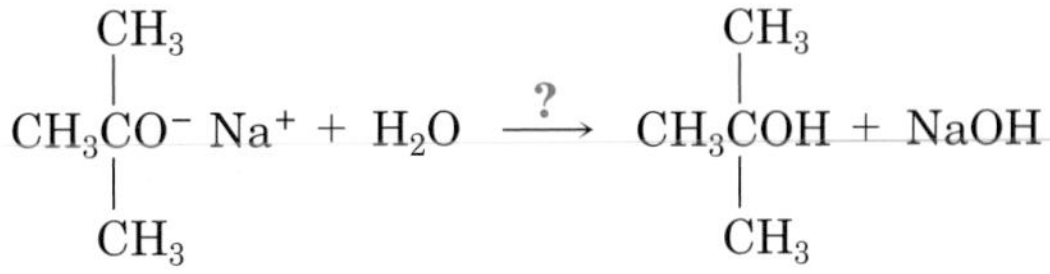

$$(CH_3)_3CO^- \; Na^+ + H_2O \xrightarrow{?} (CH_3)_3COH + NaOH$$

8.47 Sodium bicarbonate, $NaHCO_3$, is the sodium salt of carbonic acid (H_2CO_3), $pK_a = 6.4$. Which of the substances shown in Problem 8.44 will react with sodium bicarbonate?

8.48 Assume that you have two unlabeled bottles, one that contains phenol (pK_a = 9.9) and one that contains acetic acid (pK_a = 4.7). In light of your answer to Problem 8.47, propose a simple way to tell what is in each bottle.

INTEGRATED PROBLEMS

8.49 Named *bombykol*, the sex pheromone secreted by the female silkworm moth has the formula $C_{16}H_{28}O$ and the systematic name (10*E*,12*Z*)-hexadecane-10,12-dien-1-ol. Draw bombykol showing correct geometry for the two double bonds.

8.50 When 4-chlorobutane-1-thiol is treated with a strong base such as sodium hydride, NaH, tetrahydrothiophene is produced. Suggest a mechanism for this reaction.

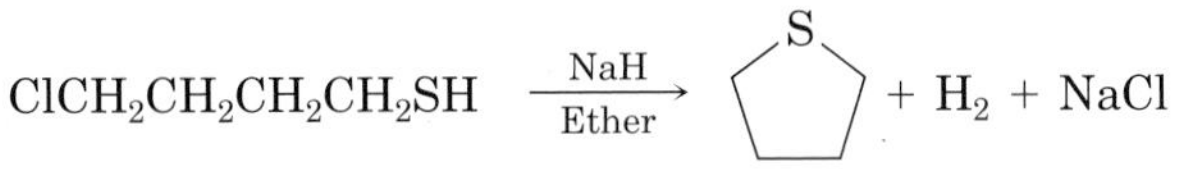

$$ClCH_2CH_2CH_2CH_2SH \xrightarrow[\text{Ether}]{\text{NaH}} \text{(tetrahydrothiophene)} + H_2 + NaCl$$

Tetrahydrothiophene

8.51 Why can't the Williamson ether synthesis be used to prepare diphenyl ether?

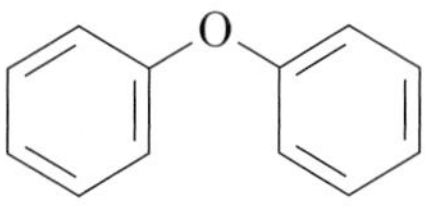

C_6H_5–O–C_6H_5 **Diphenyl ether**

8.52 Starting from benzene, how would you prepare benzyl phenyl ether? More than one step is required.

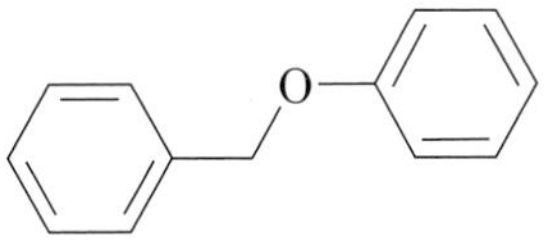

$C_6H_5CH_2$–O–C_6H_5 **Benzyl phenyl ether**

8.53 It's found experimentally that a substituted cyclohexanol with an axial –OH group reacts with CrO_3 more rapidly than its isomer with an equatorial –OH group. Draw both *cis*- and *trans*-4-*tert*-butylcyclohexanol, and predict which oxidizes faster. (The large *tert*-butyl group is equatorial in both.)

8.54 Since all hamsters look pretty much alike, pairing and mating is governed by chemical means of communication. Investigations have shown that dimethyl disulfide, CH_3SSCH_3, is secreted by female hamsters as a sex attractant for males. How would you synthesize dimethyl disulfide in the laboratory if you wanted to trick your hamster?

8.55 *tert*-Butyl ethers can be prepared by the reaction of an alcohol with 2-methylpropene in the presence of an acid catalyst. Propose a mechanism for this reaction.

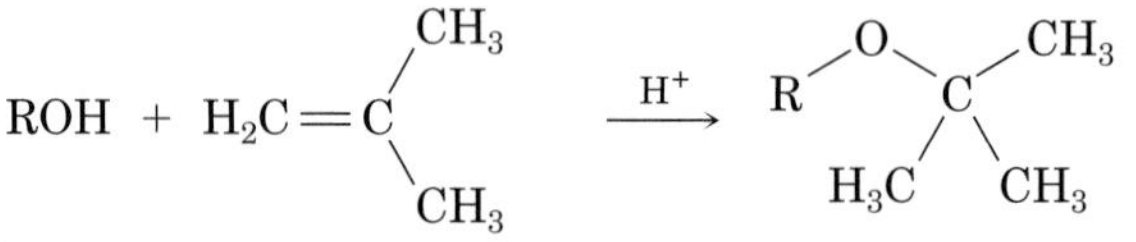

$$ROH + H_2C{=}C(CH_3)_2 \xrightarrow{H^+} R{-}O{-}C(CH_3)_3$$

8.56 *tert*-Butyl ethers react with trifluoroacetic acid, CF_3CO_2H, to yield an alcohol and 2-methylpropene. For example:

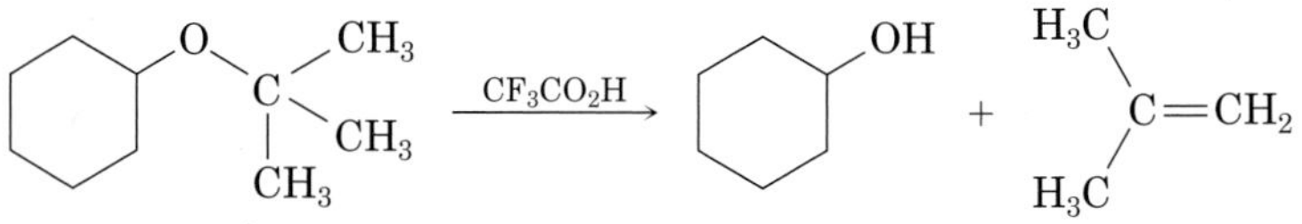

$$C_6H_{11}{-}O{-}C(CH_3)_3 \xrightarrow{CF_3CO_2H} C_6H_{11}{-}OH + (H_3C)_2C{=}CH_2$$

Tell what kind of reaction is occurring, and propose a mechanism.

8.57 How would you prepare the following ethers?

(a) $C_6H_5{-}OCH_2CH_3$ (b) epoxide: H_3C, H on one C; H, CH_3 on the other C (O bridging the two C)

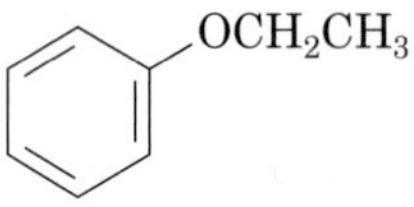

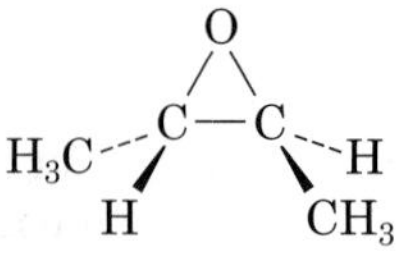

8.58 Identify the reagents a through d in the following scheme:

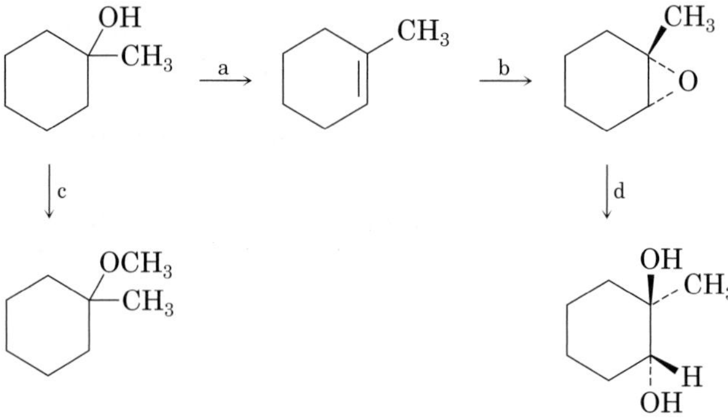

1-methylcyclohexanol (OH, CH_3) —a→ 1-methylcyclohexene (CH_3) —b→ 1-methylcyclohexene oxide (CH_3, O)

1-methylcyclohexanol —c→ 1-methoxy-1-methylcyclohexane (OCH_3, CH_3)

1-methylcyclohexene oxide —d→ 1-methylcyclohexane-1,2-diol (OH, CH_3; H, OH)

8.59 What cleavage product would you expect from reaction of tetrahydrofuran with hot aqueous HI?

Tetrahydrofuran

8.60 Methyl phenyl ether can be cleaved to yield iodomethane and lithium phenoxide when heated with LiI. Propose a mechanism for this reaction.

8.61 The *Zeisel method*, a procedure for determining the number of methoxyl groups (CH_3O–) in a compound, involves heating a weighed amount of the compound with HI. Ether cleavage occurs, and the iodomethane that forms is distilled off and passed into a solution of $AgNO_3$, where it reacts to give AgI. The silver iodide is then weighed, and the number of methoxy groups in the sample is thereby determined. For example, 1.06 g of vanillin, the material responsible for the characteristic odor of vanilla, yields 1.60 g of AgI. If vanillin has a molecular weight of 152, how many methoxyls does it contain?

8.62 Reduction of butan-2-one with $NaBH_4$ yields butan-2-ol. Explain why the product is chiral but not optically active.

$CH_3C(=O)CH_2CH_3$ **Butan-2-one**

8.63 A portion of the Nobel Prize in 1987 was awarded to Dr. Charles Pedersen for the discovery of crown ethers. These molecules are able to bind metal cations, such as Na^+ or K^+. Where and why would you expect the cations to bind? The molecule shown binds Na^+ effectively, but it does not bind Cs^+ because it is a much larger ion. How would you design a molecule that binds Cs^+?

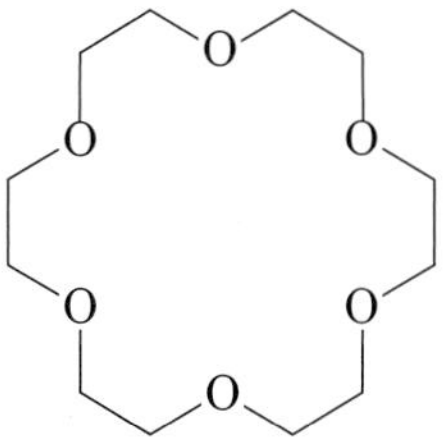

A crown ether

IN THE MEDICINE CABINET

8.64 Nonoxynol 9 is a potent spermicide made by reacting ethylene oxide with *p*-nonylphenoxide. Draw a mechanism for this multistep reaction.

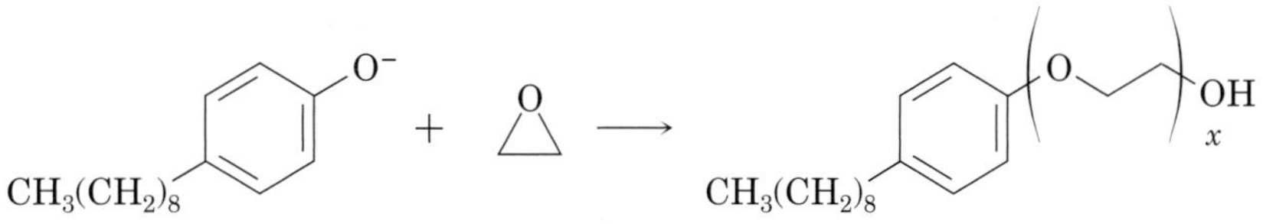

$CH_3(CH_2)_8$–C_6H_4–O^- + ethylene oxide $\longrightarrow$ $CH_3(CH_2)_8$–C_6H_4–$(OCH_2CH_2)_x$–OH

p-Nonylphenoxide **Ethylene oxide** **Nonoxynol 9**

8.65 Blood pressure increases when blood vessels constrict. Captopril is used to treat high blood pressure and heart failure because it decreases certain chemicals that constrict blood vessels.

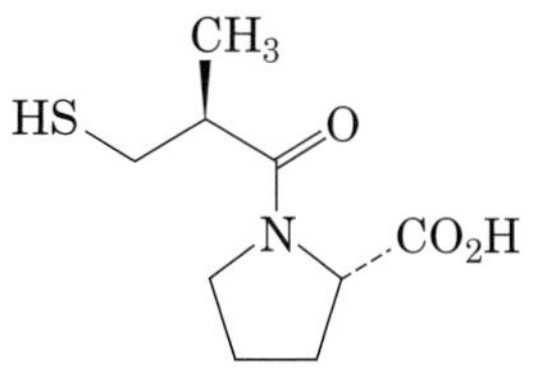

Captopril

(a) Assign the two chirality centers in captopril as *R* or *S*.
(b) What three functional groups are present in captopril?
(c) Draw the disulfide that results from oxidation of two molecules of captopril.

8.66 Synthetic dopamine can be produced by the following reaction. Draw a mechanism for this reaction. (*Hint*: Bromide ion acts as the nucleophile in these S_N2 reactions.)

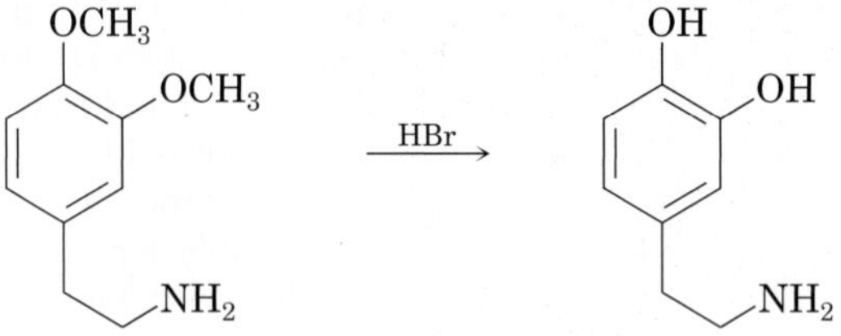

OCH_3 OCH_3 NH_2 $\xrightarrow{HBr}$ OH OH NH_2

Dopamine

In the Field with Agrochemicals

8.67 The herbicide 2,4,5-T (2,4,5-trichlorophenoxyacetic acid) can be prepared by heating a mixture of 2,4,5-trichlorophenol and $ClCH_2CO_2H$ with NaOH. Show the mechanism of the reaction.

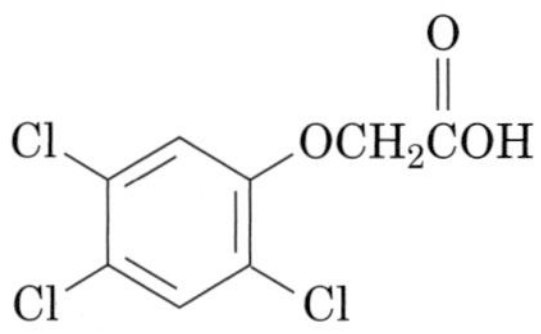

Cl Cl Cl O OCH_2COH

2,4,5-T

The ketone (*Z*)-henicos-6-en-11-one is used as an attractant by the moths responsible for defoliation of Douglas fir trees. Henicosane is $C_{21}H_{44}$. (Gregory G. Dimijian/Photo Researchers, Inc.)

Aldehydes and Ketones: Nucleophilic Addition Reactions

In this and the next two chapters, we'll discuss the most important and widely occurring functional group in both organic and biological chemistry—the **carbonyl group, C=O**. Carbonyl compounds are everywhere in nature. Most biologically important molecules contain carbonyl groups, as do many pharmaceutical agents and synthetic chemicals that touch our everyday lives. The carbonyl-containing functional groups can be divided into two categories based on what is attached to the carbon atom of the C=O group. In this chapter we'll examine aldehydes and ketones, the first of the two categories. Aldehydes have one hydrogen attached to the carbonyl carbon. Ketones have two carbons attached. Neither hydrogen nor carbon can readily stabilize a negative charge, and accordingly, these groups rarely act as leaving groups.

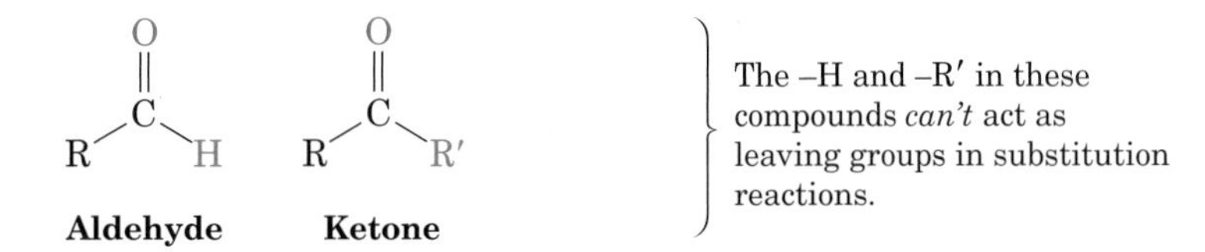

Aldehyde
Ketone
The –H and –R′ in these compounds *can't* act as leaving groups in substitution reactions.

In the next chapter, we'll examine the functional groups that have an electronegative atom, such as oxygen or nitrogen, attached to the carbonyl group. These atoms *can* stabilize a negative charge and therefore *can* act as leaving groups in substitution reactions.

Carboxylic acid **Ester** **Acid chloride**

Amide **Acid anhydride**

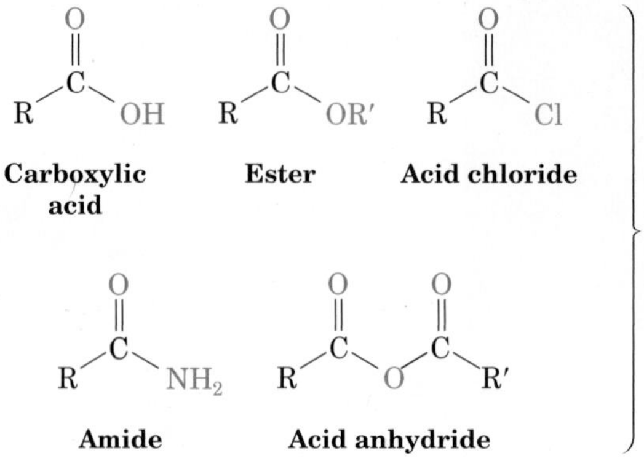

The –OH, –OR′, –Cl, $-NH_2$, and –OCOR′ in these compounds *can* act as leaving groups in substitution reactions.

9.1 Chemistry of the Carbonyl Group

The carbon–oxygen double bond of carbonyl groups is similar in some respects to the carbon–carbon double bond of alkenes (Figure 9.1). The carbonyl carbon atom is sp^2-hybridized and forms three σ bonds. The fourth valence electron remains in a carbon p orbital and forms a π bond to oxygen by overlap with an oxygen p orbital. The oxygen also has two nonbonding pairs of electrons, which occupy its remaining two orbitals. Like alkenes, carbonyl compounds are planar about the double bond and have bond angles of approximately 120°.

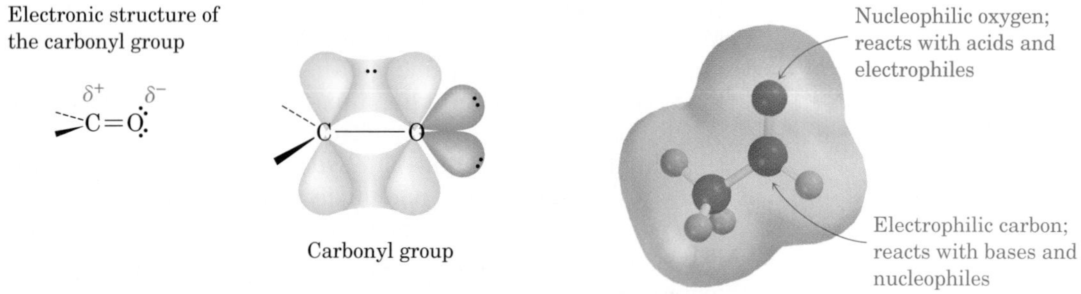

FIGURE 9.1 Electronic structure of the carbonyl group.

Electronic structure of the carbonyl group

Carbonyl group

Nucleophilic oxygen; reacts with acids and electrophiles

Electrophilic carbon; reacts with bases and nucleophiles

Carbon–oxygen double bonds are polarized because of the high electronegativity of oxygen relative to carbon. Since the carbonyl carbon is positively polarized, it is electrophilic (a Lewis acid) and reacts with nucleophiles. Conversely, the carbonyl oxygen is negatively polarized and nucleophilic (a Lewis base). We'll see in this and the next two chapters that most carbonyl-group reactions are the result of this bond polarization.

PROBLEM 9.1 Propose structures for molecules that meet the following descriptions:

(a) A ketone, $C_5H_{10}O$
(b) An aldehyde, $C_6H_{10}O$
(c) A keto aldehyde, $C_6H_{10}O_2$
(d) A cyclic ketone, C_5H_8O

9.2 Naming Aldehydes and Ketones

Aldehydes are named by replacing the terminal *-e* of the corresponding alkane name with *-al.* The parent chain must contain the –CHO group, and the –CHO carbon is always numbered as carbon 1. For example:

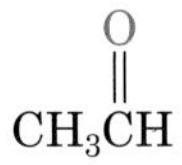

$$CH_3\overset{O}{\overset{\|}{C}}H \qquad CH_3CH_2\overset{O}{\overset{\|}{C}}H \qquad \underset{5}{C}H_3\underset{4}{\overset{CH_3}{\overset{|}{C}}}H\underset{3}{C}H_2\underset{2}{\underset{CH_2CH_3}{\underset{|}{C}}}H\underset{1}{\overset{O}{\overset{\|}{C}}}H$$

Ethanal (acetaldehyde) 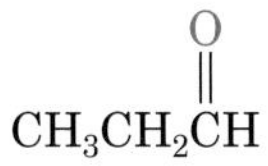 — **Propanal (propionaldehyde)** 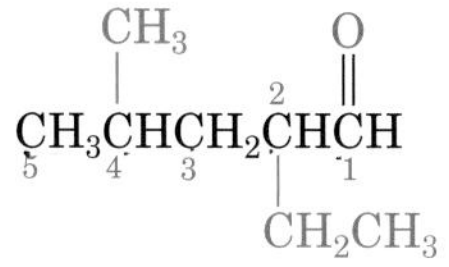 — **2-Ethyl-4-methylpentanal**

Note that the longest chain in 2-ethyl-4-methylpentanal is a hexane, but this chain does not include the –CHO group and thus is not the parent.

For more complex aldehydes in which the –CHO group is attached to a ring, the suffix *-carbaldehyde* is used:

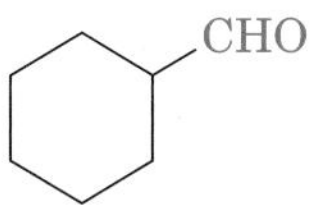

Cyclohexanecarbaldehyde 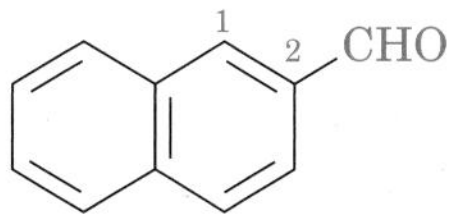 — **2-Naphthalenecarbaldehyde**

Some simple and well-known aldehydes also have common names, as indicated in Table 9.1.

Ketones are named by replacing the terminal *-e* of the corresponding alkane name with *-one.* The parent chain is the longest one that contains the ketone group, and numbering begins at the end nearer the carbonyl carbon. For example:

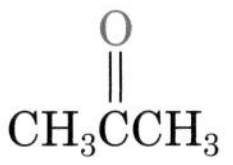

$$CH_3\overset{O}{\overset{\|}{C}}CH_3 \qquad \underset{1}{C}H_3\underset{2}{C}H_2\underset{3}{\overset{O}{\overset{\|}{C}}}\underset{4}{C}H_2\underset{5}{C}H_2\underset{6}{C}H_3 \qquad \underset{6}{C}H_3\underset{5}{C}H{=}\underset{4}{C}H\underset{3}{C}H_2\underset{2}{\overset{O}{\overset{\|}{C}}}\underset{1}{C}H_3$$

Propanone (acetone) 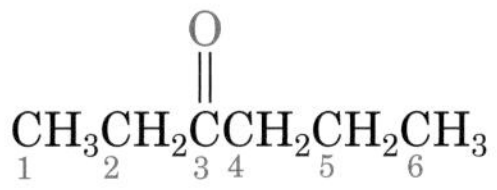 — **Hexan-3-one** 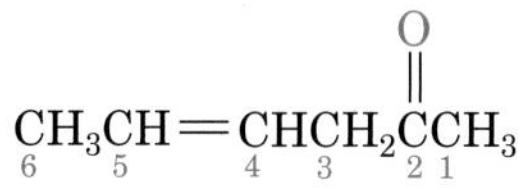 — **Hex-4-en-2-one**

TABLE 9.1 Common Names of Some Simple Aldehydes

Formula	Common name	Systematic name
HCHO	Formaldehyde	Methanal
CH_3CHO	Acetaldehyde	Ethanal
CH_3CH_2CHO	Propionaldehyde	Propanal
$CH_3CH_2CH_2CHO$	Butyraldehyde	Butanal
$CH_3CH_2CH_2CH_2CHO$	Valeraldehyde	Pentanal
$H_2C{=}CHCHO$	Acrolein	Prop-2-enal
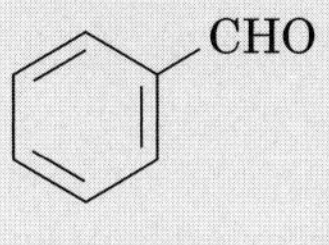 C_6H_5–CHO (benzene ring–CHO)	Benzaldehyde	Benzenecarbaldehyde

A few ketones also have common names:

CH_3COCH_3	$C_6H_5COCH_3$	$C_6H_5COC_6H_5$
Acetone	**Acetophenone**	**Benzophenone**

When it's necessary to refer to the –COR group as a substituent, the general term *acyl* is used. Similarly, –$COCH_3$ is an *acetyl* group, –CHO is a *formyl* group, –COAr is an *aroyl* group, and –COC_6H_5 is a *benzoyl* group.

R–CO–	H_3C–CO–	H–CO–	Ar–CO–	C_6H_5–CO–
An acyl group (R = alkyl, alkenyl)	Acetyl	Formyl	Aroyl (Ar = aromatic)	Benzoyl

Occasionally, the doubly bonded oxygen is considered a substituent, and the prefix *oxo-* is used. For example:

$$\underset{6}{C}H_3\underset{5}{C}H_2\underset{4}{C}H_2\underset{3}{C}(=O)\underset{2}{C}H_2\underset{1}{C}(=O)OCH_3$$

Methyl 3-oxohexanoate

PROBLEM 9.2 Name the following aldehydes and ketones:

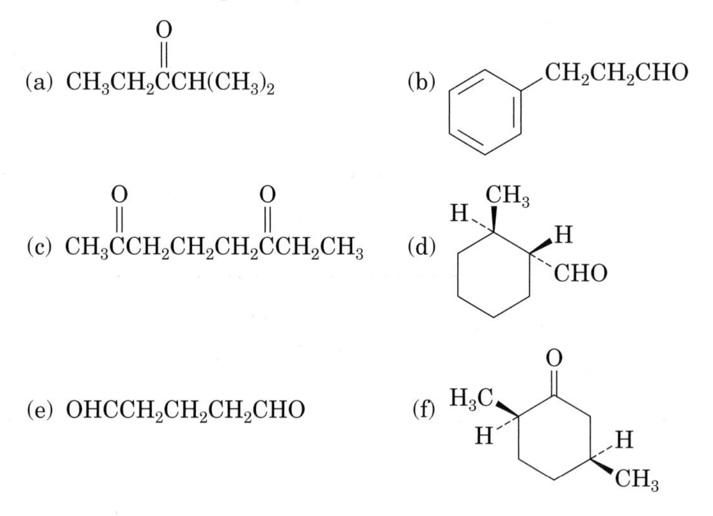

PROBLEM 9.3 Draw structures corresponding to the following IUPAC names:

(a) 3-Methylbutanal
(b) 3-Methylbut-3-enal
(c) 4-Chloropentan-2-one
(d) Phenylacetaldehyde
(e) 2,2-Dimethylcyclohexanecarbaldehyde
(f) Cyclohexane-1,3-dione

9.3 Synthesis of Aldehydes and Ketones

We've already discussed one of the best methods of preparing aldehydes and ketones—the oxidation of alcohols (see Section 8.5). Primary alcohols are oxidized to give aldehydes, and secondary alcohols are oxidized to give ketones. Pyridinium chlorochromate (PCC) in dichloromethane is usually chosen for making aldehydes, while PCC, CrO_3, and $Na_2Cr_2O_7$ are all effective for making ketones:

$$\text{Citronellol} \xrightarrow[CH_2Cl_2]{PCC} \text{Citronellal (82\%)}$$

$$(CH_3)_3C\text{-}C_6H_{10}\text{-}OH \xrightarrow[CH_2Cl_2]{PCC} (CH_3)_3C\text{-}C_6H_{10}{=}O$$

4-*tert*-Butylcyclohexanol **4-*tert*-Butylcyclohexanone (90%)**

Other methods for preparing ketones include the hydration of a terminal alkyne to yield a methyl ketone (see Section 4.12) and the Friedel–Crafts acylation of an aromatic ring to yield an alkyl aryl ketone (see Section 5.6).

$$CH_3(CH_2)_3C{\equiv}CH \xrightarrow[Hg(OAc)_2]{H_3O^+} CH_3(CH_2)_3\overset{O}{\overset{\|}{C}}{-}CH_3$$

Hex-1-yne **Hexan-2-one (78%)**

$$C_6H_6 + CH_3\overset{O}{\overset{\|}{C}}Cl \xrightarrow{AlCl_3} C_6H_5\overset{O}{\overset{\|}{C}}CH_3$$

Benzene **Acetyl chloride** **Acetophenone (95%)**

PROBLEM 9.4 How could you prepare pentanal from the following starting materials?
(a) Pentan-1-ol (b) $CH_3CH_2CH_2CH_2CO_2H$ (c) Dec-5-ene

PROBLEM 9.5 How could you prepare hexan-2-one from the following starting materials?

(a) $CH_3CH_2CH_2CH_2\overset{OH}{\overset{|}{C}}HCH_3$ (b) $CH_3CH_2CH_2CH_2C{\equiv}CH$

(c) $CH_3CH_2CH_2CH_2\overset{CH_3}{\overset{|}{C}}{=}CH_2$

PROBLEM 9.6 How would you carry out the following transformations? More than one step may be required.

(a) Hex-3-ene → Hexan-3-one (b) Benzene → 1-Phenylethanol

9.4 Oxidation of Aldehydes

Aldehydes are easily oxidized to yield carboxylic acids, RCHO → RCO_2H, but ketones are unreactive toward oxidation. This reactivity difference is a consequence of structure: aldehydes have a –CHO proton that can be removed during oxidation, but ketones do not.

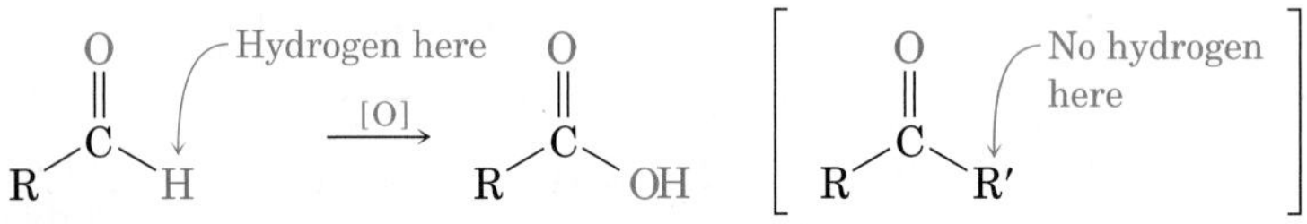

Hydrogen here

[O]

No hydrogen here

An aldehyde

A ketone

One of the simplest methods for oxidizing an aldehyde is to use silver ion, Ag^+, in dilute aqueous ammonia, a mixture called *Tollens' reagent.* As the oxidation proceeds, a shiny mirror of silver metal is deposited on the walls of the reaction flask, forming the basis of a simple test to detect the presence of an aldehyde functional group in a sample of unknown structure. A small amount of the unknown is dissolved in ethanol in a test tube, and a few drops of Tollens' reagent are added. If the test tube becomes silvery, the unknown is an aldehyde.

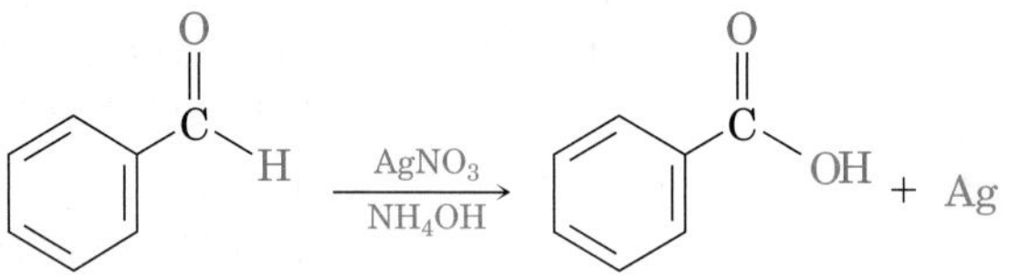

$AgNO_3$, NH_4OH; + Ag

Benzaldehyde **Benzoic acid**

Oxygen, O_2, in air can also convert aldehydes into carboxylic acids. Sunlight promotes this reaction. For both of these reasons, aldehydes such as vanilla (or vanillin) are stored in tightly sealed, dark bottles.

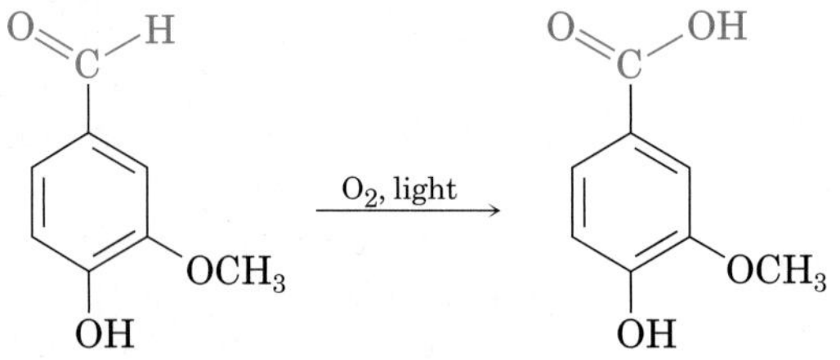

O_2, light

Vanillin (an aldehyde) **An oxidation product (a carboxylic acid)**

PRACTICE PROBLEM 9.1

What product would you obtain from the oxidation of 3-methylbutanal with Tollens' reagent?

STRATEGY Write the structure of the aldehyde, and then replace the –H bonded to the carbonyl group by –OH.

SOLUTION

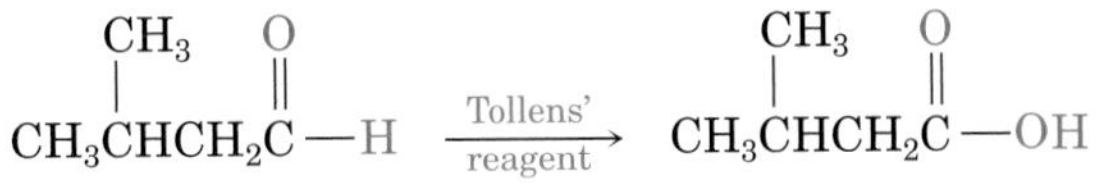

$$\underset{\textbf{3-Methylbutanal}}{CH_3\overset{CH_3}{\overset{|}{C}}HCH_2\overset{O}{\overset{\|}{C}}-H} \xrightarrow[\text{reagent}]{\text{Tollens'}} \underset{\textbf{3-Methylbutanoic acid}}{CH_3\overset{CH_3}{\overset{|}{C}}HCH_2\overset{O}{\overset{\|}{C}}-OH}$$

PROBLEM 9.7 Predict the products of the reaction of the following substances with Tollens' reagent:

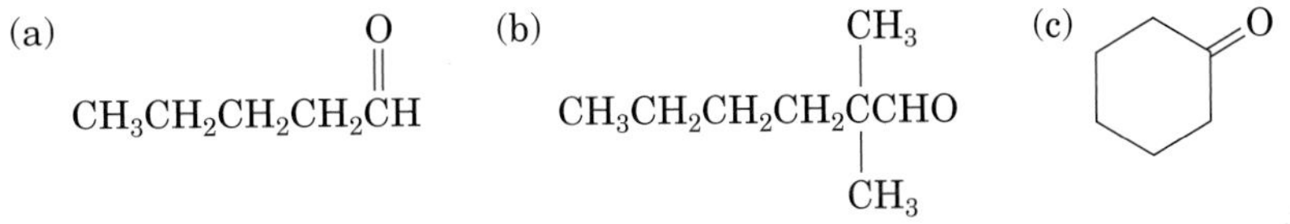

(a) $CH_3CH_2CH_2CH_2\overset{O}{\overset{\|}{C}}H$ (b) $CH_3CH_2CH_2CH_2\underset{CH_3}{\underset{|}{\overset{CH_3}{\overset{|}{C}}}}CHO$ (c) cyclohexanone (six-membered ring with C=O)

9.5 Nucleophilic Addition Reactions of Aldehydes and Ketones

The most common reaction of aldehydes and ketones is the **nucleophilic addition reaction**, in which a nucleophile adds to the electrophilic carbon of the carbonyl group. Whether it is a reduction with a source of hydride ($H:^-$), hydration with H_2O, acetal formation with ROH, Grignard addition with RMgX, or imine formation with RNH_2, the general mechanism is the same. The only variation in this mechanism is whether the conditions are acidic or basic. It is useful to commit the following guidelines to memory:

RULE 1 Under basic conditions (in the presence of ^-OH or ^-OR), negatively charged intermediates are favored; positively charged intermediates are disfavored.

RULE 2 Under acidic conditions (the presence of H^+), positively charged intermediates are favored; negatively charged intermediates are disfavored.

If we glance ahead at all the mechanisms remaining in the chapter, we will see a common trend summarized in Figure 9.2. A nucleophile attacks the carbonyl carbon. In order to maintain an octet, the π bond of the C=O bond breaks simultaneously with the C–Nu bond formation. The π electrons of the C=O bond migrate to the oxygen atom. The only variation in these mechanisms is *when* the proton appears on the carbonyl oxygen. In acidic environments, the oxygen is protonated *before* the nucleophile attacks: a protonated carbonyl is much more reactive than an unprotonated carbonyl. In basic environments, the oxygen anion forms upon nucleophile attack and is later protonated.

The remaining challenge is to recognize the nucleophiles. Table 9.2 identifies common nucleophiles. Negatively charged nucleophiles exist under basic conditions. Neutral nucleophiles exist under acidic conditions.

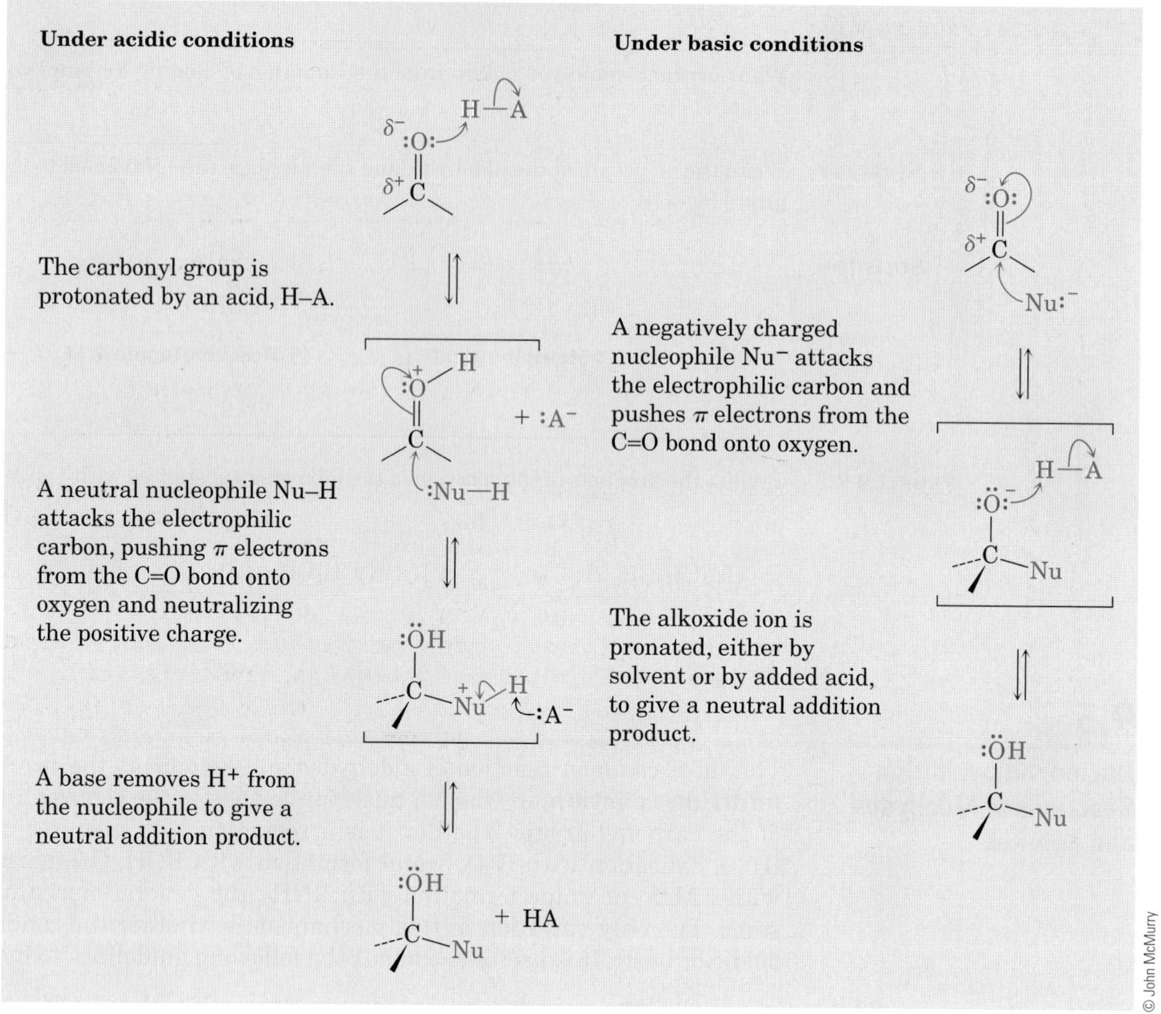

Under acidic conditions

The carbonyl group is protonated by an acid, H–A.

A neutral nucleophile Nu–H attacks the electrophilic carbon, pushing π electrons from the C=O bond onto oxygen and neutralizing the positive charge.

A base removes H^+ from the nucleophile to give a neutral addition product.

Under basic conditions

A negatively charged nucleophile Nu^- attacks the electrophilic carbon and pushes π electrons from the C=O bond onto oxygen.

The alkoxide ion is pronated, either by solvent or by added acid, to give a neutral addition product.

FIGURE 9.2 MECHANISM: The general mechanism for nucleophilic attack at a carbonyl group under basic (right) or acidic (left) conditions. The fundamental difference is when the oxygen atom receives a proton.

TABLE 9.2 Nucleophiles for Additions to Carbonyl Groups

Nu—H	Name	Reagent	Nu^-	Name	Reagent
HO—H	Water	H_2O	HO^-	Hydroxide ion	LiOH
RO—H	An alcohol	ROH	$H:^-$	Hydride ion	$NaBH_4$
H_2N—H	Ammonia	NH_3	$R_3C:^-$	A carbanion	R_3CMgX
RHN—H	An amine	RNH_2	RO^-	An alkoxide ion	NaOR
			$N{\equiv}C:^-$	A cyanide ion	KCN

PRACTICE PROBLEM 9.2

What product would you expect from nucleophilic addition of aqueous hydroxide ion to acetaldehyde?

STRATEGY The negatively charged hydroxide ion is a nucleophile, which can add to the C=O carbon atom and give an alkoxide ion intermediate. Protonation will then yield a 1,1-dialcohol.

SOLUTION

PROBLEM 9.8 What product would you expect if the nucleophile cyanide ion, CN^-, were to add to acetone and the intermediate were to be protonated?

PROBLEM 9.9 What product would you expect if the nucleophile methoxide ion, CH_3O^-, were to add to benzaldehyde and the intermediate were to be protonated?

9.6 Reduction

Throughout the text, we'll look at a number of reducing agents, that is, molecules that increase the number of C–H bonds or reduce the number of C–O bonds. Different functional groups require different reducing agents: $NaBH_4$ (sodium borohydride) is strong enough to reduce aldehydes to alcohols but not strong enough to reduce carboxylic acids. For carboxylic acids, $LiAlH_4$ (lithium aluminum hydride) is required. Unlike these simple laboratory reducing agents, nature uses more complex molecules that undergo *reversible* oxidation and reduction reactions:

Common *irreversible* reducing agents in the laboratory

Common *reversible* reducing agents in living organisms

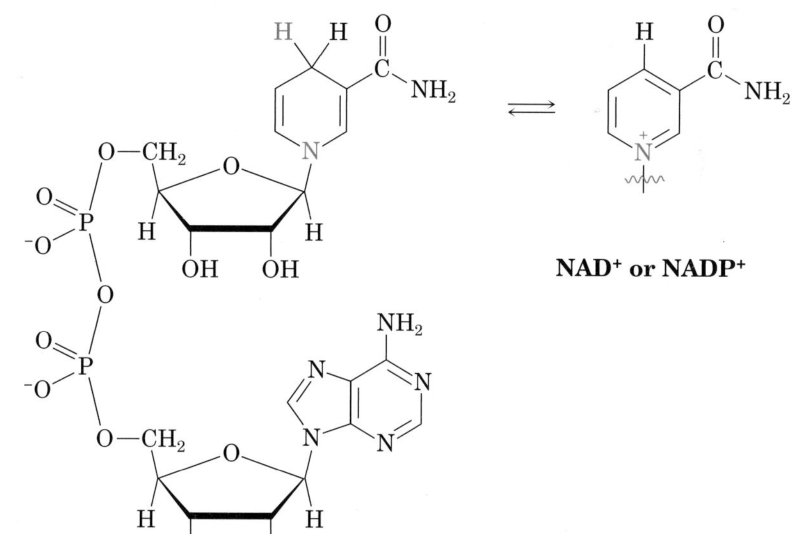

X = OH Reduced nicotinamide adenine dinucleotide (NADH)
X = OPO_3^{2-} Reduced nicotinamide adenine dinucleotide phosphate (NADPH)

The source of H:⁻ is shown in red for these molecules. By comparing the electronegativities of H to either B or Al (Figure 1.12), we can rationalize why $LiAlH_4$ is a stronger reducing agent than $NaBH_4$. Al is more electropositive than B, so the hydride of $LiAlH_4$ is more reactive than the hydride of $NaBH_4$.

But what about the source of hydride in NADH or NADPH? (The "H" denotes the reduced form.) We know that the electronegativity difference between C and H is very small. How does nature offset the cost of making an H:⁻ when breaking an unpolarized C–H bond? The answer is *aromaticity*. Donation of hydride from NADH or NADPH produces a stable aromatic ring.

All these hydride agents can be abbreviated as H:⁻ for the purpose of discussing mechanism. As shown in Figure 9.3, the hydride nucleophile uses its pair of electrons to form a bond to the carbon atom of the C=O group. At the same time, the C=O carbon atom changes hybridization from sp^2 to sp^3 and two electrons from the C=O bond move to the oxygen atom, making an alkoxide ion. Later, the addition of H^+ to the alkoxide ion yields a neutral alcohol as the product.

FIGURE 9.3 MECHANISM: General mechanism of a reduction reaction with a hydride source.

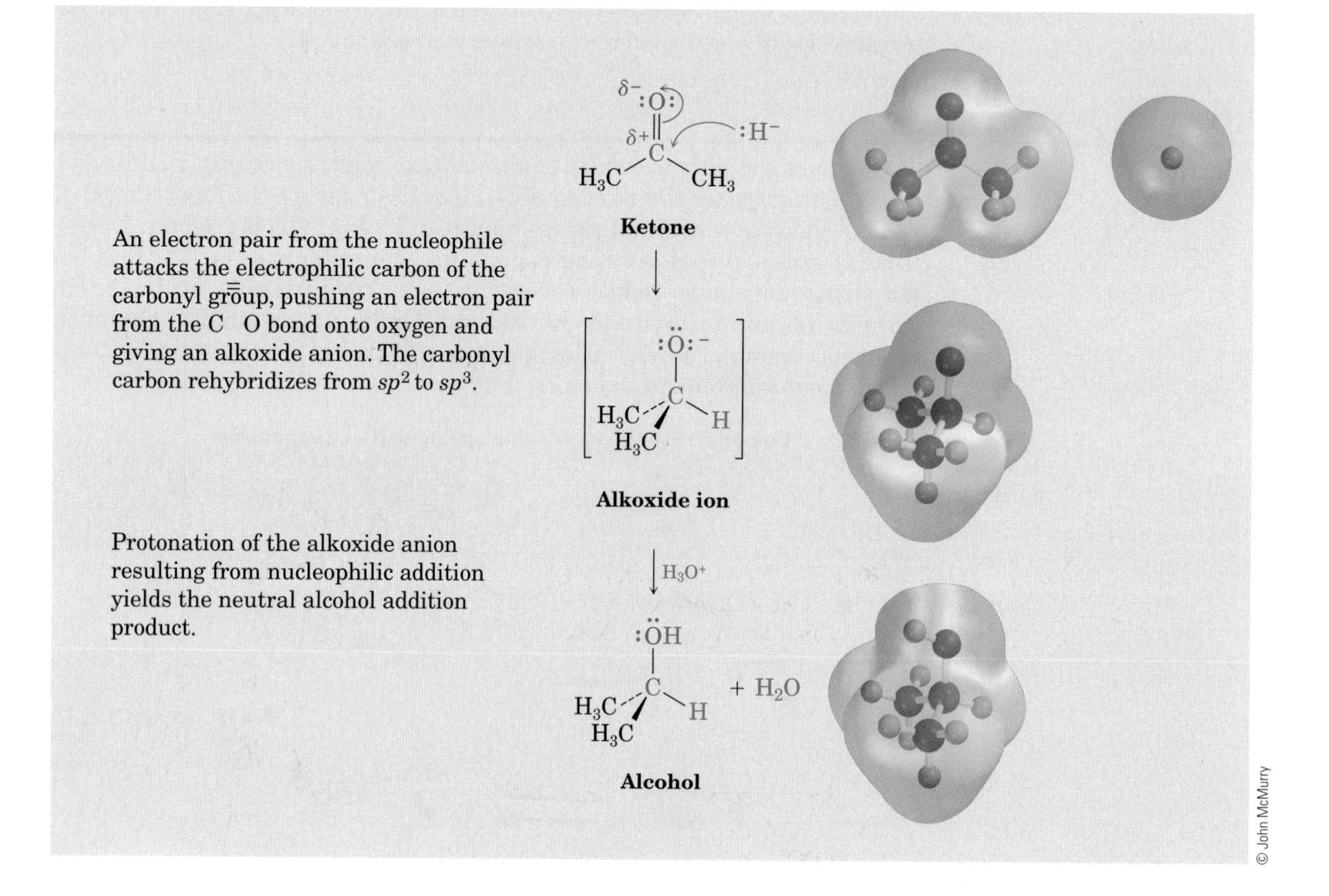
δ−
:O:
δ+
:H⁻
H₃C
CH₃
Ketone
An electron pair from the nucleophile attacks the electrophilic carbon of the carbonyl group, pushing an electron pair from the C O bond onto oxygen and giving an alkoxide anion. The carbonyl carbon rehybridizes from sp² to sp³.
:Ö:⁻
H₃C
H₃C
H
Alkoxide ion
Protonation of the alkoxide anion resulting from nucleophilic addition yields the neutral alcohol addition product.
H₃O⁺
:ÖH
+ H₂O
H₃C
H₃C
H
Alcohol
© John McMurry

9.7 Nucleophilic Addition of Water: Hydration

Before we examine the nucleophilic addition of water or alcohols to carbonyls, let's introduce the H-to-O (H_2O) rule, a mnemonic device that helps us to think about reactivity. Remember, the chemistry of addition across the C=O bond is similar when either water or an alcohol reacts with an aldehyde or ketone.

H-TO-O RULE When an RO–H group adds to a C=O bond, the *H goes to the oxygen* atom of the C=O bond and the OR goes to the carbon atom of the C=O bond.

Now, let's look at hydration.

Aldehydes and ketones undergo a nucleophilic addition reaction with water to yield 1,1-diols, called **geminal (gem) diols**. The reaction is reversible, and the diol product can eliminate water to regenerate a ketone or aldehyde:

$$H_3C-C(=O)-CH_3 + H_2O \rightleftharpoons (H_3C)_2C(OH)_2$$

Acetone **Acetone hydrate (a gem diol)**

The position of the equilibrium between gem diols and aldehydes/ketones depends on the structure of the carbonyl compound. Although the equilibrium strongly favors the carbonyl compound in most cases, the gem diol is favored for a few simple aldehydes. For example, an aqueous solution of acetone consists of about 0.1% gem diol and 99.9% ketone, whereas an aqueous solution of formaldehyde (CH_2O) consists of 99.9% gem diol and 0.1% aldehyde.

The nucleophilic addition of water to aldehydes and ketones is slow in pure water but is catalyzed by both acid and base. As always, the catalyst doesn't change the *position* of the equilibrium; it affects only the rate at which the hydration reaction occurs.

The base-catalyzed addition reaction takes place in several steps, as shown in Figure 9.4. The attacking nucleophile is the negatively charged hydroxide ion.

FIGURE 9.4 MECHANISM: Mechanism of the base-catalyzed hydration reaction of a ketone or aldehyde. Hydroxide ion is a more reactive nucleophile than neutral water.

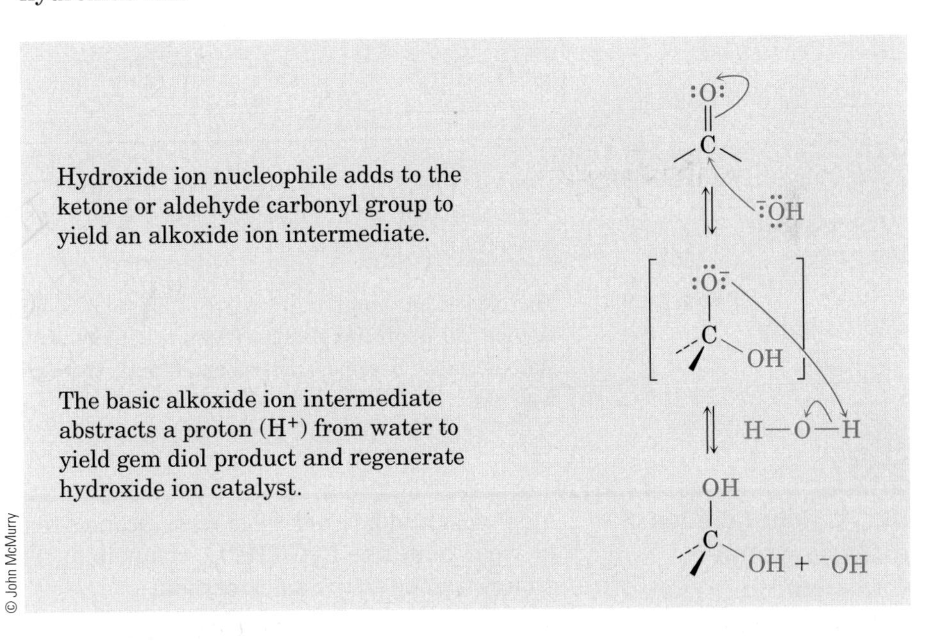

Hydroxide ion nucleophile adds to the ketone or aldehyde carbonyl group to yield an alkoxide ion intermediate.
The basic alkoxide ion intermediate abstracts a proton (H+) from water to yield gem diol product and regenerate hydroxide ion catalyst.
:O:
C
:ÖH
OH
H—O—H
OH
OH + ⁻OH
© John McMurry

The acid-catalyzed hydration reaction also takes place in several steps (Figure 9.5). The acid catalyst first protonates the basic oxygen atom of the carbonyl group, and subsequent nucleophilic addition of neutral water yields a protonated gem diol. Loss of a proton then gives the gem diol product.

Note the difference between the base-catalyzed and acid-catalyzed processes. The *base*-catalyzed reaction takes place rapidly because hydroxide ion is a much better nucleophilic *donor* than neutral water. The *acid*-catalyzed reaction takes place rapidly because the carbonyl compound is converted by protonation into a much better electrophilic *acceptor*.

FIGURE 9.5 MECHANISM: Mechanism of the acid-catalyzed hydration reaction of a ketone or aldehyde. The acid catalyst protonates the carbonyl starting material, thus making it more electrophilic and more reactive.

Acid catalyst protonates the basic carbonyl oxygen atom, making the ketone or aldehyde a much better acceptor of nucleophiles.

Nucleophilic addition of neutral water yields a protonated gem diol.

Loss of a proton regenerates the acid catalyst and gives neutral gem diol product.

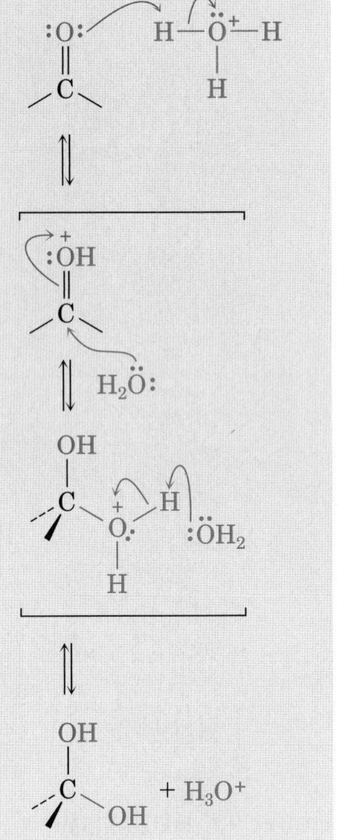

$+ H_3O^+$

PROBLEM 9.10 When dissolved in water, trichloroacetaldehyde (chloral, CCl_3CHO) exists primarily as the gem diol, chloral hydrate (better known as "knockout drops"). Show the structure of chloral hydrate.

PROBLEM 9.11 The oxygen in water is primarily (99.8%) ^{16}O, but water enriched with the heavy isotope ^{18}O is also available. When a ketone or aldehyde is dissolved in $H_2{}^{18}O$, the isotopic label becomes incorporated into the carbonyl group: $R_2C{=}O + H_2O^* \rightarrow R_2C{=}O^* + H_2O$ (where $O^* = {}^{18}O$). Explain.

9.8 Nucleophilic Addition of Alcohols: Acetal Formation

Aldehydes and ketones react with alcohols in the presence of an acid catalyst to yield **acetals, $R_2C(OR')_2$**, compounds that have two ether-like –OR′ groups bonded to the same carbon:

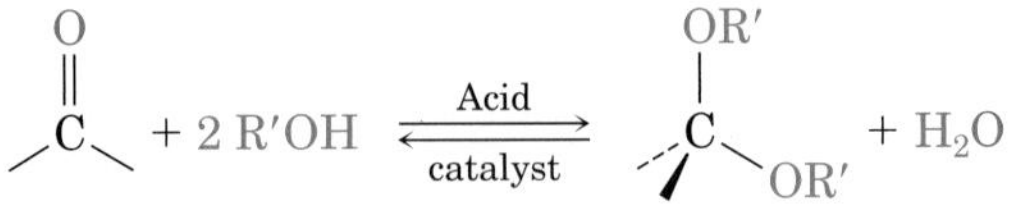

$$\text{Ketone/aldehyde} + 2\,R'OH \underset{\text{catalyst}}{\overset{\text{Acid}}{\rightleftharpoons}} \text{An acetal } (R_2C(OR')_2) + H_2O$$

Acetal formation involves the acid-catalyzed nucleophilic addition of an alcohol to a carbonyl group in a manner similar to that of acid-catalyzed hydration (see Section 9.7). The initial nucleophilic addition step yields a

hemiacetal, a carbon bearing both an OR group and an OH group. This chemistry is identical to chemistry we've seen with the addition of other nucleophiles across the C=O bond. However, this hemiacetal *can* react with a second equivalent of alcohol to yield water and an acetal, a carbon with two OR groups. How? This reactivity is made possible through the formation of an **oxonium ion**, an oxygen bearing a positive charge. The oxonium ion resembles a protonated aldehyde or ketone, and it has similar reactions.

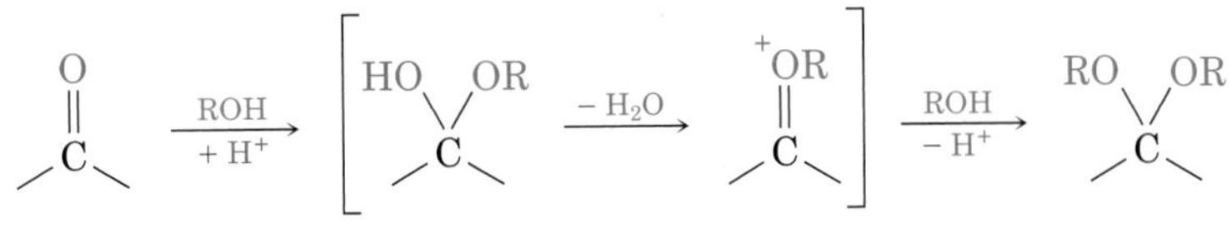

ROH, $+ H^+$ → $- H_2O$ → ROH, $- H^+$

Aldehyde or ketone **Hemiacetal** **Oxonium ion intermediate** **Acetal**

To understand the details of this reaction, let's look more closely at the first half of this mechanism to reinforce our discussions of carbonyl chemistry using cyclohexanone as an example (Figure 9.6). This mechanism is similar to that shown in Figure 9.2.

FIGURE 9.6 MECHANISM: Mechanism of the acid-catalyzed formation of a hemiacetal.

Cyclohexanone

The carbonyl group is protonated by an acid, H–A, thereby activating it for attack by an alcohol nucleophile, ROH.

$+ :A^-$

The hybridization of the carbonyl carbon changes from sp^2 to sp^3 as the π electrons from the C=O bond move onto oxygen, preventing the carbon from violating the octet rule and neutralizing the positive charge.

A base removes H^+ from the intermediate to give a neutral hemiacetal addition product and regenerate the acid catalyst.

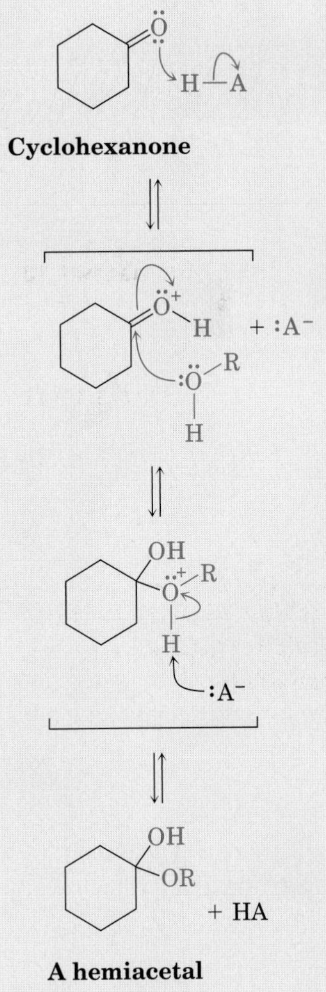

$+ HA$

A hemiacetal

Now let's see how the acetal is formed from the hemiacetal. It is necessary to substitute an –OR group for the –OH group. This process follows an S_N1 mechanism because the intermediate is actually an oxonium ion and is much more stable than a normal carbocation. It begins by protonation of the –OH group to make it a better leaving group (Figure 9.7). Then water leaves. The carbocation produced by this step is stabilized by resonance. The contribution of a lone pair from the oxygen of the –OR group produces a stabilized oxonium ion, in which all of the atoms have an octet of electrons. Next, an

FIGURE 9.7 MECHANISM: Mechanism of the acid-catalyzed formation of an acetal from a hemiacetal.

A hemiacetal

The –OH of the hemiacetal is protonated by an acid, H–A, making it a good leaving group.

+ :A⁻

An electron lone pair on the –OR moves toward carbon, expelling water as the leaving group and giving a C=O bond with a positive charge on oxygen, an oxonium ion.

Nucleophilic addition of an alcohol to the C=O bond is accompanied by movement of the π electrons to quench the positive charge on oxygen.

:A⁻

A base removes H^+ from the intermediate to give a neutral acetal addition product and regenerate the acid catalyst.

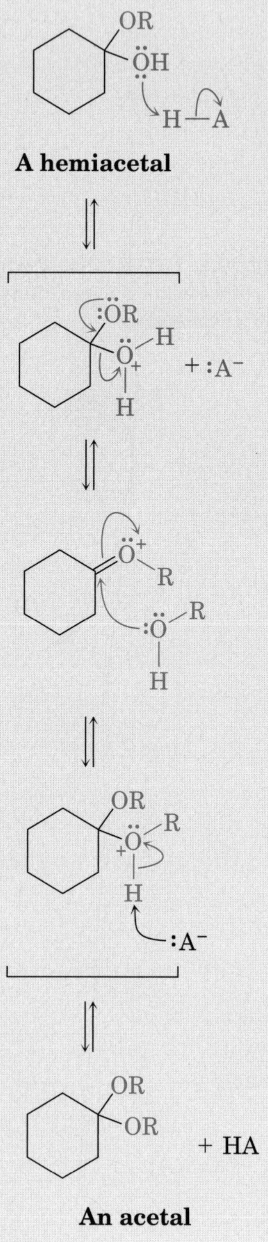

+ HA

An acetal

alcohol molecule reacts with the oxonium ion in the same manner as when it reacts with the protonated carbonyl group in Figure 9.6.

As with hydration, all of the steps during acetal formation are reversible, and the reaction can be made to go either forward (from carbonyl compound to acetal) or backward (from acetal to carbonyl compound), depending on reaction conditions. The forward reaction is favored by conditions that remove water from the medium and thus drive the equilibrium to the right. The backward reaction is favored in the presence of an excess of water, which drives the equilibrium to the left.

9.9 The Importance of Hemiacetals and Acetals: Nature and the Laboratory

Hemiacetals and acetals play important roles in nature and in the laboratory. These groups are most prevalent in sugars (or carbohydrates). Glucose, for example, contains an aldehyde group that spontaneously reacts with its own hydroxyl group to form the cyclic hemiacetal (yellow). This equilibrium lies far in favor of the cyclic hemiacetal: no aldehyde can be detected in solutions of glucose. We can polymerize these hemiacetals by making acetals: glucose hemiacetals react with other glucose hydroxyl groups (blue) to give the acetal linkages (orange). If we choose the blue hydroxyl group, two diastereomers result. One is called cellulose; the other is starch. Because they are diastereomers, they have completely different properties. We can digest starches, such as that in potatoes, but we cannot digest the cellulose of wheat grass.

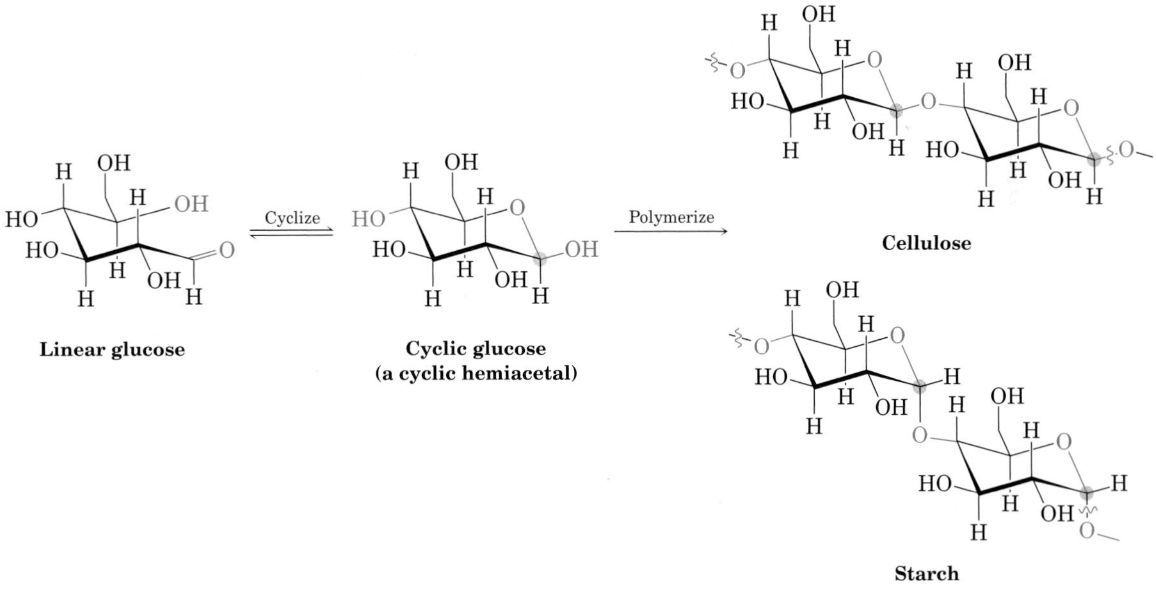

Linear glucose — Cyclize ⇌ — Cyclic glucose (a cyclic hemiacetal) — Polymerize → Cellulose / Starch

Many molecules cyclize spontaneously to form hemiacetals and acetals. Polyether antibiotics, like monensin, contain carbonyl groups that react with hydroxyl groups to form hemiacetal and acetal rings. The result of these cyclizations is a complex structure that orients functional groups in specific places in three dimensions to accomplish tasks such as binding metal ions. The ability of monesin to transport metal ions across membranes gives it potent antibacterial activity.

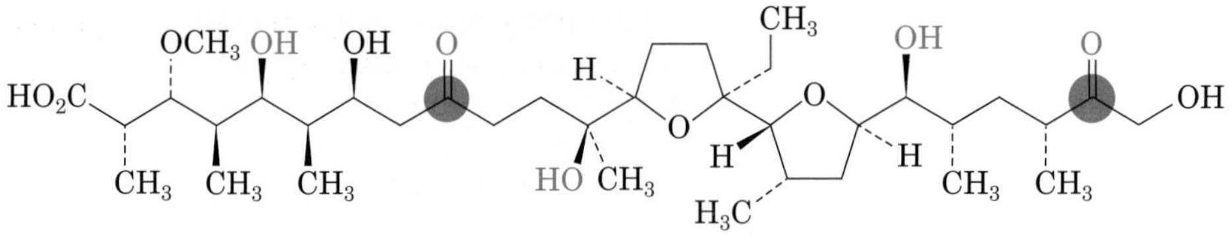

Monensin

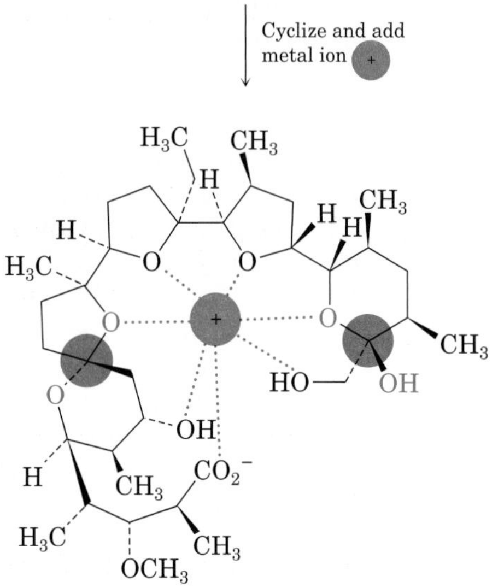

Cyclize and add metal ion

Acetals are valuable to organic chemists because they can serve as **protecting groups** for aldehydes and ketones. To see what this means, imagine that you want to reduce the ester group of methyl 4-oxopentanoate to obtain 5-hydroxypentan-2-one. This reaction can't be done in a single step because of the presence of the ketone carbonyl group in the molecule. If we were to treat methyl 4-oxopentanoate with $LiAlH_4$, both ester and ketone groups would be reduced.

$$CH_3\overset{O}{\overset{\|}{C}}CH_2CH_2\overset{O}{\overset{\|}{C}}OCH_3 \xrightarrow{?} CH_3\overset{O}{\overset{\|}{C}}CH_2CH_2CH_2OH$$

Methyl 4-oxopentanoate **5-Hydroxypentan-2-one**

This situation isn't unusual. It often happens that one functional group in a complex molecule interferes with intended chemistry on another functional group elsewhere in the molecule. In such situations, it's often possible to circumvent the problem by *protecting* the interfering functional group to render it unreactive, carrying out the desired reaction, and then removing the protecting group.

Aldehydes and ketones can be protected by converting them into acetals. Acetals, like ethers, are stable to bases, reducing agents, and various nucleophiles, but they can be cleaved by treatment with acid (see Section 8.7). Thus, you can selectively reduce the ester group in methyl 4-oxopentanoate by converting the keto group into an acetal, reducing the compound with $LiAlH_4$ in ether, and then removing the acetal protecting group by treatment with aqueous acid. The mechanism for the conversion of the acetyl group back to the ketone carbonyl group is exactly the reverse of the mechanisms of Figures 9.6 and 9.7.

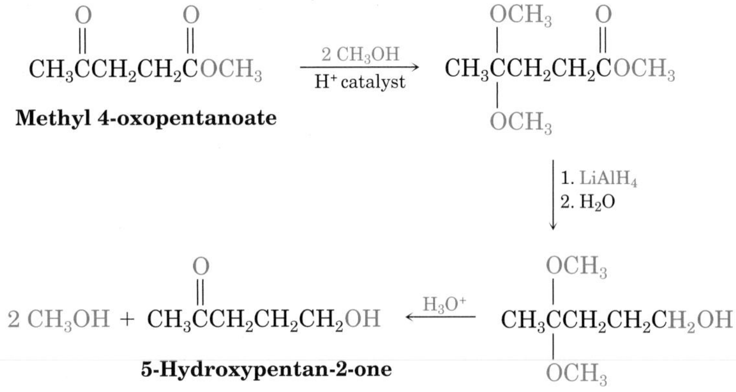

$$\underset{\textbf{Methyl 4-oxopentanoate}}{CH_3\overset{O}{\overset{\|}{C}}CH_2CH_2\overset{O}{\overset{\|}{C}}OCH_3} \xrightarrow[H^+\text{ catalyst}]{2\ CH_3OH} CH_3\underset{OCH_3}{\underset{|}{\overset{OCH_3}{\overset{|}{C}}}}CH_2CH_2\overset{O}{\overset{\|}{C}}OCH_3$$

$$\xrightarrow{1.\ LiAlH_4;\ 2.\ H_2O} CH_3\underset{OCH_3}{\underset{|}{\overset{OCH_3}{\overset{|}{C}}}}CH_2CH_2CH_2OH \xrightarrow{H_3O^+} 2\ CH_3OH + \underset{\textbf{5-Hydroxypentan-2-one}}{CH_3\overset{O}{\overset{\|}{C}}CH_2CH_2CH_2OH}$$

PRACTICE PROBLEM 9.3

What product would you obtain from the acid-catalyzed reaction of 2-methylcyclopentanone with methanol?

STRATEGY A ketone reacts with an alcohol in the presence of acid to yield an acetal. To find the product, replace the oxygen of the ketone with two $-OCH_3$ groups from the alcohol.

SOLUTION

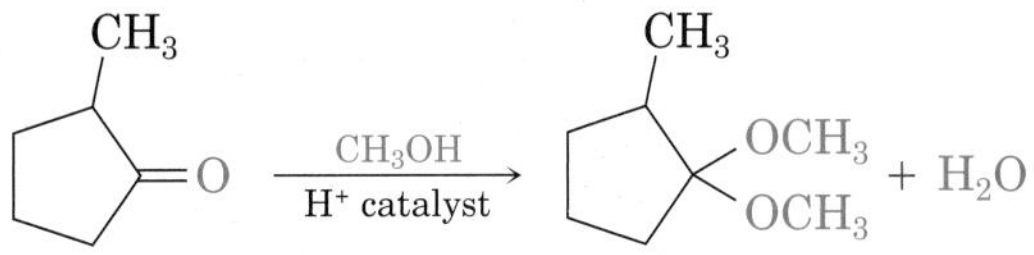

2-Methylcyclopentanone $\xrightarrow[H^+\text{ catalyst}]{CH_3OH}$ 1,1-dimethoxy-2-methylcyclopentane (CH_3, OCH_3, OCH_3) $+ H_2O$

PROBLEM 9.12 What product would you expect from the acid-catalyzed reaction of cyclohexanone and ethanol?

PROBLEM 9.13 When an aldehyde or ketone is treated with a diol such as ethylene glycol (ethane-1,2-diol) and an acid catalyst, a *cyclic* acetal is formed. Draw the structure of the product you would obtain from benzaldehyde and ethylene glycol.

PROBLEM 9.14 Show how you might carry out the following transformation. (A protection step is needed.)

$$H\overset{O}{\overset{\|}{C}}CH_2CH_2\overset{O}{\overset{\|}{C}}OCH_3 \longrightarrow H\overset{O}{\overset{\|}{C}}CH_2CH_2CH_2OH$$

9.10 Addition of Amines to Form Imines

Ammonia and primary amines, $R'NH_2$, add to aldehydes and ketones to yield **imines, $R_2C{=}NR'$**. Imines are formed by addition of the nucleophilic amine to the carbonyl group, followed by loss of water from the amino alcohol addition product.

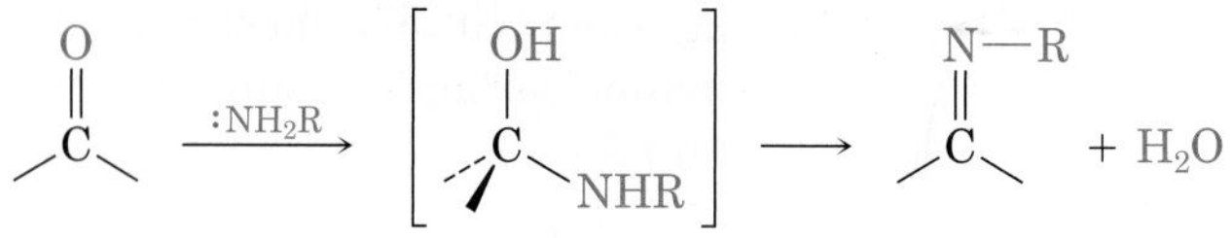

A ketone or aldehyde → (:NH_2R) [Amino alcohol intermediate] → An imine + H_2O

Imines are common intermediates in numerous biological pathways. In the context of biosynthesis, the bacterium *Bacillus subtilis* prepares alanine from pyruvic acid and ammonia. The key step is the nucleophilic addition of ammonia to the ketone group of pyruvic acid. The resulting imine C=N bond is further reduced (analogous to the reduction of C=O bonds) to the amino acid.

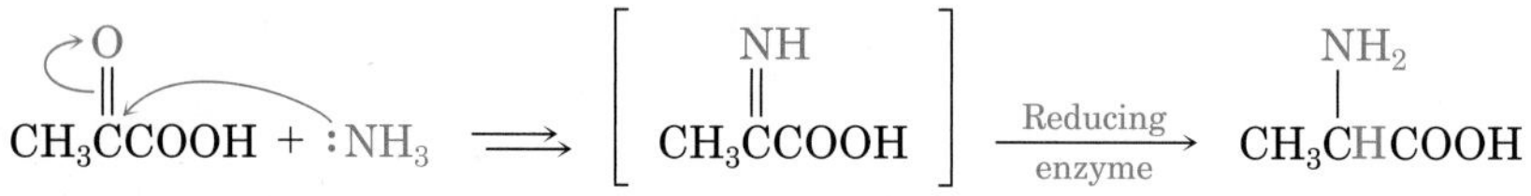

$$CH_3COCOOH + :NH_3 \rightleftharpoons [CH_3C(=NH)COOH] \xrightarrow[\text{enzyme}]{\text{Reducing}} CH_3CH(NH_2)COOH$$

Pyruvic acid — An imine — **Alanine**

In the context of degradation, the amino acid alanine, for instance, reacts with the aldehyde pyridoxal phosphate, a derivative of vitamin B_6, to yield an imine that is then further degraded. We'll see further examples in Chapter 17.

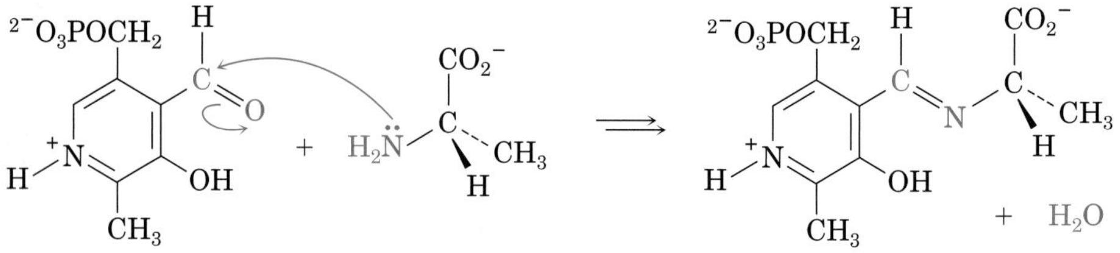

Pyridoxal phosphate + **Alanine** ⇌ **An imine** + H_2O

PRACTICE PROBLEM 9.4

What product do you expect from the reaction of butan-2-one with hydroxylamine, NH_2OH?

STRATEGY Take oxygen from the ketone and two hydrogens from the amine to form water, and then join the fragments that remain.

SOLUTION

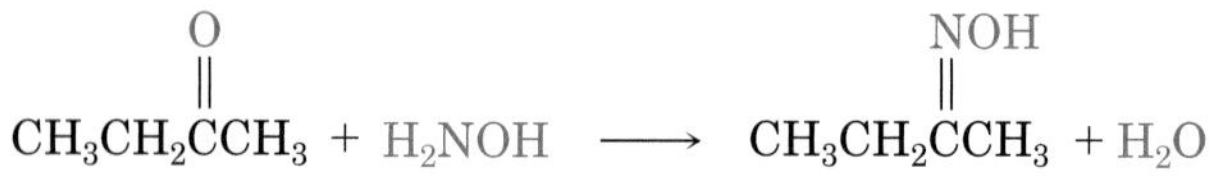

$$CH_3CH_2COCH_3 + H_2NOH \longrightarrow CH_3CH_2C(=NOH)CH_3 + H_2O$$

PROBLEM 9.15 Write the products you would obtain from treatment of cyclohexanone with the following:

(a) CH_3NH_2 (b) CH_3CH_2OH, H^+ (c) $NaBH_4$

PROBLEM 9.16 Show how the following molecule can be prepared from a carbonyl compound and an amine (blue = N):

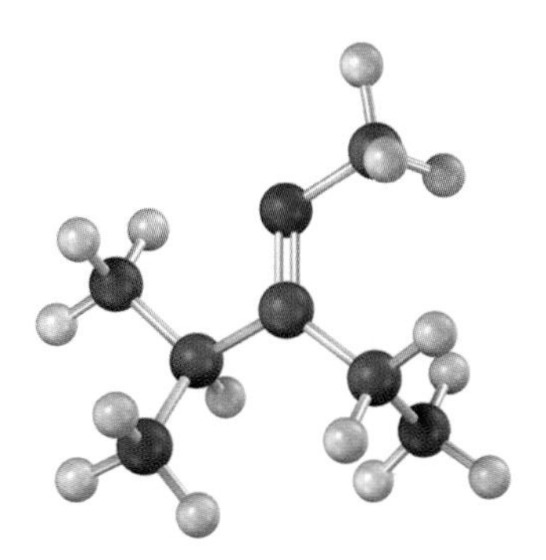

9.11 Nucleophilic Addition of Grignard Reagents: Alcohol Formation

We saw in Section 7.3 that organic halides react with magnesium metal in ether solution to give *Grignard reagents*, RMgX.

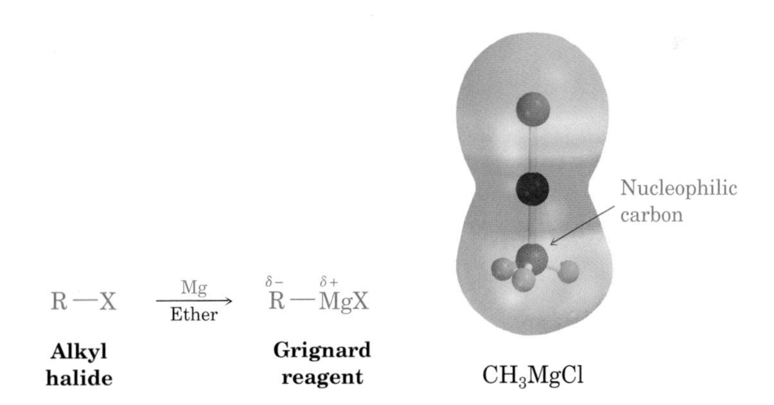

The carbon–magnesium bond of a Grignard reagent is polarized so that the carbon atom is both nucleophilic and basic. Grignard reagents therefore react as though they were carbon anions, or **carbanions**, **:R⁻**, and they undergo nucleophilic addition to aldehydes and ketones just as water and alcohols do. The reaction first produces a tetrahedral magnesium alkoxide intermediate, which is then protonated to yield the neutral alcohol on treatment with aqueous acid. Unlike the addition of water and alcohols, though, the nucleophilic addition of a Grignard reagent is irreversible.

$$\underset{\text{Ketone/aldehyde}}{\text{C=O}} \xrightarrow{:R^- \ {}^+MgX} \underset{\text{Tetrahedral intermediate}}{\left[\ :\ddot{O}:^- \ {}^+MgX \text{ on C–R} \right]} \xrightarrow{H_3O^+} \underset{\text{Alcohol}}{\text{C(OH)R}} + MgXOH$$

Ketone/ aldehyde — **Tetrahedral intermediate** — **Alcohol**

A great many alcohols can be obtained from Grignard reactions, depending on the reagents used. For example,

Formaldehyde and Grignard reagents give primary alcohols.

Cyclohexylmagnesium bromide (MgBr) + H—C(=O)—H → (1. Mix; 2. H_3O^+) → CH_2OH

Cyclohexylmagnesium bromide **Formaldehyde** **Cyclohexylmethanol (65%) (a 1° alcohol)**

Aldehydes and Grignard reagents give secondary alcohols.

$CH_3CH(CH_3)CH_2CHO$ + Phenylmagnesium bromide (MgBr) → (1. Ether solvent; 2. H_3O^+) → $CH_3CH(CH_3)CH_2CH(OH)$—Ph

3-Methylbutanal **Phenylmagnesium bromide** **3-Methyl-1-phenylbutan-1-ol (73%) (a 2° alcohol)**

Ketones and Grignard reagents give tertiary alcohols.

Cyclohexanone (=O) → (1. CH_3CH_2MgBr, ether; 2. H_3O^+) → OH, CH_2CH_3

Cyclohexanone **1-Ethylcyclohexanol (89%) (a 3° alcohol)**

Although useful, the Grignard reaction also has limitations. For example, a Grignard reagent can't be prepared from an organohalide that has other reactive functional groups in the same molecule. Some of these groups cause the Grignard reagent to react with itself. Others destroy the Grignard reagent by protonation.

—CHO, —COR, —CONR$_2$, —C≡N, —NO$_2$, —SO$_2$R } Grignard reagent reacts with these groups.

—OH, —NH, —SH, —COOH } Grignard reagent is destroyed by the acidic hydrogen.

Practice Problem 9.5

How can you use the addition of a Grignard reagent to a ketone to synthesize 2-phenylpropan-2-ol?

H_3C, OH, C, CH_3 **2-Phenylpropan-2-ol**

STRATEGY Look at the product, and identify the groups bonded to the alcohol carbon atom. In this instance, there are two methyl groups (–CH_3) and one phenyl (–C_6H_5). One of the three must come from a Grignard reagent, and the remaining two must come from a ketone. Thus, the possibilities are addition of CH_3MgBr to acetophenone and addition of C_6H_5MgBr to acetone.

SOLUTION

Acetophenone $\xrightarrow[2.\ H_3O^+]{1.\ CH_3MgBr}$ 2-Phenylpropan-2-ol $\xleftarrow[2.\ H_3O^+]{1.\ C_6H_5MgBr}$ Acetone

PROBLEM 9.17 Show the products obtained from addition of CH_3MgBr to the following compounds:

(a) Cyclopentanone (b) Benzophenone, $C_6H_5C(=O)C_6H_5$ (c) $CH_3CH_2CH_2CCH_2CH_3$ (with C=O)

PROBLEM 9.18 How might you use a Grignard addition reaction to prepare the following alcohols from an aldehyde or ketone?

(a) $CH_3C(OH)(CH_3)CH_3$ (b) 1-methylcyclohexanol (OH, CH_3 on ring carbon) (c) $CH_3CH_2C(OH)(CH_3)CH_2CH_3$

PROBLEM 9.19 How might you use a Grignard reaction to prepare the following molecule (red = O)?

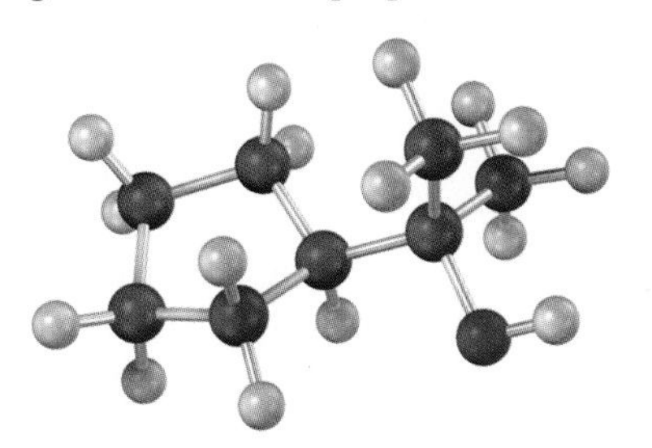

9.12 Conjugate Addition Reactions

Closely related to the direct (1,2) addition of a nucleophile to the C=O bond of an aldehyde or ketone is the **conjugate (1,4) addition** of a nucleophile to the C=C bond of an *α,β-unsaturated* aldehyde or ketone. An ***α,β*-unsaturated carbonyl compound** is one with a C=C bond between the so-called *α* carbon (the one *next to* the C=O group) and the *β* carbon (the one two carbons away from the C=O group). Thus, an *α,β*-unsaturated carbonyl compound has its C=C and C=O bonds conjugated, much as a conjugated diene does (see Section 4.10).

Direct (1,2) addition

2 O ; 1 C ; :Nu$^-$ → [O$^-$ Nu C] —H_3O^+→ HO Nu C

Conjugate (1,4) addition

4 O ; 3 C ; C 2 α ; C 1 β ; :Nu$^-$ → [:Ö:$^-$ C C C Nu ↕ :O: C C̈$^-$ C Nu] —H_3O^+→ O C C C Nu H

α, β-Unsaturated aldehyde/ketone

Enolate ion

Saturated aldehyde/ketone

The initial product of conjugate addition is a resonance-stabilized **enolate ion**, which typically undergoes protonation on the α carbon to give a saturated aldehyde or ketone product. For example, methylamine reacts with but-3-en-2-one to give an amino ketone addition product.

O H ; H_3C C C C H ; H —CH_3NH_2→ O H H ; H_3C C C C $NHCH_3$; H H

But-3-en-2-one

Conjugate addition product

Conjugate addition occurs because the electronegative oxygen atom of the α,β-unsaturated carbonyl compound withdraws electrons from the β carbon, thereby making it more electron-poor and more electrophilic than a typical alkene C=C bond.

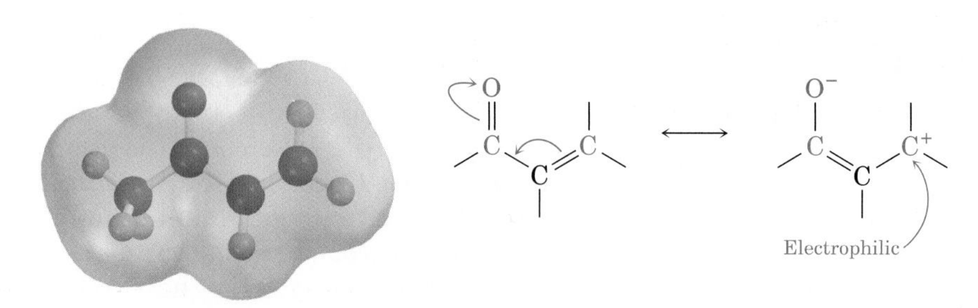

Conjugate additions are common in many biological pathways. An example is the conversion of fumarate to malate by reaction with water, a step in the citric acid cycle by which acetate is metabolized to CO_2.

$$^{-}O_2C{-}CH{=}CH{-}CO_2^{-} \xrightarrow{H_2O} {}^{-}O_2C{-}CH_2{-}CH(OH){-}CO_2^{-}$$

Fumarate **Malate**

PROBLEM 9.20 The following compound was prepared by a conjugate addition reaction between an α,β-unsaturated ketone and an alcohol. Identify the two reactants.

[Structure: cyclohexanone bearing CH_3 at C2 and $OCH(CH_3)_2$ at C3]

INTERLUDE

Carbonyl Compounds for Insect Control

© Mary M. Thacher/Photo Researchers, Inc.

Approximately 15% of the world's crops are lost each year to insects.

It has been estimated by the World Health Organization that approximately 15% of the world's crops are lost to insects each year and that more than $2 billion is spent each year on crop insecticides. Although remarkably effective, the powerful, broad-spectrum insecticides in current use are far from an ideal solution for insect control because of their potential toxicity to animals and their lack of selectivity. Beneficial as well as harmful insects are killed indiscriminately.

One new approach to insect control, which promises to be more selective and less ecologically harmful than present-day insecticides, is the use of *insect antifeedants*, substances that prevent an insect from eating but do not kill it directly. The idea of using antifeedants came from the observations of chemical ecologists that many plants have evolved elaborate and sophisticated chemical defenses against insects. Among these defenses, some plants contain chemicals that appear to block the ability or desire of an insect to feed. The insect often remains nearby, where it dies of starvation.

Most naturally occurring antifeedants are relatively complex carbonyl compounds. Polygodial, for instance, is a dialdehyde active against African army worms, and ajugarin I shows activity against locusts. Both polygodial and ajugarin I are probably too complex to be manufactured economically, but recent research has

Continued

discovered a number of substances that are structurally simpler than naturally occurring antifeedants yet retain potent activity. Among these substances synthesized in the laboratory is the following acetal.

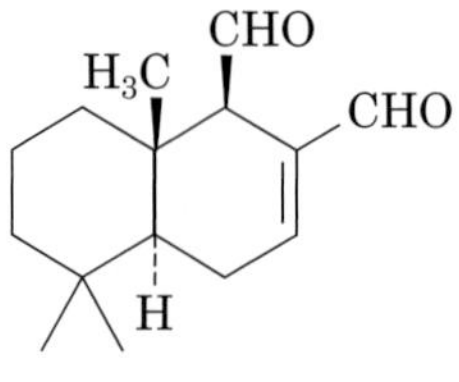

CHO, H_3C, CHO, H

Polygodial

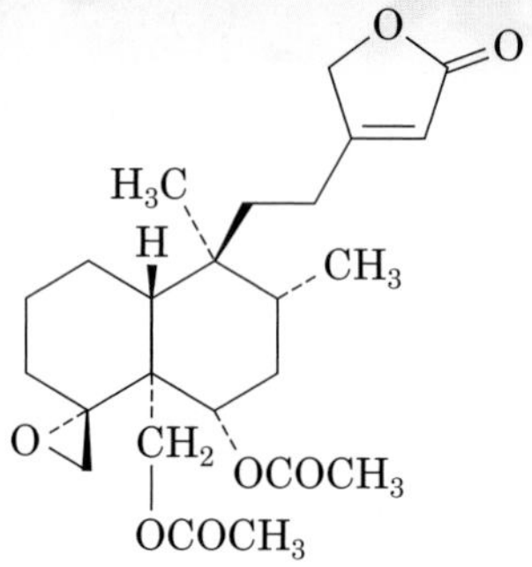

H_3C, H, CH_3, O, O, O, CH_2, $OCOCH_3$, $OCOCH_3$

Ajugarin I

O, O, H

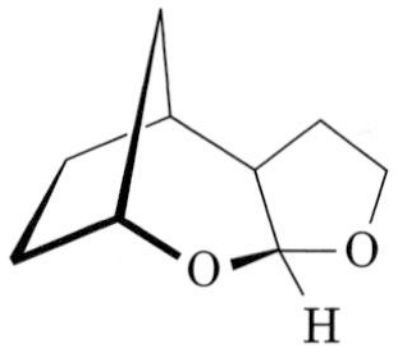

Synthetic compound

Summary and Key Words

Carbonyl compounds can be classified into two general categories:

RCHO
R_2CO

Aldehydes and **ketones** are similar in their reactivity and are distinguished by the fact that the substituents on the acyl carbon can't act as leaving groups.

RCOOH
RCOCl
RCOOCOR′
RCOOR′
$RCONH_2$

Carboxylic acids and their derivatives—**acid chlorides**, **acid anhydrides**, **esters**, and **amides**—are distinguished by the fact that the substituents on the acyl carbon *can* act as leaving groups.

A carbon–oxygen double bond is structurally similar to a carbon–carbon double bond. The carbonyl carbon atom is sp^2-hybridized and forms both a σ bond and a π bond to oxygen. Carbonyl groups are strongly polarized because of the electronegativity of oxygen.

Aldehydes are usually prepared by oxidation of primary alcohols. Ketones are similarly prepared by oxidation of secondary alcohols. Aldehydes and ketones behave similarly in much of their chemistry. Both undergo **nucleophilic addition reactions**, and a variety of products can be prepared. For example, aldehydes and ketones undergo a reversible addition reaction with water to yield 1,1-dialcohols, or **gem diols**, according to the H-to-O rule. Similarly, they react with alcohols to yield **acetals, $R_2C(OR')_2$**, which are valuable as carbonyl **protecting groups**. Primary amines add to aldehydes and ketones to give **imines, $R_2C{=}NR'$**. In addition, aldehydes and ketones are reduced by $NaBH_4$ to yield primary and secondary alcohols, respectively, and they react with Grignard reagents to give alcohols.

Closely related to the direct (1,2) addition of nucleophiles to aldehydes and ketones is the **conjugate (1,4) addition** of nucleophiles to the C=C bond of ***α,β*-unsaturated aldehydes and ketones**. Both direct and conjugate addition reactions are common in numerous biological pathways.

Summary of Reactions

1. Reaction of aldehydes and ketones with alcohols to yield acetals (Section 9.8)

$$>C{=}O + ROH \underset{\text{catalyst}}{\overset{H^+}{\rightleftharpoons}} \left[>C(OH)(OR) \right] \rightleftharpoons >C(OR)(OR) + H_2O$$

2. Reaction of aldehydes and ketones with amines to yield imines (Section 9.10)

$$>C{=}O + RNH_2 \longrightarrow >C{=}NR + H_2O$$

3. Reaction of aldehydes and ketones with Grignard reagents to yield alcohols (Section 9.11)

$$>C{=}O \xrightarrow[\text{2. } H_3O^+]{\text{1. RMgX}} >C(OH)R$$

4. Conjugate (1,4) nucleophilic addition reaction (Section 9.12)

$$-C(=O)-C{=}C< \xrightarrow{\text{Nu—H}} -C(=O)-C(H)-C-Nu$$

EXERCISES

Visualizing Chemistry

9.21 Identify the kinds of carbonyl groups in the following molecules (red = O, blue = N):

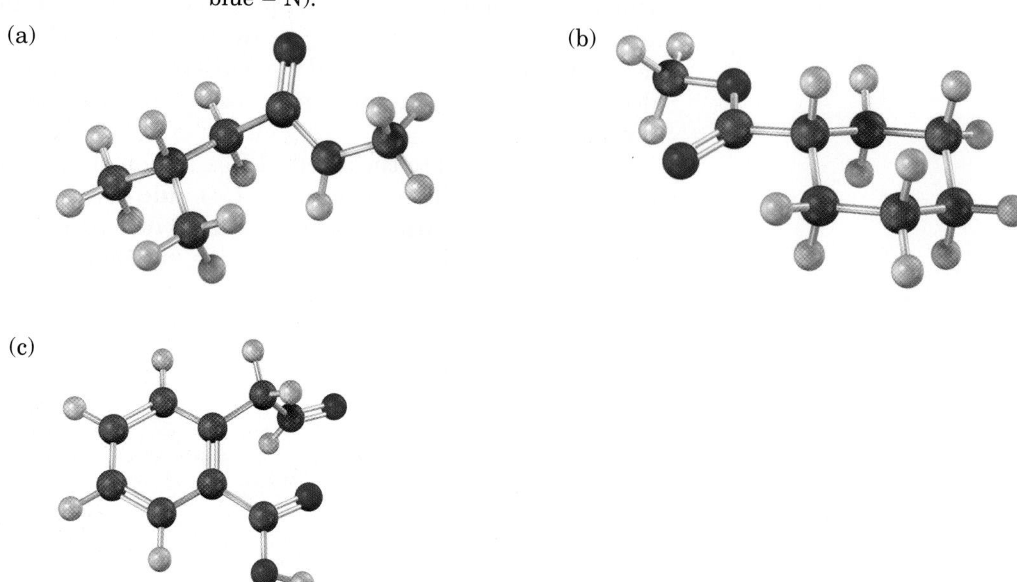

9.22 Identify the reactants from which the following molecules were prepared. If an acetal, identify the carbonyl compound and the alcohol; if an imine, identify the carbonyl compound and the amine; if an alcohol, identify the carbonyl compound and the Grignard reagent (red = O, blue = N):

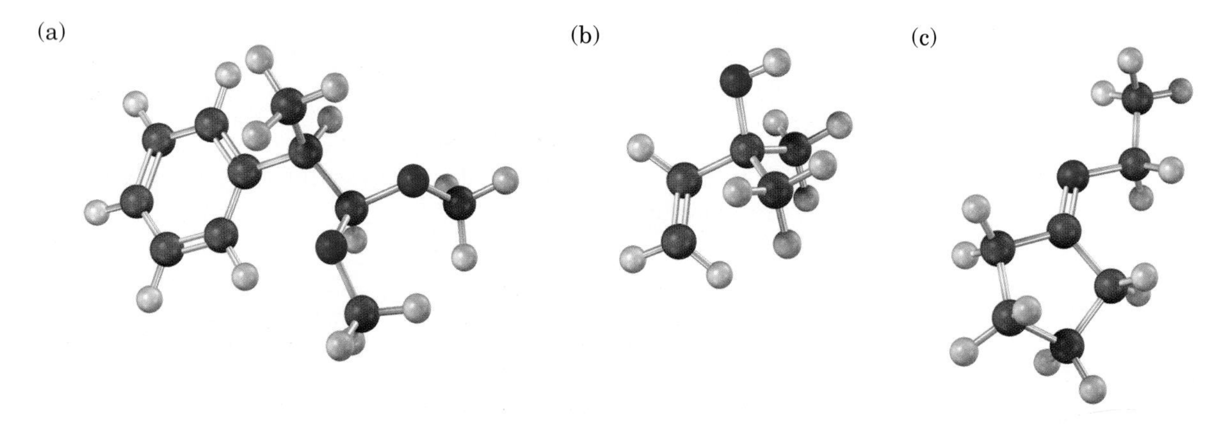

9.23 Compounds called *cyanohydrins* result from the nucleophilic addition of HCN to an aldehyde or ketone. Draw and name the carbonyl compound that the following cyanohydrin was prepared from (red = O, blue = N):

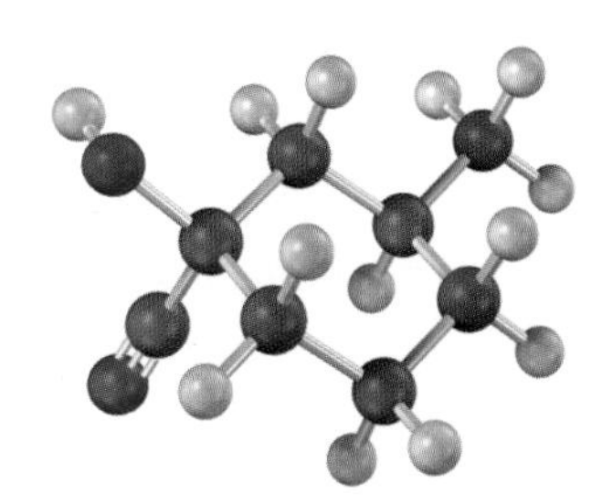

9.24 The following model represents the product resulting from addition of a nucleophile to an aldehyde or ketone. Identify the reactants, and write the reaction (red = O, blue = N).

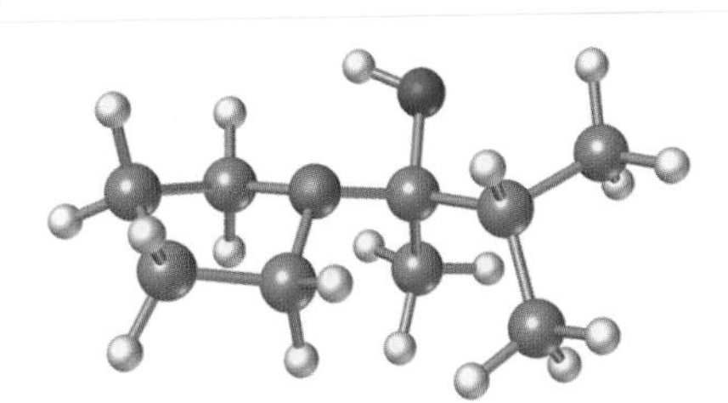

Additional Problems

NOMENCLATURE AND ISOMERISM

9.25 Identify the different kinds of carbonyl functional groups in the following molecules:

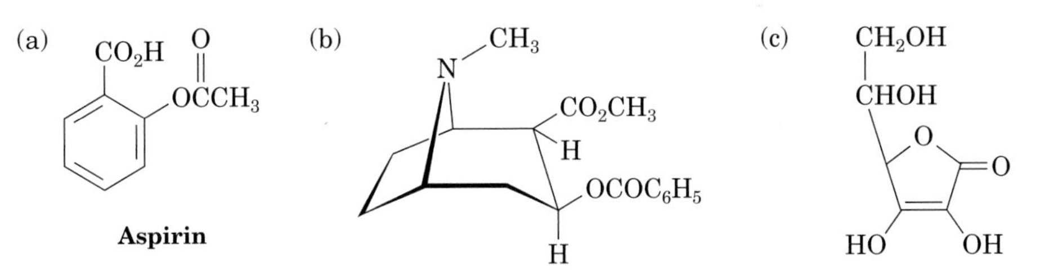

9.26 Draw structures corresponding to the following names:
(a) Bromoacetone
(b) 3-Methylbutan-2-one
(c) 3,5-Dinitrobenzaldehyde
(d) 3,5-Dimethylcyclohexanone
(e) 2,2,4,4-Tetramethylpentan-3-one
(f) Butanedial
(g) (*S*)-2-Hydroxypropanal
(h) 3-Phenylprop-2-enal

9.27 Draw and name the seven aldehydes and ketones with the formula $C_5H_{10}O$.

9.28 Which of the compounds you identified in Problem 9.27 are chiral?

9.29 Draw structures of molecules that meet the following descriptions:
(a) A cyclic ketone, C_6H_8O
(b) A diketone, $C_6H_{10}O_2$
(c) An aryl ketone, $C_9H_{10}O$
(d) A 2-bromo aldehyde, C_5H_9BrO

9.30 Give IUPAC names for the following structures:

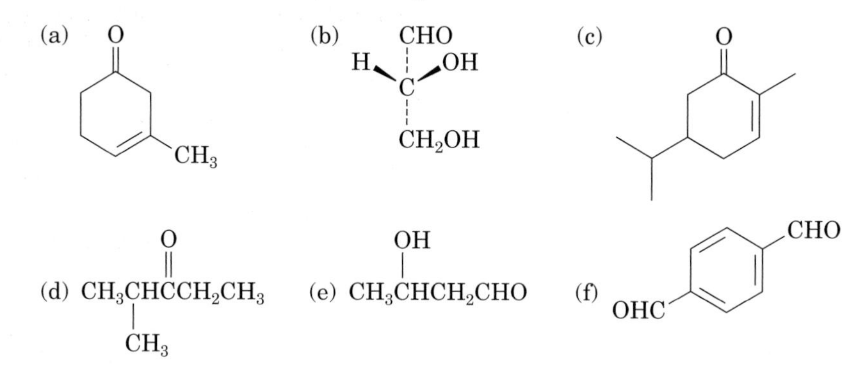

9.31 Give an example of each of the following:
(a) An acetal (b) A gem diol (c) An imine (d) A hemiacetal

REACTIONS

9.32 Predict the products of the reaction of phenylacetaldehyde, $C_6H_5CH_2CHO$, with the following reagents:
(a) $NaBH_4$, then H_3O^+ (b) Tollens' reagent (c) NH_2OH
(d) CH_3MgBr, then H_3O^+ (e) CH_3OH, H^+ catalyst

9.33 Answer Problem 9.32 for the reaction of acetophenone, $C_6H_5COCH_3$, with the same reagents.

9.34 Identify the nucleophile that has added to acetone to give the following products:

(a) $CH_3CH(OH)CH_3$ (b) $CH_3C(OH)(CH_3)CH_2CH_3$ (c) $CH_3C(=NCH_3)CH_3$ (d) $CH_3C(OH)(SCH_3)CH_3$

9.35 Reaction of butan-2-one with HCN yields a *cyanohydrin* product [$R_2C(OH)CN$] having a new stereocenter. Explain why the product is not optically active.

9.36 In light of your answer to Problem 9.35, what stereochemistry would you expect the product from the reaction of phenylmagnesium bromide with butan-2-one to have?

SYNTHESIS

9.37 Starting from cyclohex-2-enone and any other reagents needed, how would you prepare the following substances? (More than one step may be required.)

(a) (b) OH CH_3 (c) OH (d) OH

9.38 How can you explain the observation that the S_N2 reaction of (dibromomethyl)-benzene with NaOH yields benzaldehyde rather than (dihydroxymethyl)-benzene?

$CHBr_2$ $\xrightarrow[H_2O]{NaOH}$ O C H

(Dibromomethyl)benzene **Benzaldehyde**

9.39 Use a Grignard reaction on an aldehyde or ketone to synthesize the following compounds from an aldehyde or ketone:
(a) Pentan-2-ol (b) 2-Phenylbutan-2-ol
(c) 1-Ethylcyclohexanol (d) Diphenylmethanol

9.40 Show the products that result from the reaction of phenylmagnesium bromide with the following reagents:
(a) CH_2O (b) Benzophenone ($C_6H_5COC_6H_5$) (c) Pentan-3-one

9.41 Show how you could make the following alcohols using a Grignard reaction of an aldehyde or ketone. Show all possibilities.

(a) $CH_3CH(CH_3)CH_2CH_2CH_2OH$ (b) 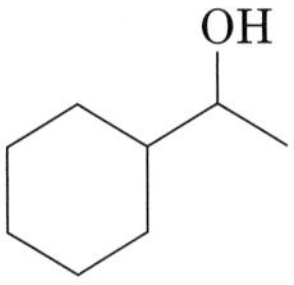 [cyclohexyl–CH(OH)–CH₃ structure] (c) $CH_3CH_2CH(OH)CH{=}CHCH_3$

9.42 Show how you could make the following alcohols using a Grignard reaction of an aldehyde or ketone. Show all possibilities.

(a) 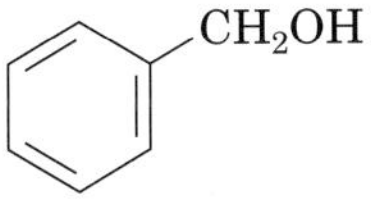 [benzene ring]–CH_2OH (b) 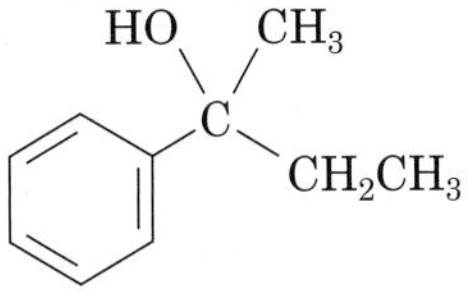 [benzene ring]–C(HO)(CH_3)(CH_2CH_3) (c) [cyclohexane ring]–C(HO)(CH_3)(CH_3)

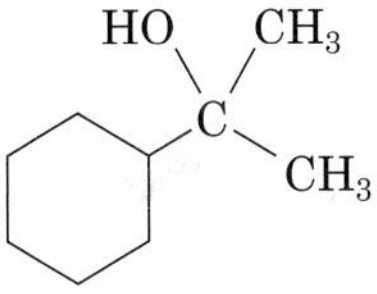

(d) $CH_3CH_2CH(CH_3)CH_2CH(OH)CH_3$ (e) $CH_3CH_2C(CH_2CH_3)(CH_2CH_3)OH$ (f) 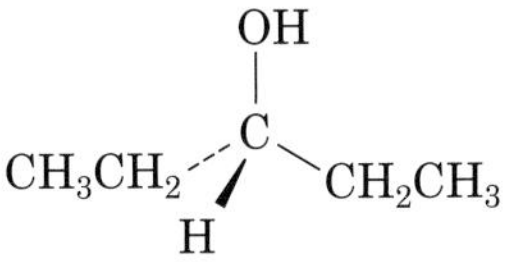 CH_3CH_2–C(OH)(H)–CH_2CH_3

9.43 How could you convert bromobenzene into benzoic acid, $C_6H_5CO_2H$? (More than one step is required.)

Hemiacetals and Acetals

9.44 Show the structures of the intermediate hemiacetals and the final acetals that result from the following reactions:

(a) 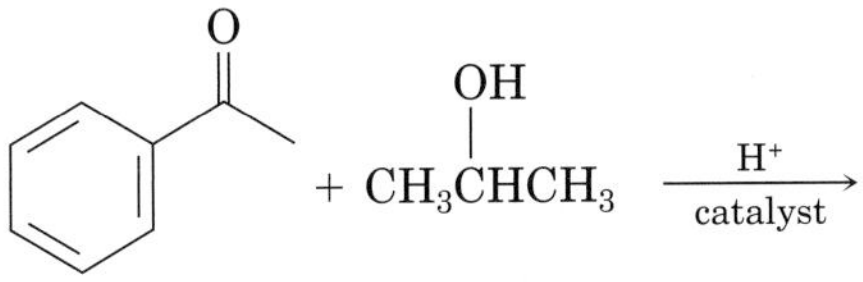 [benzene ring]–C(=O)–CH_3 + $CH_3CH(OH)CH_3$ $\xrightarrow[\text{catalyst}]{H^+}$ (b) 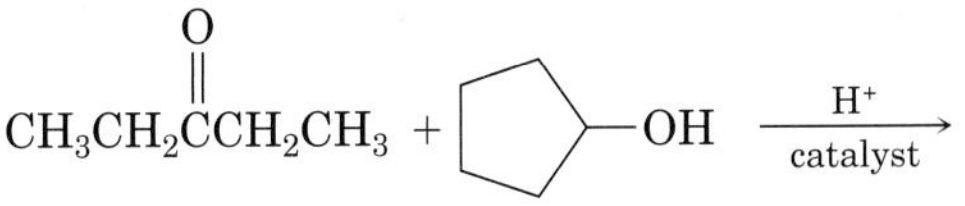 $CH_3CH_2C(=O)CH_2CH_3$ + [cyclopentane ring]–OH $\xrightarrow[\text{catalyst}]{H^+}$

9.45 Show the structures of the alcohols and aldehydes or ketones you would use to make the following acetals:

(a) $CH_3CH_2CH(CH_3)CH_2CH(OCH_3)OCH_3$ (b) [benzene ring]–C(OCH_2CH_3)(OCH_2CH_3)–CH_3 (c) [cyclopentane ring spiro-fused to 1,3-dioxolane, O, O]

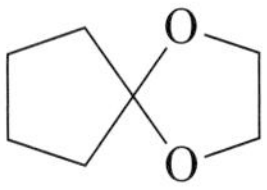

Integrated Problems

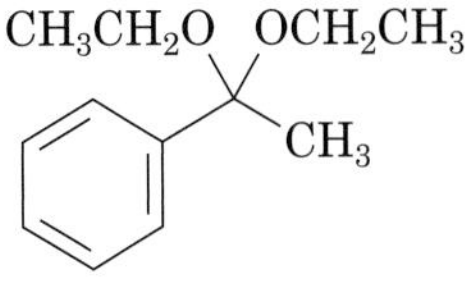

9.46 Show the products from the reaction of pentan-2-one with the following reagents:

(a) NH_2OH (b) 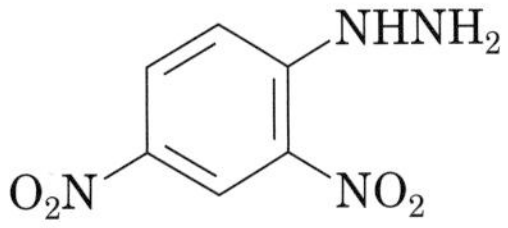 [2,4-dinitrophenyl ring with $NHNH_2$, O_2N, NO_2] (c) CH_3CH_2OH, H^+

9.47 How would you synthesize the following compounds from cyclohexanone?

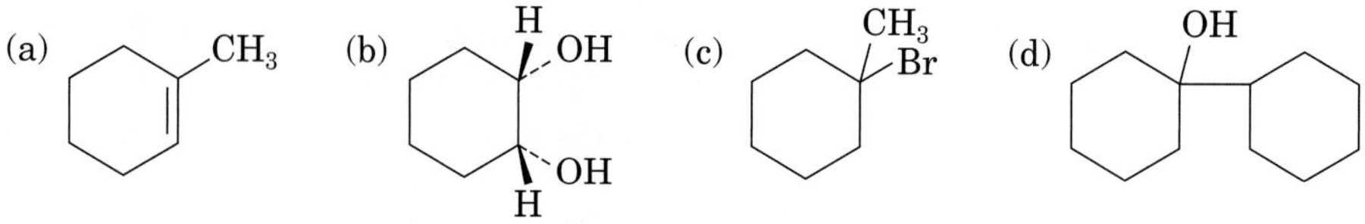

(a) CH_3 (b) H, OH, OH, H (c) CH_3, Br (d) OH

9.48 Draw the product(s) obtained by conjugate addition of the following reagents to cyclohex-2-enone:
(a) H_2O (b) NH_3 (c) CH_3OH (d) CH_3CH_2SH

9.49 Carvone is the major constituent of spearmint oil. What products would you expect from the reaction of carvone with the following reagents?
(a) $LiAlH_4$, then H_3O^+ (b) C_6H_5MgBr, then H_3O^+
(c) H_2, Pd catalyst (d) CH_3OH, H^+

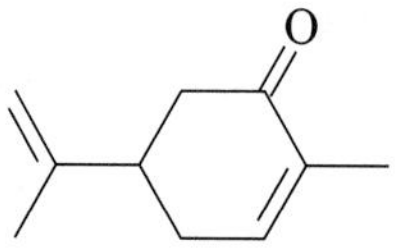

O

Carvone

9.50 Treatment of a ketone or aldehyde with a thiol in the presence of an acid catalyst yields a *thioacetal*, $R_2C(SR')_2$. To what other reaction is this thioacetal formation analogous? Show how the reaction occurs.

9.51 Treatment of a ketone or aldehyde with hydrazine, H_2NNH_2, yields an *azine*, $R_2C{=}N{-}N{=}CR_2$. Show how the reaction occurs.

9.52 When glucose is treated with $NaBH_4$, reaction occurs to yield *sorbitol*, a commonly used food additive. Show how this reduction occurs

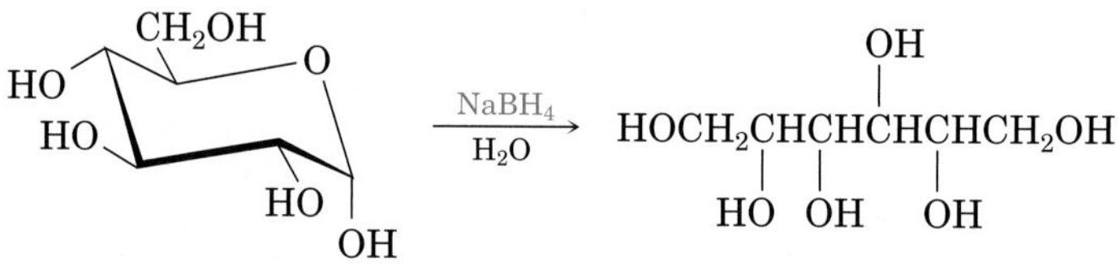

CH_2OH, O, HO, HO, HO, OH $\xrightarrow[H_2O]{NaBH_4}$ $HOCH_2CH(OH)CH(OH)CH(OH)CH(OH)CH_2OH$

Glucose **Sorbitol**

9.53 Ketones react with dimethylsulfonium methylide to yield epoxides by a mechanism that involves an initial nucleophilic addition followed by an intramolecular S_N2 substitution. Show the mechanism.

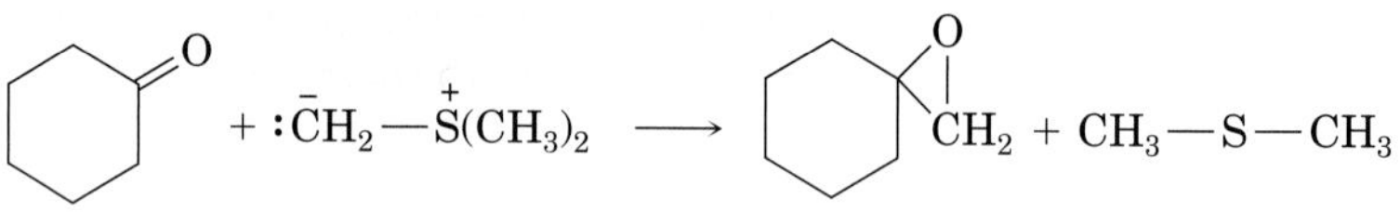

O $+ :\overset{-}{C}H_2{-}\overset{+}{S}(CH_3)_2 \longrightarrow$ O, $CH_2 + CH_3{-}S{-}CH_3$

Dimethylsulfonium methylide

9.54 Identify the reagents a through d in the following scheme:

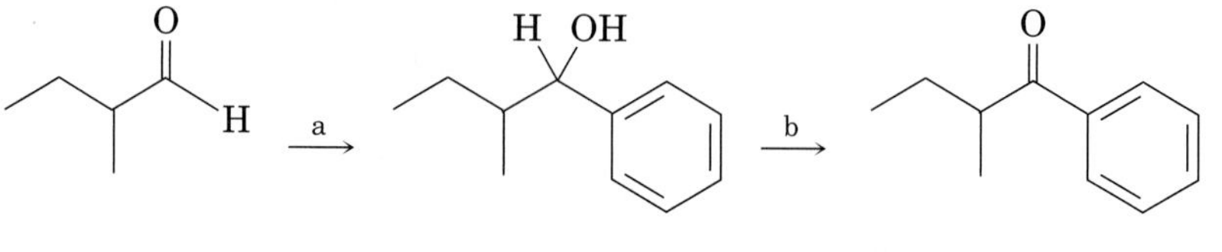

O, H $\xrightarrow{a}$ H OH $\xrightarrow{b}$ O

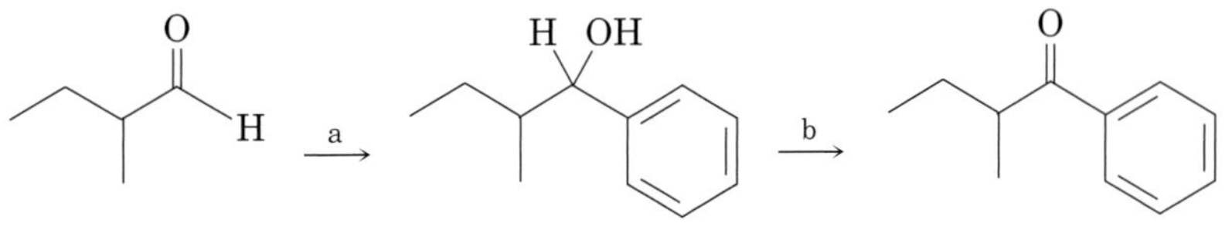

O H a H OH b O

IN THE MEDICINE CABINET

9.55 How would you prepare tamoxifen, a drug used in the treatment of breast cancer, from benzene, the following ketone, and any other reagents needed?

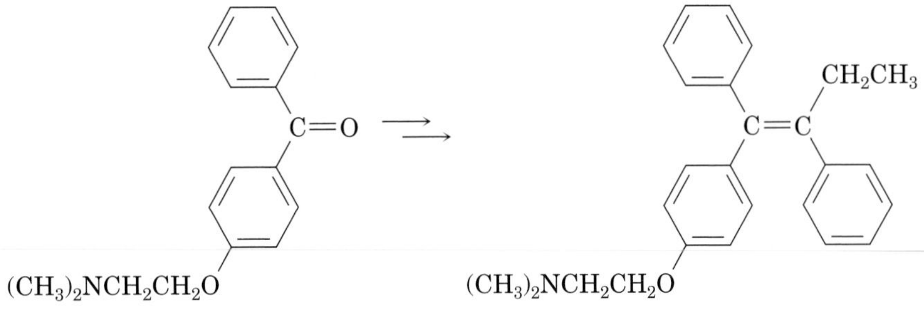

$C=O$ ⟹ CH_2CH_3 $C=C$

$(CH_3)_2NCH_2CH_2O$ $(CH_3)_2NCH_2CH_2O$

Tamoxifen

9.56 Pralidoxime iodide is a general antidote for many compounds that target the protein acetylcholine esterase, including some insecticides and nerve toxins such as sarin. The drug is made in two steps, starting with pyridine-2-carbaldehyde.

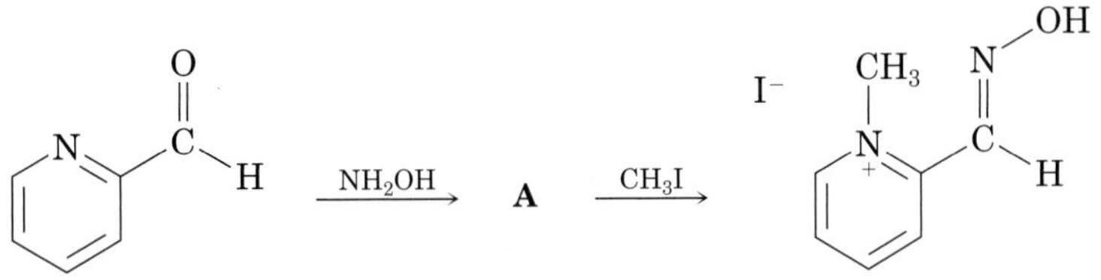

NH_2OH → **A** CH_3I → I^- CH_3 OH N H

Pyridine-2-carbaldehyde **Pralidoxime iodide**

(a) Draw a mechanism for the nucleophilic substitution reaction of hydroxylamine, NH_2OH, with pyridine-2-carbaldehyde to determine the structure of intermediate **A**.

(b) Reaction of intermediate **A** with methyl iodide by an S_N2 reaction produces pralidoxime iodide. Draw this mechanism.

(c) What can you say about the relative nucleophilicity of a pyridine nitrogen versus an *N*-hydroxyl group toward methyl iodide?

9.57 The production of vitamin C (ascorbic acid) highlights the importance of oxidation and reduction chemistry of oxygen-containing functional groups.

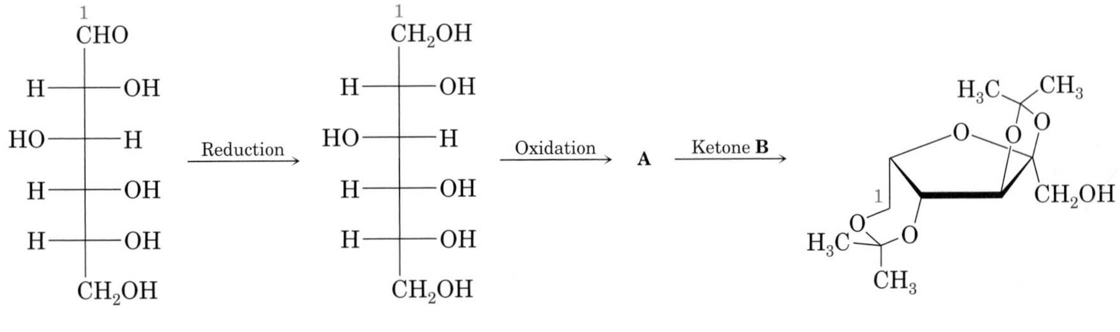

CHO, H—OH, HO—H, H—OH, H—OH, CH_2OH Reduction → CH_2OH, H—OH, HO—H, H—OH, H—OH, CH_2OH Oxidation → **A** Ketone **B** → H_3C CH_3 CH_2OH H_3C CH_3

D-Glucose

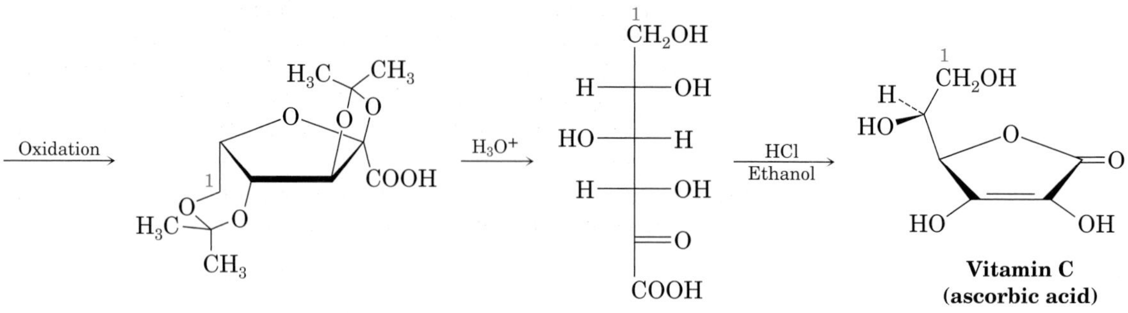

Oxidation → (diacetone intermediate with COOH) → H_3O^+ → (open-chain acid) → HCl, Ethanol → **Vitamin C (ascorbic acid)**

(a) The synthesis of vitamin C starts with D-glucose, shown here in its open-chain form. How many chirality centers are in D-glucose?
(b) The reduction of D-glucose involves the conversion of what functional group into what type of alcohol?
(c) The critical step in vitamin C production is the oxidation step to intermediate **A**. Intermediate **A** contains a single carbonyl group. Draw the eight possible oxidation products, and identify them as aldehydes, ketones, or carboxylic acids.
(d) The oxidation is carried out by an enzyme from a bacterium. Why would an enzyme be used instead of a chemical oxidizing agent such as $KMnO_4$?
(e) The subsequent steps also rely on oxidation chemistry; however, before the second oxidation is executed, intermediate **A** is reacted with an excess of a ketone to make acetals. What ketone is used?
(f) These acetals are cleaved with H_3O^+ after the second oxidation step. What purpose do these acetals serve? What could these groups be called?
(g) Concentrated HCl promotes an esterification reaction (see Chapter 10) to yield vitamin C, which has a single chiral center. Is the configuration *R* or *S*?

In the Field with Agrochemicals

9.58 The early steps of the synthesis of the herbicide Metolachlor rely on oxidation and imine formation.

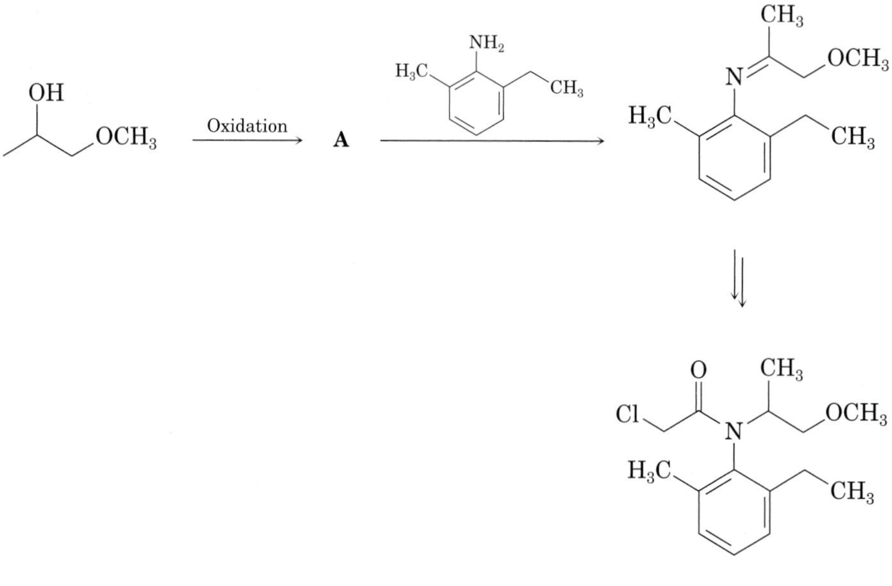

(1-methoxy-2-propanol) → Oxidation → **A** → (2-ethyl-6-methylaniline) → imine → **Metolachlor**

(a) The starting ether can be obtained by reacting an epoxide with $-OCH_3$. Propose a structure for the epoxide commonly called propylene oxide.

(b) Propose the structure of intermediate **A**.
(c) What oxidizing agent would you use for the oxidation?
(d) Draw a mechanism that describes the reaction of ketone **A** with the aniline derivative (an Ar—NH_2) to form the imine.
(e) Two imine products are obtained based on cis–trans isomerism. Draw these two isomers.

9.59 Interfering with ecdysone, the molting hormone for insects, is a common mechanism of action for some insecticides.

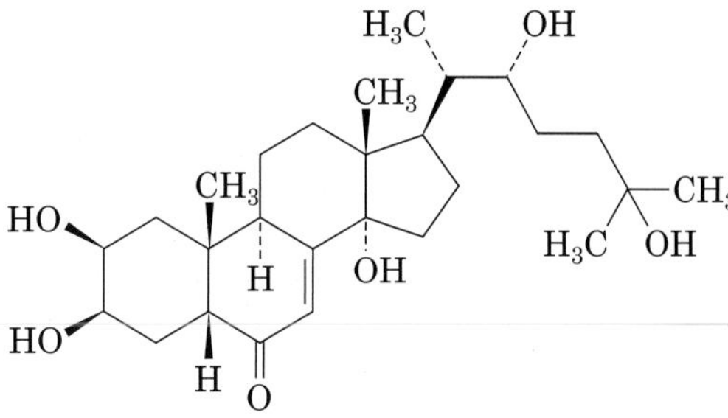

Ecdysone

(a) Categorize each of the hydroxyl groups as primary, secondary, or tertiary.
(b) Oxidation of which hydroxyl groups of ecdysone will provide aldehydes?
(c) Oxidation of which hydroxyl groups of ecdysone will provide ketones?
(d) Oxidation of which hydroxyl groups will be unsuccessful?
(e) How many chirality centers are in ecdysone?
(f) Addition of a reducing agent to ecdysone can produce two different classes of products. Draw the mechanisms for a 1,4-addition of hydride to produce a ketone after protonation and a 1,2-addition of hydride to produce an alcohol with an α-alkene.
(g) From where does the “one” of ecdysone originate?

9.60 Both the 1,4- and 1,2-addition of hydride to ecdysone (Problem 9.59) produce two products depending on which side of the molecule the hydride attacks.
(a) Draw these stereoisomers.
(b) What term describes the relationship between the 1,4-addition products? Between the 1,2-addition products?
(c) Do you expect that the products of either 1,4- or 1,2-addition will be in equal 50:50 ratios? Why or why not?

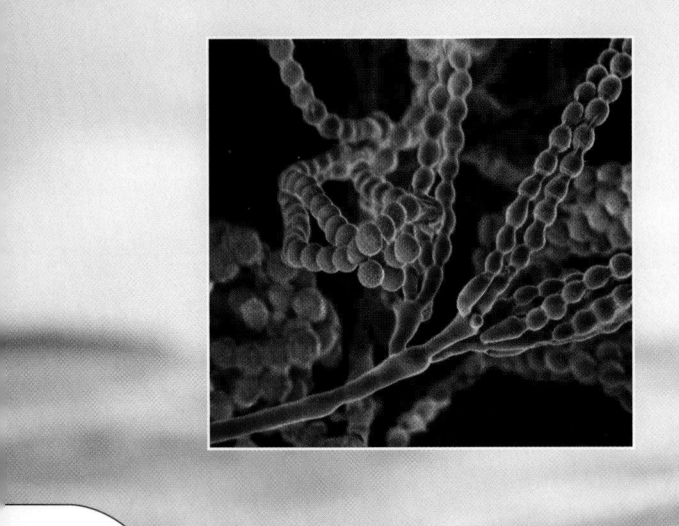

The penicillins are a large group of carboxylic acid derivatives used commonly as antibiotics. Shown here is a colored scanning electron micrograph of *Penicillium chrysogenum* fungus, the source of penicillin G. (SciMAT/Photo Researchers, Inc.)

10 CHAPTER Carboxylic Acids and Derivatives

Carboxylic acids and their derivatives are among the most abundant of organic compounds, both in the laboratory and in living organisms. Although there are many different kinds of carboxylic acid derivatives, we'll be concerned only with some of the most common ones: *acid halides*, *acid anhydrides*, *esters*, *amides*, and related compounds called *nitriles*. In addition, *acyl phosphates* and *thioesters* are acid derivatives of great importance in numerous biological processes, but we'll save our coverage of these latter two classes until Chapter 17. The common structural feature of all these compounds is that they contain an acyl group bonded to an electronegative atom or substituent that can act as a leaving group in substitution reactions.

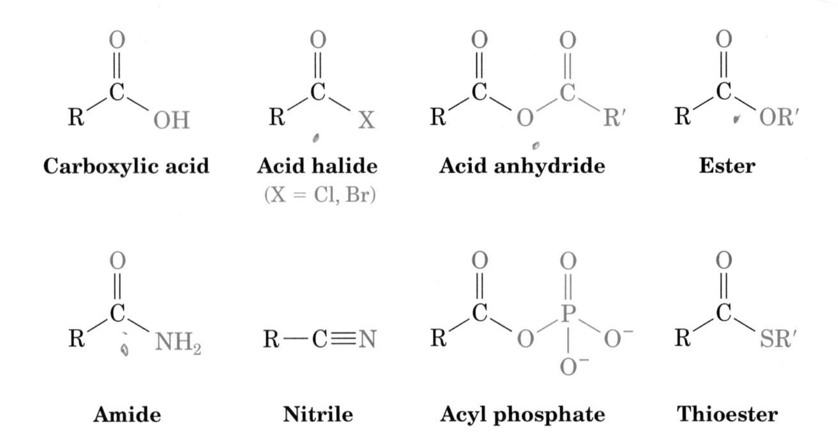

10.1

Naming Carboxylic Acids and Derivatives

Carboxylic Acids: RCO_2H

Simple open-chain carboxylic acids are named by replacing the terminal *-e* of the alkane name with *-oic acid.* The $–CO_2H$ carbon (the **carboxyl group** carbon) is always numbered C1.

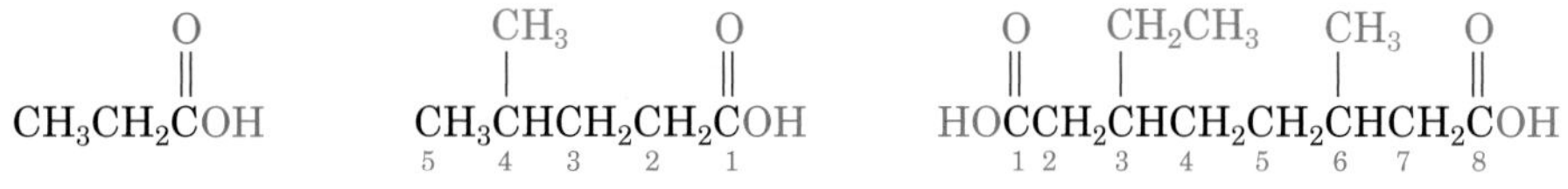

CH_3CH_2COH (C=O)

Propanoic acid

$CH_3CHCH_2CH_2COH$ (CH_3 on C4; C=O on C1; numbered 5 4 3 2 1)

4-Methyl**pentan**oic acid

$HOCCH_2CHCH_2CH_2CHCH_2COH$ (CH_2CH_3 on C3; CH_3 on C6; C=O on C1 and C8; numbered 1 2 3 4 5 6 7 8)

3-Ethyl-6-methyl**octanedi**oic acid

Alternatively, compounds that have a $–CO_2H$ group bonded to a ring are named by using the suffix *-carboxylic acid.* In this alternative system, the carboxylic acid carbon is *attached to* C1 on the ring but is not itself numbered.

COOH (ring positions 1–6; Br on C3)

3-Bromo**cyclohexane**carboxylic acid

COOH (ring positions 1–5; double bond C1=C2)

Cyclopent-1-encarboxylic acid

Because many carboxylic acids were among the first organic compounds to be isolated and purified, there are a large number of acids with common names (Table 10.1). We'll use systematic names in this book, with the exception of formic (methanoic) acid, HCO_2H, and acetic (ethanoic) acid, CH_3CO_2H, whose names are so well known that it makes little sense to refer to them in any other way. Also listed in Table 10.1 are the names for acyl groups derived from the parent acids.

TABLE 10.1 Some Common Names of Carboxylic Acids and Acyl Groups

Carboxylic acid		Acyl group	
Structure	Name	Name	Structure
HCO_2H	Formic	Formyl	$HCO—$
CH_3CO_2H	Acetic	Acetyl	$CH_3CO—$
$CH_3CH_2CO_2H$	Propionic	Propionyl	$CH_3CH_2CO—$
$CH_3CH_2CH_2CO_2H$	Butyric	Butyryl	$CH_3(CH_2)_2CO—$
HO_2CCO_2H	Oxalic	Oxalyl	$—OCCO—$
$HO_2CCH_2CO_2H$	Malonic	Malonyl	$—OCCH_2CO—$
$HO_2CCH_2CH_2CO_2H$	Succinic	Succinyl	$—OC(CH_2)_2CO—$
$H_2C{=}CHCO_2H$	Acrylic	Acryloyl	$H_2C{=}CHCO—$
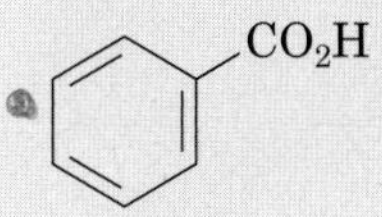 (benzene ring)–CO_2H	Benzoic	Benzoyl	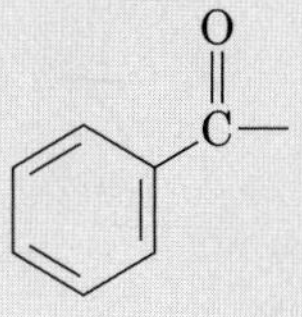 (benzene ring)–C(=O)—

Acid Halides: RCOX

Acid halides are named by identifying first the acyl group and then the halide. The acyl group name is derived from the acid name by replacing the *-ic acid* ending with *-yl,* or the *-carboxylic acid* ending with *-carbonyl.* For example

$CH_3\overset{O}{\overset{\|}{C}}Cl$ — **Acetyl chloride (from acetic acid)**

Br, C, O — **Benzoyl bromide (from benzoic acid)**

Cl, C, O — **Cyclohexanecarbonyl chloride (from cyclohexanecarboxylic acid)**

Acid Anhydrides: RCO_2COR'

Anhydrides from simple carboxylic acids and cyclic anhydrides from dicarboxylic acids are named by replacing the word *acid* with *anhydride:*

H_3C, C, O, O, C, O, CH_3 — **Acetic anhydride**

O, C, O, C, O — **Benzoic anhydride**

O, O, O — **Succinic anhydride**

Amides: $RCONH_2$

Amides with an unsubstituted —NH_2 group are named by replacing the *-oic acid* or *-ic acid* ending with *-amide*, or by replacing the *-carboxylic acid* ending with *-carboxamide:*

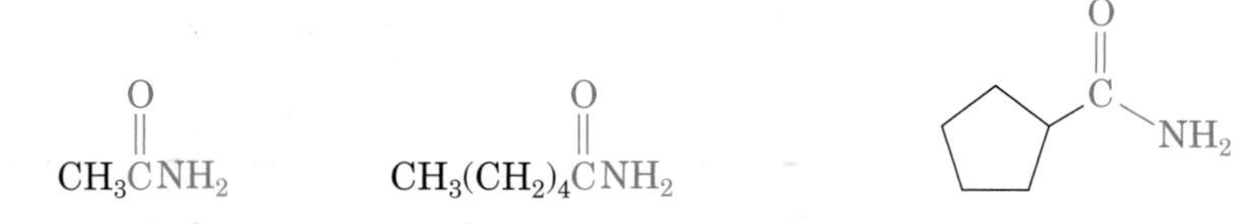

Acetamide (from acetic acid) **Hexanamide (from hexanoic acid)** **Cyclopentanecarboxamide (from cyclopentanecarboxylic acid)**

If the nitrogen atom is substituted, the amide is named by first identifying the substituent group and then the parent. The substituents are preceded by the letter *N* to identify them as being directly attached to nitrogen.

$CH_3CH_2\overset{O}{\overset{\|}{C}}NHCH_3$ — ***N*-Methylpropanamide**

O, C, N, CH_2CH_3, CH_2CH_3 — ***N,N*-Diethylcyclohexanecarboxamide**

Esters: RCO_2R'

Systematic names for esters are derived by first giving the name of the alkyl group attached to oxygen and then identifying the carboxylic acid. In so doing, the *-ic acid* ending is replaced by *-ate:*

$$CH_3\overset{\overset{\displaystyle O}{\|}}{C}OCH_2CH_3$$

Ethyl acetate
(the ethyl ester of acetic acid)

$$CH_3O\overset{\overset{\displaystyle O}{\|}}{C}CH_2\overset{\overset{\displaystyle O}{\|}}{C}OCH_3$$

Dimethyl malonate
(the dimethyl ester of malonic acid)

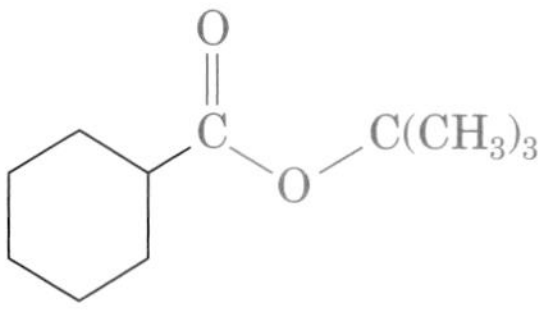

***tert*-Butyl cyclohexanecarboxylate**
(the *tert*-butyl ester of cyclohexanecarboxylic acid)

Nitriles: R—C≡N

Compounds containing the —C≡N functional group are called **nitriles.** Simple acyclic nitriles are named by adding *-nitrile* as a suffix to the alkane name, with the nitrile carbon numbered C1.

$$\underset{5}{CH_3}\underset{4}{\overset{\overset{\displaystyle CH_3}{|}}{C}H}\underset{3}{CH_2}\underset{2}{CH_2}\underset{1}{CN}$$

4-Methylpentanenitrile

More complex nitriles are named as derivatives of carboxylic acids by replacing the *-ic acid* or *-oic acid* ending with *-onitrile*, or by replacing the *-carboxylic acid* ending with *-carbonitrile.* In this system, the nitrile carbon atom is attached to C1 but is not itself numbered:

$$CH_3C{\equiv}N$$

Acetonitrile
(from acetic acid)

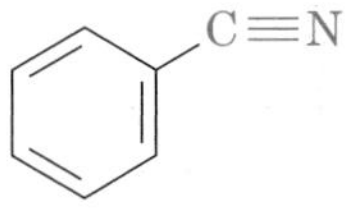

Benzonitrile
(from benzoic acid)

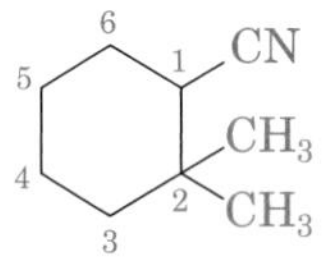

2,2-Dimethylcyclohexanecarbonitrile
(from 2,2-dimethylcyclohexanecarboxylic acid)

PROBLEM 10.1 Give IUPAC names for the following carboxylic acids:

(a) $CH_3\overset{\overset{\displaystyle CH_3}{|}}{C}HCH_2COOH$

(b) $CH_3\overset{\overset{\displaystyle Br}{|}}{C}HCH_2CH_2COOH$

(c) $CH_3CH{=}CHCH_2CH_2COOH$

(d) $CH_3CH_2\overset{\overset{\displaystyle COOH}{|}}{C}HCH_2CH_2CH_3$

(e)

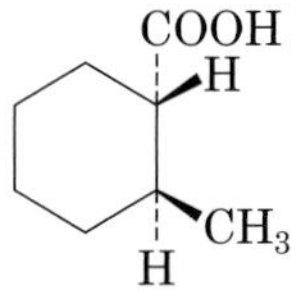

PROBLEM 10.2 Draw structures corresponding to the following names:

(a) 2,3-Dimethylhexanoic acid (b) 4-Methylpentanoic acid
(c) *o*-Hydroxybenzoic acid (d) *trans*-Cyclobutane-1,2-dicarboxylic acid

PROBLEM 10.3 Give IUPAC names for the following structures:

(a) 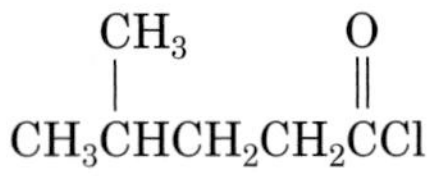 CH_3, O; $CH_3CHCH_2CH_2CCl$

(b) 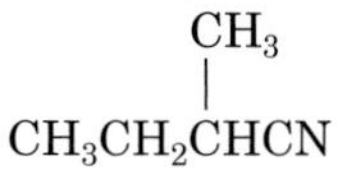 CH_3; CH_3CH_2CHCN

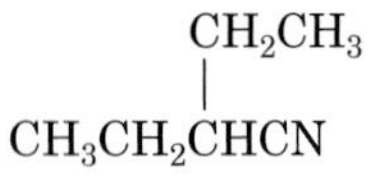

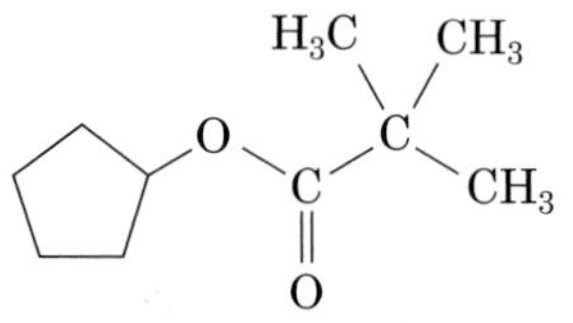

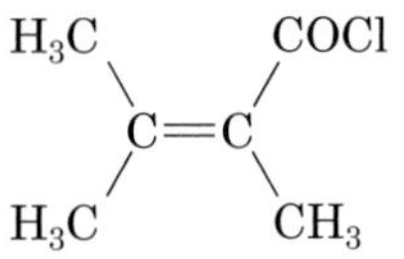

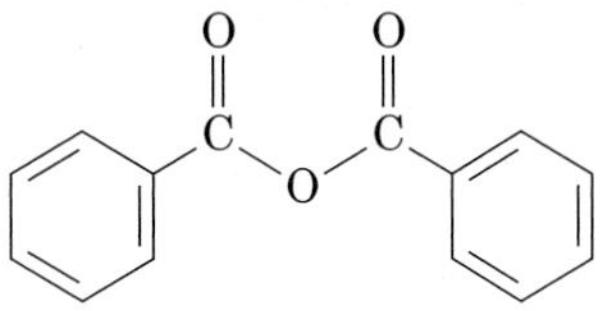

(c)  O; $H_2C{=}CHCH_2CH_2CNH_2$

(d) CH_2CH_3; CH_3CH_2CHCN

(e) H_3C, CH_3, O, C, C, CH_3, O

(f) H_3C, COCl, C=C, H_3C, CH_3

(g) O, O, C, O, C

(h) O, C, $OCH(CH_3)_2$

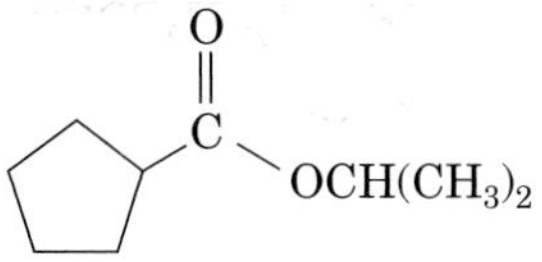

PROBLEM 10.4 Draw structures corresponding to the following names:

(a) 2,2-Dimethylpropanoyl chloride (b) *N*-Methylbenzamide
(c) 5,5-Dimethylhexanenitrile (d) *tert*-Butyl butanoate
(e) *trans*-2-Methylcyclohexanecarboxamide (f) *p*-Methylbenzoic anhydride
(g) *cis*-3-Methylcyclohexanecarbonyl bromide (h) *p*-Bromobenzonitrile

10.2 Occurrence and Properties of Carboxylic Acids and Derivatives

Carboxylic acids are abundant in nature. Acetic acid, CH_3CO_2H, for example, is the principal organic component of vinegar; butanoic acid, $CH_3CH_2CH_2CO_2H$, is responsible for the rancid odor of sour butter; and hexanoic acid (caproic acid), $CH_3(CH_2)_4CO_2H$, is partially responsible for the unmistakable aroma of goats (Latin *caper*, meaning "goat") and dirty sneakers.

Like alcohols, carboxylic acids form strong intermolecular hydrogen bonds. Most carboxylic acids, in fact, exist as *dimers* held together by two hydrogen bonds:

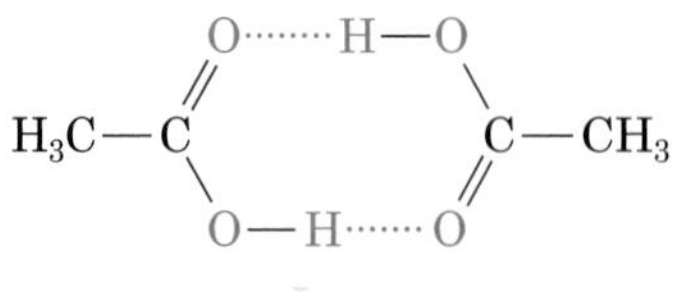

O·······H—O
H_3C—C C—CH_3
O—H·······O

Acetic acid dimer

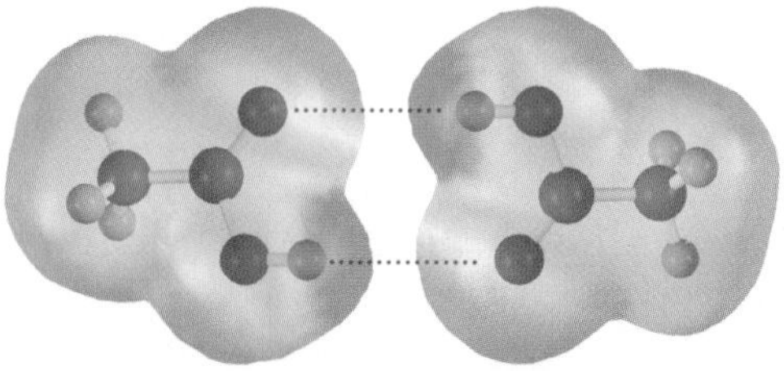

This strong hydrogen bonding has a noticeable effect on boiling points; carboxylic acids normally boil at much higher temperatures than alkanes or alkyl halides of similar molecular weight. Acetic acid, for example, boils at 118 °C, whereas chloropropane boils at 46.6 °C.

Esters are carboxylic acid derivatives. Unlike carboxylic acids, which are often foul in odor, many simple esters are pleasant-smelling liquids, which are responsible for the fragrant aromas of fruits and flowers. Methyl butanoate, for example, has been isolated from pineapple oil, and isopentyl acetate has been found in banana oil. The ester group is also present in animal fats and other biologically important molecules. The lack of strong intermolecular forces (like the hydrogen bonds of carboxylic acids) gives esters lower boiling points and makes them more volatile at room temperature.

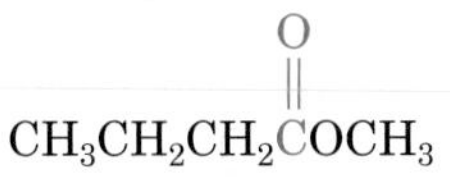

$CH_3CH_2CH_2COCH_3$ (C=O)

Methyl butanoate (from pineapples)

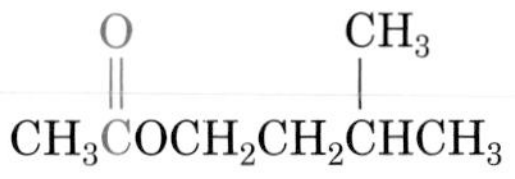

$CH_3COCH_2CH_2CHCH_3$ (C=O; CH_3 branch)

Isopentyl acetate (from bananas)

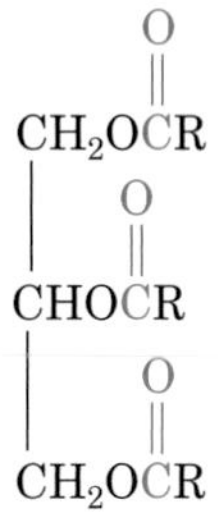

CH_2OCR / $CHOCR$ / CH_2OCR (each C=O)

A fat ($R = C_{11-17}$ chains)

Amides are less reactive than esters; this stability make amides ideal linkages in peptides and proteins, the topic of Chapter 15. Hydrogen bonding between amides increases their boiling points. A diverse range of biological events—from protein folding to the action of drugs like the antibiotic vancomycin—depend on hydrogen bonding between amides.

A hydrogen bond

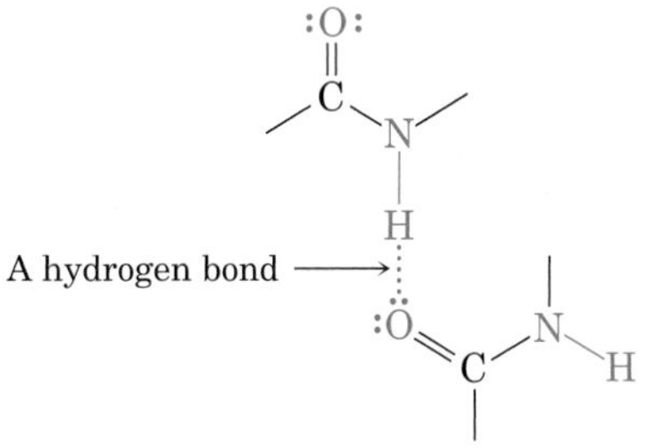

Acyl chlorides and anhydrides are commonly used in the chemical and pharmaceutical industries. With few exceptions, these groups are not found in nature due to their reactivity. We'll examine some uses of these synthetic molecules later in the chapter.

10.3 Acidity of Carboxylic Acids

The most obvious property of carboxylic acids is implied by their name—their *acidity*. Although much weaker than mineral acids, carboxylic acids are nevertheless much stronger acids than alcohols; acetic acid, for example, has $K_a = 1.76 \times 10^{-5}$ ($pK_a = 4.75$), while ethanol has $K_a = 10^{-16}$ ($pK_a = 16$). In practical terms, a K_a value near 10^{-5} means that only about 1% of the molecules in a 0.1 M aqueous solution are dissociated. Because of their acidity, carboxylic acids react with bases such as NaOH to give water-soluble metal **carboxylates**, $RCO_2^-\ Na^+$.

$$\underset{\text{A carboxylic acid (water-insoluble)}}{RCO_2H} + NaOH \xrightarrow{H_2O} \underset{\text{A carboxylic acid salt (water-soluble)}}{RCO_2^- Na^+} + H_2O$$

A carboxylic acid (water-insoluble)

A carboxylic acid salt (water-soluble)

As indicated by the list of K_a values in Table 10.2, there is a considerable range in the strengths of various carboxylic acids. Trichloroacetic acid ($K_a = 0.23$), for example, is more than 12,000 times as strong as acetic acid ($K_a = 1.76 \times 10^{-5}$).

How can we account for the large differences in pK_a listed in Table 10.2? Because acid dissociation is an equilibrium process, anything that stabilizes the carboxylate anion favors increased dissociation and increased acidity. Thus, the presence of an electron-withdrawing chlorine atom spreads out the negative charge on the anion and makes chloroacetic acid stronger than acetic acid by a factor of 75. Introducing two electronegative chlorine atoms makes dichloroacetic acid some 3000 times as strong as acetic acid, and introducing three makes trichloroacetic acid more than 12,000 times as strong.

CH_3CO_2H pK_a = 4.75 — $ClCH_2CO_2H$ pK_a = 2.85 — Cl_2CHCO_2H pK_a = 1.48 — Cl_3CCO_2H pK_a = 0.64

Weaker acid → Stronger acid

Why are carboxylic acids so much more acidic than alcohols even though both contain O–H groups? To answer this question, compare the relative stabilities of carboxylate anions versus alkoxide anions (Figure 10.1). In an alkoxide ion, the negative charge is localized on the one oxygen atom. In a carboxylate ion, however, the negative charge is shared by both oxygen atoms. In other words, a carboxylate anion is a stabilized resonance hybrid

TABLE 10.2 Acid Strengths of Some Carboxylic Acids

Name	K_a	pK_a	
HCl (hydrochloric acid)[a]	(10^7)	(−7)	Stronger acid
CCl_3CO_2H	0.23	0.64	↑
$CHCl_2CO_2H$	3.3×10^{-2}	1.48	
CH_2ClCO_2H	1.4×10^{-3}	2.85	
HCO_2H	1.77×10^{-4}	3.75	
$C_6H_5CO_2H$	6.46×10^{-5}	4.19	
$H_2C{=}CHCO_2H$	5.6×10^{-5}	4.25	
CH_3CO_2H	1.76×10^{-5}	4.75	
CH_3CH_2OH (ethanol)[a]	(10^{-16})	(16)	Weaker acid

[a]Values for HCl and ethanol are shown for reference.

(see Section 4.11) of two equivalent structures. Since a carboxylate ion is more stable than an alkoxide ion, it is lower in energy and is present in greater amount at equilibrium.

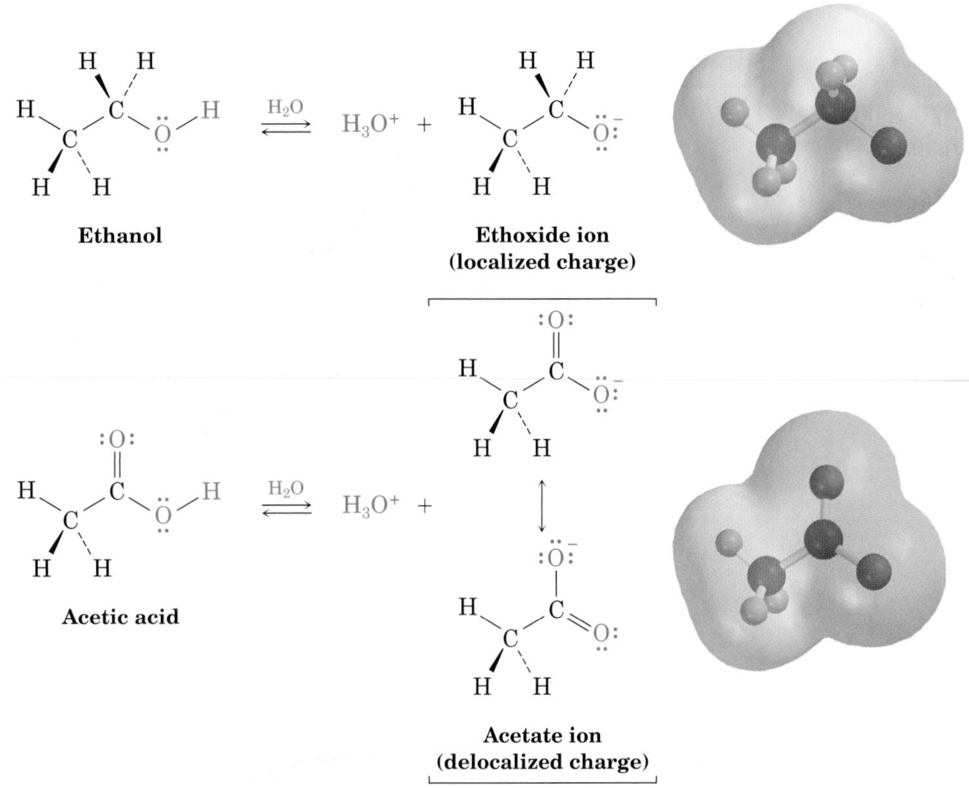

FIGURE 10.1 An alkoxide ion has its charge localized on one oxygen atom and is less stable, while a carboxylate ion has the charge spread equally over both oxygens by two resonance forms and is therefore more stable.

Ethanol $\underset{}{\overset{H_2O}{\rightleftharpoons}}$ H_3O^+ + Ethoxide ion (localized charge)

Acetic acid $\underset{}{\overset{H_2O}{\rightleftharpoons}}$ H_3O^+ + Acetate ion (delocalized charge)

Practice Problem 10.1

Which would you expect to be the stronger acid, benzoic acid or *p*-nitrobenzoic acid?

Solution The more stabilized the carboxylate anion, the stronger the acid. We know from its effect on aromatic substitution (see Section 5.8) that a nitro group is electron-withdrawing and can stabilize a negative charge. Thus, a *p*-nitrobenzoate ion is more stable than a benzoate ion, and *p*-nitrobenzoic acid is stronger than benzoic acid.

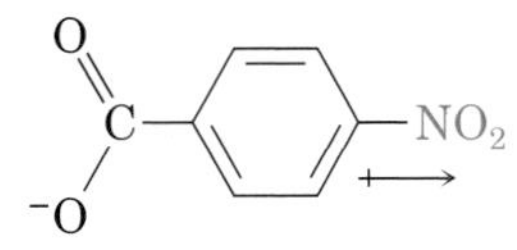

Nitro group withdraws electrons from ring and stabilizes negative charge.

Problem 10.5 Draw structures for the products of the following reactions:

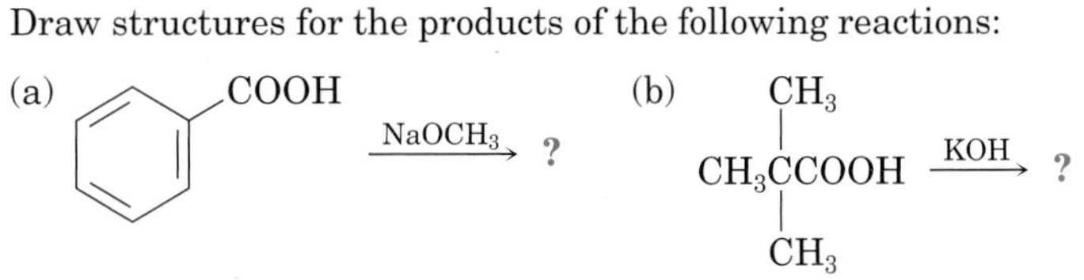

(a) $C_6H_5COOH \xrightarrow{NaOCH_3}$?

(b) $(CH_3)_3CCOOH \xrightarrow{KOH}$?

Problem 10.6 Rank the following compounds in order of increasing acidity: sulfuric acid, methanol, phenol, *p*-nitrophenol, acetic acid.

PROBLEM 10.7 Rank the following compounds in order of increasing acidity:
(a) $CH_3CH_2CO_2H$, $BrCH_2CO_2H$, $BrCH_2CH_2CO_2H$
(b) Benzoic acid, ethanol, *p*-cyanobenzoic acid

10.4 Synthesis of Carboxylic Acids

We've already seen several methods for preparing carboxylic acids:

- A substituted alkylbenzene can be oxidized with $KMnO_4$ to give a substituted benzoic acid (see Section 5.9).

$$O_2N-C_6H_4-CH_3 \xrightarrow[H_2O,\ 95\ °C]{KMnO_4} O_2N-C_6H_4-COOH$$

***p*-Nitrotoluene** → ***p*-Nitrobenzoic acid (88%)**

- Primary alcohols and aldehydes can be oxidized to give carboxylic acids (see Sections 8.5 and 9.4). A primary alcohol is often oxidized with CrO_3 or $Na_2Cr_2O_7$; an aldehyde is oxidized with Tollens' reagent ($AgNO_3$ in NH_4OH).

$$CH_3(CH_2)_8CH_2OH \xrightarrow[H_2O,\ H_2SO_4]{CrO_3} CH_3(CH_2)_8COOH$$

Decan-1-ol → **Decanoic acid (93%)**

$$CH_3CH_2CH_2CH_2CH_2CHO \xrightarrow[NH_4OH]{AgNO_3} CH_3CH_2CH_2CH_2CH_2COOH$$

Hexanal → **Hexanoic acid (85%)**

In addition to the preceding two methods, there are numerous other ways to prepare carboxylic acids. For instance, carboxylic acids can be prepared from nitriles, R—C≡N, by a *hydrolysis* reaction with aqueous acid or base. Since nitriles themselves are usually prepared by an S_N2 reaction between an alkyl halide and cyanide ion, CN^-, the two-step sequence of cyanide ion displacement followed by nitrile hydrolysis is an excellent method for converting an alkyl halide into a carboxylic acid: RBr → RC≡N → RCO_2H. As with all S_N2 reactions, the method works best with primary alkyl halides, although secondary alkyl halides can sometimes be used (see Section 7.5).

A good example of the reaction occurs in the commercial synthesis of the antiarthritis drug fenoprofen, a nonsteroidal anti-inflammatory agent marketed under the name Nalfon.

$$C_6H_5-O-C_6H_4-CH(Br)CH_3 \xrightarrow[\text{2. }^-OH/H_2O\ \ \text{3. }H_3O^+]{\text{1. NaCN}} C_6H_5-O-C_6H_4-CH(CH_3)COOH$$

Fenoprofen
(an antiarthritic agent)

PROBLEM 10.8 Show the steps in the conversion of iodomethane to acetic acid by the nitrile hydrolysis route. Would a similar route work for the conversion of iodobenzene to benzoic acid? Explain.

10.5 Nucleophilic Acyl Substitution Reactions

We saw in Chapter 9 that the addition of nucleophiles to the polar C=O bond is a general feature of aldehyde and ketone chemistry. Carboxylic acids and their derivatives also react with nucleophiles, but the ultimate product is different from that of the aldehyde/ketone reaction. Instead of undergoing protonation to yield an alcohol, the initially formed alkoxide intermediate expels one of the substituents originally bonded to the carbonyl carbon, leading to the formation of a new carbonyl compound by a **nucleophilic acyl substitution reaction** (Figure 10.2).

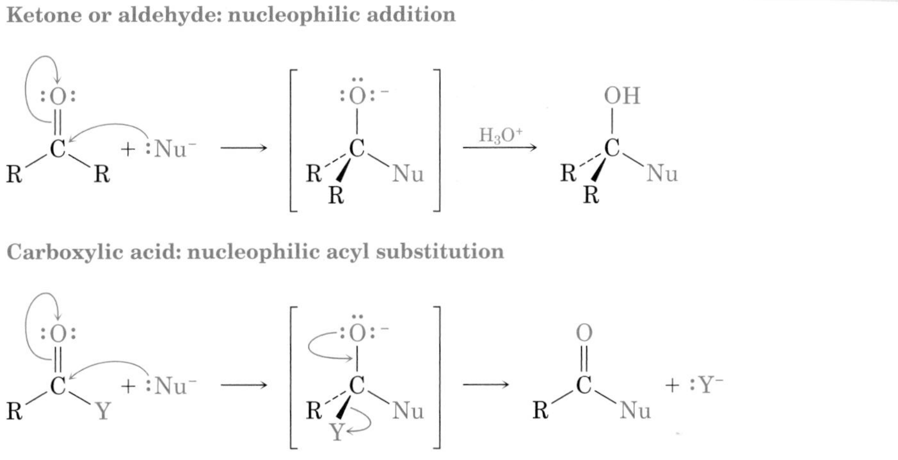

Ketone or aldehyde: nucleophilic addition

Carboxylic acid: nucleophilic acyl substitution

FIGURE 10.2 The general mechanisms of nucleophilic addition and nucleophilic acyl substitution reactions. Both reactions begin with the addition of a nucleophile to a polar C=O bond to give a tetrahedral, alkoxide ion intermediate. The intermediate formed from an aldehyde or ketone is protonated to give an alcohol, but the intermediate formed from a carboxylic acid derivative expels a leaving group to give a new carbonyl compound.

The different behavior toward nucleophiles of aldehydes/ketones and carboxylic acid derivatives is a consequence of structure. Carboxylic acid derivatives have an acyl carbon bonded to a group that can leave as a stable anion. As soon as addition of a nucleophile occurs, the group leaves and a new carbonyl compound forms. Aldehydes and ketones have no such leaving group, however, and therefore don't undergo substitution.

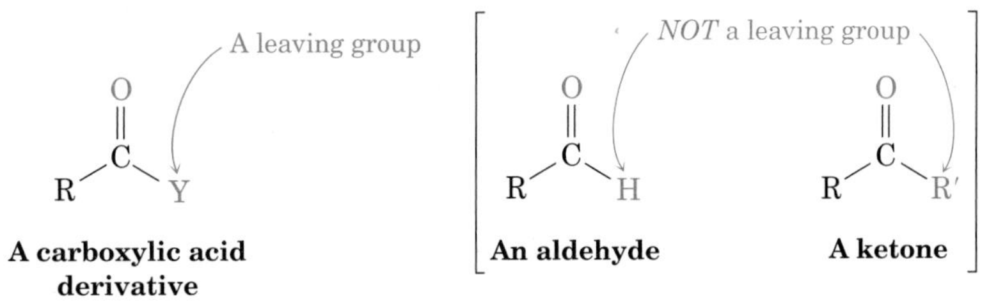

A leaving group

NOT a leaving group

A carboxylic acid derivative

An aldehyde

A ketone

In comparing the reactivity of different acyl derivatives, the more electron-poor the C=O carbon, the more readily the compound reacts with nucleophiles. Thus, acid chlorides are the most reactive compounds because the electronegative chlorine atom strongly withdraws electrons from the carbonyl carbon, whereas amides are the least reactive compounds. Electrostatic potential maps indicate these differences among various carboxylic acid derivatives by the relative blueness on the C=O carbons.

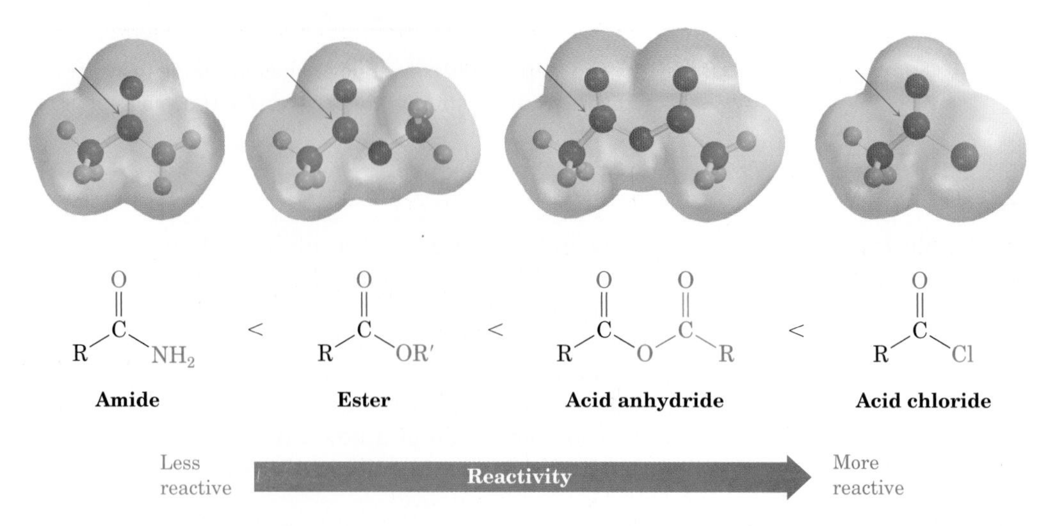

A consequence of these reactivity differences is that it's usually possible to convert a more reactive acid derivative into a less reactive one. Acid chlorides, for example, can be converted into esters and amides, but amides and esters can't be converted into acid chlorides. Remembering the reactivity order is therefore a useful way to keep track of a large number of reactions (Figure 10.3).

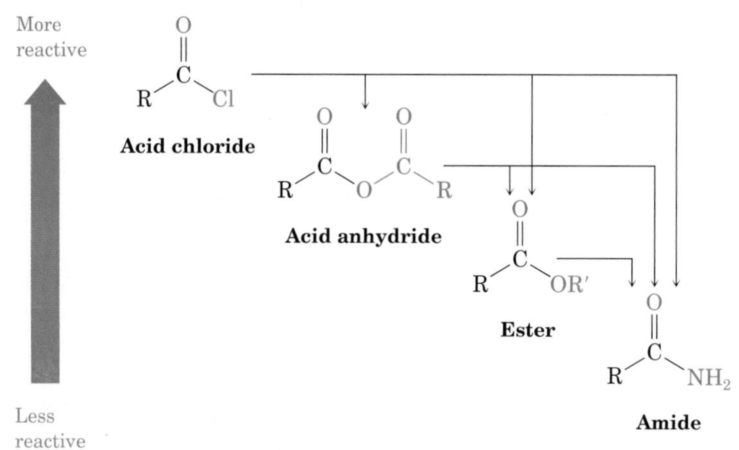

FIGURE 10.3 Interconversions of carboxylic acid derivatives. More reactive compounds can be converted into less reactive ones, but not vice versa.

PRACTICE PROBLEM 10.2

Which is more reactive in a nucleophilic acyl substitution reaction with hydroxide ion, CH_3CONH_2 or CH_3COCl?

STRATEGY Identify the functional groups and determine which is more reactive based on Figure 10.3.

SOLUTION The acid chloride CH_3COCl is more reactive than the amide CH_3CONH_2.

PROBLEM 10.9 Which compound in each of the following sets is more reactive in nucleophilic acyl substitution reactions?

(a) CH_3COCl or $CH_3CO_2CH_3$
(b) $(CH_3)_2CHCONH_2$ or $CH_3CH_2CO_2CH_3$
(c) $CH_3CO_2CH_3$ or $CH_3CO_2COCH_3$
(d) $CH_3CO_2CH_3$ or CH_3CHO

PROBLEM 10.10 Methyl trifluoroacetate, $CF_3CO_2CH_3$, is more reactive than methyl acetate, $CH_3CO_2CH_3$, in nucleophilic acyl substitution reactions. Explain.

10.6

The Tetrahedral Intermediate

Let's look closely at one example of a nucleophilic acyl substitution reaction called the **Fischer esterification reaction**. Here, an ester is produced by heating a carboxylic acid with an acid catalyst in an alcohol solvent. As shown in Figure 10.4, the acid catalyst first protonates the carbonyl oxygen of the $-CO_2H$ group, giving the carboxylic acid a positive charge and making it more reactive toward nucleophiles. An alcohol molecule then adds to the protonated carboxylic acid, and subsequent loss of water yields the ester product.

FIGURE 10.4 MECHANISM: Mechanism of the Fischer esterification reaction of a carboxylic acid. The reaction is an acid-catalyzed nucleophilic acyl substitution.

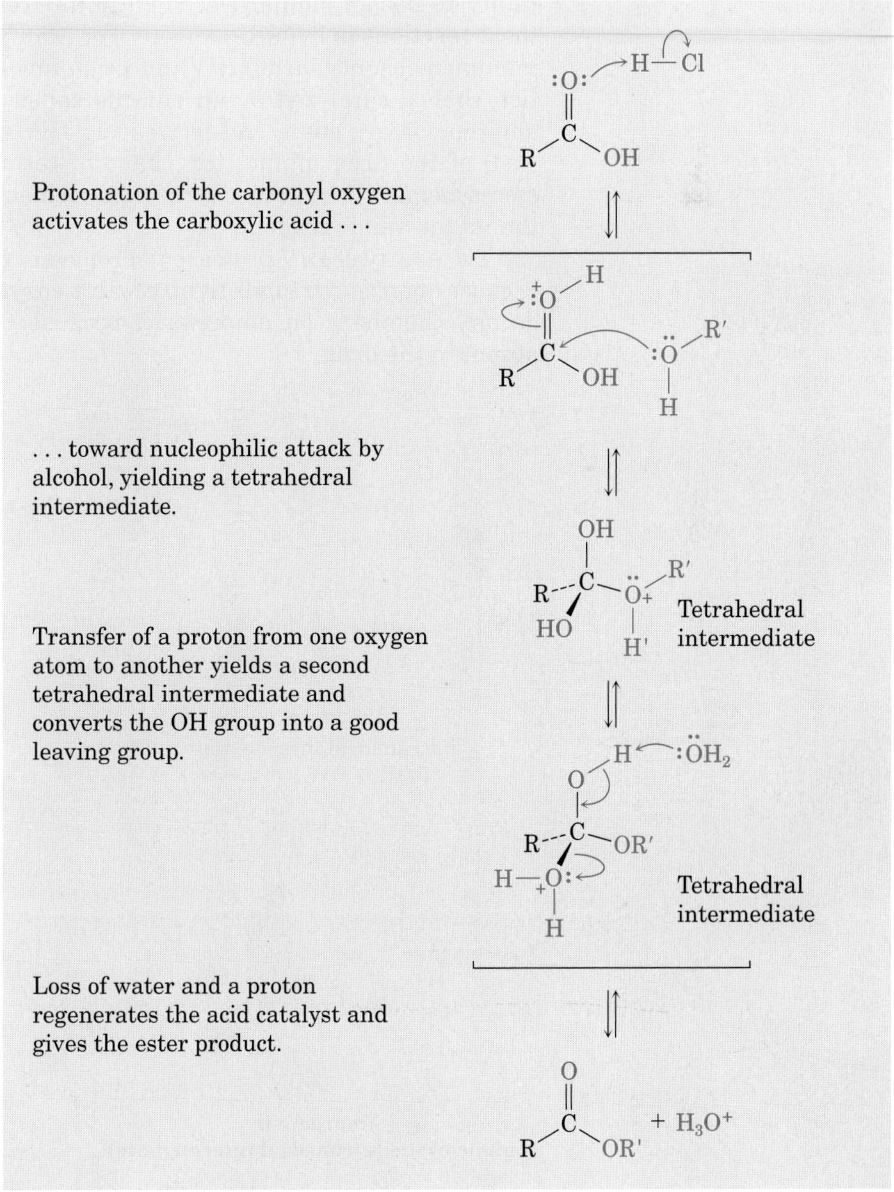

Protonation of the carbonyl oxygen activates the carboxylic acid . . .

. . . toward nucleophilic attack by alcohol, yielding a tetrahedral intermediate.

Transfer of a proton from one oxygen atom to another yields a second tetrahedral intermediate and converts the OH group into a good leaving group.

Loss of water and a proton regenerates the acid catalyst and gives the ester product.

All steps in the Fischer esterification reaction are reversible, and the position of the equilibrium depends on the reaction conditions. Ester formation is favored when alcohol is used as the solvent, but a carboxylic acid is favored when the solvent is water.

As this reaction proceeds, the hybridization of the carbonyl carbon changes from sp^2 to sp^3 upon addition of the nucleophile, and then from sp^3 back to sp^2 upon loss of the leaving group. The intermediate with the sp^3-hybridized carbon atom is called the *tetrahedral intermediate*: sp^3 carbons have a tetrahedral shape. Tetrahedral intermediates are common to the reactions of carboxylic acids, esters, amides, acyl halides, and anhydrides. Clearly, a fundamental understanding of reaction mechanisms reduces a diverse number of reactions with varied reactants, reagents, and products into a single chemical theme that will emerge from this chapter.

The well-defined tetrahedral shape of this intermediate offers an opportunity to design similar, yet slightly different molecules that will inhibit these reactions in living organisms. For example, a critical step of the human immunodeficiency virus (HIV) infection involves an amide hydrolysis reaction that is catalyzed by an enzyme called HIV protease. Drugs such as amprenavir are successful in treating HIV because the secondary alcohol (red) of the drug mimics the shape of the tetrahedral intermediate. The stereochemistry and other structural elements (blue) of amprenavir match that of the amide reagent.

The ability of HIV protease to propagate the infection is greatly reduced because amprenavir binds tightly to this enzyme catalyst. HIV protease can't do any chemistry on amprenavir because the amide functional group is absent in the drug.

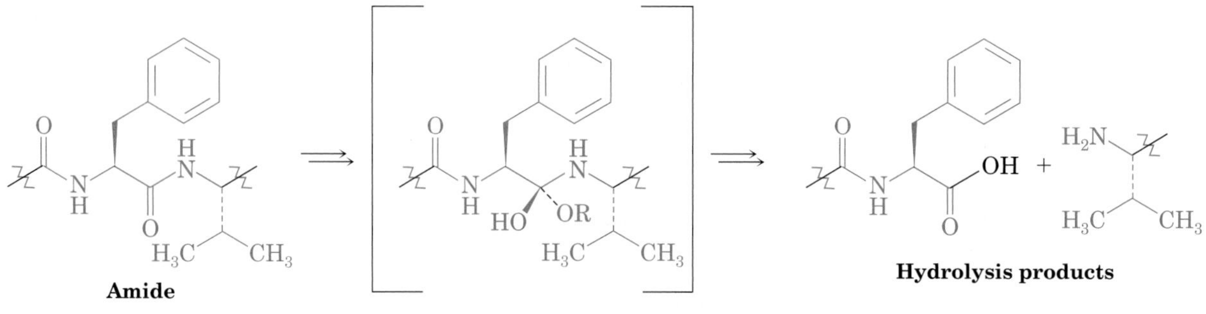

Amide

Tetrahedral intermediate
(ROH is HIV protease)

Hydrolysis products

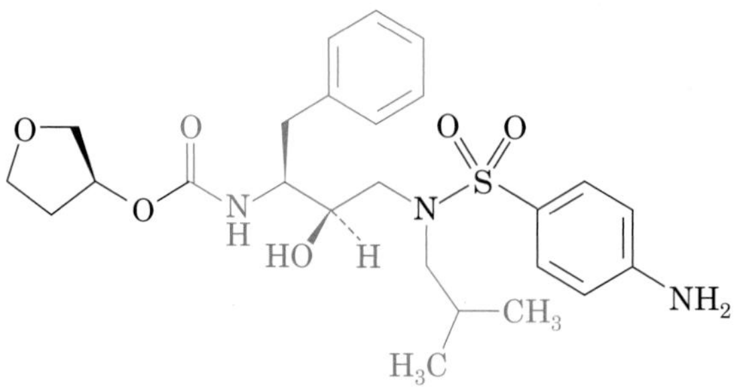

Amprenavir
(mimic of the tetrahedral intermediate)

10.7 Overview of Reactions

With the exception of the C—C bond-forming reactions that we will treat separately in Chapter 11, our basic road map for the reactions of functional groups is now complete. Figure 10.5 shows how the reactions that we will discuss in the rest of this chapter link the functional groups together. While superficially it appears that there are a lot of new reactions to learn in this chapter, in reality there are only a few. These few reactions can be grouped into three categories, indicated by the blue (hydrolysis reactions), green (nucleophilic addition reactions), and red (oxidation–reduction reactions) lines. Spend a few minutes contemplating the chart in Figure 10.5. Make flash cards for all the reactions. The "big picture" should be emerging. We'll need this big picture as we move into the final chapters of this book, which focus on the chemistry of life . . . processes that are much more beautiful and orchestrated than these boxes suggest.

The following sections break this road map into smaller pieces.

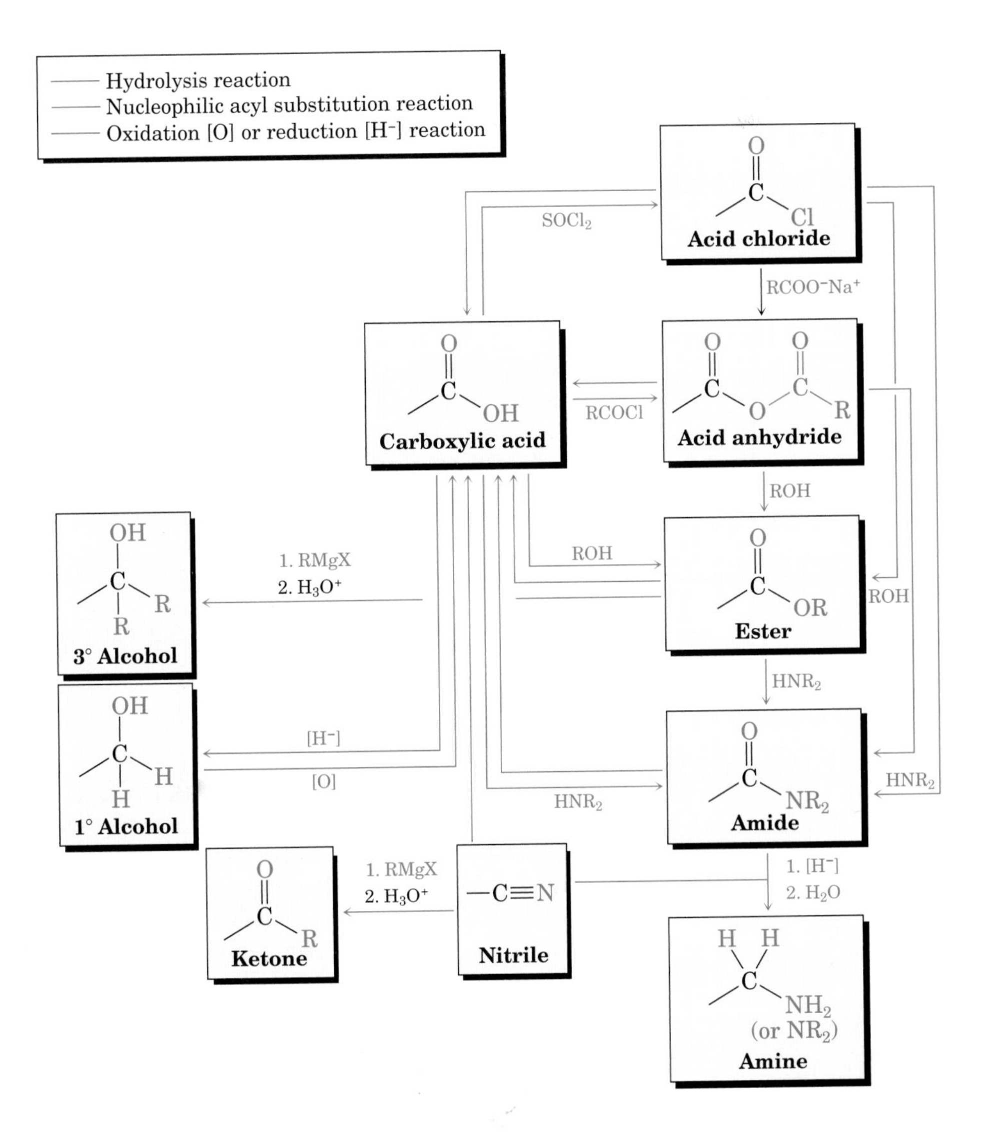

FIGURE 10.5 Road map for the reactions of functional groups in this chapter.

10.8

Reactions of Carboxylic Acids

Conversion of Acids into Alcohols by Reduction ($RCO_2H \rightarrow RCH_2OH$)

We saw in Section 8.4 that carboxylic acids are reduced by lithium aluminum hydride ($LiAlH_4$) to yield primary alcohols. The net effect is substitution of the acid –OH group by –H, giving an aldehyde that is further reduced to the alcohol. For example:

$$CH_3(CH_2)_7CH{=}CH(CH_2)_7\overset{O}{\overset{\|}{C}}OH \xrightarrow[2.\ H_3O^+]{1.\ LiAlH_4} CH_3(CH_2)_7CH{=}CH(CH_2)_7CH_2OH$$

Oleic acid **_cis_-Octadec-9-en-1-ol (87%)**

Conversion of Acids into Acid Chlorides ($RCO_2H \rightarrow RCOCl$)

Carboxylic acids are converted into acid chlorides by treatment with thionyl chloride, $SOCl_2$. The net effect is substitution of the –OH group by –Cl. For example:

2,4,6-Trimethylbenzoic acid $\xrightarrow[CHCl_3]{SOCl_2}$ 2,4,6-Trimethylbenzoyl chloride (90%) $+ HCl + SO_2$

2,4,6-Trimethylbenzoic acid **2,4,6-Trimethylbenzoyl chloride (90%)**

Conversion of Acids into Esters ($RCO_2H \rightarrow RCO_2R'$)

Perhaps the most useful reaction of carboxylic acids is their conversion into esters by reaction with an alcohol—the substitution of –OH by –OR. Called the Fischer esterification reaction, the simplest method involves heating the carboxylic acid with an acid catalyst in an alcohol solvent.

$$C_6H_5COOH + CH_3CH_2OH \xrightarrow[\text{catalyst}]{H_2SO_4} C_6H_5COOCH_2CH_3 + H_2O$$

Benzoic acid **Ethyl benzoate (91%)**

Conversion of Acids into Amides ($RCO_2H \rightarrow RCONH_2$)

Amides are carboxylic acid derivatives in which the acid –OH group has been replaced by a nitrogen substituent, $-NH_2$, –NHR, or $-NR_2$. Amides are difficult to prepare directly from acids by substitution with an amine because amines are bases, which convert acidic carboxyl groups into their carboxylate anions. Because the carboxylate anion has a negative charge, it no longer undergoes attack by nucleophiles, except at high temperatures.

$$R\overset{O}{\overset{\|}{C}}OH + :NH_3 \rightleftharpoons R\overset{O}{\overset{\|}{C}}O^-\ NH_4^+$$

We'll see a good method for converting an acid into an amide in Section 15.4 in connection with the synthesis of proteins from amino acids.

PRACTICE PROBLEM 10.3

How might you prepare the following ester using a Fischer esterification reaction?

$$C_6H_5-\overset{O}{\overset{\|}{C}}-OCH_2CH_2CH_3$$

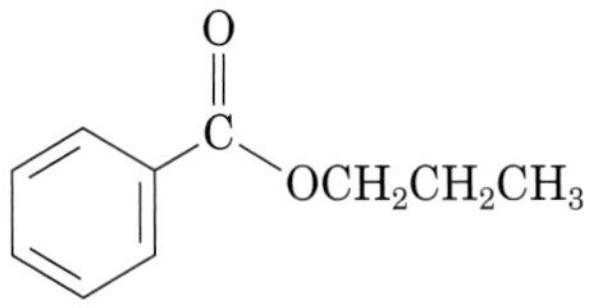

STRATEGY The trick is to identify the two parts of the ester. The acyl part comes from the carboxylic acid, and the –OR part comes from the alcohol. In this case the target molecule is propyl benzoate, so it can be prepared by treating benzoic acid with propan-1-ol.

SOLUTION

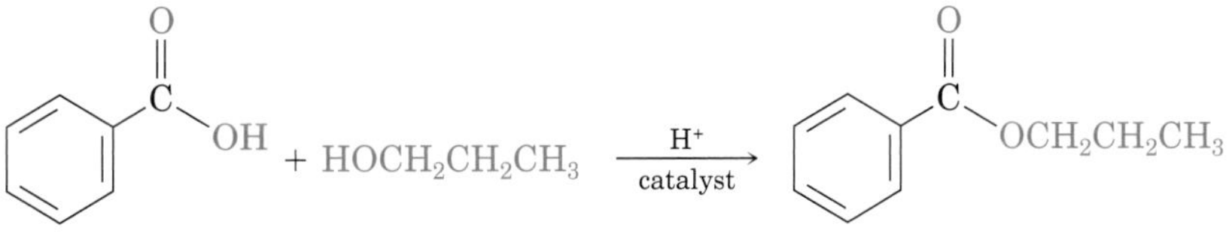

$$C_6H_5-\overset{O}{\overset{\|}{C}}-OH + HOCH_2CH_2CH_3 \xrightarrow[\text{catalyst}]{H^+} C_6H_5-\overset{O}{\overset{\|}{C}}-OCH_2CH_2CH_3$$

Benzoic acid **Propan-1-ol** **Propyl benzoate**

PROBLEM 10.11 What products would you obtain by treating benzoic acid with the following reagents? Formulate the reactions.

(a) $SOCl_2$ (b) CH_3OH, HCl (c) 1. $LiAlH_4$, 2. H_3O^+ (d) NaOH

PROBLEM 10.12 Show how you might prepare the following esters using a Fischer esterification reaction:

(a) $CH_3\overset{O}{\overset{\|}{C}}OCH_2CH_2CH_2CH_3$ (b) $CH_3CH_2CH_2\overset{O}{\overset{\|}{C}}OCH_3$ (c) $C_6H_5-\overset{O}{\overset{\|}{C}}-OCH(CH_3)CH_3$

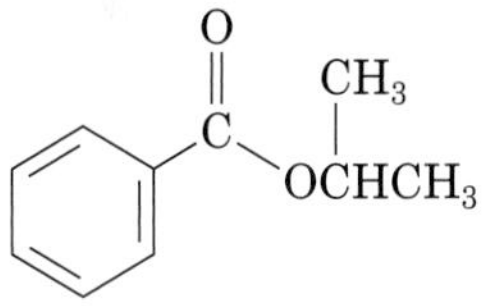

10.9 Chemistry of Acid Halides

Acid chlorides are prepared from carboxylic acids by reaction with thionyl chloride, $SOCl_2$, as we saw in the previous section:

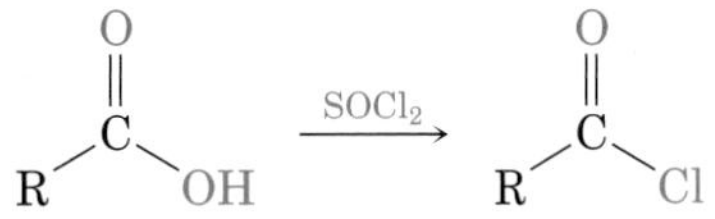

$$R-\overset{O}{\overset{\|}{C}}-OH \xrightarrow{SOCl_2} R-\overset{O}{\overset{\|}{C}}-Cl$$

Acid halides are among the most reactive of the various carboxylic acid derivatives and can be converted into many other kinds of substances. The halogen can be replaced by –OH to yield an acid, by –OR to yield an ester, or by $-NH_2$ to yield an amide (Figure 10.6). Although illustrated only for acid chlorides, similar processes take place with other acid halides.

FIGURE 10.6 Reactions of acid chlorides.

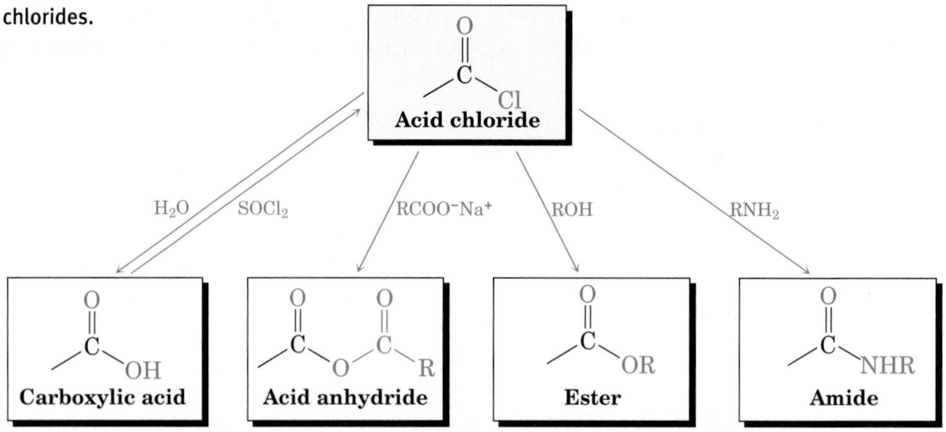

Acid chloride

H_2O / $SOCl_2$ → Carboxylic acid

$RCOO^-Na^+$ → Acid anhydride

ROH → Ester

RNH_2 → Amide

Conversion of Acid Chlorides into Acids ($RCOCl \rightarrow RCO_2H$)

Acid chlorides react with water to yield carboxylic acids—the substitution of –Cl by –OH. This hydrolysis reaction is a typical nucleophilic acyl substitution process and is initiated by attack of the nucleophile water on the acid chloride carbonyl group. The initially formed tetrahedral intermediate undergoes loss of HCl to yield the product:

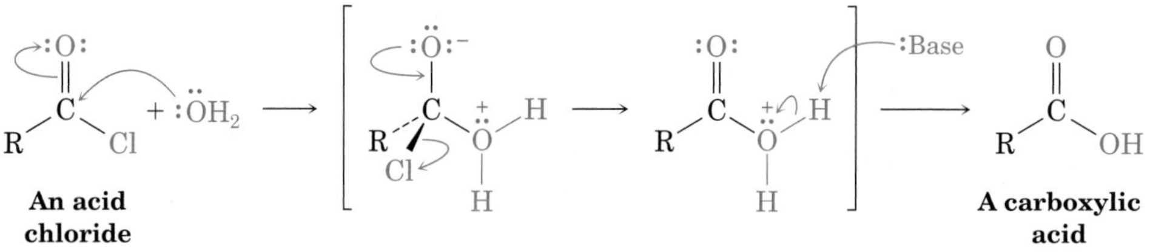

An acid chloride + $:\ddot{O}H_2$ → [tetrahedral intermediate] → :Base → **A carboxylic acid**

Conversion of Acid Chlorides into Esters ($RCOCl \rightarrow RCO_2R'$)

Acid chlorides react with alcohols to yield esters in a reaction analogous to their reaction with water to yield acids:

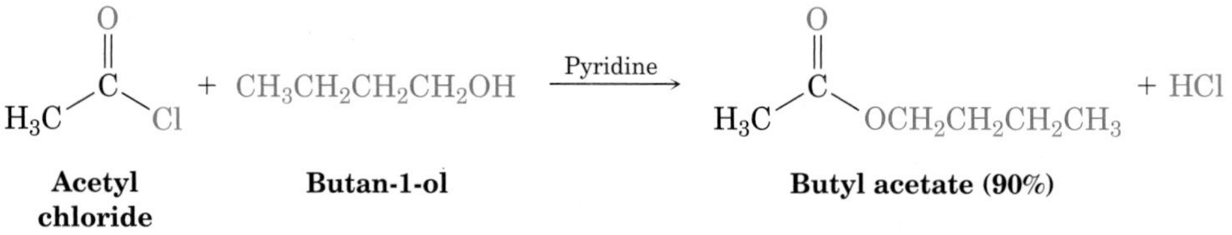

$$H_3CCOCl + CH_3CH_2CH_2CH_2OH \xrightarrow{\text{Pyridine}} H_3CCOOCH_2CH_2CH_2CH_3 + HCl$$

Acetyl chloride **Butan-1-ol** **Butyl acetate (90%)**

Since HCl is generated as a by-product, the reaction is usually carried out in the presence of an amine base such as pyridine (see Section 12.3), which reacts with the HCl as it's formed and prevents it from causing side reactions.

Conversion of Acid Chlorides into Amides ($RCOCl \rightarrow RCONH_2$)

Acid chlorides react rapidly with ammonia and with amines to give amides—the substitution of –Cl by $–NR_2$. Both monosubstituted and disubstituted amines can be used. For example, 2-methylpropanamide is prepared by reaction of 2-methylpropanoyl chloride with ammonia. Note that one extra equivalent of ammonia is added to react with the HCl generated.

$$(CH_3)_2CHCOCl + 2\,NH_3 \longrightarrow (CH_3)_2CHCONH_2 + NH_4Cl$$

2-Methylpropanoyl chloride **2-Methylpropanamide**

PRACTICE PROBLEM 10.4

Show how you could prepare ethyl benzoate by reaction of an acid chloride with an alcohol.

STRATEGY As its name implies, ethyl benzoate can be made by reaction of *ethyl* alcohol with the acid chloride of *benzoic* acid.

SOLUTION

$$C_6H_5COCl + CH_3CH_2OH \xrightarrow[\text{solvent}]{\text{Pyridine}} C_6H_5COOCH_2CH_3$$

Benzoyl chloride **Ethanol** **Ethyl benzoate**

PRACTICE PROBLEM 10.5

Show how you would prepare *N*-methylpropanamide by reaction of an acid chloride with an amine.

STRATEGY The name of the product gives a hint as to how it can be prepared. Reaction of *methyl*-amine with *propanoyl* chloride gives *N*-methylpropanamide.

SOLUTION

$$CH_3CH_2\overset{O}{\overset{\|}{C}}Cl + 2\,CH_3NH_2 \longrightarrow CH_3CH_2\overset{O}{\overset{\|}{C}}NHCH_3 + CH_3NH_3^+\,Cl^-$$

Propanoyl chloride **Methylamine** ***N*-Methylpropanamide**

PROBLEM 10.13 How could you prepare the following esters using the reaction of an acid chloride with an alcohol?

(a) $CH_3CH_2CO_2CH_3$ (b) $CH_3CO_2CH_2CH_3$ (c) Cyclohexyl acetate

PROBLEM 10.14 Write the steps in the mechanism of the reaction between ammonia and 2-methylpropanoyl chloride to yield 2-methylpropanamide.

PROBLEM 10.15 What amines would react with what acid chlorides to give the following amide products?

(a) $CH_3CH_2CONH_2$ (b) $(CH_3)_2CHCH_2CONHCH_3$
(c) *N,N*-Dimethylpropanamide (d) *N,N*-Diethylbenzamide

10.10 Chemistry of Acid Anhydrides

The best method for preparing acid anhydrides is by a nucleophilic acyl substitution reaction of an acid chloride with a carboxylic acid anion. Both symmetrical and unsymmetrical acid anhydrides can be prepared in this way.

$$RCOO^- Na^+ + ClCOCH_3 \xrightarrow[25\ ^\circ C]{Ether} HCOOCOCH_3$$

Sodium formate **Acetyl chloride** **Acetic formic anhydride (64%)**

The chemistry of acid anhydrides is similar to that of acid chlorides. Thus, acid anhydrides react with water to form acids, with alcohols to form esters, and with amines to form amides (Figure 10.7).

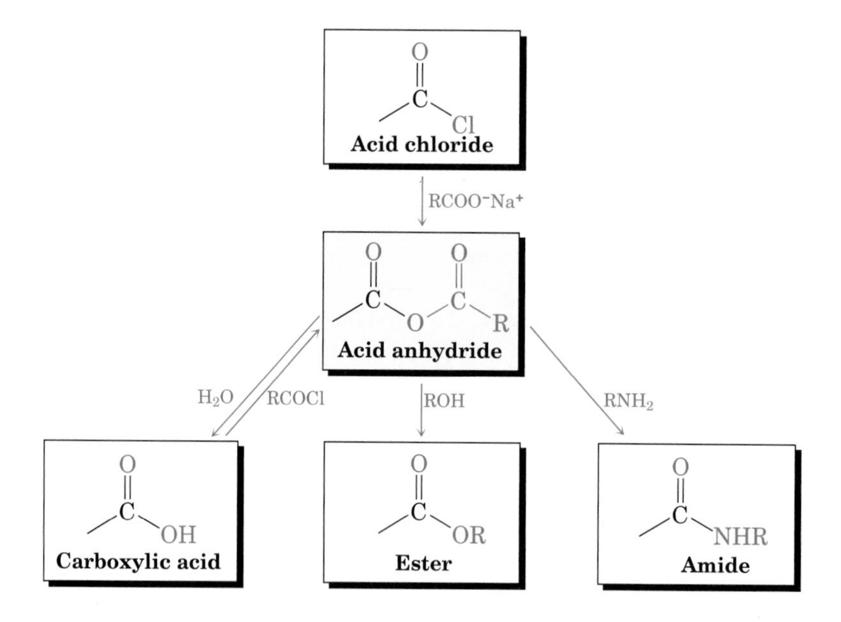

FIGURE 10.7 Some reactions of acid anhydrides.

Acetic anhydride is often used to prepare acetate esters of complex alcohols and to prepare substituted acetamides from amines. For example, aspirin (an ester) is prepared by reaction of acetic anhydride with *o*-hydroxybenzoic acid. Similarly, acetaminophen (an amide; the active ingredient in Tylenol) is prepared by reaction of acetic anhydride with *p*-hydroxyaniline.

$$\text{Salicylic acid } (o\text{-Hydroxybenzoic acid}) + \underset{\text{Acetic anhydride}}{CH_3\overset{O}{\overset{\|}{C}}O\overset{O}{\overset{\|}{C}}CH_3} \xrightarrow{\text{Pyridine}} \text{Aspirin (an ester)} + CH_3\overset{O}{\overset{\|}{C}}O^-$$

Salicylic acid
(o-Hydroxybenzoic acid)

Acetic anhydride

Aspirin (an ester)

$$p\text{-Hydroxyaniline} + \underset{\text{Acetic anhydride}}{CH_3\overset{O}{\overset{\|}{C}}O\overset{O}{\overset{\|}{C}}CH_3} \xrightarrow{\text{Pyridine}} \text{Acetaminophen} + CH_3\overset{O}{\overset{\|}{C}}O^-$$

p-Hydroxyaniline

Acetic anhydride

Acetaminophen

Notice in both of these examples that only "half" of the anhydride molecule is used; the other half acts as the leaving group during the nucleophilic acyl substitution step and produces carboxylate anion as a by-product. Thus, anhydrides are inefficient to use, and acid chlorides are normally used instead.

Practice Problem 10.6

What is the product of the following reaction?

$$\text{Cyclohexanol (OH)} + CH_3\overset{O}{\overset{\|}{C}}O\overset{O}{\overset{\|}{C}}CH_3 \xrightarrow{\text{Pyridine}} ?$$

Strategy You have to remember that acid anhydrides undergo a nucleophilic acyl substitution reaction with alcohols to give esters. Reaction of cyclohexanol with acetic anhydride yields cyclohexyl acetate by nucleophilic acyl substitution of the $-OCOCH_3$ group of the anhydride by the $-OR$ group of the alcohol.

Solution

$$\underset{\text{Cyclohexanol}}{\text{C}_6\text{H}_{11}\text{OH}} + CH_3\overset{O}{\overset{\|}{C}}O\overset{O}{\overset{\|}{C}}CH_3 \xrightarrow{\text{Pyridine}} \underset{\text{Cyclohexyl acetate}}{\text{C}_6\text{H}_{11}O\overset{O}{\overset{\|}{C}}CH_3}$$

Cyclohexanol

Cyclohexyl acetate

Problem 10.16 Write the steps in the mechanism of the reaction between *p*-hydroxyaniline and acetic anhydride to prepare acetaminophen.

PROBLEM 10.17 What product would you expect to obtain from the reaction of 1 equivalent of methanol with a cyclic anhydride such as phthalic anhydride?

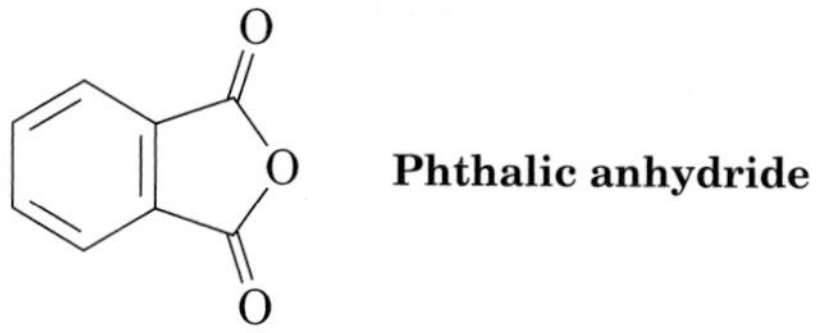

Phthalic anhydride

10.11 Chemistry of Esters

Esters are usually prepared either from acids or acid chlorides by the methods already discussed. Thus, carboxylic acids are converted directly into esters by Fischer esterification with an alcohol (see Section 10.5), and acid chlorides are converted into esters by reaction with an alcohol in the presence of pyridine (see Section 10.8).

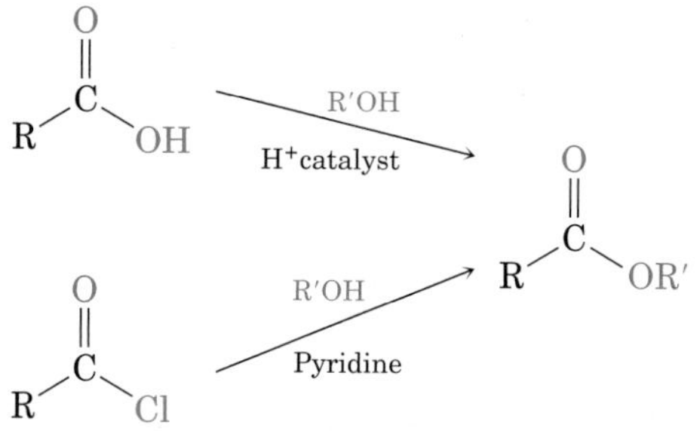

$$RCOOH \xrightarrow[\text{H}^+\text{catalyst}]{R'OH} RCOOR'$$

$$RCOCl \xrightarrow[\text{Pyridine}]{R'OH} RCOOR'$$

Esters show the same kinds of chemistry we've seen for other acyl derivatives, but they're less reactive toward nucleophiles than acid chlorides or anhydrides. Figure 10.8 shows some general reactions of esters.

FIGURE 10.8 Some reactions of esters.

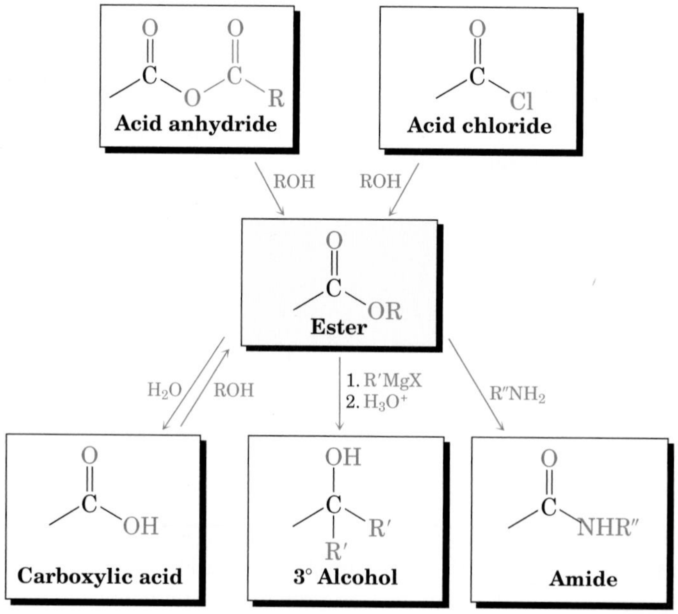

Acid anhydride

Acid chloride

ROH

ROH

Ester

H_2O

ROH

1. R′MgX
2. H_3O^+

R″NH_2

Carboxylic acid

3° Alcohol

Amide

Conversion of Esters into Acids ($RCO_2R' \rightarrow RCO_2H$)

Esters are hydrolyzed either by aqueous base or by aqueous acid to yield a carboxylic acid plus an alcohol:

$$RCOOR' \xrightarrow[\text{NaOH or } H_3O^+]{H_2O} RCOOH + R'OH$$

Ester **Acid**

Hydrolysis in basic solution is called *saponification*, after the Latin word *sapo*, meaning "soap." (As we'll see in Section 16.3, soap is made by the base-induced ester hydrolysis of animal fat.) Ester hydrolysis occurs by a typical nucleophilic acyl substitution pathway in which OH^- nucleophile adds to the ester carbonyl group, yielding a tetrahedral intermediate. Loss of OR'^- then gives a carboxylic acid, which is deprotonated to give the acid carboxylate:

$$RCOOR' + :\ddot{O}H^- \longrightarrow [R(R'O)C(O^-)OH] \longrightarrow RCOOH + R'O^- \longrightarrow RCOO^- + R'OH$$

Ester **Tetrahedral intermediate** **Acid** **Acid salt**

Conversion of Esters into Alcohols by Reduction ($RCO_2R' \rightarrow RCH_2OH$)

Esters are reduced by treatment with $LiAlH_4$ to yield primary alcohols (see Section 8.4):

$$CH_3CH_2CH{=}CHCOCH_2CH_3 \xrightarrow[2.\ H_3O^+]{1.\ LiAlH_4,\ \text{ether}} CH_3CH_2CH{=}CHCH_2OH + CH_3CH_2OH$$

Ethyl pent-2-enoate **Pent-2-en-1-ol (91%)**

The reaction involves an initial nucleophilic acyl substitution reaction, as hydride ion first adds to the carbonyl group followed by elimination of an alkoxide ion to yield an aldehyde intermediate. Further reduction of the aldehyde gives the primary alcohol.

$$RCOOR' + :H^- \xrightarrow[\text{Ether}]{LiAlH_4} [R(R'O)C(:\ddot{O}:^-)H \longrightarrow RCHO + R'O^-]$$

$$[RCHO + :H^- \xrightarrow[\text{Ether}]{LiAlH_4} RCH_2(:\ddot{O}:^-)] \xrightarrow{H_3O^+} RCH_2OH$$

PRACTICE PROBLEM 10.7

Write the products of the following saponification reaction:

$$\underset{\text{Ethyl 3-methylbutanoate}}{CH_3\overset{CH_3}{\overset{|}{C}}HCH_2\overset{O}{\overset{\|}{C}}OCH_2CH_3} \xrightarrow[2.\ H_3O^+]{1.\ NaOH,\ H_2O} ?$$

STRATEGY Treatment of an ester with aqueous base cleaves the ester into its acid and alcohol components by breaking the bond between the carbonyl carbon and the alcohol oxygen.

SOLUTION

$$\underset{\text{Ethyl 3-methylbutanoate}}{CH_3\overset{CH_3}{\overset{|}{C}}HCH_2\overset{O}{\overset{\|}{C}}OCH_2CH_3} \xrightarrow[2.\ H_3O^+]{1.\ NaOH,\ H_2O} \underset{\text{3-Methylbutanoic acid}}{CH_3\overset{CH_3}{\overset{|}{C}}HCH_2\overset{O}{\overset{\|}{C}}OH} + \underset{\text{Ethanol}}{CH_3CH_2OH}$$

PRACTICE PROBLEM 10.8

What products would you obtain by reduction of propyl benzoate with $LiAlH_4$?

STRATEGY Reduction of an ester with $LiAlH_4$ yields two molecules of alcohol product, one from the acyl part of the ester and one from the alkoxy part. Thus, reduction of propyl benzoate yields benzyl alcohol (from the acyl group) and propan-1-ol (from the alkoxyl group).

SOLUTION

$$\underset{\text{Propyl benzoate}}{C_6H_5-\overset{O}{\overset{\|}{C}}OCH_2CH_2CH_3} \xrightarrow[2.\ H_3O^+]{1.\ LiAlH_4} \underset{\text{Benzyl alcohol}}{C_6H_5-CH_2OH} + \underset{\text{Propan-1-ol}}{HOCH_2CH_2CH_3}$$

PROBLEM 10.18 Show the products of hydrolysis of the following esters:

(a) $CH_3\overset{O}{\overset{\|}{C}}O\overset{CH_3}{\overset{|}{C}}HCH_3$

(b) Cyclohexyl–CO_2CH_3

PROBLEM 10.19 Why do you suppose the saponification of esters is not reversible? In other words, why doesn't treatment of a carboxylic acid with an alkoxide ion give an ester?

PROBLEM 10.20 Show the products you would obtain by reduction of the following esters with $LiAlH_4$:

(a) $CH_3CH_2CH_2\overset{H_3C}{\overset{|}{C}}H\overset{O}{\overset{\|}{C}}OCH_3$

(b) $C_6H_5-\overset{O}{\overset{\|}{C}}-O-C_6H_5$

PROBLEM 10.21 What product would you expect from the reaction of a cyclic ester such as butyrolactone with $LiAlH_4$ followed by treatment with H_3O^+?

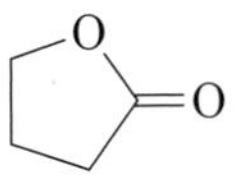

Butyrolactone

Conversion of Esters into Alcohols by Reaction with Grignard Reagents

Grignard reagents react with esters to yield tertiary alcohols in which two of the substituents on the hydroxyl-bearing carbon are identical. For example, methyl benzoate reacts with 2 equivalents of CH_3MgBr to yield 2-phenylpropan-2-ol. The reaction occurs by addition of a Grignard reagent to the ester, elimination of alkoxide ion to give an intermediate ketone, and further addition to the ketone to yield the tertiary alcohol (Figure 10.9):

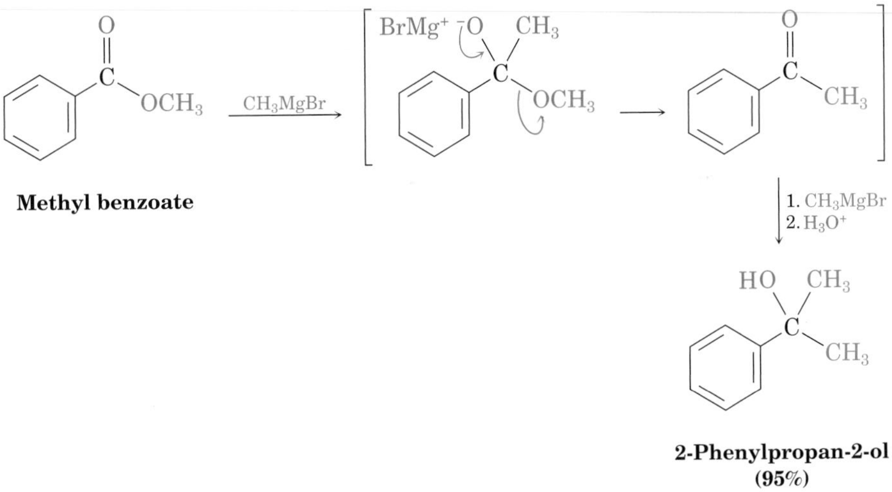

FIGURE 10.9 Mechanism of the reaction of a Grignard reagent with an ester to yield a tertiary alcohol. A ketone intermediate is involved.

Methyl benzoate $\xrightarrow{CH_3MgBr}$ [$BrMg^+\ ^-O$, CH_3, OCH_3 intermediate $\longrightarrow$ ketone ($C_6H_5COCH_3$)] $\xrightarrow{1.\ CH_3MgBr;\ 2.\ H_3O^+}$ **2-Phenylpropan-2-ol (95%)**

PRACTICE PROBLEM 10.9

How could you use the reaction of a Grignard reagent with an ester to prepare 1,1-diphenylpropan-1-ol?

STRATEGY The product of the reaction between a Grignard reagent and an ester is a tertiary alcohol in which the alcohol carbon and one of the attached groups have come from the ester and the remaining two groups bonded to the alcohol carbon have come from the Grignard reagent. Since 1,1-diphenylpropan-1-ol has two phenyl groups and one ethyl group bonded to the alcohol carbon, it must be prepared from reaction of a phenylmagnesium halide with an ester of propanoic acid.

SOLUTION

$$2\ C_6H_5MgBr + CH_3CH_2\overset{O}{\overset{\|}{C}}OCH_3 \xrightarrow[2.\ H_3O^+]{1.\ \text{Ether solvent}} CH_3CH_2-\overset{OH}{\underset{C_6H_5}{C}}-C_6H_5$$

1,1-Diphenylpropan-1-ol

PROBLEM 10.22 What ester and what Grignard reagent might you use to prepare the following alcohols?

(a) H_3C, CH_3, C, OH (b) H_3C, OH, C (c) OH, $CH_3CH_2CH_2CH_2CCH_2CH_3$, CH_2CH_3

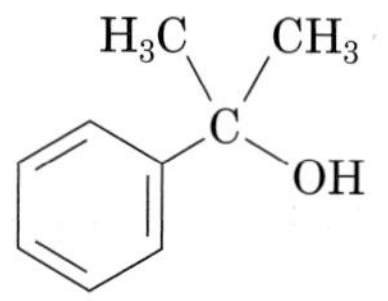

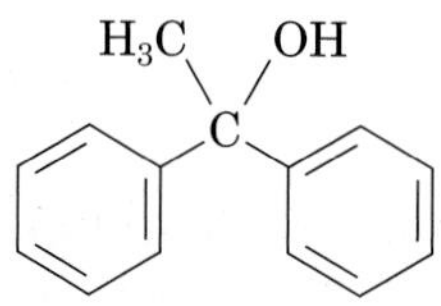

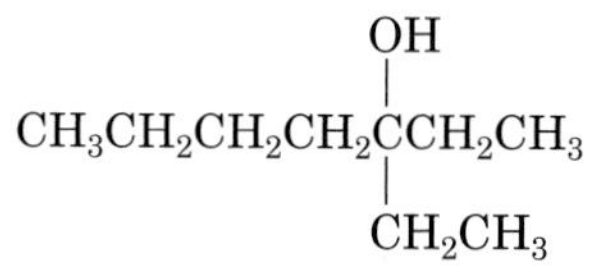

10.12 Chemistry of Amides

Amides are usually prepared by reaction of an acid chloride with an amine, as we saw in Section 10.9. Ammonia, monosubstituted amines, and disubstituted amines all undergo this reaction (Figure 10.10).

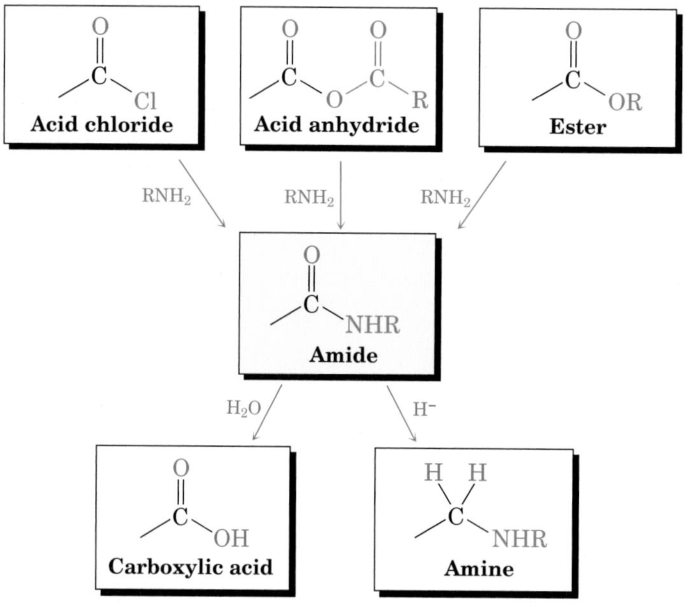

FIGURE 10.10 Some reactions of amides.

Acid chloride, Acid anhydride, Ester —(RNH_2)→ Amide; Amide —(H_2O)→ Carboxylic acid; Amide —(H^-)→ Amine

Amides are much less reactive than acid chlorides, acid anhydrides, and esters. We'll see in Chapter 15, for example, that the amide linkage is stable enough to serve as the basic unit from which proteins are made.

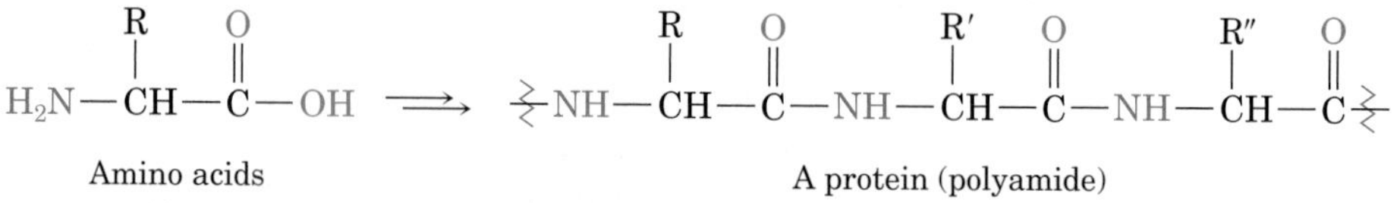

$$H_2N{-}CH(R){-}C(=O){-}OH \Longrightarrow {-}NH{-}CH(R){-}C(=O){-}NH{-}CH(R'){-}C(=O){-}NH{-}CH(R''){-}C(=O){-}$$

Amino acids A protein (polyamide)

Conversion of Amides into Acids ($RCONH_2 \rightarrow RCO_2H$)

Amides undergo hydrolysis to yield carboxylic acids plus amine on heating in either aqueous acid or base. Although the reaction is slow and requires prolonged heating, the overall transformation is a typical nucleophilic acyl substitution of –OH for $-NH_2$.

$$RCONH_2 \xrightarrow[\text{Heat}]{H_3O^+ \text{ or } HO^-,\ H_2O} RCOOH + NH_3$$

Amide **Acid**

Conversion of Amides into Amines by Reduction ($RCONH_2 \rightarrow RCH_2NH_2$)

Like other carboxylic acid derivatives, amides are reduced by $LiAlH_4$. The product of this reduction, however, is an *amine* rather than an alcohol:

$$C_6H_5CONH_2 \xrightarrow[2.\ H_2O]{1.\ LiAlH_4,\ \text{ether}} C_6H_5CH_2NH_2$$

Benzamide **Benzylamine (93%)**

The effect of amide reduction is to convert the amide carbonyl group into a methylene group ($C{=}O \rightarrow CH_2$). This kind of reaction is specific for amides and does not occur with other carboxylic acid derivatives.

Practice Problem 10.10

How could you prepare *N*-ethylaniline by reduction of an amide with $LiAlH_4$?

Strategy Reduction of an amide with $LiAlH_4$ yields an amine. To find the starting material for synthesis of *N*-ethylaniline, look for a CH_2 position next to the nitrogen atom and replace that CH_2 by $C{=}O$. In this case the amide is *N*-phenylacetamide.

Solution

$$C_6H_5NHCOCH_3 \xrightarrow[2.\ H_2O]{1.\ LiAlH_4,\ \text{ether}} C_6H_5NHCH_2CH_3 + H_2O$$

***N*-Phenylacetamide** ***N*-Ethylaniline**

Problem 10.23 How would you convert *N*-ethylbenzamide into the following substances?

(a) Benzoic acid

(b) Benzyl alcohol

(c) *N*-Ethylbenzylamine, $C_6H_5CH_2NHCH_2CH_3$

$$C_6H_5CONHCH_2CH_3$$

***N*-Ethylbenzamide**

PROBLEM 10.24 The reduction of an amide with $LiAlH_4$ followed by protonation with water to yield an amine is effective with both acyclic and cyclic amides (*lactams*). What product would you obtain from reduction of 5,5-dimethylpyrrolidin-2-one with $LiAlH_4$?

H_3C, H_3C, N, H, O

5,5-Dimethylpyrrolidin-2-one
(a lactam)

10.13

Chemistry of Nitriles

Nitriles, R—C≡N, are not related to carboxylic acids in the same sense that acyl derivatives are, but the chemistries of nitriles and carboxylic acids are so similar that the two classes of compounds can be considered together. Both functional groups have a carbon atom with three bonds to an electronegative atom, and both contain a multiple bond.

R—C≡N

A nitrile—three bonds to nitrogen

R—C(=O)OH

An acid—three bonds to two oxygens

The simplest method of preparing nitriles is by the S_N2 reaction of cyanide ion with a primary alkyl halide (see Section 7.5):

$$RCH_2Br + Na^+\ CN^- \xrightarrow[\text{reaction}]{S_N2} RCH_2CN + NaBr$$

Reactions of Nitriles

The chemistry of nitriles is similar in many respects to the chemistry of carbonyl compounds. Thus, nitriles undergo many of the same kinds of reactions as carboxylic acid derivatives (Figure 10.11).

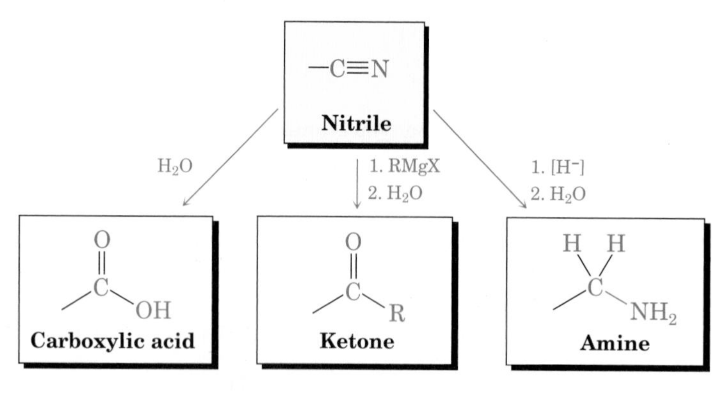

FIGURE 10.11 Some reactions of nitriles.

Like carbonyl groups, nitriles are strongly polarized. The nitrile carbon atom is electrophilic and undergoes attack by nucleophiles to yield an sp^2-hybridized intermediate imine anion that is analogous to the sp^3-hybridized intermediate alkoxide anion formed by addition of a nucleophile to a carbonyl group. Once formed, the intermediate imine anion can go on to yield further products.

Carbonyl compound

Alkoxide ion

Nitrile

Imine anion

Conversion of Nitriles into Carboxylic Acids ($RCN \rightarrow RCO_2H$)

Nitriles are hydrolyzed in either acidic or basic solution to yield carboxylic acids and ammonia (or an amine).

$$R-C\equiv N \xrightarrow[\text{or NaOH, } H_2O]{H_3O^+} RCOOH + NH_3$$

Conversion of Nitriles into Amines by Reduction ($RCN \rightarrow RCH_2NH_2$)

Reduction of a nitrile with $LiAlH_4$ gives a primary amine, just as reduction of an ester gives a primary alcohol.

$$\text{o-Methylbenzonitrile} \xrightarrow[\text{2. } H_2O]{\text{1. } LiAlH_4\text{, ether}} \text{o-Methylbenzylamine (88\%)}$$

o-Methylbenzonitrile **o-Methylbenzylamine (88%)**

Conversion of Nitriles into Ketones by Reaction with Grignard Reagents

Grignard reagents, RMgX, add to nitriles to give intermediate imine anions that can be hydrolyzed to yield ketones:

$$R-C\equiv N: + R'^{-}\ \overset{+}{Mg}X \longrightarrow \left[R-\underset{}{C}(=N^{-}\ \overset{+}{Mg}X)-R'\right] \xrightarrow{H_3O^+} R-C(=O)-R' + NH_3$$

Nitrile Imine anion **Ketone**

For example, benzonitrile reacts with ethylmagnesium bromide to give propiophenone:

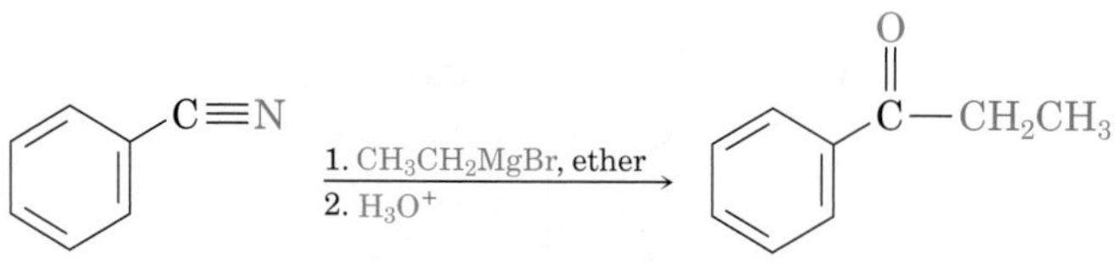

$$\text{Benzonitrile } (C_6H_5\text{–}C{\equiv}N) \xrightarrow[2.\ H_3O^+]{1.\ CH_3CH_2MgBr,\ ether} \text{Propiophenone } (C_6H_5\text{–}\overset{O}{\overset{\|}{C}}\text{–}CH_2CH_3)\ (89\%)$$

Benzonitrile **Propiophenone (89%)**

PRACTICE PROBLEM 10.11

Show how you could prepare hexan-2-one by reaction of a Grignard reagent with a nitrile.

STRATEGY Look at the structure of the target ketone. The C=O carbon comes from the C≡N carbon, one of the two attached groups comes from the Grignard reagent, and the other attached group was present in the nitrile. Thus, there are two ways to prepare a ketone from a nitrile by Grignard addition.

SOLUTION

$$CH_3C{\equiv}N + CH_3CH_2CH_2CH_2MgBr \longrightarrow CH_3\overset{O}{\overset{\|}{C}}CH_2CH_2CH_2CH_3$$

$$CH_3MgBr + N{\equiv}CCH_2CH_2CH_2CH_3 \longrightarrow CH_3\overset{O}{\overset{\|}{C}}CH_2CH_2CH_2CH_3$$

PROBLEM 10.25 How would you prepare the following ketones by reaction of a Grignard reagent and a nitrile?

(a) $CH_3CH_2COCH_2CH_3$

(b) $CH_3CH_2COCH(CH_3)_2$

(c) Acetophenone (methyl phenyl ketone)

(d) Dicyclohexyl ketone (structure: two cyclohexane rings joined by C=O)

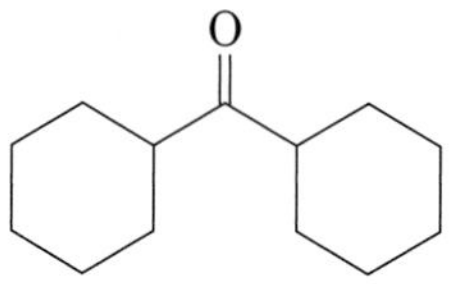

10.14 Polymers from Carbonyl Compounds: Polyamides and Polyesters

Now that we've seen the main classes of carboxylic acid derivatives, it's interesting to note how some of these compounds are used in daily life. Surely the most important such use is as *polymers*, particularly as polyamides (*nylons*) and polyesters.

There are two main classes of synthetic polymers: *chain-growth polymers* and *step-growth polymers.* Polyethylene and other alkene polymers like those we saw in Section 4.8 are called **chain-growth polymers** because they are prepared in chain-reaction processes. An initiator first adds to the double bond of an alkene monomer to produce a reactive intermediate, which then adds to a second alkene monomer unit, and so on. The polymer chain lengthens as more monomer units add successively to the end of the growing chain.

Step-growth polymers are prepared by polymerization reactions between two difunctional molecules, with each new bond formed in a discrete

TABLE 10.3 Some Important Step-Growth Polymers and Their Uses

Monomer name	Formula	Trade or common name of polymer	Uses
Hexamethylene-diamine	$H_2N(CH_2)_6NH_2$	Nylon 66	Fibers, clothing, tire cord, bearings
Adipic acid	$HO_2C(CH_2)_4CO_2H$		
Ethylene glycol	$HOCH_2CH_2OH$	Dacron, Terylene, Mylar	Fibers, clothing, tire cord, film
Dimethyl terephthalate	CH_3O_2C–C_6H_4–CO_2CH_3 (para)		
Caprolactam	(seven-membered cyclic amide, C=O adjacent to N—H)	Nylon 6, Perlon	Fibers, large cast articles

step, independent of all other bonds in the polymer. The key bond-forming step is often a nucleophilic acyl substitution of a carboxylic acid derivative. Some commercially important step-growth polymers are shown in Table 10.3.

Nylons

The best-known step-growth polymers are the *polyamides*, or **nylons**, first prepared by Wallace Carothers at the DuPont Company. Nylons are usually prepared by reaction between a diacid and a diamine. For example, nylon 66 is prepared by heating the six-carbon adipic acid (hexanedioic acid) with the six-carbon hexamethylenediamine (1,6-hexanediamine) at 280 °C:

$$HO\overset{O}{\overset{\|}{C}}CH_2CH_2CH_2CH_2\overset{O}{\overset{\|}{C}}OH + H_2NCH_2CH_2CH_2CH_2CH_2CH_2NH_2$$

Adipic acid **Hexamethylenediamine (hexane-1,6-diamine)**

$$\downarrow \text{Heat}$$

$$\left(\overset{O}{\overset{\|}{C}}CH_2CH_2CH_2CH_2\overset{O}{\overset{\|}{C}}—NHCH_2CH_2CH_2CH_2CH_2CH_2NH \right)_n + 2n\ H_2O$$

Nylon 66

Nylons are used both in engineering applications and in making fibers. A combination of high-impact strength and abrasion resistance makes nylon an excellent metal substitute for bearings and gears. As fibers, nylon is used in a variety of applications, from clothing to tire cord to mountaineering ropes.

Polyesters

Just as a polyamide is made by reaction between a diacid and a diamine, a **polyester** is made by reaction between a diacid and a dialcohol. The most generally useful polyester is made by a nucleophilic acyl substitution reaction between dimethyl terephthalate (dimethyl benzene-1,4-dicarboxylic acid) and ethylene glycol. The product is used under the trade name Dacron to make clothing fiber and tire cord and under the name Mylar to make plastic film and recording tape. The tensile strength of polyester film is nearly equal to that of steel.

$CH_3OC(=O)-C_6H_4-C(=O)OCH_3$ + $HOCH_2CH_2OH$

Dimethyl terephthalate **Ethylene glycol**

↓ 200 °C

$\left[OCH_2CH_2O-C(=O)-C_6H_4-C(=O) \right]_n$ + $2n\ CH_3OH$

Polyester, Dacron, Mylar

PRACTICE PROBLEM 10.12

Draw the structure of Qiana, a polyamide made by high-temperature reaction of hexanedioic acid with cyclohexane-1,4-diamine.

SOLUTION

$HOC(=O)(CH_2)_4C(=O)OH$

Hexanedioic acid

+

$H_2N-C_6H_{10}-NH_2$

Cyclohexane-1,4-diamine

→ $\left[HN-C_6H_{10}-NH-C(=O)(CH_2)_4C(=O) \right]_n$

Qiana

PROBLEM 10.26 Kevlar, a nylon polymer used in bulletproof vests, is made by reaction of benzene-1,4-dicarboxylic acid with benzene-1,4-diamine. Show the structure of Kevlar.

10.15 Enzymes in Organic Synthesis

Nature provides examples of almost all of the reactions described in this book. We've seen some examples of the selectivity of nature: recall the reduction of a *cis*-alkene with fumarase but not the *trans*-alkene (see Section 4.4). Although we'll discuss nature's catalysts in depth in Chapter 15, let's look at some examples of reactions from this chapter that are accomplished *selectively* by nature.

How does nature achieve selectivity? The catalysts that nature uses have complex three-dimensional shapes that "organize" the substrate (reactant) very specifically before doing chemistry. We call these catalysts **enzymes** when they comprise specific building blocks called amino acids. Enzymes are now being used in reaction flasks to do chemistry just like HCl or NaOH is. One example of the use of enzymes in organic chemistry is the hydrolysis of esters. Often, these esters are of great importance to the pharmaceutical and chemical industries. The partial hydrolysis of the diester shown using NaOH produces a racemate because the two monoesters are enantiomers. The separation of these enantiomers is difficult because they have similar chemical properties.

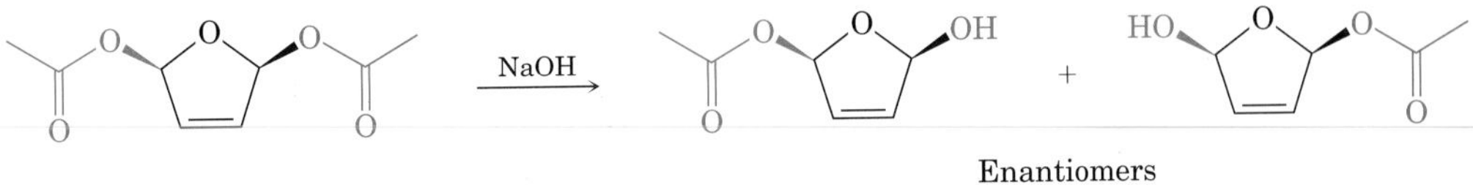

NaOH

Enantiomers

What if only one of these two materials is useful? A strategy that produces only the desired monoester would be ideal. One solution to this problem is to use a chiral hydrolysis catalyst. Enzymes called **esterases** (because they do chemistry on esters) provide the necessary *chiral environment* to allow selective hydrolysis. Esterases from different organisms have different three-dimensional structures and, as a result, perform different chemistry. It so happens that acetylcholine esterase from electric eels cleaves one ester, whereas the lipase (a lipid esterase) from a pig's pancreas cleaves the other.

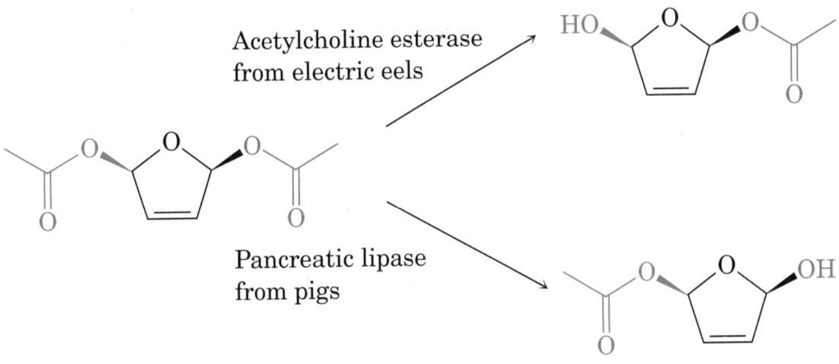

Acetylcholine esterase from electric eels

Pancreatic lipase from pigs

Another example of how enzymes are used in organic synthesis is in the commercial production of vitamin C, also called ascorbic acid. A key step in the synthesis of vitamin C is the oxidation of a hydroxyl group of sorbitol to a ketone. The microorganism *Acetobacter suboxydans* performs this chemistry with exquisite selectivity, which cannot be matched by any catalyst or reagent designed in the laboratory.

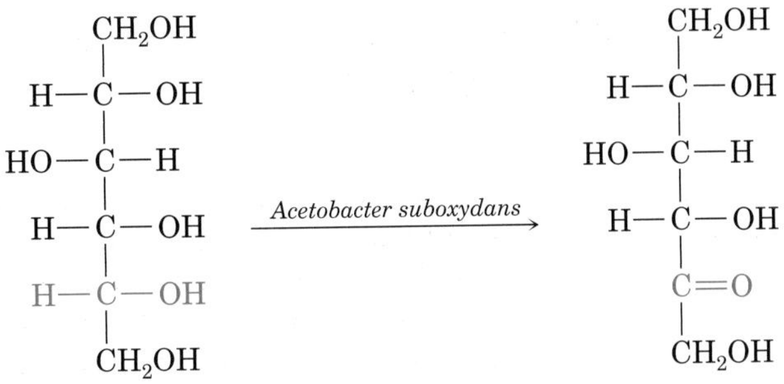

Acetobacter suboxydans

INTERLUDE

β-Lactam Antibiotics—Part I

Penicillium mold growing in a petri dish.

Although you should never underestimate the value of hard work and logical thinking, sheer luck often plays a role in most scientific breakthroughs. What has been called "the supreme example [of luck] in all scientific history" occurred in the late summer of 1928, when the Scottish bacteriologist Alexander Fleming went on vacation, leaving in his lab a culture plate recently inoculated with the bacterium *Staphylococcus aureus*.

While Fleming was away, an extraordinary chain of events occurred. First, a 9-day cold spell lowered the laboratory temperature to a point where the *Staphylococcus* on the plate could not grow. During this time, spores from a colony of the mold *Penicillium notatum*, which were being grown on the floor below, wafted up into Fleming's lab and landed in the culture plate. The temperature then rose, and both *Staphylococcus* and *Penicillium* began to grow. On returning from vacation, Fleming discarded the plate into a tray of antiseptic, intending to sterilize it. Evidently, though, the plate did not sink deeply enough into the antiseptic, because when Fleming glanced at the plate a few days later, what he saw changed the course of human history. He noticed that the growing *Penicillium* mold appeared to dissolve the colonies of staphylococci.

Fleming realized that the mold must be producing a chemical that killed bacteria, and he spent several years trying to isolate the substance. Finally, in 1939, the Australian pathologist Howard Florey and the German refugee Ernst Chain managed to isolate the active substance, called *penicillin*. Penicillin's dramatic ability to cure infections in mice was soon demonstrated, and successful tests in humans followed shortly thereafter. By 1943, penicillin was being produced on a large scale for military use, and by 1944, it was being used in civilians. Since then, untold millions of lives have been saved. In 1945, Fleming, Florey, and Chain shared the Nobel Prize in Medicine.

Now called *benzylpenicillin*, or penicillin G, the substance discovered by Fleming is just one member of a large class of so-called β-lactam antibiotics. A lactam is a cyclic amide. The designation "β" signifies that the amide appears in a four-membered ring.

β-Lactam

Benzylpenicillin (penicillin G)

From the chemistry we have learned so far, we know that amides are relatively unreactive compared with acid chlorides or esters. Hydrolysis of amides to carboxylic acids requires harsh conditions. From these observations, we would conclude that nucleophilic substitution at the amide carbon doesn't occur readily. Should it come as a surprise, then, to find out that the β-lactam family of antibiotics fight bacteria by employing a reactive amide? Perhaps not. Ring strain makes this amide reactive. We know that the ideal bond angle is 109.5° for sp^3-hybridized atoms and 120° for sp^2-hybridized atoms. When atoms are confined into a four-membered ring,

Continued

the bond angles are squeezed closer to 90°. This strain makes the amide reactive toward some nucleophiles.

β-Lactam antibiotics react with a nucleophile on an active site of the enzyme transpeptidase. Transpeptidase is responsible for building the cell wall of the bacteria. A nucleophilic hydroxyl group reacts with the β-lactam antibiotic to form a stable ester. This ester prevents transpeptidase from doing any additional chemistry. The bacterial cell wall is never finished, and the bacteria ultimately die.

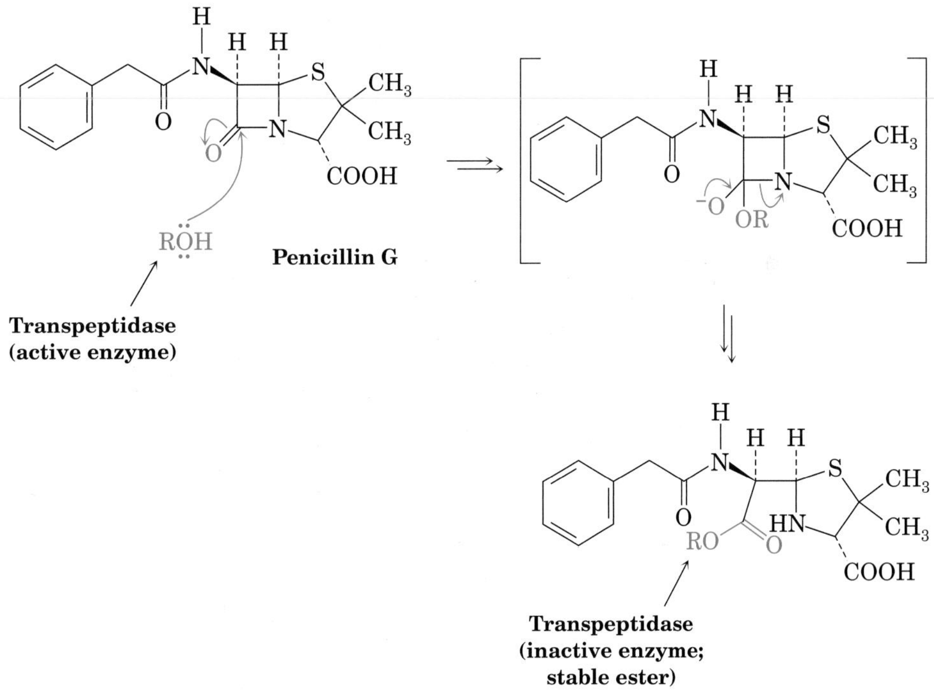

ROH

Penicillin G

Transpeptidase (active enzyme)

Transpeptidase (inactive enzyme; stable ester)

Unfortunately, bacteria have become resistant to penicillin by evolving enzymes similar to transpeptidase that bind penicillin and efficiently hydrolyze the amide. The hydrolyzed penicillin has no antibacterial activity. We'll look at how chemists have countered this advance in Chapter 11.

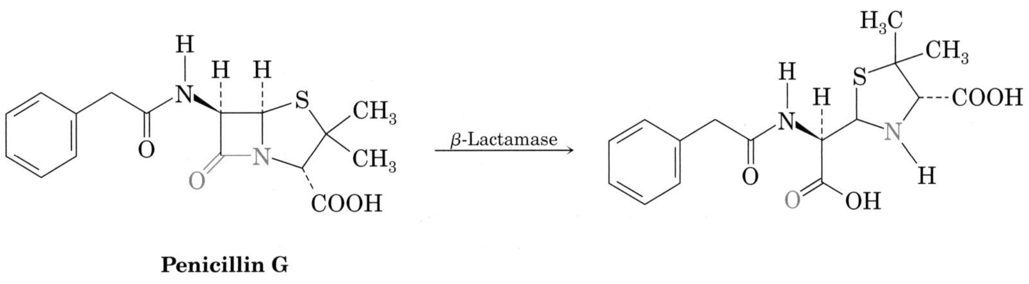

β-Lactamase

Penicillin G

Summary and Key Words

Carboxylic acids are useful building blocks for synthesizing other molecules, both in nature and in the chemical laboratory. The distinguishing characteristic of carboxylic acids is their acidity. Although weaker than mineral acids like HCl, carboxylic acids are much more acidic than alcohols because carboxylate ions are stabilized by resonance. Most carboxylic acids have pK_a values near 5, but the exact acidity of an acid depends on its structure. Carboxylic acids substituted by an electron-withdrawing group are more acidic (have a lower pK_a) because their carboxylate ions are more stable.

Carboxylic acids can be transformed into a variety of **carboxylic acid derivatives** in which the acid —OH group has been replaced by another substituent. **Acid chlorides**, **acid anhydrides**, **esters**, and **amides** are the most common. The chemistry of all these derivatives is similar and is dominated by a single general reaction type: the **nucleophilic acyl substitution reaction**. These substitutions take place by addition of a nucleophile to the polar carbonyl group of the acid derivative, followed by expulsion of the leaving group. Carboxylic acid derivatives can undergo reaction with many different nucleophiles. Among the most important are reaction with water, alcohols, amines, hydride ion, and Grignard reagents.

$$RC(=O)Y + :Nu^- \longrightarrow [R-C(O^-)(Y)-Nu] \longrightarrow RC(=O)Nu + :Y^-$$

where Y = Cl, Br, I (acid halide); OR (ester); OCOR (anhydride); or NH_2 (amide)

Nitriles, R—C≡N, can also be considered as carboxylic acid derivatives because they undergo nucleophilic additions to the polar C≡N bond in the same way carbonyl compounds do. The most important reactions of nitriles are their hydrolysis to yield carboxylic acids, their reduction to yield primary amines, and their reaction with Grignard reagents to yield ketones.

Summary of Reactions

1. Reactions of carboxylic acids (Section 10.8)
   (a) Conversion into acid chlorides

$$RC(=O)OH \xrightarrow{SOCl_2} RC(=O)Cl$$

   (b) Conversion into esters (Fischer esterification)

$$RC(=O)OH + R'OH \xrightarrow[\text{catalyst}]{H^+} RC(=O)OR'$$

2. Reactions of acid halides (Section 10.9)
   (a) Conversion into carboxylic acids

$$RCOCl + H_2O \longrightarrow RCOOH$$

   (b) Conversion into esters

$$RCOCl + R'OH \xrightarrow{\text{Pyridine}} RCOOR'$$

   (c) Conversion into amides

$$RCOCl + NH_3 \longrightarrow RCONH_2$$

3. Reactions of acid anhydrides (Section 10.10)
   (a) Conversion into esters

$$RCOOCOR + R'OH \xrightarrow{\text{Pyridine}} RCOOR'$$

   (b) Conversion into amides

$$RCOOCOR + NH_3 \xrightarrow{\text{Pyridine}} RCONH_2$$

4. Reactions of esters (Section 10.11)
   (a) Conversion into acids

$$RCOOR' + H_2O \xrightarrow[\text{NaOH}]{H^+ \text{ or}} RCOOH$$

   (b) Conversion into primary alcohols by reduction

$$RCOOR' \xrightarrow[2.\ H_3O^+]{1.\ LiAlH_4} RCH_2OH$$

   (c) Conversion into tertiary alcohols by Grignard reaction

$$RCOOR' \xrightarrow[2.\ H_3O^+]{1.\ R''MgX} R-C(OH)(R'')-R''$$

5. Reactions of amides (Section 10.12)
   (a) Conversion into carboxylic acids

$$RC(=O)NH_2 + H_2O \xrightarrow[NaOH]{H^+ \text{ or}} RC(=O)OH$$

   (b) Conversion into amines by reduction

$$RC(=O)NH_2 \xrightarrow[2.\ H_2O]{1.\ LiAlH_4} RCH_2NH_2$$

6. Reactions of nitriles (Section 10.13)
   (a) Conversion into carboxylic acids

$$R{-}C{\equiv}N + H_2O \xrightarrow[NaOH]{H^+ \text{ or}} RC(=O)OH$$

   (b) Conversion into amines by reduction

$$R{-}C{\equiv}N \xrightarrow[2.\ H_2O]{1.\ LiAlH_4} RCH_2NH_2$$

   (c) Conversion into ketones by Grignard reaction

$$R{-}C{\equiv}N \xrightarrow[2.\ H_3O^+]{1.\ R'MgX} RC(=O)R'$$

EXERCISES

Visualizing Chemistry

10.27 Name the following compounds (red = O, blue = N):

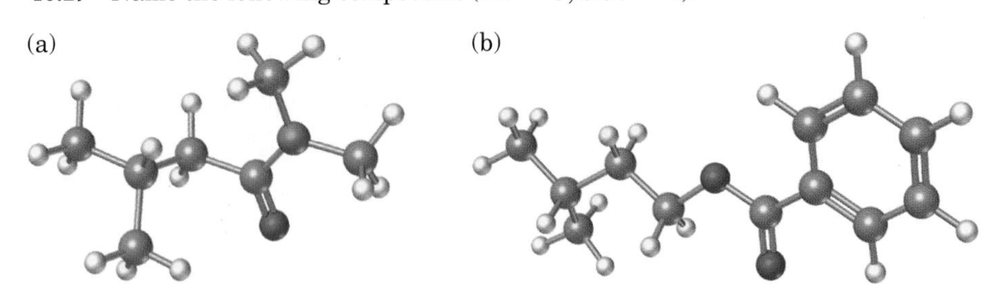

10.28 Show how you could prepare each of the following compounds starting with an appropriate carboxylic acid and any other reagents needed (red = O, blue = N, reddish brown = Br):

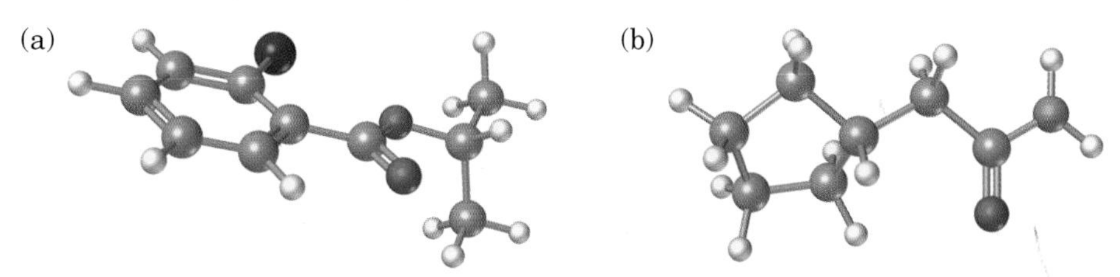

10.29 The following structure represents a tetrahedral alkoxide ion intermediate formed by addition of a nucleophile to a carboxylic acid derivative. Identify the nucleophile, the leaving group, the reactant, and the ultimate product (red = O, blue = N, yellow-green = Cl).

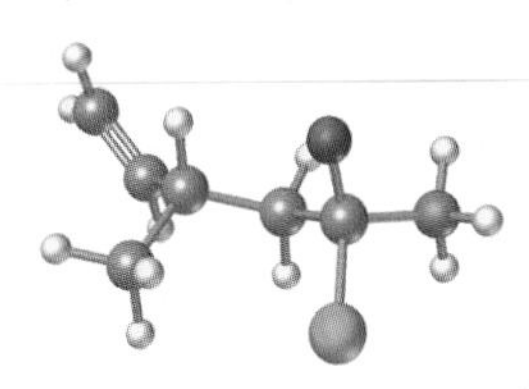

10.30 The following structure represents a tetrahedral alkoxide ion intermediate formed by addition of a nucleophile to a carboxylic acid derivative. Identify the nucleophile, the leaving group, the reactant, and the ultimate product (red = O, blue = N).

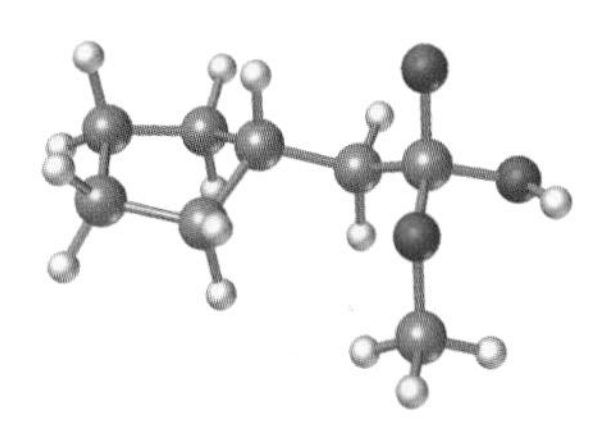

10.31 Electrostatic potential maps of methyl thioacetate and methyl acetate are shown. Which of the two do you think is more reactive in nucleophilic acyl substitution reactions? Explain.

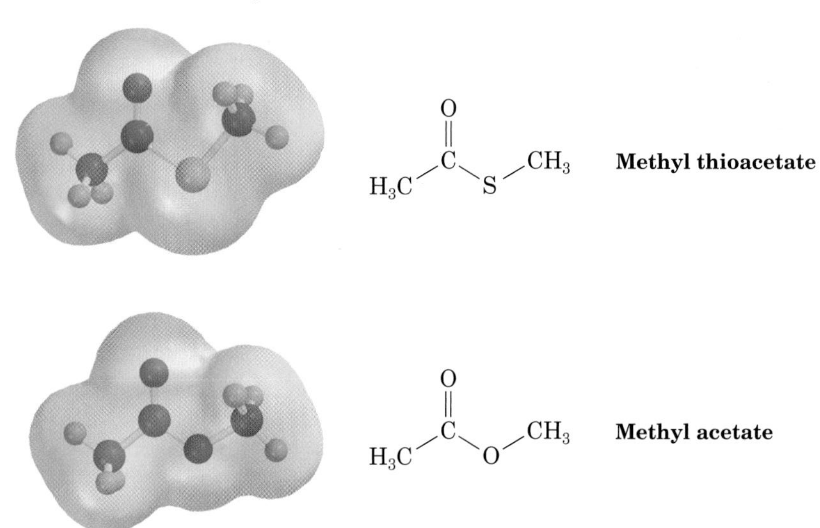

Additional Problems

NOMENCLATURE

10.32 Give IUPAC names for the following carboxylic acids:

(a) $CH_3CH(COOH)CH_2CH_2CH(COOH)CH_3$

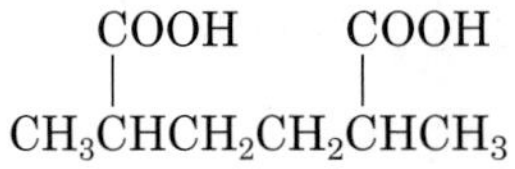

(b) $(CH_3)_3CCOOH$

(c) $CH_3CH_2CH_2CH(CH_2CH_2CH_3)CH_2COOH$

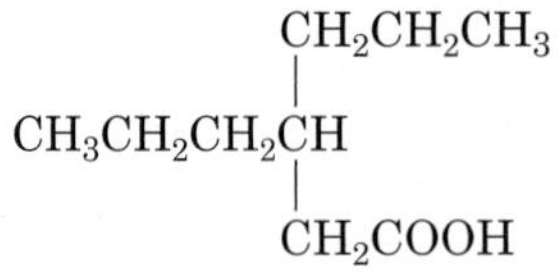

(d) COOH on a benzene ring, with NO_2 para

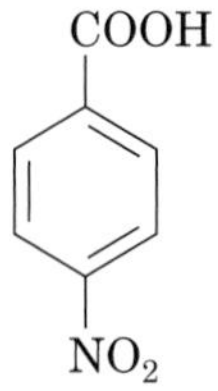

(e) COOH on a cyclodecene ring (at the double-bond carbon)

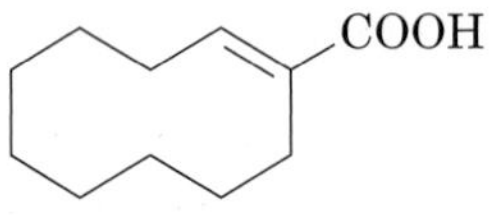

(f) $BrCH_2CH(Br)CH_2CH_2COOH$

10.33 Give IUPAC names for the following carboxylic acid derivatives:

(a) $CONH_2$ on a benzene ring, with H_3C para

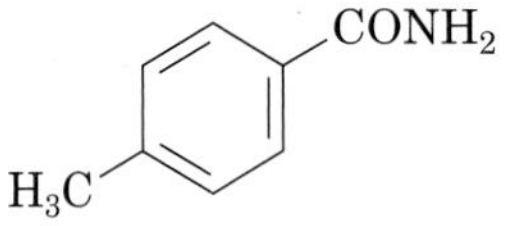

(b) $(CH_3CH_2)_2CHCH{=}CHCN$

(c) $CH_3O_2CCH_2CH_2CO_2CH_3$

(d) $CH_2CH_2CO_2CH(CH_3)_2$ on a benzene ring

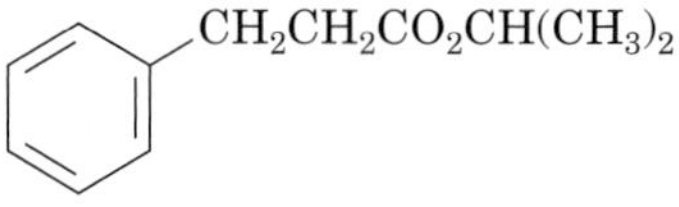

(e) C_6H_5–C(=O)–O–C_6H_5

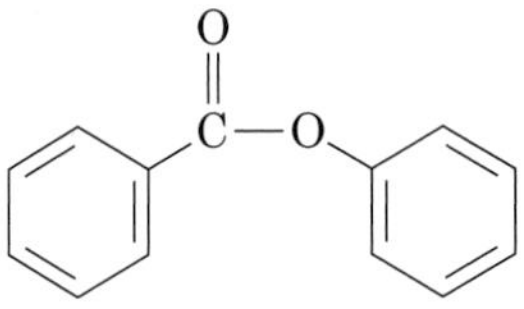

(f) $CH_3CH(Br)CH_2CONHCH_3$

(g) C(=O)–Cl on a benzene ring, with Br at the two meta positions

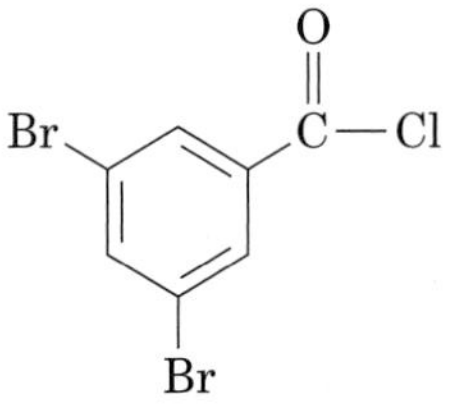

(h) CN on a cyclopentene ring (at the double-bond carbon)

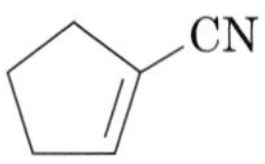

10.34 Draw structures corresponding to the following IUPAC names:

(a) 4,5-Dimethylheptanoic acid
(b) *cis*-Cyclohexane-1,2-dicarboxylic acid
(c) Heptanedioic acid
(d) Triphenylacetic acid
(e) 2,2-Dimethylhexanamide
(f) Phenylacetamide
(g) Cyclobut-2-enecarbonitrile
(h) Ethyl cyclohexanecarboxylate

10.35 Draw and name the eight carboxylic acids with formula $C_6H_{12}O_2$. Which are chiral?

10.36 Draw and name compounds that meet the following descriptions:

(a) Three acid chlorides, C_6H_9ClO
(b) Three amides, $C_7H_{11}NO$
(c) Three nitriles, C_5H_7N
(d) Three esters, $C_5H_8O_2$

PHYSICAL PROPERTIES

10.37 Acetic acid boils at 118 °C, but its ethyl ester boils at 77 °C. Why is the boiling point of the acid so much higher, even though it has a lower molecular weight?

REACTIVITY

10.38 The following reactivity order has been found for the saponification of alkyl acetates by aqueous NaOH:

$$CH_3CO_2CH_3 > CH_3CO_2CH_2CH_3 > CH_3CO_2CH(CH_3)_2 > CH_3CO_2C(CH_3)_3$$

How can you explain this reactivity order?

10.39 Rank the following compounds in order of their reactivity toward nucleophilic acyl substitution:
(a) $CH_3CO_2CH_3$ (b) CH_3COCl (c) CH_3CONH_2 (d) $CH_3CO_2COCH_3$

ACIDITY

10.40 Citric acid has $pK_a = 3.14$, and tartaric acid has $pK_a = 2.98$. Which acid is stronger?

10.41 Order the compounds in each of the following sets with respect to increasing acidity:
(a) Acetic acid, chloroacetic acid, trifluoroacetic acid
(b) Benzoic acid, *p*-bromobenzoic acid, *p*-nitrobenzoic acid
(c) Acetic acid, phenol, cyclohexanol

10.42 How can you explain the fact that 2-chlorobutanoic acid has $pK_a = 2.86$, 3-chlorobutanoic acid has $pK_a = 4.05$, 4-chlorobutanoic acid has $pK_a = 4.52$, and butanoic acid itself has $pK_a = 4.82$?

REACTIONS

10.43 Predict the product(s) of the following reactions:

(a) Cyclohexyl–$CO_2CH_2CH_3$ $\xrightarrow[2.\ H_3O^+]{1.\ CH_3CH_2MgBr}$?

(b) $CH_3CH(CH_3)CH_2CH_2C{\equiv}N$ $\xrightarrow[2.\ H_2O]{1.\ LiAlH_4}$?

(c) Cyclopentyl–$COCl$ $\xrightarrow{CH_3NH_2}$?

(d) Cyclohexane with CO_2H and H on one carbon and CH_3 and H on the adjacent carbon $\xrightarrow[H_2SO_4]{CH_3OH}$?

(e) $H_2C{=}CHCH(CH_3)CH_2CO_2CH_3$ $\xrightarrow[2.\ H_3O^+]{1.\ LiAlH_4}$?

(f) Cyclohexyl–OH $\xrightarrow[Pyridine]{CH_3CO_2COCH_3}$?

(g) Benzene ring with $CONH_2$ and ortho CH_3 $\xrightarrow[2.\ H_2O]{1.\ LiAlH_4}$?

(h) Benzene ring with CH_2CO_2H and ortho Br $\xrightarrow{SOCl_2}$?

10.44 Predict the product of the reaction of *p*-methylbenzoic acid with each of the following reagents:
(a) 1. $LiAlH_4$, 2. H_3O^+ (b) CH_3OH, HCl (c) $SOCl_2$ (d) NaOH, then CH_3I

10.45 A chemist in need of 2,2-dimethylpentanoic acid decided to synthesize some by reaction of 2-chloro-2-methylpentane with NaCN, followed by hydrolysis of the product. After carrying out the reaction sequence, however, none of the desired product could be found. What do you suppose went wrong?

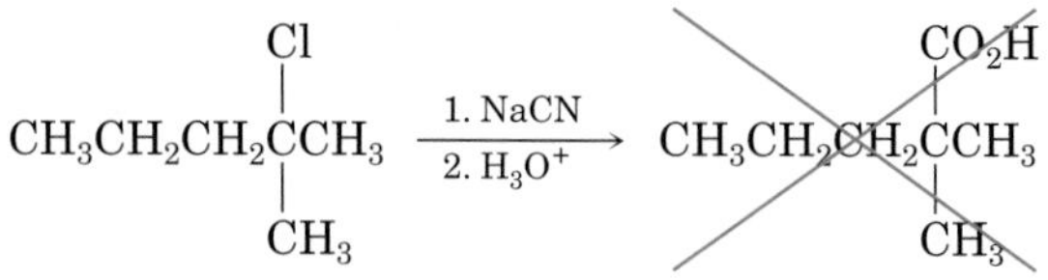

$$CH_3CH_2CH_2\overset{Cl}{\underset{CH_3}{C}}CH_3 \xrightarrow[2.\ H_3O^+]{1.\ NaCN} \cancel{CH_3CH_2CH_2\overset{CO_2H}{\underset{CH_3}{C}}CH_3}$$

10.46 If 5-hydroxypentanoic acid is treated with an acid catalyst, an intramolecular esterification reaction occurs. What is the structure of the product? (*Intramolecular* means within the same molecule.)

$HOCH_2CH_2CH_2CH_2CO_2H$ **5-Hydroxypentanoic acid**

10.47 How can you explain the observation that an attempted Fischer esterification of 2,4,6-trimethylbenzoic acid with methanol/HCl is unsuccessful? No ester is obtained, and the starting acid is recovered unchanged.

10.48 Acid chlorides undergo reduction with $LiAlH_4$ in the same way that esters do to yield primary alcohols. What are the products of the following reactions?

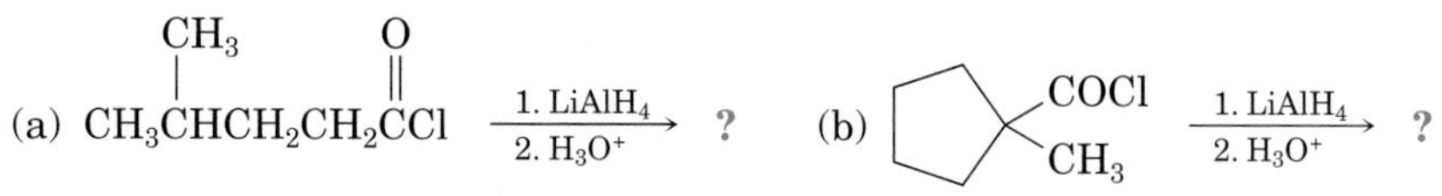

(a) $CH_3\overset{CH_3}{CH}CH_2CH_2\overset{O}{\overset{\|}{C}}Cl \xrightarrow[2.\ H_3O^+]{1.\ LiAlH_4}$?

(b) 1-methylcyclopentane-1-carbonyl chloride (cyclopentane ring bearing COCl and CH_3) $\xrightarrow[2.\ H_3O^+]{1.\ LiAlH_4}$?

MECHANISMS

10.49 The reaction of an acid chloride with $LiAlH_4$ to yield a primary alcohol (Problem 10.48) takes place in two steps. The first step is a nucleophilic acyl substitution of H^- for Cl^- to yield an aldehyde, and the second step is nucleophilic addition of H^- to the aldehyde to yield an alcohol. Write the mechanism of the reduction of CH_3COCl.

10.50 One method for preparing acid anhydrides is by treatment of an acid chloride with a carboxylate ion. For example:

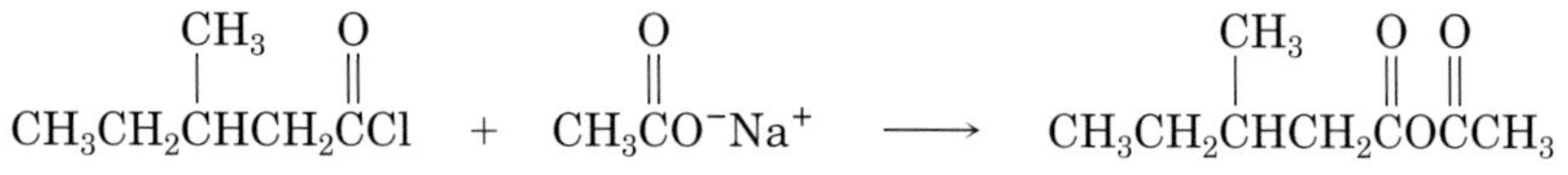

$$CH_3CH_2\overset{CH_3}{CH}CH_2\overset{O}{\overset{\|}{C}}Cl + CH_3\overset{O}{\overset{\|}{C}}O^-Na^+ \longrightarrow CH_3CH_2\overset{CH_3}{CH}CH_2\overset{O}{\overset{\|}{C}}O\overset{O}{\overset{\|}{C}}CH_3$$

Propose a mechanism for this reaction.

10.51 Acid chlorides undergo reaction with Grignard reagents at −78°C to yield ketones. Propose a mechanism for the reaction.

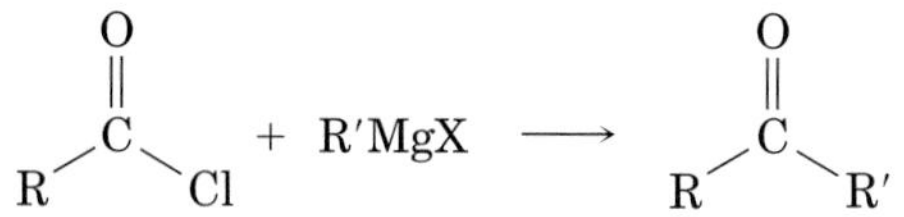

$$R\overset{O}{\overset{\|}{C}}Cl + R'MgX \longrightarrow R\overset{O}{\overset{\|}{C}}R'$$

10.52 If the reaction of an acid chloride with a Grignard reagent (Problem 10.51) is carried out at room temperature, a tertiary alcohol is formed.

(a) Propose a mechanism for this reaction.

(b) What are the products of the reaction of CH_3MgBr with the acid chlorides given in Problem 10.48?

SYNTHESIS

10.53 How can you prepare acetophenone (methyl phenyl ketone) from the following starting materials? (More than one step may be required.)

(a) Benzonitrile (b) Bromobenzene

(c) Methyl benzoate (d) Benzene

10.54 How might you prepare the following products from butanoic acid? (More than one step may be required.)
(a) $CH_3CH_2CH_2CH_2OH$ (b) $CH_3CH_2CH_2CHO$ (c) $CH_3CH_2CH_2CH_2Br$
(d) $CH_3CH_2CH_2CH_2CN$ (e) $CH_3CH_2CH{=}CH_2$ (f) $CH_3CH_2CH_2CH_2NH_2$

10.55 Show how you might prepare the anti-inflammatory agent ibuprofen starting from isobutylbenzene. More than one step is needed.

Isobutylbenzene $\Rightarrow$ Ibuprofen (CO_2H)

INTEGRATED PROBLEMS

10.56 When dimethyl carbonate, $CH_3OCO_2CH_3$, is treated with phenylmagnesium bromide, triphenylmethanol is formed. Explain.

10.57 Predict the product, if any, of reaction between propanoyl chloride and the following reagents. (See Problems 10.48 and 10.52.)
(a) 1. Excess CH_3MgBr in ether, 2. H_3O^+ (b) NaOH in H_2O
(c) Methylamine (d) 1. $LiAlH_4$, 2. H_3O^+
(e) Cyclohexanol (f) Sodium acetate

10.58 Answer Problem 10.57 for reaction between methyl propanoate and the listed reagents.

10.59 What esters and what Grignard reagents would you use to make the following alcohols? Show all possibilities.

(a) $CH_3CH_2CH_2C(OH)(CH_3)CH_3$ (b) HO CH$_3$ (1,1-diphenylethanol)

10.60 Show two ways to make the following esters:

(a) $CH_3CH(CH_3)CH_2CH_2C(=O)OCH_2CH_3$ (b) cyclopentyl$-CH_2OC(=O)CH_3$

10.61 How would you prepare the following substances from the indicated starting materials? More than one step is required in each case.
(a) $(CH_3)_2CHCH_2CH_2NH_2$ from $(CH_3)_2CHCH_2I$
(b) 1-Phenylbutan-2-one from benzyl bromide, $C_6H_5CH_2Br$

10.62 What products would you obtain on saponification of the following esters?

(a)
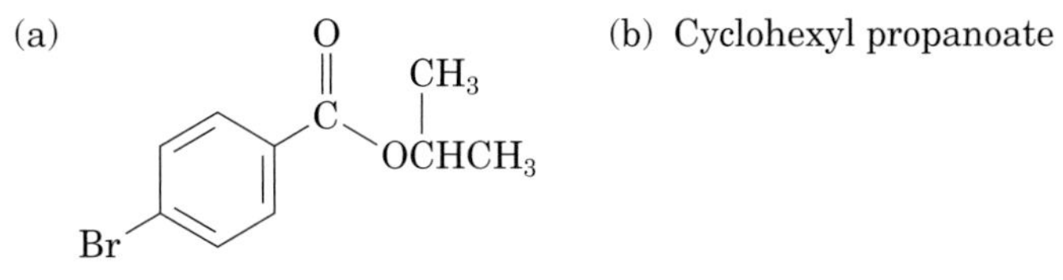

(b) Cyclohexyl propanoate

10.63 When *methyl* acetate is heated in pure ethanol containing a small amount of HCl catalyst, *ethyl* acetate results. Explain.

10.64 *tert*-Butoxycarbonyl azide, an important reagent used in protein synthesis, is prepared by treating *tert*-butoxycarbonyl chloride with sodium azide. Propose a mechanism for this reaction.

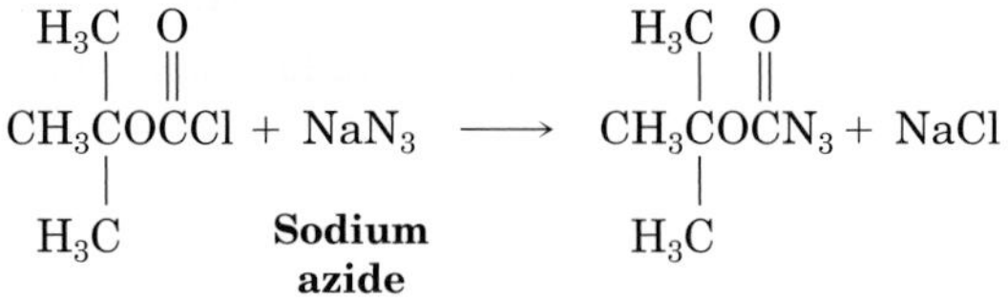

$$(CH_3)_3COCOCl + NaN_3 \longrightarrow (CH_3)_3COCON_3 + NaCl$$

tert-Butoxycarbonyl chloride **Sodium azide**

10.65 What product would you expect to obtain on treatment of the cyclic ester butyrolactone with excess phenylmagnesium bromide?

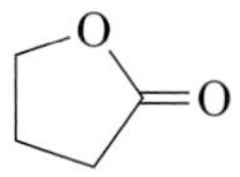

Butyrolactone

10.66 *N,N*-Diethyl-*m*-toluamide (DEET) is the active ingredient in many insect repellents. How might you synthesize DEET from *m*-bromotoluene?

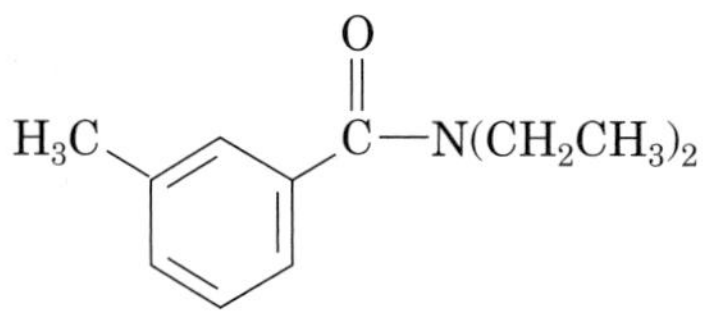

H_3C–C_6H_4–$C(=O)$–$N(CH_2CH_3)_2$ **N, N-Diethyl-m-toluamide**

10.67 In the *iodoform reaction*, a triiodomethyl ketone reacts with aqueous NaOH to yield a carboxylate ion and iodoform (triiodomethane). Propose a mechanism for this reaction.

$$R-\overset{O}{\overset{\|}{C}}-CI_3 \xrightarrow{NaOH,\ H_2O} R-\overset{O}{\overset{\|}{C}}-O^- + CHI_3$$

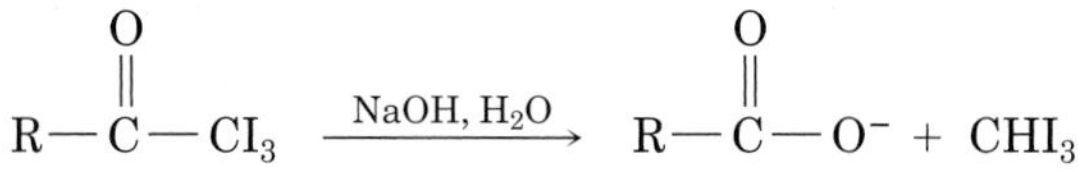

10.68 The K_a for bromoacetic acid is approximately 1×10^{-3}. What percentage of the acid is dissociated in a 0.10 M aqueous solution?

10.69 The step-growth polymer called nylon 6 is prepared from caprolactam. The reaction involves initial reaction of caprolactam with water to give an intermediate amino acid, followed by heating to form the polymer. Propose mechanisms for both steps, and show the structure of nylon 6.

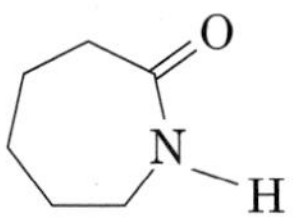

Caprolactam

10.70 Draw a representative segment of the polyester that would result from reaction of pentanedioic acid ($HO_2CCH_2CH_2CH_2CO_2H$) and pentane-1,5-diol.

10.71 The marine natural product and potential chemotherapeutic agent spongistatin is rich in functional groups. How many of the 15 oxygen-containing functional groups are alcohols? Ethers? Aldehydes? Ketones? Esters? Carboxylic acids? Hemiacetals? Acetals?

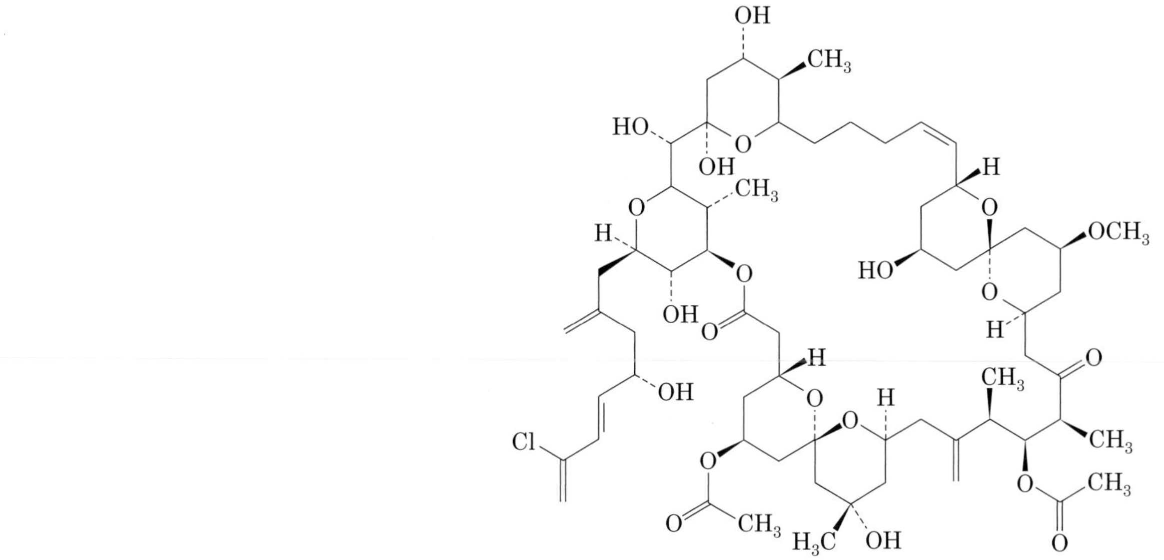

Spongistatin

10.72 Dendrimers, unlike linear polymers, are perfectly branched, tree-shaped molecules. The most widely studied dendrimer is the PAMAM dendrimer based on poly(amidoamine). Its synthesis starts with a diamine core like ethylenediamine. Four conjugate additions (see Section 9.11) occur (two on each N) between the diamine and methyl acrylate. Draw the conjugate addition product, a tetraester, of this reaction:

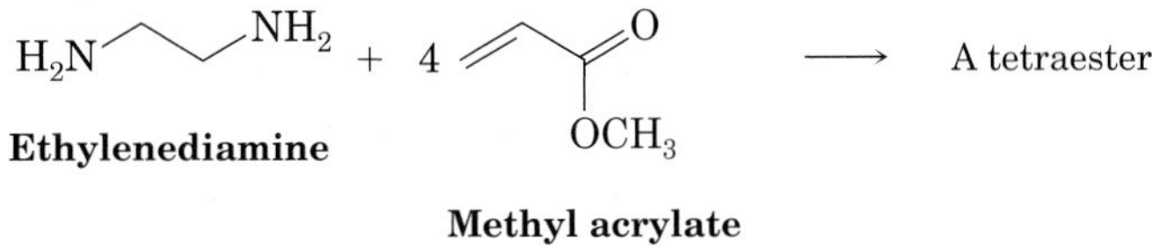

H_2N–CH_2CH_2–NH_2 + 4 $CH_2{=}CH{-}C(=O)OCH_3$ ⟶ A tetraester

Ethylenediamine **Methyl acrylate**

(a) What does this reaction tell you about the relative rates of 1,2-addition versus 1,4-addition in this reaction?

(b) To continue the dendrimer synthesis, the tetraester obtained is reacted with an excess of ethylenediamine to yield the tetraamine. Draw the product of this reaction.

(c) If the tetramine is reacted with an excess of methyl acrylate, how many esters will the product have (assuming each primary amine reacts twice)?

In the Medicine Cabinet

10.73 Yesterday's drug is often today's poison. Cocaine enjoyed a much better reputation 100 years ago, when it was used as a stimulant in products such as "Forced March," which helped Shackleton explore Antarctica, as well as in drops to treat toothaches and depression. What three molecules are produced on hydrolysis of cocaine?

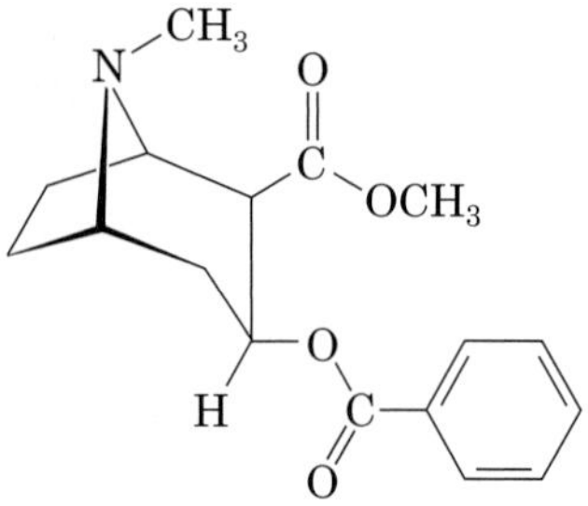

Cocaine

10.74 Because of cocaine's addictive properties, researchers have looked for less addictive alternatives to relieve pain. Lidocaine and Novocain preserve many of the common structural features of cocaine (red) but do not have the same risk factors.

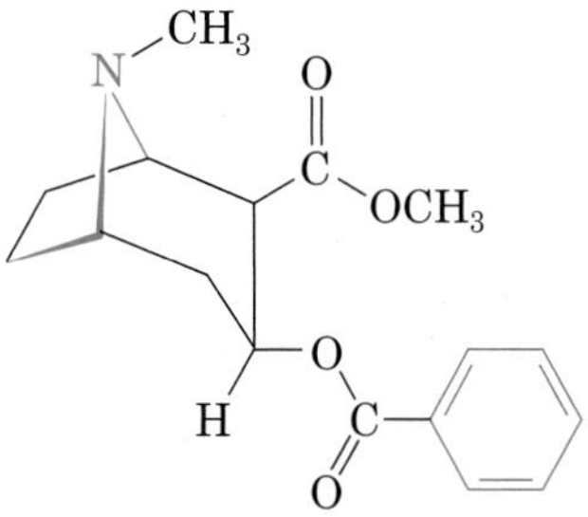

Cocaine 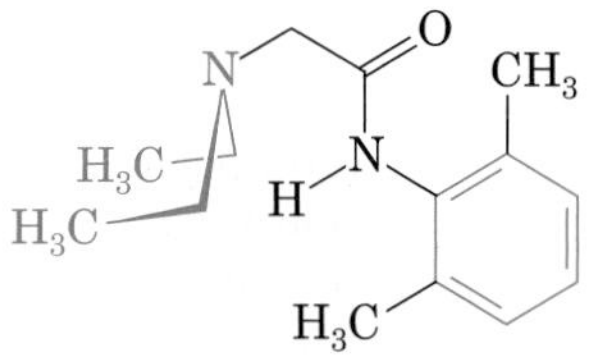 **Lidocaine** 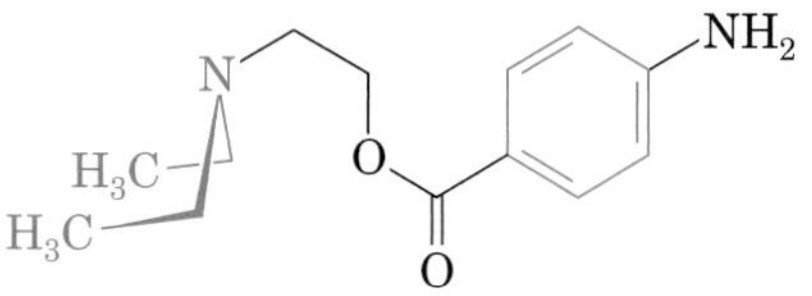 **Novocain**

Lidocaine is prepared by the reaction sequence shown. For each step, indicate the type of reaction and draw a mechanism.

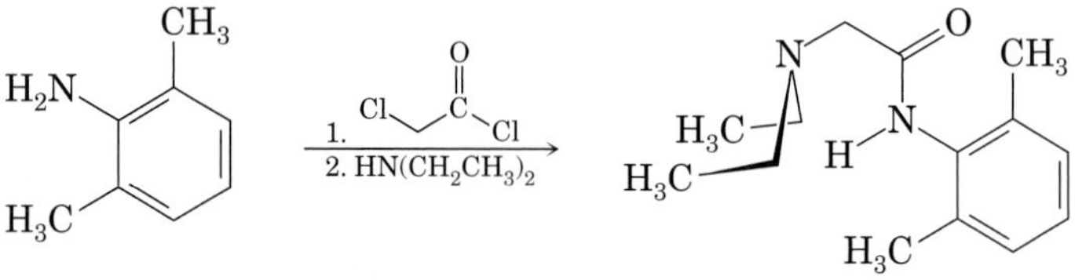

1. $ClCH_2C(=O)Cl$
2. $HN(CH_2CH_3)_2$

Lidocaine

10.75 Consumption of wheat contaminated with certain fungi (that appear like brown kernels) produces hallucinations and a condition called St. Anthony's fire. Long-term exposure causes gangrene due to the potent vasoconstrictive action of a molecule present, a naturally occurring derivative of lysergic acid.

LSD, or lysergic acid diethylamide, is synthesized by the hydrolysis of the amide bond indicated and reaction with diethylamine.

Tom Volk, University of Wisconsin–La Crosse, TomVolkFungi.net

(a) Draw the product of hydrolysis of the indicated bond.
(b) Mixing lysergic acid with diethylamine does not produce LSD. What is the product of this acid–base reaction?
(c) Reaction of lysergic acid with NaOH, followed by acetyl chloride, provides a mixed anhydride. What is the structure of the anhydride?
(d) Reaction of this anhydride with a twofold excess of diethylamine provides LSD. Why is a twofold excess required?

10.76 Derivatives of lysergic acid (Problem 10.75) cause vasoconstriction. By mimicking LSD, a number of vasoconstrictors have been designed and marketed to treat migraines, caused by pressure exerted by dilated blood vessels in the head. Sumatriptan is one example. The structural features common to lysergic acid and the triptan family of vasoconstrictors are highlighted in blue.

Sumatriptan

Sumatriptan contains a sulfonamide group. These groups can be prepared from amines and sulfonyl chlorides. Draw a mechanism for the reaction of CH_3NH_2 with $ArCH_2SO_2Cl$ (Ar signifies an *ar*omatic group).

IN THE FIELD WITH AGROCHEMICALS

10.77 Acetylcholine is a neurotransmitter that is constantly synthesized and hydrolyzed at nerve endings. Poisoning the enzyme acetylcholine esterase, which is responsible for hydrolysis of acetylcholine, kills organisms. A nucleophilic hydroxyl group on the enzyme, abbreviated ROH, reacts with acetylcholine to produce choline and an acetylated enzyme intermediate, $ROCOCH_3$. In a subsequent step, a water molecule hydrolyzes the acetyl group from the enzyme to regenerate ROH and produce acetic acid. Show a mechanism for both steps of this reaction.

ROH = Acetylcholine esterase

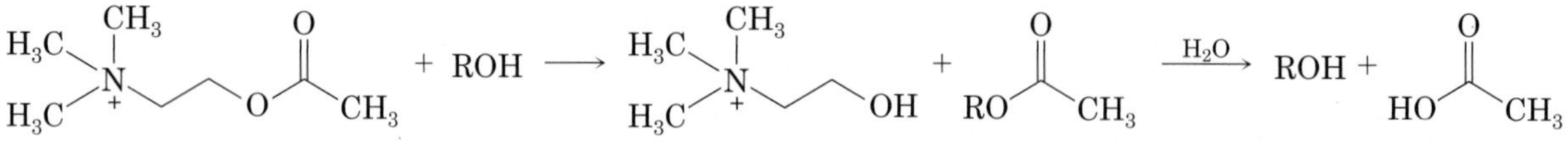

Acetylcholine + ROH (**Enzyme**) ⟶ **Choline** + **Acetylated enzyme** $\xrightarrow{H_2O}$ ROH (**Enzyme**) + **Acetic acid**

10.78 Parathion was one of the first organophosphate insecticides available. The P=S group shows the same chemistry as a C=O group. *p*-Nitrophenoxide is an excellent leaving group. Use a mechanism to show how parathion forms a stable enzyme intermediate analogous to the acetylated enzyme intermediate from Problem 10.77.

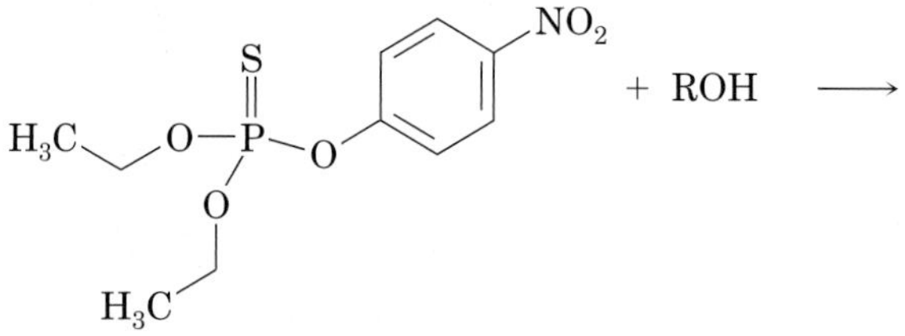

Parathion + ROH ⟶

10.79 Carbaryl, named for its *carb*amate (–NHCOOR) and *ar*omatic functional groups and marketed under the name Sevin, works by poisoning acetylcholine esterase (Problem 10.77). If ArO^- (Ar signifies an *ar*omatic group) is a better leaving group than $-NHCH_3$, propose the structure of the modified acetylcholine esterase intermediate using ROH to represent the enzyme.

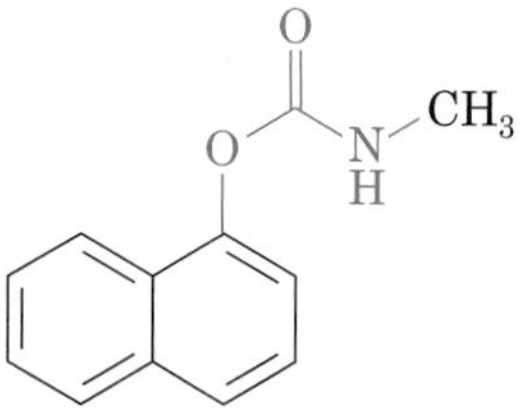

Carbaryl (a carbamate)

10.80 Carbaryl is prepared by reacting an aromatic alcohol nucleophile with methyl isocyanate, $H_3C{-}N{=}C{=}O$. Propose a mechanism for this reaction. Methyl isocyanate is the poisonous gas responsible for the accident that led to the 1984 tragedy in Bhopal, India.

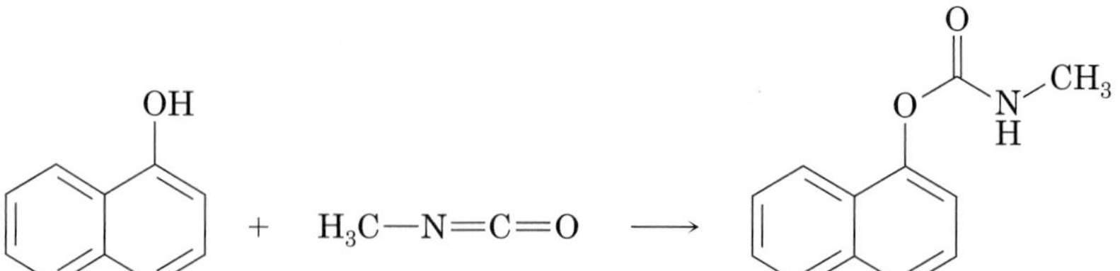

1-Naphthol + $H_3C{-}N{=}C{=}O$ (**Methyl isocyanate**) ⟶ **Carbaryl**

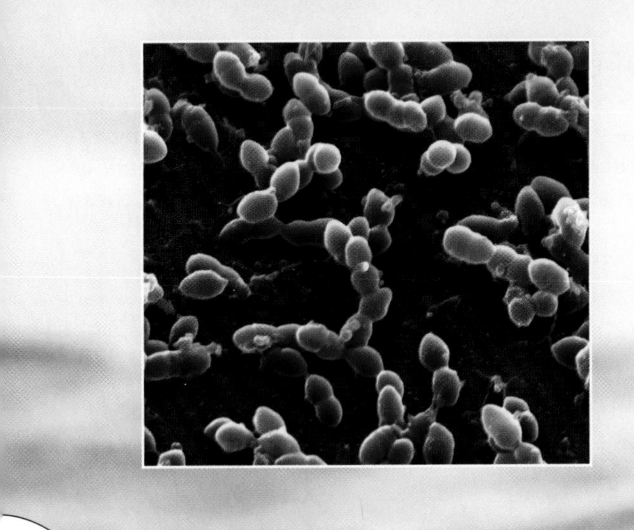

Augmentin, a mixture of a β-lactam antibiotic and a β-lactamase inhibitor, is used to treat bacterial infections. (Eye of Science/Photo Researchers, Inc.)

CHAPTER 11

Carbonyl Alpha-Substitution Reactions and Condensation Reactions

Much of the chemistry of carbonyl compounds can be explained by just four fundamental reactions. We've already looked in detail at two of the four: the nucleophilic addition reaction of aldehydes and ketones (Chapter 9) and the nucleophilic acyl substitution reaction of carboxylic acid derivatives (Chapter 10). In this chapter, we'll look at the remaining two: the *alpha-substitution reaction* and the *carbonyl condensation reaction.*

Alpha-substitution reactions occur at the position *next to* the carbonyl group—the alpha (α) position—and result in the substitution of an α hydrogen atom by some other electrophilic group, E^+:

$$\text{R–C(=O)–C}_\alpha\text{–H} \xrightarrow{E^+} \text{R–C(=O)–C–E} + H^+$$

Carbonyl condensation reactions take place when *two* carbonyl compounds react with each other in such a way that the α carbon of one partner becomes bonded to the carbonyl carbon of the second partner:

$$2\ \text{H–C–C(=O)} \longrightarrow \text{H–C–C(OH)–C–C(=O)}$$

The key feature of both α-substitution reactions and carbonyl condensation reactions is that they take place through the formation of either *enol* or *enolate ion* intermediates. Let's begin our study by learning more about these two species.

An enol   An enolate ion

11.1 Keto–Enol Tautomerism

A carbonyl compound that has a hydrogen atom on its α carbon rapidly interconverts with its corresponding **enol** (*ene* + *ol*; unsaturated alcohol) isomer. This interconversion between what are called the *keto* and *enol* forms is a special kind of isomerism called *tautomerism* (Greek *tauto*, meaning "the same," and *meros*, meaning "part"). The individual isomers are called **tautomers**. The interconversion of tautomers involves the movement of atoms, so they are different from resonance forms. We use equilibrium arrows to indicate the chemical reaction that interconverts tautomers.

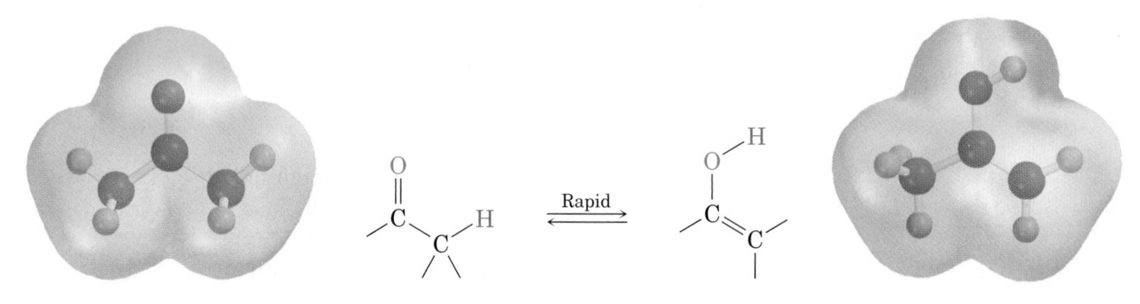

Keto tautomer   **Enol tautomer**

Note that two isomers must interconvert *rapidly* to be considered tautomers. Thus, keto and enol isomers of carbonyl compounds *are* tautomers, but two alkene isomers such as but-1-ene and but-2-ene are not because they don't interconvert rapidly.

Most carbonyl compounds exist almost entirely in the keto form at equilibrium, and it's usually difficult to isolate the pure enol. Cyclohexanone, for example, contains only about 0.000 1% of its enol tautomer at room temperature, and acetone contains only about 0.000 001% enol. The amount of enol tautomer is even less for carboxylic acids, esters, and amides. Even though enols are difficult to isolate and are present to only a small extent at equilibrium, they are nevertheless critically important intermediates in the chemistry of carbonyl compounds.

99.999 9%   0.000 1%

Cyclohexanone

$H_3C-C(=O)-CH_3 \rightleftharpoons H_3C-C(OH)=CH_2$

99.999 999%   0.000 001%

Acetone

Keto–enol tautomerism of carbonyl compounds is catalyzed by both acids and bases. Acid catalysis involves protonation of the carbonyl oxygen atom (a Lewis base) to give an intermediate cation that then loses H^+ from the α carbon to yield the enol (Figure 11.1). This proton loss from the positively charged intermediate is analogous to what occurs during an E1 reaction when a carbocation loses H^+ from the neighboring carbon to form an alkene (see Section 7.8).

FIGURE 11.1 MECHANISM: Mechanism of acid-catalyzed enol formation.

Keto tautomer

Protonation of the carbonyl oxygen atom by an acid catalyst HA yields a cation that can be represented by two resonance structures.

Loss of H^+ from the α position by reaction with a base A^- then yields the enol tautomer and regenerates HA catalyst.

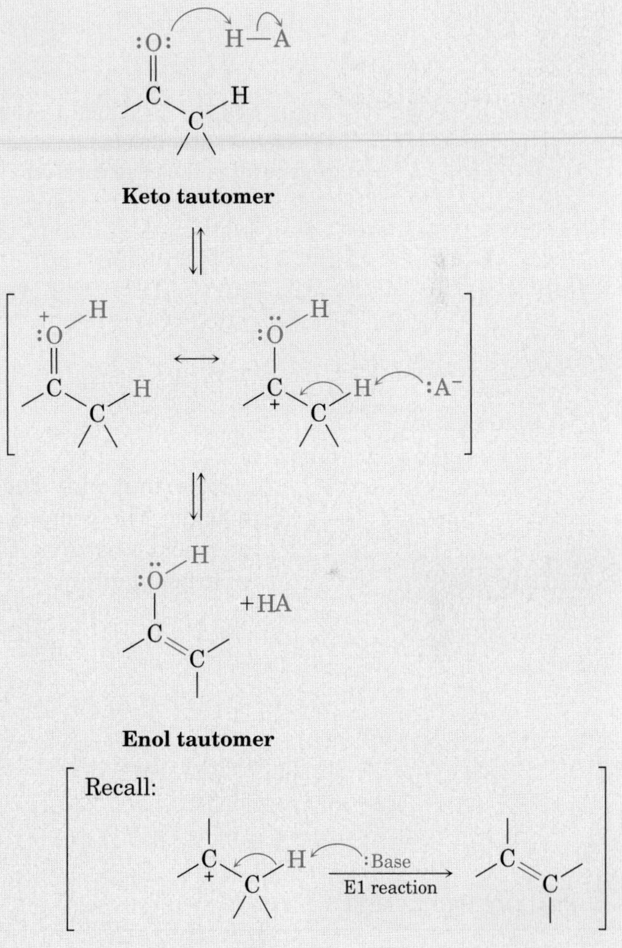

+HA

Enol tautomer

Recall:

:Base

E1 reaction

Base-catalyzed enol formation occurs because the presence of a carbonyl group makes the hydrogens on the α carbon weakly acidic. Thus, a carbonyl compound can act as an acid and donate one of its α hydrogens to the base. The resultant resonance-stabilized anion, an **enolate ion**, is then protonated to yield a neutral compound. If protonation of the enolate ion takes place on the α carbon, the keto tautomer is regenerated and no net change occurs. If, however, protonation takes place on the oxygen atom, then an enol tautomer is formed (Figure 11.2).

FIGURE 11.2 MECHANISM: Mechanism of base-catalyzed enol formation. The intermediate enolate anion, a resonance hybrid of two forms, can be protonated on oxygen to give an enol.

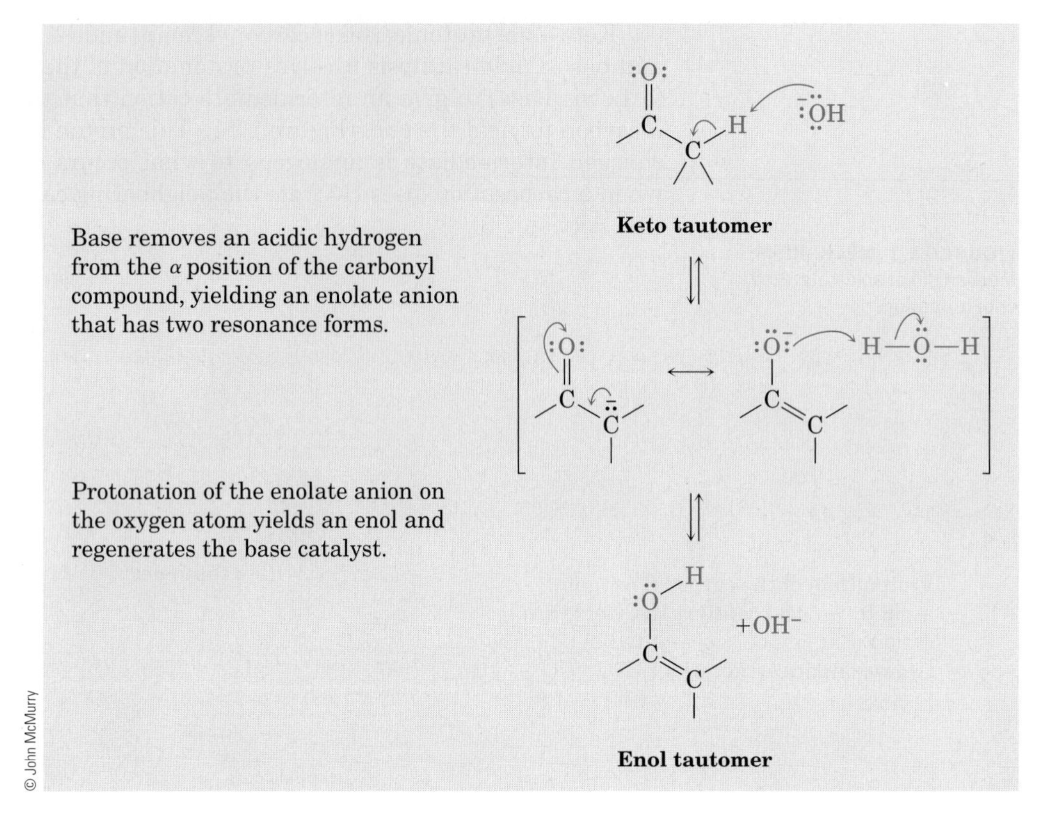

Keto tautomer
Base removes an acidic hydrogen from the α position of the carbonyl compound, yielding an enolate anion that has two resonance forms.
Protonation of the enolate anion on the oxygen atom yields an enol and regenerates the base catalyst.
+OH⁻
Enol tautomer
© John McMurry

Note that only the protons on the α position of carbonyl compounds are acidic. The protons at beta (β), gamma (γ), delta (δ), and other positions aren't acidic because the resulting anions can't be resonance-stabilized by the carbonyl group.

Acidic — Not acidic

$$-\overset{\overset{\displaystyle O}{\|}}{C}-\underset{\alpha}{\overset{\overset{\displaystyle H}{|}}{C}}-\underset{\beta}{\overset{\overset{\displaystyle H}{|}}{C}}-\underset{\gamma}{\overset{\overset{\displaystyle H}{|}}{C}}-\underset{\delta}{\overset{\overset{\displaystyle H}{|}}{C}}-$$

PRACTICE PROBLEM 11.1

Show the structure of the enol tautomer of butanal.

STRATEGY Form the enol by removing a hydrogen from the carbon next to the carbonyl carbon. Then draw a resonance structure that has a double bond between the two carbons, and replace the hydrogen on the carbonyl oxygen.

SOLUTION

Acidic α hydrogens

$$CH_3CH_2-\underset{\underset{\displaystyle H}{|}}{\overset{\overset{\displaystyle H}{|}}{C}}-\overset{\overset{\displaystyle O}{\|}}{C}-H \rightleftharpoons CH_3CH_2-\overset{\overset{\displaystyle H}{|}}{C}=\overset{\overset{\displaystyle OH}{|}}{C}-H$$

Enol tautomer

PROBLEM 11.1 Draw structures for the enol tautomers of the following compounds:

(a) Cyclopentanone (ring C=O)

(b) CH_3CCl (with C=O)

(c) $CH_3COCH_2CH_3$ (with C=O)

(d) CH_3COH (with C=O)

(e) Acetophenone (phenyl ketone, C=O)

PROBLEM 11.2 How many acidic hydrogens does each of the molecules listed in Problem 11.1 have? Identify them.

PROBLEM 11.3 2-Methylcyclohexanone can form two enol tautomers. Show the structures of both.

11.2 Reactivity of Enols: The Mechanism of Alpha-Substitution Reactions

What kind of chemistry might we expect of enols? Since their double bonds are electron-rich, enols behave as nucleophiles and react with electrophiles in much the same way alkenes do (see Section 4.1). Because of electron donation from the oxygen lone-pair electrons, however, enols are even more reactive than alkenes.

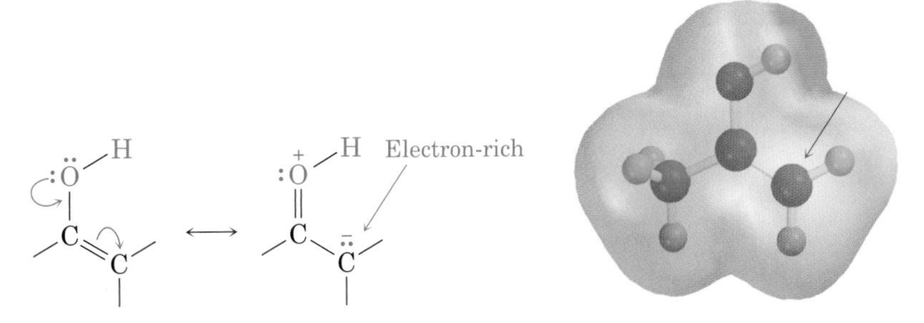

Enol tautomer

When an *alkene* reacts with an electrophile, such as Br_2, addition of Br^+ occurs to give an intermediate carbocation that reacts with Br^- to give the addition product. When an *enol* reacts with an electrophile, however, the addition step is the same but the intermediate cation loses the –OH proton to regenerate a carbonyl compound. The net result of the reaction is α substitution by the mechanism shown in Figure 11.3.

11.3 Alpha Halogenation of Aldehydes and Ketones

Aldehydes and ketones are halogenated at their α positions by reaction with Cl_2, Br_2, or I_2 in acidic solution. Bromine is most often used, and acetic acid is often employed as solvent.

$$\text{H–C–C(=O)} + X_2 \xrightarrow{CH_3COOH} \text{X–C–C(=O)} + HX$$

X = Cl, Br, I

What evidence do we have for this mechanism? Mixing an aldehyde or ketone in *isotopically labeled* acid, D_3O^+, reveals that the proton at the α-position can undergo exchange. As an isotope of hydrogen, deuterium has

FIGURE 11.3 MECHANISM: The general mechanism of a carbonyl α-substitution reaction with an electrophile, E^+.

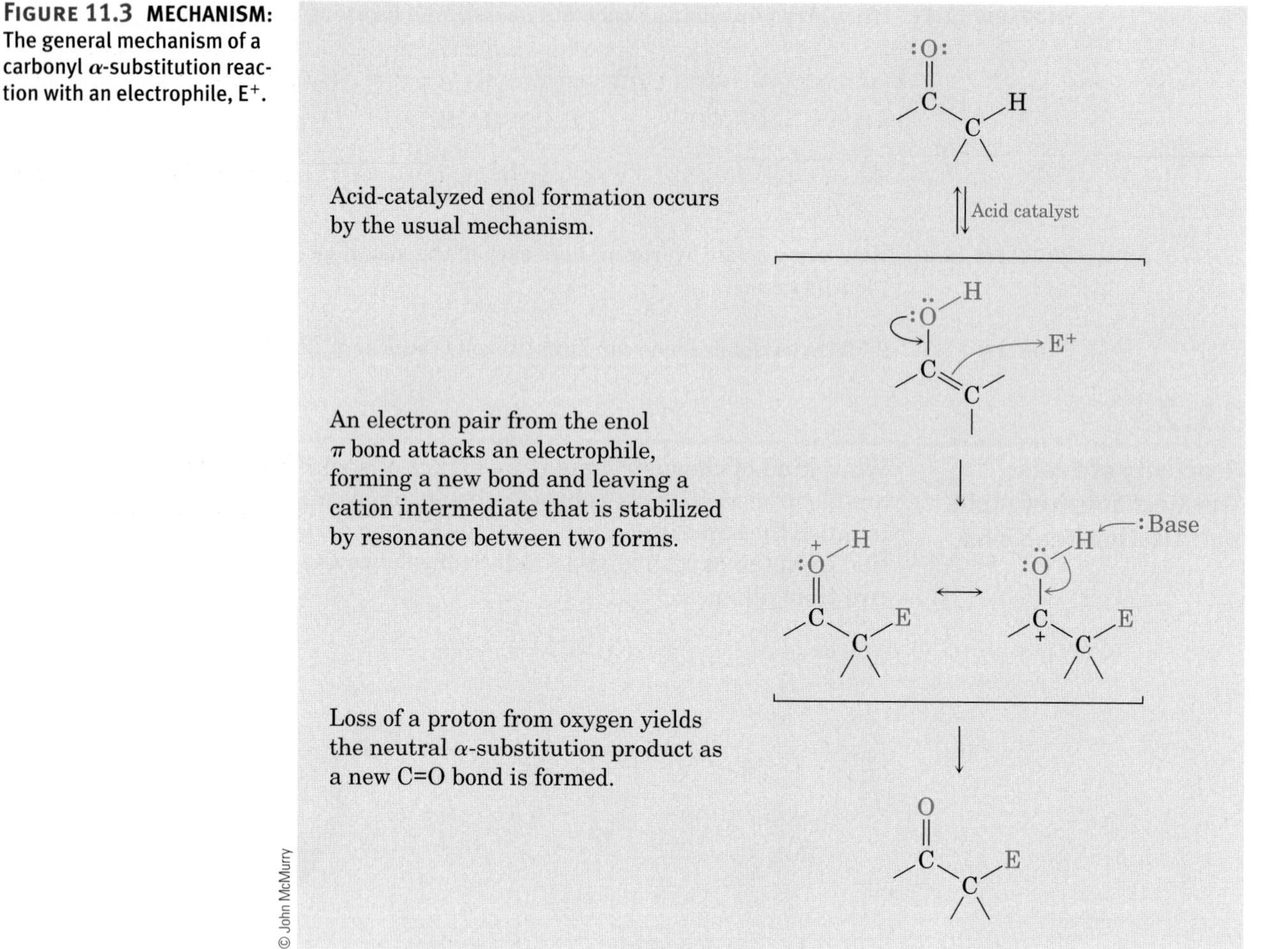
Acid-catalyzed enol formation occurs by the usual mechanism.
Acid catalyst
An electron pair from the enol π bond attacks an electrophile, forming a new bond and leaving a cation intermediate that is stabilized by resonance between two forms.
E+
:Base
Loss of a proton from oxygen yields the neutral α-substitution product as a new C=O bond is formed.

very similar chemical properties. Deuterium differs from hydrogen in mass, however, making it experimentally detectable and making deuterium–hydrogen exchange a useful tool for probing mechanism.

D_3O^+ ⇌ D_3O^+

α-Bromo ketones are useful because they undergo elimination of HBr on treatment with base to yield α,β-unsaturated ketones. For example, 2-bromo-2-methylcyclohexanone gives 2-methylcyclohex-2-enone when heated in the organic base pyridine. The reaction takes place by an E2 elimination pathway (see Section 7.7) and is a good way to introduce a C=C bond into a molecule.

$\xrightarrow[CH_3COOH]{Br_2}$ $\xrightarrow{Pyridine}$ + HBr

2-Methylcyclohexanone **2-Bromo-2-methylcyclohexanone** **2-Methylcyclohex-2-en-1-one (62%)**

PRACTICE PROBLEM 11.2

What product would you obtain from the reaction of cyclopentanone with Br_2 in acetic acid?

STRATEGY Locate the α hydrogens in the starting ketone and replace one of them by Br to carry out an α-substitution reaction.

SOLUTION

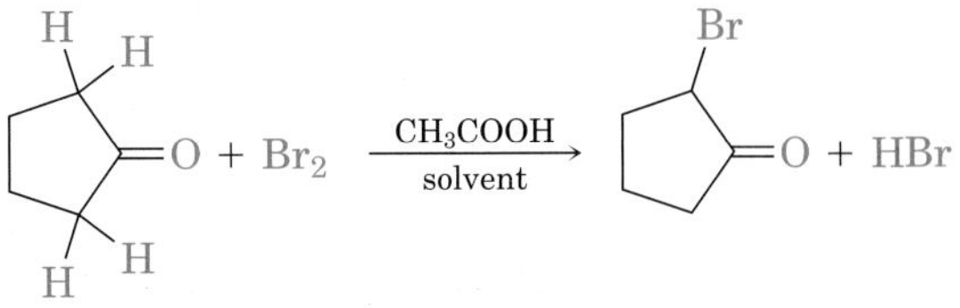

$$\text{Cyclopentanone} + Br_2 \xrightarrow[\text{solvent}]{CH_3COOH} \text{2-Bromocyclopentanone} + HBr$$

Cyclopentanone **2-Bromocyclopentanone**

PROBLEM 11.4 Show the products of the following reactions:

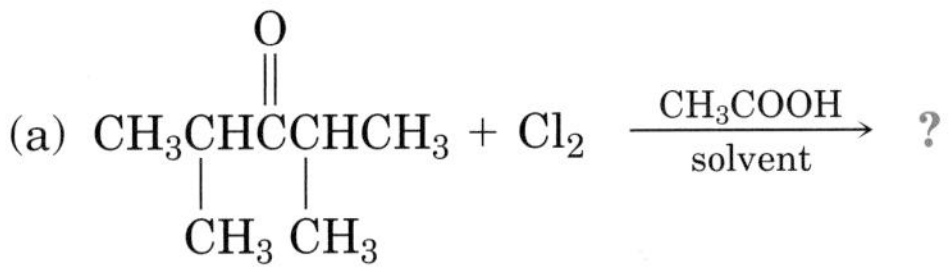

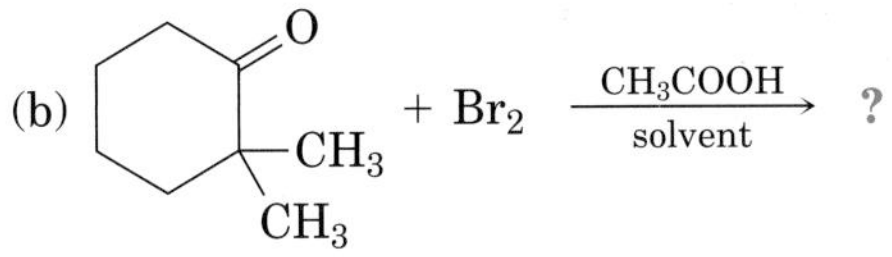

(a) $CH_3CH(CH_3)C(=O)CH(CH_3)CH_3 + Cl_2 \xrightarrow[\text{solvent}]{CH_3COOH}$?

(b) 2,2-Dimethylcyclohexanone $+ Br_2 \xrightarrow[\text{solvent}]{CH_3COOH}$?

PROBLEM 11.5 Show how you might prepare pent-1-en-3-one from pentan-3-one:

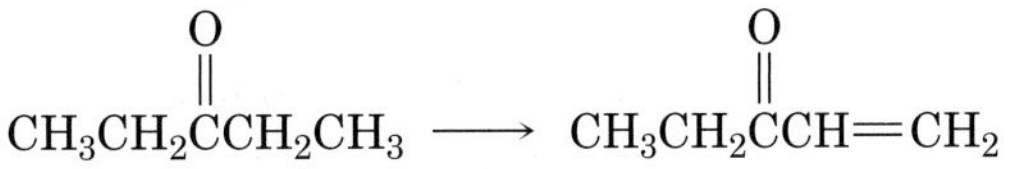

$$CH_3CH_2C(=O)CH_2CH_3 \longrightarrow CH_3CH_2C(=O)CH{=}CH_2$$

Pentan-3-one **Pent-1-en-3-one**

11.4 Acidity of Alpha Hydrogen Atoms: Enolate Ion Formation

During the discussion of base-catalyzed enol formation in Section 11.1, we said that carbonyl compounds are weak acids. Strong bases can abstract an acidic α proton from a carbonyl compound to form a resonance-stabilized enolate ion:

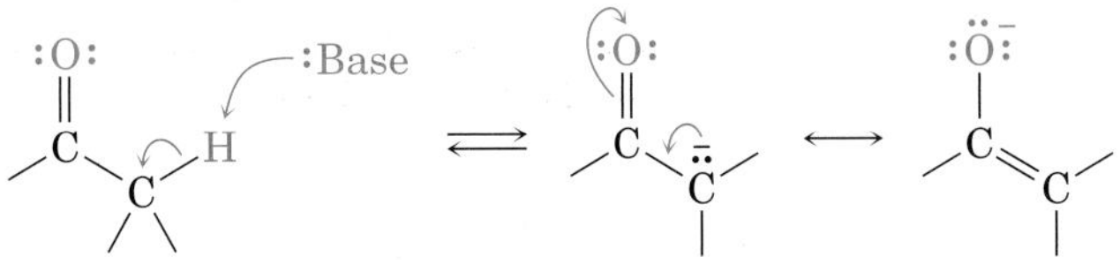

:Base

A carbonyl compound **An enolate ion**

Why are carbonyl compounds weakly acidic? If we compare acetone, $pK_a = 19.3$, with ethane, $pK_a \approx 60$, we find that the presence of the carbonyl group increases the acidity of the neighboring C–H by a factor of 10^{40}.

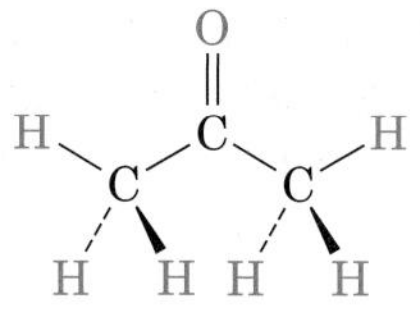

Acetone
(pK_a = 19.3)

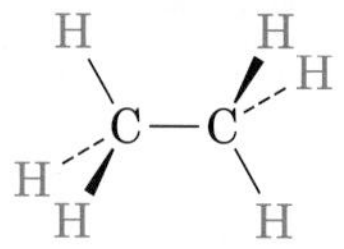

Ethane
($pK_a \approx 60$)

The reason for the acidity of carbonyl compounds can be seen by looking at an orbital picture of an enolate ion (Figure 11.4). Proton abstraction from a carbonyl compound occurs when the α C–H bond is oriented parallel to the *p* orbitals of the carbonyl group. The α carbon of the enolate ion is sp^2-hybridized and has a *p* orbital that overlaps the carbonyl *p* orbitals. Thus, the negative charge is shared by the electronegative oxygen atom, and the enolate ion is stabilized by resonance between two forms.

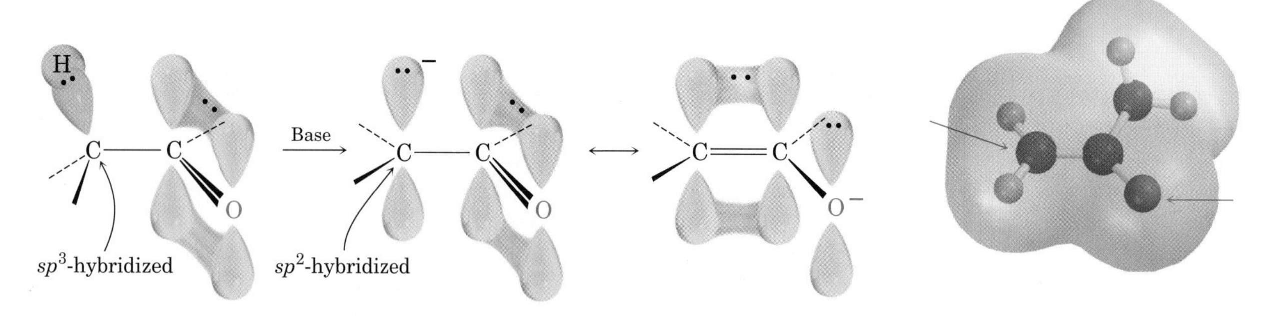

FIGURE 11.4 Mechanism of enolate ion formation by abstraction of an acidic α hydrogen from a carbonyl compound. The enolate ion is stabilized by resonance, and the negative charge is shared by the oxygen and the α carbon, as indicated by the electrostatic potential map.

Carbonyl compounds are more acidic than alkanes for the same reason that carboxylic acids are more acidic than alcohols (Section 10.3). In both cases, the anions are stabilized by resonance. Enolate ions differ from carboxylate ions, though, because their two resonance forms aren't equivalent. The resonance form with the negative charge on the enolate oxygen atom is lower in energy than the form with the charge on carbon.

CH_3CH_3 versus

Ethane
($pK_a \approx 60$)

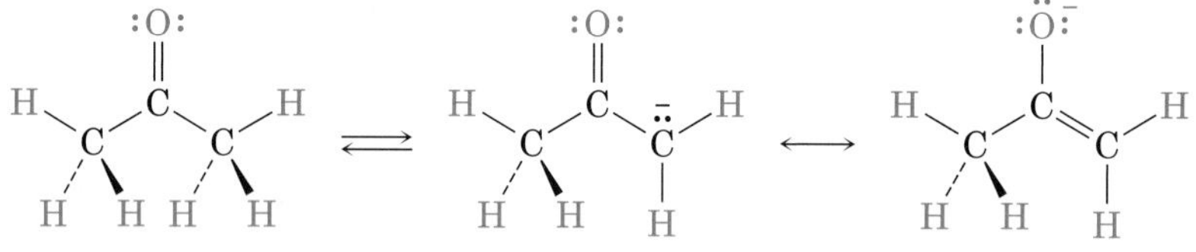

Acetone
(pK_a = 19.3)

Nonequivalent resonance forms

CH_3OH versus

Methanol
(pK_a = 15.5)

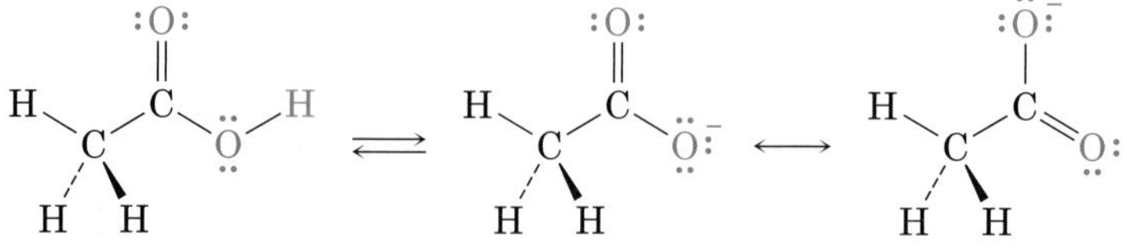

Acetic acid
(pK_a = 4.7)

Equivalent resonance forms

Because α hydrogen atoms of carbonyl compounds are only weakly acidic, strong bases are needed to form enolate ions. If an alkoxide ion, such as sodium ethoxide, is used, ionization of acetone takes place only to the extent of about 0.1% because acetone ($pK_a = 19.3$) is a weaker acid than ethanol ($pK_a = 16$). If, however, a more powerful base such as sodium amide ($NaNH_2$, the sodium salt of ammonia) or sodium hydride (NaH, the sodium salt of H_2) is used, then a carbonyl compound is completely converted into its enolate ion.

Table 11.1 lists the approximate pK_a values of different kinds of carbonyl compounds and shows how these values compare with other common acids.

When a C–H bond is flanked by *two* carbonyl groups, its acidity is enhanced even more. Thus, Table 11.1 shows that 1,3-diketones (called *β-diketones*), 3-keto esters (*β-keto esters*), and 1,3-diesters are more acidic than water. The enolate ions derived from these β-dicarbonyl compounds are stabilized by sharing of the negative charge onto *two* neighboring carbonyl

TABLE 11.1 Acidity Constants for Some Organic Compounds

Compound type	Compound	pK_a	
Carboxylic acid	CH_3COOH	5	Stronger acid
1,3-Diketone	$CH_2(COCH_3)_2$	9	↑
3-Keto ester	$CH_3COCH_2CO_2CH_3$	11	
1,3-Diester	$CH_2(CO_2CH_3)_2$	13	
Water	HOH	15.74	
Primary alcohol	CH_3CH_2OH	16	
Acid chloride	CH_3COCl	16	
Aldehyde	CH_3CHO	17	
Ketone	CH_3COCH_3	19	
Ester	$CH_3CO_2CH_3$	25	
Nitrile	CH_3CN	25	
Dialkylamide	$CH_3CON(CH_3)_2$	30	
Ammonia	NH_3	35	Weaker acid

oxygens. For example, there are three resonance forms for the enolate ion from pentane-2,4-dione:

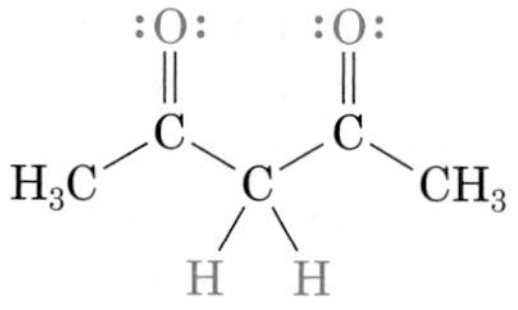

Pentane-2,4-dione (pK_a = 9)

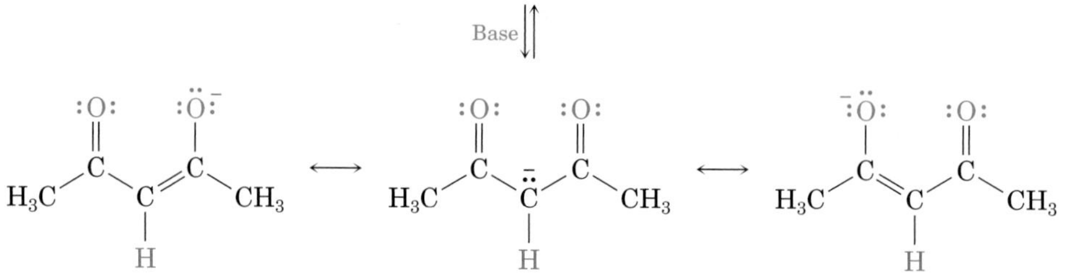

Base

Practice Problem 11.3

Draw structures of the two enolate ions you could obtain by deprotonation of 3-methylcyclohexanone.

Strategy Locate the acidic hydrogens, and then remove them one at a time to generate the possible enolate ions. In this case, 3-methylcyclohexanone can be deprotonated either at C2 or at C6.

Solution

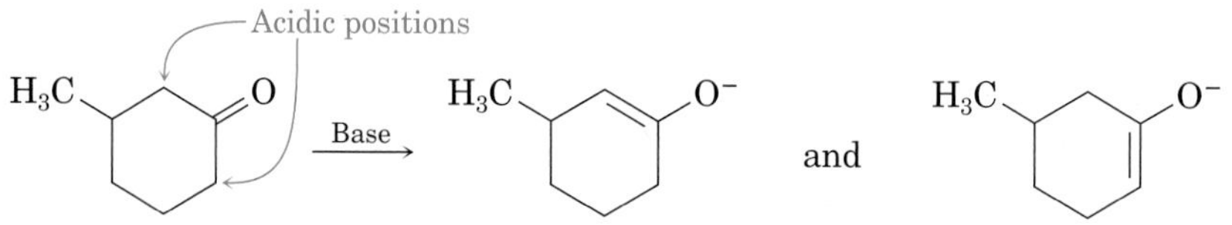

Acidic positions

Base

and

Problem 11.6 Identify all acidic hydrogens in the following molecules:

(a) CH_3CH_2CHO (b) $(CH_3)_3CCOCH_3$ (c) CH_3CO_2H
(d) $CH_3CH_2CH_2C{\equiv}N$ (e) Cyclohexane-1,3-dione

Problem 11.7 Show the enolate ions you would obtain by deprotonation of the following carbonyl compounds:

(a) 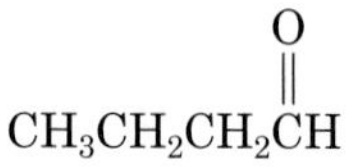 $CH_3CH_2CH_2CH{=}O$ (b) 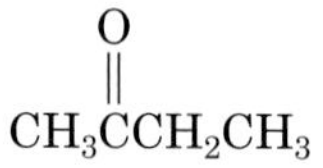 $CH_3C(=O)CH_2CH_3$ (c) 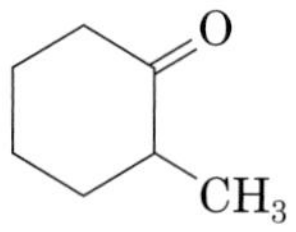 2-methylcyclohexanone

Problem 11.8 Draw three resonance forms for the most stable enolate ion you would obtain by deprotonation of methyl 3-oxobutanoate.

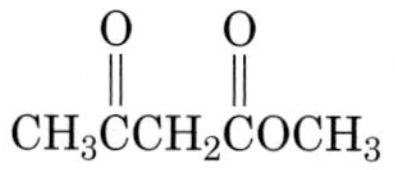

$CH_3C(=O)CH_2C(=O)OCH_3$ **Methyl 3-oxobutanoate**

11.5 Reactivity of Enolate Ions

Enolate ions are more useful than enols for two reasons. First, pure enols normally can't be isolated but are instead generated only as transient intermediates in low concentration. By contrast, stable solutions of pure enolate ions are easily prepared from many carbonyl compounds by treatment with a strong base. Second, enolate ions are more reactive than enols. Whereas enols are neutral, enolate ions have a negative charge that makes them much better nucleophiles. Thus, the α position of an enolate ion is electron-rich and highly reactive toward electrophiles.

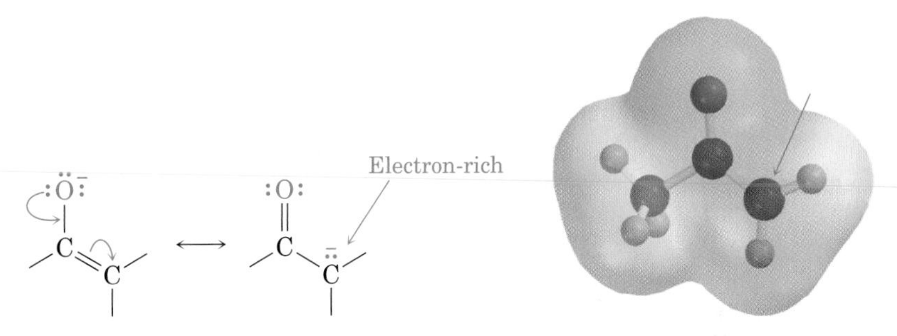

An enolate ion

As resonance hybrids, enolate ions can be thought of either as α-keto carbanions ($^-$C—C=O) or as vinylic alkoxides (C=C—O$^-$). Thus, enolate ions can react with electrophiles either on oxygen or on carbon. Reaction on oxygen yields an enol derivative, whereas reaction on carbon yields an α-substituted carbonyl compound (Figure 11.5). Both kinds of reactivity are known, but reaction on carbon is more common.

FIGURE 11.5 Two modes of enolate ion reactivity. Reaction on carbon to yield an α-substituted carbonyl product is more common.

Reaction here OR Reaction here

α-Keto carbanion **Vinylic alkoxide**

E^+ E^+

An α-substituted carbonyl compound **An enol derivative**

11.6 Alkylation of Enolate Ions

Perhaps the most useful reaction of enolate ions is their **alkylation** by treatment with an alkyl halide. The reaction forms a new C–C bond, thereby joining two smaller pieces into one larger molecule. Alkylation occurs when the nucleophilic enolate ion reacts with the electrophilic alkyl halide in an S_N2 reaction, displacing the halide ion in the usual way.

S_N2 reaction

Enolate ion **Alkyl halide**

Like many S_N2 reactions (see Section 7.5), alkylations are successful only when a primary alkyl halide (RCH_2X) or methyl halide (CH_3X) is used, because a competing E2 elimination occurs if a secondary or tertiary halide is used. The leaving group X can be chloride, bromide, or iodide.

One of the best-known carbonyl alkylation reactions, a process called the **malonic ester synthesis**, is an excellent method for preparing a substituted acetic acid from an alkyl halide:

$$R{-}X \xrightarrow[\text{ester synthesis}]{\text{Via malonic}} R{-}CH_2\overset{O}{\overset{\|}{C}}OH$$

Alkyl halide α-Substituted acetic acid

Diethyl propanedioate, commonly called diethyl malonate or *malonic ester*, is relatively acidic ($pK_a = 13$) because its α hydrogen atoms are flanked by two carbonyl groups. Thus, malonic ester is easily converted into its enolate ion by reaction with sodium ethoxide in ethanol. The enolate ion, in turn, is readily alkylated by treatment with an alkyl halide, yielding an α-substituted malonic ester. Note in the following examples that the abbreviation "Et" is used for an ethyl group, $—CH_2CH_3$.

$$H{-}C(CO_2Et)(H){-}CO_2Et \xrightarrow[\text{EtOH}]{Na^+\ ^-OEt} \left[Na^+\ ^-{:}C(CO_2Et)(H){-}CO_2Et\right] \xrightarrow{RX} R{-}C(CO_2Et)(H){-}CO_2Et$$

Diethyl propanedioate (malonic ester) **Sodio malonic ester** **An alkylated malonic ester**

The product of a malonic ester alkylation has one acidic α hydrogen remaining, so the alkylation process can be repeated a second time to yield a dialkylated malonic ester:

$$R{-}C(CO_2Et)(H){-}CO_2Et \xrightarrow[\text{2. R'X}]{\text{1. } Na^+\ ^-OEt} R{-}C(CO_2Et)(R'){-}CO_2Et$$

An alkylated malonic ester **A dialkylated malonic ester**

On heating with aqueous hydrochloric acid, the alkylated (or dialkylated) malonic ester undergoes hydrolysis and **decarboxylation** (loss of CO_2) to yield a substituted monocarboxylic acid. Note that decarboxylation is not a general reaction of carboxylic acids but is a unique feature of compounds like malonic acids that have a second carbonyl group two atoms away from the $—CO_2H$.

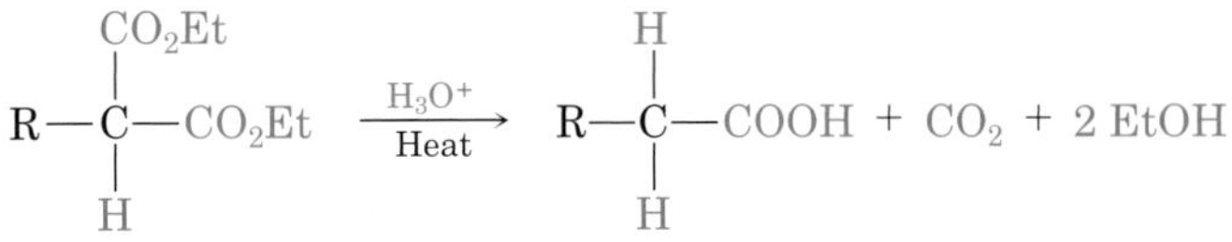

$$R—C(H)(CO_2Et)—CO_2Et \xrightarrow[\text{Heat}]{H_3O^+} R—CH_2—COOH + CO_2 + 2\ EtOH$$

The overall result of the malonic ester synthesis is to convert an alkyl halide into a carboxylic acid and to lengthen the carbon chain by two atoms ($RX \rightarrow RCH_2CO_2H$).

$$CH_3CH_2CH_2CH_2Br + Na^+\ ^-{:}CH(CO_2Et)_2 \longrightarrow CH_3CH_2CH_2CH_2CH(CO_2Et)_2$$

1-Bromobutane **Sodio diethylmalonate** **(84%)**

$$\downarrow H_3O^+$$

$$CH_3CH_2CH_2CH_2CH_2COH\ (C{=}O)$$

Hexanoic acid (75%)

Practice Problem 11.4

How would you prepare heptanoic acid by a malonic ester synthesis?

Strategy The malonic ester synthesis converts an alkyl halide into a carboxylic acid having two more carbon atoms. Thus, a seven-carbon acid chain must be derived from a five-carbon alkyl halide such as 1-bromopentane.

Solution

$$CH_3CH_2CH_2CH_2CH_2Br + CH_2(CO_2Et)_2 \xrightarrow[2.\ H_3O^+,\ heat]{1.\ Na^+\ ^-OEt} CH_3CH_2CH_2CH_2CH_2CH_2COH\ (C{=}O)$$

Problem 11.9 What alkyl halide would you use to prepare the following compounds by a malonic ester synthesis?

(a) $CH_3CH_2CH_2COH$ (C=O)

(b) $C_6H_5CH_2CH_2COH$ (C=O)

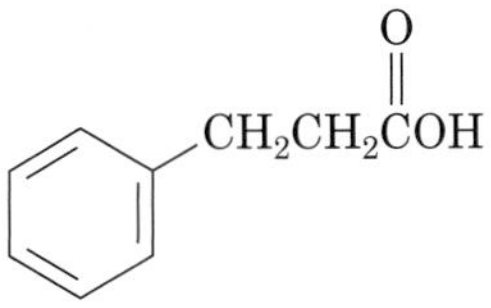

(c) $CH_3CH(CH_3)CH_2CH_2CH_2COH$ (C=O)

Problem 11.10 Show how you could use a malonic ester synthesis to prepare the following compounds:
(a) 4-Methylpentanoic acid (b) 2-Methylpentanoic acid

Problem 11.11 Show how you could use a malonic ester synthesis to prepare the following compound:

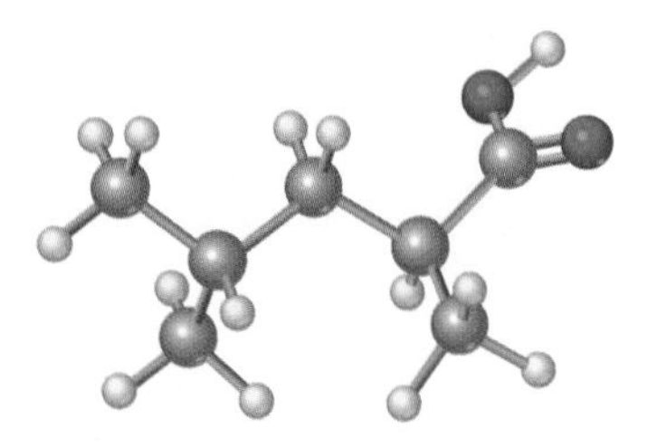

11.7

Enolate Alkylation and the Dawn of Modern Medicine

Although using herbal remedies to treat illness and disease goes back thousands of years, the prescription of chemicals prepared in the laboratory has a much shorter history. The barbiturates, a class of potent sedatives, constitute an early example of medicinal chemistry. The synthesis and medical use of these molecules traces back to the early 1900s and relies on reactions that are now quite familiar—enolate alkylations and the chemistry of esters. Starting with diethyl malonate, enolate alkylations with simple alkyl halides provide a wealth of potential derivatives (Figure 11.6). Conversion to the barbiturate occurs on heating the diester with urea in the presence of a base. While the NH_2 groups of urea are not strong nucleophiles, deprotonation generates a more reactive anion.

Each barbiturate comes as a tablet of regulated size, shape, and color. The street names used when the drugs are trafficked illegally are equally colorful and are often derived from the color of the pill. Although still used today, most of these drugs have been replaced with a safer class of sedative-hypnotics, including diazepam (Valium), compounds with markedly different structures.

11.8

Carbonyl Condensation Reactions

We've seen now that carbonyl compounds can behave as either electrophiles or nucleophiles. In a nucleophilic addition reaction or a nucleophilic acyl substitution reaction, the carbonyl group behaves as an electrophile by accepting electrons from an attacking nucleophile. In an α-substitution reaction, however, the carbonyl compound behaves as a nucleophile after conversion into an enol or enolate ion.

$\delta-$:O: $\delta+$ C :Nu^-

Electrophilic carbonyl group is attacked by nucleophiles.

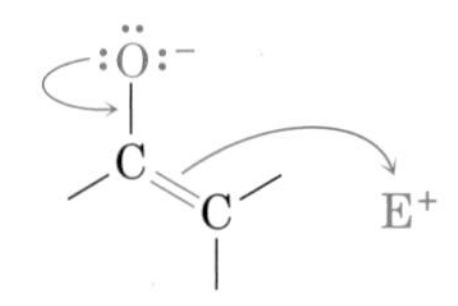

Nucleophilic enolate ion attacks electrophiles.

FIGURE 11.6 The synthesis of barbiturates relies on enolate alkylations and nucleophilic substitution reactions of carbonyl compounds.

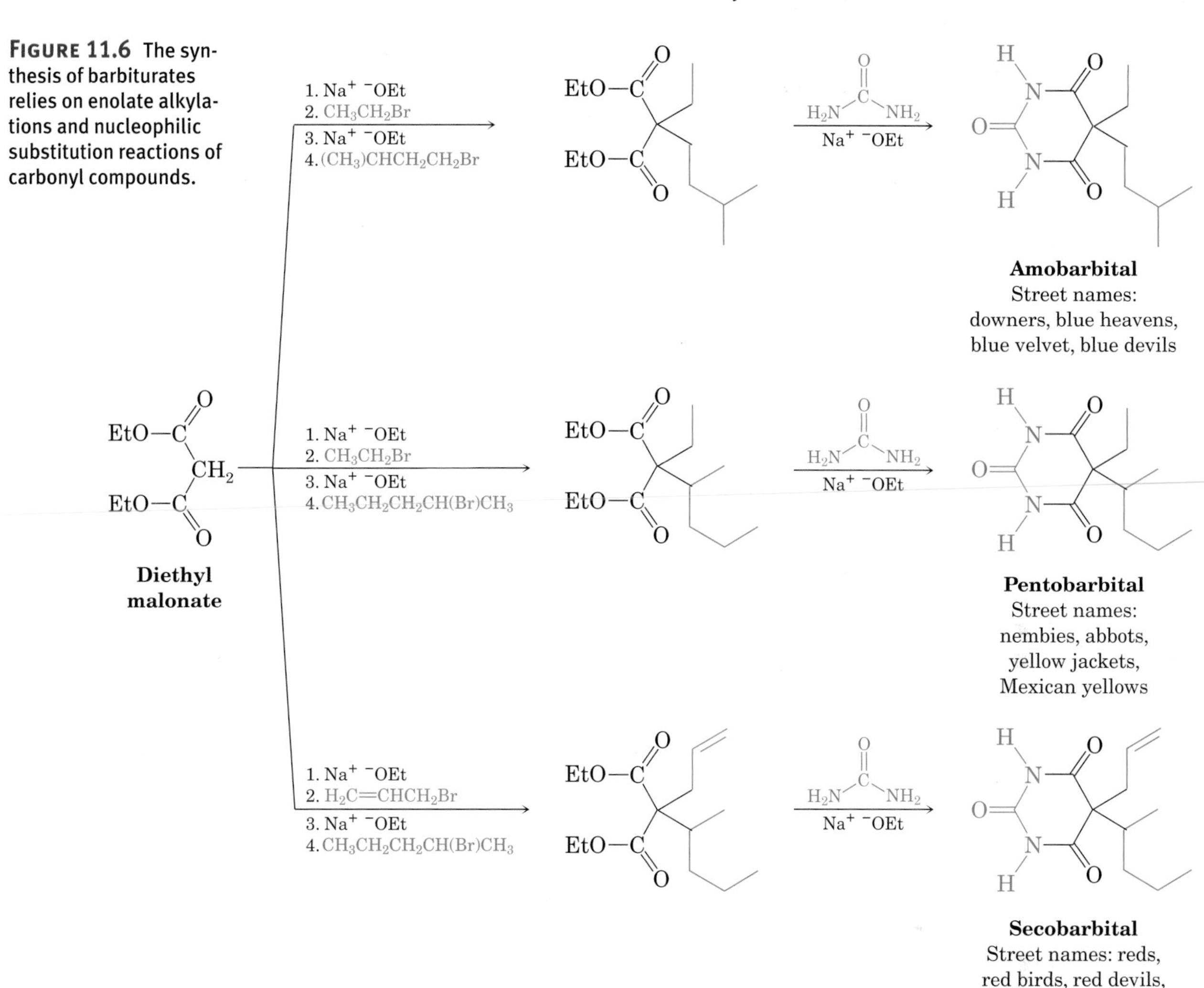

Carbonyl condensation reactions, the fourth and last general category of carbonyl-group reactions we'll study, involve *both* kinds of reactivity. These reactions take place between two carbonyl partners and involve a combination of nucleophilic addition and α-substitution steps. One partner (the nucleophilic *donor*) is converted into an enolate ion and undergoes an α-substitution reaction, while the other partner (the electrophilic *acceptor*) undergoes a nucleophilic addition reaction. There are numerous variations of carbonyl condensation reactions, depending on the two carbonyl partners, but the general mechanism remains the same (Figure 11.7).

11.9 Condensations of Aldehydes and Ketones: The Aldol Reaction

When acetaldehyde is dissolved in an alcohol solvent and treated with a base, such as sodium hydroxide or sodium ethoxide, a rapid and reversible condensation reaction occurs. The product is a β-hydroxy aldehyde product known commonly as *aldol* (*ald*ehyde + alcoh*ol*). Called the **aldol reaction**,

FIGURE 11.7 MECHANISM: The general mechanism of a carbonyl condensation reaction. One partner (the donor) acts as a nucleophile, while the other partner (the acceptor) acts as an electrophile.

One carbonyl partner with an α hydrogen atom is converted by base into its enolate ion.

This enolate ion acts as a nucleophilic donor and adds to the electrophilic carbonyl group of the acceptor partner.

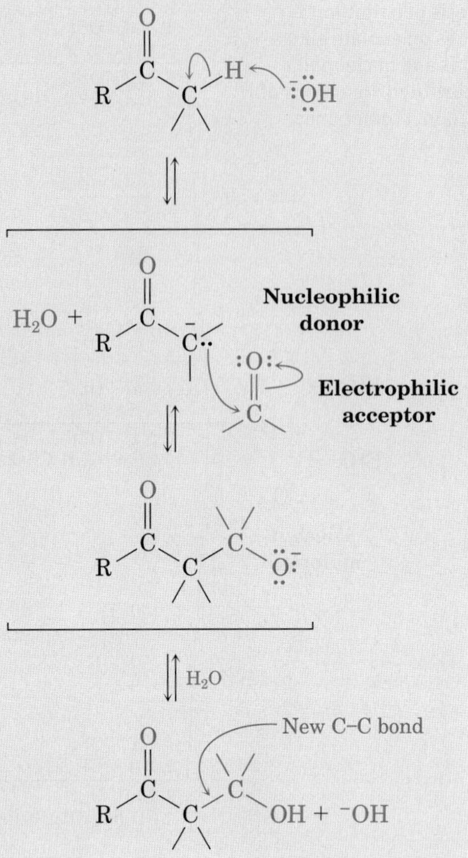

Nucleophilic donor

Electrophilic acceptor

Protonation of the tetrahedral alkoxide ion intermediate gives the neutral condensation product.

New C–C bond

base-catalyzed condensation is a general reaction of all aldehydes and ketones with α hydrogen atoms. If the aldehyde or ketone does not have an α hydrogen atom, aldol condensation can't occur.

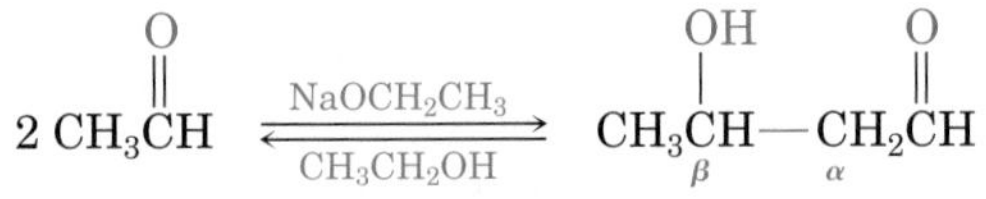

$$2\ CH_3CHO \xrightleftharpoons[CH_3CH_2OH]{NaOCH_2CH_3} CH_3CH(OH)-CH_2CHO$$

Acetaldehyde → **Aldol (a β-hydroxy aldehyde)**

The exact position of the aldol equilibrium depends both on reaction conditions and on substrate structure. The condensation product is generally favored for reaction of monosubstituted acetaldehydes (RCH_2CHO), but the starting material is favored for reaction of disubstituted acetaldehydes (R_2CHCHO) and for ketones.

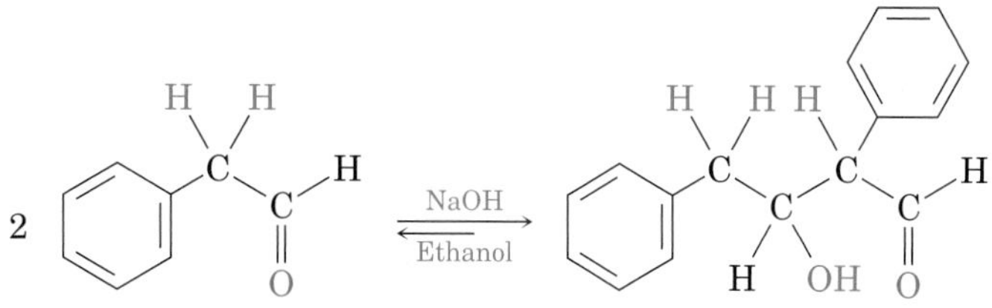

$$2\ C_6H_5CH_2CHO \underset{\text{Ethanol}}{\overset{\text{NaOH}}{\rightleftharpoons}} C_6H_5CH_2CH(OH)CH(C_6H_5)CHO$$

Phenylacetaldehyde 90%

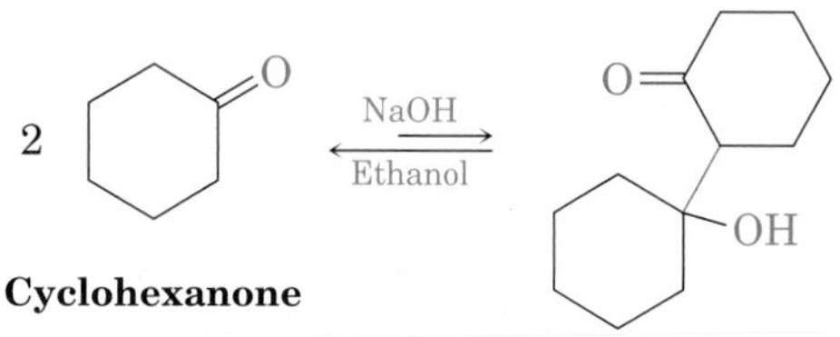

$$2\ \text{Cyclohexanone} \underset{\text{Ethanol}}{\overset{\text{NaOH}}{\rightleftharpoons}} \text{2-(1-hydroxycyclohexyl)cyclohexanone}$$

Cyclohexanone

PRACTICE PROBLEM 11.5

What is the structure of the aldol product derived from propanal?

STRATEGY An aldol reaction combines two molecules of starting material, forming a bond between the α carbon of one partner and the carbonyl carbon of the second partner.

SOLUTION

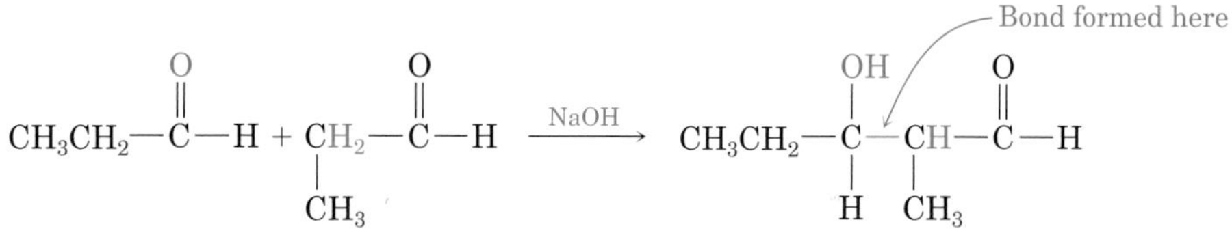

$$CH_3CH_2-\overset{O}{\overset{\|}{C}}-H + \underset{CH_3}{\underset{|}{CH}_2}-\overset{O}{\overset{\|}{C}}-H \xrightarrow{\text{NaOH}} CH_3CH_2-\underset{H}{\underset{|}{\overset{OH}{\overset{|}{C}}}}-\underset{CH_3}{\underset{|}{CH}}-\overset{O}{\overset{\|}{C}}-H$$

Bond formed here

PROBLEM 11.12 Which of the following compounds can undergo the aldol reaction, and which cannot? Explain.

(a) Cyclohexanone (b) Benzaldehyde
(c) 2,2,6,6-Tetramethylcyclohexanone (d) Formaldehyde

PROBLEM 11.13 Show the product of the aldol reaction of the following compounds:

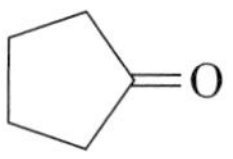

(a)  $CH_3CH_2CH_2\overset{O}{\overset{\|}{C}}H$ (b) Cyclopentanone (c) Acetophenone ($C_6H_5COCH_3$)

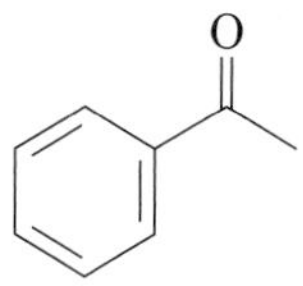

11.10 Dehydration of Aldol Products: Synthesis of Enones

The β-hydroxy ketones and β-hydroxy aldehydes formed in aldol reactions are easily dehydrated to yield conjugated (α,β-unsaturated) **enones** (*ene* + *one*). In fact, it's this loss of water that gives the aldol *condensation* its name, since water condenses out of the reaction.

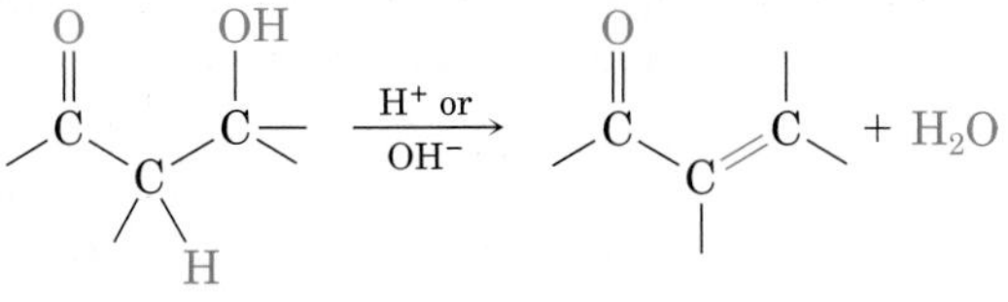

A β-hydroxy ketone or aldehyde $\xrightarrow[OH^-]{H^+ \text{ or}}$ A conjugated enone $+ H_2O$

Most alcohols are resistant to dehydration by dilute acid or base (see Section 8.5), but –OH groups that are two carbons away from a carbonyl group are special. Under *basic* conditions, an α hydrogen is abstracted and the resultant enolate ion expels the nearby OH^- leaving group. Under *acidic* conditions, the –OH group of the enol is protonated and H_2O is then expelled as the leaving group.

Base-catalyzed $\xrightleftharpoons{OH^-}$ [Enolate ion] $\rightleftharpoons$ $+ OH^-$

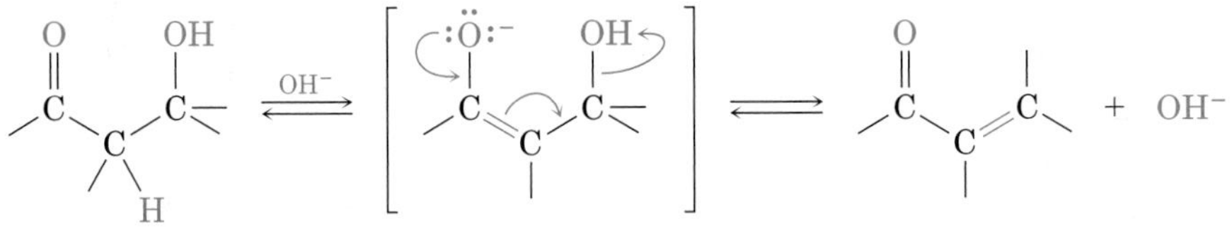

Enolate ion

Acid-catalyzed $\xrightleftharpoons{H^+}$ [Enol] $\rightleftharpoons$ $+ H_3O^+$

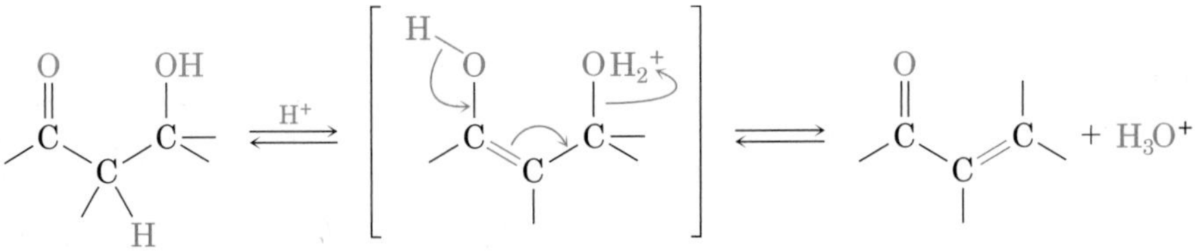

Enol

The reaction conditions needed for aldol dehydration are often only a bit more vigorous (slightly higher temperature, for instance) than the conditions needed for the aldol condensation itself. As a result, conjugated enones are often obtained directly from aldol reactions without ever isolating the intermediate β-hydroxy carbonyl compounds.

PRACTICE PROBLEM 11.6

What is the structure of the enone obtained from aldol condensation of acetaldehyde?

STRATEGY In the aldol reaction, H_2O is eliminated and a double bond is formed by removing two hydrogens from the acidic α position of one partner and the oxygen from the second partner.

SOLUTION

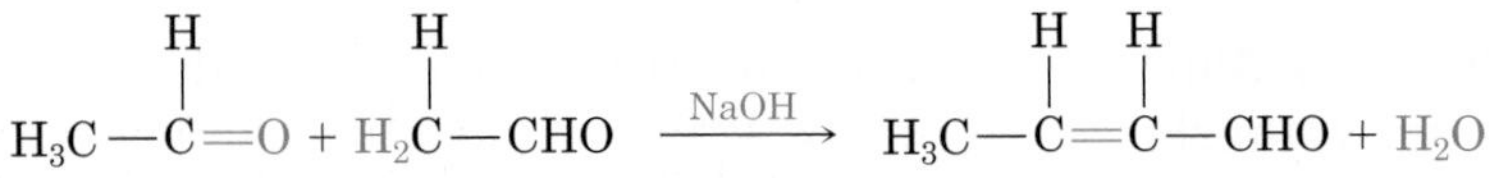

$$H_3C—CH{=}O + H_2CH—CHO \xrightarrow{NaOH} H_3C—CH{=}CH—CHO + H_2O$$

But-2-enal

PROBLEM 11.14 Write the structures of the enone products you would obtain from aldol condensation of the following compounds:

(a) CH_3CCH_3 (C=O)

(b) cyclopentanone (ring =O)

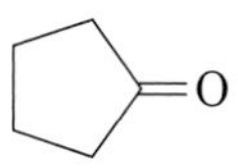

(c) CH_3CH_2CH (C=O)

PROBLEM 11.15 Aldol condensation of butan-2-one leads to a mixture of two enones (ignoring double-bond stereochemistry). Draw them.

PROBLEM 11.16 From what aldehyde or ketone was the following molecule prepared by aldol reaction?

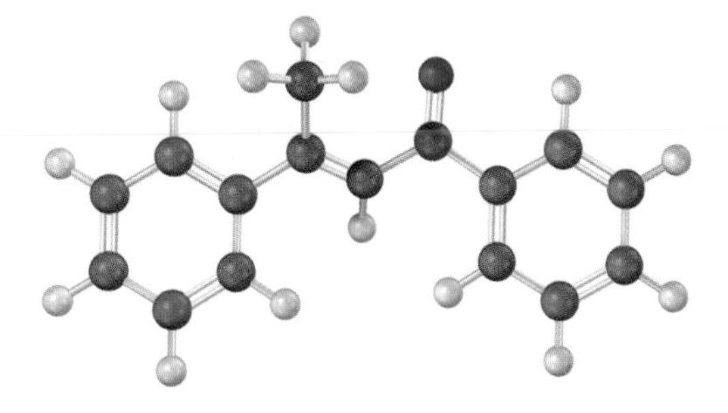

11.11 Condensations of Esters: The Claisen Condensation Reaction

Esters, like aldehydes and ketones, are weakly acidic. When an ester with an α hydrogen is treated with 1 equivalent of a base such as sodium ethoxide, a reversible condensation reaction occurs to yield a β-keto ester product. For example, ethyl acetate yields ethyl acetoacetate on treatment with base. This reaction between two ester components is known as the **Claisen condensation reaction**.

$$2\ CH_3\overset{O}{\overset{\|}{C}}OCH_2CH_3 \xrightarrow[2.\ H_3O^+]{1.\ Na^+\ ^-OEt,\ ethanol} CH_3\overset{O}{\overset{\|}{C}}-CH_2\overset{O}{\overset{\|}{C}}OCH_2CH_3 + CH_3CH_2OH$$

(β and α labels under the product carbons)

Ethyl acetate → **Ethyl acetoacetate a β-keto ester (75%)**

The mechanism of the Claisen reaction is similar to that of the aldol reaction. As shown in Figure 11.8, the reaction involves the nucleophilic addition of an ester enolate ion to the carbonyl group of a second ester molecule. From the point of view of the enolate ion partner, the Claisen condensation is simply an α-substitution reaction. From the point of view of the other partner, the Claisen condensation is a nucleophilic acyl substitution reaction.

The only difference between an aldol condensation and a Claisen condensation involves the fate of the initially formed tetrahedral intermediate. The tetrahedral intermediate in the aldol reaction is protonated to give a stable alcohol—exactly the behavior previously seen for aldehydes and ketones (see Section 9.5). The tetrahedral intermediate in the Claisen reaction, however, expels a leaving group to yield an acyl substitution product—exactly the behavior previously seen for esters (see Section 10.5).

FIGURE 11.8 MECHANISM: The mechanism of the Claisen condensation reaction.

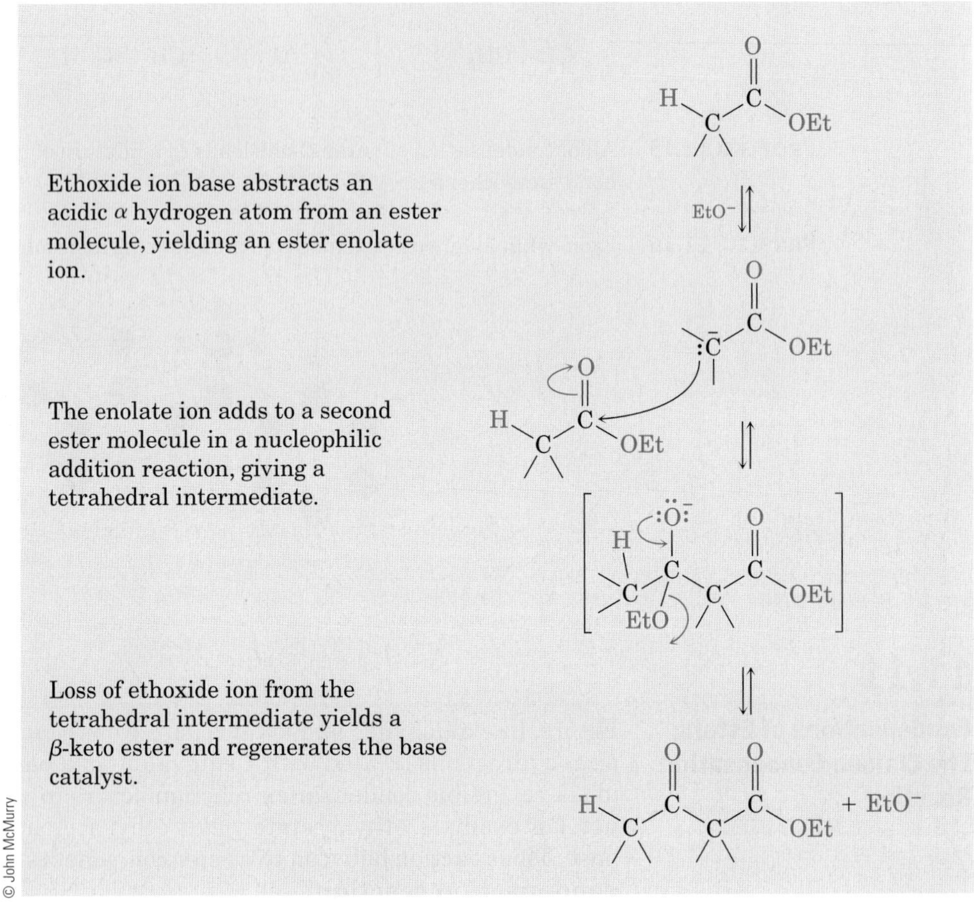

Ethoxide ion base abstracts an acidic α hydrogen atom from an ester molecule, yielding an ester enolate ion.

The enolate ion adds to a second ester molecule in a nucleophilic addition reaction, giving a tetrahedral intermediate.

Loss of ethoxide ion from the tetrahedral intermediate yields a β-keto ester and regenerates the base catalyst.

PRACTICE PROBLEM 11.7

What product would you obtain from Claisen condensation of methyl propanoate?

STRATEGY The Claisen condensation of an ester results in the loss of one molecule of alcohol and the formation of a product in which an acyl group of one partner bonds to the α carbon of the second partner.

SOLUTION

$$CH_3CH_2\overset{O}{\overset{\|}{C}}-OCH_3 + H-\underset{CH_3}{\underset{|}{C}}H\overset{O}{\overset{\|}{C}}OCH_3 \xrightarrow[2.\ H_3O^+]{1.\ NaOCH_3} CH_3CH_2\overset{O}{\overset{\|}{C}}-\underset{CH_3}{\underset{|}{C}}H\overset{O}{\overset{\|}{C}}OCH_3 + CH_3OH$$

2 Methyl propanoate → **Methyl 2-methyl-3-oxopentanoate**

PROBLEM 11.17 Which of the following esters can't undergo Claisen condensation? Explain.

(a) $H\overset{O}{\overset{\|}{C}}OCH_3$ (b) $H_2C{=}CH\overset{O}{\overset{\|}{C}}OCH_3$ (c) $CH_3CH_2\overset{O}{\overset{\|}{C}}OCH_3$

PROBLEM 11.18 Show the products you would obtain by Claisen condensation of the following esters:

(a) 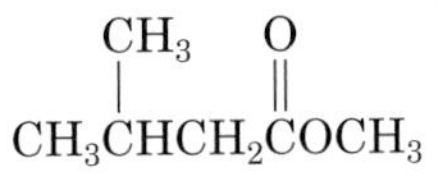 $CH_3CH(CH_3)CH_2COOCH_3$ (CH$_3$ and O above)

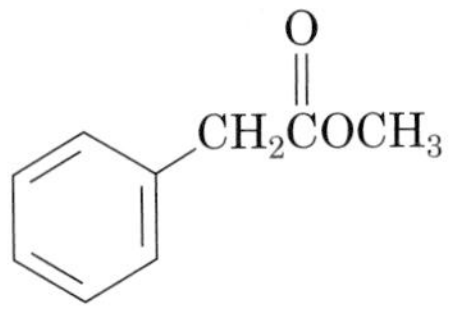

(b) $C_6H_5CH_2COOCH_3$

(c) $C_6H_{11}CH_2COOCH_3$

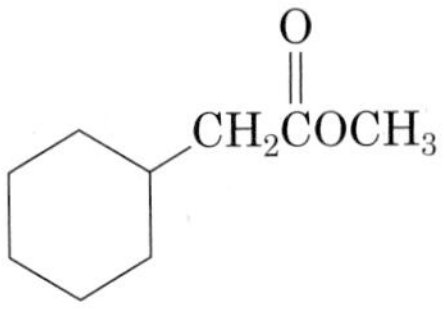

11.12 Biological Carbonyl Reactions

Biochemistry *is* carbonyl chemistry. Almost all metabolic processes used by living organisms involve one or more of the four fundamental carbonyl-group reactions we've seen in the last three chapters. The digestion and metabolic breakdown of all the major classes of food molecules—fats, carbohydrates, and proteins—take place by nucleophilic addition reactions, nucleophilic acyl substitutions, α substitutions, and carbonyl condensations. Similarly, hormones and other crucial biological molecules are built up from smaller precursors by these same carbonyl-group reactions.

Take *glycolysis*, for example, the metabolic pathway by which organisms convert glucose to pyruvate as the first step in extracting energy from carbohydrates:

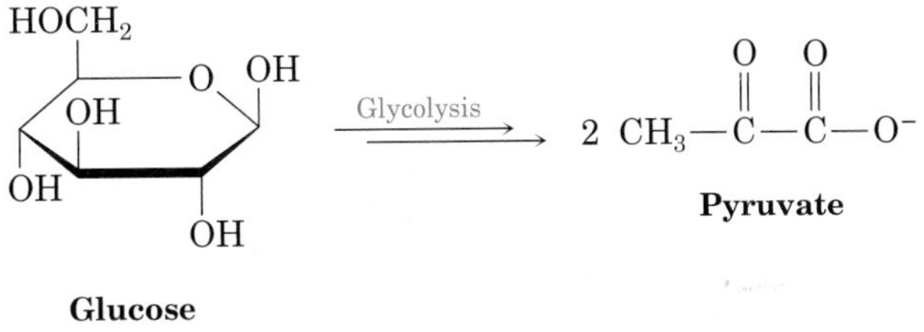

$$\text{Glucose} \xrightarrow{\text{Glycolysis}} 2\ CH_3{-}\overset{O}{\overset{\|}{C}}{-}\overset{O}{\overset{\|}{C}}{-}O^-$$

Glucose　　　　Pyruvate

Glycolysis is a ten-step process that begins with conversion of glucose from its cyclic hemiacetal form to its open-chain aldehyde form—a retro nucleophilic addition reaction. The aldehyde then undergoes tautomerization to yield an enol, which undergoes yet another tautomerization to give the ketone fructose.

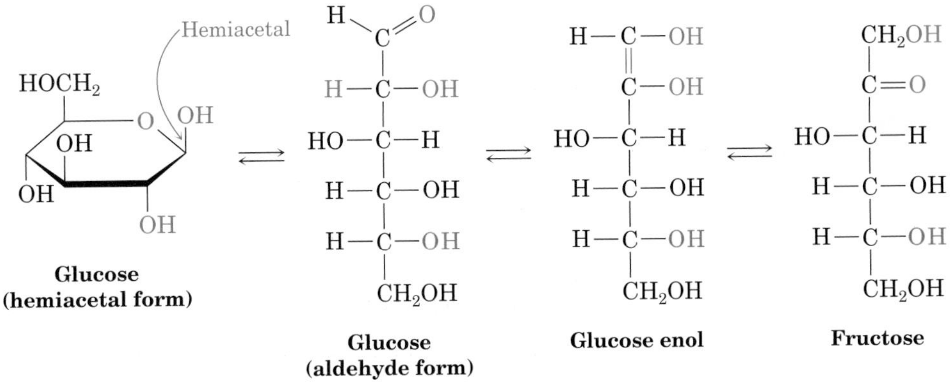

Glucose (hemiacetal form) ⇌ Glucose (aldehyde form) ⇌ Glucose enol ⇌ Fructose

Glucose (aldehyde form)	Glucose enol	Fructose
H–C=O	H–C–OH	CH_2OH
H–C–OH	C–OH	C=O
HO–C–H	HO–C–H	HO–C–H
H–C–OH	H–C–OH	H–C–OH
H–C–OH	H–C–OH	H–C–OH
CH_2OH	CH_2OH	CH_2OH

Hemiacetal

Fructose, a β-hydroxy ketone, is then cleaved into two three-carbon molecules—one ketone and one aldehyde—by a retro aldol reaction. Still further carbonyl-group reactions then occur until pyruvate results.

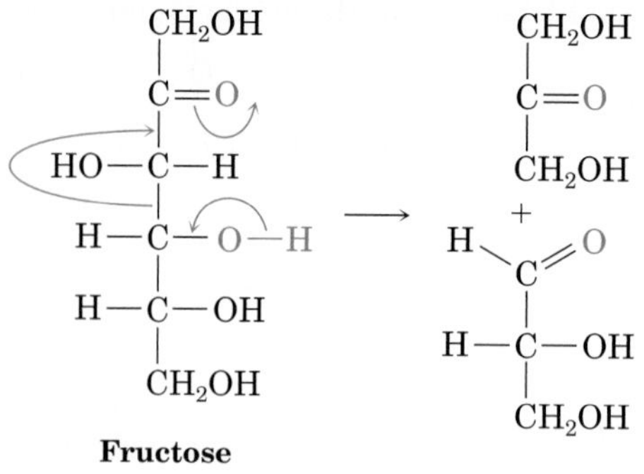

$$\underset{\textbf{Fructose}}{HOCH_2-CO-CH(OH)-CH(OH)-CH(OH)-CH_2OH} \longrightarrow HOCH_2-CO-CH_2OH + H-C(=O)-CH(OH)-CH_2OH$$

As another example of a biological carbonyl reaction, nature uses the two-carbon acetate fragment of acetyl CoA as the major building block for synthesis. Acetyl CoA can act not only as an electrophilic acceptor, being attacked by nucleophiles at the carbonyl group, but also as a nucleophilic donor by loss of its acidic α hydrogen. Once formed, the enolate ion of acetyl CoA can add to another carbonyl group in a condensation reaction. For example, citric acid is biosynthesized by nucleophilic addition of acetyl CoA to the ketone carbonyl group of oxaloacetic acid (2-oxobutanedioic acid) in a kind of mixed aldol reaction.

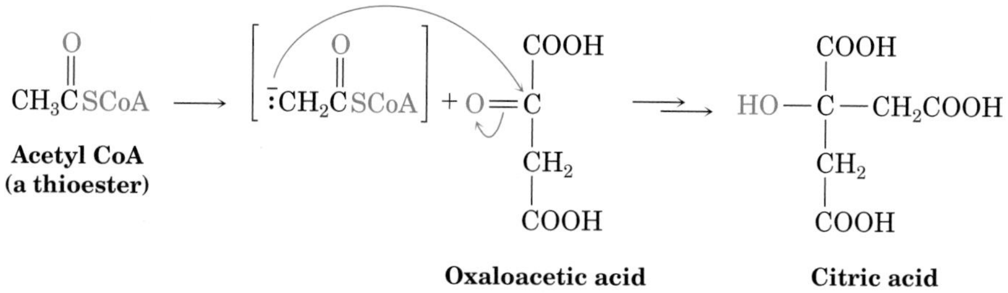

$$\underset{\textbf{Acetyl CoA (a thioester)}}{CH_3\overset{O}{\overset{\|}{C}}SCoA} \longrightarrow \left[:\bar{C}H_2\overset{O}{\overset{\|}{C}}SCoA\right] + \underset{\textbf{Oxaloacetic acid}}{O=C(COOH)CH_2COOH} \longrightarrow \underset{\textbf{Citric acid}}{HO-C(COOH)(CH_2COOH)-CH_2COOH}$$

The few examples just given are only an introduction; we'll look at several of the major metabolic pathways and see many more carbonyl-group reactions in Chapter 17. A good grasp of carbonyl chemistry is crucial to an understanding of biochemistry.

INTERLUDE

β-Lactam Antibiotics—Part II

In Chapter 10 we introduced the story of the β-lactam antibiotics. Penicillin G has saved countless lives since its introduction in the 1940s through its ability to interfere with the action of transpeptidase, the bacterial enzyme critical for cell wall synthesis. Unfortunately, bacteria have evolved resistance strategies, including the production of enzymes called β-lactamases that hydrolyze the critical β-lactam ring, rendering the transpeptidase drugs useless.

Continued

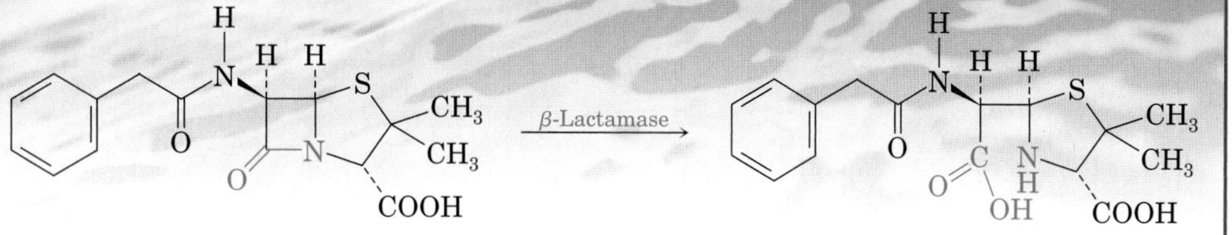

β-Lactamase

Active drug (penicillin G)

Inactive product

Chemistry, however, lets us fight back. One way is to change the structure of the drug slightly, with the expectation that the β-lactamase will no longer recognize it. The acylamino side chain can be varied in the laboratory to provide literally hundreds of penicillin analogs with different biological activity profiles. Ampicillin, for instance, has an extra amine group. Closely related to the penicillins are the *cephalosporins*, a group of β-lactam antibiotics that contain an unsaturated six-membered sulfur-containing ring. Cephalexin, marketed under the trade name Keflex, is one example. Cephalosporins generally have much greater antibacterial activity than do penicillins, particularly against resistant strains of bacteria.

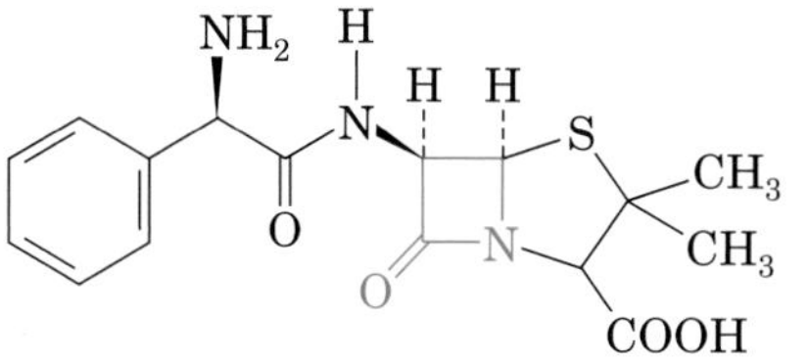

Ampicillin

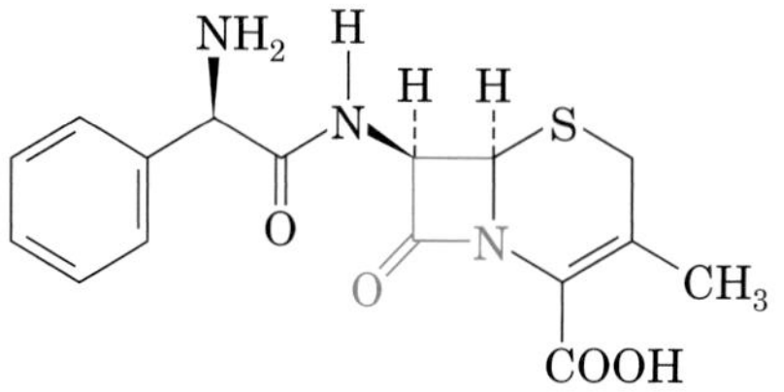

Cephalexin (a cephalosporin)

Another strategy is to protect the transpeptidase drug by adding a second drug that inhibits the β-lactamase enzyme. Clavulanic acid has a β-lactam ring and a shape that is recognized by β-lactamase; however, the sulfur atom of the penicillins is replaced with an oxygen. What special property does the end ether group have?

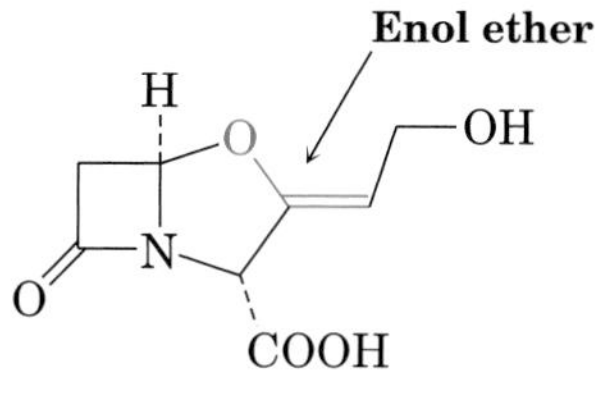

Enol ether

Clavulanic acid

This ether group is part of an enolate leaving group. When the tetrahedral intermediate collapses to cleave the lactam ring, the electrons in the C–N bond migrate to form a stable imine and expel the enolate. Intuition tells us that producing an enolate with the negative charge on oxygen is more favorable than leaving the electrons on the nitrogen atom because oxygen is more electronegative than nitrogen. After protonation, the enolate has accomplished its role, but the story is not over.

Continued

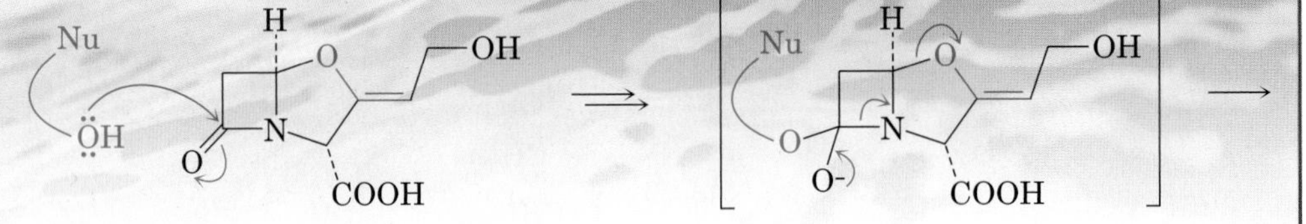

β-Lactamase **Clavulanic acid**

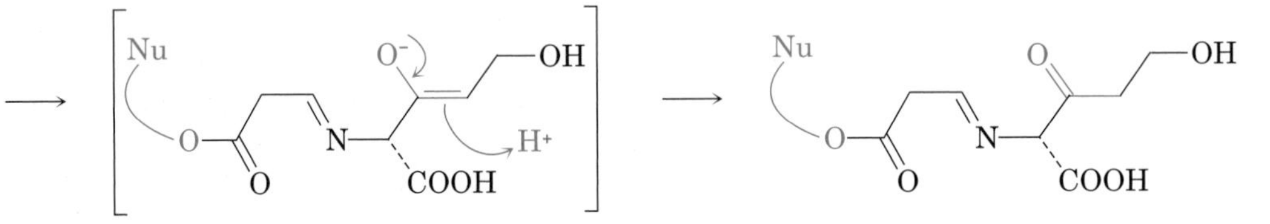

Let's turn our attention back to the imine. The C=N double bond migrates to form a C=C double bond through a simple proton transfer reaction. This C=C bond is special because it is part of an electrophilic enone. It so happens that a second nucleophilic group on the β-lactamase enzyme reacts with the enone by conjugate addition. The covalent drug–enzyme conjugate that results from these reactions blocks the β-lactamase from binding and hydrolyzing other β-lactams, including the drugs that target transpeptidase.

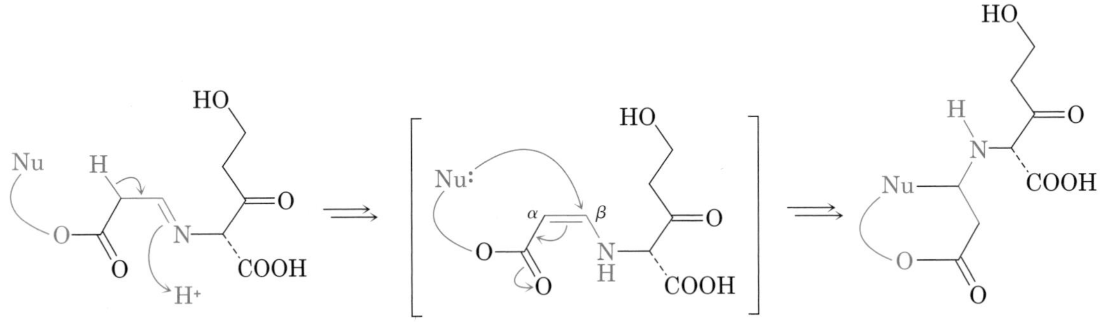

An imine **An enone** **A trapped β-lactamase**

Augmentin is one of the most widely used two-drug mixtures. Augmentin contains both a penicillin-like amoxicillin that targets the transpeptidase responsible for bacterial cell wall production and clavulanic acid that targets β-lactamase. Amazingly, the simple reactions we have been studying are used in drugs to win the day-to-day, life-or-death fight against bacterial infection.

Summary and Key Words

Aldol reaction, p. 357
Alkylation reaction, p. 353
Alpha-substitution reaction, p. 343

Alpha substitutions and **carbonyl condensations** are two of the four fundamental reaction types in carbonyl-group chemistry. Alpha-substitution reactions take place via **enol** or **enolate ion** intermediates and result in the replacement of an α hydrogen atom by another substituent.

Carbonyl compounds are in rapid equilibrium with their enols, a process known as *tautomerism.* Enol **tautomers** are normally present to only a small extent, and pure enols usually can't be isolated. Nevertheless, enols

Carbonyl condensation reaction, p. 343
Claisen condensation reaction, p. 361
Decarboxylation, p. 355
Enol, p. 344
Enolate ion, p. 345
Enone, p. 359
Malonic ester synthesis, p. 354
Tautomers, p. 344

react rapidly with a variety of electrophiles. For example, aldehydes and ketones are halogenated by reaction with Cl_2, Br_2, or I_2 in acetic acid solution.

Alpha hydrogen atoms in carbonyl compounds are acidic and can be abstracted by bases to yield enolate ions. Ketones, aldehydes, esters, amides, and nitriles can all be deprotonated. The most important reaction of enolate ions is their S_N2 **alkylation** by reaction with alkyl halides. The nucleophilic enolate ion attacks an alkyl halide, displacing the leaving halide group and yielding an α-alkylated product. The **malonic ester synthesis**, which involves alkylation of diethyl malonate with an alkyl halide, is a good method for preparing a monoalkylated or dialkylated acetic acid.

A carbonyl condensation reaction takes place between two carbonyl components and involves a combination of nucleophilic addition and α-substitution steps. One carbonyl partner (the donor) is converted into its enolate ion, which then adds to the carbonyl group of the second partner (the acceptor).

The **aldol reaction** is a carbonyl condensation that occurs between two aldehyde or ketone components. Aldol reactions are reversible, leading first to a β-hydroxy ketone and then to an α,β-unsaturated ketone, or **enone**. The **Claisen condensation reaction** is a carbonyl condensation reaction that occurs between two ester components and leads to a β-keto ester product.

Summary of Reactions

1. Halogenation of aldehydes and ketones (Section 11.3)

$$\text{C(=O)–C–H} + X_2 \xrightarrow{CH_3COOH} \text{C(=O)–C–X} \qquad \text{where } X = \text{Cl, Br, or I}$$

2. Malonic ester synthesis (Section 11.6)

$$\text{H–CH(CO}_2\text{R)–CO}_2\text{R} \xrightarrow[\text{2. R'X}]{\text{1. Base}} \text{R'–CH(CO}_2\text{R)–CO}_2\text{R} \xrightarrow[\text{Heat}]{H_3O^+} R'CH_2COOH$$

3. Aldol reaction of aldehydes and ketones (Section 11.9)

$$2\ \text{H–C–C(=O)} \xrightarrow{\text{NaOH}} \text{H–C–C(HO)–C–C(=O)}$$

4. Claisen condensation reaction of esters (Section 11.11)

$$2\ \text{H–C–C(=O)OR} \xrightarrow[\text{2. } H_3O^+]{\text{1. } NaOCH_2CH_3} \text{H–C–C(=O)–C–C(=O)OR}$$

EXERCISES

Visualizing Chemistry

11.19 What aldehydes or ketones might the following enones have been prepared from by aldol reaction?

(a) (b)

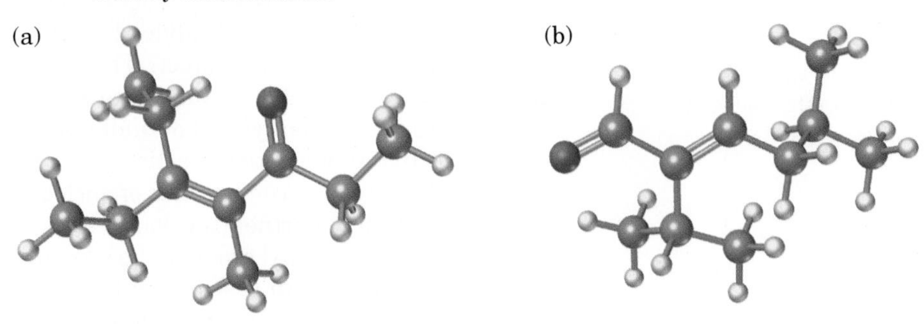

11.20 The following structure represents an intermediate formed by addition of an ester enolate ion to a second ester molecule. Identify the reactant, the leaving group, and the product.

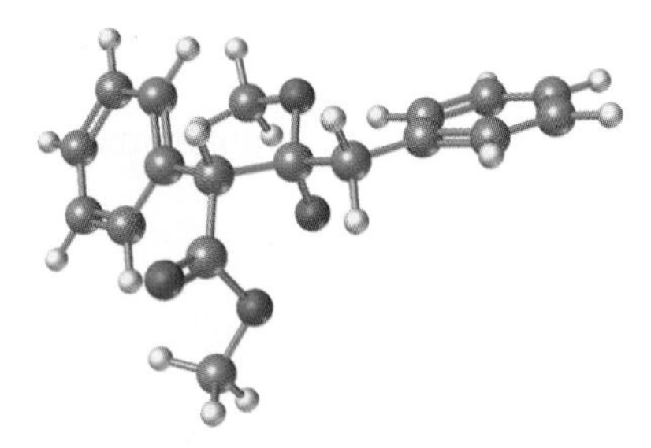

11.21 The following molecule was formed by an *intramolecular* aldol reaction of a *dicarbonyl* compound. Show the structure of the dicarbonyl reactant.

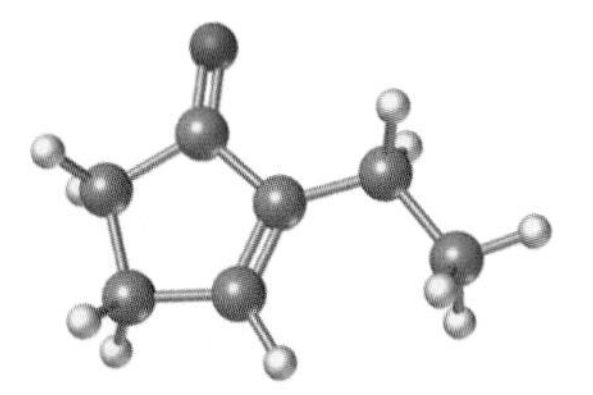

11.22 Show the steps in preparing the following molecule using a malonic ester synthesis:

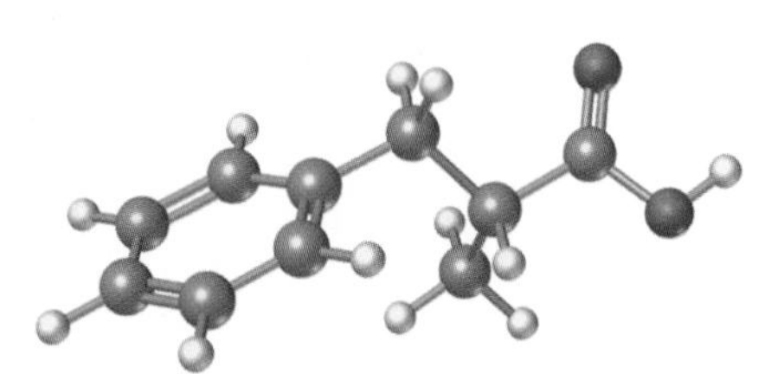

Additional Problems

Acidity, Tautomerism, and Reactivity

11.23 Indicate all acidic hydrogen atoms in the following molecules:

(a) $HOCH_2\overset{O}{\overset{\|}{C}}CH_3$ (b) $HOCH_2CH_2\overset{O}{\overset{\|}{C}}C(CH_3)_3$ (c) Cyclopentane-1,3-dione

(d) $CH_3CH{=}CHCHO$

11.24 Draw structures for the monoenol tautomers of cyclohexane-1,3-dione. How many enol forms are possible, and which would you expect to be most stable? Explain.

11.25 Rank the following compounds in order of increasing acidity:

$$CH_3CH_2\overset{O}{\overset{\|}{C}}OH,\ CH_3\overset{O}{\overset{\|}{C}}CH_3,\ CH_3CH_2OH,\ CH_3\overset{O}{\overset{\|}{C}}CH_2\overset{O}{\overset{\|}{C}}CH_3$$

11.26 Why do you suppose pentane-2,4-dione is 76% enolized at equilibrium although acetone is enolized only to the extent of about 0.0001%?

11.27 Write resonance forms for the following anions:

(a) $CH_3\overset{O}{\overset{\|}{C}}\overset{\cdot\cdot\,-}{C}H\overset{O}{\overset{\|}{C}}CH_3$ (b) $:\overset{-}{C}H_2C{\equiv}N$ (c) $CH_3CH{=}CH\overset{\cdot\cdot\,-}{C}H\overset{O}{\overset{\|}{C}}CH_3$

(d) $N{\equiv}C\overset{\cdot\cdot\,-}{C}HCO_2C_2H_5$

11.28 Why is an enolate ion generally more reactive than a neutral enol?

Mechanisms

11.29 How do the mechanisms of base-catalyzed enolization and acid-catalyzed enolization differ?

11.30 Nonconjugated β,γ-unsaturated ketones, such as 3-cyclohexenone, are in an acid-catalyzed equilibrium with their conjugated α,β-unsaturated isomers. Propose a mechanism for the acid-catalyzed interconversion.

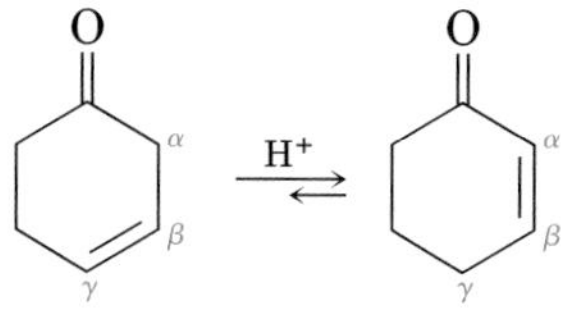

O α β γ $\underset{\longleftarrow}{\overset{H^+}{\longrightarrow}}$ O α β γ

11.31 The α,β to β,γ interconversion of unsaturated ketones (see Problem 11.30) is catalyzed by base as well as by acid. Propose a mechanism.

11.32 One consequence of the base-catalyzed α,β to β,γ isomerization of unsaturated ketones (see Problem 11.31) is that C5-substituted cyclopent-2-enones can be interconverted with C2-substituted cyclopent-2-enones. Propose a mechanism for this isomerization.

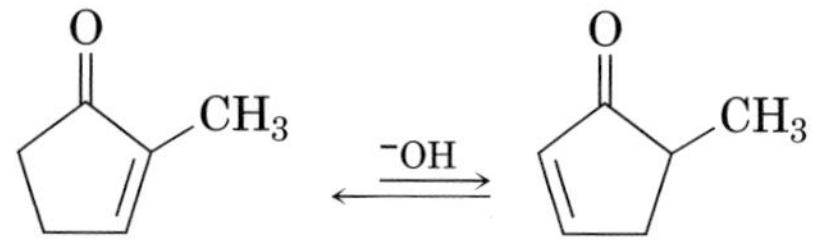

O CH_3 $\overset{^-OH}{\longleftrightarrow}$ O CH_3

11.33 How can you account for the fact that *cis*- and *trans*-4-*tert*-butyl-2-methylcyclohexanone are interconverted by base treatment? Which of the two isomers is more stable, and why? (See Section 2.10.)

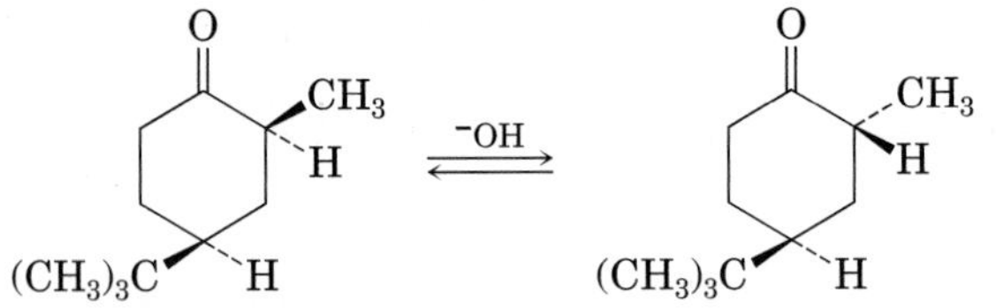

O, CH_3, H, $(CH_3)_3C$, H $\underset{}{\overset{^-OH}{\rightleftharpoons}}$ O, CH_3, H, $(CH_3)_3C$, H

11.34 When acetone is treated with acid in deuterated water, D_2O, deuterium becomes incorporated into the molecule. Propose a mechanism.

$$CH_3\overset{O}{\overset{\|}{C}}CH_3 + D_2O \overset{DCl}{\rightleftharpoons} CH_3\overset{O}{\overset{\|}{C}}CH_2D$$

11.35 When optically active (*R*)-2-methylcyclohexanone is treated with aqueous HCl or NaOH, racemic 2-methylcyclohexanone is produced. Explain.

11.36 When optically active (*R*)-3-methylcyclohexanone is treated with aqueous HCl or NaOH, no racemization occurs. Instead, the optically active ketone is recovered unchanged. How can you reconcile this observation with your answer to Problem 11.35?

ADOL CONDENSATION

11.37 Which of the following compounds would you expect to undergo aldol condensation? Draw the product in each case.

(a) $CH_3-\overset{H_3C}{\underset{H_3C}{C}}-\overset{O}{\overset{\|}{C}}H$ (b) 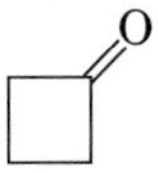 (c)

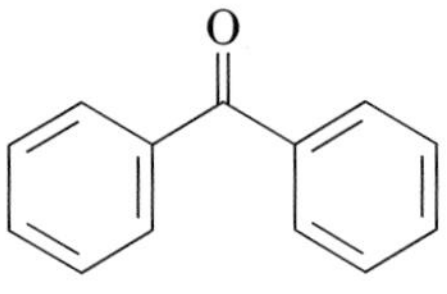

(d) $CH_3(CH_2)_8\overset{O}{\overset{\|}{C}}H$ (e) 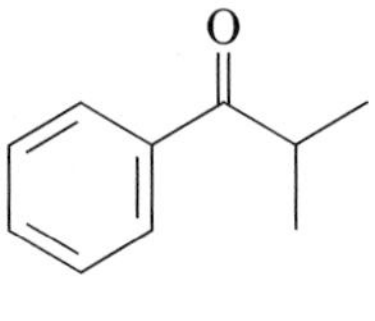 (f) $H\overset{O}{\overset{\|}{C}}H$

11.38 How might you prepare the following compounds using an aldol condensation reaction?

(a) 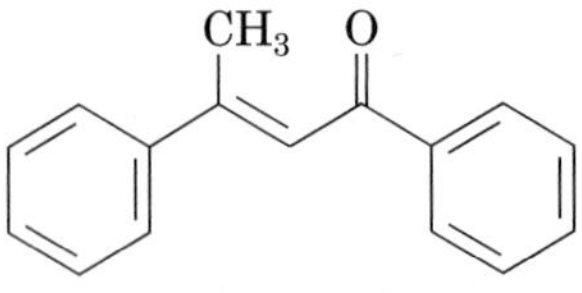 CH_3 O (b) $CH_3\overset{CH_3}{\overset{|}{C}}{=}CH\overset{O}{\overset{\|}{C}}CH_3$

(c) $CH_3CH_2CH_2CH{=}\underset{CH_2CH_3}{\underset{|}{\overset{O}{\overset{\|}{C}}}}H$ (d) 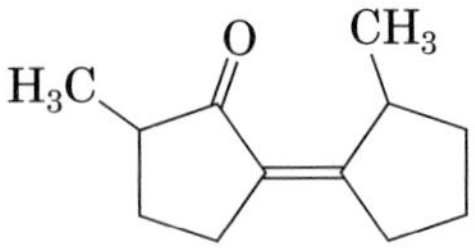 H_3C O CH_3

11.39 The aldol condensation reaction can be carried out intramolecularly by treatment of a diketone with base. What diketone would you start with to prepare 3-methylcyclohex-2-enone? Show the reaction.

11.40 What product would you expect to obtain from aldol condensation of hexanedial, $OHCCH_2CH_2CH_2CH_2CHO$? (See Problem 11.39.)

MALONIC ESTER SYNTHESIS

11.41 Which of the following esters can be prepared by a malonic ester synthesis? Show what reagents you would use.

(a) $CH_3CH_2CH_2CH_2C(=O)OCH_2CH_3$

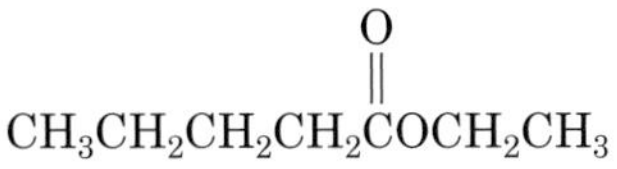

(b) $CH_3CH(CH_3)CH_2C(=O)OCH_2CH_3$

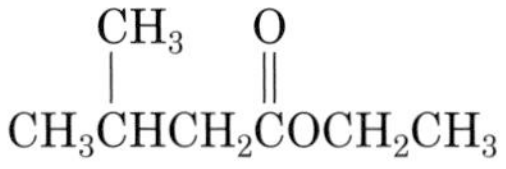

(c) $CH_3CH_2CH(CH_3)C(=O)OCH_2CH_3$

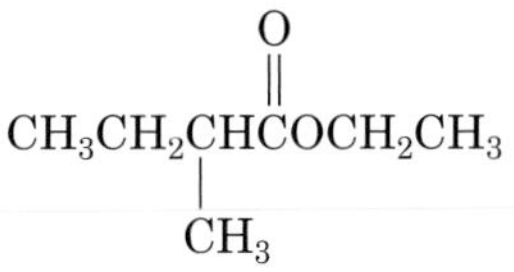

(d) $CH_3{-}C(CH_3)_2{-}C(=O)OCH_2CH_3$

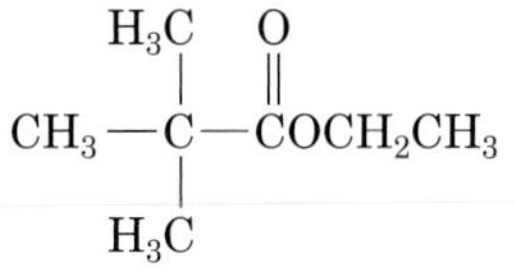

11.42 By starting with a *dihalide*, cyclic compounds can be prepared using the malonic ester synthesis. What product would you expect to obtain from the reaction of diethyl malonate, 1,4-dibromobutane, and 2 equivalents of base?

11.43 Show how you might convert geraniol, the chief constituent of rose oil, into ethyl geranylacetate.

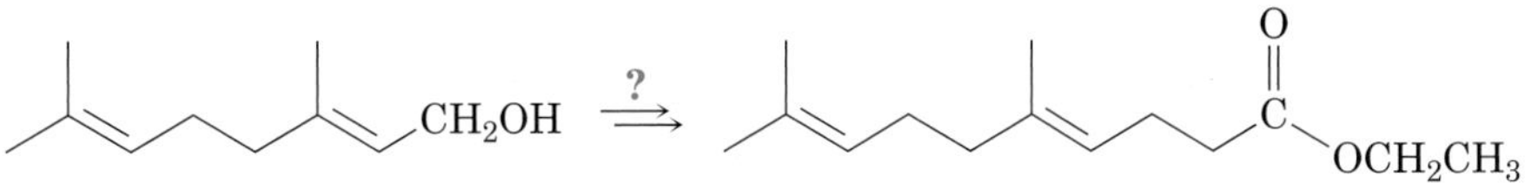

$$\text{Geraniol } (\ldots CH_2OH) \xrightarrow{?} \text{Ethyl geranylacetate } (\ldots C(=O)OCH_2CH_3)$$

Geraniol **Ethyl geranylacetate**

CLAISEN CONDENSATION

11.44 If a 1:1 mixture of ethyl acetate and ethyl propanoate is treated with base under Claisen condensation conditions, a mixture of four β-keto ester products is obtained. Show their structures, and explain.

11.45 If a mixture of ethyl acetate and ethyl benzoate is treated with base, a mixture of two Claisen condensation products is obtained. Show their structures, and explain.

INTEGRATED PROBLEMS

11.46 The *acetoacetic ester synthesis* is closely related to the malonic ester synthesis but involves alkylation with the anion of ethyl acetoacetate rather than diethyl malonate. Treatment of the ethyl acetoacetate anion with an alkyl halide, followed by decarboxylation, yields a ketone product:

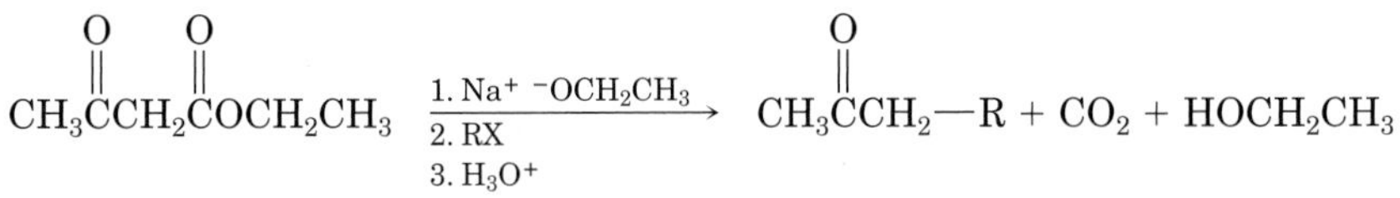

$$CH_3C(=O)CH_2C(=O)OCH_2CH_3 \xrightarrow[\text{2. RX} \quad \text{3. } H_3O^+]{\text{1. } Na^+ \; ^-OCH_2CH_3} CH_3C(=O)CH_2{-}R + CO_2 + HOCH_2CH_3$$

How would you prepare the following compounds using an acetoacetic ester synthesis?

(a) $C_6H_5CH_2CH_2C(=O)CH_3$

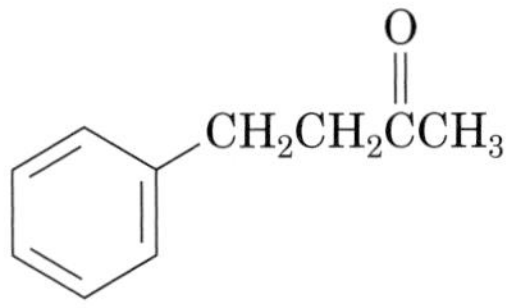

(b) $CH_3CH(CH_3)CH_2CH_2C(=O)CH_3$

(c) $CH_3CH_2CH_2CH(CH_3)C(=O)CH_3$

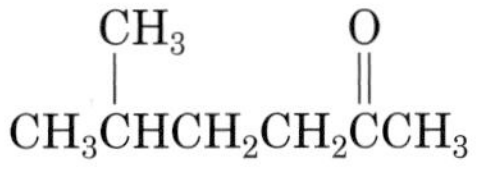

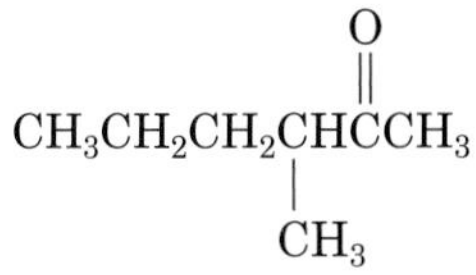

11.47 Which of the following compounds can't be prepared by an acetoacetic ester synthesis (see Problem 11.46)? Explain.

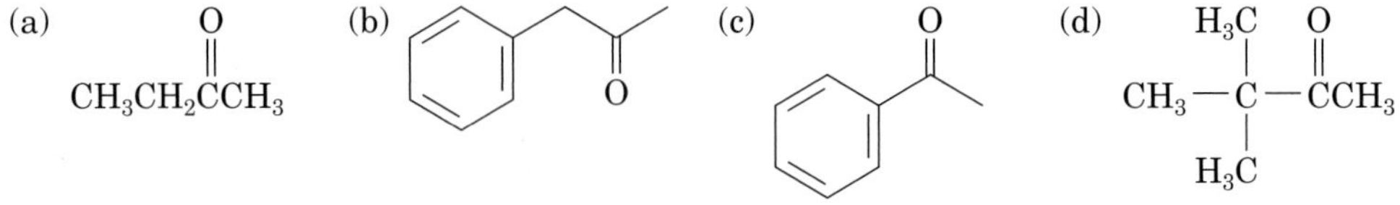

(a) $CH_3CH_2\overset{O}{\overset{\|}{C}}CH_3$ (b) (c) (d) $CH_3-\overset{H_3C}{\overset{|}{\underset{H_3C}{\underset{|}{C}}}}-\overset{O}{\overset{\|}{C}}CH_3$

11.48 Cinnamaldehyde, the aromatic constituent of cinnamon oil, can be synthesized by a mixed aldol-like reaction between benzaldehyde and acetaldehyde. Formulate the reaction. What other product would you expect to obtain?

CHO

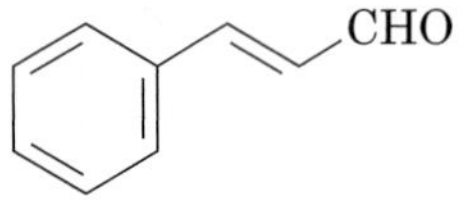

Cinnamaldehyde

11.49 Butan-1-ol is synthesized commercially from acetaldehyde by a three-step route that involves an aldol reaction followed by two reductions. Show how you might carry out this transformation.

11.50 Monoalkylated acetic acids (RCH_2CO_2H) and dialkylated acetic acids (R_2CHCO_2H) can be prepared by malonic ester synthesis, but trialkylated acetic acids (R_3CCO_2H) can't be prepared. Explain.

11.51 Just as the aldol condensation can be carried out intramolecularly on a diketone (Problem 11.39), so too can the Claisen condensation be carried out intramolecularly on a diester. Called the *Dieckmann cyclization*, reaction of a diester with base yields a cyclic β-keto ester product. What product would you expect to obtain in the following reaction?

$CH_3O\overset{O}{\overset{\|}{C}}$ ... $\overset{O}{\overset{\|}{C}}OCH_3$ $\xrightarrow[CH_3OH]{Na^+ \ ^-OCH_3}$?

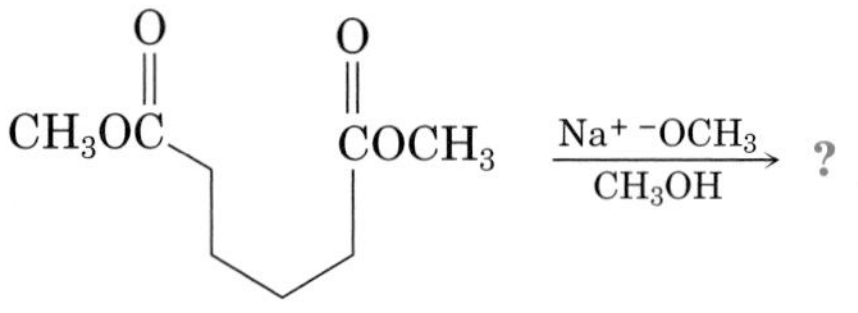

11.52 The cyclic β-keto ester formed in a Dieckmann cyclization reaction (Problem 11.51) can be converted by treatment with base into an anion and alkylated in a process much like that of the acetoacetic ester synthesis (Problem 11.46). Predict the product of the following reaction:

O, H, CO_2CH_3 $\xrightarrow{\text{1. } Na^+ \ ^-OCH_3 \quad \text{2. } CH_3Br \quad \text{3. } H_3O^+\text{, heat}}$?

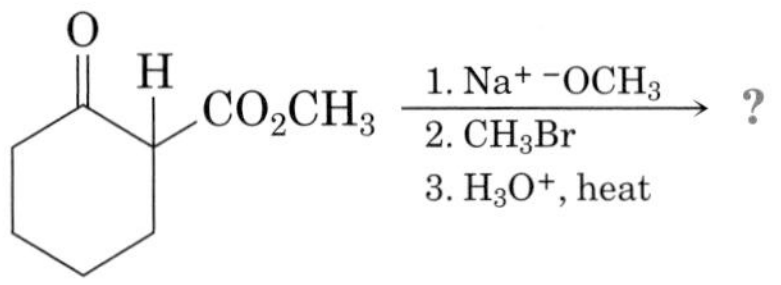

11.53 The Claisen condensation is reversible. That is, a β-keto ester can be cleaved by base into two fragments. Show the mechanism by which the following cleavage occurs:

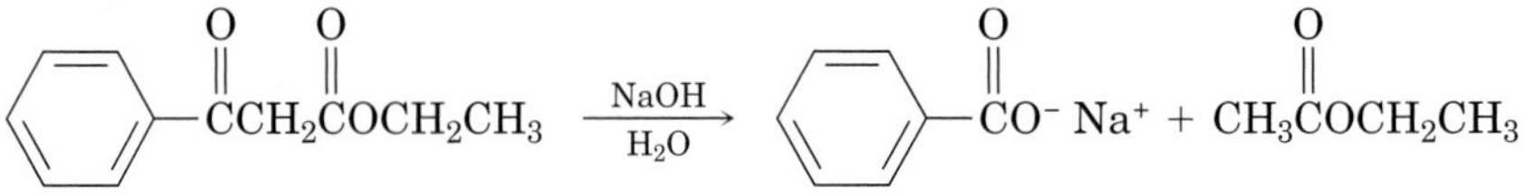

$C_6H_5-\overset{O}{\overset{\|}{C}}CH_2\overset{O}{\overset{\|}{C}}OCH_2CH_3 \xrightarrow[H_2O]{NaOH} C_6H_5-\overset{O}{\overset{\|}{C}}O^- \ Na^+ + CH_3\overset{O}{\overset{\|}{C}}OCH_2CH_3$

11.54 Treatment of an α,β-unsaturated carbonyl compound with base yields an anion by removal of H^+ from the γ-carbon. Why are hydrogens on the γ-carbon atom acidic?

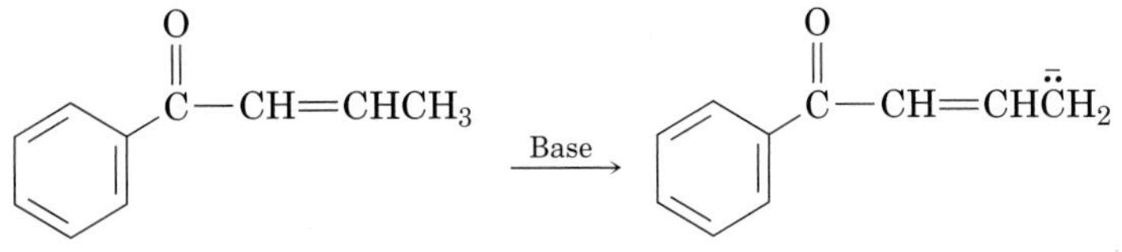

$C_6H_5C(=O)-CH=CHCH_3 \xrightarrow{\text{Base}} C_6H_5C(=O)-CH=CH\ddot{C}H_2$

11.55 Amino acids can be prepared by reaction of alkyl halides with diethyl acetamidomalonate, followed by heating the initial alkylation product with aqueous HCl. Show how you would prepare alanine, $CH_3CH(NH_2)CO_2H$, one of the 20 amino acids found in proteins.

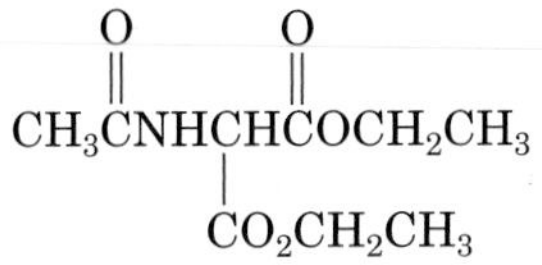

$CH_3C(=O)NHCH(CO_2CH_2CH_3)C(=O)OCH_2CH_3$ **Diethyl acetamidomalonate**

IN THE MEDICINE CABINET

11.56 Tazobactam is a β-lactam that inhibits the enzyme that bacteria secrete to resist penicillins. Draw a mechanism for the formation of the tazobactam-trapped β-lactamase conjugate that proceeds through the following steps:

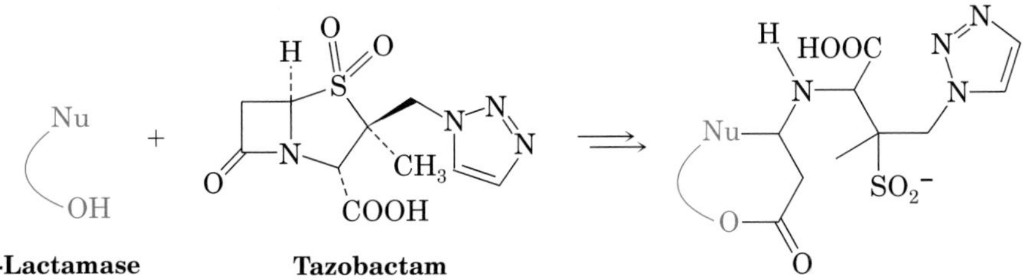

β-Lactamase + **Tazobactam** ⟹ **A trapped β-lactamase**

(a) Attack by the hydroxyl group of β-lactamase on the β-lactam to form a tetrahedral intermediate and collapse to form the nitrogen anion
(b) Formation of the imine by expulsion of the SO_2R leaving group (remember to conserve charge)
(c) Rearrangement of the imine to an α,β-unsaturated carbonyl by proton migration
(d) Conjugate addition by the nucleophilic group onto the α,β-unsaturated carbonyl and formation of the product depicted

11.57 Typically, esters are less stable than amides. Why does the hydroxyl group attack the lactam of tazobactam (Problem 11.56a) to form an ester from an amide?

11.58 Collapse of the tetrahedral intermediate leads to imine formation (Problem 11.56b and c). Why is $-SO_2R$ a good leaving group?

11.59 Most barbituric acids are made by first alkylating the diethyl ester of malonic acid, $CH_3CH_2OOCCH_2COOCH_2CH_3$. Draw the mechanism for the reaction of $CH_3CH_2OOCCH_2COOCH_2CH_3$ with ethyl iodide and sodium ethoxide.

11.60 Pentobarbital, also known as Nembutal, is a barbiturate used in the treatment of insomnia. It is synthesized by the following reaction sequence. Draw mechanisms for these reactions. Sodium ethoxide is strong enough to remove a proton from urea, H_2NCONH_2.

EtO—C(=O)—CH(CH₂CH₃)—C(=O)—OEt → (1. Na⁺ ⁻OEt; 2. $CH_3CH_2CH_2CH(Br)CH_3$) → EtO—C(=O)/EtO—C(=O) → (H_2N—C(=O)—NH_2, Na⁺ ⁻OEt) →

Pentobarbital (Nembutal)

11.61 Nembutal (Problem 11.60) is administered as a racemate. Identify the chiral center and draw the *R*- and *S*-enantiomers. Do you expect these enantiomers to have identical biological activity?

IN THE FIELD WITH AGROCHEMICALS

11.62 Carboxin, or Vitavax, is a fungicide used on corn and wheat fields. One synthesis of carboxin is shown here.

HOOC → ($(CH_3CO)_2O$) → **A** → ($C_6H_5NH_2$) → **B** → (Cl_2, Acetic acid) → **C** → ($HSCH_2CH_2OH$) → **D** → (H^+) → [**E**] → **Carboxin (a fungicide)**

(a) Draw a mechanism that produces the mixed-anhydride intermediate, **A**.
(b) Draw a mechanism for the conversion of **A** to **B**.
(c) Draw a mechanism for the acid-catalyzed chlorination, which yields **C** from **B**.
(d) Reaction of **C** with $HSCH_2CH_2OH$ occurs by an S_N2 mechanism. Show this mechanism starting with the (*R*)-enantiomer of **C**. How would you describe the relative nucleophilicity of a thiol versus a hydroxyl group?
(e) Formation of carboxin from **D** occurs through intermediate **E**. What new functional group appears in this intermediate? Show a mechanism for formation of **E**.
(f) The conversion of **E** to carboxin occurs through an enol intermediate. Show a mechanism for this reaction.
(g) The O=C—C=C group of carboxin (red) has a specific name and chemistry. Describe both.

11.63 Some of the intermediates produced during the synthesis of carboxin (Problem 11.62) are chiral. Identify the chiral intermediates. Why is carboxin achiral?

Many naturally occurring amines produce significant biological responses. Nicotine triggers the flight-or-fight response. (Getty Images/Travis Heying/AFP)

12 CHAPTER Amines

Amines are composed of nitrogen atoms bearing alkyl or aromatic groups. The lone pair of electrons on the nitrogen makes amines basic and nucleophilic. Of all organic compounds, those containing amines produce some of the most interesting effects. Many amines are "bioactive" in that they produce profound biological responses ranging from mood alteration to hallucinations to death upon ingestion.

Amines occur widely in both plants and animals. Trimethylamine, for instance, occurs in animal tissues and is partially responsible for the distinctive odor of fish, nicotine is found in tobacco, and cocaine is a stimulant found in the South American coca bush.

CH_3 H_3C N CH_3 — **Trimethylamine**

H N N CH_3 — **Nicotine**

CH_3 :N CO_2CH_3 H O H O — **Cocaine**

12.1

Naming Amines

Amines are classed as **primary (RNH_2)**, **secondary (R_2NH)**, or **tertiary (R_3N)**, depending on the number of organic substituents attached to nitrogen. For example, methylamine, CH_3NH_2, is a primary amine, and trimethylamine $(CH_3)_3N$, is a tertiary amine. Note that this usage of the terms *primary*, *secondary*, and *tertiary* is different from our previous usage. When we speak of a tertiary alcohol or alkyl halide, we refer to the degree of substitution at the alkyl *carbon* atom, but when we speak of a tertiary amine, we refer to the degree of substitution at the *nitrogen* atom.

$(CH_3)_3C{-}OH$ — ***tert*-Butyl alcohol (a tertiary alcohol)**

$(CH_3)_3N$ — **Trimethylamine (a tertiary amine)**

$(CH_3)_3C{-}NH_2$ — ***tert*-Butylamine (a primary amine)**

Compounds with four groups attached to nitrogen are also known, but the nitrogen atom must carry a positive charge. Such compounds are called **quaternary ammonium salts.**

$R_4N^+\ X^-$ A quaternary ammonium salt

Amines can be either alkyl-substituted (**alkylamines**) or aryl-substituted (**arylamines**). Although much of the chemistry of the two classes is similar, we'll soon see that there are also important differences.

$CH_3CH_2\ddot{N}H_2$ — **Ethylamine (an alkylamine)**

$C_6H_5\ddot{N}H_2$ — **Aniline (an arylamine)**

$C_6H_5CH_2\ddot{N}H_2$ — **Benzylamine (an alkylamine)**

Primary amines, RNH_2, are named in the IUPAC system by adding the suffix *-amine* to the name of the organic substituent:

$(CH_3)_3C{-}NH_2$ — ***tert*-Butylamine**

$C_6H_{11}{-}NH_2$ — **Cyclohexylamine**

$H_2NCH_2CH_2CH_2CH_2NH_2$ — **Butane-1,4-diamine**

Amines that have additional functional groups are named by considering the $-NH_2$ as an *amino* substituent on the parent molecule:

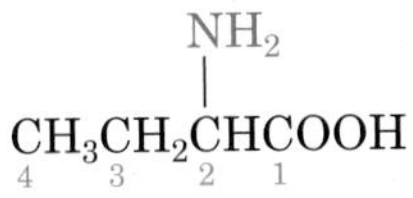

$CH_3CH_2CH(NH_2)COOH$ (numbered 4 3 2 1)

2-Aminobutanoic acid

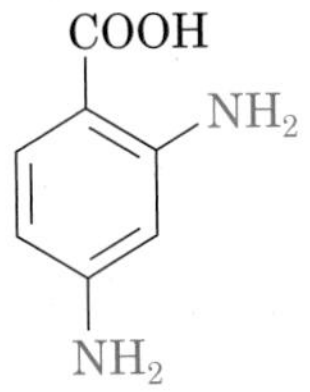

2,4-Diaminobenzoic acid

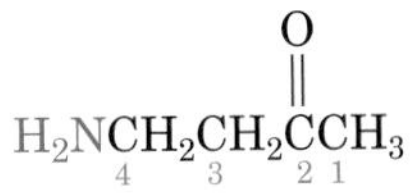

$H_2NCH_2CH_2C(=O)CH_3$ (numbered 4 3 2 1)

4-Aminobutan-2-one

Symmetrical secondary and tertiary amines are named by adding the prefix *di-* or *tri-* to the alkyl group:

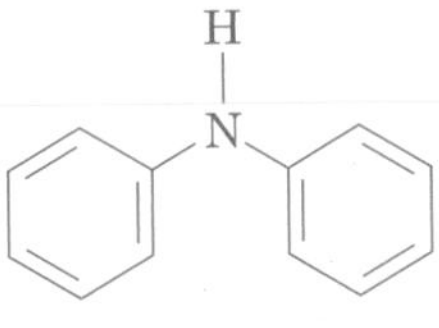

Diphenylamine

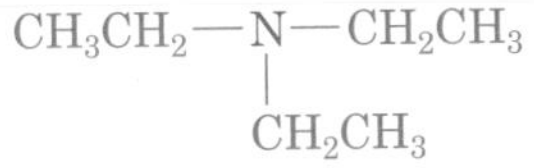

$CH_3CH_2-N(CH_2CH_3)-CH_2CH_3$

Triethylamine

Unsymmetrically substituted secondary and tertiary amines are named as *N*-substituted primary amines. The largest organic group is chosen as the parent, and the other groups are considered as *N*-substituents on the parent (*N* because they're attached to nitrogen).

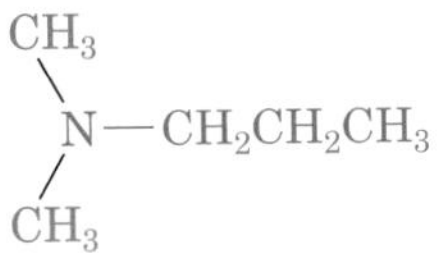

$(CH_3)_2N-CH_2CH_2CH_3$

***N,N*-Dimethylpropylamine**
(propylamine is the parent name; the two methyl groups are substituents on nitrogen)

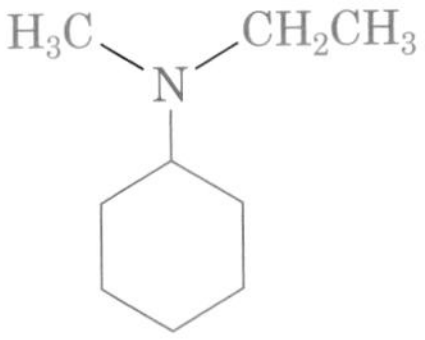

$H_3C-N(CH_2CH_3)-C_6H_{11}$

***N*-Ethyl-*N*-methylcyclohexylamine**
(cyclohexylamine is the parent name; methyl and ethyl are *N*-substituents)

There are few common names for simple amines, although phenylamine is usually called *aniline.*

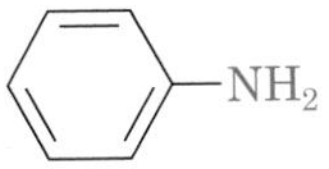

$C_6H_5-NH_2$ **Aniline**

Heterocyclic amines—compounds in which the nitrogen atom occurs as part of a ring—are also common, and each different heterocyclic ring system has its own parent name. In all cases, the nitrogen atom is numbered as position 1.

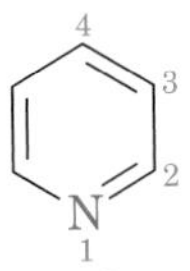

Pyridine 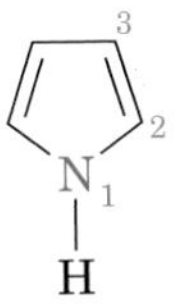 **Pyrrole** 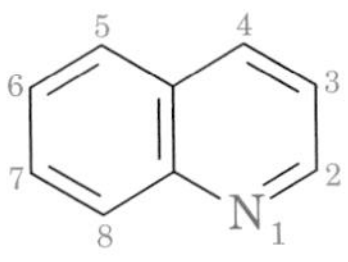 **Quinoline** 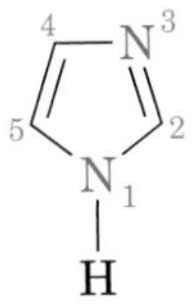 **Imidazole** 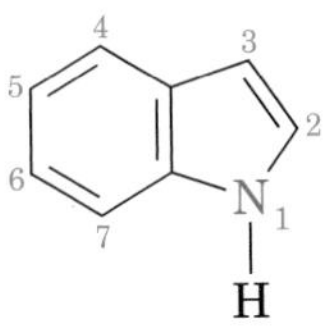 **Indole** 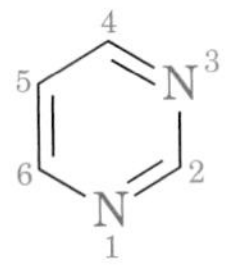 **Pyrimidine**

Practice Problem 12.1

Classify the following amines as primary, secondary, or tertiary:

(a) $CH_3CH_2CH(CH_3)NH_2$ (b) pyrrolidine (N—H) (c) $C_6H_5N(CH_3)_2$

Solution Amine (a) has one organic group attached to nitrogen and is primary, (b) has two organic groups attached to nitrogen and is secondary, and (c) is tertiary.

Problem 12.1 Classify each of the following compounds as either a primary, secondary, or tertiary amine, or as a quaternary ammonium salt:

(a) $(CH_3)_2CHNH_2$ (b) $(CH_3CH_2)_2NH$

(c) $C_6H_{11}N(CH_3)_2$ (cyclohexyl) (d) $C_6H_5CH_2\overset{+}{N}(CH_3)_3\ I^-$

Problem 12.2 Draw structures of compounds that meet the following descriptions:

(a) A secondary amine with one isopropyl group
(b) A tertiary amine with one phenyl group and one ethyl group
(c) A quaternary ammonium salt with four different groups bonded to nitrogen

Problem 12.3 Name the following compounds by IUPAC rules:

(a) $CH_3NHCH_2CH_3$ (b) $(C_6H_{11})_3N$ (c) N—CH_3 (pyrrole ring)

(d) $C_6H_{11}N(CH_3)CH_2CH_2CH_3$ (e) $H_2NCH_2CH_2CH(CH_3)NH_2$

Problem 12.4 Draw structures corresponding to the following IUPAC names:

(a) Triethylamine (b) *N*-Methylaniline
(c) Tetraethylammonium bromide (d) *p*-Bromoaniline
(e) *N*-Ethyl-*N*-methylcyclopentylamine

12.2 Structure and Properties of Amines

The bonding in amines is similar to the bonding in ammonia. The nitrogen atom is sp^3-hybridized, with the three substituents occupying three corners of a regular tetrahedron and the lone pair of electrons occupying the fourth corner. As expected, the C–N–C bond angles are very close to the 109° tetrahedral value. For trimethylamine, the C–N–C angle is 108° and the C–N bond length is 147 pm.

sp^3-hybridized

H_3C, N, CH_3, H_3C

Trimethylamine

Like alcohols, amines are highly polar, and those with fewer than five carbon atoms are generally water-soluble. Also like alcohols, primary and secondary amines form hydrogen bonds and therefore have higher boiling points than alkanes of similar molecular weight.

One other characteristic property of amines is their *odor.* Low-molecular-weight amines such as trimethylamine have a distinctive fishlike aroma, while diamines such as putrescine (butane-1,4-diamine) have odors as putrid as their names suggest.

12.3 Basicity of Amines

The chemistry of amines is dominated by the lone pair of electrons on nitrogen. Because of this lone pair, amines are both basic and nucleophilic, as shown by the following electrostatic potential map. They therefore react with acids to form acid–base salts, and they react with electrophiles in many of the polar reactions seen in previous chapters.

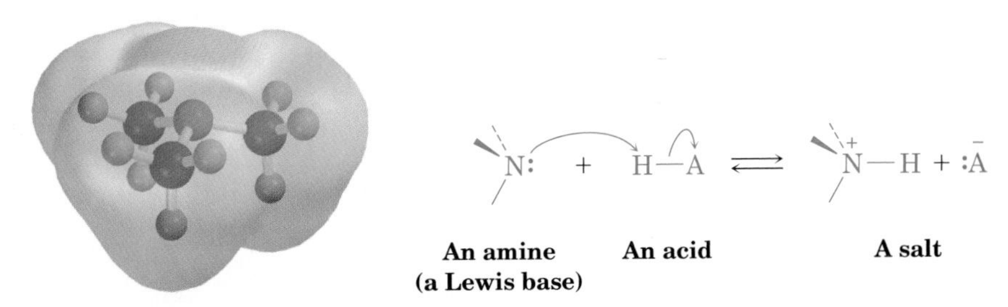

Amines are much stronger bases than alcohols, ethers, or water. When an amine is dissolved in water, an equilibrium is established in which water acts as an acid and donates H^+ to the amine. By finding the equilibrium constant for the reaction, we can define a **basicity constant, K_b,** that measures the ability of an amine to accept a proton, and we can thereby establish a relative order of base strength. The larger the K_b (and the smaller the pK_b), the more favorable the equilibrium and the stronger the base; the smaller the K_b (and the larger the pK_b), the weaker the base.

For the reaction: $RNH_2 + H_2O \rightleftharpoons RNH_3^+ + OH^-$

$$K_b = \frac{[RNH_3^+][OH^-]}{[RNH_2]}$$

$$pK_b = -\log K_b$$

Table 12.1 gives the pK_b values of some common amines. As indicated, substitution has relatively little effect on alkylamine basicity. Most simple alkylamines have pK_b's in the narrow range 3 to 4, regardless of their exact structure.

The most important conclusion from Table 12.1 is that *arylamines*, such as aniline, are weaker bases than alkylamines by a factor of about 10^6. The nitrogen lone-pair electrons in an arylamine are shared by orbital overlap with the π orbitals of the aromatic ring, and they are therefore less available for bonding to an acid. In resonance terms, an arylamine is more stable than an alkylamine because of its five resonance structures:

$:NH_2$ ⟷ $:NH_2$ ⟷ $^+NH_2$ ⟷ $^+NH_2$ ⟷ $^+NH_2$

In contrast to amines, *amides* ($RCONH_2$) are nonbasic. Amides don't react with acids, and their aqueous solutions are neutral. The main reason for the decreased basicity of amides relative to amines is that the nitrogen lone-pair electrons are shared by orbital overlap with the neighboring carbonyl-group π orbital. The electrons are therefore much less available for bonding to an acid. In resonance terms, amides are more stable and less

TABLE 12.1 Basicity of Some Common Amines

Name	Structure	pK_b	
Triethylamine	$(CH_3CH_2)_3N$	2.99	More basic
Ethylamine	$CH_3CH_2NH_2$	3.19	
Dimethylamine	$(CH_3)_2NH$	3.27	
Methylamine	CH_3NH_2	3.34	
Diethylamine	$(CH_3CH_2)_2NH$	3.51	
Trimethylamine	$(CH_3)_3N$	4.19	
Ammonia	NH_3	4.74	
Pyridine	N (ring)	8.75	
Aniline	C_6H_5–NH_2	9.37	Less basic

reactive than amines because they are hybrids of two resonance forms. Electrostatic potential maps show clearly this decreased electron density on the amide nitrogen.

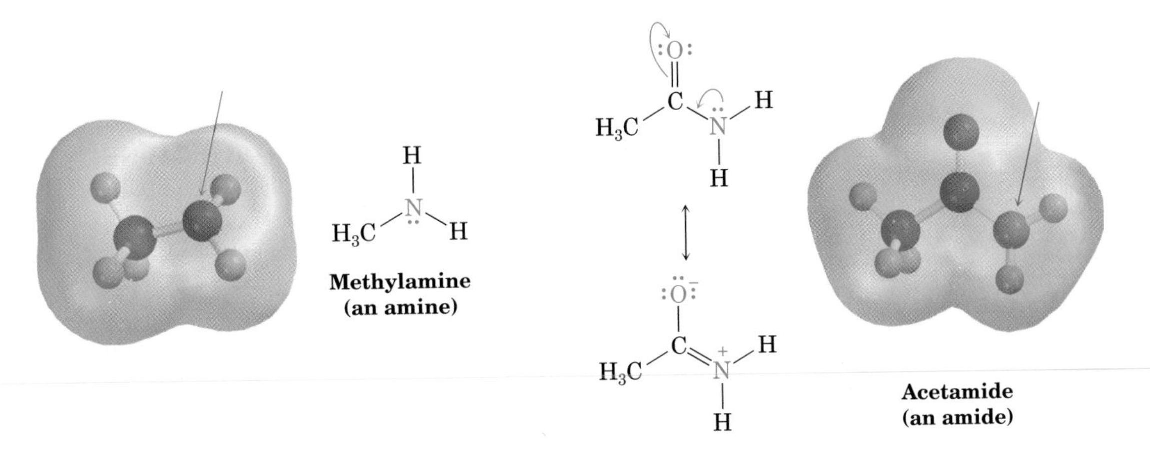

It's often possible to take advantage of its basicity to purify an amine. For example, if a mixture of an amine (basic) and a ketone (neutral) is dissolved in an organic solvent and aqueous HCl is added, the basic amine dissolves in the acidic water as its ammonium ion, while the ketone remains in the organic solvent. Separation of the water layer and neutralization of the ammonium ion by addition of NaOH then provides the pure amine (Figure 12.1).

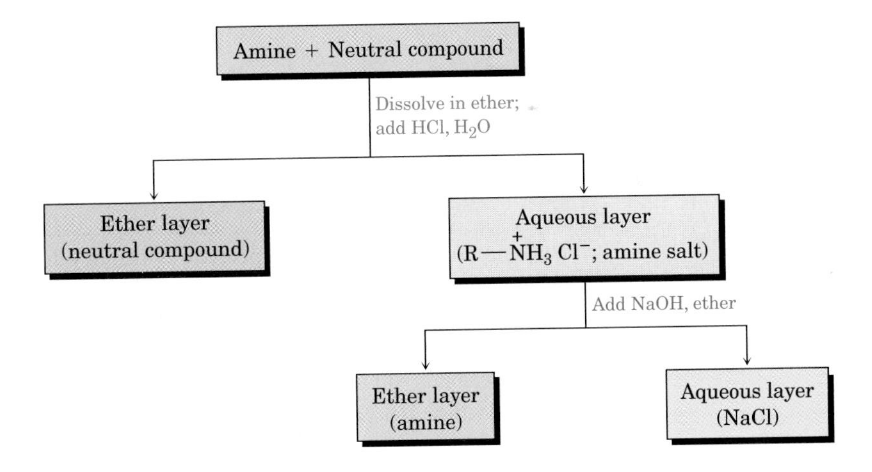

FIGURE 12.1 Separation and purification of an amine from a mixture.

PRACTICE PROBLEM 12.2

Predict the product of the following reaction:

$$CH_3CH_2NHCH_3 + HCl \rightarrow ?$$

STRATEGY Amines are protonated by acids to yield ammonium salts.

SOLUTION

$$CH_3CH_2NHCH_3 + HCl \longrightarrow CH_3CH_2N^+H_2CH_3\ Cl^-$$

PROBLEM 12.5 Predict the product of the following reaction:

$$\text{C}_5\text{H}_9\text{–N(CH}_3\text{)H} + HBr \longrightarrow ?$$

PROBLEM 12.6 Which compound in each of the following pairs is more basic?

(a) $CH_3CH_2NH_2$ or $CH_3CH_2CONH_2$ (b) NaOH or $C_6H_5NH_2$
(c) CH_3NHCH_3 or $CH_3NHC_6H_5$ (d) CH_3OCH_3 or $(CH_3)_3N$

12.4 Synthesis of Amines

Reduction of Nitriles and Amides

We've already seen how amines can be prepared by reduction of amides (Section 10.12) and nitriles (Section 10.13) with $LiAlH_4$. The two-step sequence of S_N2 reaction of an alkyl halide with cyanide ion, followed by reduction, is a good method for converting an alkyl halide into a primary amine having one more carbon atom than the original halide. Amide reduction provides a method for converting a carboxylic acid into an amine having the same number of carbon atoms.

$$\underset{\textbf{Alkyl halide}}{RX} \xrightarrow{NaCN} RCN \xrightarrow[2.\ H_2O]{1.\ LiAlH_4,\ ether} \underset{\textbf{1° amine}}{RCH_2NH_2}$$

$$\underset{\textbf{Carboxylic acid}}{R\text{—}\overset{O}{\overset{\|}{C}}\text{—}OH} \xrightarrow[2.\ NH_3]{1.\ SOCl_2} R\text{—}\overset{O}{\overset{\|}{C}}\text{—}NH_2 \xrightarrow[2.\ H_2O]{1.\ LiAlH_4,\ ether} \underset{\textbf{1° amine}}{RCH_2NH_2}$$

PRACTICE PROBLEM 12.3

What amide would you use to prepare *N*-ethylcyclohexylamine?

STRATEGY Reduction of an amide with $LiAlH_4$ yields an amine in which the amide carbonyl group has been replaced by a methylene (–CH_2–) unit, $RCONR_2 \rightarrow RCH_2NR_2$. Since *N*-ethylcyclohexylamine has only one –CH_2– carbon attached to nitrogen (the ethyl group), the product must come from reduction of *N*-cyclohexylacetamide.

SOLUTION

$$\underset{\textbf{\textit{N}-Cyclohexylacetamide}}{C_6H_{11}\text{—}NH\text{—}\overset{O}{\overset{\|}{C}}\text{—}CH_3} \xrightarrow[2.\ H_2O]{1.\ LiAlH_4} \underset{\textbf{\textit{N}-Ethylcyclohexylamine}}{C_6H_{11}\text{—}NHCH_2CH_3}$$

PROBLEM 12.7 Propose structures for amides that might be precursors of the following amines:

(a) $CH_3CH_2CH_2NH_2$ (b) $NH(CH_2CH_2CH_3)_2$ (c) $C_6H_5CH_2NH_2$ (benzene ring with CH_2NH_2)

PROBLEM 12.8 Propose structures for nitriles that might be precursors of the following amines:

(a) $CH_3CH(CH_3)CH_2CH_2NH_2$ (b) Benzylamine, $C_6H_5CH_2NH_2$

S_N2 Alkylation Reactions of Alkyl Halides

Perhaps the simplest method of amine synthesis is by an S_N2 alkylation reaction of ammonia or an alkylamine with an alkyl halide (see Section 7.5). If ammonia is used, a primary amine results; if a primary amine is used, a secondary amine results; and so on. Even tertiary amines react with alkyl halides to yield quaternary ammonium salts, $R_4N^+ X^-$.

S_N2 reaction

Ammonia	$\ddot{N}H_3 + R{-}X \longrightarrow$	$RNH_3^+ X^-$	$\xrightarrow{NaOH}$	RNH_2	Primary
Primary	$R\ddot{N}H_2 + R{-}X \longrightarrow$	$R_2NH_2^+ X^-$	$\xrightarrow{NaOH}$	R_2NH	Secondary
Secondary	$R_2\ddot{N}H + R{-}X \longrightarrow$	$R_3NH^+ X^-$	$\xrightarrow{NaOH}$	R_3N	Tertiary
Tertiary	$R_3\ddot{N} + R{-}X \longrightarrow$	$R_4N^+ X^-$	Quaternary ammonium salt		

Unfortunately, none of these reactions stops cleanly after a single alkylation has occurred. Because primary, secondary, and tertiary amines all have similar reactivity, the initially formed monoalkylated amine often undergoes further reaction to yield a mixture of products. For example, treatment of 1-bromooctane with a twofold excess of ammonia leads to a mixture containing only a 45% yield of octylamine. A nearly equal amount of dioctylamine is produced by double alkylation, along with smaller amounts of trioctylamine and tetraoctylammonium bromide.

$$CH_3(CH_2)_6CH_2Br + :NH_3 \longrightarrow CH_3(CH_2)_6CH_2\ddot{N}H_2 + [CH_3(CH_2)_6CH_2]_2\ddot{N}H$$

1-Bromooctane **Octylamine (45%)** **Dioctylamine (43%)**

$$+ [CH_3(CH_2)_6CH_2]_3N: + [CH_3(CH_2)_6CH_2]_4\overset{+}{N}\ \overset{-}{Br}$$

Trace Trace

PRACTICE PROBLEM 12.4

How could you prepare diethylamine from ammonia and an alkyl halide?

STRATEGY Look at the starting material (NH_3) and the product ($(CH_3CH_2)_2NH$), and note the difference. Since two ethyl groups have become bonded to the nitrogen atom, the reaction must involve ammonia and 2 equivalents of an ethyl halide.

SOLUTION

$$2\ CH_3CH_2Br + NH_3 \longrightarrow (CH_3CH_2)_2NH$$

PROBLEM 12.9 How could you prepare the following amines from ammonia and appropriate alkyl halides?

(a) Triethylamine (b) Tetramethylammonium bromide

PROBLEM 12.10 The following molecule can be prepared by reaction between a primary amine and a *dihalide.* Identify the two reactants, and write the reaction.

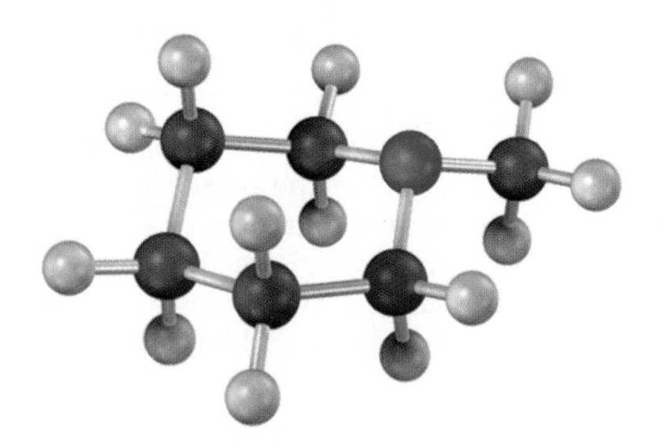

Reductive Amination of Aldehydes and Ketones

Amines can be synthesized in a single step by treatment of an aldehyde or ketone with ammonia or an amine in the presence of a reducing agent, a process called **reductive amination**. For example, amphetamine, a central nervous system stimulant, is prepared commercially by reductive amination of 1-phenylpropan-2-one with ammonia, using hydrogen gas over a nickel catalyst as the reducing agent.

$$C_6H_5CH_2C(=O)CH_3 \xrightarrow[H_2/Ni]{:NH_3} C_6H_5CH_2CH(:NH_2)CH_3 + H_2O$$

1-Phenylpropan-2-one **Amphetamine**

Reductive amination takes place by the pathway shown in Figure 12.2. An imine intermediate is first formed by a nucleophilic addition reaction (see Section 9.10), and the C=N bond of the imine is then reduced.

Ammonia, primary amines, and secondary amines can all be used in the reductive amination reaction, yielding primary, secondary, and tertiary amines, respectively.

$$RC(=O)R' \xrightarrow[H_2/cat.]{NH_3} RCH(NH_2)R' \qquad RC(=O)R' \xrightarrow[H_2/cat.]{R''NH_2} RCH(NHR'')R' \qquad RC(=O)R' \xrightarrow[H_2/cat.]{R''_2NH} RCH(NR''_2)R'$$

Primary amine **Secondary amine** **Tertiary amine**

We'll see in Section 17.5 that a process closely related to reductive amination occurs frequently in the biological pathways by which amino acids

FIGURE 12.2 MECHANISM: Mechanism of reductive amination of a ketone to yield an amine.

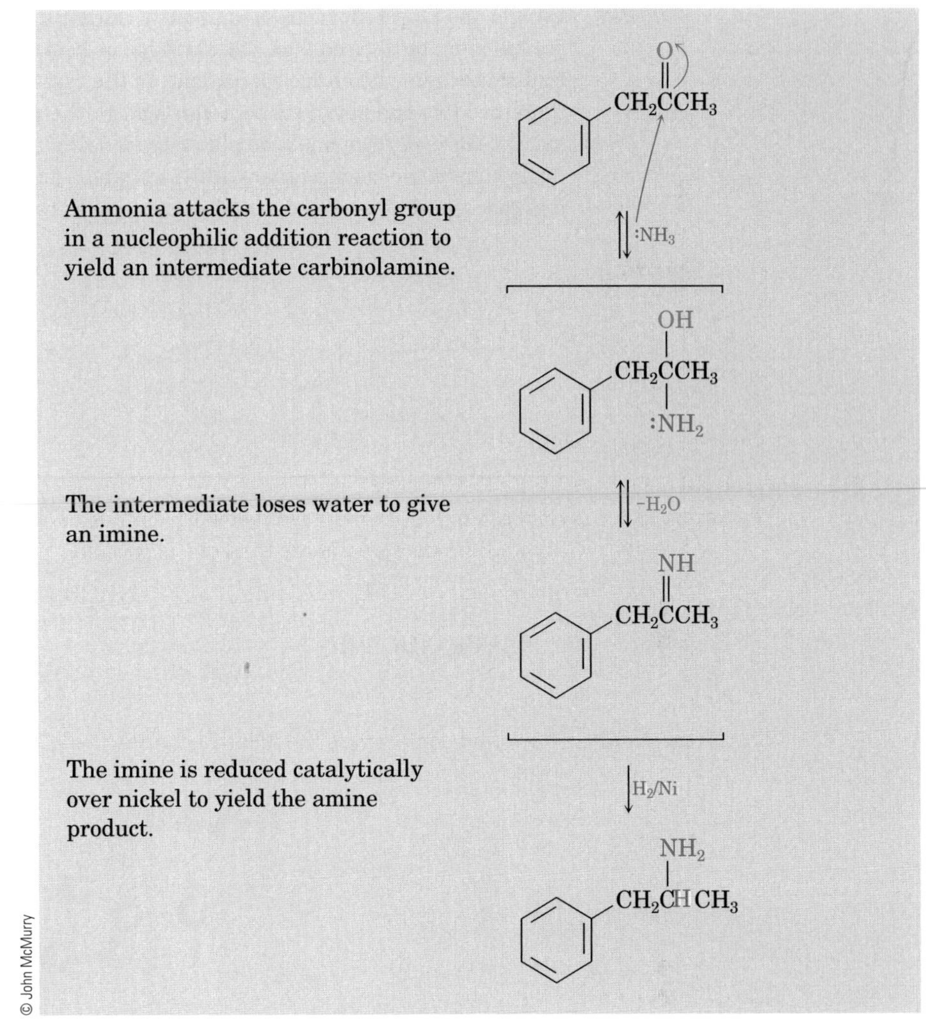

Ammonia attacks the carbonyl group in a nucleophilic addition reaction to yield an intermediate carbinolamine.
The intermediate loses water to give an imine.
The imine is reduced catalytically over nickel to yield the amine product.
:NH3
OH
CH2CCH3
:NH2
-H2O
NH
H2/Ni
NH2
CH2CHCH3
© John McMurry

are synthesized. The amino acid alanine, for instance, arises by reaction of pyruvic acid and ammonia to give an intermediate imine that is then reduced.

$$\underset{\textbf{Pyruvic acid}}{CH_3\overset{O}{\overset{\|}{C}}COOH} \xrightarrow{NH_3} \left[\underset{\textbf{Imine}}{CH_3\overset{NH}{\overset{\|}{C}}COOH}\right] \xrightarrow[\text{enzymes}]{\text{Reducing}} \underset{\textbf{Alanine}}{CH_3\overset{NH_2}{\overset{|}{C}}HCOOH}$$

PRACTICE PROBLEM 12.5

How might you prepare *N*-methyl-2-phenylethylamine using a reductive amination reaction?

$NHCH_3$ **_N_-Methyl-2-phenylethylamine**

STRATEGY Look at the target molecule, and identify the groups attached to nitrogen. One of the groups must be derived from the aldehyde or ketone component, and the other must be derived from the amine component. In the case of *N*-methyl-2-phenylethylamine, there are two combinations that can lead to the product: phenylacetaldehyde plus methylamine or formaldehyde plus 2-phenylethylamine. In general, it's usually better to choose the combination with the simpler amine component—methylamine in this case—and to use an excess of that amine as reactant.

SOLUTION

Phenylacetaldehyde (Ph–CH₂–CHO) + CH_3NH_2 $\xrightarrow{H_2/catalyst}$ Ph–CH₂CH₂–NHCH₃ $\xleftarrow{H_2/catalyst}$ Ph–CH₂CH₂–NH₂ + CH_2O

PROBLEM 12.11 How could you prepare the following amines using reductive amination reactions? Show all precursors if more than one is possible.

(a) $CH_3CH_2NHCH(CH_3)CH_3$ (b) Ph–CH₂–$NHCH_2CH_3$ (c) Cyclopentyl–$NHCH_3$

PROBLEM 12.12 How could you prepare the following amine using a reductive amination reaction?

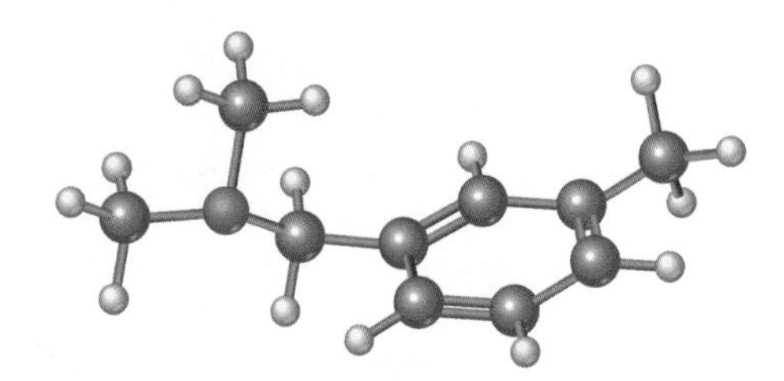

Reduction of Nitrobenzenes

Arylamines are prepared by nitration of an aromatic starting material (see Section 5.5), followed by reduction of the nitro group. The reduction step can be carried out in different ways, depending on the circumstances. Catalytic hydrogenation over platinum works well but is sometimes incompatible with the presence elsewhere in the molecule of other reducible groups, such as C=C bonds. Iron, zinc, tin, and stannous chloride ($SnCl_2$) in aqueous acid are also effective.

$(CH_3)_3C$–C_6H_4–NO_2 $\xrightarrow{H_2,\ Pt\ catalyst}$ or $\xrightarrow{1.\ Fe,\ H_3O^+;\ 2.\ NaOH}$ $(CH_3)_3C$–C_6H_4–NH_2

***p-tert*-Butylnitrobenzene** → ***p-tert*-Butylaniline**

PRACTICE PROBLEM 12.6

How could you synthesize *p*-methylaniline from benzene? More than one step is required.

STRATEGY A methyl group is introduced onto a benzene ring by a Friedel–Crafts reaction with $CH_3Cl/AlCl_3$ (see Section 5.6), and an amino group is introduced onto a ring by nitration and reduction. Since a methyl group is ortho- and para-directing (see Section 5.7), it would be best to introduce the methyl group first followed by nitration and reduction.

SOLUTION

Benzene $\xrightarrow[AlCl_3]{CH_3Cl}$ Toluene $\xrightarrow[H_2SO_4]{HNO_3}$ *p*-Nitrotoluene $\xrightarrow[2.\ NaOH]{1.\ Fe,\ H_3O^+}$ *p*-Methylaniline

Benzene **Toluene** ***p*-Nitrotoluene** ***p*-Methylaniline**

PROBLEM 12.13 How could you synthesize the following amines from benzene? More than one step is required in each case.

(a) H_2N–C_6H_4–CO_2H (meta) (b) 2,4,6-tribromoaniline (Br, Br, Br, NH_2)

12.5 Reactions of Amines

We've already seen the two most important reactions of alkylamines: *alkylation* and *acylation*. As we saw in the previous section, primary, secondary, and tertiary amines can be alkylated by reaction with alkyl halides. Primary and secondary (but not tertiary) amines can also be acylated by nucleophilic acyl substitution reactions with acid chlorides or acid anhydrides (see Sections 10.9 and 10.10) to give amides.

$$RCOCl + NH_3 \xrightarrow[\text{solvent}]{\text{Pyridine}} RCONH_2 + HCl$$

$$RCOCl + R'NH_2 \xrightarrow[\text{solvent}]{\text{Pyridine}} RCONHR' + HCl$$

$$RCOCl + R'_2NH \xrightarrow[\text{solvent}]{\text{Pyridine}} RCONR'_2 + HCl$$

12.6 Heterocyclic Amines

Cyclic organic compounds can be either *carbocycles* or *heterocycles.* Carbocycles contain only carbon atoms in their rings, while **heterocycles** contain one or more different atoms in addition to carbon. Heterocyclic amines are particularly common in organic chemistry, and many have important biological properties. For example, the antiulcer agent cimetidine and the sedative phenobarbital are both heterocycles but display very different nitrogen-containing functional groups. Only cimetidine has amine groups.

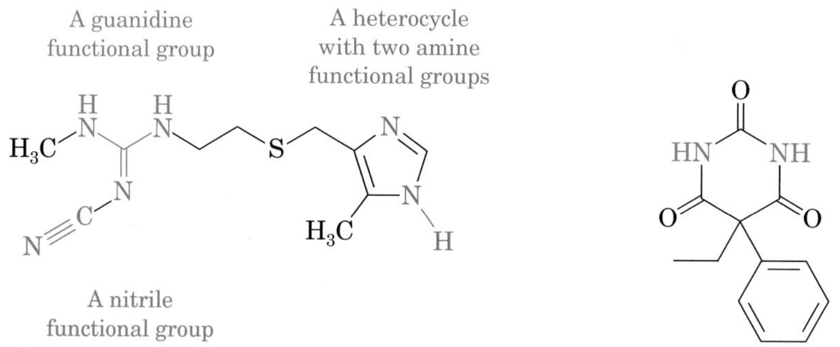

A guanidine functional group

A heterocycle with two amine functional groups

A nitrile functional group

Cimetidine (an antiulcer agent), with multiple nitrogen-containing functional groups

Phenobarbital (a sedative), a heterocycle

For the most part, heterocyclic amines have the same chemistry as their open-chain counterparts. Similarly, the chemistry of phenobarbital is analogous to open-chain amides. In certain cases, particularly when the ring is unsaturated, heterocyclic amines have unique and interesting properties. Let's look at several examples.

Pyrrole, a Five-Membered Aromatic Heterocycle

Pyrrole (two "r's"; one "l") is a five-membered heterocyclic amine with two double bonds and one nitrogen. Although pyrrole is both an amine and a conjugated diene, its chemistry is not consistent with either of these structural features. Unlike most amines, pyrrole is not basic; unlike most conjugated dienes, pyrrole doesn't undergo electrophilic addition reactions. How can we explain these observations?

In fact, pyrrole is *aromatic.* Even though it has a five-membered ring, pyrrole has six π electrons in a cyclic, conjugated π orbital system, just as benzene does (see Section 5.2). Each of the four carbon atoms contributes one π electron, and the sp^2-hybridized nitrogen atom contributes two more (its lone pair). The six π electrons occupy p orbitals with lobes above and below the plane of the flat ring, as shown in Figure 12.3. Because the lone-pair electrons on nitrogen are shared in the aromatic ring, they are not available for donation to an acid and pyrrole is nonbasic. Note in Figure 12.3 how the nitrogen atom is neutral (green) rather than electron-rich (red).

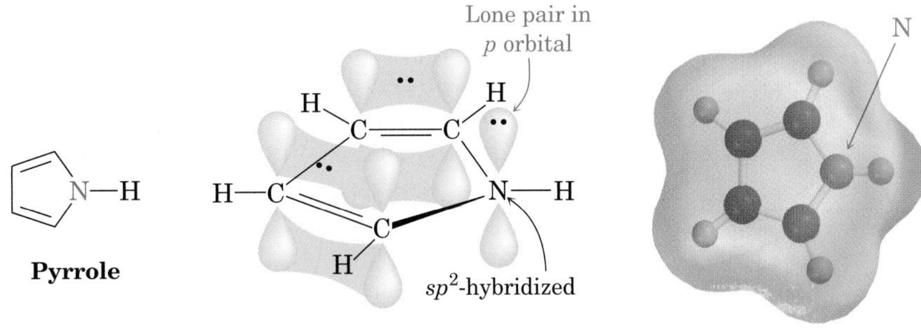

FIGURE 12.3 Pyrrole, an aromatic heterocycle, has a π electron structure similar to that of benzene. The nitrogen atom is nonbasic.

Like benzene, pyrrole undergoes substitution of a ring hydrogen atom on reaction with an electrophile. Substitution normally occurs at the position next to nitrogen, as the following nitration shows:

$$\text{Pyrrole} + HNO_3 \xrightarrow[\text{anhydride}]{\text{Acetic}} \text{2-Nitropyrrole} + H_2O$$

Pyrrole **2-Nitropyrrole** **(83)%**

Substituted pyrrole rings form the basic building blocks from which many important plant and animal pigments are constructed. Among these is *heme*, an iron-containing tetrapyrrole found in blood.

H_3C $CH{=}CH_2$ H_3C CH_3 N—Fe—N $HOOCCH_2CH_2$ $CH{=}CH_2$ $HOOCCH_2CH_2$ CH_3

Heme

PROBLEM 12.14 Pyrrole undergoes other typical electrophilic substitution reactions in addition to nitration. What products would you expect to obtain from reaction of *N*-methylpyrrole with the following reagents?

(a) Br_2 (b) $CH_3Cl, AlCl_3$ (c) $CH_3COCl, AlCl_3$

PROBLEM 12.15 Review the mechanism of the nitration of benzene (see Section 5.5), and then propose a mechanism for the nitration of pyrrole.

Pyridine, a Six-Membered Aromatic Heterocycle

Pyridine is a nitrogen-containing heterocyclic analog of benzene. Like benzene, pyridine is a flat molecule with bond angles of approximately 120° and with C–C bond lengths of 139 pm, intermediate between normal single and double bonds. Also like benzene, pyridine is aromatic with six π electrons in a cyclic, conjugated π orbital system. The sp^2-hybridized nitrogen atom and the five carbon atoms each contribute one π electron to the cyclic, conjugated *p* orbitals of the ring. Unlike the situation in pyrrole, however, the lone-pair electrons on the pyridine nitrogen atom are not part of the π orbital system but instead occupy an sp^2 orbital in the plane of the ring (Figure 12.4). As a result, the pyridine lone-pair electrons are available for donation to an acid and pyridine is therefore basic. Compare the electrostatic potential maps of pyrrole (Figure 12.3) and pyridine (Figure 12.4) to see this difference in basicity.

FIGURE 12.4 Electronic structure of pyridine, a nitrogen-containing analog of benzene.

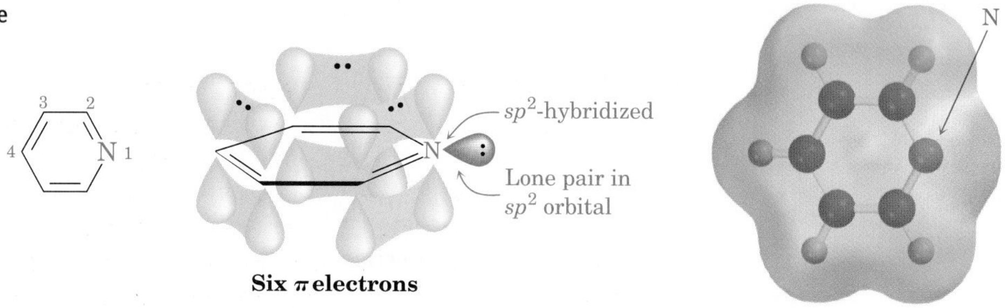

Although less basic than typical alkylamines, pyridine (pK_b = 8.75) is nevertheless used in a variety of organic reactions when a base catalyst is required. Recall, for instance, that the reaction of an acid chloride with an alcohol to yield an ester is commonly done in the presence of pyridine (see Section 10.9).

Substituted pyridines, such as the B_6 complex vitamins pyridoxal and pyridoxine, are important biologically. Present in yeast, cereal, and other foodstuffs, the B_6 vitamins are necessary for the synthesis of some amino acids.

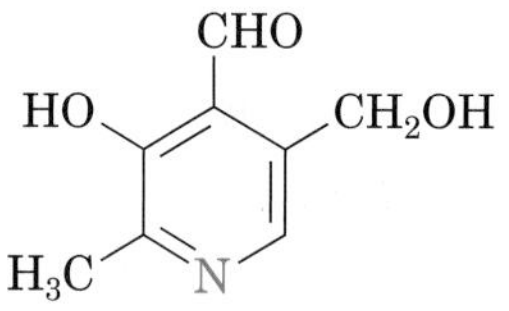

Pyridoxal

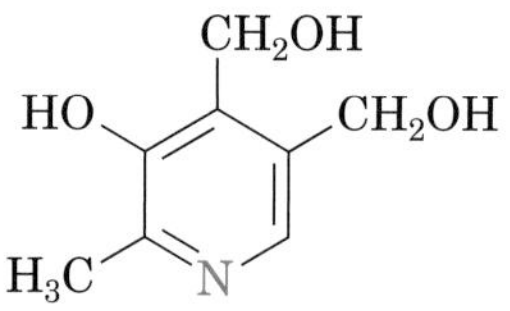

Pyridoxine

PROBLEM 12.16 The five-membered heterocycle imidazole contains two nitrogen atoms, one “pyrrole-like” and one “pyridine-like.” Draw an orbital picture of imidazole, and indicate the orbital in which each nitrogen has its electron lone pair.

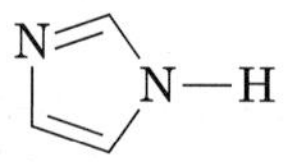

Imidazole

PROBLEM 12.17 Which nitrogen atom in imidazole (Problem 12.16) is more basic according to the following electrostatic potential map? Why?

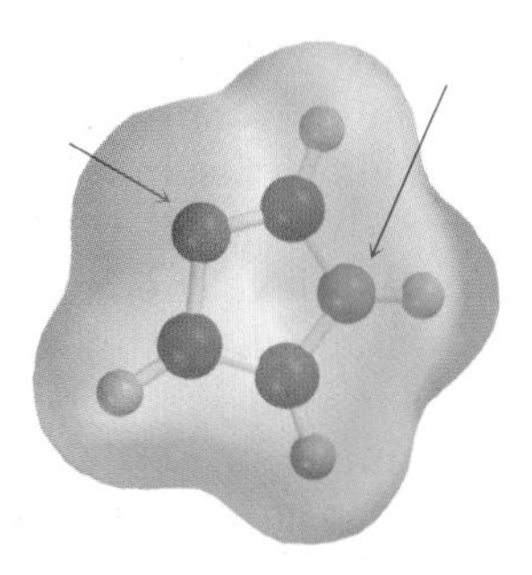

Imidazole

Fused-Ring Aromatic Heterocycles

Fused-ring heterocycles like quinoline, isoquinoline, and indole are more complex than simple monocyclic compounds. All three have a benzene ring and a heterocyclic ring sharing a common bond. These and many other fused-ring systems occur widely in nature, and many members of the class have useful biological properties. Quinine, a quinoline derivative found in the bark of the South American cinchona tree, is an important antimalarial drug. *N,N*-Dimethyltryptamine, which contains an indole ring, is a powerful hallucinogen.

Quinoline **Isoquinoline** **Indole**

Quinine, an antimalarial drug (a quinoline alkaloid)

***N, N*-Dimethyltryptamine, a hallucinogen (an indole alkaloid)**

12.7 Alkaloids: Naturally Occurring Amines

Naturally occurring amines derived from plant sources were once known as "vegetable alkali" because their aqueous solutions are slightly basic, but they are now referred to as **alkaloids**. The study of alkaloids provided much of the impetus for the growth of organic chemistry in the 19th century and remains today a fascinating area of research.

Alkaloids vary widely in structure, from the simple to the enormously complex. The odor of rotting fish, for example, is caused by the simplest alkaloid, methylamine. (The common use of acidic lemon juice to mask fish odors is just an acid–base reaction that forms the nonvolatile methylammonium salt.)

Many alkaloids have pronounced biological properties, and many of the pharmaceutical agents used today are alkaloids from natural sources. As only a few of many thousand examples, atropine, an antispasmodic agent used for the treatment of colitis, is obtained from the flowering plant *Atropa belladonna*, commonly called the deadly nightshade. Cocaine, both an anesthetic and a central nervous system stimulant, is obtained from the coca bush *Erythroxylon coca*, endemic to upland rainforest areas of Colombia, Ecuador, Peru, Bolivia, and western Brazil. Reserpine, a tranquilizer and antihypertensive, comes from powdered roots of the semitropical plant *Rauwolfia serpentina*. Ephedrine, a bronchodilator and decongestant, is obtained from the Chinese plant *Ephedra sinica*.

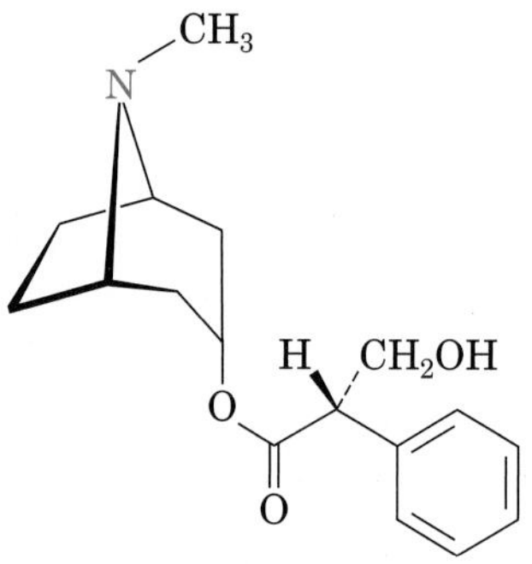

Atropine

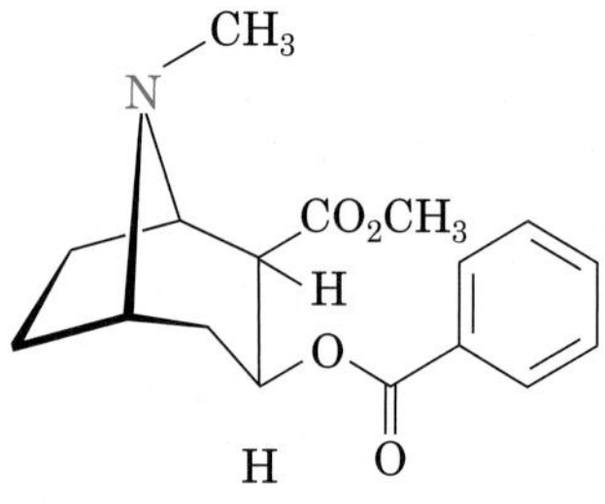

Cocaine

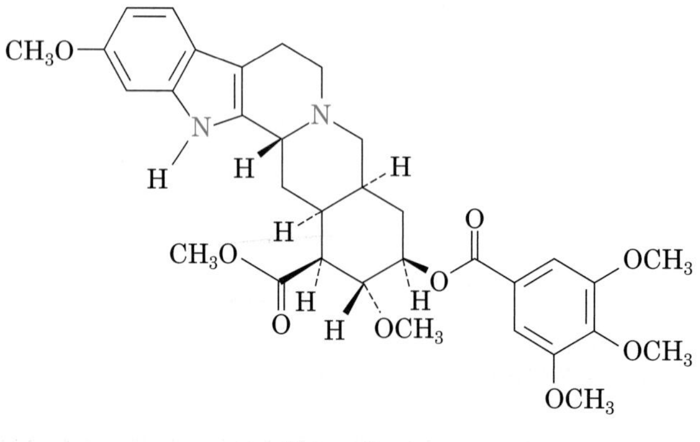

Reserpine

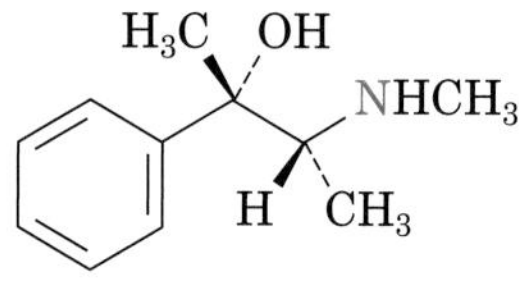

Ephedrine

INTERLUDE

Opium and Opiates

Medical uses of the poppy, *Papaver somniferum*, have been known at least since the 17th century, when crude extracts—called *opium*—were used for the relief of pain. Morphine was the first pure alkaloid to be isolated from opium, but its close relative, codeine, also occurs naturally. Codeine, which is simply the methyl ether of morphine, is used in prescription cough medicines and as an analgesic. Heroin, another close relative of morphine, does not occur naturally but is synthesized by diacetylation of morphine.

Morphine **Codeine** **Heroin**

Chemical investigations into the structure of morphine occupied some of the finest chemical minds of the 19th and early 20th centuries, and it was not until 1924 that the puzzle was finally solved by Robert Robinson.

Morphine and its relatives are extremely useful pharmaceutical agents, yet they also pose an enormous social problem because of their addictive properties. Much effort has therefore gone into understanding how morphine works and into developing modified morphine analogs that retain the analgesic activity but don't cause physical dependence. Our present understanding is that morphine functions by binding to so-called mu opioid receptor sites both in the spinal cord, where it interferes with the transmission of pain signals, and in brain neurons, where it changes the brain's reception of the signal.

Morphine and several of its relatives are isolated from the opium poppy, *Papaver somniferum*.

Hundreds of morphine-like molecules have been synthesized and tested for their analgesic properties. Research has shown that only part of the complex framework of morphine is necessary for biological activity. According to the "morphine rule," biological activity requires (1) an aromatic ring attached to (2) a quaternary carbon atom and (3) a tertiary amine situated (4) two carbon atoms farther away. Meperidine (Demerol), a widely used analgesic, and methadone, a substance used in the treatment of heroin addiction, are two of many compounds that fit the morphine rule.

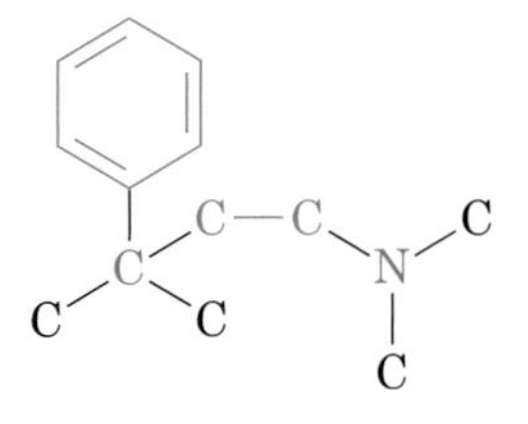

The morphine rule: an aromatic ring, attached to a quaternary carbon, attached to two more carbons, attached to a tertiary amine

Methadone **Meperidine**

Summary and Key Words

Alkaloid, p. 392
Alkylamine, p. 376
Amine, p. 375
Arylamine, p. 376
Basicity constant, K_b, p. 379
Heterocycle, p. 388
Primary amine (RNH_2), p. 376
Quaternary ammonium salt, p. 376
Reductive amination, p. 384
Secondary amine (R_2NH), p. 376
Tertiary amine (R_3N), p. 376

Amines are organic derivatives of ammonia. They are named in the IUPAC system either by adding the suffix *-amine* to the name of the alkyl substituent or by considering the amino group as a substituent on a more complex parent molecule. Bonding in amines is similar to that in ammonia: the nitrogen atom is sp^3-hybridized, the three substituents are directed to three corners of a regular tetrahedron, and the lone pair of electrons occupies the fourth corner of the tetrahedron.

The chemistry of amines is dominated by the presence of the lone-pair electrons on nitrogen, which make amines both basic and nucleophilic. **Arylamines** are generally weaker bases than **alkylamines** because their lone-pair electrons are shared by orbital overlap with the aromatic π electron system.

The simplest method of amine synthesis involves S_N2 reaction of ammonia or an amine with an alkyl halide. Alkylation of ammonia yields a primary amine, alkylation of a primary amine yields a secondary amine, and so on. Amines can also be prepared from amides and nitriles by reduction with $LiAlH_4$ and from aldehydes and ketones by **reductive amination** with ammonia and a reducing agent. Arylamines are prepared by nitration of an aromatic ring followed by reduction of the nitro group. Many of the reactions that amines undergo are familiar from previous chapters. Thus, amines react with alkyl halides in S_N2 reactions and with acid chlorides in nucleophilic acyl substitution reactions.

Heterocyclic amines, compounds in which the nitrogen atom is in a ring, have a great diversity in their structures and properties. Pyrrole, pyridine, indole, and quinoline all show aromatic properties.

Summary of Reactions

1. Synthesis of amines (Section 12.4)
   (a) Alkylamines by S_N2 reaction

$$NH_3 + RX \longrightarrow RNH_2$$
$$RNH_2 + RX \longrightarrow R_2NH$$
$$R_2NH + RX \longrightarrow R_3N$$
$$R_3N + RX \longrightarrow R_4N^+X^-$$

   (b) Reductive amination of aldehydes and ketones

$$RC(=O)R' + NH_3 \xrightarrow{H_2,\ Ni} RCH(NH_2)R'$$

   (c) Arylamines by reduction of nitrobenzenes

$$C_6H_5NO_2 \xrightarrow{Fe,\ H_3O^+} C_6H_5NH_2$$

EXERCISES

Visualizing Chemistry

12.18 Name the following amines, and identify each as primary, secondary, or tertiary:

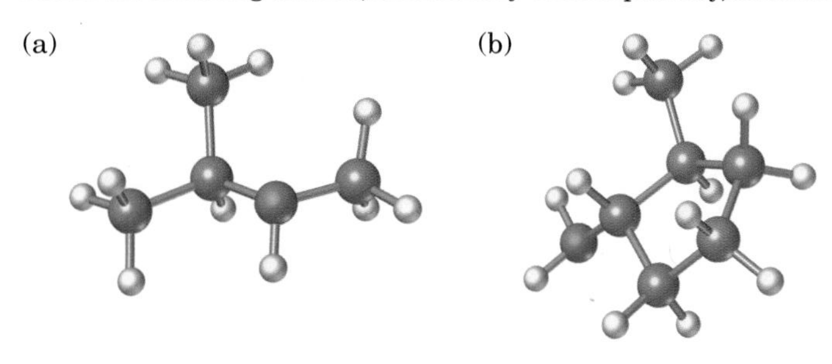

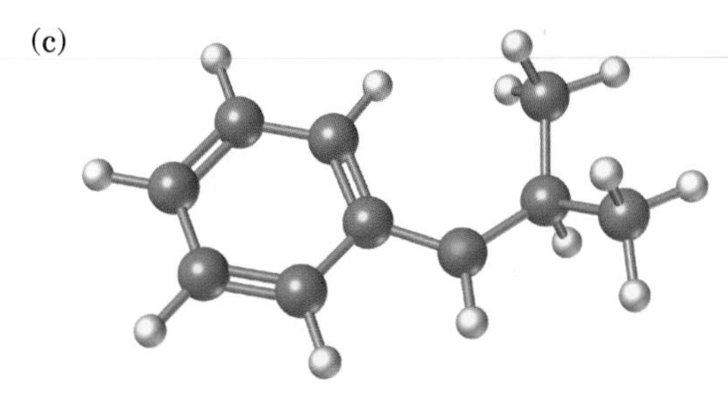

12.19 The following amine contains three nitrogen atoms. Rank them in order of expected increasing basicity.

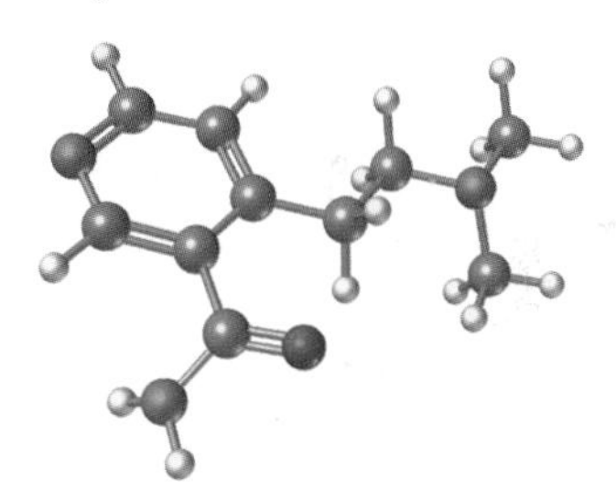

12.20 Which nitrogen atom in the alkaloid tryptamine is more basic? Explain.

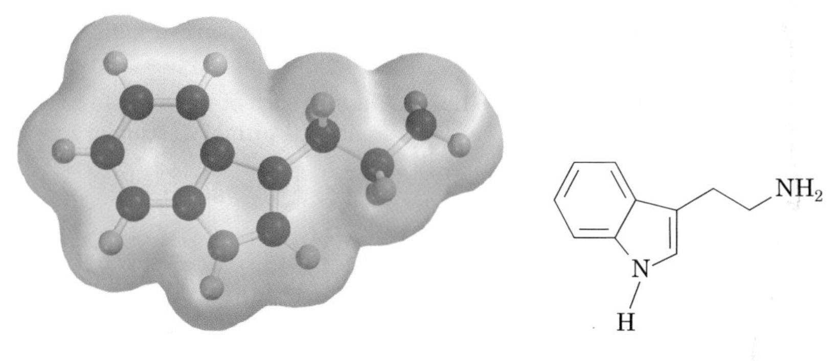

Tryptamine

12.21 Name the following amine, including *R,S* stereochemistry, and draw the product of its reaction with (i) CH_3CH_2Br and (ii) CH_3COCl:

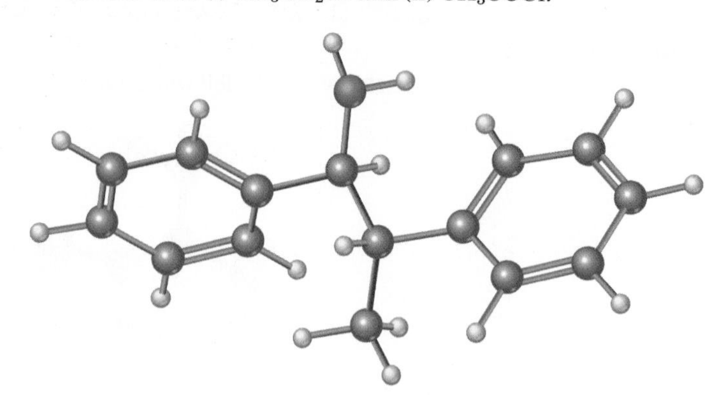

Additional Problems

NOMENCLATURE

12.22 Draw structures corresponding to the following IUPAC names:
(a) *N,N*-Dimethylaniline
(b) *N*-Methylcyclohexylamine
(c) (Cyclohexylmethyl)amine
(d) (2-Methylcyclohexyl)amine
(e) 3-(*N,N*-Dimethylamino)propanoic acid

12.23 Name the following compounds according to IUPAC rules:

(a) NH_2, Br, Br
(b) $CH_2CH_2NH_2$
(c) $NHCH_2CH_3$
(d) CH_3, N, CH_3
(e) $N—CH_2CH_2CH_3$
(f) $H_2NCH_2CH_2CH_2CN$

12.24 There are eight isomeric amines with the formula $C_4H_{11}N$. Draw them; name them; and classify each as primary, secondary, or tertiary.

12.25 Mescaline, a powerful hallucinogen derived from the peyote cactus, has the systematic name 2-(3,4,5-trimethoxyphenyl)ethylamine. Draw its structure.

12.26 Classify each of the amine (not amide) nitrogen atoms in the following substances as primary, secondary, or tertiary:

(a) O, $(C_2H_5)_2N—C$, $N—CH_3$, N, H

Lysergic acid diethylamide

(b) O, CH_3, H_3C, N, N, N, O, N, CH_3

Caffeine

12.27 Propose structures for amines that fit the following descriptions:
(a) A secondary arylamine
(b) A 1,3,5-trisubstituted arylamine
(c) An achiral quaternary ammonium salt
(d) A five-membered heterocyclic amine

PHYSICAL PROPERTIES

12.28 How can you explain the fact that trimethylamine (bp 3 °C) has a lower boiling point than dimethylamine (bp 7 °C) even though it has a higher molecular weight?

REACTIONS AND SYNTHESIS

12.29 Show the products of the following reactions:

(a) $CH_3CH_2CH_2NH_2 \xrightarrow{CH_3Br} ?$

(b) Cyclohexyl–$NH_2 \xrightarrow{HBr} ?$

(c) $CH_3CH_2\overset{O}{\overset{\|}{C}}NH_2 \xrightarrow[2.\ H_2O]{1.\ LiAlH_4} ?$

(d) Phenyl–$C{\equiv}N \xrightarrow[2.\ H_2O]{1.\ LiAlH_4} ?$

12.30 How might you prepare the following amines from ammonia and any alkyl halides needed?

(a) $CH_3CH_2CH_2CH_2CH_2CH_2NH_2$

(b) $(CH_3)_4N^+\ I^-$

(c) Phenyl–CH_2NH_2

(d)

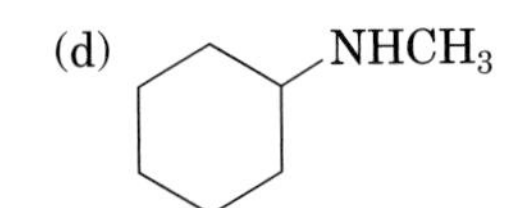

12.31 How might you prepare the following amines from 1-bromobutane?
(a) Butylamine (b) Dibutylamine (c) Pentylamine

12.32 How might you prepare each of the amines in Problem 12.31 from butan-1-ol?

12.33 How would you prepare benzylamine, $C_6H_5CH_2NH_2$, from each of the following starting materials?

(a) Phenyl–$CONH_2$ (b) Phenyl–CO_2H (c) Benzene

12.34 How might you prepare pentylamine from the following starting materials?
(a) Pentanamide (b) Pentanenitrile (c) Pentanoic acid

BASICITY

12.35 Which compound is more basic, $CH_3CH_2NH_2$ or $CF_3CH_2NH_2$? Explain.

12.36 Which compound is more basic, *p*-aminobenzaldehyde or aniline?

12.37 Which compound is more basic, triethylamine or aniline? Does the following reaction proceed as written?

$$(CH_3CH_2)_3NH^+\ Cl^- + C_6H_5NH_2 \longrightarrow C_6H_5NH_3^+\ Cl^- + (CH_3CH_2)_3N$$

INTEGRATED PROBLEMS

12.38 Suppose that you were given a mixture of toluene, aniline, and phenol. Describe how you would separate the mixture into its three pure components.

12.39 Would you expect diphenylamine to be more basic or less basic than aniline? Explain.

12.40 Hexane-1,6-diamine, one of the starting materials used for the manufacture of nylon 66, can be synthesized by a route that begins with the addition of Cl_2 to buta-1,3-diene (Section 4.9). How would you carry out the complete synthesis?

12.41 Another method for making hexane-1,6-diamine (see Problem 12.40) starts from adipic acid (hexanedioic acid). How would you carry out the synthesis?

12.42 Give the structures of the major organic products you would obtain from the reaction of *m*-methylaniline with the following reagents:
(a) Br_2 (1 mol) (b) CH_3I (excess) (c) CH_3COCl, pyridine

12.43 Draw structures for the following amines:
(a) 2-Ethylpyrrole (b) 2,3-Dimethylaniline (c) 3-Methylindole

12.44 Furan, the oxygen-containing analog of pyrrole, is aromatic in the same way that pyrrole is. Draw an orbital picture of furan, and show how it has six electrons in its cyclic conjugated π orbitals.

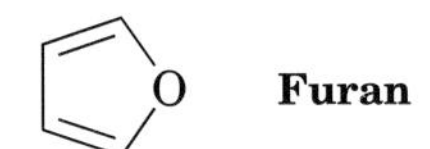

Furan

12.45 By analogy with the chemistry of pyrrole, what product would you expect from the reaction of furan with Br_2 (see Problem 12.44)?

12.46 How can you explain the observation that *p*-nitroaniline is less basic than aniline by a factor of 40,000? (See Section 5.8.)

12.47 In light of your answer to Problem 12.46, which would you expect to be more basic, aniline or *p*-methoxyaniline? Explain.

12.48 We've seen that amines are basic and amides are neutral. *Imides*, compounds with two carbonyl groups flanking an N–H, are actually acidic. Show by drawing resonance structures of the anion why imides are acidic.

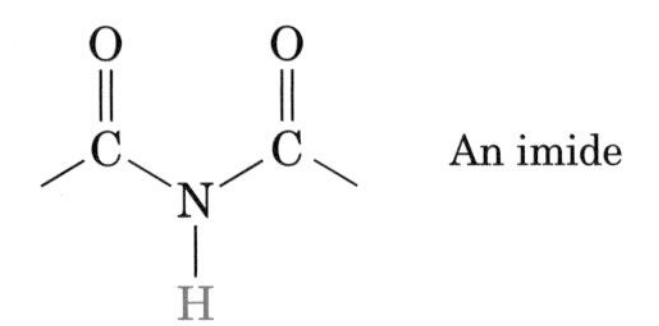

An imide

12.49 Atropine, $C_{17}H_{23}NO_3$, is a poisonous alkaloid isolated from the leaves and roots of the deadly nightshade, *Atropa belladonna*. In small doses, atropine acts as a muscle relaxant: 0.5 ng (1 nanogram = 10^{-9} g) is sufficient to cause pupil dilation. On reaction with aqueous NaOH, atropine yields tropic acid, $C_6H_5CH(CH_2OH)COOH$, and tropine, $C_8H_{15}NO$. Tropine, an optically inactive alcohol, yields tropidene on dehydration. Propose a structure for atropine.

N CH_3

Tropidene

12.50 Choline, a component of the phospholipids in cell membranes, can be prepared by S_N2 reaction of trimethylamine with ethylene oxide. Show the structure of choline, and propose a mechanism for the reaction.

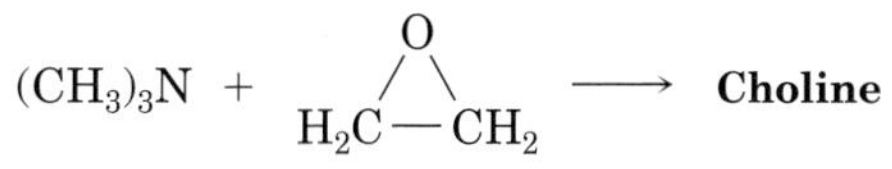

$$(CH_3)_3N + H_2C\overset{O}{—}CH_2 \longrightarrow \textbf{Choline}$$

12.51 The following transformation involves a conjugate nucleophilic addition reaction (Section 9.11) followed by an intramolecular nucleophilic acyl substitution reaction (Section 10.5). Show the mechanism.

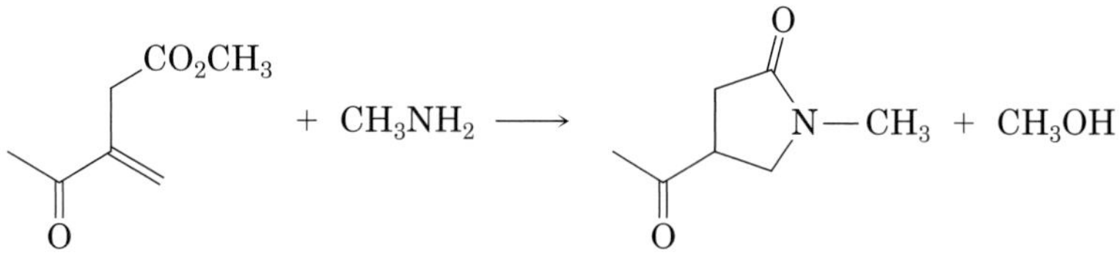

$$+ CH_3NH_2 \longrightarrow N—CH_3 + CH_3OH$$

12.52 Fill in the missing reagents a–f in the following scheme:

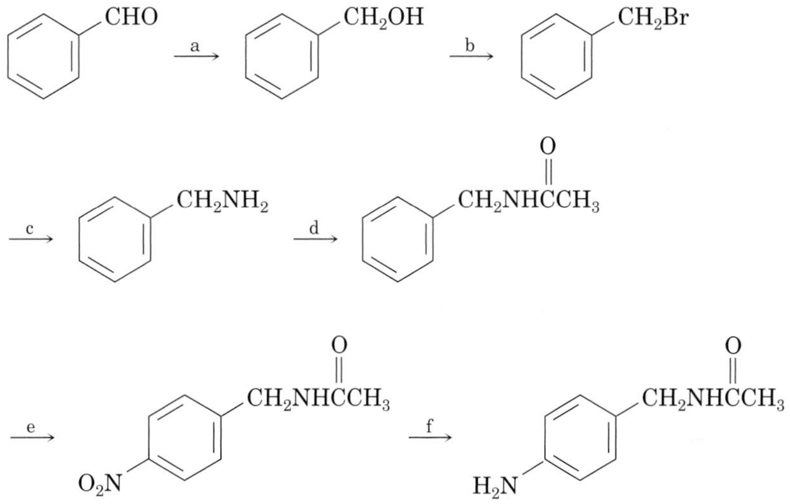

$$C_6H_5CHO \xrightarrow{a} C_6H_5CH_2OH \xrightarrow{b} C_6H_5CH_2Br \xrightarrow{c} C_6H_5CH_2NH_2 \xrightarrow{d} C_6H_5CH_2NHCCH_3 \xrightarrow{e} O_2N{-}C_6H_4CH_2NHCCH_3 \xrightarrow{f} H_2N{-}C_6H_4CH_2NHCCH_3$$

In the Medicine Cabinet

12.53 Mitomycin is a naturally occurring molecule that is effective at killing cancerous cells. Classify the amines in mitomycin as primary, secondary, and tertiary. The carbamate group (red) of mitomycin is different from an amine because it has different chemistry.

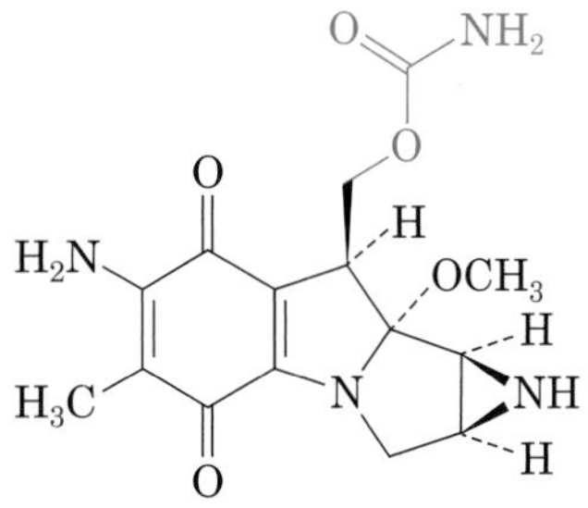

Mitomycin C

12.54 Identify the chirality centers in mitomycin (Problem 12.53).

12.55 Mitomycin (Problem 12.53) is toxic to cells because it cross-links DNA. Draw mechanisms for each step of this process.

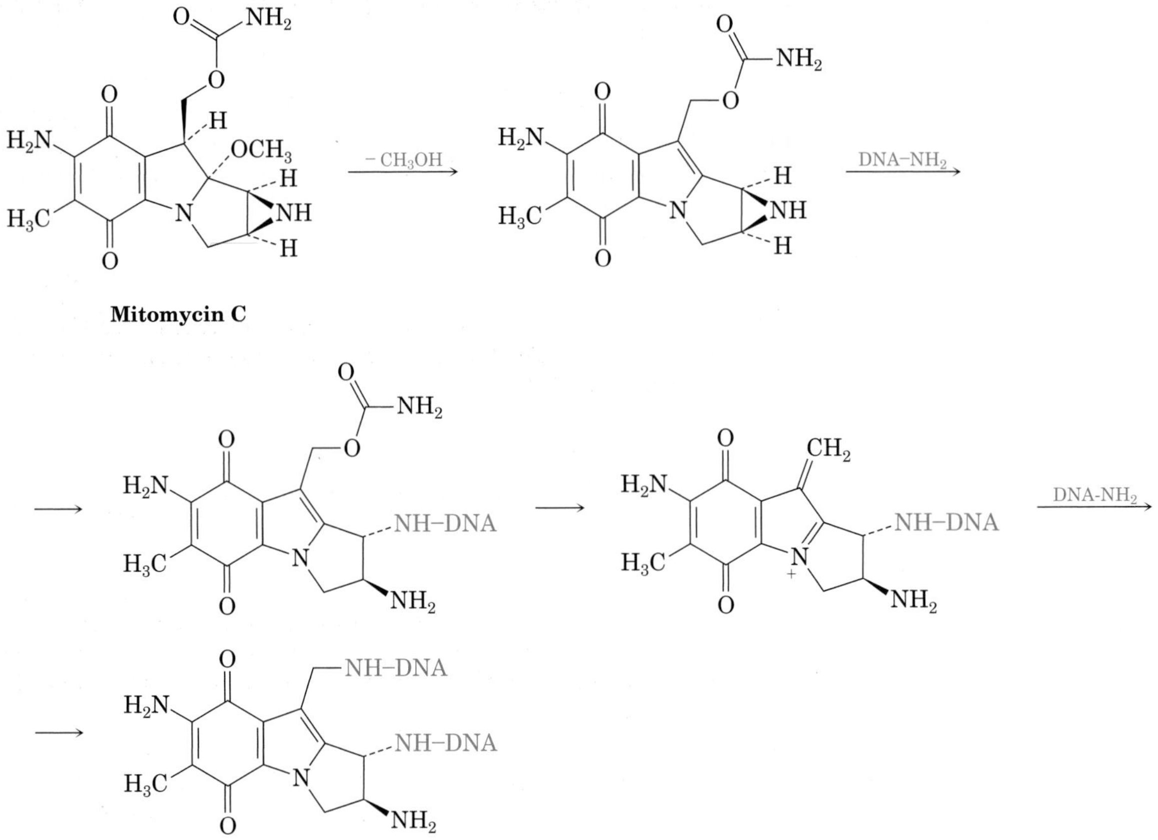

Mitomycin C

$- CH_3OH$

DNA–NH_2

NH–DNA

DNA-NH_2

NH–DNA

12.56 In Problem 12.55, why is the three-membered ring containing nitrogen (called an aziridine group) so reactive?

12.57 In Problem 12.55, classify the steps as a conjugate addition, elimination, or S_N2 reaction.

12.58 The anti-inflammatory drug celecoxib recently was withdrawn from the market and is now being considered for readmission. Draw a mechanism for the heterocycle-forming step in this synthesis of celecoxib.

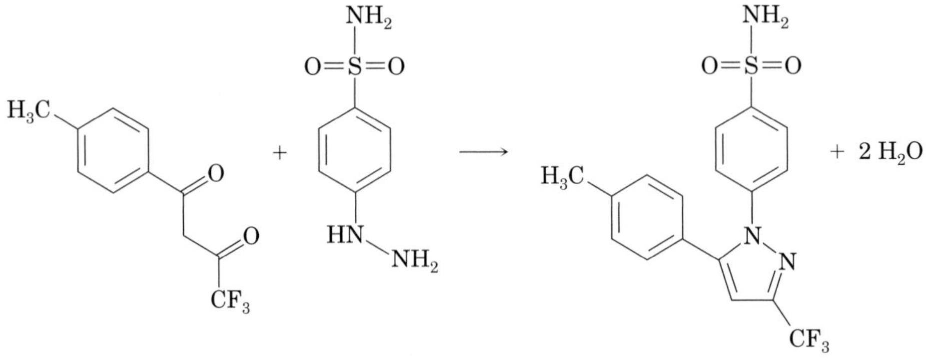

$+ 2\,H_2O$

Celecoxib

IN THE FIELD WITH AGROCHEMICALS

12.59 Marketed under names such as Stampede and Chem-Rice, propanil is commonly used to prevent weeds in rice fields. Draw the mechanism for the synthesis of propanil shown. Identify the key functional groups involved in the reaction.

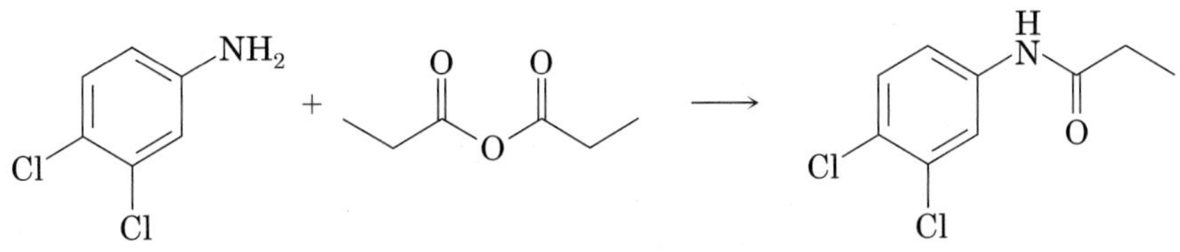

NH_2 Cl Cl + O O O → H N O Cl Cl

Propanil

12.60 DEET is the active ingredient in most mosquito repellants. DEET is prepared by a simple reaction that produces water. What two building blocks are used to prepare DEET?

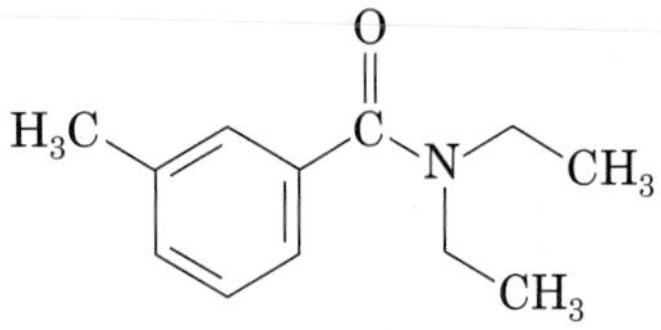

O H_3C C N CH_3 CH_3 **DEET**

12.61 Paraquat is a broad-spectrum weed killer. The following structure of paraquat shown is *incorrect*: the reaction is not charge-balanced. Fill in the missing charges on paraquat to correct the structure, and draw a mechanism for its synthesis.

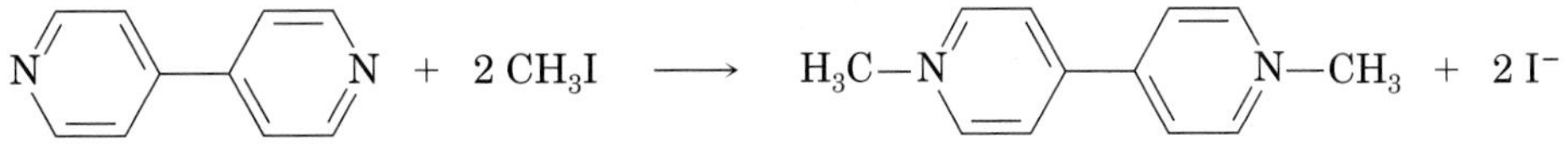

N N + 2 CH_3I ⟶ H_3C—N N—CH_3 + 2 I^-

Paraquat

INTERLUDE

The Road Map of Chemical Reactions

The reactions that we have covered in Chapters 1–12 link all the functional groups. Modern chemistry has produced many more reaction pathways than are shown here. Understanding these reactions enables us to appreciate the chemistry of life and its molecules. These are the topics we'll address in the remaining chapters of the text. Keeping fundamental principles in mind—chirality, carbocation stablility, Markovnikov's rule, the C–C bond-forming reactions of Chapter 12—will help us integrate this material into our emerging chemical intuition.

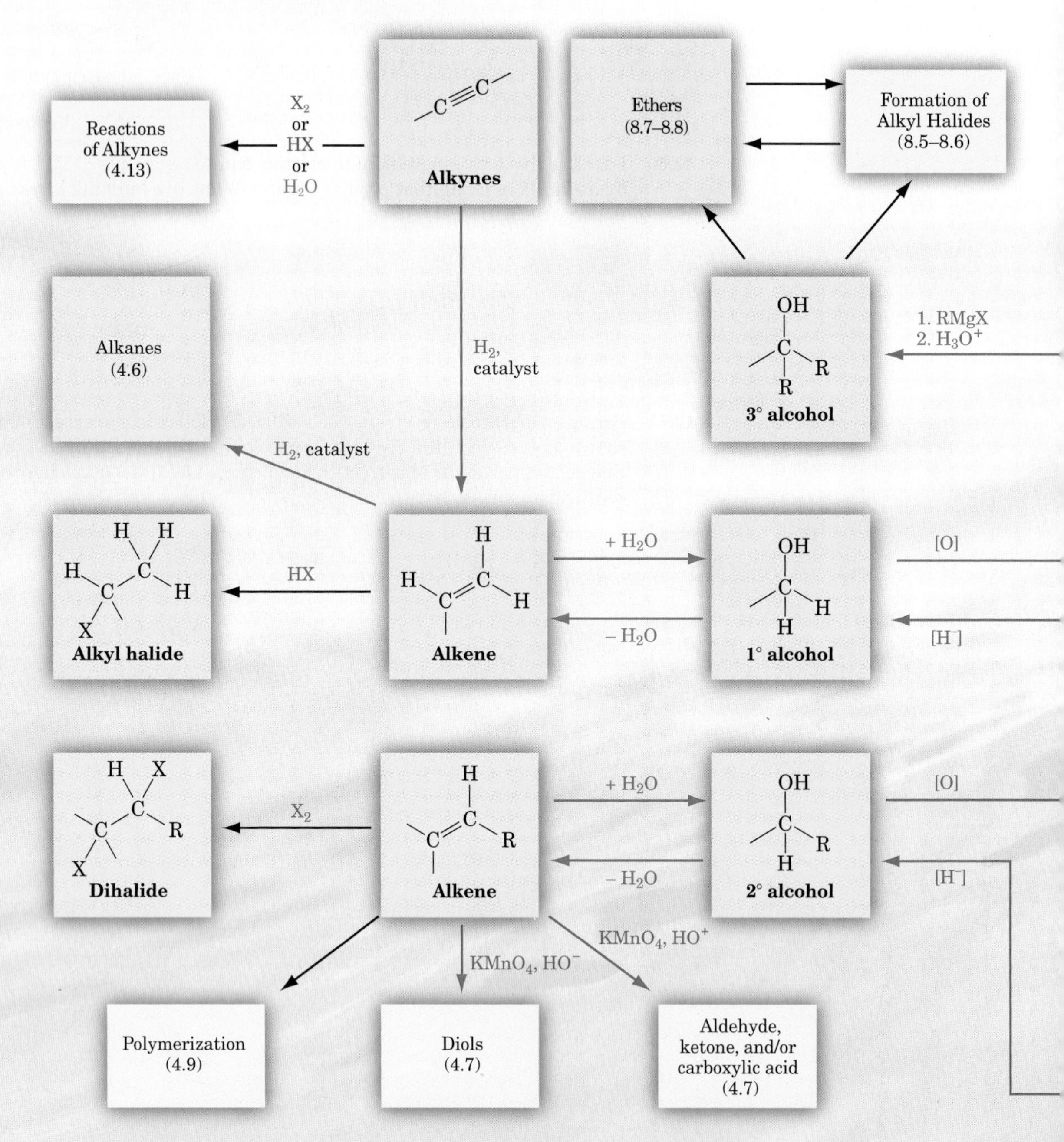

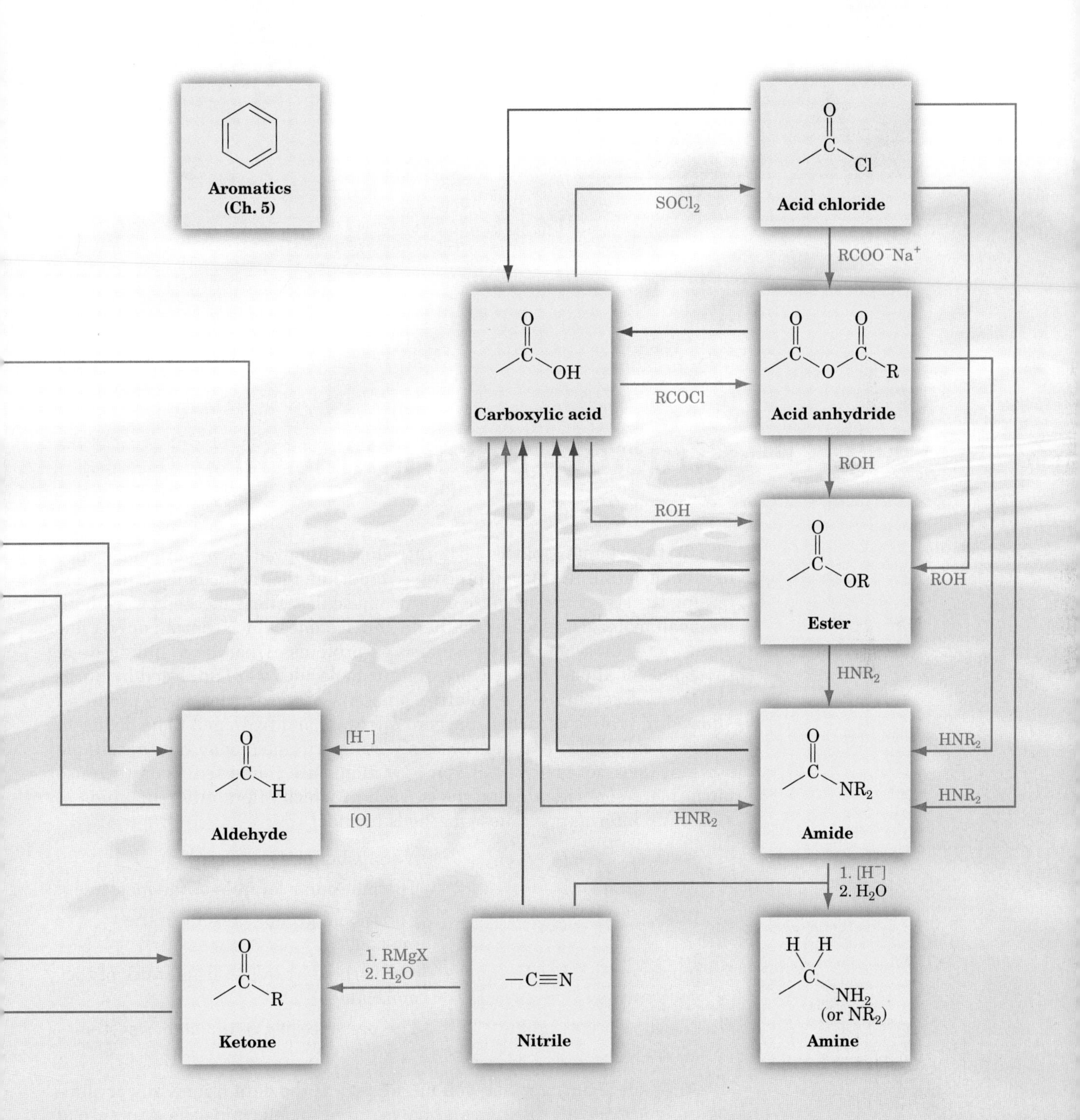

Aromatics
(Ch. 5)
Cl
Acid chloride
$SOCl_2$
$RCOO^-Na^+$
OH
Carboxylic acid
RCOCl
O
R
Acid anhydride
ROH
ROH
OR
Ester
ROH
HNR_2
H
Aldehyde
$[H^-]$
[O]
NR_2
Amide
HNR_2
HNR_2
HNR_2
1. $[H^-]$
2. H_2O
R
Ketone
1. RMgX
2. H_2O
—C≡N
Nitrile
H H
NH_2
(or NR_2)
Amine

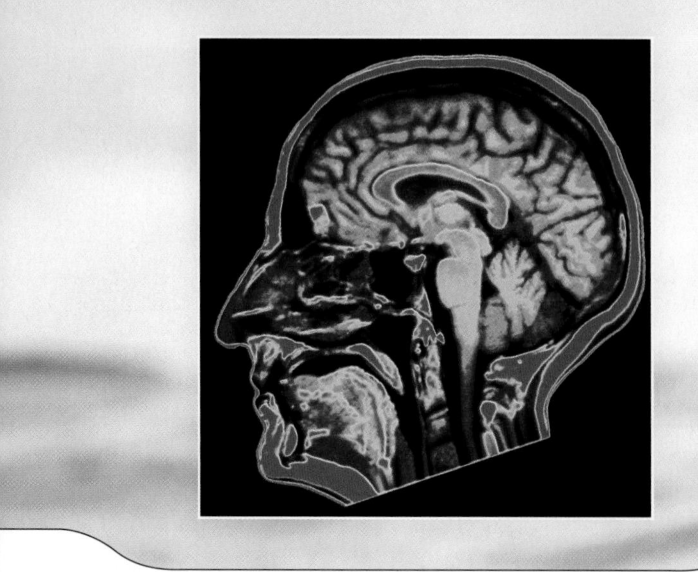

MRI is an application of NMR spectroscopy. Contrast agents like the gadolinium chelate can enhance the images. (Scott Camazine/Photo Researchers, Inc.)

13 CHAPTER Structure Determination

The first 12 chapters of this book introduced hundreds of organic molecules in atomic detail. In this chapter we will outline the methods used in determining the structures of these organic molecules. These methods work both for small molecules, such as methane and acetone, which contain only a few atoms, and for large molecules, such as proteins, which have hundreds of thousands of atoms. Historically, organic molecules were analyzed by burning them. Examining the relative amounts of gases that were produced allowed the ratios of carbon, oxygen, nitrogen, and hydrogen to be deduced. These so-called elemental analyses are still performed today, but this simple strategy does not give us data on how atoms are connected. Today, we use much more sophisticated methods, each of which offers different kinds of structural information. These methods include:

X-ray crystallography	A snapshot of the molecule
Mass spectrometry	The molecular formula of the molecule
Infrared spectroscopy	Functional groups present in the molecule
Nuclear magnetic resonance spectroscopy	The carbon–hydrogen framework of the molecule
Ultraviolet and visible spectroscopy	The presence of a conjugated π system

After surveying this list, you might ask, "If the data from X-ray crystallography provide all the information required to determine the structure of a molecule, why bother with the other techniques?" As it turns out, while X-ray crystallography data alone can determine a molecule's structure, it is rarely easy or quick to perform. (We'll discuss why later.) In addition, the other methods—with the right equipment—can be performed in a matter of minutes. Sometimes the equipment will fit into a suitcase; other times, it requires an entire room. These practical issues often dictate where and when a particular method is used. Most research laboratories have access to all these techniques and commonly use as many as possible to accurately determine a molecule's structure.

13.1 Electromagnetic Radiation: A Common Probe of Structure

Common to most of the characterization techniques that we will discuss is electromagnetic radiation. Electromagnetic radiation is light: not only the light we see, but the light we cannot see as well. Even though the **electromagnetic spectrum** (Figure 13.1) represents a continuous range of energy, it can be divided into regions such as visible light, X-rays, and microwaves.

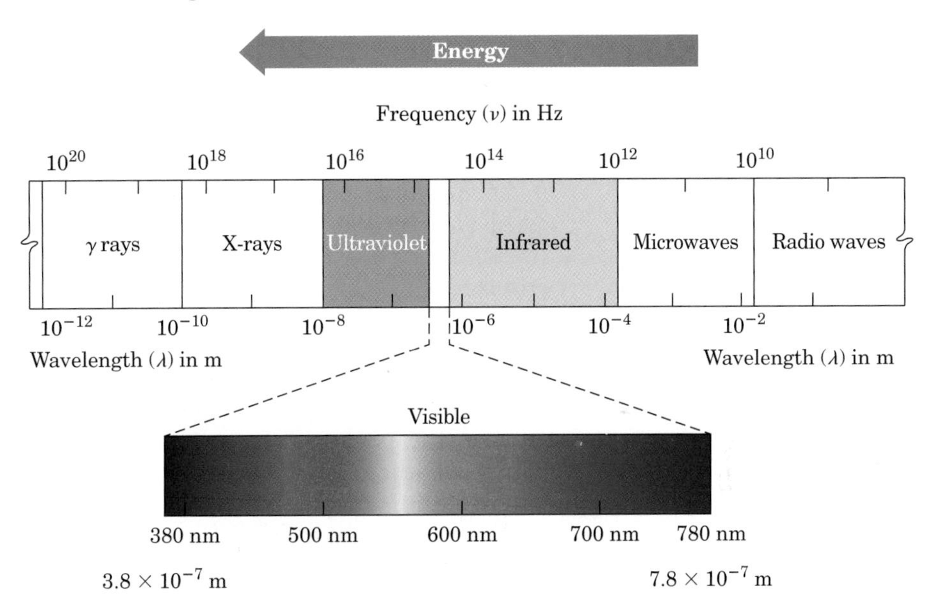

FIGURE 13.1 The electromagnetic spectrum.

Electromagnetic radiation is characterized by a frequency, a wavelength, and an amplitude (Figure 13.2). The energy of light is proportional to its **frequency**, ν, which is the number of waves that pass a fixed point per unit time. More waves mean more energy. Therefore, light of shorter wavelength is higher in energy because more waves can pass a fixed point per unit time. Frequency, which is measured in **hertz** (Hz), corresponds to a "per second," or s^{-1}, as in "900,000,000 waves per second," or 900 MHz, the operating frequency of most cellular phones.

Just as matter comes only in discrete units called *atoms*, electromagnetic energy is transmitted only in discrete amounts called *quanta*. The amount of energy, ϵ, corresponding to 1 quantum of energy (1 photon) of a given frequency, ν, is expressed by the Planck equation:

$$\epsilon = h\nu$$

where h = Planck's constant (6.62×10^{-34} J · s = 1.58×10^{-34} cal · s).

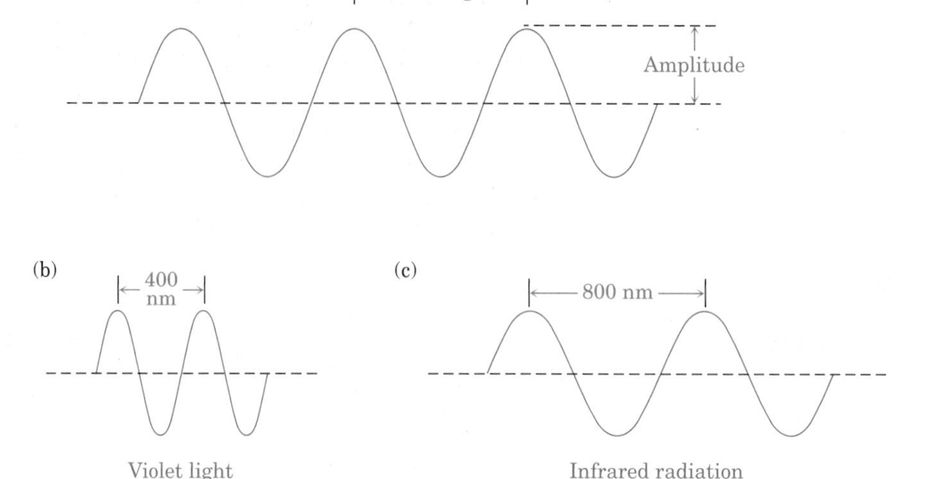

FIGURE 13.2 (a) Wavelength (λ) is the distance between two successive wave maxima. Amplitude is the height of the wave measured from the center. (b)–(c) What we perceive as different kinds of electromagnetic radiation are simply waves with different wavelengths and frequencies.

The length of a single wave of electromagnetic radiation, or **wavelength**, can be calculated because all waves travel at the same speed. Waves move at the speed of electromagnetic radiation, or as we more commonly say, the speed of light. We abbreviate the speed of light as c, a constant equal to 3×10^8 m/s. Wavelength, λ, can be calculated from the frequency or energy with the following equations:

$$\lambda = \frac{c}{\upsilon} \quad \text{or} \quad \lambda = \frac{hc}{\epsilon}$$

The **amplitude** is the height of the wave, measured from the midpoint between the peak and trough to the maximum. The intensity of radiant energy, whether it is a feeble beam or blinding glare, is proportional to the square of the wave's amplitude.

We take advantage of different portions of the electromagnetic spectrum to obtain different types of information about the structure of a molecule. To summarize:

- **X-ray crystallography** High-energy, short-wavelength X rays are passed through a periodic lattice (or crystal) of organic molecules. This light diffracts off the planes of atoms in the crystal, and from this scattering, the three-dimensional structure can be calculated.
- **Mass spectrometry** Electromagnetic radiation is not required. We'll discuss this technique later.
- **Ultraviolet (UV) and visible spectroscopy** A broad spectrum (the UV-visible range) of light is passed through a sample of molecules. Conjugated networks of π bonds can absorb this moderate-energy light. Differences in the conjugated π bond networks lead to the absorption of different regions of the spectrum, and from this, the structure can be identified. Often, a single broad and characteristic absorption is diagnostic.
- **IR spectroscopy** **IR spectroscopy** works like UV-visible spectroscopy except that a lower-energy region of the electromagnetic spectrum is used. The portions of the spectrum absorbed by the sample correspond to

vibrations of specific bonds in the molecule. Unlike the UV-visible spectrum, the IR spectrum of a molecule contains multiple absorptions that correspond to all the possible vibrations in the molecule. Only a small amount of this information is needed to identify specific functional groups.

- **Nuclear magnetic resonance spectroscopy** When organic molecules containing carbon and hydrogen are placed in a strong magnetic field, radio waves can be used to probe the connectivity of these atoms. From the nuclear magnetic resonance (NMR) spectrum, the carbon and hydrogen arrangements can be determined.

PRACTICE PROBLEM 13.1

Which is higher in energy, FM radio waves with a frequency of 1.015×10^8 Hz (101.5 MHz) or visible light with a frequency of 5×10^{14} Hz?

STRATEGY Remember the equation $\epsilon = h\nu$, which says that energy increases as frequency increases and as wavelength decreases.

SOLUTION Since visible light has a higher frequency than radio waves, it is higher in energy.

PRACTICE PROBLEM 13.2

What is the wavelength in meters of visible light with a frequency of 4.5×10^{14} Hz?

STRATEGY Frequency and wavelength are related by the equation $\lambda = c/\nu$, where c is the speed of light (3.0×10^8 m/s).

SOLUTION

$$\lambda = \frac{3.0 \times 10^8 \text{ m/s}}{4.5 \times 10^{14} \text{ s}^{-1}} = 6.7 \times 10^{-7} \text{ m}$$

PROBLEM 13.1 How does the energy of infrared radiation with $\lambda = 1.0 \times 10^{-6}$ m compare with that of an X ray having $\lambda = 3.0 \times 10^{-9}$ m?

PROBLEM 13.2 Which is higher in energy, radiation with $\nu = 4.0 \times 10^9$ Hz or radiation with $\lambda = 9.0 \times 10^{-6}$ m?

PROBLEM 13.3 Why are we concerned more about potential health risks associated with X rays and sunlight than those associated with cellular phones and radios? What are the typical energies associated with a 900 MHz cellular phone, a transistor radio, sunlight, and a medical X-ray film? Use Figure 13.1 to choose representative frequencies on which to base your calculations. How do these energies compare with the strength of a typical C–C bond?

13.2 X-Ray Crystallography

Recent advances in the speed of computers have led to a revolution in chemistry, physics, and biology. With fast computers, the opportunity to process the data obtained from **X-ray crystallographic** studies has led to the elucidation of the structures of small molecules and large proteins alike. A snapshot of four organic molecules is shown in Figure 13.3. Two molecules of arachidonic acid, $C_{20}H_{32}O_2$, a molecule involved in the inflammatory response, are bound to the dimer of cyclooxygenase (blue), the enzyme that processes arachidonic acid in cells. The exact connectivity of atoms, and thus the structure of this molecule, is easily determined from this data. While the size at which this structure is printed does not allow us to readily see the atom-to-atom connectivity, the data can be seen in the expanded portion. Twenty years ago, such detail was just a dream, and most proteins were depicted as colored circles in textbooks. Now, the three-dimensional structures of thousands of proteins are known.

FIGURE 13.3 The structure of arachidonic acid and cyclooxygenase as determined from X-ray crystallography. (Dr. Lisa M. Pérez, Texas A&M University. Used with permission.)

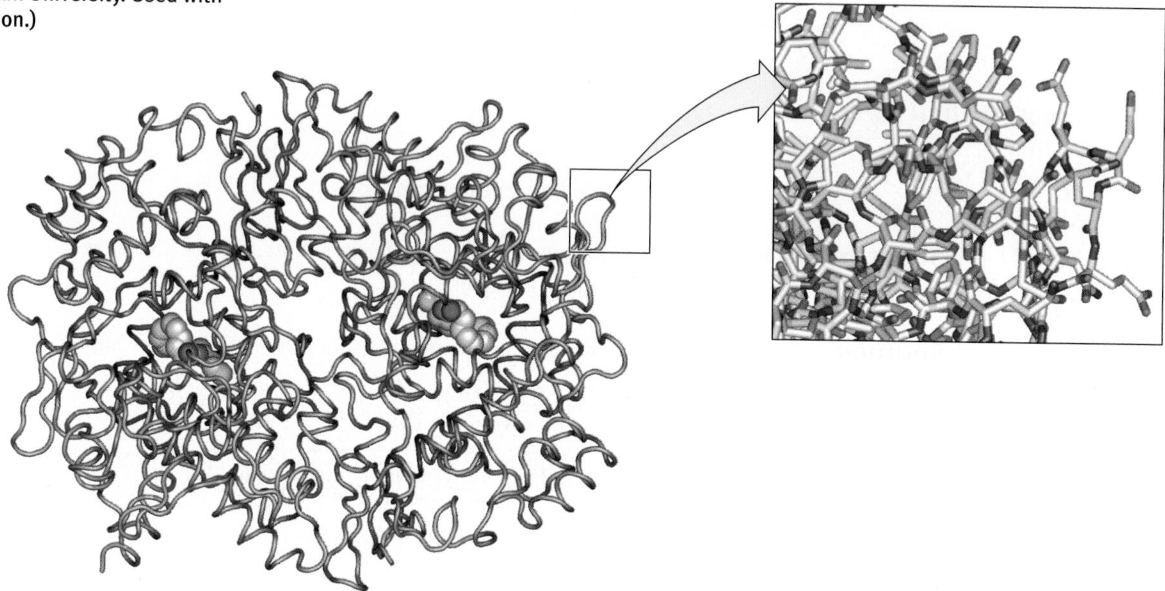

The details of X-ray crystallography constitute an entire course, but the basic principles behind this method can be simply understood (Figure 13.4). As its name implies, X-ray crystallography requires a source of X rays, which are typically generated by passing large amounts of energy through a filament and then focusing the light that is produced toward a target. The target in this case is a crystal—a periodic arrangement of molecules of the compound in question. With periodicity, the atoms of the molecules form planes that diffract the X rays. This diffraction is measured on a detector as a series of spots, which are given coordinates. If enough spots are collected, a computer can reconstruct the three-dimensional network of the atoms' electron density. This electron density is then interpreted to provide the three-dimensional structure.

FIGURE 13.4 The steps required to perform a structure determination using X-ray crystallography include obtaining crystals, collecting X-ray diffraction data, solving the data set to generate an electron density map, and interpreting the electron density map.

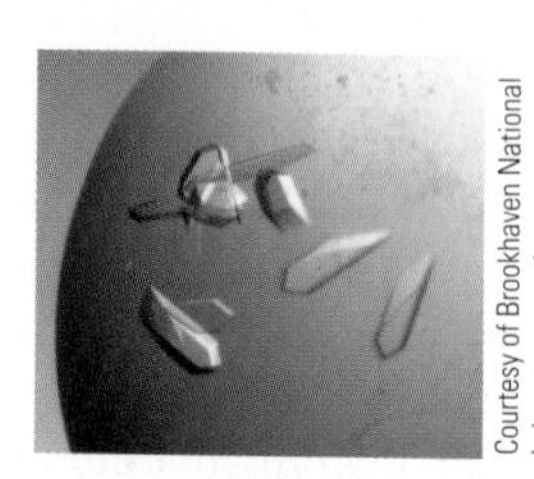

Courtesy of Brookhaven National Laboratory

Crystals

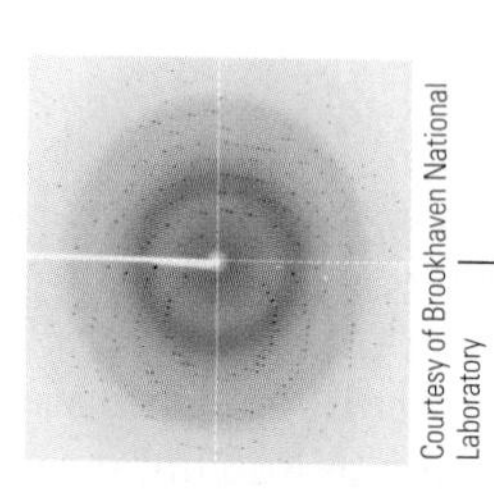

Courtesy of Brookhaven National Laboratory

Diffraction spots

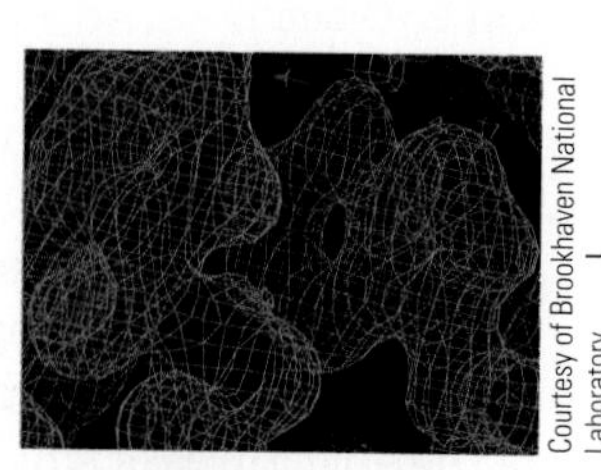

Courtesy of Brookhaven National Laboratory

Electron density map

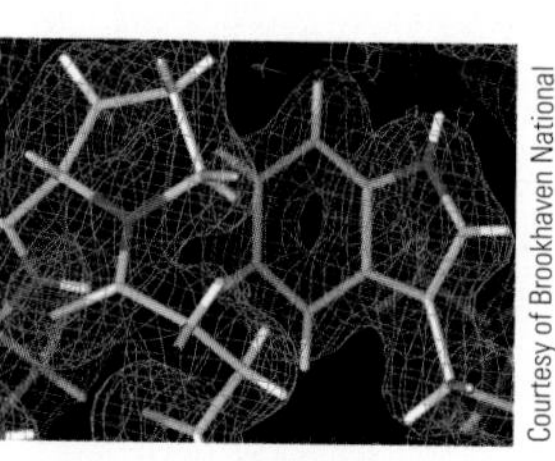

Courtesy of Brookhaven National Laboratory

Molecular structure

While some organic molecules like table sugar form perfect crystals quickly, it can take months or years to obtain crystals of other organic molecules. To obtain enough data to determine the structure, sometimes a special X ray–generating facility is needed. Data for proteins are usually collected at only a handful of sites in the world. Even with the data in hand, the amount of time required to process the data, an exercise that requires both a scientist and a computer, can vary. Despite these hurdles, it is now likely that the three-dimensional structures of most human proteins will be determined in our lifetimes. Often, proteins are co-crystallized with small molecules, including potential drugs, to aid in the drug-discovery process or to understand how the protein works to do chemistry. Both Figure 13.3 and our discussions of atrazine in Chapter 5 are examples of these uses.

13.3 Mass Spectrometry

Mass spectrometry provides the molecular weight and formula of a molecule. This method takes anywhere from seconds to minutes and is routinely used to verify the presence or absence of a molecule in a sample. You have probably seen a mass spectrometer at work. Airport security personnel will swab luggage and place the swab under a rubber flap. What happens next is mass spectrometry. A vacuum pulls molecules from the swab and sends them flying into the spectrometer, which provides a charge to (or *ionizes*) each molecule. Timing how long the newly charged molecules take to transit part of the spectrometer (their time of flight) allows the molecular weight and formula of the molecule to be calculated. In our example, if the time measured matches that of a known chemical hazard, such as an explosive, the airport machine will signal that a threat was detected.

Mass spectrometers differ in a number of ways, including how the samples are introduced into the machine, how the molecules are given charge, and the ways that the samples are separated (time of flight or path through a magnetic field). Regardless of the instrumentation, the net result is the same. A mass spectrometer provides the molecular weights of small organic molecules in a sample, often to four decimal places. From this accuracy, the molecular formula can be determined unambiguously.

Notice that the data are not reported in molecular weight (g/mol) but rather as a ratio of **mass to charge (m/z)**. This reporting scheme is necessary because only ions are detected in mass spectrometry. The x-axis of the mass spectrum gives the ion's mass in amu (atomic mass units) divided by its charge. In most cases the charge is 1, so the mass of the ion can be obtained. The y-axis of a mass spectrum indicates the percent relative abundance of the ion, the distribution of ions measured by the instrument. For example, the mass spectrum of benzene (C_6H_6) shows that the predominant ion measured has an m/z ratio of 78, which is the same as its molecular weight of benzene in grams per mole.

Data from mass spectrometry can provide more information than just molecular weight. If the molecule decomposes, the fragments can provide structural data. For benzene, these major fragments correspond to C_2H_3 (m/z 27), C_3H_3 (m/z 39), and C_4H_4 (m/z 52).

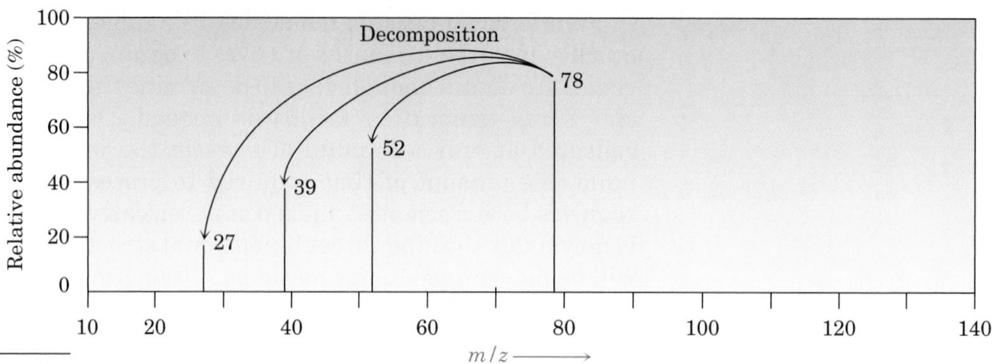

TABLE 13.1 Common Isotopic Ratios Observed in Mass Spectometry

Isotopes	Relative population
$^{12}C:^{13}C$	100:1
$^{32}S:^{34}S$	25:1
$^{35}Cl:^{37}Cl$	3:1
$^{79}Br:^{81}Br$	1:1

Because mass spectrometry looks at individual molecules, individual isotopes are observable. One out of every 100 carbons is ^{13}C instead of ^{12}C, and the mass spectrum of methane shows two major peaks; the first peak, corresponding to 16 mass units ($^{12}CH_4$), is 100 times greater in abundance than the second peak, which corresponds to 17 mass units ($^{13}CH_4$). Other common isotopes are shown in Table 13.1.

PRACTICE PROBLEM 13.3

Assign the peaks in the spectrum of chlorobenzene:

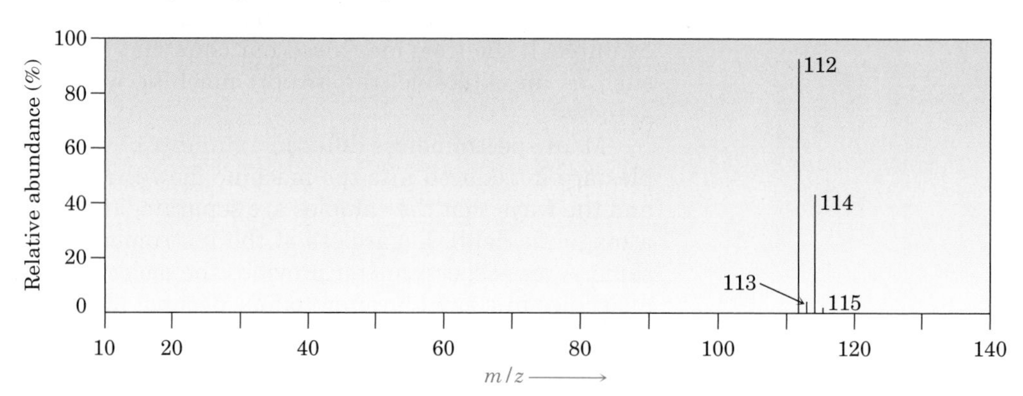

SOLUTION Table 13.1 shows that the ratio of ^{12}C:^{13}C is 100:1 and the ratio of ^{35}Cl:^{37}Cl is 3:1. The molecular weight of $^{12}C_6H_5{}^{35}Cl$ is 112 amu. This peak is three times higher than the peak at 114 *m/z* corresponding to $^{12}C_6H_5{}^{37}Cl$. The chance that one of the carbon atoms is ^{13}C is only 1%, so the tiny peaks at 113 *m/z* and 115 *m/z* correspond to chlorobenzene molecules with a single ^{13}C. We don't see chlorobenzene molecules with more than one ^{13}C because the peaks are too small.

PROBLEM 13.4 Sketch the mass spectrum of the fumigant methylbromide and explain the relative heights of the peaks.

PROBLEM 13.5 The mass spectrum recorded for the halogenation of but-2-ene with X_2 showed one peak with an *m/z* of 310. What halogen was used?

PROBLEM 13.6 The oxidation of butan-1-ol proceeded to give the desired major product, butanal, and a trace impurity. Based on the mass spectrum and your chemical intuition, what is the impurity?

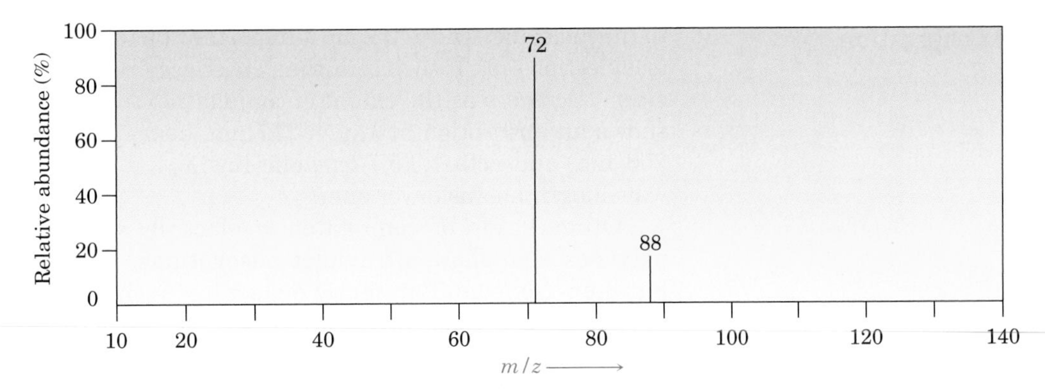

13.4 Ultraviolet Spectroscopy

The **ultraviolet (UV)** region of the electromagnetic spectrum extends from 10^{-8} m to the low-wavelength end of the visible region (4×10^{-7} m). The portion of greatest interest to organic chemists, though, is the narrow range from 2×10^{-7} m to 4×10^{-7} m. Absorptions in this region are measured in *nanometers* (nm), where 1 nm = 10^{-9} m = 10^{-7} cm. Thus, the ultraviolet range of interest is from 200 to 400 nm.

The energy of UV radiation corresponds to the amount necessary to raise the energy level of a π electron in an unsaturated molecule.

Ultraviolet spectra are recorded by irradiating a sample with UV light of continuously changing wavelength. When the wavelength of light corresponds to the amount of energy required to promote a π electron in an unsaturated molecule to a higher level, energy is absorbed. The absorption is detected and displayed on a chart that plots wavelength versus the amount of radiation absorbed.

A typical UV spectrum—that of buta-1,3-diene—is shown in Figure 13.5. Unlike IR spectra (see Section 13.6), which generally have many peaks, UV spectra are usually quite simple. Often, there is only a single broad peak, which is identified by noting the wavelength at the very top (λ_{max}). For buta-1,3-diene, λ_{max} = 217 nm.

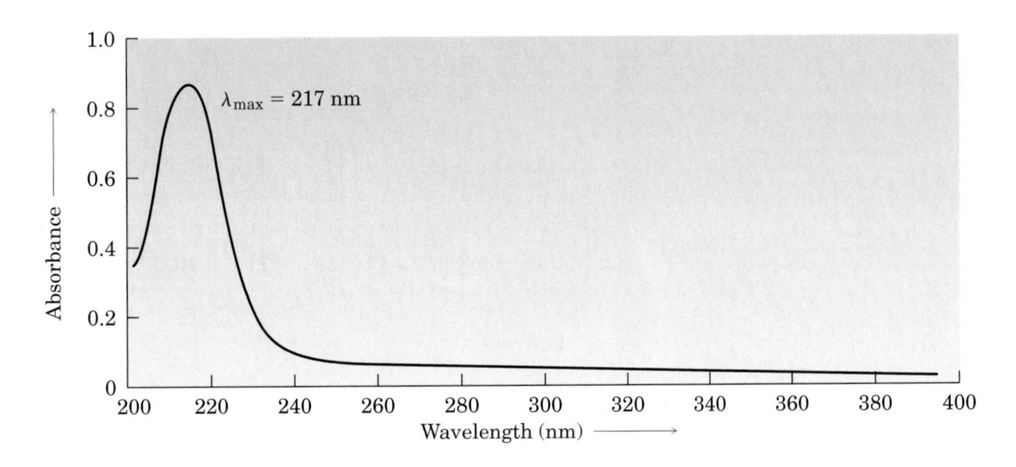

FIGURE 13.5 Ultraviolet spectrum of buta-1,3-diene.

13.5 Interpreting Ultraviolet Spectra: The Effect of Conjugation

The wavelength of radiation necessary to raise the energy of a π electron in an unsaturated molecule depends on the nature of the π electron system in the molecule. One of the most important factors is the extent of conjugation (see Section 4.9). It turns out that the energy required for an electronic transition *decreases* as the extent of conjugation *increases*. Thus, buta-1,3-*di*ene shows an absorption at λ_{max} = 217 nm, hexa-1,3,5-*tri*ene absorbs at λ_{max} = 258 nm, and octa-1,3,5,7-*tetra*ene has λ_{max} = 290 nm. (Remember: longer wavelength means lower energy.)

Other kinds of conjugated π electron systems besides dienes and polyenes also show ultraviolet absorptions. Conjugated enones, such as but-3-en-2-one, and aromatic molecules, such as benzene, also have characteristic UV absorptions that aid in structure determination. The UV absorption maxima of some representative conjugated molecules are given in Table 13.2.

Ultraviolet-visible spectroscopy is a common technique for measuring the concentration of species in solution. Beer's law tells us that the absorbance, A, of a sample is proportional to the concentration of molecules in solution.

$$A = \epsilon c l$$

where ϵ = the molar absorptivity of the molecule, a molecule-specific constant that reflects how well it absorbs light in units of L mol^{-1} cm^{-1} (usually given for the λ_{max} of the compound)

c = concentration of molecules in solution in mol/L

l = the distance the light travels through the sample, the pathlength, in cm

For example, β-carotene, the pigment responsible for the color of carrots, absorbs light very well (ϵ = 100,000 L mol^{-1} cm^{-1}). If a sample of β-carotene

TABLE 13.2 Ultraviolet Absorption Maxima of Some Conjugated Molecules

Name	Structure	λ_{max} (nm)
Ethylene	$H_2C{=}CH_2$	171
2-Methylbuta-1,3-diene	$H_2C{=}C(CH_3){-}CH{=}CH_2$	220
Cyclohexa-1,3-diene	(cyclohexa-1,3-diene ring)	256
Hexa-1,3,5-triene	$H_2C{=}CH{-}CH{=}CH{-}CH{=}CH_2$	258
But-3-en-2-one	$H_2C{=}CH{-}C(CH_3){=}O$	219
Benzene	(benzene ring)	254

is placed in a holder that has a pathlength of 1 cm and the absorbance reads 0.2, we can calculate the concentration of β-carotene to be

$$A = \epsilon c l$$

$$c = A/\epsilon l = 0.2/(100{,}000)(1) = 2 \times 10^{-6}\ \mathrm{M}$$

PRACTICE PROBLEM 13.5

Hexa-1,5-diene and hexa-1,3-diene are isomers. How can you distinguish them by UV spectroscopy?

$H_2C{=}CHCH_2CH_2CH{=}CH_2$ — **Hexa-1,5-diene**

$CH_3CH_2CH{=}CHCH{=}CH_2$ — **Hexa-1,3-diene**

SOLUTION Hexa-1,3-diene is a conjugated diene, but hexa-1,5-diene is nonconjugated. Only the conjugated isomer shows a UV absorption above 200 nm.

PROBLEM 13.7 Which of the following compounds show UV absorptions in the range 200 to 400 nm?

(a)

(b) 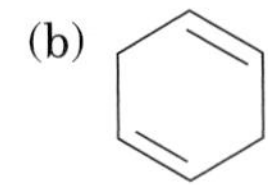

(c) 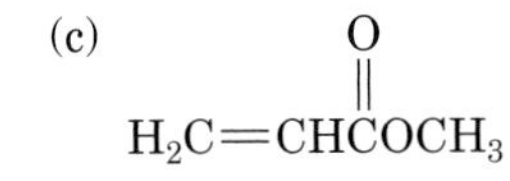

(d) Br, CH3

(e) O, CH3

(f) O, CH3

PROBLEM 13.8 How can you distinguish between hexa-1,3-diene and hexa-1,3,5-triene by UV spectroscopy?

PROBLEM 13.9 The molar absorptivity for the base cytosine is 6,100 $M^{-1}\ cm^{-1}$ when measured at 220 nm. If the absorbance of a sample of cytosine is measured to be 0.2 for a 1 cm pathlength, what is the concentration?

PROBLEM 13.10 A 0.018 μM solution of hexa-1,3,5-triene has an absorbance of $A = 0.63$ when the pathlength is 1 cm. What is the molar absorptivity of hexa-1,3,5-triene?

13.6 Infrared Spectroscopy of Organic Molecules

The infrared region of the electromagnetic spectrum covers the range from just above the visible (7.8×10^{-7} m) to approximately 10^{-4} m, but only the middle of the region is used by organic chemists. This midportion extends from 2.5×10^{-5} to 2.5×10^{-6} cm, and wavelengths are usually given in *micrometers* (μm; 1 μm $= 10^{-6}$ m). Frequencies are usually given in **wavenumbers ($\tilde{\nu}$)**, rather than in hertz. The wavenumber is equal to the reciprocal of the wavelength in centimeters and is thus expressed in units of reciprocal centimeters (cm^{-1}):

$$\text{Wavenumber:}\quad \tilde{\nu}\ (\mathrm{cm}^{-1}) = \frac{1}{\lambda\ (\mathrm{cm})}$$

Why does a molecule absorb some wavelengths of infrared energy but not others? All molecules have a certain amount of energy, which causes bonds to stretch and contract, atoms to wag back and forth, and other molecular motions to occur. Following are some of the kinds of allowed vibrations:

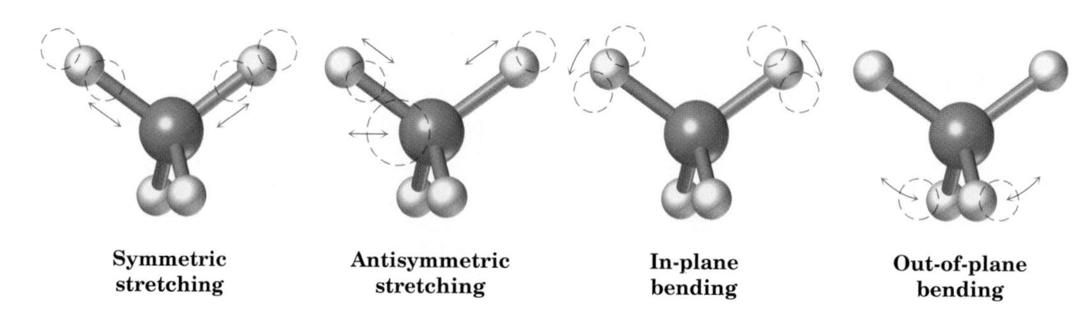

The amount of energy a molecule contains is not continuously variable but is *quantized.* That is, a molecule can vibrate only at specific frequencies corresponding to specific energy levels. Take bond stretching, for example. Although we usually speak of bond lengths as if they were fixed, the numbers given are really averages because bonds are constantly stretching and bending, lengthening and contracting. Thus, a typical C–H bond with an average bond length of 110 pm is actually vibrating at a specific frequency, alternately stretching and compressing as if there were a spring connecting the two atoms.

When the molecule is irradiated with electromagnetic radiation, *energy is absorbed if the frequency of the radiation matches the frequency of the vibration.* The result of energy absorption is an increased amplitude for the vibration; in other words, the "spring" connecting the two atoms stretches and compresses a bit further. Since each frequency absorbed by a molecule corresponds to a specific molecular motion, we can find what kinds of motions a molecule has by measuring its IR spectrum. By then interpreting those motions, we can find out what kinds of bonds (functional groups) are present in the molecule.

IR spectrum ⟶ What molecular motions? ⟶ What functional groups?

The full interpretation of an IR spectrum is difficult because most organic molecules are so large that they have dozens of different bond stretching and bending motions. Thus, an IR spectrum usually contains dozens of absorptions. Fortunately, we don't need to interpret an IR spectrum fully to get useful information because *functional groups have characteristic IR absorptions that don't change from one compound to another.* The C=O absorption of a ketone is almost always in the range 1680 to 1750 cm^{-1}, the O–H absorption of an alcohol is almost always in the range 3400 to 3650 cm^{-1}, the C=C absorption of an alkene is almost always in the range 1640 to 1680 cm^{-1}, and so forth. By learning to recognize where characteristic functional-group absorptions occur, it's possible to get structural information from IR spectra.

Look at the IR spectra of cyclohexanol and cyclohexanone in Figure 13.6 to see how they can be used. Although both spectra contain many peaks, the

FIGURE 13.6 Infrared spectra of (a) cyclohexanol and (b) cyclohexanone. Such spectra are easily obtained in minutes with milligram amounts of material.

characteristic absorptions of the different functional groups allow the compounds to be distinguished. Cyclohexanol shows a characteristic alcohol O–H absorption at 3300 cm^{-1} and a C–O absorption at 1060 cm^{-1}; cyclohexanone shows a characteristic ketone C=O peak at 1715 cm^{-1}.

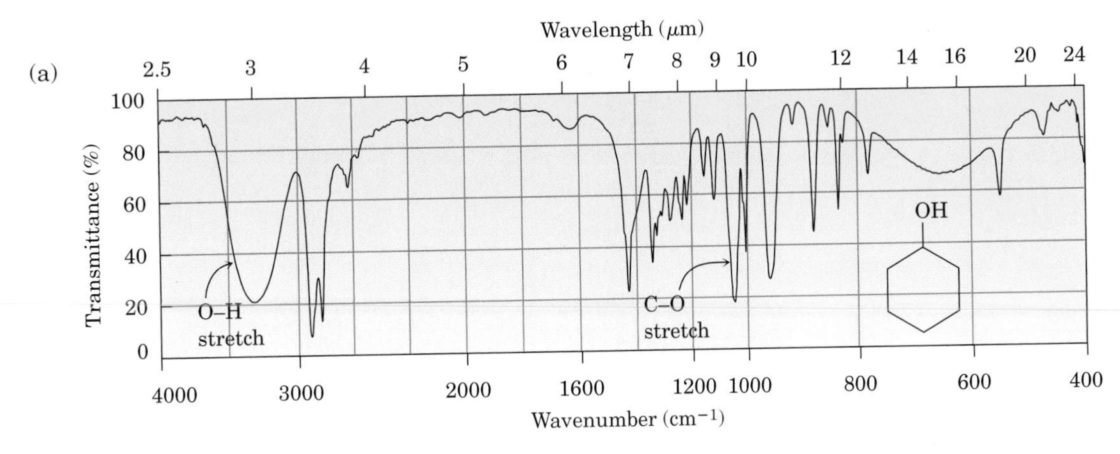

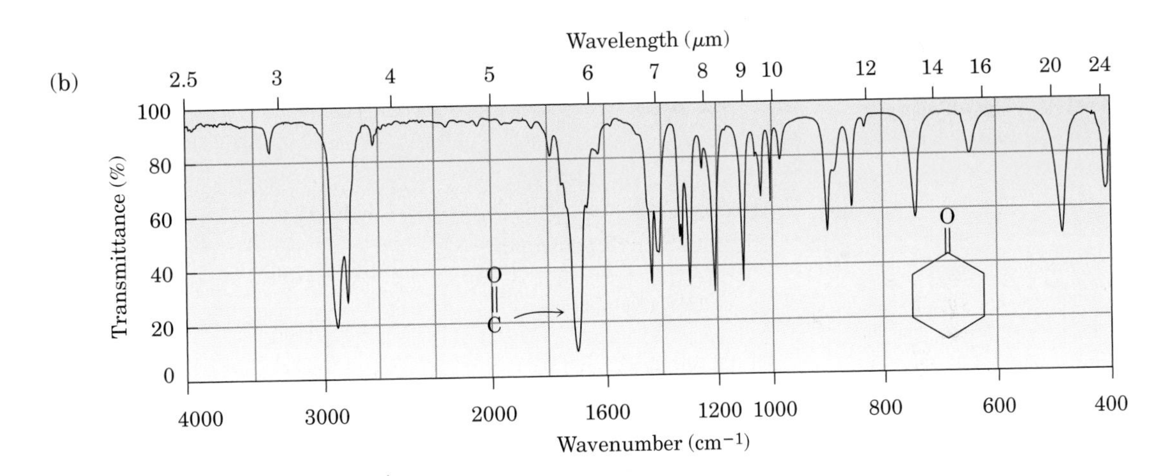

One further point about infrared spectroscopy: it's also possible to obtain structural information from an IR spectrum by noticing which absorptions are *not* present. If the spectrum of an unknown does *not* have an absorption near 3400 cm^{-1}, the unknown is not an alcohol; if the spectrum does not have an absorption near 1715 cm^{-1}, the unknown is not a ketone; and so on. Table 13.3 lists characteristic IR absorption frequencies of some common functional groups.

It helps in remembering the positions of various IR absorptions to divide the infrared range from 4000 to 200 cm^{-1} into four parts, as shown in Figure 13.7.

- The region from 4000 to 2500 cm^{-1} corresponds to N–H, C–H, and O–H bond-stretching motions. Both N–H and O–H bonds absorb in the 3300 to 3600 cm^{-1} range, whereas C–H bond-stretching occurs near 3000 cm^{-1}. Since almost all organic compounds have C–H bonds, almost all IR spectra have an intense absorption in this region.

TABLE 13.3 Characteristic Infrared Absorptions of Some Functional Groups

Functional group class	Band position (cm^{-1})	Intensity of absorption
Alkanes; alkyl groups		
C—H	2850–2960	Medium to strong
Alkenes		
=C—H	3020–3100	Medium
C=C	1640–1680	Medium
Alkynes		
≡C—H	3300	Strong
—C≡C—	2100–2260	Medium
Alkyl halides		
C—Cl	600–800	Strong
C—Br	500–600	Strong
C—I	500	Strong
Alcohols		
O—H	3400–3650	Strong, broad
C—O	1050–1150	Strong
Aromatics		
C—H	3030	Weak
(aromatic ring C=C)	1660–2000	Weak
	1450–1600	Medium
Amines		
N—H	3300–3500	Medium
C—N	1030–1230	Medium
Carbonyl compounds[a]		
C=O	1670–1780	Strong
Carboxylic acids		
O—H	2500–3100	Strong, very broad
Nitriles		
C≡N	2210–2260	Medium
Nitro compounds		
NO_2	1540	Strong

[a]Carboxylic acids, esters, aldehydes, and ketones.

- The region from 2500 to 2000 cm^{-1} is where triple-bond stretching occurs. Both nitriles (RC≡N) and alkynes (RC≡CR′) absorb here.
- The region from 2000 to 1500 cm^{-1} is where C=O, C=N, and C=C bonds absorb. Carbonyl groups generally absorb from 1670 to 1780 cm^{-1}, and alkene stretching normally occurs in the narrow range from 1640 to 1680 cm^{-1}. The exact position of a C=O absorption is often diagnostic of the exact kind of carbonyl group in the molecule. Esters usually absorb at 1735 cm^{-1}; aldehydes, at 1725 cm^{-1}; and open-chain ketones, at 1715 cm^{-1}.
- The region below 1500 cm^{-1} is the so-called fingerprint region. A large number of absorptions due to various C–O, C–C, and C–N single-bond vibrations occur here, forming a unique pattern that acts as an identifying "fingerprint" of each organic molecule.

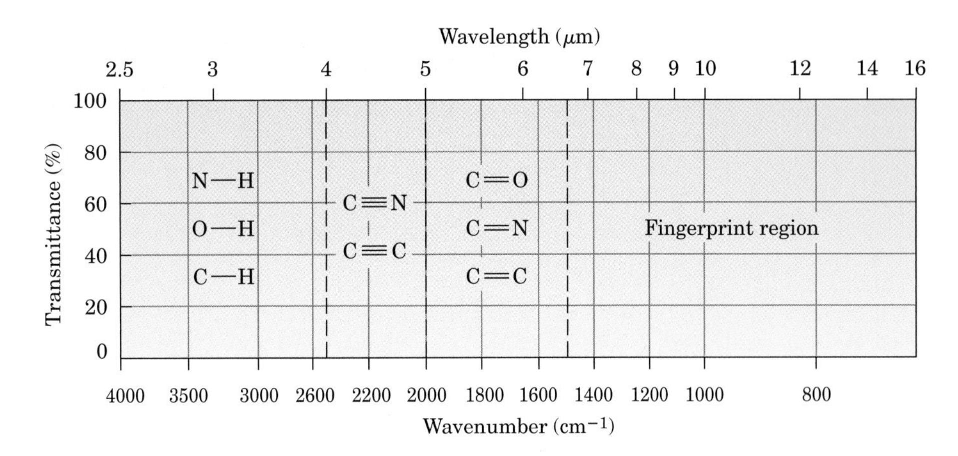

FIGURE 13.7 Single-bond, double-bond, triple-bond, and fingerprint regions in the infrared spectrum.

PRACTICE PROBLEM 13.4

Refer to Table 13.3 and make educated guesses about the functional groups that cause the following IR absorptions:
(a) 1735 cm^{-1} (b) 3500 cm^{-1}

SOLUTION (a) An absorption at 1735 cm^{-1} is in the carbonyl-group region of the IR spectrum, probably an ester.
(b) An absorption at 3500 cm^{-1} is in the –OH (alcohol) region.

PRACTICE PROBLEM 13.5

Acetone and 2-propen-1-ol ($H_2C{=}CHCH_2OH$) are isomers. How could you distinguish them by IR spectroscopy?

STRATEGY Identify the functional groups in each molecule, and refer to Table 13.3.

SOLUTION Table 13.3 shows that acetone has a strong C=O absorption at 1715 cm^{-1}, while prop-2-en-1-ol has an –OH absorption at 3500 cm^{-1} and a C=C absorption at 1660 cm^{-1}.

PROBLEM 13.11 What functional groups might molecules contain if they show IR absorptions at the following frequencies?
(a) 1715 cm^{-1} (b) 1540 cm^{-1}
(c) 2210 cm^{-1} (d) 1720 and 2500–3100 cm^{-1}
(e) 3500 and 1735 cm^{-1}

PROBLEM 13.12 How might you use IR spectroscopy to help distinguish between the following pairs of isomers?

(a) CH_3CH_2OH and CH_3OCH_3

(b) 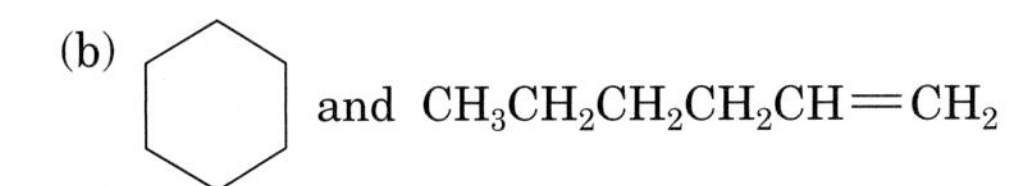

(c) 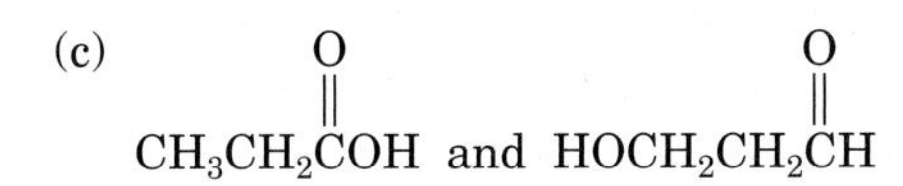

PROBLEM 13.13 Where might the following molecule have IR absorptions?

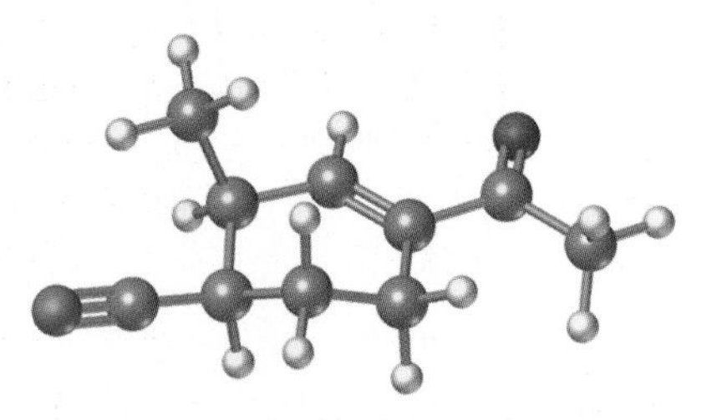

13.7 Nuclear Magnetic Resonance Spectroscopy

Nuclear magnetic resonance (NMR) spectroscopy provides a "map" of the carbon–hydrogen framework in an organic molecule. How does NMR spectroscopy work? Many kinds of nuclei, including ^{1}H and ^{13}C, behave like a child's top spinning about an axis. Since they're positively charged, these spinning nuclei act like tiny magnets and interact with an external magnetic field (denoted $\boldsymbol{B_0}$). In the absence of an external magnetic field, the nuclear spins of magnetic nuclei are oriented randomly. When a sample containing these nuclei is placed between the poles of a strong magnet, however, the nuclei adopt specific orientations, much as a compass needle orients itself in the earth's magnetic field.

A spinning ^{1}H or ^{13}C nucleus can orient so that its own tiny magnetic field is aligned either with (parallel to) or against (antiparallel to) the external field. The two orientations don't have the same energy and therefore aren't equally likely. The parallel orientation is slightly lower in energy, making this spin state slightly favored over the antiparallel orientation (Figure 13.8).

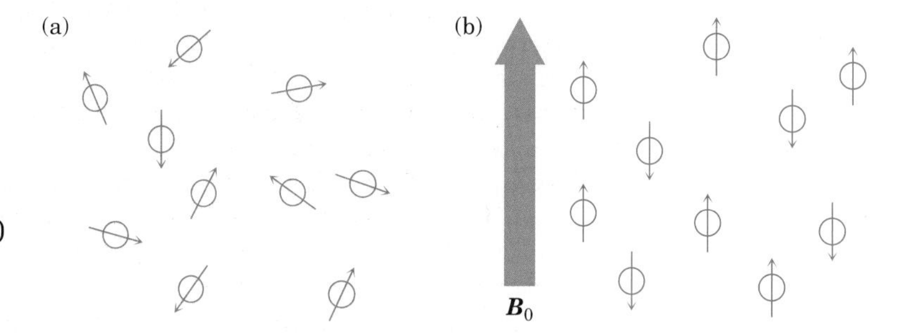

FIGURE 13.8 (a) Nuclear spins are oriented randomly in the absence of an external magnetic field but (b) have a specific orientation in the presence of an external field B_0. Note that some of the spins (red) are aligned parallel to the external field and others (blue) are antiparallel. The parallel spin state is lower in energy.

If the oriented nuclei are now irradiated with electromagnetic radiation of the right frequency, energy absorption occurs and the lower-energy state "spin-flips" to the higher-energy state. When this spin-flip occurs, the nuclei are said to be in resonance with the applied radiation—hence the name *nuclear magnetic resonance.*

The exact frequency necessary for resonance depends both on the strength of the external magnetic field and on the identity of the nuclei. If a very strong external field is applied, the energy difference between the two spin states is large and higher-energy (higher-frequency) radiation is required. If a weaker magnetic field is applied, less energy is required to effect the transition between nuclear spin states.

In practice, superconducting magnets that produce enormously powerful fields up to 14.1 tesla (T) are sometimes used, but field strengths in the range 2.62 to 4.7 T are more common. At a magnetic field strength of 2.62 T, so-called radiofrequency (rf) energy in the 100 MHz range (1 MHz = 10^6 Hz) is required to bring a ^1H nucleus into resonance, and rf energy of 25 MHz is required to bring a ^{13}C nucleus into resonance.

PROBLEM 13.14 NMR spectroscopy uses electromagnetic radiation with a frequency of 1×10^8 Hz. Is this a greater or lesser amount of energy than that used by IR spectroscopy?

13.8 The Nature of NMR Absorptions

From the description thus far, you might expect all ^1H nuclei in a molecule to absorb energy at the same frequency and all ^{13}C nuclei to absorb at the same frequency. If this were true, we would observe only a single NMR absorption in the ^1H or ^{13}C spectrum of a molecule, a situation that would be of little use for structure determination. In fact, the absorption frequency is not the same for all ^1H or all ^{13}C nuclei.

All nuclei are surrounded by circulating electrons. When an external magnetic field is applied to a molecule, the moving electrons around nuclei set up tiny local magnetic fields of their own. These local fields act in opposition to the applied field so that the *effective* field actually felt by the nucleus is a bit weaker than the applied field:

$$\boldsymbol{B}_{\text{effective}} = \boldsymbol{B}_{\text{applied}} - \boldsymbol{B}_{\text{local}}$$

In describing this effect of local fields, we say that the carbon and hydrogen nuclei are **shielded** from the full effect of the applied field by their surrounding electrons. Since each specific ^1H or ^{13}C nucleus in a molecule is in a slightly different electronic environment, each specific nucleus is shielded to a slightly different extent and the effective magnetic field felt by each is not the same. These slight differences can be detected, and we therefore see different NMR signals for each chemically distinct ^1H or ^{13}C nucleus.

Figure 13.9 shows both the ^1H and the ^{13}C NMR spectra of methyl acetate, $CH_3CO_2CH_3$. In both spectra, the horizontal axis tells the effective field strength felt by the nuclei and the vertical axis indicates the intensity of absorption of rf energy. Each peak in the NMR spectrum corresponds to a chemically distinct hydrogen or carbon in the molecule. Note, though, that ^1H and ^{13}C spectra can't be observed at the same time on the same spectrometer because different amounts of energy are required to spin-flip the different kinds of nuclei. The two spectra must be recorded separately.

The ^{13}C spectrum of methyl acetate shown in Figure 13.9b has three peaks, one for each of the three chemically distinct carbons in the molecule. The ^1H spectrum shows only *two* peaks, however, even though methyl acetate has *six* hydrogens. One peak is due to the $CH_3C{=}O$ hydrogens and the other to the $–OCH_3$ hydrogens. Because the three hydrogens in each methyl group have the same chemical (and magnetic) environment, they are shielded to the same extent and are said to be *equivalent. Chemically equivalent nuclei show a single absorption.* The two methyl groups themselves, however, are nonequivalent, so the two sets of hydrogens absorb at different positions.

FIGURE 13.9 (a) The ¹H NMR spectrum and (b) the ^{13}C NMR spectrum of methyl acetate, $CH_3CO_2CH_3$. The peak at the far right of each spectrum marked TMS is a calibration peak, explained in Section 13.9.

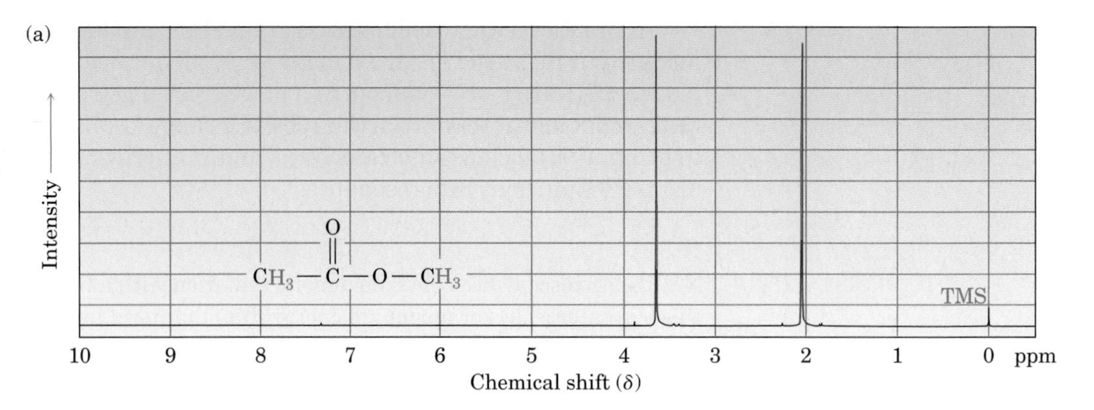

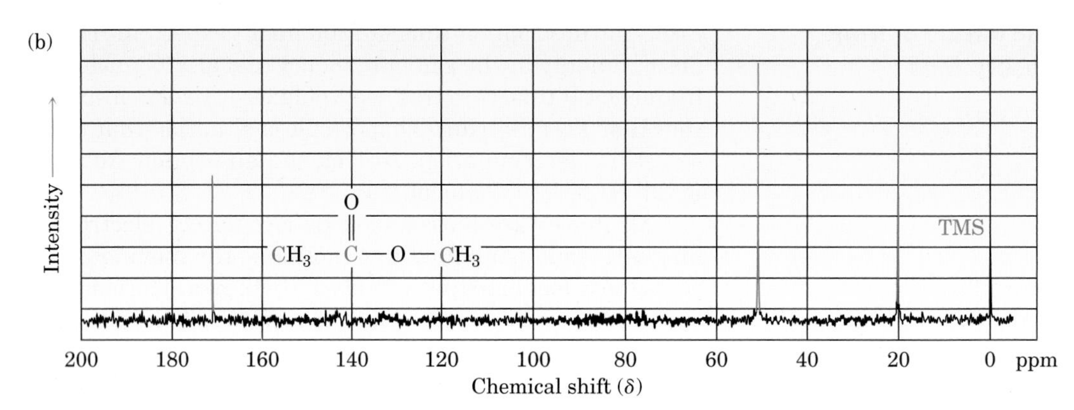

Practice Problem 13.6

How many signals would you expect *p*-dimethylbenzene to show in its ^{1}H and ^{13}C NMR spectra?

Strategy Look at the structure of the molecule, and count the number of kinds of chemically distinct ^{1}H and ^{13}C nuclei.

Solution Because of the molecule's symmetry, the two methyl groups in *p*-dimethylbenzene are equivalent and all four ring hydrogens are equivalent. Thus, there are only two absorptions in the ^{1}H NMR spectrum. Also because of symmetry, there are only three absorptions in the ^{13}C NMR spectrum: one for the two equivalent methyl-group carbons, one for the four equivalent C–H ring carbons, and one for the two equivalent ring carbons bonded to the methyl groups.

H H
H_3C CH_3
H H

Two kinds of hydrogens

Three kinds of carbons

***p*-Dimethylbenzene**

PROBLEM 13.15 How many absorptions would you expect each of the following compounds to show in its ^{1}H and ^{13}C NMR spectra?

(a) Methane (b) Ethane (c) Propane
(d) Cyclohexane (e) Dimethyl ether (f) Benzene
(g) $(CH_3)_3COH$ (h) Chloroethane (i) $(CH_3)_2C{=}C(CH_3)_2$

PROBLEM 13.16 2-Chloropropene shows signals for three kinds of hydrogens in its ^{1}H NMR spectrum. Explain.

PROBLEM 13.17 How many signals would you expect the following molecule to show in its ^{1}H and ^{13}C NMR spectra?

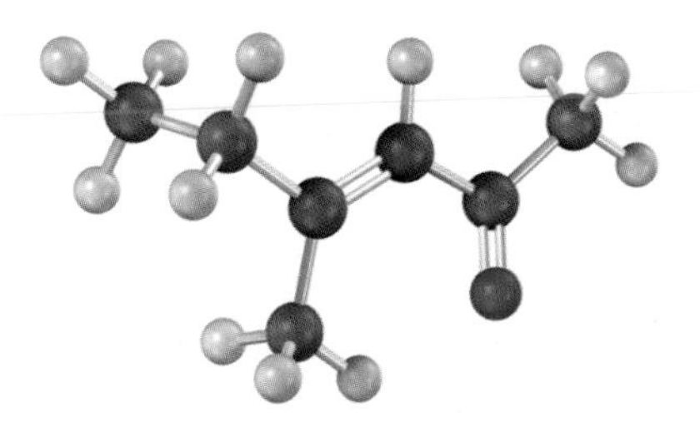

13.9 Chemical Shifts

NMR spectra are displayed on charts that show the applied field strength increasing from left to right (Figure 13.9). Thus, the left side of the chart is the low-field (or **downfield**) side, and the right side is the high-field (or **upfield**) side. Nuclei that absorb on the downfield side of the chart require a lower field strength for resonance, implying that they have relatively little shielding. Nuclei that absorb on the upfield side require a higher field strength for resonance, implying that they are strongly shielded.

To define the position of an absorption, the NMR chart is calibrated and a reference point is used. In practice, a small amount of tetramethylsilane [TMS, $(CH_3)_4Si$] is added to the sample so that a reference absorption line is produced when the spectrum is run. TMS is used as a reference for both ^{1}H and ^{13}C spectra, because in both kinds of spectra it produces a single peak that occurs upfield (farther right on the chart) of other absorptions normally found in organic molecules.

The position on the chart at which a nucleus absorbs is called its **chemical shift**. By convention, the chemical shift of TMS is set as the zero point, and other peaks normally occur downfield, to the left on the chart. NMR charts are calibrated in units of frequency using an arbitrary scale called the **delta (δ) scale**, where 1 delta unit is equal to 1 part per million (ppm) of the spectrometer operating frequency. For example, if we were using a 100 MHz instrument to measure the ^{1}H NMR spectrum of a substance, 1 δ would be 1 ppm of 100,000,000 Hz, or 100 Hz. If we were measuring the spectrum with a 300 MHz instrument, however, then 1 δ would be 300 Hz.

Although this method of calibrating NMR charts may seem complex, there's a good reason for it. As we saw earlier, the rf frequency required to bring a given nucleus into resonance depends on the spectrometer's magnetic field strength. But because there are many different kinds of spectrometers with many different magnetic field strengths available, chemical shifts given in frequency units (Hz) vary from one instrument to another. A resonance that occurs at 120 Hz downfield from TMS on one spectrometer might occur at 600 Hz downfield from TMS on another spectrometer with a more powerful magnet.

By using a system of measurement in which NMR absorptions are expressed in *relative* terms (parts per million relative to spectrometer frequency) rather than in absolute terms (Hz), comparisons of spectra obtained on different instruments are possible. *The chemical shift of an NMR absorption in δ units is constant, regardless of the operating frequency of the instrument.* A ^{1}H nucleus that absorbs at 2.0 δ on a 100 MHz instrument also absorbs at 2.0 δ on a 300 MHz instrument.

PRACTICE PROBLEM 13.7

Cyclohexane shows an absorption at 1.43 δ in its ^{1}H NMR spectrum. How many hertz away from TMS is this on a spectrometer operating at 100 MHz? On a spectrometer operating at 220 MHz?

SOLUTION On a 100 MHz spectrometer, 1 δ = 100 Hz. Thus, 1.43 δ = 143 Hz away from the TMS reference peak. On a 220 MHz spectrometer, 1 δ = 220 Hz and 1.43 δ = 315 Hz.

PROBLEM 13.18 When the ^{1}H NMR spectrum of acetone is recorded on a 100 MHz instrument, a single sharp resonance line at 2.1 δ is observed.

(a) How far away from TMS (in hertz) does the acetone absorption occur?
(b) What is the position of the acetone absorption in δ units on a 220 MHz instrument?
(c) How many hertz away from TMS does the absorption in the 220 MHz spectrum correspond to?

PROBLEM 13.19 The following ^{1}H NMR resonances were recorded on a spectrometer operating at 100 MHz. Convert each into δ units.

(a) $CHCl_3$, 727 Hz (b) CH_3Cl, 305 Hz
(c) CH_3OH, 347 Hz (d) CH_2Cl_2, 530 Hz

13.10 Chemical Shifts in ^{1}H NMR Spectra

Everything we've said thus far about NMR spectroscopy applies to both ^{1}H and ^{13}C spectra. Now, let's focus only on ^{1}H NMR spectroscopy to see how it can be used in organic structure determination. Most ^{1}H NMR absorptions occur in the range 0 to 10 δ, which can be divided into five regions as shown in Table 13.4. By remembering the positions of these regions, it's possible to tell at a glance what kinds of protons a molecule contains. (In speaking about NMR, the ^{1}H nucleus is often referred to as a *proton.*)

Table 13.4 also shows the correlation of ^{1}H chemical shift with electronic environment in more detail. In general, protons bonded to sp^3 carbons absorb at higher fields (right side of the spectrum), whereas protons bonded to sp^2 carbons absorb at lower fields (left side of the spectrum). Protons on carbons that are bonded to electronegative atoms such as N, O, or halogen also absorb at lower fields.

TABLE 13.4 Correlation of ^{1}H Chemical Shift with Environment

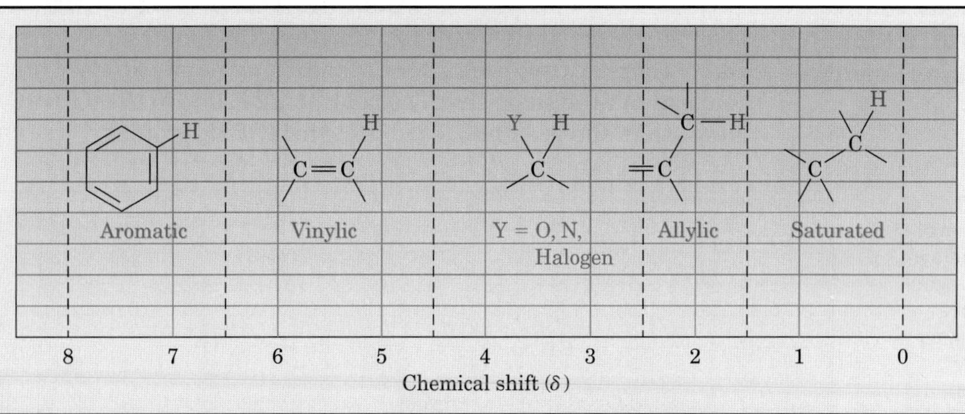

Type of hydrogen		Chemical shift (δ)	Type of hydrogen		Chemical shift (δ)
Reference	$(CH_3)_4Si$	0	Alkyl halide X = Cl, Br, I	X–C–H	2.5–4.0
Saturated primary	$—CH_3$	0.7–1.3	Alcohol	C–O–H	2.5–5.0 (Variable)
Saturated secondary	$—CH_2—$	1.2–1.6	Alcohol, ether	O–C–H	3.3–4.5
Saturated tertiary	H–C	1.4–1.8	Vinylic	H–C=C–H	4.5–6.5
Allylic	C=C–C–H	1.6–2.2	Aromatic	Ar—H	6.5–8.0
Methyl ketone	O=C–CH_3	2.0–2.4	Aldehyde	O=C–H	9.7–10.0
Aromatic methyl	Ar—CH_3	2.4–2.7	Carboxylic acid	O=C–O–H	11.0–12.0
Alkynyl	—C≡C—H	2.5–3.0			

PRACTICE PROBLEM 13.8

Methyl 2,2-dimethylpropanoate, $(CH_3)_3CCO_2CH_3$, has two peaks in its ^{1}H NMR spectrum. At what approximate chemical shifts do they come?

SOLUTION See Table 13.4. The $CH_3O–$ protons absorb around 3.5 to 4.0 δ because they are on carbon bonded to oxygen. The $(CH_3)_3C–$ protons absorb around 1.0 δ because they are typical alkane protons.

PROBLEM 13.20 Each of the following compounds exhibits a single ^{1}H NMR peak. Approximately where would you expect each to absorb?

(a) Ethane (b) Acetone (c) Benzene (d) Trimethylamine

13.11

Integration of ¹H NMR Spectra: Proton Counting

Look at the ¹H NMR spectrum of methyl 2,2-dimethylpropanoate in Figure 13.10. There are two peaks, corresponding to the two kinds of protons, but the peaks aren't the same size. The peak at 1.2 δ, due to the $(CH_3)_3C$– protons, is larger than the peak at 3.7 δ, due to the $-OCH_3$ protons.

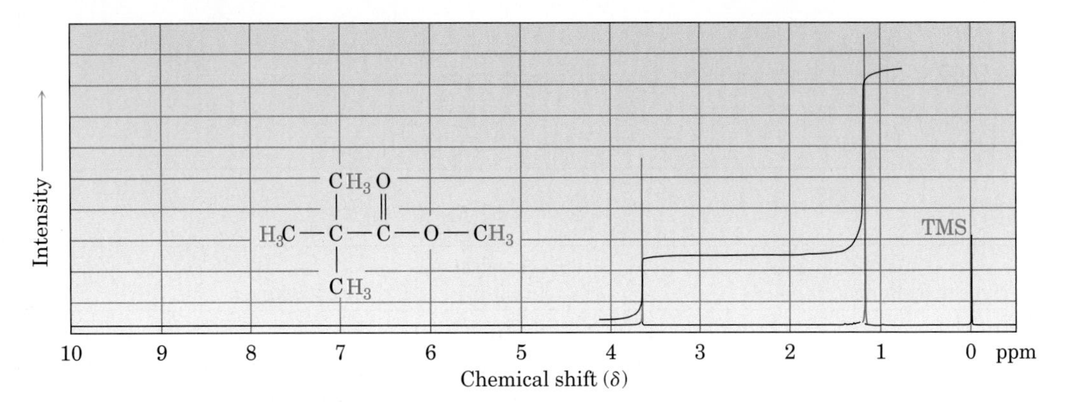

FIGURE 13.10 The ¹H NMR spectrum of methyl 2,2-dimethylpropanoate. Integrating the peaks in a "stair-step" manner shows that they have a 1:3 ratio, corresponding to the ratio of the numbers of protons (3:9) responsible for each peak.

The area under each peak is proportional to the number of protons causing that peak. By electronically measuring, or **integrating**, the area under each peak, it's possible to measure the relative numbers of each kind of proton in a molecule. Integrated peak areas can be either reported digitally or superimposed over the spectrum in a "stair-step" manner, with the height of each step proportional to the area of the peak and therefore proportional to the relative number of protons causing the peak. To compare the size of one peak with that of another, we simply measure the heights of the various steps. For example, the two peaks in methyl 2,2-dimethylpropanoate are found to have a 1:3 (or 3:9) height ratio when integrated—exactly what we expect because the three $-OCH_3$ protons are equivalent and the nine $(CH_3)_3C$– protons are equivalent.

PROBLEM 13.21 How many peaks would you expect in the ¹H NMR spectrum of *p*-dimethylbenzene (*p*-xylene)? What ratio of peak areas would you expect to find on integration of the spectrum? Refer to Table 13.4 for approximate chemical shift values, and sketch what the spectrum might look like.

13.12

Spin–Spin Splitting in ¹H NMR Spectra

In the ¹H NMR spectra we've seen thus far, each chemically distinct proton in a molecule has given rise to a single peak. It often happens, though, that a given absorption splits into *multiple* peaks, or **multiplets**. For example, the ¹H NMR spectrum of bromoethane in Figure 13.11 indicates that the $-CH_2Br$ protons appear as four peaks (a *quartet*) at 3.42 δ and the $-CH_3$ protons appear as a *triplet* at 1.68 δ.

Called **spin–spin splitting**, the phenomenon of multiple absorptions is caused by the interaction, or **coupling**, of the nuclear spins of neighboring atoms. In other words, the tiny magnetic field of one nucleus affects the magnetic field felt by a neighboring nucleus.

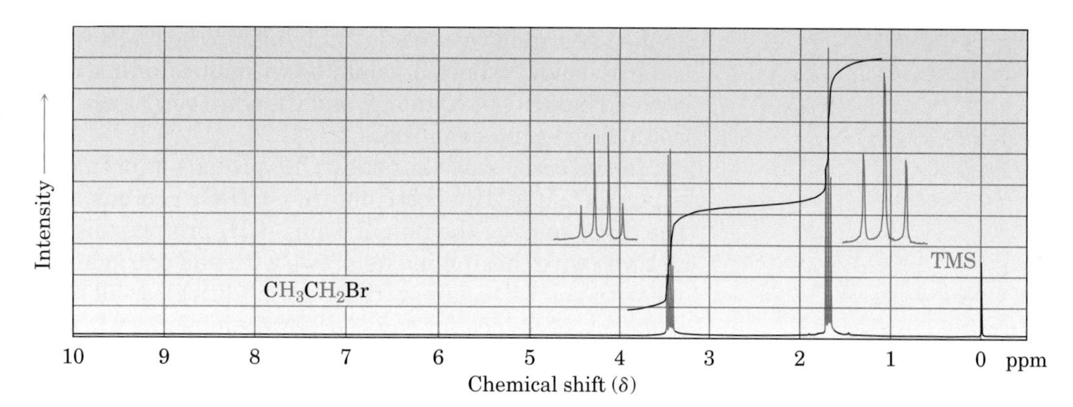

FIGURE 13.11 The ¹H NMR spectrum of bromoethane, CH_3CH_2Br. The $-CH_2Br$ protons appear as a quartet at 3.42 δ, and the $-CH_3$ protons appear as a triplet at 1.68 δ.

To understand the reasons for spin–spin splitting, look at the $-CH_3$ protons in bromoethane. The three equivalent $-CH_3$ protons are neighbored by two other magnetic nuclei, the $-CH_2Br$ protons. Each of the $-CH_2Br$ protons has its own nuclear spin, which can align either with or against the applied magnetic field, producing a tiny effect that is felt by the neighboring $-CH_3$ protons.

There are three ways in which the two $-CH_2Br$ protons can align, as shown in Figure 13.12. If both protons align *with* the applied magnetic field, the total effective field felt by the neighboring $-CH_3$ protons is slightly larger than it would otherwise be. Consequently, the applied field necessary to cause resonance is slightly reduced. Alternatively, if one $-CH_2Br$ proton aligns *with* and one aligns *against* the applied field (two possible ways), there is no effect on the neighboring $-CH_3$ protons. Finally, if both $-CH_2Br$ protons align *against* the applied field, the effective field felt by the $-CH_3$ protons is slightly smaller than it would otherwise be and the applied field needed for resonance must be slightly increased.

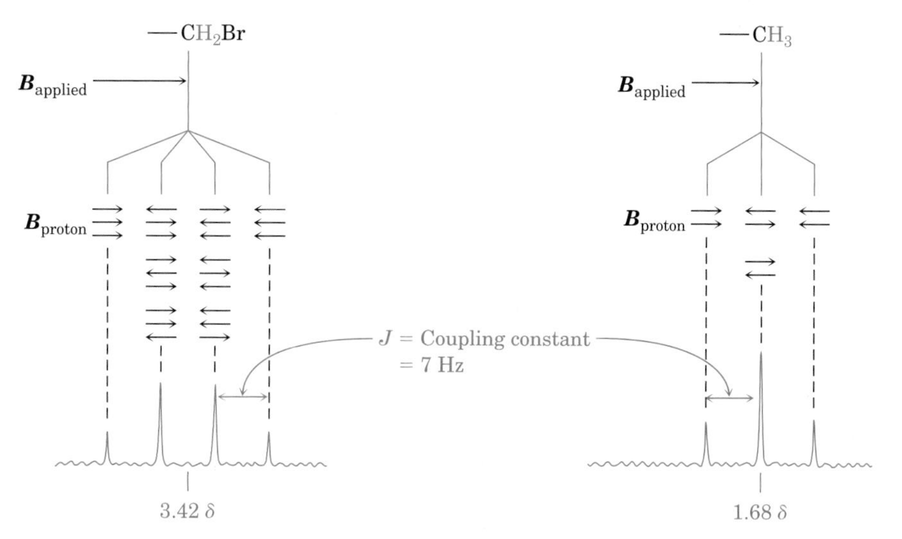

FIGURE 13.12 The origin of spin–spin splitting in bromoethane. The nuclear spins of neighboring protons, indicated by horizontal arrows, align either with or against the applied field, causing the splitting of absorptions into multiplets.

Any given molecule adopts only one of the three possible alignments of $-CH_2Br$ spins, but in a large collection of molecules, all three spin states are represented in a 1:2:1 statistical ratio. We therefore find that the neighboring $-CH_3$ protons come into resonance at three slightly different values of the applied field, and we see a 1:2:1 triplet in the NMR spectrum. One resonance is a little above where it would be without coupling, one is at the same place it would be without coupling, and the third resonance is a little below where it would be without coupling.

In the same way that the –CH_3 protons of bromoethane are split into a triplet in the NMR spectrum, the –CH_2Br protons are split into a quartet. The three spins of the neighboring –CH_3 protons align in four combinations: all three with the applied field, two with and one against (three possibilities), one with and two against (three possibilities), or all three against. Thus, four peaks are produced for the –CH_2Br protons in a 1:3:3:1 ratio.

As a general rule, called the ***n* + 1 rule**, protons that have n equivalent neighboring protons split into $n + 1$ peaks in their NMR absorption. For example, the spectrum of 2-bromopropane in Figure 13.13 shows a doublet at 1.71 δ and a seven-line multiplet, or *septet*, at 4.28 δ. The septet is caused by splitting of the –CHBr– proton signal by six equivalent neighboring protons on the two methyl groups ($n + 1 = 7$ when $n = 6$). The doublet results from splitting of the six equivalent –CH_3 protons by the single –CHBr– proton.

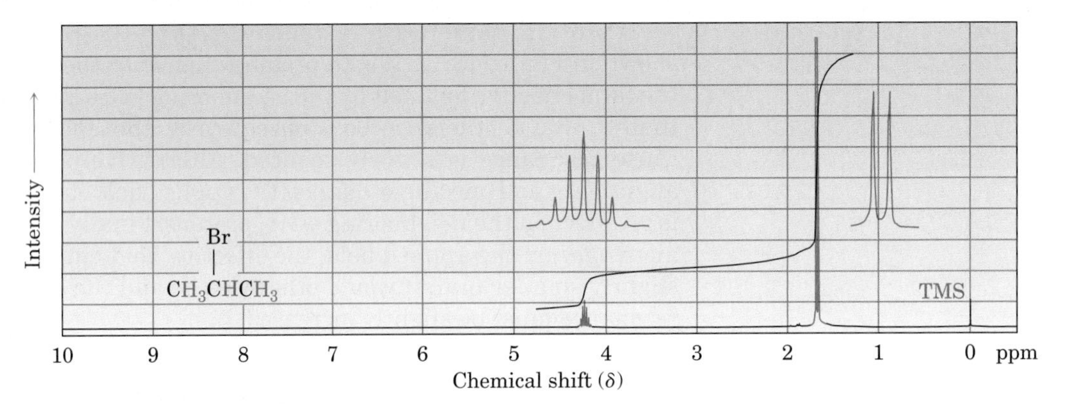

FIGURE 13.13 The ¹H NMR spectrum of 2-bromopropane. The –CH_3 proton signal at 1.71 δ is split into a doublet, and the –CHBr– proton signal 4.28 δ is split into a septet.

The distance between peaks in a multiplet is called the **coupling constant, *J*.** Coupling constants are measured in hertz and generally fall in the range 0 to 18 Hz. The exact value of J depends on the geometry of the molecule, but a typical value for an open-chain alkane is J = 6–8 Hz. Note that the same coupling constant is shared by both groups of hydrogens whose spins are coupled. In bromoethane, for instance, the –CH_2Br protons are coupled to the –CH_3 protons with coupling constant J = 7 Hz. The –CH_3 protons are similarly coupled to the –CH_2Br protons with the same J = 7 Hz coupling constant.

Three important points about spin–spin splitting are illustrated by the spectra of bromoethane in Figure 13.11 and 2-bromopropane in Figure 13.13:

- **Chemically equivalent protons don't show spin–spin splitting.** The equivalent protons can be on the same carbon or on different carbons, but their signals still appear as singlets and don't split.

Three C–H protons are chemically equivalent; no splitting occurs.

Four C–H protons are chemically equivalent; no splitting occurs.

- **The signal of a proton with *n* equivalent neighboring protons is split into a multiplet of *n* + 1 peaks with coupling constant *J*.** Protons that are more than two carbon atoms apart usually don't split each other's signals.

```
  H     H              H           H
   \   /                \    |    /
  —C — C—              —C — C — C—
   /   \                /    |    \
```

Splitting observed   Splitting not usually observed

- **Two groups of protons coupled to each other have the same coupling constant *J*.**

PRACTICE PROBLEM 13.9

Predict the splitting pattern for each kind of hydrogen in isopropyl propanoate, $CH_3CH_2CO_2CH(CH_3)_2$.

STRATEGY First, find how many different kinds of protons are present (there are four). Then, find out how many neighboring protons each kind has, and apply the $n + 1$ rule.

SOLUTION

```
   (Quartet)              (Septet)
(Triplet)        O        H
                 ‖        |
   CH3 — CH2 — C — O — C — CH3
                          |
                          CH3    (Doublet)
```

Isopropyl propanoate

PROBLEM 13.22 Predict the splitting patterns for each proton in the following molecules:

(a) $(CH_3)_3CH$ (b) CH_3CHBr_2 (c) $CH_3OCH_2CH_2Br$
(d) $CH_3CH_2CO_2CH_3$ (e) $ClCH_2CH_2CH_2Cl$ (f) $(CH_3)_2CHCO_2CH_3$

PROBLEM 13.23 Propose structures for compounds that show the following ¹H NMR spectra:

(a) C_2H_6O; one singlet (b) $C_3H_6O_2$; two singlets
(c) C_3H_7Cl; one doublet and one septet

PROBLEM 13.24 Predict the splitting patterns for each kind of chemically distinct proton in the following molecule:

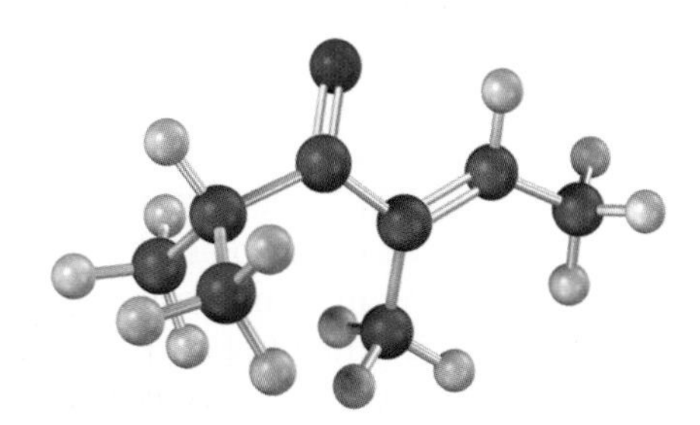

13.13

Uses of ^{1}H NMR Spectra

^{1}H NMR spectroscopy is used to help identify the products of nearly every reaction run in the laboratory. For example, we said in Section 4.4 that acid-catalyzed addition of H_2O to an alkene occurs with Markovnikov orientation; that is, the more highly substituted alcohol is formed. With the help of ^{1}H NMR, we can now prove this statement.

Does addition of H_2O to 1-methylcyclohexene yield 1-methylcyclohexanol or 2-methylcyclohexanol?

CH_3 $\xrightarrow[H_2SO_4]{H_2O}$ CH_3, OH or CH_3, OH

1-Methylcyclohexene **1-Methylcyclohexanol** **2-Methylcyclohexanol**

The ^{1}H NMR spectrum of the reaction product is shown in Figure 13.14. Although many of the ring protons overlap into a broad, poorly defined multiplet centered around 1.6 δ, the spectrum also shows a large singlet absorption in the saturated methyl region at 1.2 δ, indicating that the product has a methyl group with no neighboring hydrogens (R_3C—$\mathbf{CH_3}$). Furthermore, the spectrum shows no absorptions around 4 δ, where we would expect the signal of an $R_2\mathbf{CH}$OH proton to occur. Thus, the reaction product must be 1-methylcyclohexanol.

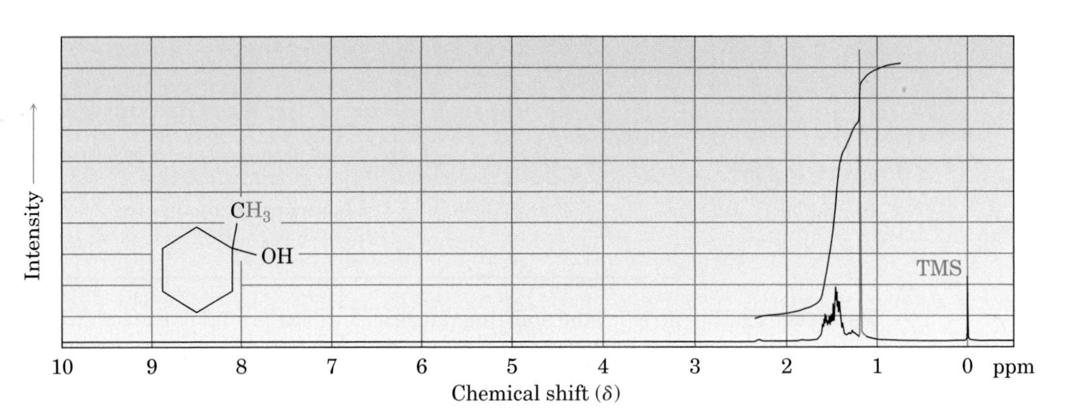

FIGURE 13.14 The ^{1}H NMR spectrum of the reaction product from H_2O and 1-methylcyclohexene. The presence of the $-CH_3$ absorption at 1.2 δ and the absence of any absorptions near 4 δ identify the product as 1-methylcyclohexanol.

13.14

^{13}C NMR Spectroscopy

Having now looked at ^{1}H NMR spectroscopy, let's take a brief look at ^{13}C NMR before ending. In some ways, it's surprising that carbon NMR is even possible. After all, ^{12}C, the most abundant carbon isotope, has no nuclear spin and is not observable by NMR. The only naturally occurring carbon isotope with a magnetic moment is ^{13}C, but its natural abundance is only 1.1%. Thus, only about 1 of every 100 carbon atoms in an organic molecule is observable by NMR. Fortunately, the technical problems caused by this low abundance have been overcome by computer techniques, and ^{13}C NMR is a routine structural tool.

At its simplest, ^{13}C NMR makes it possible to count the number of carbon atoms in a molecule. In addition, it's possible to get information about the environment of each carbon by observing its chemical shift. As illustrated by the ^{13}C NMR spectrum of methyl acetate shown earlier (Figure 13.9b), we normally observe a single, sharp resonance line for each kind of carbon atom in a molecule. Thus, methyl acetate has three nonequivalent carbon atoms and three peaks in its ^{13}C NMR spectrum. (Coupling between adjacent carbon atoms isn't seen, because the low natural abundance of ^{13}C makes it unlikely that two such nuclei will be adjacent in a molecule.)

Most ^{13}C resonances are between 0 and 220 δ downfield from the TMS reference line, with the exact chemical shift dependent on a carbon's environment in the molecule. Figure 13.15 shows how environment and chemical shift are correlated.

FIGURE 13.15 Chemical shift correlations for ^{13}C NMR.

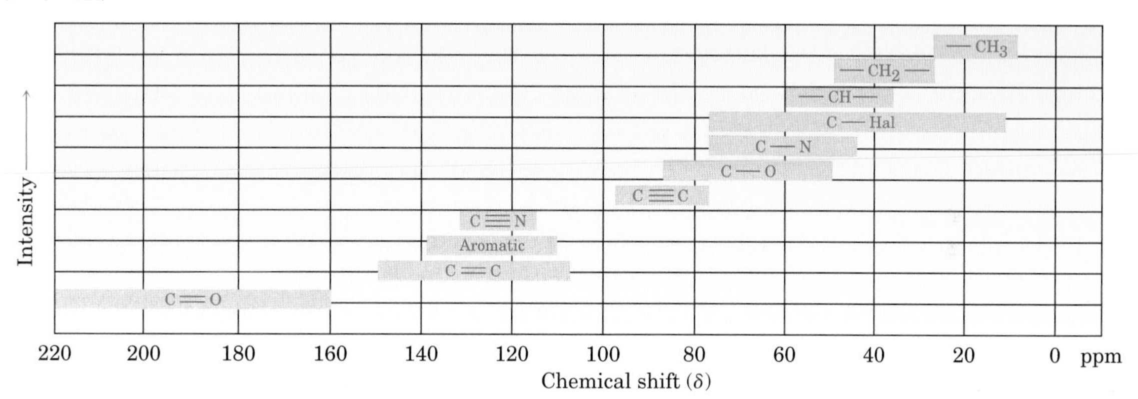

The factors that determine chemical shifts are complex, but it's possible to make some generalizations. One trend is that a carbon bonded to an electronegative atom like oxygen, nitrogen, or halogen absorbs downfield of a typical alkane carbon. Another trend is that sp^3-hybridized carbons absorb in the range 0 to 100 δ, and sp^2 carbons absorb in the range 100 to 220 δ. Carbonyl carbons (C=O) are particularly distinct in the ^{13}C NMR spectrum and are easily observed at the extreme low-field side of the chart, in the range 170 to 220 δ. For example, the ^{13}C NMR spectrum of *p*-bromoacetophenone in Figure 13.16 shows an absorption for the carbonyl carbon at 197 δ.

The ^{13}C NMR spectrum of *p*-bromoacetophenone is interesting for another reason as well. Note that only six absorptions are observed even though the molecule has eight carbons. *p*-Bromoacetophenone has a symmetry plane that makes carbons 4 and 4′, and carbons 5 and 5′, equivalent. Thus, the six ring carbons show only four absorptions in the range 128 to 137 δ. In addition, the $-CH_3$ carbon absorbs at 26 δ.

FIGURE 13.16 The ^{13}C NMR spectrum of *p*-bromoacetophenone, $BrC_6H_4COCH_3$.

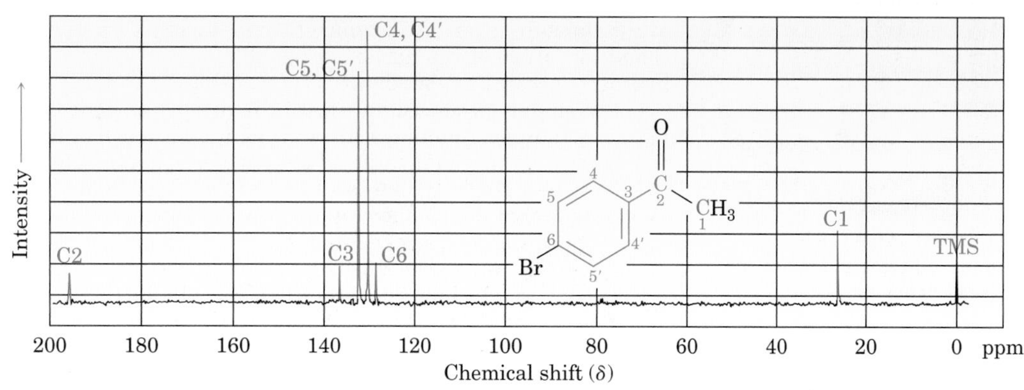

PRACTICE PROBLEM 13.10

How many peaks would you expect in the ^{13}C NMR spectrum of methylcyclopentane?

SOLUTION Methylcyclopentane has a symmetry plane. Thus, it has only four kinds of carbons and only four peaks in its ^{13}C NMR spectrum.

3, 4, 2, 1, CH_3 ----- Symmetry plane, 4, 3

PROBLEM 13.25 How many peaks would you expect in the ^{13}C NMR spectra of the following compounds?

(a) cyclopentane (b) H_3C, CH_3 (m-xylene) (c) CH_3, CH_3 (o-xylene) (d) CH_3 (1-methylcyclohexene)

PROBLEM 13.26 Propose structures for compounds whose ^{13}C NMR spectra fit the following descriptions:

(a) A hydrocarbon with seven peaks in its spectrum
(b) A six-carbon compound with only five peaks in its spectrum
(c) A four-carbon compound with three peaks in its spectrum

INTERLUDE

Magnetic Resonance Imaging (MRI)

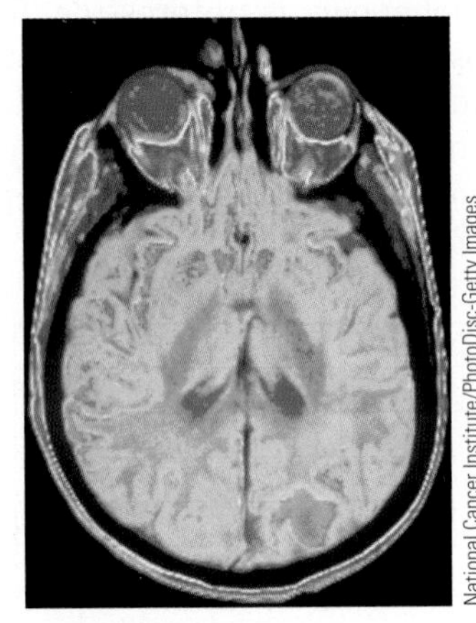

National Cancer Institute/PhotoDisc-Getty Images

This MRI image of a human brain indicates a tumor on the lower right side.

As practiced by organic chemists, NMR spectroscopy is a powerful method of structure determination. A small amount of sample, typically a few milligrams or less, is dissolved in approximately 1 mL of a suitable solvent, the solution is placed in a thin glass tube, and the tube is placed into the narrow (1–2 cm) gap between the poles of a strong magnet. Imagine, though, that a much larger NMR instrument were available. Instead of a few milligrams, the sample size could be tens of kilograms; instead of a narrow gap between magnet poles, the gap could be large enough for a person to climb into so that an NMR spectrum of body parts could be obtained. What you've just imagined is an instrument for *magnetic resonance imaging (MRI),* a diagnostic technique of enormous importance in medicine because of its advantages over X-ray or radioactive imaging methods.

Like NMR spectroscopy, MRI takes advantage of the magnetic properties of certain nuclei, typically hydrogen, and of the signals emitted when those nuclei are stimulated by radiofrequency energy. Unlike what happens in NMR spectroscopy, though, MRI instruments use powerful computers and data manipulation techniques to look at the three-dimensional *location* of magnetic nuclei in the body rather than at the chemical nature of the nuclei. As noted, most MRI instruments currently look at hydrogen, present in abundance wherever there is water or fat in the body.

The signals produced vary with the density of hydrogen atoms and with the nature of their surroundings, allowing identification of different types of tissue and even allowing the visualization of motion. For example, the volume of blood leaving the heart in a single stroke can be measured, and heart motion can be observed. Soft tissues that do not show up well on X rays can be seen clearly, allowing diagnosis of brain tumors, strokes, and other conditions. The technique is also valuable in diagnosing damage to knees or other joints and is a painless alternative to arthroscopy, in which an endoscope is physically introduced into the knee joint.

Several types of atoms in addition to hydrogen can be detected by MRI, and the applications of images based on ^{31}P atoms are being explored. The technique holds great promise for studies of metabolism.

Summary and Key Words

Amplitude, p. 406
Chemical shift, p. 421
Coupling, p. 424
Coupling constant (J), p. 426
Delta (δ) scale, p. 421
Downfield, p. 421
Electromagnetic spectrum, p. 405
Frequency (ν), p. 405
Hertz (Hz), p. 405
Infrared (IR) spectroscopy, p. 406
Integration, p. 424
Mass spectrometry, p. 409
Mass-to-charge ratio (m/z), p. 409
Multiplet, p. 424
$n + 1$ rule, p. 426
Nuclear magnetic resonance (NMR) spectroscopy, p. 418
Shielding, p. 419
Spin–spin splitting, p. 424
Ultraviolet (UV) spectroscopy, p. 411
Upfield, p. 421
Wavelength (λ), p. 406
Wavenumber ($\tilde{\nu}$), p. 413
X-ray crystallography, p. 407

Five main spectroscopic methods are used to determine the structures of organic molecules. Each gives a different kind of information:

X-ray crystallography	What is the structure of the molecule?
Mass spectrometry	What is the molecular weight and formula?
Infrared spectroscopy	What functional groups are present?
Nuclear magnetic resonance spectroscopy	What is the carbon–hydrogen framework?
Ultraviolet and visible spectroscopy	Is a conjugated π electron system present?

Using **X-ray crystallography**, the structure of the molecule in three-dimensions can be obtained, but only if the molecule forms crystals, and sometimes only after much effort. **Mass spectrometry** provides the molecular weight and formula of ionized organic molecules. Recorded as a ratio of **mass to charge (m/z)**, the isotopic composition of the molecule is also provided.

When an organic molecule is irradiated with **infrared (IR)** energy, frequencies of light corresponding to the energies of molecular stretching and bending motions are absorbed. Each kind of functional group has a characteristic set of IR absorptions that allows the group to be identified. For example, an alkene C=C bond absorbs in the range 1640 to 1680 cm^{-1}, a saturated ketone absorbs near 1715 cm^{-1}, and a nitrile absorbs near 2230 cm^{-1}. By observing which frequencies of IR radiation are absorbed by a molecule *and which are not*, the functional groups in a molecule can be identified.

Ultraviolet (UV) spectroscopy is applicable to conjugated π electron systems. When a conjugated molecule is irradiated with ultraviolet light, energy absorption occurs, leading to excitation of π electrons to higher energy levels. The greater the extent of conjugation, the longer the wavelength needed for excitation.

Nuclear magnetic resonance (NMR) spectroscopy is the most valuable of the common spectroscopic techniques for organic chemists. When ^{1}H and ^{13}C nuclei are placed in a magnetic field, their spins orient either with or against the field. On irradiation with radiofrequency (rf) waves, energy is absorbed and the nuclear spins flip from the lower-energy state to the higher-energy state. This absorption of energy is detected, amplified, and displayed as an NMR spectrum. NMR spectra display four general features:

- **Number of peaks.** Each nonequivalent kind of ^{1}H or ^{13}C nucleus in a molecule gives rise to a different peak.
- **Chemical shift.** The exact position of each peak is called its chemical shift and is correlated to the chemical environment of each ^{1}H or ^{13}C nucleus. Most ^{1}H absorptions fall in the range 0 to 10 δ downfield from the TMS reference signal.
- **Integration.** The area under each peak can be electronically integrated to determine the relative number of protons responsible for each peak.
- **Spin–spin splitting.** Neighboring nuclear spins can **couple**, splitting an NMR absorption into a **multiplet**. The NMR signal of a ^{1}H nucleus neighbored by n adjacent protons splits into $n + 1$ peaks with coupling constant J.

EXERCISES

Visualizing Chemistry

13.27 Into how many peaks would you expect the ^1H NMR signal of the indicated protons to be split? (Yellow-green = Cl.)

(a) (b)

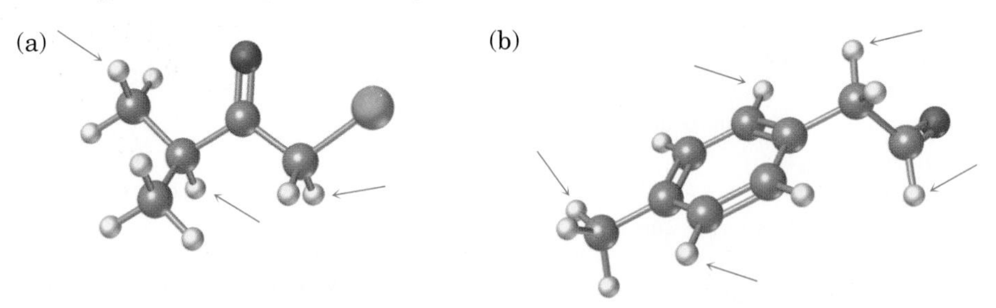

13.28 Where in the infrared spectrum would you expect each of the following compounds to absorb?

(a) (b)

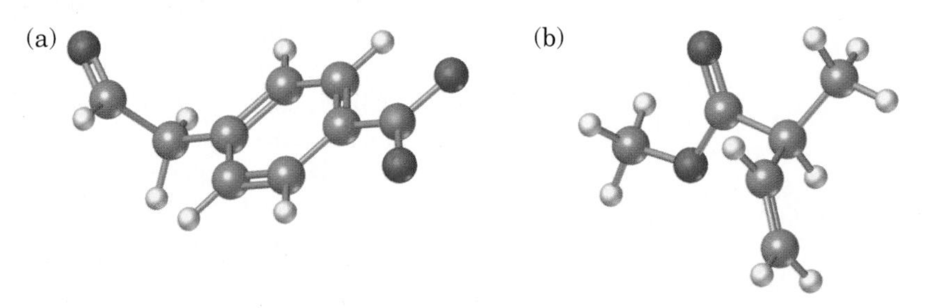

(c)

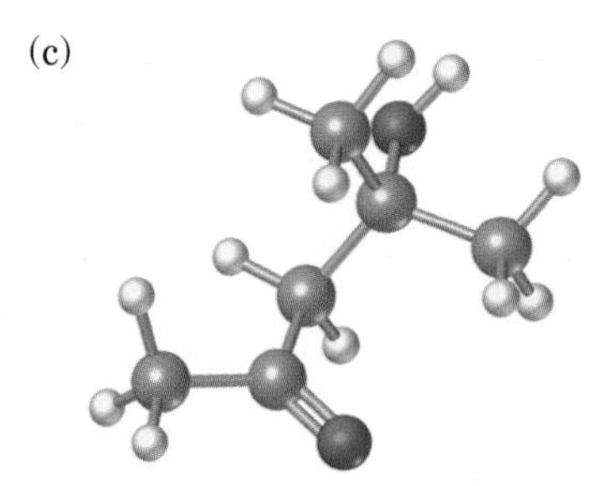

13.29 How many absorptions would you expect the following compound to have in its ^{13}C NMR spectrum?

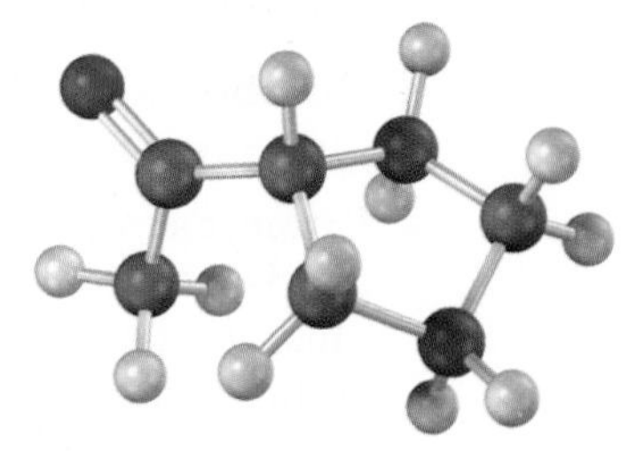

13.30 Sketch what you might expect the ^{1}H and ^{13}C NMR spectra of the following compound to look like (yellow-green = Cl):

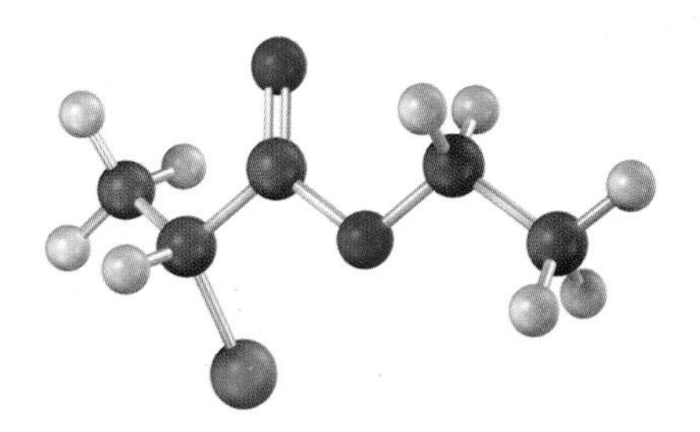

Additional Problems

GENERAL

13.31 Tell what is meant by each of the following terms:
(a) Chemical shift (b) Coupling constant (c) λ_{max}
(d) Spin–spin splitting (e) Wavenumber (f) Applied magnetic field
(g) m/z

13.32 The energy E of electromagnetic radiation, expressed in units of kilojoules per mole (kJ/mol), can be determined by the following formula:

$$E = \frac{1.20 \times 10^{-2}}{\lambda \text{ (in cm)}} \text{ kJ/mol}$$

What is the energy of infrared radiation of wavelength 6.55×10^{-6} m?

13.33 Using the formula given in Problem 13.32, calculate the energy required to effect the electronic excitation of buta-1,3-diene (λ_{max} = 217 nm).

13.34 Using the equation given in Problem 13.32, calculate the amount of energy required to spin-flip a proton in a spectrometer operating at 100 MHz. Does increasing the spectrometer frequency from 100 to 220 MHz increase or decrease the amount of energy necessary for resonance?

UV SPECTROSCOPY

13.35 The active metabolite of the antiviral drug abacavir is a triphosphate derivative. The molar absorptivity at 253 nm is ϵ_{253} = 13,260 M^{-1} cm^{-1}. What absorbance would you expect for a 1 cm pathlength sample at a concentration of 42 μM?

13.36 What is the concentration of abacavir oral suspension (in mg/mL) if the absorbance measured through 1 cm of solution obtained on dilution of the suspension by 1000 times is A = 0.93, assuming a molar absorptivity of ϵ = 13,260 M^{-1} cm^{-1}?

13.37 Metabolism of drugs by the body often depends on the species. The molecule coumarin, an anticoagulant, metabolizes to a toxic compound to rats but is safe in humans. How would the UV spectra differ between these different liver metabolites?

HO O O ← Liver enzymes in humans — O O — Liver enzymes in rats → O O O

(Human metabolite) **Coumarin** **(Rat metabolite)**

13.38 The mosquito attractant oct-1-en-3-ol can be prepared by reduction of oct-1-en-3-one. How can UV spectroscopy be used to monitor this reaction?

IR SPECTROSCOPY

13.39 What kinds of functional groups might compounds contain if they show the following IR absorptions?

(a) 1670 cm^{-1} (b) 1735 cm^{-1}

(c) 1540 cm^{-1} (d) 1715 cm^{-1} and 2500–3100 cm^{-1} (broad)

13.40 At what approximate positions might the following compounds show IR absorptions?

(a) CO_2H

(b)

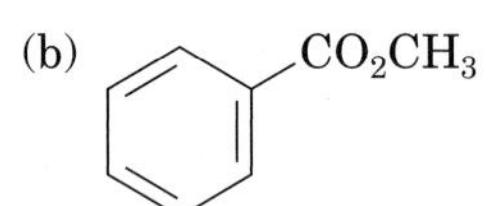

(c)

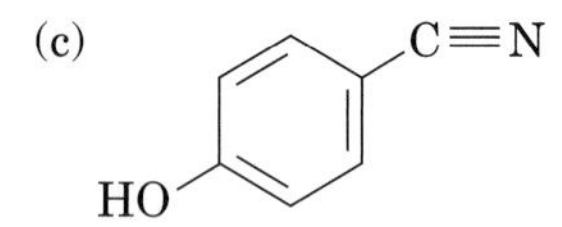

(d) O

(e)

$$\mathrm{CH_3\overset{O}{\overset{\|}{C}}CH_2CH_2\overset{O}{\overset{\|}{C}}OCH_3}$$

13.41 If C–O single-bond stretching occurs at 1000 cm^{-1} and C=O double-bond stretching occurs at 1700 cm^{-1}, which of the two requires more energy? How does your answer correlate with the relative strengths of single and double bonds?

13.42 Propose structures for compounds that meet the following descriptions:

(a) C_5H_8, with IR absorptions at 3300 and 2150 cm^{-1}

(b) C_4H_8O, with a strong IR absorption at 3400 cm^{-1}

(c) C_4H_8O, with a strong IR absorption at 1715 cm^{-1}

(d) C_8H_{10}, with IR absorptions at 1600 and 1500 cm^{-1}

MASS SPECTROMETRY

13.43 Hair analysis by mass spectrometry can provide evidence of drug use. What mass spectral peaks do you expect to see for the following drugs?

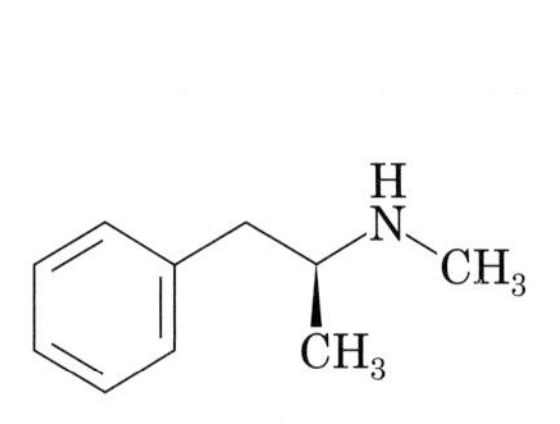

Methamphetamine

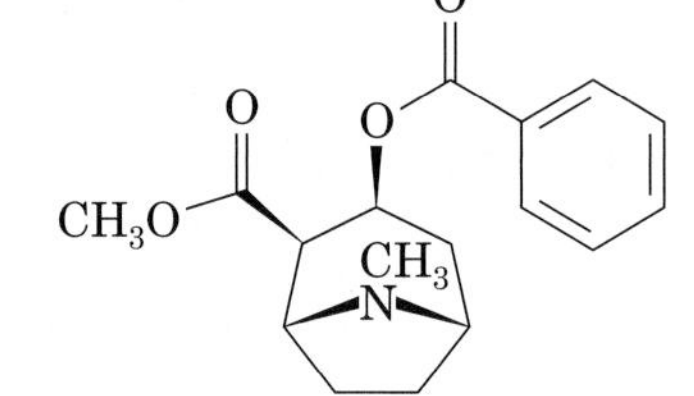

Cocaine

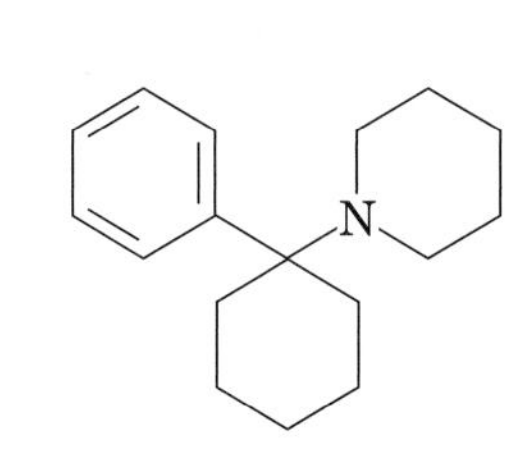

Phencyclidine

HO (CH2)4CH3 H H3C O H H3C CH3

Tetrahydrocannabinol (THC)

O O H N CH3 CH3

Ecstasy

13.44 After prolonged storage in a humid environment, a mass spectrum of benzyl bromide gave the following peaks. Assign structures to all these species.

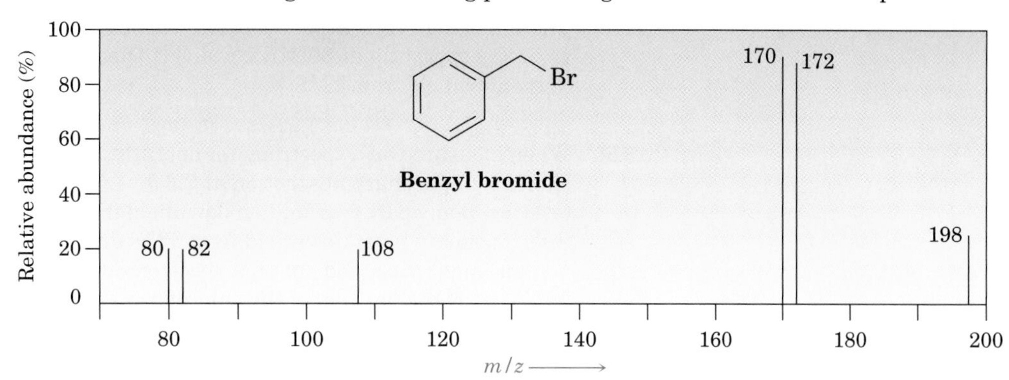

13.45 Benzoylecgonine is commonly used to screen for cocaine (see Problem 13.43) because it appears when cocaine is metabolized in the body. The molecular weight of benzoylecgonine is 289 g/mol. Propose a structure of this molecule.

13.46 Analysis of a mixture of steroids by an expert revealed that the sample contained, among other things, testosterone but not the banned drug clostebol. Are the data in the following mass spectrum consistent with this conclusion?

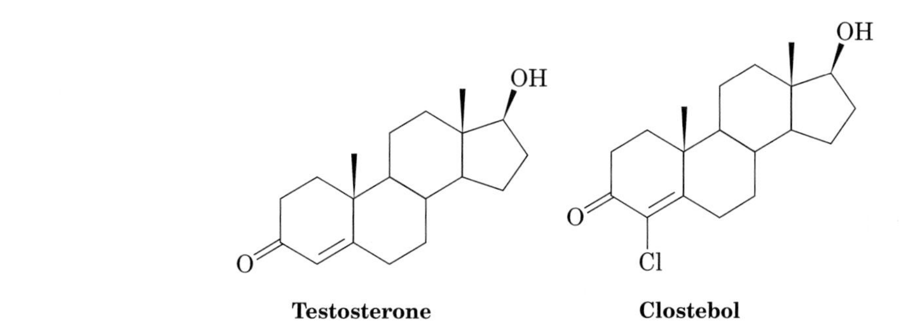

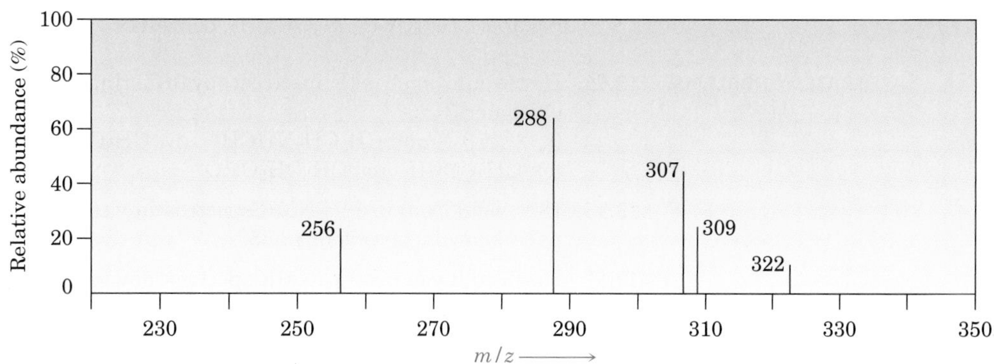

NMR SPECTROSCOPY

13.47 The following ^1H NMR absorptions, determined on a spectrometer operating at 100 MHz, are given in hertz downfield from the TMS standard. Convert the absorptions to δ units.

(a) 218 Hz (b) 478 Hz (c) 751 Hz

13.48 At what positions, in hertz downfield from TMS standard, would the NMR absorptions in Problem 13.47 appear on a spectrometer operating at 220 MHz?

13.49 The following NMR absorptions, given in δ units, were obtained on a spectrometer operating at 80 MHz. Convert the chemical shifts from δ units into hertz downfield from TMS.
(a) 2.1 δ (b) 3.45 δ (c) 6.30 δ

13.50 When measured on a spectrometer operating at 100 MHz, chloroform ($CHCl_3$) shows a single sharp absorption at 7.3 δ.
(a) How many parts per million downfield from TMS does chloroform absorb?
(b) How many hertz downfield from TMS does chloroform absorb if the measurement is carried out on a spectrometer operating at 360 MHz?
(c) What is the position of the chloroform absorption in δ units measured on a 360 MHz spectrometer?

13.51 How many absorptions would you expect in the ^{13}C NMR spectra of the following compounds?

(a) CH_3, CH_3 (on cyclohexane ring) (b) $CH_3CH_2OCH_3$ (c)

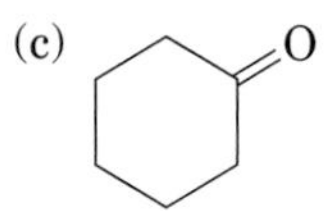

(d) $CH_3CH{=}C(CH_3)CH_3$ (e) $CH_3CH_2(H)C{=}C(H)CH_3$ (f) $H(CH_3CH_2)C{=}C(CH_3)H$

13.52 How many types of nonequivalent protons are in each of the molecules listed in Problem 13.51?

13.53 Describe the ^1H NMR spectra you would expect for the following compounds:
(a) CH_3CHCl_2 (b) $CH_3CO_2CH_2CH_3$ (c) $(CH_3)_3CCH_2CH_3$

13.54 The following compounds all show a single peak in their ^1H NMR spectra. List them in order of expected increasing chemical shift: CH_4, CH_2Cl_2, cyclohexane, CH_3COCH_3, $H_2C{=}CH_2$, benzene.

INTEGRATED PROBLEMS

13.55 How would you use IR spectroscopy to distinguish between the following pairs of isomers?
(a) $(CH_3)_3N$ and $CH_3CH_2NHCH_3$ (b) CH_3COCH_3 and $CH_2{=}CHCH_2OH$
(c) CH_3COCH_3 and CH_3CH_2CHO

13.56 How would you use ^1H NMR spectroscopy to distinguish between the isomer pairs shown in Problem 13.55?

13.57 How could you use ^{13}C NMR spectroscopy to distinguish between the isomer pairs shown in Problem 13.55?

13.58 Assume that you're carrying out the dehydration of 1-methylcyclohexanol to yield 1-methylcyclohexene. How could you use IR spectroscopy to determine when the reaction is complete? What characteristic absorptions would you expect for both starting material and product?

13.59 How would you expect the mass spectrum of the starting material and product to differ for Problem 13.58?

13.60 The IR spectrum of a compound with the formula C_7H_6O is shown. Propose a likely structure.

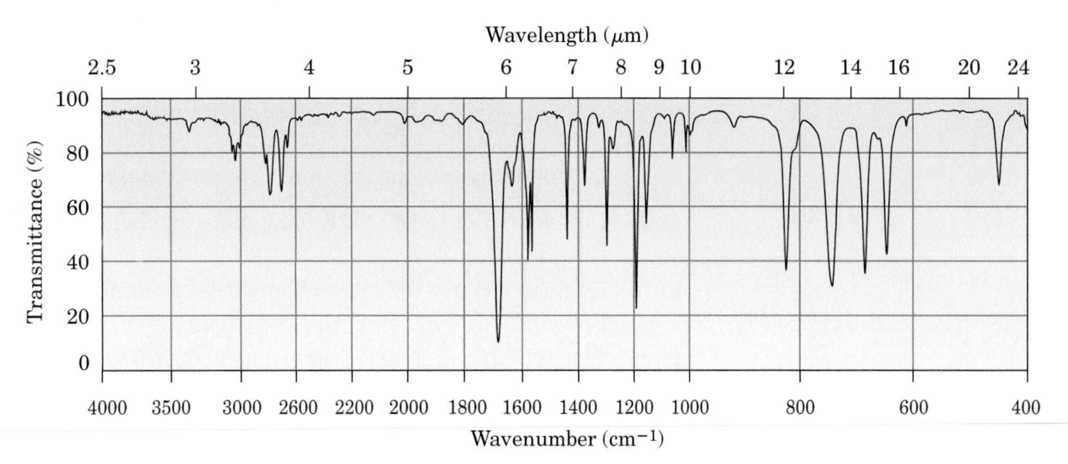

13.61 Dehydration of 1-methylcyclohexanol might lead to either of two isomeric alkenes, 1-methylcyclohexene or methylenecyclohexane. How could you use NMR spectroscopy (both ^1H and ^{13}C) to determine the structure of the product?

13.62 3,4-Dibromohexane can undergo base-induced double dehydrobromination to yield either hex-3-yne or hexa-2,4-diene. How could you use UV spectroscopy to help identify the product? How could you use ^1H NMR spectroscopy?

13.63 Describe the ^1H and ^{13}C NMR spectra you expect for the following compounds:
(a) $ClCH_2CH_2CH_2Cl$ (b) $CH_3COCH_2CH_2Cl$

13.64 Propose structures for compounds with the following formulas that show only one peak in their ^1H NMR spectra:
(a) C_5H_{12} (b) C_5H_{10} (c) $C_4H_8O_2$

13.65 Assume that you have a compound with formula C_3H_6O.
(a) Propose as many structures as you can that fit the molecular formula (there are seven).
(b) If your compound has an IR absorption at 1715 cm^{-1}, what can you conclude?
(c) If your compound has a single ^1H NMR absorption at 2.1 δ, what is its structure?

13.66 Propose structures for compounds that fit the following ^1H NMR data:

(a) $C_5H_{10}O$
6 H doublet at 0.95 δ, $J = 7$ Hz
3 H singlet at 2.10 δ
1 H multiplet at 2.43 δ

(b) C_3H_5Br
3 H singlet at 2.32 δ
1 H singlet at 5.25 δ
1 H singlet at 5.54 δ

13.67 How can you use ^1H and ^{13}C NMR to help distinguish among the following four isomers?

$CH_2—CH_2$
| |
$CH_2—CH_2$

$H_2C{=}CHCH_2CH_3$

$CH_3CH{=}CHCH_3$

CH_3
|
$CH_3C{=}CH_2$

13.68 How can you use ^{1}H NMR to help distinguish between the following isomers?

3-Methylcyclohex-2-enone **Cyclopent-4-enyl methyl ketone**

13.69 How can you use ^{13}C NMR to help distinguish between the isomers in Problem 13.68?

13.70 How can you use UV spectroscopy to help distinguish between the isomers in Problem 13.68?

13.71 The ^{1}H NMR spectrum of compound A, $C_3H_6Br_2$, is shown. Propose a structure for A, and explain how the spectrum fits your structure.

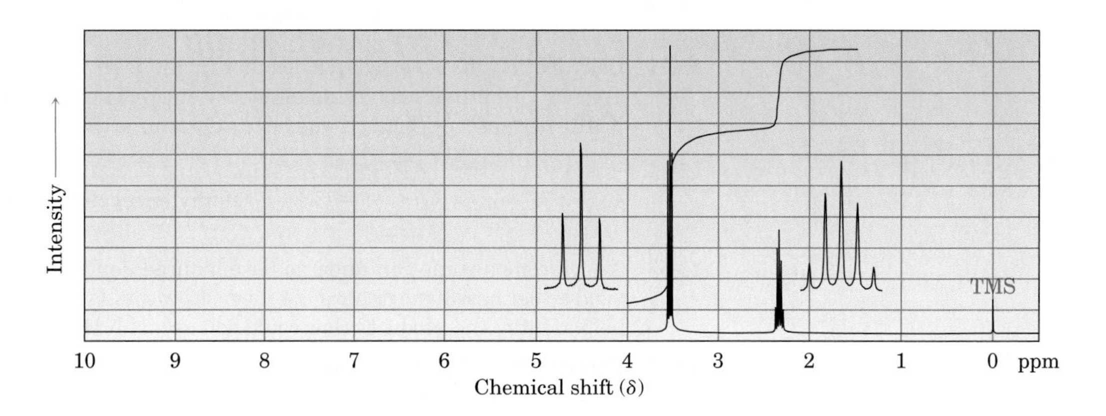

13.72 The compound whose ^{1}H NMR spectrum is shown has the formula $C_4H_7O_2Cl$ and has an IR absorption peak at 1740 cm^{-1}. Propose a structure.

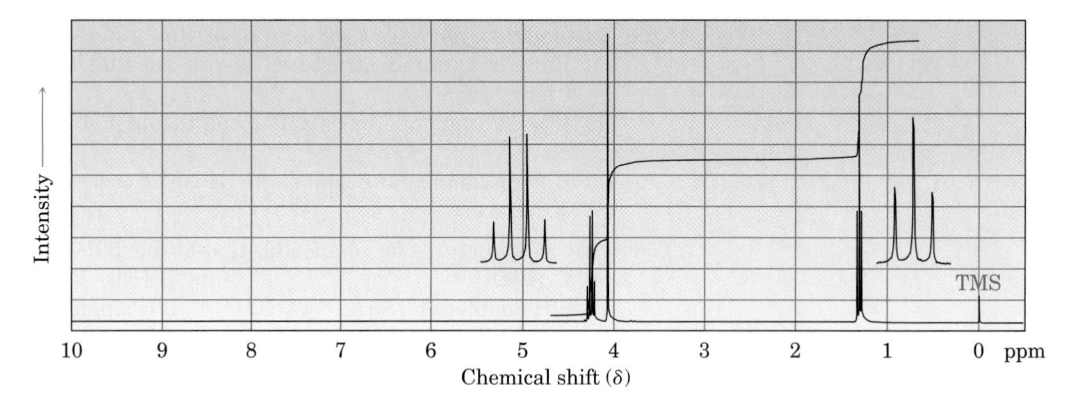

13.73 Propose a structure for a compound with formula C_4H_9Br that has the following ^1H NMR spectrum:

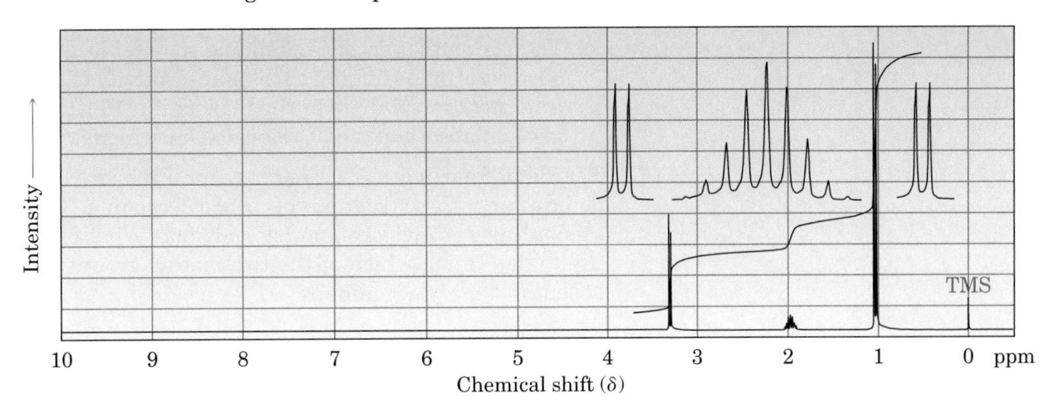

13.74 Propose structures for compounds that fit the following ^1H NMR data:

(a) $C_4H_6Cl_2$
- 3 H singlet at 2.18 δ
- 2 H doublet at 4.16 δ, $J = 7$ Hz
- 1 H triplet at 5.71 δ, $J = 7$ Hz

(b) $C_{10}H_{14}$
- 9 H singlet at 1.30 δ
- 5 H singlet at 7.30 δ

13.75 Nitriles (RC≡N) react with Grignard reagents (RMgBr). The reaction product from 2-methylpropanenitrile with methylmagnesium bromide has the following spectroscopic properties. Propose a structure.

Molecular weight = 86 IR: 1715 cm^{-1}

^1H NMR: 1.05 δ (6 H doublet, $J = 7$ Hz); 2.12 δ (3 H singlet); 2.67 δ (1 H septet, $J = 7$ Hz)

^{13}C NMR: 18.2, 27.2, 41.6, 211.2 δ

$$CH_3CH(CH_3)C{\equiv}N \xrightarrow[2.\ H_3O^+]{1.\ CH_3MgBr} ?$$

2-Methylpropanenitrile

IN THE MEDICINE CABINET

13.76 Lamivudine, a drug used in the management of acquired immunodeficiency syndrome (AIDS), greatly resembles cytidine. Lamivudine is also referred to as 3TC, an acronym for 3-thiocytidine.

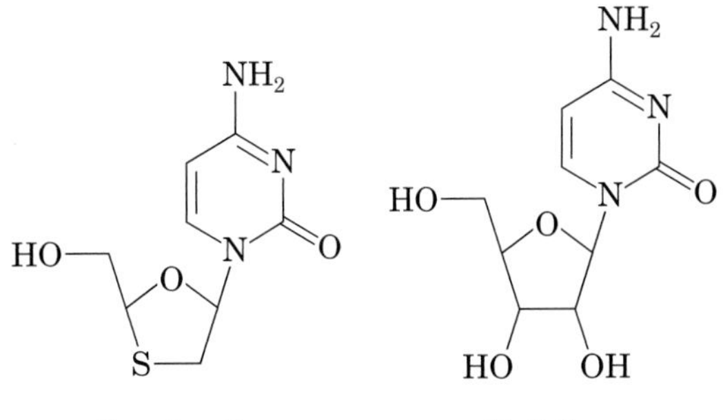

Lamivudine **Cytidine**

(a) How many chiral centers are in lamivudine?

(b) Cytidine has one glycosidic (acetal-like) bond to the nitrogen of the heterocylic ring. Identify the anomeric carbon of cytidine.

(c) Lamivudine has two acid-sensitive acetal-like groups that will hydrolyze with acid. Identify these groups.
(d) What three products arise from the acid hydrolysis of lamivudine?
(e) The concentration of lamivudine in the blood can be measured by UV spectroscopy. Assuming a molar absorptivity of 8600 $M^{-1}cm^{-1}$, what is the concentration of drug in the plasma if a 10 times *dilution* of the sample gave an absorbance of 0.195 using a pathlength of 1 cm?

13.77 The data from mass spectrometry for Problem 13.45 are inconclusive for determining the structure of benzoylecgonine. Cocaine has two methyl groups, but benzoylecgonine has only one. What other techniques could you use to determine which of the two methyl groups that appear in cocaine is absent in benzoylecgonine?

IN THE FIELD WITH AGROCHEMICALS

13.78 Although the U.S. Environmental Protection Agency (EPA) banned the use of chlordane more than 30 years ago, it remains a common environmental contaminant, especially in termite-infested areas, where it was widely used to control their populations. Mass spectrometry of chlordane is complicated for two reasons. First, chlordane exists as a mixture of many different chemicals. Second, it contains multiple chlorine atoms. The mass spectrum of the molecule shown appears as a range of peaks separated by two mass units. What are the minimum and maximum molecular weights of this molecule based on chlorine isotopes (assume ^{1}H and ^{12}C)?

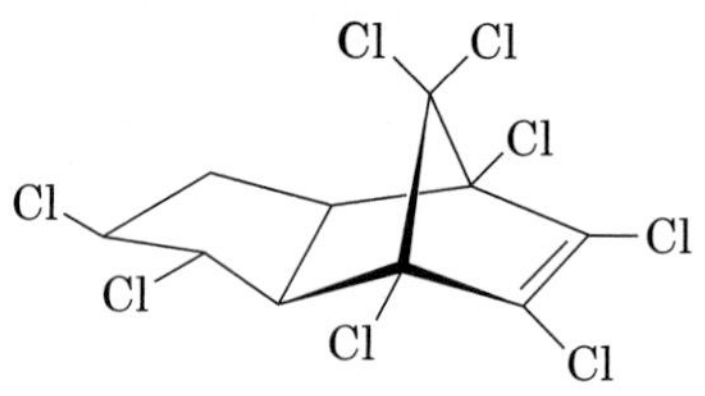

Cl Cl Cl Cl Cl Cl Cl Cl

Chlordane

The anti-influenza agent zanamivir, marketed under the name Relenza, was designed to mimic the cell-surface carbohydrate that a flu virus binds to during infection.

14 CHAPTER Biomolecules: Carbohydrates

Carbohydrates are found in every living organism. The sugar and starch in food and the cellulose in wood, paper, and cotton are nearly pure carbohydrate. Modified carbohydrates form part of the coating around living cells; other carbohydrates are found in the DNA that carries genetic information, and still others are used as medicines.

The word **carbohydrate** derives historically from the fact that glucose, the first carbohydrate to be obtained pure, has the molecular formula $C_6H_{12}O_6$ and was originally thought to be a "hydrate of carbon," $C_6(H_2O)_6$. This view was soon abandoned, but the name persisted. Today, the term *carbohydrate* is used to refer loosely to the broad class of polyhydroxylated aldehydes and ketones commonly called *sugars.*

```
 H
   \
    C=O
    |
H — C — OH
    |
HO— C — H
    |
H — C — OH
    |
H — C — OH
    |
   CH2OH
```

Glucose (also called dextrose), a pentahydroxyhexanal

Carbohydrates are made by green plants during photosynthesis, a complex process in which sunlight provides the energy to convert CO_2 and H_2O

into glucose. Many molecules of glucose are then chemically linked for storage by the plant in the form of either cellulose or starch. It has been estimated that more than 50% of the dry weight of the earth's biomass—all plants and animals—consists of glucose polymers. When eaten and then metabolized, carbohydrates provide the major source of energy required by organisms. Thus, carbohydrates act as the chemical intermediaries by which solar energy is stored and used to support life.

$$6\ CO_2 + 6\ H_2O \xrightarrow{\text{Sunlight}} 6\ O_2 + \underset{\textbf{Glucose}}{C_6H_{12}O_6} \longrightarrow \text{Cellulose, starch}$$

14.1 Classification of Carbohydrates

Carbohydrates are generally classed into two groups: *simple* and *complex*. **Simple sugars**, or **monosaccharides**, are carbohydrates like glucose and fructose that can't be converted into smaller sugars by hydrolysis. **Complex carbohydrates** are composed of two or more simple sugars linked together. Sucrose (table sugar), for example, is a *disaccharide* made up of one glucose molecule linked to one fructose molecule. Similarly, cellulose is a *polysaccharide* made up of several thousand glucose molecules linked together. Hydrolysis of polysaccharides breaks them down into their constituent monosaccharide units.

$$\text{1 Sucrose} \xrightarrow{H_3O^+} \text{1 Glucose + 1 Fructose}$$

$$\text{Cellulose} \xrightarrow{H_3O^+} \sim\text{3000 Glucose}$$

Monosaccharides are further classified as either **aldoses** or **ketoses**. The *-ose* suffix is used as the family name ending for carbohydrates, and the *aldo-* and *keto-* prefixes identify the nature of the carbonyl group, whether aldehyde or ketone. The number of carbon atoms in the monosaccharide is indicated by using *tri-*, *tetr-*, *pent-*, *hex-*, and so forth, in the name. For example, glucose is an *aldohexose*, a six-carbon aldehydo sugar; fructose is a *ketohexose*, a six-carbon keto sugar; and ribose is an *aldopentose*, a five-carbon aldehydo sugar. Most of the commonly occurring simple sugars are either aldopentoses or aldohexoses.

$$\begin{array}{c}
\text{H}-\text{C}=\text{O} \\
| \\
\text{H}-\text{C}-\text{OH} \\
| \\
\text{HO}-\text{C}-\text{H} \\
| \\
\text{H}-\text{C}-\text{OH} \\
| \\
\text{H}-\text{C}-\text{OH} \\
| \\
CH_2OH
\end{array}
\qquad
\begin{array}{c}
CH_2OH \\
| \\
\text{C}=\text{O} \\
| \\
\text{HO}-\text{C}-\text{H} \\
| \\
\text{H}-\text{C}-\text{OH} \\
| \\
\text{H}-\text{C}-\text{OH} \\
| \\
CH_2OH
\end{array}
\qquad
\begin{array}{c}
\text{H}-\text{C}=\text{O} \\
| \\
\text{H}-\text{C}-\text{OH} \\
| \\
\text{H}-\text{C}-\text{OH} \\
| \\
\text{H}-\text{C}-\text{OH} \\
| \\
CH_2OH
\end{array}$$

Glucose (an aldohexose) **Fructose (a ketohexose)** **Ribose (an aldopentose)**

PRACTICE PROBLEM 14.1

Classify the following monosaccharide:

```
 H   O
  \ //
   C
   |
H—C—OH
   |
H—C—OH
   |
H—C—OH
   |
H—C—OH
   |
  CH2OH
```

Allose

SOLUTION Since allose has six carbons and an aldehyde carbonyl group, it's an aldohexose.

PROBLEM 14.1 Classify each of the following monosaccharides:

(a)
```
  H   O
   \ //
    C
    |
HO—C—H
    |
 H—C—OH
    |
   CH2OH
```
Threose

(b)
```
   CH2OH
    |
    C=O
    |
 H—C—OH
    |
 H—C—OH
    |
   CH2OH
```
Ribulose

(c)
```
   CH2OH
    |
    C=O
    |
HO—C—H
    |
HO—C—H
    |
 H—C—OH
    |
   CH2OH
```
Tagatose

(d)
```
  H   O
   \ //
    C
    |
 H—C—H
    |
 H—C—OH
    |
 H—C—OH
    |
   CH2OH
```
2-Deoxyribose

14.2 Configurations of Monosaccharides: Fischer Projections

Since all carbohydrates are chiral and have stereocenters (see Section 6.2), it was recognized long ago that a standard method of representation is needed to describe carbohydrate stereoisomers. In 1891, Emil Fischer suggested a method based on the projection of a tetrahedral carbon atom onto a flat surface. These **Fischer projections** were soon adopted and are now a standard means of depicting stereochemistry.

A tetrahedral carbon atom is represented in a Fischer projection by two perpendicular lines. The horizontal lines indicate bonds coming out of the page, and the vertical lines indicate bonds going into the page:

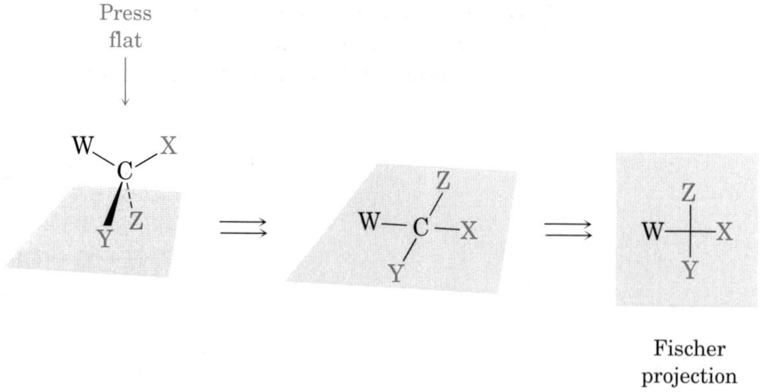

Press flat

Fischer projection

By convention, the carbonyl carbon is placed at or near the top in Fischer projections of carbohydrates. Thus, (*R*)-glyceraldehyde, the simplest monosaccharide, is represented as shown in Figure 14.1.

FIGURE 14.1 Fischer projection of (*R*)-glyceraldehyde.

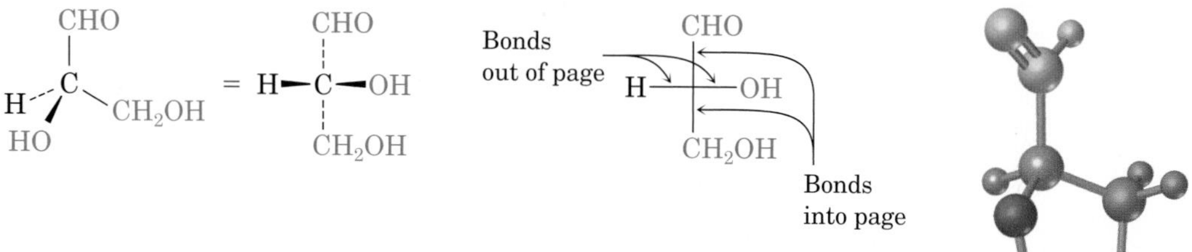

Bonds out of page

Bonds into page

Fischer projection of (*R*)-glyceraldehyde

Carbohydrates with more than one stereocenter are shown by stacking the atoms, one on top of the other, with the carbonyl carbon at or near the top of the Fischer projection. Glucose, for example, has four stereocenters stacked on top of one another in a Fischer projection:

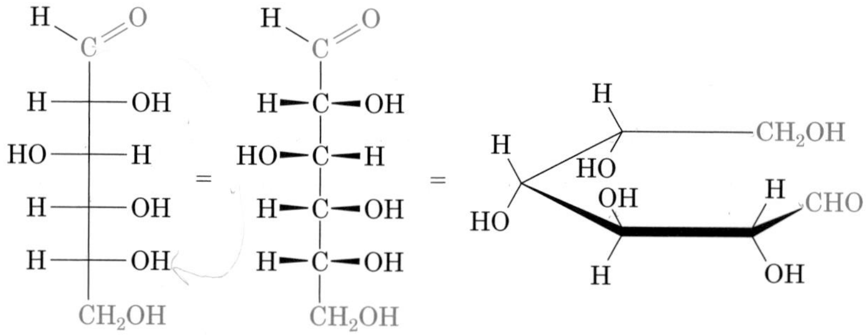

Glucose (carbonyl group at top)

PRACTICE PROBLEM 14.2

Convert the following tetrahedral representation of (*R*)-butan-2-ol into a Fischer projection:

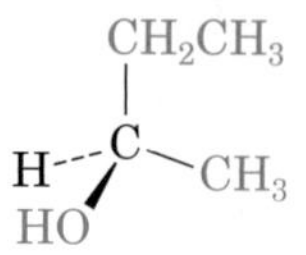

(*R*)-Butan-2-ol

STRATEGY Orient the molecule so that two horizontal bonds are facing you and two vertical bonds are receding away from you. Then press the molecule flat into the paper, indicating the stereocenter as the intersection of two crossed lines.

SOLUTION

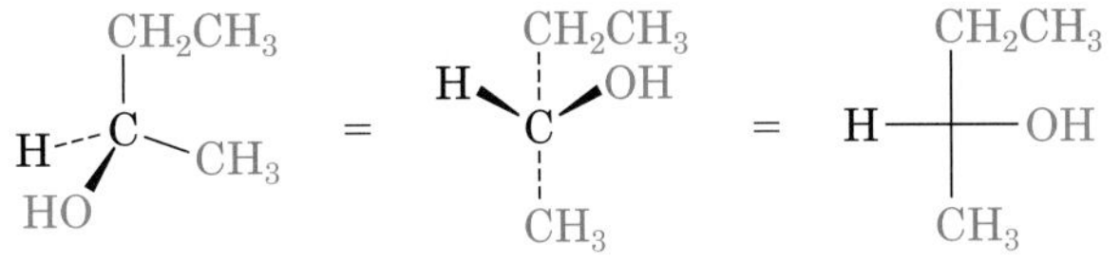

(*R*)-Butan-2-ol

PRACTICE PROBLEM 14.3

Convert the following Fischer projection of lactic acid into a tetrahedral representation, and indicate whether the molecule is (*R*) or (*S*):

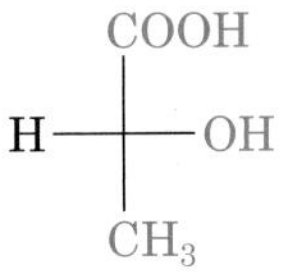

Lactic acid

STRATEGY Place a carbon atom at the intersection of the two crossed lines, and imagine that the two horizontal bonds are coming toward you and the two vertical bonds are receding away from you. The projection represents (*R*)-lactic acid.

SOLUTION

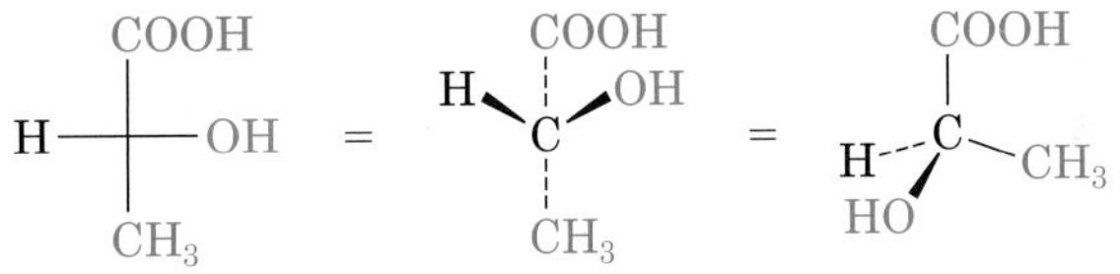

(*R*)-Lactic acid

PROBLEM 14.2 Convert the following tetrahedral representation of (*S*)-glyceraldehyde into a Fischer projection:

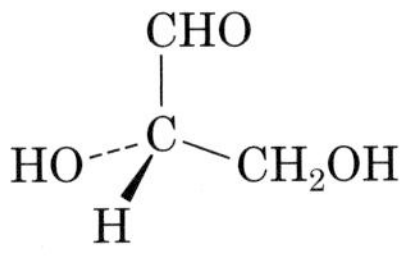

(*S*)-Glyceraldehyde

PROBLEM 14.3 Draw Fischer projections of both (*R*)-2-chlorobutane and (*S*)-2-chlorobutane.

PROBLEM 14.4 Convert the following Fischer projections into tetrahedral representations, and assign *R* or *S* stereochemistry to each:

(a) (b) 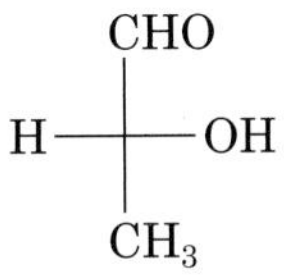 (c)

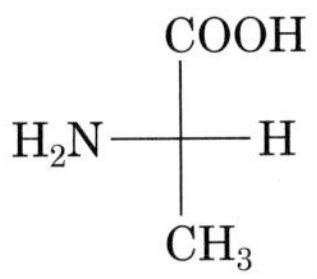

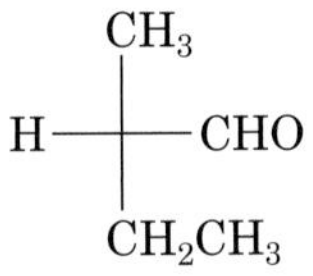

PROBLEM 14.5 Redraw the following molecule as a Fischer projection, and assign *R* or *S* configuration to the stereocenter (yellow-green = Cl):

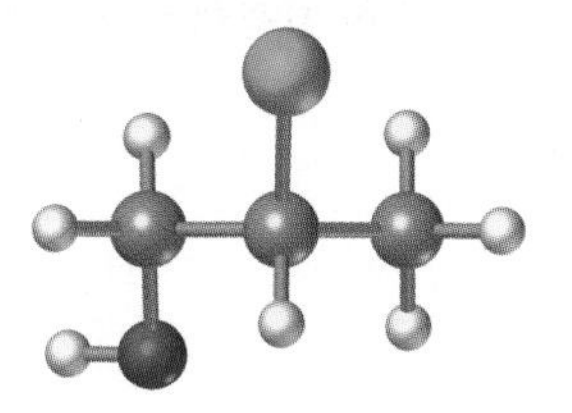

14.3 D,L Sugars

Glyceraldehyde has only one stereocenter and therefore has two enantiomeric (mirror-image) forms. Only the dextrorotatory enantiomer occurs naturally, however. That is, a sample of naturally occurring glyceraldehyde placed in a polarimeter (see Section 6.4) rotates plane-polarized light in a clockwise direction, denoted (+). Since (+)-glyceraldehyde is known to have the *R* configuration at C2, it can be represented as in Figure 14.2. For historical reasons dating from long before the adoption of the *R,S* system, (*R*)-(+)-glyceraldehyde is also referred to as *D-glyceraldehyde* (D for dextrorotatory). The other enantiomer, (*S*)-(−)-glyceraldehyde, is known as *L-glyceraldehyde* (L for levorotatory).

Because of the way that monosaccharides are synthesized in nature, glucose, fructose, ribose, and most other naturally occurring monosaccharides have the same *R* stereochemical configuration as D-glyceraldehyde at the stereocenter farthest from the carbonyl group. In Fischer projections, therefore, most naturally occurring sugars have the –OH group at the bottom stereocenter pointing to the *right* (Figure 14.2). Such compounds are referred to as **D sugars.**

FIGURE 14.2 Some naturally occurring D sugars. The –OH at the bottom stereocenter (the one farthest from the carbonyl group) is on the right in Fischer projections.

D-Glyceraldehyde [(*R*)-(+)-glyceraldehyde]: CHO; H—C—OH; CH_2OH

D-Ribose: CHO; H—C—OH; H—C—OH; H—C—OH; CH_2OH

D-Glucose: CHO; H—C—OH; HO—C—H; H—C—OH; H—C—OH; CH_2OH

D-Fructose: CH_2OH; C=O; HO—C—H; H—C—OH; H—C—OH; CH_2OH

In contrast to D sugars, all **L sugars** have an *S* configuration at the lowest stereocenter, with the bottommost –OH group pointing to the *left* in Fischer projections. Thus, an L sugar is the mirror image (enantiomer) of the corresponding D sugar and has the opposite configuration at all stereocenters. Note that the D and L notations have no relation to the direction in which a given sugar rotates plane-polarized light. A D sugar may be either dextrorotatory or levorotatory. The prefix D indicates only that the stereochemistry of the bottommost stereocenter is the same as that of D-glyceraldehyde and is to the right in a Fischer projection when the molecule is drawn in the standard way with the carbonyl group at or near the top. Note also that the D,L system of carbohydrate nomenclature describes the configuration at only *one* stereocenter and says nothing about the configuration of other stereocenters that may be present.

L-Glyceraldehyde
[(*S*)-(–)-glyceraldehyde]

L-Glucose
(not naturally occurring)

Mirror

D-Glucose

Practice Problem 14.4

Look at the Fischer projection of D-fructose in Figure 14.2, and draw a Fischer projection of L-fructose.

Strategy Since L-fructose is the enantiomer (mirror image) of D-fructose, we simply take the structure of D-fructose and reverse the configuration at each stereocenter.

Solution

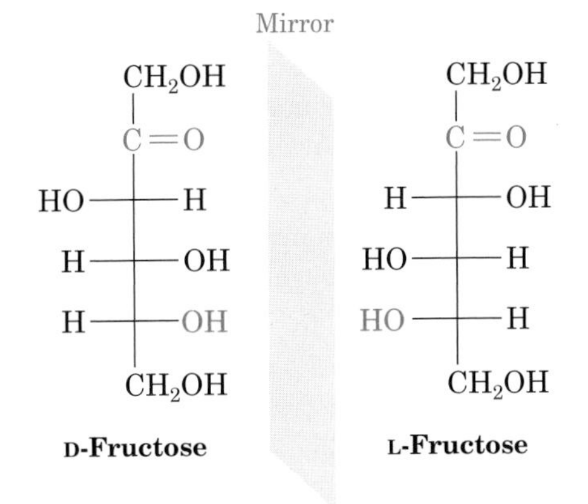

Problem 14.6

Which of the following are L sugars, and which are D sugars?

(a)

(b)

(c)

PROBLEM 14.7 Draw the enantiomers (mirror images) of the carbohydrates shown in Problem 14.6, and identify each as a D sugar or an L sugar.

14.4 Configurations of Aldoses

Aldotetroses are four-carbon sugars with two stereocenters. Thus, there are $2^2 = 4$ possible stereoisomeric aldotetroses, or two D,L pairs of enantiomers, named *erythrose* and *threose.*

Aldopentoses have three stereocenters and a total of $2^3 = 8$ possible stereoisomers, or four D,L pairs of enantiomers. These four pairs are named *ribose*, *arabinose*, *xylose*, and *lyxose.* All except lyxose occur widely in nature. D-Ribose is an important part of RNA (ribonucleic acid), L-arabinose is found in many plants, and D-xylose is found in wood.

Aldohexoses have four stereocenters, for a total of $2^4 = 16$ possible stereoisomers, or eight D,L pairs of enantiomers. The names of the eight are *allose*, *altrose*, *glucose*, *mannose*, *gulose*, *idose*, *galactose*, and *talose.* Only D-glucose, from starch and cellulose, and D-galactose, from gums and fruit pectins, are widely distributed in nature. D-Mannose and D-talose also occur naturally, but in lesser abundance.

Fischer projections of the four-, five-, and six-carbon aldoses are shown in Figure 14.3 for the D series. Starting from D-glyceraldehyde, we can imagine constructing the two D aldotetroses by inserting a new stereocenter just below the aldehyde carbon. Each of the two D aldotetroses then leads to two D aldopentoses (four total), and each of the four D aldopentoses leads to two D aldohexoses (eight total).

PROBLEM 14.8 Write Fischer projections for the following L sugars. Remember that an L sugar is the mirror image of the corresponding D sugar shown in Figure 14.3.

(a) L-Arabinose (b) L-Threose (c) L-Galactose

PROBLEM 14.9 How many aldoheptoses are possible? How many of them are D sugars, and how many are L sugars?

PROBLEM 14.10 Draw Fischer projections for the two D aldoheptoses (Problem 14.9) whose stereochemistry at C3, C4, C5, and C6 is the same as that of glucose at C2, C3, C4, and C5.

PROBLEM 14.11 The following model is that of an aldopentose. Draw a Fischer projection of the sugar, and identify it. Is it a D sugar or an L sugar?

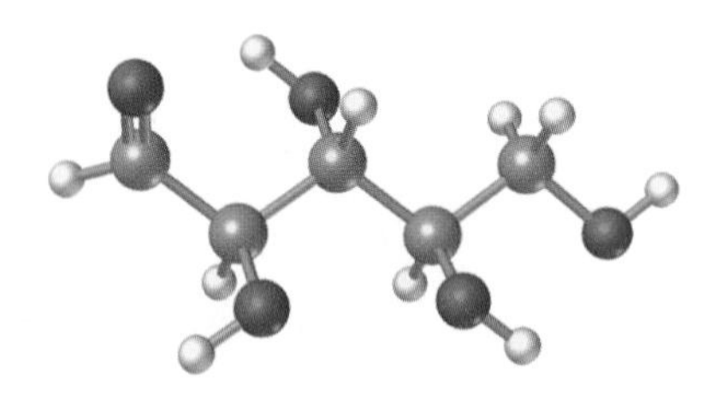

FIGURE 14.3 Configurations of D aldoses. The structures are arranged from left to right so that the –OH groups on C2 alternate right/left (R/L) in going across a series. Similarly, the –OH groups at C3 alternate two right/two left (2R/2L), the –OH groups at C4 alternate 4R/4L, and the –OH groups at C5 are to the right in all eight (8R).

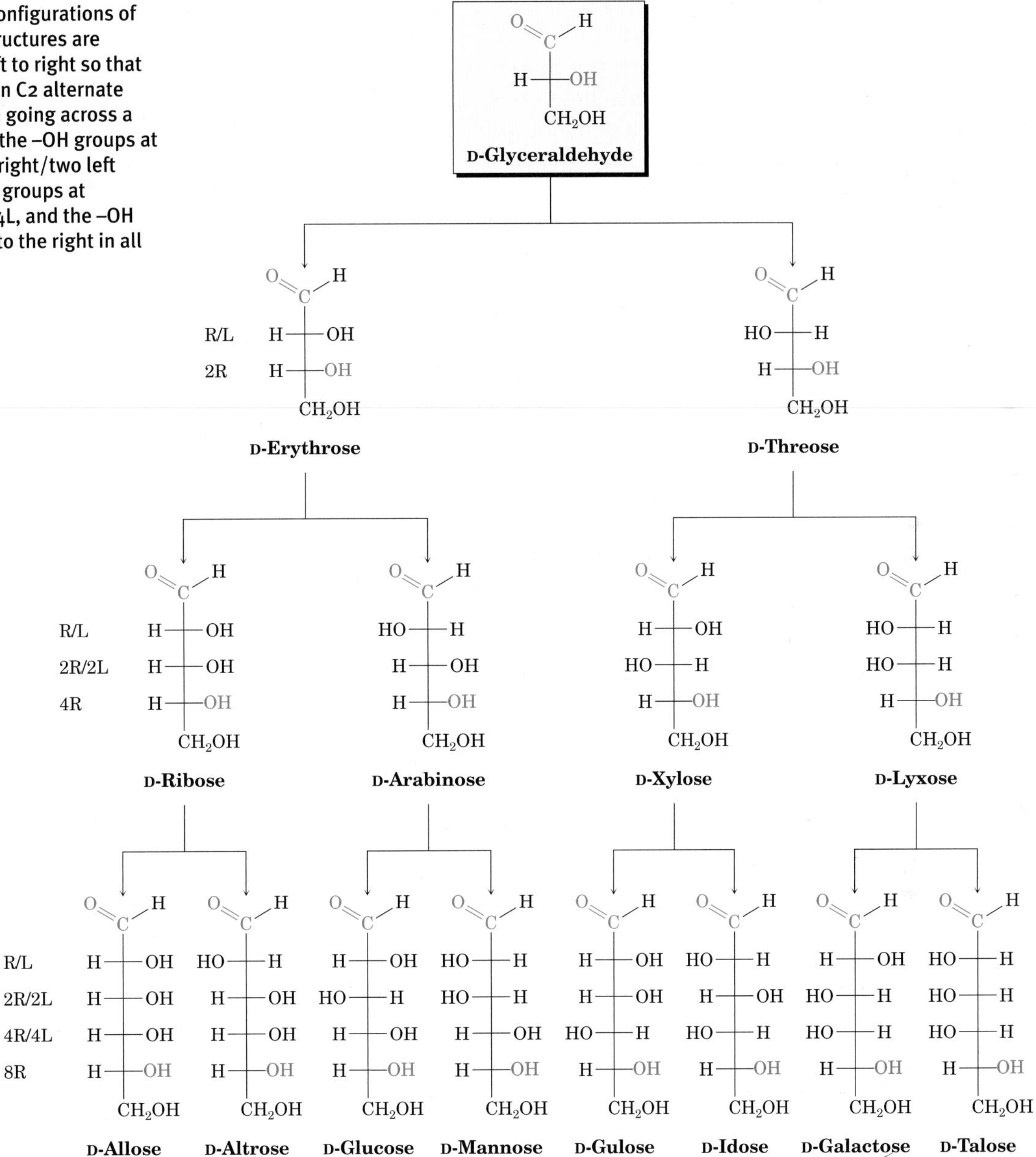

14.5

Cyclic Structures of Monosaccharides: Hemiacetal Formation

During the discussion of carbonyl-group chemistry in Section 9.8, we said that alcohols undergo a rapid and reversible nucleophilic addition reaction with aldehydes and ketones to form hemiacetals:

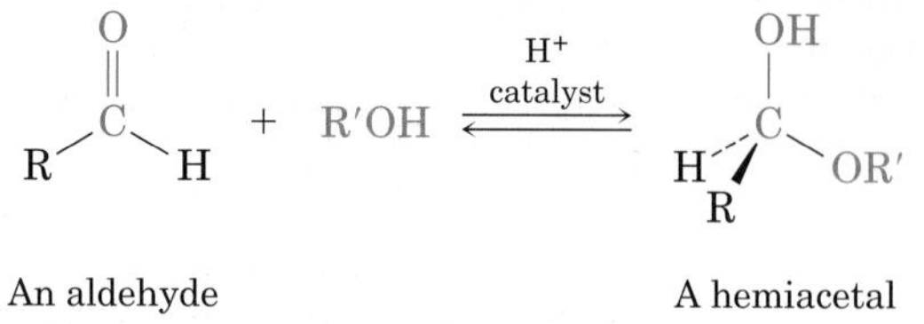

$$\text{RCHO} + \text{R'OH} \underset{}{\overset{\text{H}^+ \text{ catalyst}}{\rightleftharpoons}} \text{R(H)C(OH)OR'}$$

An aldehyde — A hemiacetal

If both the hydroxyl and the carbonyl group are in the same molecule, an *intramolecular* nucleophilic addition can take place, leading to the formation of a *cyclic* hemiacetal. Five- and six-membered cyclic hemiacetals form particularly easily, and many carbohydrates therefore exist in an equilibrium between open-chain and cyclic hemiacetal forms. For example, glucose exists in aqueous solution primarily in the six-membered **pyranose** form resulting from intramolecular nucleophilic addition of the –OH group at C5 to the C1 aldehyde group. Fructose, on the other hand, exists to the extent of about 72% in the pyranose form and about 28% in the five-membered **furanose** form resulting from addition of the –OH group at C5 to the C2 ketone. (The names *pyranose* for a six-membered ring and *furanose* for a five-membered ring are derived from the names of the simple cyclic ethers pyran and furan.) The cyclic forms of glucose and fructose are shown in Figure 14.4.

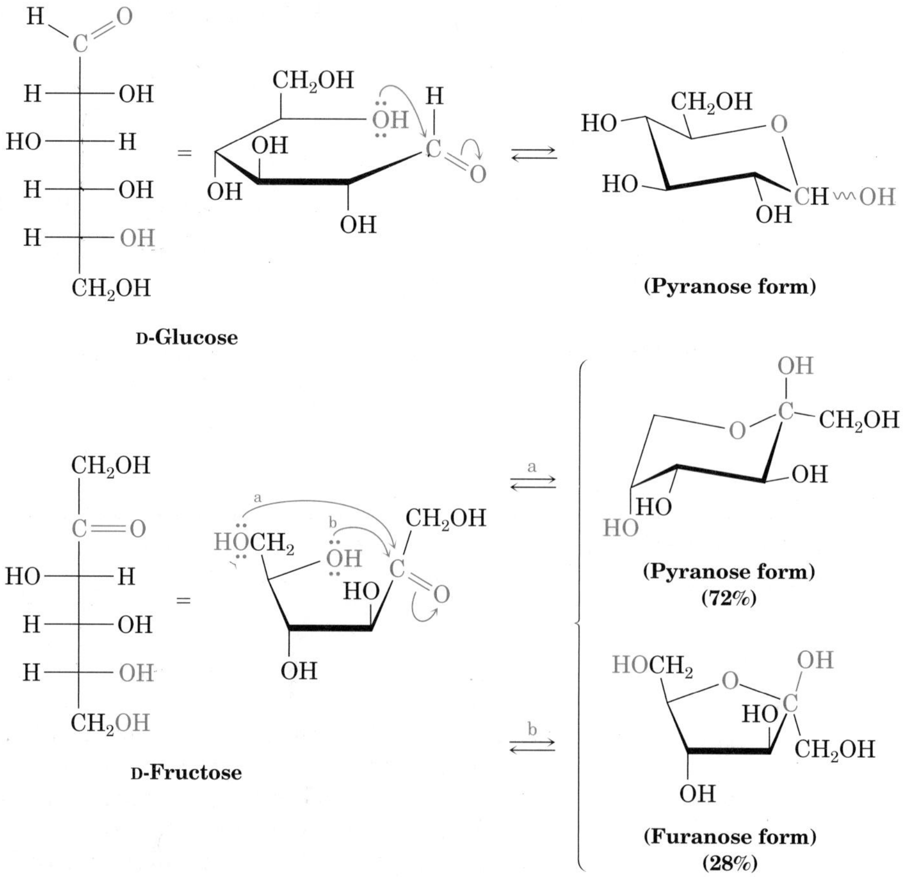

FIGURE 14.4 Glucose and fructose in their cyclic pyranose and furanose forms.

D-Glucose

(Pyranose form)

D-Fructose

(Pyranose form) (72%)

(Furanose form) (28%)

Like cyclohexane rings (see Section 2.9), pyranose rings have a chairlike geometry with axial and equatorial substituents. By convention, the rings are usually drawn by placing the hemiacetal oxygen atom at the right rear, as shown in Figure 14.4. Note that an –OH group on the *right* in a Fischer projection is on the *bottom* face of the pyranose ring, and an –OH group on the *left* in a Fischer projection is on the *top* face of the ring. For D sugars, the terminal $-CH_2OH$ group is on the top of the ring, whereas for L sugars, the $-CH_2OH$ group is on the bottom.

PRACTICE PROBLEM 14.5

D-Mannose differs from D-glucose in its stereochemistry at C2. Draw D-mannose in its pyranose form.

STRATEGY First draw a Fischer projection of D-mannose. Then lay it on its side, and curl it around so that the –CHO group (C1) is on the right front and the $-CH_2OH$ group (C6) is toward the left rear. Now, connect the –OH at C5 to the C1 carbonyl group to form the pyranose ring. In drawing the chair form, raise the leftmost carbon (C4) up and drop the rightmost carbon (C1) down.

SOLUTION

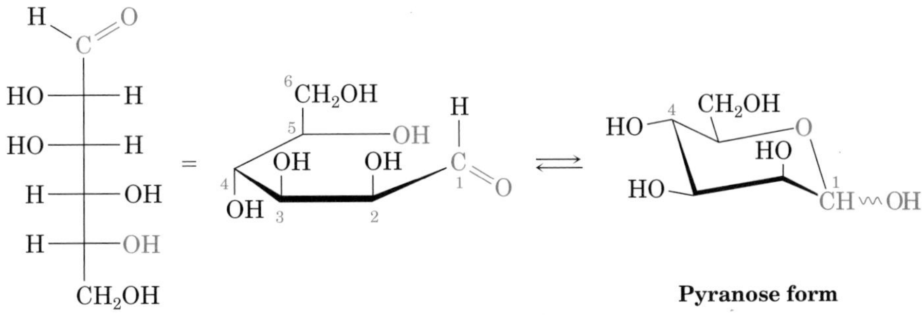

D-Mannose **Pyranose form**

PROBLEM 14.12 D-Galactose differs from D-glucose in its stereochemistry at C4. Draw D-galactose in its pyranose form.

PROBLEM 14.13 Ribose exists largely in a furanose form, produced by addition of the C4 –OH group to the C1 aldehyde. Find the structure of D-ribose in Figure 14.3, and draw it in its furanose form.

14.6 Monosaccharide Anomers: Mutarotation

When an open-chain monosaccharide cyclizes to a pyranose or furanose form, a new stereocenter forms at the former carbonyl carbon. Two diastereomers, called **anomers**, are produced, with the hemiacetal carbon referred to as the **anomeric center**. For example, glucose cyclizes reversibly in aqueous solution to yield a 37:63 mixture of two anomers (Figure 14.5). The minor anomer, which has the C1 –OH group trans to the $-CH_2OH$ substituent at C5, is called the **alpha (*α*) anomer**; its full name is *α*-D-glucopyranose. The major anomer, which has the C1 –OH group cis to the $-CH_2OH$ substituent at C5, is called the **beta (*β*) anomer**; its full name is *β*-D-glucopyranose.

Both anomers of D-glucopyranose can be crystallized and purified. Pure *α*-D-glucopyranose has a melting point of 146 °C and a specific rotation

$[\alpha]_D = +112.2°$; pure β-D-glucopyranose has a melting point of 148 to 155 °C and a specific rotation $[\alpha]_D = +18.7°$. When a sample of either pure α-D-glucopyranose or pure β-D-glucopyranose is dissolved in water, however, the optical rotation slowly changes and ultimately reaches a constant value of +52.6°. The specific rotation of the α anomer solution decreases from +112.2° to +52.6°, and the specific rotation of the β anomer solution increases from +18.7° to +52.6°. Called **mutarotation**, this spontaneous change in optical rotation is caused by the slow conversion of the pure α and β enantiomers into the 37:63 equilibrium mixture.

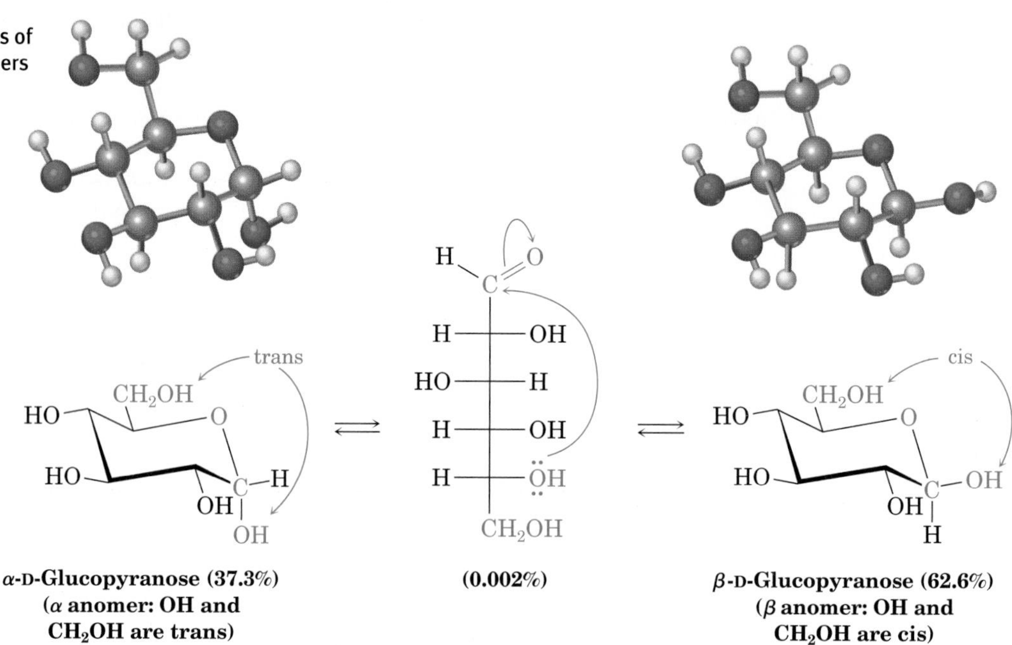

FIGURE 14.5 Structures of the alpha and beta anomers of glucose.

Mutarotation occurs by a reversible ring opening of each anomer to the open-chain aldehyde form, followed by reclosure. Although equilibration is slow at neutral pH, it is catalyzed by both acid and base.

α-D-Glucopyranose (37%)
$[\alpha]_D = +112.2°$

D-Glucose

β-D-Glucopyranose (63%)
$[\alpha]_D = +18.7°$

PRACTICE PROBLEM 14.6

Draw the two pyranose anomers of D-galactose, and identify each as α or β.

SOLUTION The α anomer has the –OH group at C1 pointing down, trans to the CH_2OH, and the β anomer has the –OH group at C1 pointing up, cis to the CH_2OH.

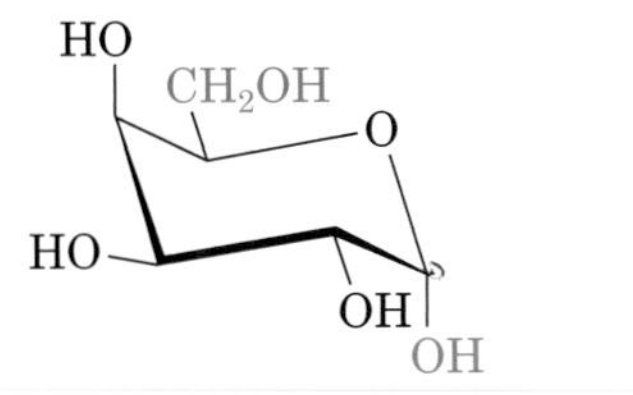

α-D-Galactopyranose

β-D-Galactopyranose

PROBLEM 14.14 At equilibrium in aqueous solution, D-fructose consists of 70% β pyranose, 2% α pyranose, 23% β furanose, and 5% α furanose forms. Draw all four.

PROBLEM 14.15 Draw β-D-mannopyranose in its chair conformation, and label all substituents as axial or equatorial. Which would you expect to be more stable, mannose or galactose (Practice Problem 14.6)?

14.7 Reactions of Monosaccharides

Since monosaccharides contain only two kinds of functional groups, carbonyls and hydroxyls, most of the chemistry of monosaccharides is the now-familiar chemistry of these two groups.

Ester and Ether Formation

Monosaccharides behave as simple alcohols in much of their chemistry. For example, carbohydrate –OH groups can be converted into esters and ethers, which are often easier to work with than the free sugars. Because of their many hydroxyl groups, monosaccharides are usually soluble in water but insoluble in organic solvents such as ether. They are also difficult to purify and have a tendency to form syrups rather than crystals when water is removed. Ester and ether derivatives, however, are soluble in organic solvents and are easily purified and crystallized.

Esterification is carried out by treating the carbohydrate with an acid chloride or acid anhydride in the presence of a base. All the –OH groups react, including the anomeric one. For example, D-glucopyranose is converted into its pentaacetate by treatment with acetic anhydride in pyridine solution:

$(CH_3CO)_2O$, Pyridine, 0 °C

D-Glucopyranose

Penta-*O*-acetyl-D-glucopyranose (91%)

Carbohydrates are converted into ethers by treatment with an alkyl halide in the presence of base (the Williamson ether synthesis; see Section 8.5). Silver oxide is a particularly mild and useful base for this reaction, since hydroxide and alkoxide bases tend to degrade the sensitive sugar molecules. For example, α-D-glucopyranose is converted into its pentamethyl ether in 85% yield on reaction with iodomethane and silver oxide:

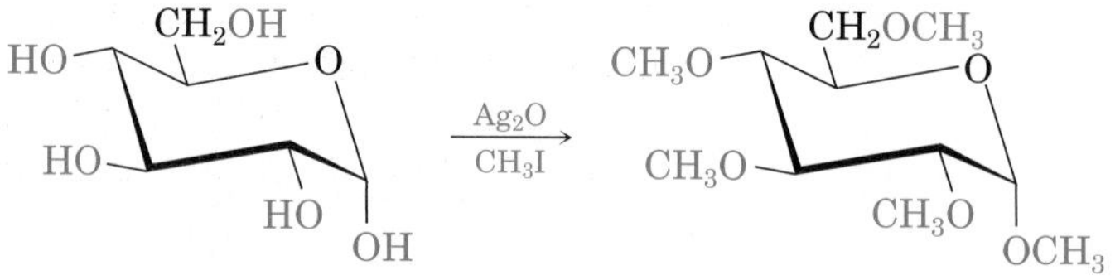

$\xrightarrow[CH_3I]{Ag_2O}$

α-D-Glucopyranose **α-D-Glucopyranose pentamethyl ether (85%)**

PROBLEM 14.16 Draw the products you would obtain by reaction of β-D-ribofuranose with the following:
(a) CH_3I, Ag_2O (b) $(CH_3CO)_2O$, pyridine

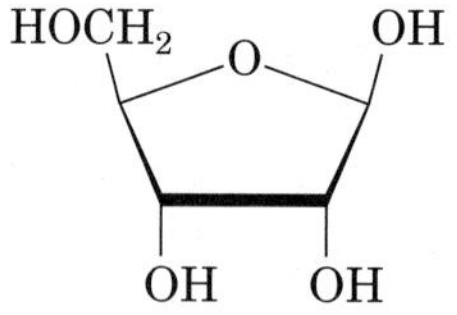

β-D-Ribofuranose

Glycoside Formation

We saw in Section 9.8 that treatment of a hemiacetal with an alcohol and an acid catalyst yields an acetal:

$$\text{hemiacetal} + ROH \underset{}{\overset{HCl}{\rightleftharpoons}} \text{acetal} + H_2O$$

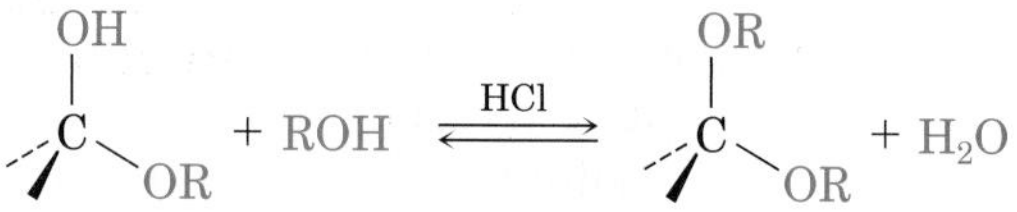

In the same way, treatment of a monosaccharide hemiacetal with an alcohol and an acid catalyst yields an acetal in which the anomeric –OH group has been replaced by an –OR group. For example, reaction of glucose with methanol gives a mixture of α and β methyl D-glucopyranosides:

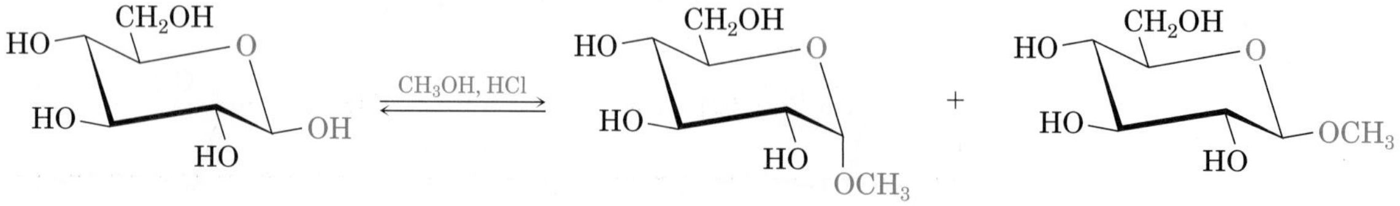

$\overset{CH_3OH,\ HCl}{\rightleftharpoons}$

β-D-Glucopyranose (a cyclic hemiacetal) **Methyl α-D-glucopyranoside (66%)** + **Methyl β-D-glucopyranoside (33%)**

Called **glycosides**, carbohydrate acetals are named by first citing the alkyl group and then replacing the *-ose* ending of the sugar with *-oside.* Like all acetals, glycosides are stable to water. They aren't in equilibrium with an open-chain form, and they don't show mutarotation. They can, however, be converted back to the free monosaccharide by hydrolysis with aqueous acid.

Glycosides are widespread in nature, and many biologically active molecules contain glycosidic linkages. For example, digitoxin, the active component of the digitalis preparations used for treatment of heart disease, is a glycoside consisting of a complex steroid alcohol linked to a trisaccharide. Note that the three sugars are also linked by glycoside bonds.

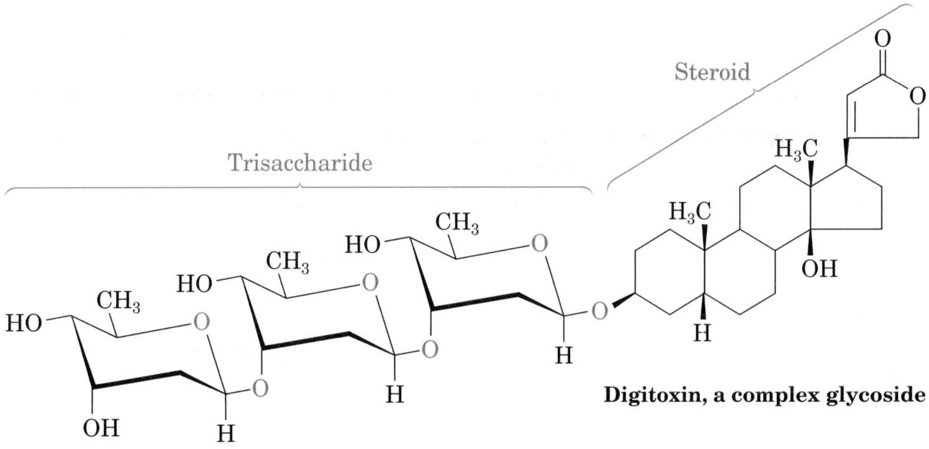

Trisaccharide

Steroid

Digitoxin, a complex glycoside

PRACTICE PROBLEM 14.7

What product would you expect from the acid-catalyzed reaction of β-D-ribofuranose with methanol?

STRATEGY The acid-catalyzed reaction of a monosaccharide with an alcohol yields a glycoside in which the anomeric –OH group is replaced by the –OR group of the alcohol:

SOLUTION

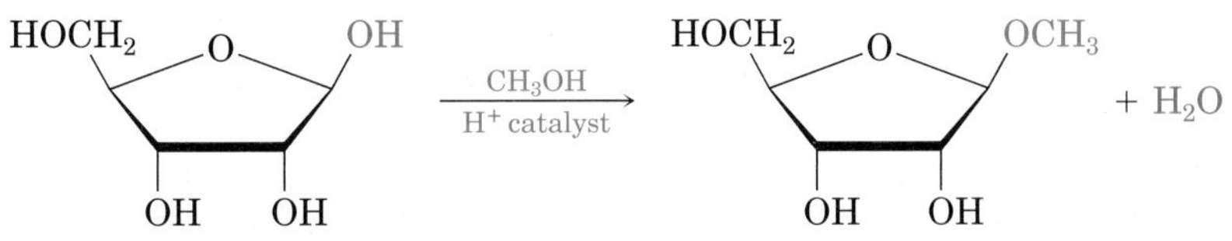

$\xrightarrow[\text{H}^+ \text{ catalyst}]{CH_3OH}$ + H_2O

β-D-Ribofuranose **Methyl β-D-ribofuranoside**

PROBLEM 14.17 Draw the product you would obtain from the acid-catalyzed reaction of β-D-galactopyranose with ethanol.

Reduction of Monosaccharides

Treatment of an aldose or a ketose with $NaBH_4$ reduces it to a polyalcohol called an **alditol**. The reaction occurs by reaction of the open-chain form present in the aldehyde $\rightleftharpoons$ hemiacetal equilibrium.

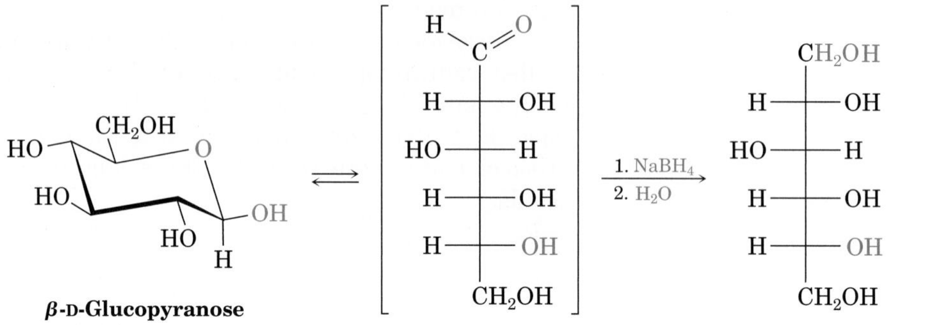

β-D-Glucopyranose $\rightleftharpoons$ [**D-Glucose**] $\xrightarrow[2.\ H_2O]{1.\ NaBH_4}$ **D-Glucitol (D-sorbitol), an alditol**

D-Glucitol, the alditol produced on reduction of D-glucose, is itself a naturally occurring substance that has been isolated from many fruits and berries. It is used under the name D-sorbitol as a sweetener and sugar substitute in many foods.

PRACTICE PROBLEM 14.8

Show the structure of the alditol you would obtain from reduction of D-galactose.

STRATEGY First draw D-galactose in its open-chain form. Then convert the –CHO group at C1 into a $-CH_2OH$ group.

SOLUTION

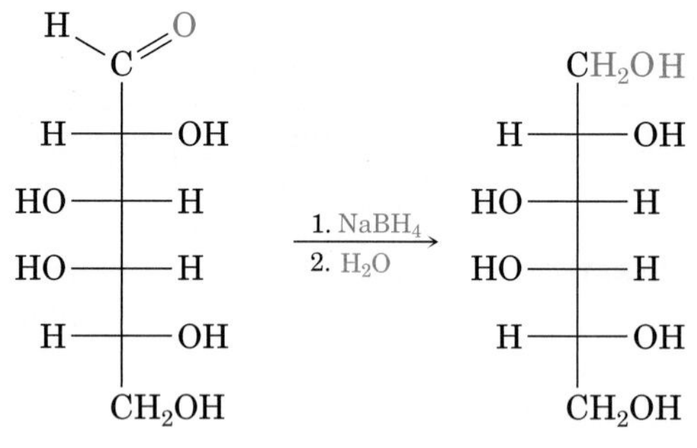

D-Galactose $\xrightarrow[2.\ H_2O]{1.\ NaBH_4}$ **D-Galactitol**

PROBLEM 14.18 How can you account for the fact that reduction of D-glucose leads to an optically active alditol (D-glucitol), whereas reduction of D-galactose leads to an optically inactive alditol (see Section 6.8)?

PROBLEM 14.19 Reduction of L-gulose with $NaBH_4$ leads to the same alditol (D-glucitol) as reduction of D-glucose. Explain.

Oxidation of Monosaccharides

Like other aldehydes, an aldose is easily oxidized to yield the corresponding carboxylic acid, called an **aldonic acid**. Aldoses react with Tollens' reagent (Ag^+ in aqueous ammonia), Fehling's reagent (Cu^{2+} with aqueous sodium tartrate), and Benedict's reagent (Cu^{2+} with aqueous sodium citrate) to yield the oxidized sugar and a reduced metallic species. All three reactions serve as simple chemical tests for what are called **reducing sugars** (*reducing* because the sugar reduces the metallic oxidizing agent).

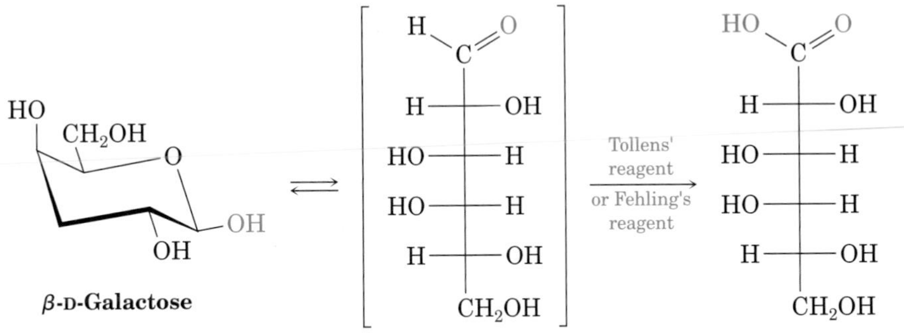

β-D-Galactose

D-Galactonic acid (an aldonic acid)

If Tollens' reagent is used, metallic silver is produced as a shiny mirror on the walls of the reaction flask or test tube (see Section 9.4). If Fehling's or Benedict's reagent is used, a reddish precipitate of Cu_2O signals a positive result. Some simple diabetes self-test kits sold in drugstores for home use employ Benedict's test. As little as 0.1% glucose in urine gives a positive test.

All aldoses are reducing sugars because they contain aldehyde carbonyl groups, but glycosides are nonreducing. Glycosides don't react with Tollens' or Fehling's reagents because the acetal group can't open to an aldehyde under basic conditions.

If warm dilute HNO_3 (nitric acid) is used as the oxidizing agent, an aldose is oxidized to a dicarboxylic acid called an **aldaric acid**. Both the aldehyde carbonyl and the terminal $-CH_2OH$ group are oxidized in this reaction.

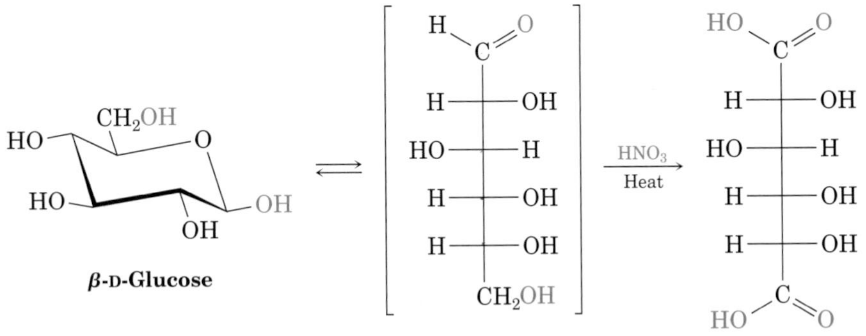

β-D-Glucose

D-Glucaric acid (an aldaric acid)

PROBLEM 14.20 D-Glucose yields an optically active aldaric acid on treatment with nitric acid, but D-allose yields an optically inactive aldaric acid. Explain.

PROBLEM 14.21 Which of the other six D aldohexoses yield optically active aldaric acids, and which yield optically inactive aldaric acids? (See Problem 14.20.)

14.8 Disaccharides

We saw in the previous section that reaction of a monosaccharide hemiacetal yields a glycoside in which the anomeric –OH group is replaced by an –OR substituent. If the alcohol is itself a sugar, the glycoside product is a disaccharide.

Maltose and Cellobiose

Disaccharides can contain a glycosidic acetal bond between the anomeric carbon (the carbonyl carbon) of one sugar and an –OH group at *any* position on the other sugar. A glycosidic link between C1 of the first sugar and C4 of the second sugar is particularly common. Such a bond is called a **1,4′ link**, where the "prime" superscript indicates that the 4′ position is on a different sugar than the 1 position.

A glycosidic bond can be either α or β. Maltose, the disaccharide obtained by partial hydrolysis of starch, consists of two D-glucopyranoses joined by a 1,4′-α-glycoside bond. Cellobiose, the disaccharide obtained by partial hydrolysis of cellulose, consists of two D-glucopyranoses joined by a 1,4′-β-glycoside bond.

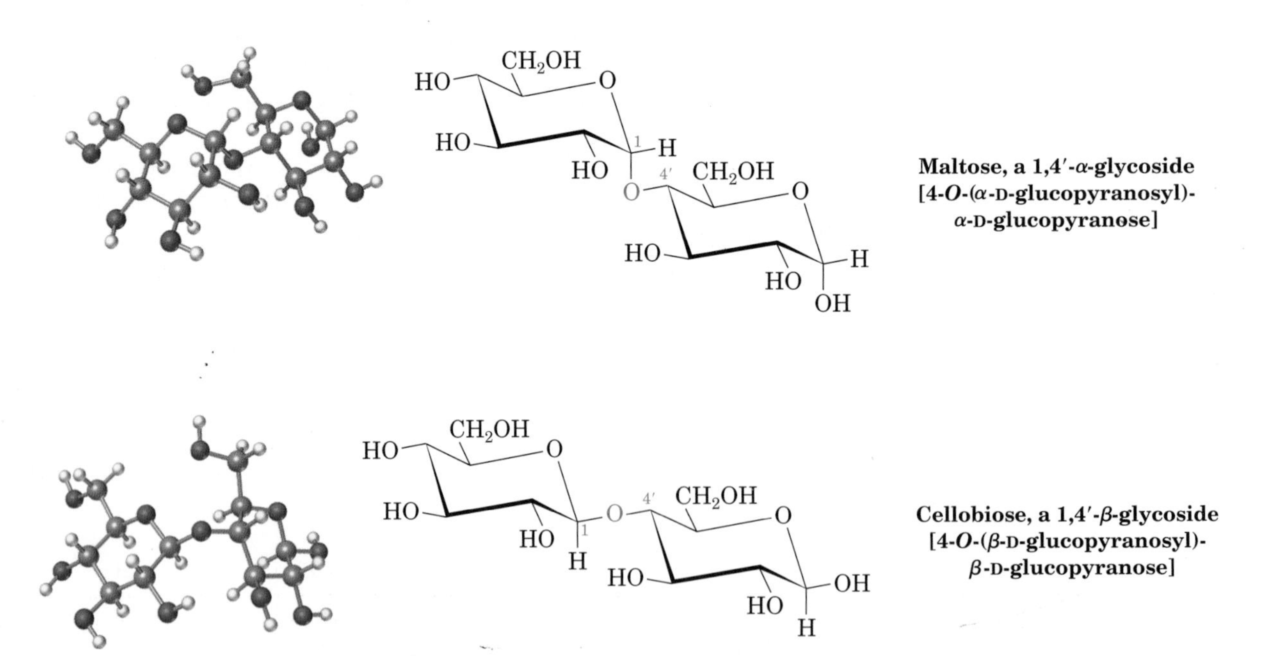

Maltose, a 1,4′-α-glycoside [4-*O*-(α-D-glucopyranosyl)-α-D-glucopyranose]

Cellobiose, a 1,4′-β-glycoside [4-*O*-(β-D-glucopyranosyl)-β-D-glucopyranose]

Maltose and cellobiose are both reducing sugars because the right-hand saccharide unit in each has a hemiacetal group. Both are therefore in equilibrium with aldehyde forms, which can reduce Tollens' or Fehling's reagent. For a similar reason, both maltose and cellobiose show mutarotation.

Despite the similarities of their structures, maltose and cellobiose are dramatically different biologically. Cellobiose can't be digested by humans and can't be fermented by yeast. Maltose, however, is digested without difficulty and is readily fermented.

PROBLEM 14.22 Draw the structures of the products obtained from reaction of cellobiose with the following:

(a) $NaBH_4$ (b) $AgNO_3$, H_2O, NH_3

Sucrose

Sucrose, or ordinary table sugar, is probably the most abundant pure organic chemical in the world. Whether from sugar cane (20% by weight) or from sugar beets (15% by weight), and whether raw or refined, all table sugar is sucrose.

Sucrose is a disaccharide that yields 1 equivalent of glucose and 1 equivalent of fructose on hydrolysis. This 1:1 mixture of glucose and fructose is often referred to as *invert sugar* because the sign of optical rotation changes (inverts) during the hydrolysis from sucrose, $[\alpha]_D = +66.5°$, to a glucose/fructose mixture, $[\alpha]_D = -22°$. Insects such as honeybees have enzymes called *invertases* that catalyze the hydrolysis of sucrose to glucose + fructose. Honey, in fact, is primarily a mixture of glucose, fructose, and sucrose.

Unlike most other disaccharides, sucrose is not a reducing sugar and does not exhibit mutarotation. These observations imply that sucrose has no hemiacetal group and that the glucose and fructose units must *both* be glycosides. This can happen only if the two sugars are joined by a glycoside link between the anomeric carbons of both sugars—C1 of glucose and C2 of fructose.

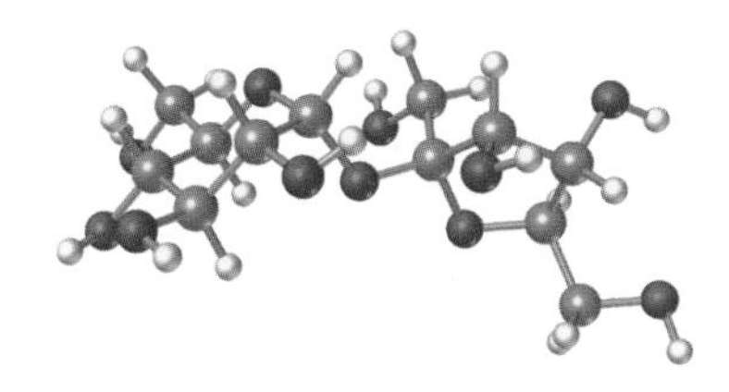

Sucrose, a 1,2′-glycoside
[2-*O*-(α-D-glucopyranosyl)-β-D-fructofuranoside]

14.9 Polysaccharides

Polysaccharides are carbohydrates in which tens, hundreds, or even thousands of simple sugars are linked by glycoside bonds. Since these compounds have no free anomeric –OH groups (except for one at the end of the chain), they aren't reducing sugars and don't show mutarotation. Cellulose and starch are the two most widely occurring polysaccharides.

Cellulose

Cellulose consists of several thousand D-glucose units linked by 1,4′-β-glycoside bonds like those in cellobiose. Different molecules can then interact to form a large aggregate structure held together by hydrogen bonds:

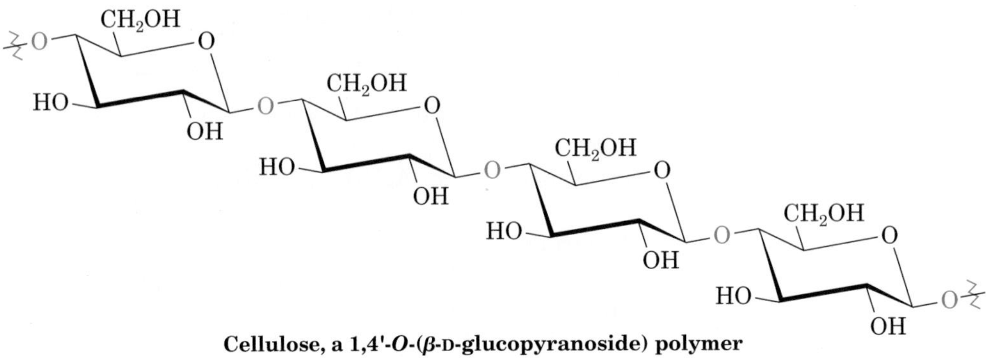

Cellulose, a 1,4′-*O*-(β-D-glucopyranoside) polymer

Nature uses cellulose primarily as a structural material to impart strength and rigidity to plants. Wood, leaves, grasses, and cotton are primarily cellulose. Cellulose also serves as a raw material for the manufacture of cellulose acetate, known commercially as *rayon*, and cellulose nitrate, known as *guncotton.* Guncotton is the major ingredient in smokeless powder, the explosive propellant used in artillery shells and in ammunition for firearms.

Starch and Glycogen

Potatoes, corn, and cereal grains contain large amounts of *starch*, a polymer of glucose in which the monosaccharide units are linked by 1,4′-α-glycoside bonds like those in maltose. Starch can be separated into two fractions: *amylose*, which is insoluble in cold water, and *amylopectin*, which *is* soluble in cold water. Amylose accounts for about 20% by weight of starch and consists of several hundred glucose molecules linked together by 1,4′-α-glycoside bonds.

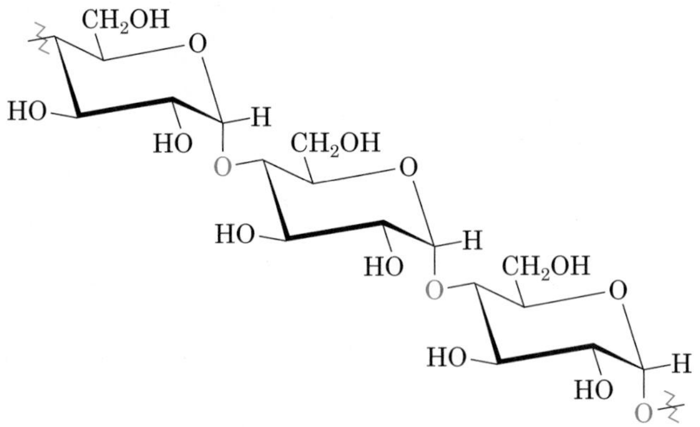

Amylose, a 1,4′-*O*-(α-D-glucopyranoside) polymer

Amylopectin, which accounts for the remaining 80% of starch, is more complex in structure than amylose. Unlike cellulose or amylose, which are linear polymers, amylopectin contains 1,6′-α-glycoside *branches* approximately every 25 glucose units. As a result, amylopectin has an exceedingly complex three-dimensional structure.

Amylopectin

Nature uses starch as the medium by which plants store energy for later use. When eaten, starch is digested in the mouth and stomach by enzymes called *glycosidases*, which catalyze the hydrolysis of glycoside bonds and release individual molecules of glucose. Like most enzymes, glycosidases are highly selective in their action. They hydrolyze only the α-glycoside links in starch and leave the β-glycoside links in cellulose untouched. Thus, humans can digest potatoes and grains but not grass.

Glycogen is a polysaccharide that serves the same energy-storage function in animals that starch serves in plants. Dietary carbohydrate not needed for immediate energy is converted by the body to glycogen for long-term storage. Like the amylopectin found in starch, glycogen has a complex three-dimensional structure with both 1,4′ and 1,6′ links (Figure 14.6). Glycogen molecules are larger than those of amylopectin—up to 100,000 glucose units—and contain even more branches.

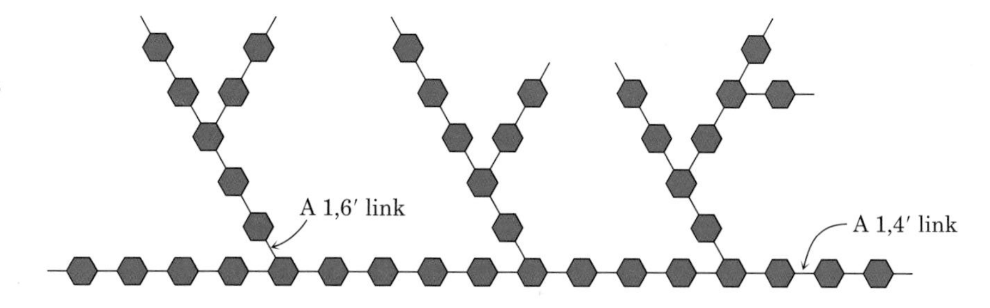

FIGURE 14.6 A representation of the structure of glycogen. The hexagons represent glucose units linked by 1,4′ and 1,6′ acetal bonds.

14.10 Other Important Carbohydrates

In addition to the common carbohydrates mentioned in previous sections, there are a variety of important carbohydrate-derived materials whose structures have been chemically modified. Their structural resemblance to sugars is clear, but they aren't simple aldoses or ketoses.

Deoxy sugars have one of their oxygen atoms "missing." That is, an –OH group is replaced by an –H. The most common deoxy sugar is 2-deoxyribose, a monosaccharide found in DNA (deoxyribonucleic acid). Note that 2-deoxyribose exists in water solution as a complex equilibrium mixture of both furanose and pyranose forms.

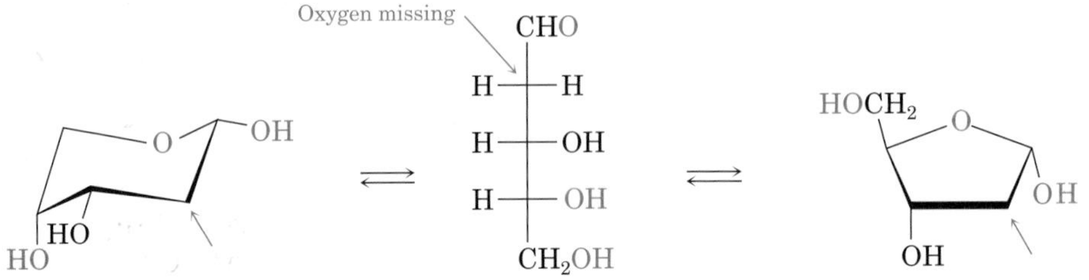

Oxygen missing

α-D-2-Deoxyribopyranose (40%) (+ 35% β anomer) ⇌ **(0.7%)** ⇌ **α-D-2-Deoxyribofuranose (13%) (+ 12% β anomer)**

Amino sugars, such as D-glucosamine, have one of their –OH groups replaced by an $-NH_2$. The *N*-acetyl amide derived from D-glucosamine is the monosaccharide unit from which *chitin*, the hard crust around insects and shellfish, is made. Antibiotics such as streptomycin and gentamicin contain still other amino sugars.

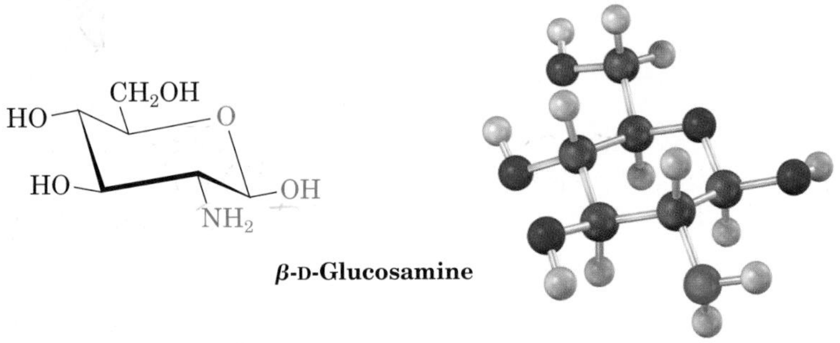

β-D-Glucosamine

14.11 Cell-Surface Carbohydrates and Carbohydrate Vaccines

It was once thought that carbohydrates were useful in nature only as structural materials and energy sources. Although carbohydrates do indeed serve these purposes, they also have many other important biochemical functions. For example, polysaccharides are centrally involved in cell–cell recognition, the critical process by which one type of cell distinguishes another. Small polysaccharide chains, covalently bound by glycosidic links to hydroxyl groups on proteins (*glycoproteins*), act as biochemical markers on cell surfaces, as illustrated by the human blood-group antigens.

It has been known for more than a century that human blood can be classified into four blood-group types (A, B, AB, and O) and that blood from a donor of one type can't be transfused into a recipient with another type unless the two types are compatible (Table 14.1). Should an incompatible mix be made, the red blood cells clump together, or *agglutinate.*

The agglutination of incompatible red blood cells, which indicates that the body's immune system has recognized the presence of foreign cells in the

TABLE 14.1 Human Blood-Group Compatibilities

Donor blood type	Acceptor blood type			
	A	B	AB	O
A	o	×	o	×
B	×	o	o	×
AB	×	×	o	×
O	o	o	o	o

o = Compatible; × = Incompatible.

body and has formed antibodies against them, results from the presence of polysaccharide markers on the surface of the cells. Types A, B, and O red blood cells each have characteristic markers, called *antigenic determinants*; type AB cells have both type A and type B markers. The structures of all three blood-group determinants are shown in Figure 14.7.

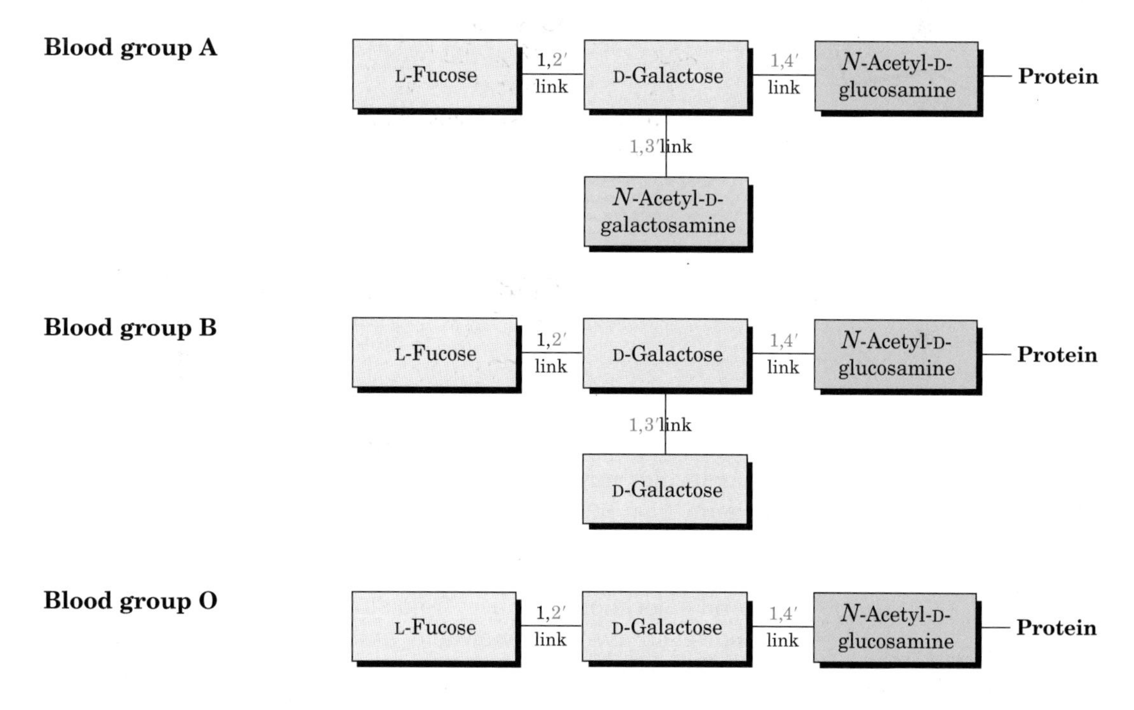

FIGURE 14.7 Structures of the A, B, and O blood-group antigenic determinants.

Note that some unusual carbohydrates are involved. All three blood-group antigenic determinants contain *N*-acetyl amino sugars as well as the unusual monosaccharide L-fucose.

HO, CH_2OH, O, HO, OH, $NHCOCH_3$

β-D-*N*-Acetylglucosamine (D-2-acetamino-2-deoxyglucose)

HO, CH_2OH, O, HO, OH, $NHCOCH_3$

β-D-*N*-Acetylgalactosamine (D-2-acetamino-2-deoxygalactose)

OH, H_3C, O, OH, OH, HO

α-L-Fucose (L-6-deoxygalactose)

Elucidation of the role of carbohydrates in cell recognition is a vigorous area of current research that offers hope of breakthroughs in the understanding of a wide range of diseases from bacterial infections to cancer. Particularly exciting is the possibility of developing useful anticancer vaccines to help mobilize the body's immune system against tumor cells. Recent advances along these lines have included a laboratory synthesis of the so-called globo H antigen, found on the surface of human breast, prostate, colon, and pancreatic cancer cells. Mice treated with the synthetic globo H hexasaccharide linked to a carrier protein developed large amounts of antibodies, which then recognized tumor cells.

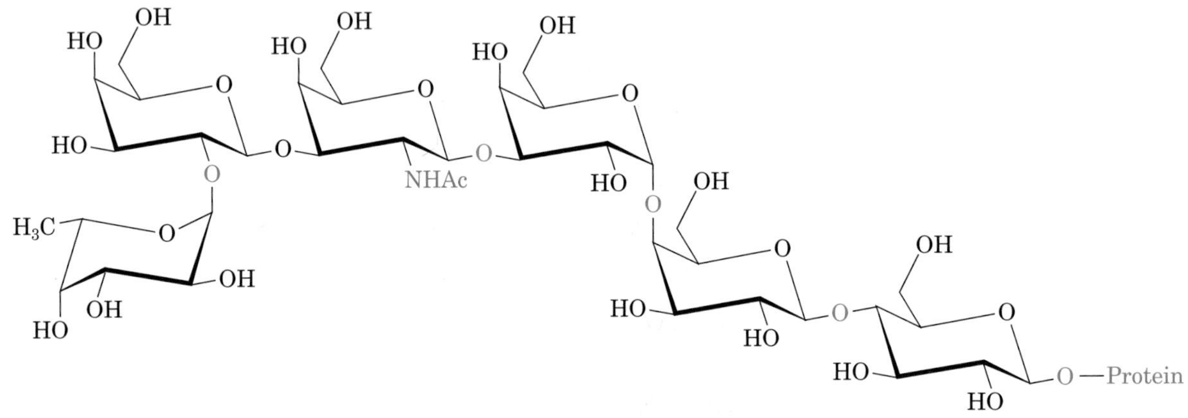

Globo H antigen

14.12 Plants: An Alternative to Petroleum?

Carbohydrates are the most abundant molecules on earth. Because carbohydrates constitute a renewable resource, chemists are studying how these molecules might be used instead of petrochemicals. The impact that petroleum has on chemistry and civilization cannot be overstated. Petroleum powers our machines and automobiles. It serves as the basis for plastics and most specialty chemicals that ultimately become our medicines, clothing, and objects of daily use: from surfboards to ballpoint pens. Petrochemicals, however, are not renewable, and alternatives will ultimately be required. Carbohydrates from plants, particularly glucose, have attracted the most attention.

Glucose is obtained commercially through hydrolysis of starches derived most often from corn, rice, or wheat. Glucose can then be used to synthesize a variety of useful compounds. The vitamin C we consume starts with glucose. Sorbitol, which is made by reducing glucose, is diuretic. More exotic chemicals can also be prepared from glucose using chemical and enzymatic strategies. These chemicals include monomers for polymer synthesis, aromatics, and specialty chemicals that serve as the basis for more elaborate syntheses, like the antiviral drug Tamiflu (oseltamivir). Because microbes commonly use glucose, chemists are not limited to traditional, one-step reactions carried out in laboratory glassware. Chemists are also exploring the use of microbes to do chemistry.

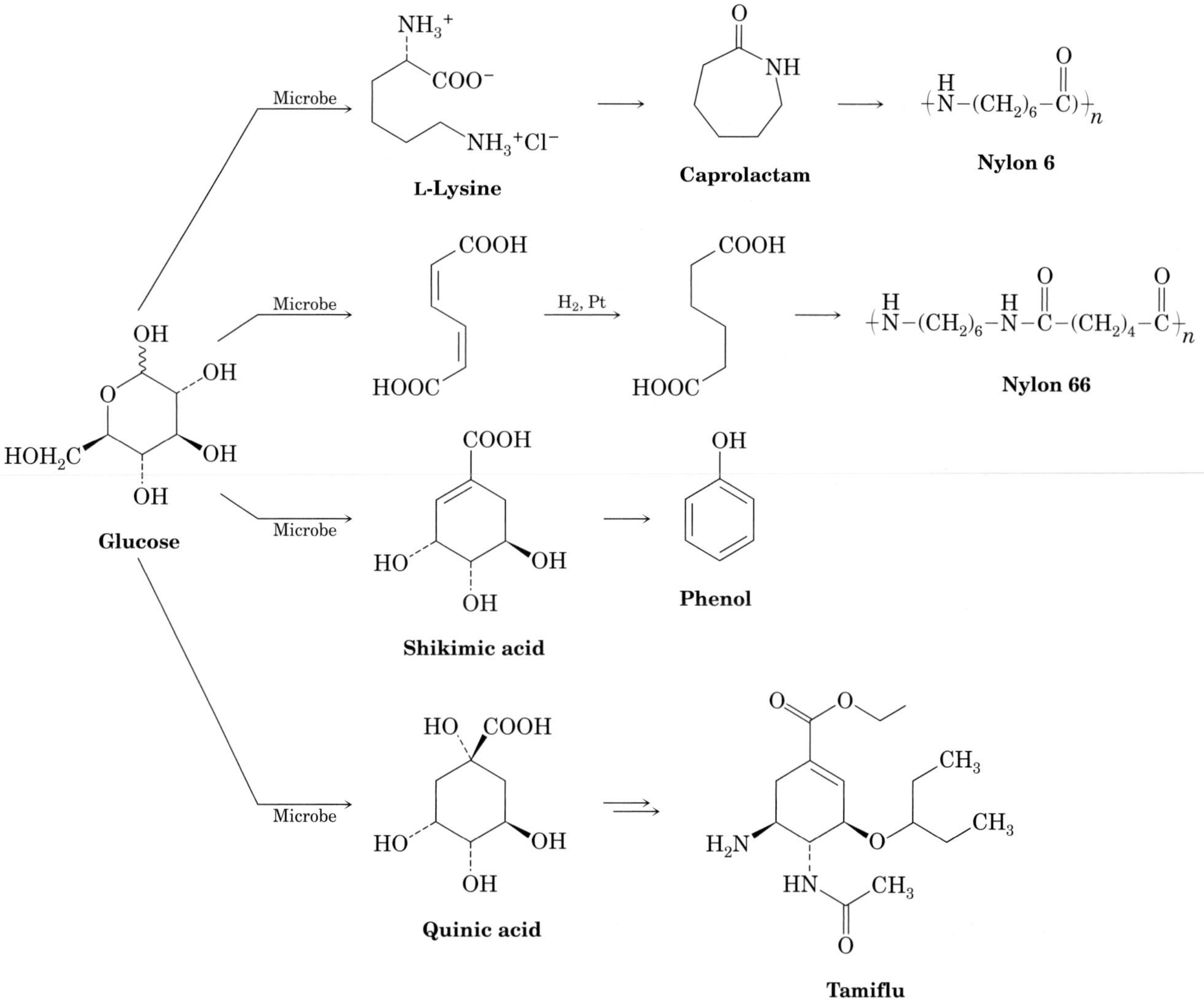

Microbe → L-Lysine → Caproloctam → Nylon 6

Glucose

Microbe → H₂, Pt → Nylon 66

Microbe → Shikimic acid → Phenol

Microbe → Quinic acid → Tamiflu

INTERLUDE

Sweetness

Say the word *sugar* and most people immediately think of sweet-tasting candies, desserts, and such. In fact, most simple carbohydrates *do* taste sweet, but the degree of sweetness varies greatly from one sugar to another. With sucrose (table sugar) as a reference point, fructose is nearly twice as sweet, but lactose is only about one-sixth as sweet. Comparisons are difficult, though, because sweetness is a matter of taste and the ranking of sugars is a matter of personal opinion. Nevertheless, the ordering in the following table is generally accepted.

Continued

Sweetness of Some Sugars and Sugar Substitutes

Name	Type	Sweetness
Lactose	Disaccharide	0.16
Glucose	Monosaccharide	0.75
Sucrose	Disaccharide	1.00
Fructose	Monosaccharide	1.75
Aspartame	Synthetic	180
Acesulfame-K	Synthetic	200
Saccharin	Synthetic	350

"Dietary disaster!"

The desire of many people to cut their caloric intake has led to the development of synthetic sweeteners such as saccharin, aspartame, and acesulfame. All are far sweeter than natural sugars, so the choice of one or another depends on personal taste, government regulations, and (for baked goods) heat stability. Saccharin, the oldest synthetic sweetener, has been used for more than a century, although it has a somewhat metallic aftertaste. Doubts about its safety and potential carcinogenicity were raised in the early 1970s, but it has now been cleared of suspicion. Acesulfame potassium, the most recently approved sweetener, is proving to be extremely popular in soft drinks because it has little aftertaste. None of the three synthetic sweeteners has any structural resemblance to a carbohydrate.

Saccharin **Aspartame** **Acesulfame potassium**

Summary and Key Words

Aldaric acid, p. 457
Alditol, p. 456
Aldonic acid, p. 457
Aldose, p. 442
Alpha (α) anomer, p. 451
Amino sugar, p. 462

Carbohydrates are polyhydroxy aldehydes and ketones. They are classified according to the number of carbon atoms and the kind of carbonyl group they contain. Thus, glucose is an *aldohexose*, a six-carbon aldehydo sugar. **Monosaccharides** are further classified as either **D** or **L sugars**, depending on the stereochemistry of the stereocenter farthest from the carbonyl group. Most naturally occurring sugars are in the D series.

Monosaccharides normally exist as cyclic hemiacetals rather than as open-chain aldehydes or ketones. The hemiacetal linkage results from reaction of the carbonyl group with an –OH group three or four carbon atoms away. A five-membered ring hemiacetal is a **furanose**, and a six-membered ring hemiacetal is a **pyranose**. Cyclization leads to the formation of a new stereocenter (the **anomeric center**) and the production of two diastereomeric hemiacetals called **alpha (α)** and **beta (β) anomers**. Stereoisomers of monosaccharides are portrayed as **Fischer projections**, which display stereocenters as a pair of crossed lines.

Much of the chemistry of monosaccharides is the familiar chemistry of alcohol and carbonyl functional groups. Thus, the –OH groups of carbohydrates form esters and ethers in the normal way. The carbonyl group of a monosaccharide can be reduced with $NaBH_4$ to yield an **alditol**, can be oxidized with Tollens' or Fehling's reagent to yield an **aldonic acid**, can be oxidized with warm HNO_3 to yield an **aldaric acid**, and can be treated with an alcohol in the presence of acid catalyst to yield a **glycoside**.

Disaccharides are complex carbohydrates in which two simple sugars are linked by a glycoside bond between the anomeric carbon of one unit and an –OH of the second unit. The two sugars can be the same, as in maltose and cellobiose, or different, as in sucrose. The glycoside bond can be either α (maltose) or β (cellobiose) and can involve any –OH of the second sugar. A **1,4′ link** is most common (cellobiose, maltose), but other links, such as 1,2′ (sucrose), also occur. **Polysaccharides**, such as cellulose, starch, and glycogen, are used in nature both as structural materials and for long-term energy storage.

EXERCISES

Visualizing Chemistry

14.23 Identify the following aldoses (see Figure 14.3), and indicate whether each is a D or L sugar.

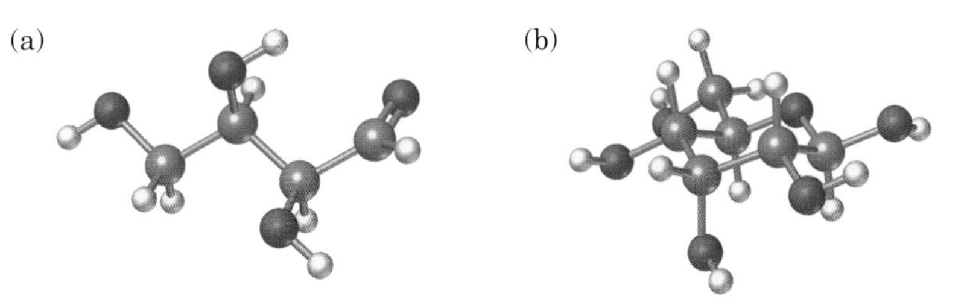

14.24 Draw Fischer projections of the following molecules, placing the carbonyl group at the top in the usual way. Identify each as a D or L sugar.

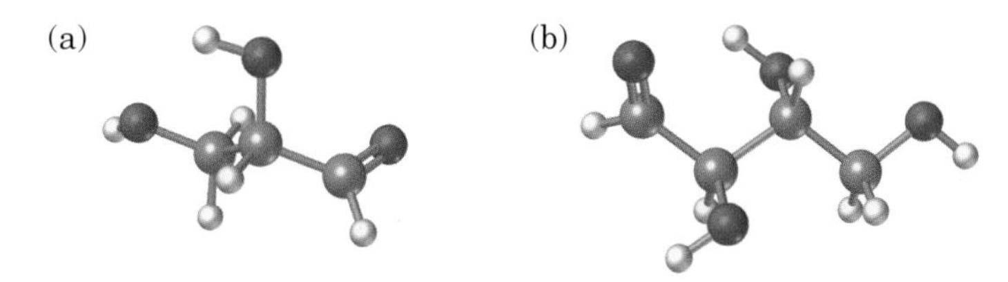

14.25 The following structure is that of an L aldohexose in its pyranose form. Identify it (see Figure 14.3).

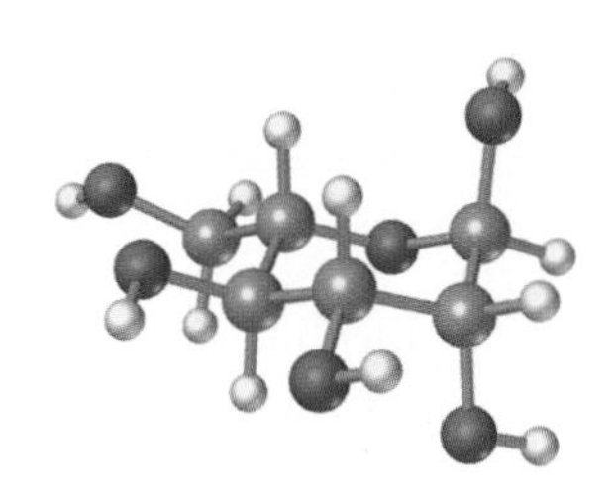

14.26 The following model is that of an aldohexose:

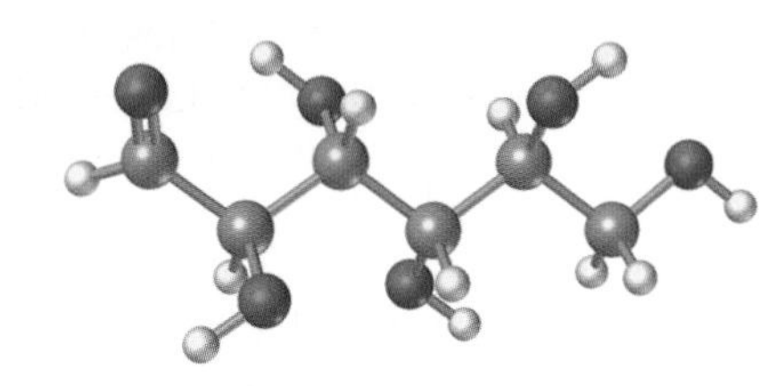

(a) Draw Fischer projections of the sugar, its enantiomer, and a diastereomer.
(b) Is this a D sugar or an L sugar? Explain.
(c) Draw the β anomer of the sugar in its furanose form.

Additional Problems

NOMENCLATURE

14.27 Classify the following sugars by type (for example, glucose is an aldohexose):

(a)
CH_2OH
C=O
CH_2OH

(b)
CH_2OH
H—OH
C=O
H—OH
CH_2OH

(c)
CHO
H—OH
HO—H
H—OH
HO—H
H—OH
CH_2OH

14.28 Write open-chain structures for a ketotetrose and a ketopentose.

14.29 Write an open-chain structure for a deoxyaldohexose.

14.30 Write an open-chain structure for a five-carbon amino sugar.

14.31 Define the following terms, and give an example of each:

(a) Monosaccharide (b) Anomeric center (c) Fischer projection
(d) Glycoside (e) Reducing sugar (f) Pyranose form
(g) 1,4′ Link (h) D-Sugar

ISOMERISM

14.32 The structure of ascorbic acid (vitamin C) is shown. Does ascorbic acid have a D or L configuration?

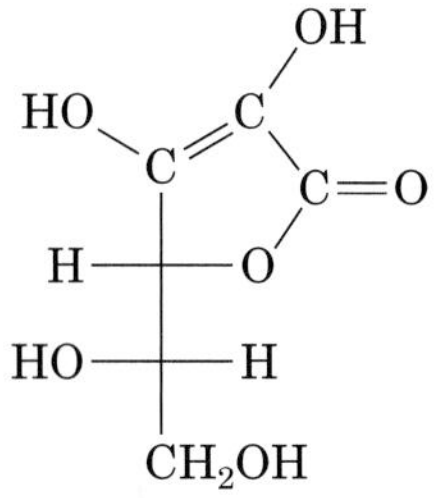

OH
HO C
C C=O
H O
HO H
CH_2OH

Ascorbic acid

14.33 Assign *R* or *S* stereochemistry to each stereocenter in ascorbic acid (Problem 14.32).

14.34 The following cyclic structure is that of gulose. Is this a furanose or pyranose form? Is it an α or β anomer? Is it a D sugar or L sugar?

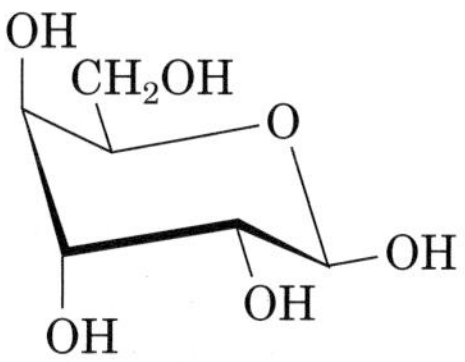

OH
CH_2OH
O
OH
OH
OH

Gulose

14.35 Uncoil gulose (see Problem 14.34), and write it in its open-chain form.

14.36 Draw D-ribulose in its five-membered cyclic β hemiacetal form.

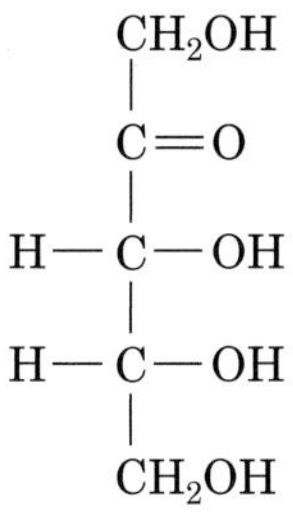

CH_2OH
C=O
H—C—OH
H—C—OH
CH_2OH

Ribulose

14.37 Look up the structures of maltose and sucrose in Section 14.8, and explain why maltose is reduced by $NaBH_4$ but sucrose is not.

14.38 Look up the structure of D-talose in Figure 14.3, and draw the β anomer in its pyranose form. Identify the ring substituents as axial or equatorial.

14.39 What is the stereochemical relationship of D-allose to L-allose? What generalizations can you make about the following properties of the two sugars?
(a) Melting point (b) Solubility in water
(c) Specific rotation (d) Density

14.40 What is the stereochemical relationship of D-ribose to L-xylose? What generalizations can you make about the following properties of the two sugars?
(a) Melting point (b) Solubility in water
(c) Specific rotation (d) Density

14.41 How many D-2-ketohexoses are there? Draw them.

REACTIONS

14.42 Draw structures for the products you would expect to obtain from the reaction of β-D-talopyranose (see Problem 14.38) with each of the following reagents:
(a) $NaBH_4$ (b) Warm dilute HNO_3 (c) $AgNO_3$, NH_3, H_2O
(d) CH_3CH_2OH, H^+ (e) CH_3I, Ag_2O (f) $(CH_3CO)_2O$, pyridine

14.43 One of the D-2-ketohexoses (see Problem 14.41) is called *sorbose*. On treatment with $NaBH_4$, sorbose yields a mixture of gulitol and iditol. What is the structure of D-sorbose? (Gulitol and iditol are the alditols obtained by reduction of gulose and idose.)

14.44 Another D-2-ketohexose, *psicose*, yields a mixture of allitol and altritol when reduced with $NaBH_4$ (see Problems 14.41 and 14.43). What is the structure of psicose?

INTEGRATED PROBLEMS

14.45 Draw Fischer projections of the following substances:
(a) (*R*)-2-Methylbutanoic acid (b) (*S*)-3-Methylpentan-2-one

14.46 Convert the following Fischer projections into tetrahedral representations:

(a) Br (top), H (left), OCH_3 (right), CH_3 (bottom)

(b) CH_3 (top), H (left), NH_2 (right), CH_2CH_3 (bottom)

14.47 Which of the eight D aldohexoses yield optically inactive (meso) alditols on reduction with $NaBH_4$?

14.48 What other D aldohexose gives the same alditol as D-talose? (See Problem 14.47.)

14.49 Which of the eight D aldohexoses give the same aldaric acids as their L enantiomers?

14.50 Which of the other three D aldopentoses gives the same aldaric acid as D-lyxose?

14.51 The *Ruff degradation* is a method used to shorten an aldose chain by one carbon atom. The original C1 carbon atom is cut off, and the original C2 carbon atom becomes the aldehyde of the chain-shortened aldose. For example, D-glucose, an aldohexose, is converted by Ruff degradation into D-arabinose, an aldopentose. What other D aldohexose would also yield D-arabinose on Ruff degradation?

14.52 D-Galactose and D-talose yield the same aldopentose on Ruff degradation (Problem 14.51). What does this tell you about the stereochemistry of galactose and talose? Which D aldopentose is obtained?

14.53 The aldaric acid obtained by nitric acid oxidation of D-erythrose, one of the D aldotetroses, is optically inactive. The aldaric acid obtained from oxidation of the other D aldotetrose, D-threose, however, is optically active. How does this information allow you to assign structures to the two D aldotetroses?

14.54 Gentiobiose is a rare disaccharide found in saffron and gentian. It is a reducing sugar and forms only glucose on hydrolysis with aqueous acid. If gentiobiose contains a 1,6′-β-glycoside link, what is its structure?

14.55 Many other sugars besides glucose exhibit mutarotation. For example, α-D-galactopyranose has $[\alpha]_D = +150.7°$, and β-D-galactopyranose has $[\alpha]_D = +52.8°$. If either anomer is dissolved in water and allowed to reach equilibrium, the specific rotation of the solution is $+80.2°$. What are the percentages of each anomer at equilibrium?

14.56 Raffinose, a trisaccharide found in sugar beets, is formed by a 1,6′ α linkage of D-galactose to the glucose unit of sucrose. Draw the structure of raffinose.

14.57 Is raffinose (see Problem 14.56) a reducing sugar? Explain.

14.58 Glucose and fructose can be interconverted by treatment with dilute aqueous NaOH. Propose a mechanism (see Section 11.1).

IN THE MEDICINE CABINET

14.59 Erythromycin is a broad-spectrum antibiotic with a number of functional groups. Identify the functional groups indicated. How many chiral centers exist in erythromycin?

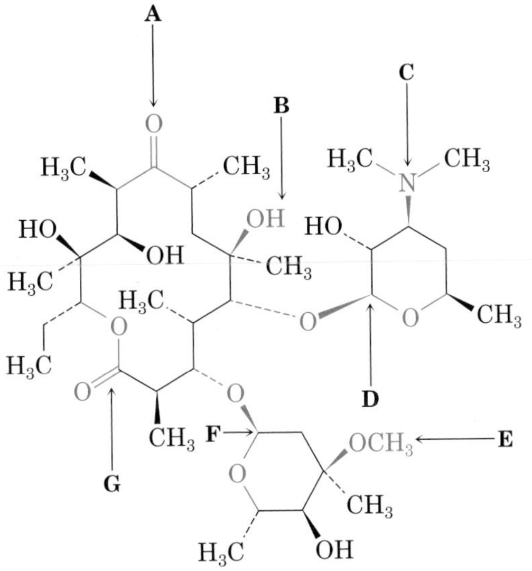

Erythromycin

14.60 Erythromycin (Problem 14.59) is made in two stages by bacteria. In the first stage, the macrocycle is prepared as a linear molecule and cyclized through a condensation reaction. Which bond do you predict is formed during the cyclization reaction?

14.61 After cyclization (Problem 14.60), the two rare sugars are added; without these sugars, erythromycin is not a useful drug. The biosynthesis of one of these sugars, desosamine, is shown starting with a glucose molecule derivatized with thymidine diphosphate (TDP; see Chapter 16).

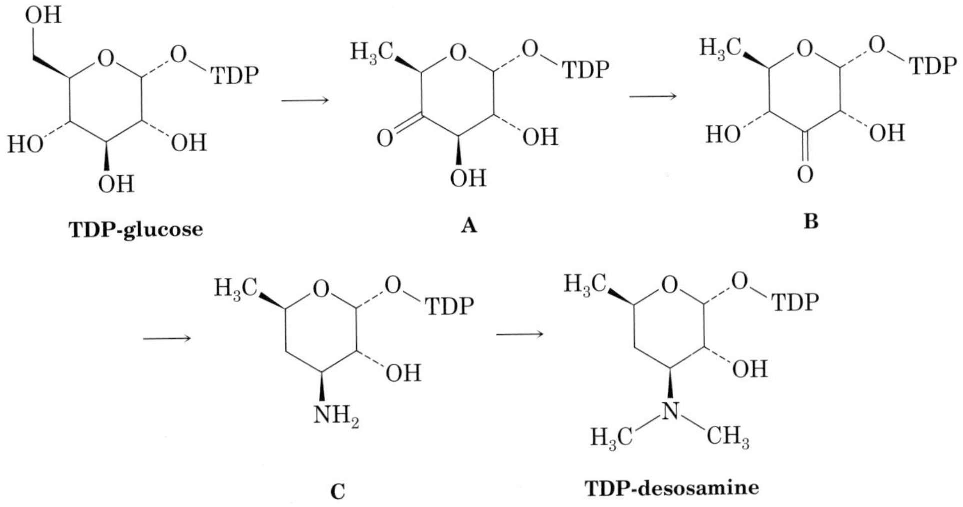

TDP-glucose → **A** → **B** → **C** → **TDP-desosamine**

(a) During the conversion of TDP-glucose to **A**, one carbon is oxidized while another is reduced. Describe these changes in terms of functional groups.
(b) The conversion of **A** to **B** is the result of isomerization through keto–enol tautomerism. Draw a mechanism that produces the enol intermediate using H^+ as an acid catalyst.
(c) The production of amine **C** from **B** is formally considered a reductive amination. Draw mechanism for this reaction using ammonia, NH_3, and the hydride source, H^-.
(d) Methylation to produce TDP-desosamine from **C** involves two molecules of *S*-adenosylmethionine. Draw a mechanism for these methylation reactions.

***S*-Adenosylmethionine**

14.62 The accumulation of glycolipids called glucocerebrosides in different organs can lead to severe neurological problems and a life-threatening condition called Gaucher's disease.

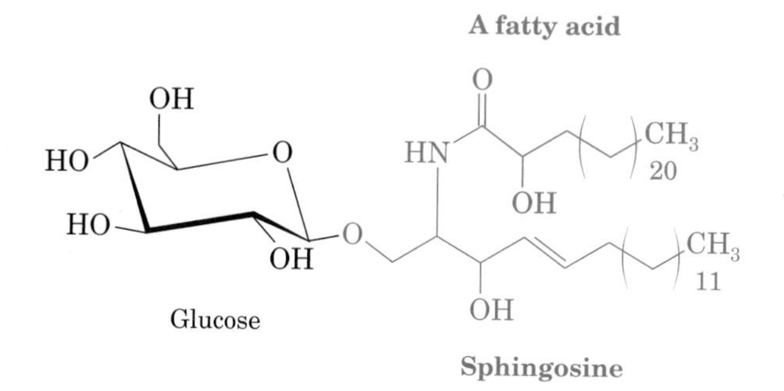

A glucocerebroside

(a) Identify the functional groups in this glucocerebroside.
(b) Hydrolysis of glucocerebrosides in acidic water produces glucose, sphingosine, and a fatty acid. Draw mechanisms for these hydrolysis reactions.
(c) One treatment for Gaucher's disease is enzyme replacement therapy. Cerezyme is a protein that catalyzes the hydrolysis of glycosidic bonds. What are the structures of the products?

14.63 The flu virus attacks cells by recognizing a terminal sialic acid on a cell surface carbohydrate and subsequently hydrolyzing it off the cell surface. Draw a mechanism for the hydrolysis of the sialic acid using H_3O^+. Why is zanamivir (Relenza) a good mimic of the intermediate oxonium ion? Remember, an oxonium ion has a positively charged oxygen atom.

R = rest of the cell

+ ROH

Oxonium ion

A sialic acid

Zanamivir (Relenza)

IN THE FIELD WITH AGROCHEMICALS

14.64 Trees have provided us with a wealth of medicines. Aspirin finds its origins in the bark of willow trees. Future drugs may come from trees as well; betulinic acid is found in the bark of a white birch tree and is useful against melanomas, dangerous forms of skin cancer.

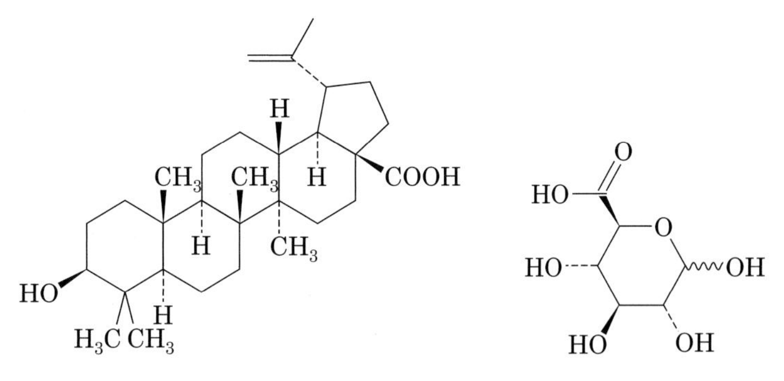

Betulinic acid **Glucuronic acid**

(a) How many chiral centers appear in betulinic acid?
(b) How many stereoisomers could exist?
(c) Identify three functional groups in the molecule, and describe from where part of the name derives.
(d) To increase the solubility of betulinic acid, a number of carbohydrates could be added to it, including glucuronic acid. Draw a mechanism for the formation of a glycosidic bond between betulinic acid and glucuronic acid.

14.65 "Super slurper" is a natural carbohydrate polymer backbone onto which multiple synthetic polymer "grafts" are grown. "Super slurper" derives its name from its ability to absorb 2000 times its weight in water.

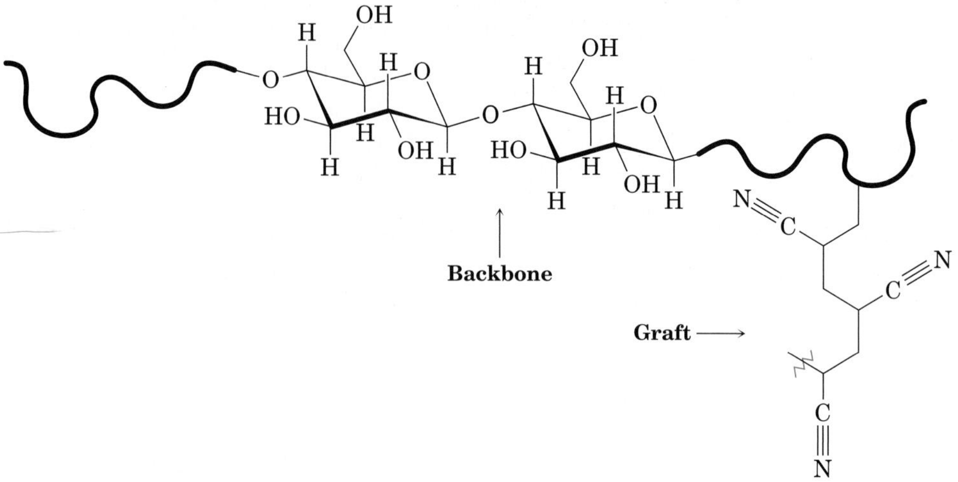

Backbone

Graft

(a) What is the carbohydrate monomer of this polymer?
(b) Describe the stereochemistry of the glycosidic bonds. What is one natural source of this polymer?
(c) The structure is an intermediate in the synthesis of "super slurper." What alkene is polymerized to form the graft?
(d) The actual structure of "super slurper" relies on hydrolysis of the nitrile groups to generate a mixture of amides and carboxylate groups using NaOH. Draw one possible structure of "super slurper."
(e) Draw mechanisms for the hydrolysis of a nitrile with NaOH to form first an amide and, ultimately, a carboxylate. You can abbreviate the graft using RCN.

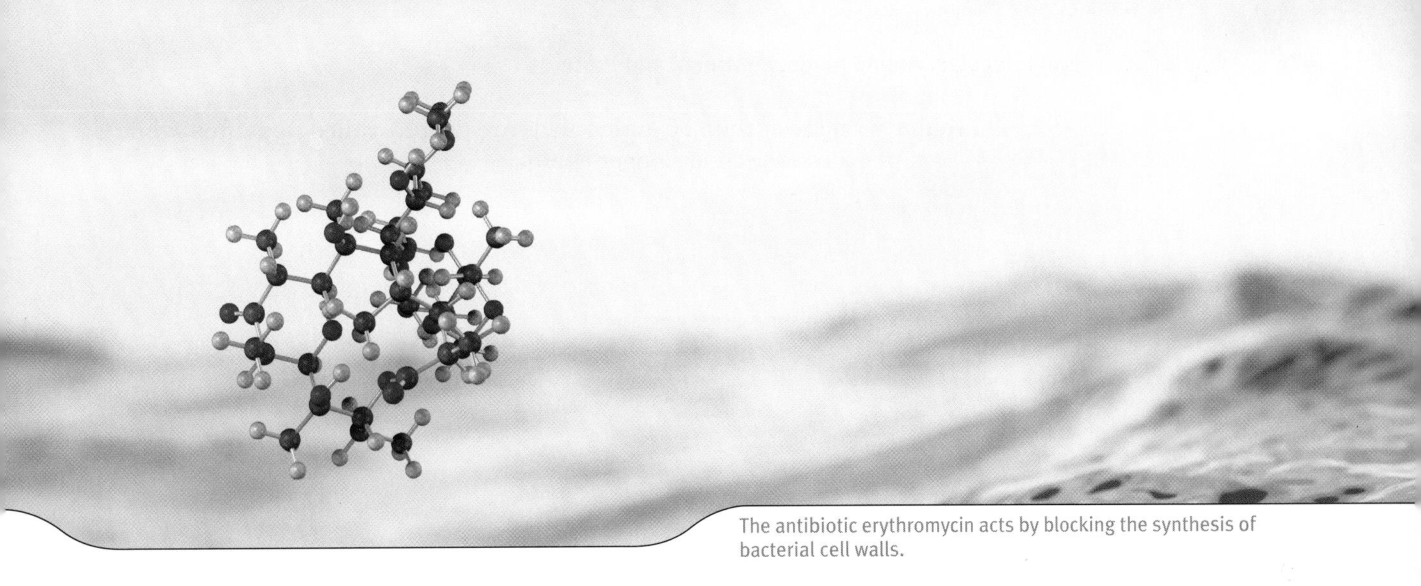

The antibiotic erythromycin acts by blocking the synthesis of bacterial cell walls.

CHAPTER 15

Biomolecules: Amino Acids, Peptides, and Proteins

Proteins are large biomolecules that occur in every living organism. They are of many types and have many biological functions. The keratin of skin and fingernails, the insulin that regulates glucose metabolism in the body, and the DNA polymerase that catalyzes the synthesis of DNA in cells are all proteins. Regardless of their appearance or function, all proteins are chemically similar. All are made up of many *amino acid* units linked together by amide bonds in a long chain.

Amino acids, as their name implies, are difunctional. They contain both a basic amino group and an acidic carboxyl group:

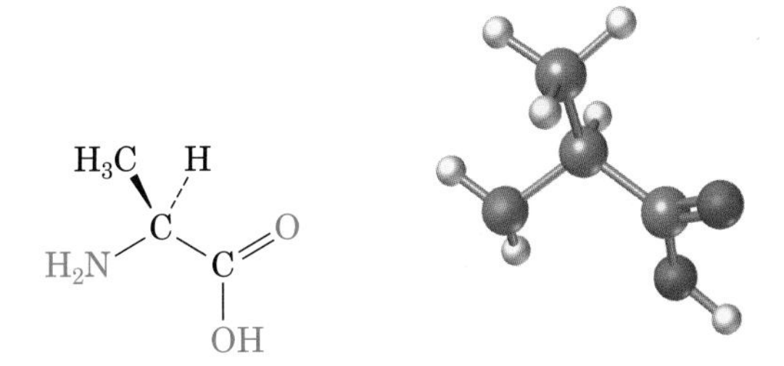

Alanine, an amino acid

Their value as building blocks for proteins stems from the fact that amino acids can link together into long chains by forming amide bonds between the $-NH_2$ of one amino acid and the $-CO_2H$ of another. For classification purposes, chains with fewer than 50 amino acids are usually called **peptides**, while the term **protein** is used for longer chains.

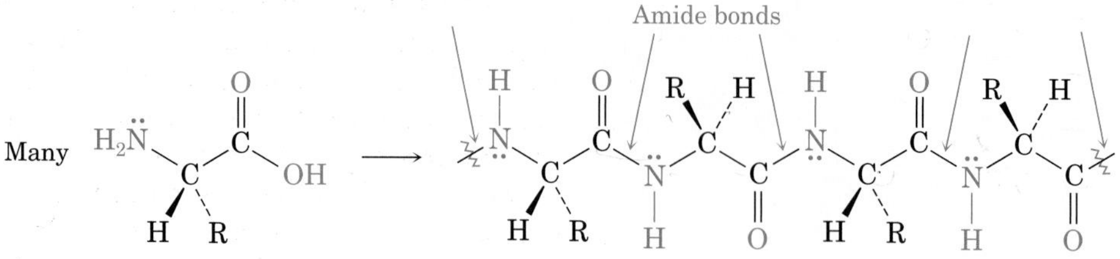

Amide bonds

Many

15.1

Structures of Amino Acids

Since amino acids contain both an acidic and a basic group, they undergo an intramolecular acid–base reaction and exist primarily in the form of a dipolar ion, or **zwitterion** (German *zwitter*, meaning "hybrid"):

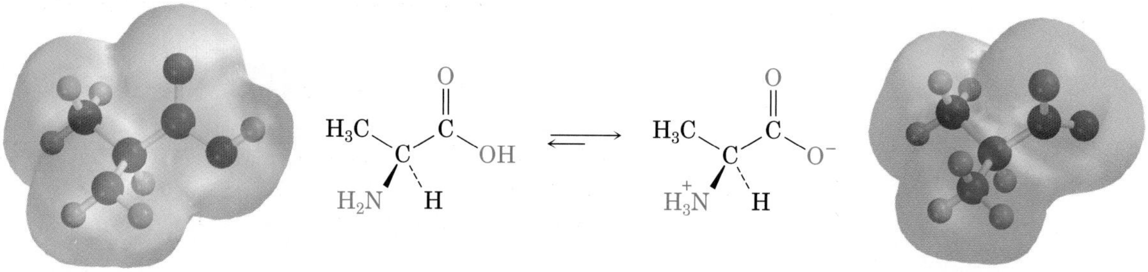

(uncharged)

(zwitterion)

Alanine

Amino acid zwitterions are salts and therefore have many of the physical properties associated with salts. They are soluble in water but insoluble in hydrocarbons and are crystalline substances with high melting points. In addition, amino acids are *amphoteric*: they can react either as acids or as bases, depending on the circumstances. In aqueous acid solution, an amino acid zwitterion is a base that *accepts* a proton to yield a cation; in aqueous base solution, the zwitterion is an acid that *loses* a proton to form an anion.

In acid solution

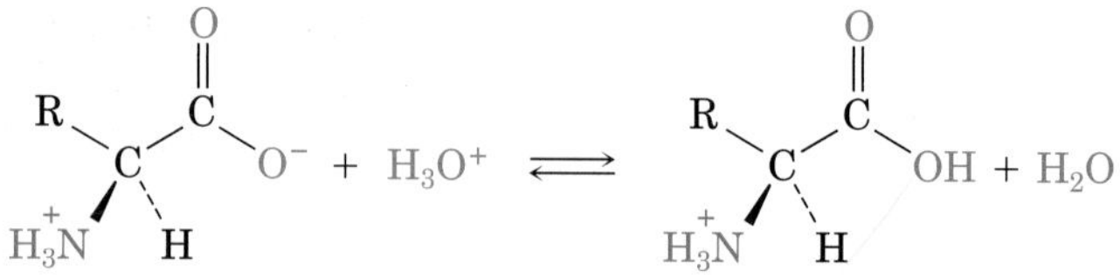

In base solution

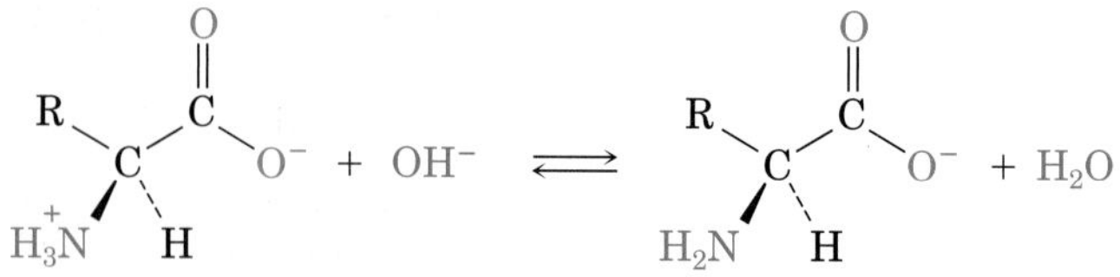

Note that it's the carboxylate, $-CO_2^-$, that acts as the basic site and accepts a proton in acid solution, and it's the ammonium cation, $-NH_3^+$, that acts as the acidic site and donates a proton in base solution.

The structures, abbreviations, and pK_a values of the 20 amino acids commonly found in proteins are shown in Table 15.1 in the form that predominates within cells at a physiological pH of 7.3. All 20 are ***α*-amino acids**, meaning that the amino group is a substituent on the *α* carbon—the one next to the carbonyl group. Nineteen of the twenty are primary amines ($-NH_2$) and differ only in the identity of the **side chain**—the substituent attached to the *α* carbon. Proline, however, is a secondary amine whose nitrogen and *α* carbon atoms are part of a five-membered pyrrolidine ring.

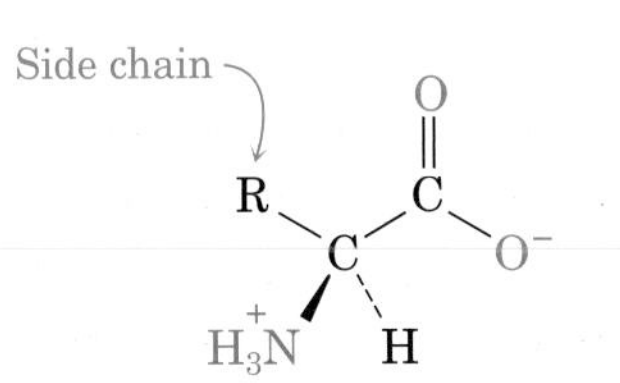

Side chain

A primary *α*-amino acid

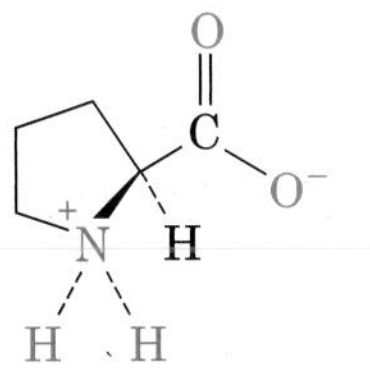

Proline, a secondary *α*-amino acid

In addition to the 20 amino acids found in proteins, there are a number of other biologically important amino acids. *γ*-Aminobutyric acid (GABA), for instance, is found in the brain and acts as a neurotransmitter; homocysteine is found in blood and is linked to coronary heart disease; and thyroxine is found in the thyroid gland, where it acts as a hormone.

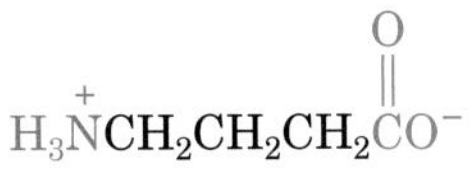

$H_3\overset{+}{N}CH_2CH_2CH_2CO^-$ (with C=O)

γ-Amino-butyric acid

$HSCH_2CH_2CH(\overset{+}{N}H_3)CO^-$ (with C=O)

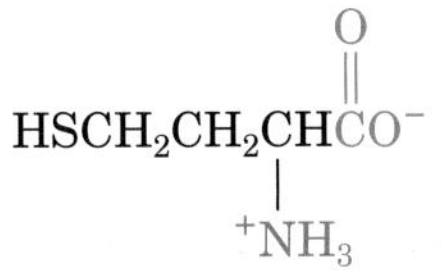

Homocysteine

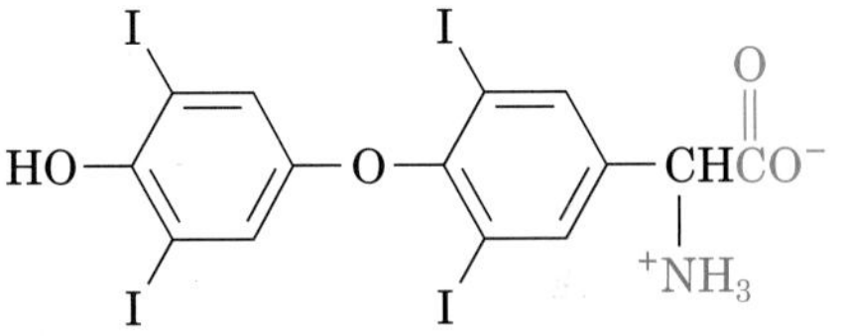

Thyroxine

With the exception of glycine, $H_2NCH_2CO_2H$, the *α* carbons of the amino acids are stereocenters. Two enantiomeric forms are therefore possible, but nature uses only a single enantiomer to build proteins. In Fischer projections, naturally occurring amino acids are represented by placing the carboxyl group at the top as if drawing a carbohydrate (see Section 14.2) and then placing the amino group on the left. Because of their stereochemical similarity to L sugars (see Section 14.3), the naturally occurring *α*-amino acids are often referred to as L-amino acids.

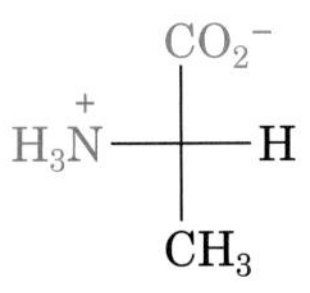

L-Alanine
(*S*)-Alanine

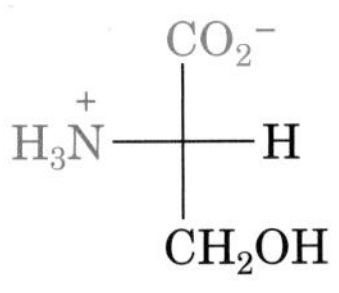

L-Serine
(*S*)-Serine

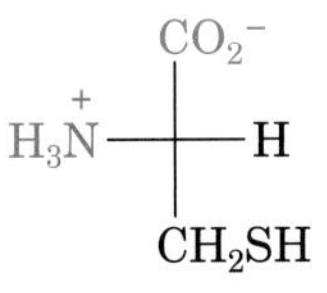

L-Cysteine
(*R*)-Cysteine

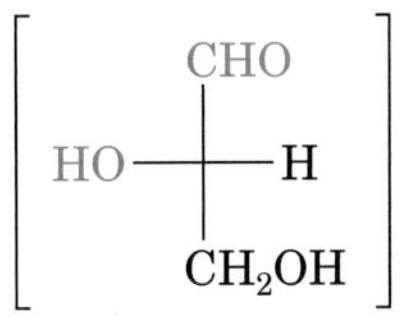

L-Glyceraldehyde

TABLE 15.1 Structures of the 20 Common Amino Acids Found in Proteins (Names of the amino acids essential to the human diet are shown in red.)

Name	Abbreviations	MW	Structure	pK_{a1} α-COOH	pK_{a2} α-NH_3^+	pK_a side chain	Isoelectric point
Neutral amino acids							
Alanine	Ala (A)	89	$CH_3CH(NH_3^+)CO^-$ (C=O)	2.34	9.69	—	6.01
Asparagine	Asn (N)	132	$H_2NC(=O)CH_2CH(NH_3^+)C(=O)O^-$	2.02	8.80	—	5.41
Cysteine	Cys (C)	121	$HSCH_2CH(NH_3^+)C(=O)O^-$	1.96	10.28	8.18	5.07
Glutamine	Gln (Q)	146	$H_2NC(=O)CH_2CH_2CH(NH_3^+)C(=O)O^-$	2.17	9.13	—	5.65
Glycine	Gly (G)	75	$CH_2(NH_3^+)C(=O)O^-$	2.34	9.60	—	5.97
Isoleucine	Ile (I)	131	$CH_3CH_2CH(CH_3)CH(NH_3^+)C(=O)O^-$	2.36	9.60	—	6.02
Leucine	Leu (L)	131	$CH_3CH(CH_3)CH_2CH(NH_3^+)C(=O)O^-$	2.36	9.60	—	5.98
Methionine	Met (M)	149	$CH_3SCH_2CH_2CH(NH_3^+)C(=O)O^-$	2.28	9.21	—	5.74
Phenylalanine	Phe (F)	165	C_6H_5–$CH_2CH(NH_3^+)C(=O)O^-$	1.83	9.13	—	5.48
Proline	Pro (P)	115	pyrrolidine ring with $\overset{+}{N}$–H, H; ring C bearing $C(=O)O^-$	1.99	10.60	—	6.30
Serine	Ser (S)	105	$HOCH_2CH(NH_3^+)C(=O)O^-$	2.21	9.15	—	5.68

Name	Abbreviations	MW	Structure	pK_{a1} α-COOH	pK_{a2} α-NH_3^+	pK_a side chain	Isoelectric point
Threonine	Thr (T)	119	$CH_3CH(OH)CH(NH_3^+)C(=O)O^-$	2.09	9.10	—	5.60
Tryptophan	Trp (W)	204	(indol-3-yl, N–H)–$CH_2CH(NH_3^+)C(=O)O^-$	2.83	9.39	—	5.89
Tyrosine	Tyr (Y)	181	HO–C_6H_4–$CH_2CH(NH_3^+)C(=O)O^-$	2.20	9.11	10.07	5.66
Valine	Val (V)	117	$CH_3CH(CH_3)CH(NH_3^+)C(=O)O^-$	2.32	9.62	—	5.96
Acidic amino acids							
Aspartic acid	Asp (D)	133	$^-OC(=O)CH_2CH(NH_3^+)C(=O)O^-$	1.88	9.60	3.65	2.77
Glutamic acid	Glu (E)	147	$^-OC(=O)CH_2CH_2CH(NH_3^+)C(=O)O^-$	2.19	9.67	4.25	3.22
Basic amino acids							
Arginine	Arg (R)	174	$H_2NC(=NH_2^+)NHCH_2CH_2CH_2CH(NH_3^+)C(=O)O^-$	2.17	9.04	12.48	10.76
Histidine	His (H)	155	(imidazol-4-yl, N–H)–$CH_2CH(NH_3^+)C(=O)O^-$	1.82	9.17	6.00	7.59
Lysine	Lys (K)	146	$H_3\overset{+}{N}CH_2CH_2CH_2CH_2CH(NH_3^+)C(=O)O^-$	2.18	8.95	10.53	9.74

The 20 common amino acids can be further classified as either neutral, acidic, or basic, depending on the structure of their side chain. Of the 20 amino acids, 15 have neutral side chains, 2 (aspartic acid and glutamic acid) have an extra carboxylic acid function in their side chains, and 3 (lysine, arginine, and histidine) have basic amino groups in their side chains. Note, however, that both cysteine (a thiol) and tyrosine (a phenol), although classified as neutral amino acids, have weakly acidic side chains and can be deprotonated in strongly basic solution.

At the pH of 7.3 found within cells, the side-chain carboxyl groups of aspartic acid and glutamic acid are dissociated and exist as carboxylate ions, $-CO_2^-$. Similarly, the basic side-chain nitrogens of lysine and arginine are protonated at pH 7.3 and exist as ammonium ions, $-NH_3^+$. Histidine, however, which contains a heterocyclic imidazole ring in its side chain, is not quite basic enough to be protonated at pH 7.3. Note that only the pyridine-like doubly bonded nitrogen in histidine is basic. The pyrrole-like singly bonded nitrogen is nonbasic because its lone pair of electrons is part of the aromatic imidazole ring (see Section 12.6).

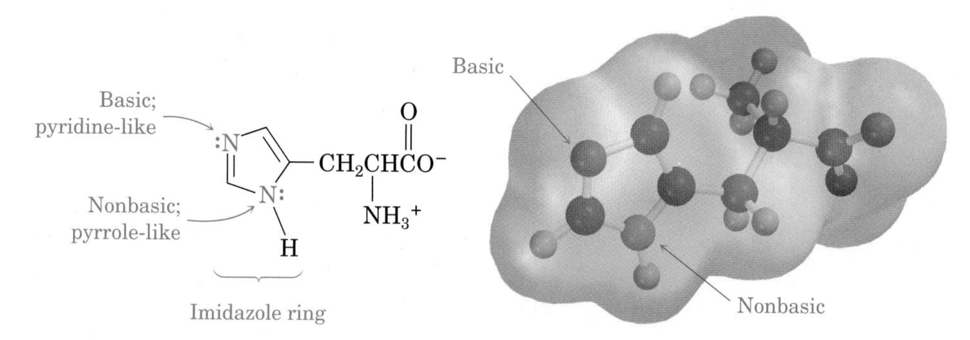

Histidine

All 20 of the amino acids are necessary for protein synthesis, but humans can synthesize only 10 of the 20. The other 10 are called *essential amino acids* because they must be obtained from food. Failure to include an adequate dietary supply of any of these essential amino acids leads to poor growth and general failure to thrive.

PRACTICE PROBLEM 15.1

Write an equation for the reaction of glycine hydrochloride with (a) 1 equivalent of NaOH and (b) 2 equivalents of NaOH.

SOLUTION Glycine hydrochloride has the structure

$$Cl^- \ H_3\overset{+}{N}CH_2\overset{O}{\overset{\|}{C}}OH$$

(a) Reaction with the first equivalent of NaOH removes the acidic $-CO_2H$ proton:

$$Cl^- \ H_3\overset{+}{N}CH_2\overset{O}{\overset{\|}{C}}OH + NaOH \longrightarrow H_3\overset{+}{N}CH_2\overset{O}{\overset{\|}{C}}O^- + H_2O + NaCl$$

(b) Reaction with a second equivalent of NaOH removes the $-NH_3^+$ proton:

$$H_3\overset{+}{N}CH_2\overset{\overset{O}{\|}}{C}O^- + NaOH \longrightarrow H_2NCH_2\overset{\overset{O}{\|}}{C}O^-\ Na^+ + H_2O$$

PROBLEM 15.1 Look at the 20 amino acids in Table 15.1. How many contain aromatic rings? How many contain sulfur? How many are alcohols? How many have hydrocarbon side chains?

PROBLEM 15.2 Of the 19 L-amino acids, 18 have the *S* configuration at the α carbon. Cysteine is the only L-amino acid that has an *R* configuration. Explain.

PROBLEM 15.3 Draw L-alanine in the standard three-dimensional format using solid, wedged, and dashed lines.

PROBLEM 15.4 Write the products of the following reactions:

(a) Phenylalanine + 1 equiv NaOH $\rightarrow$?
(b) Product of (a) + 1 equiv HCl $\rightarrow$?
(c) Product of (a) + 2 equiv HCl $\rightarrow$?

15.2 Isoelectric Points

In acid solution (low pH), an amino acid is protonated and exists primarily as a cation. In base solution (high pH), an amino acid is deprotonated and exists primarily as an anion. Thus, at some intermediate pH, the amino acid must be exactly balanced between anionic and cationic forms and exist primarily as the neutral, dipolar zwitterion. This pH is called the amino acid's **isoelectric point, p*I*.**

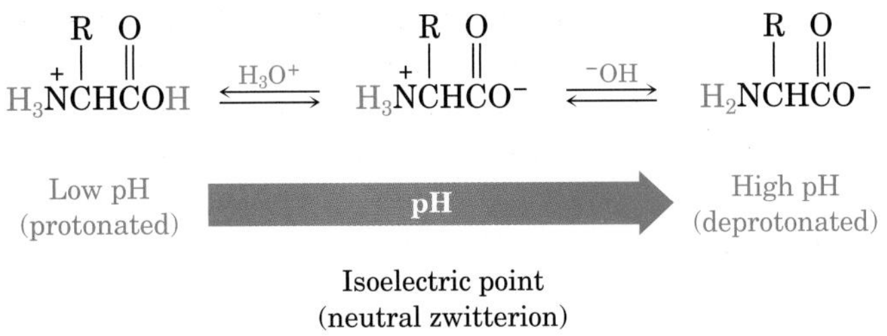

$$H_3\overset{+}{N}\overset{\overset{R}{|}}{C}H\overset{\overset{O}{\|}}{C}OH \underset{H_3O^+}{\rightleftharpoons} H_3\overset{+}{N}\overset{\overset{R}{|}}{C}H\overset{\overset{O}{\|}}{C}O^- \underset{^-OH}{\rightleftharpoons} H_2N\overset{\overset{R}{|}}{C}H\overset{\overset{O}{\|}}{C}O^-$$

Low pH (protonated) — pH → — High pH (deprotonated)

Isoelectric point (neutral zwitterion)

The isoelectric point of an amino acid depends on its structure, with values for the 20 common amino acids given in Table 15.1. The 15 amino acids with neutral side chains have isoelectric points near neutrality, in the pH range 5.0 to 6.5. The two acidic amino acids have isoelectric points at lower pH so that deprotonation of the side-chain $-CO_2H$ is suppressed, and the three basic amino acids have isoelectric points at higher pH so that protonation of the side-chain amino group is suppressed.

More specifically, the p*I* of any amino acid is the average of the two acid-dissociation constants that involve the neutral zwitterion. For the 13 amino acids with a neutral side chain, p*I* is the average of pK_{a1} and pK_{a2}. For the four amino acids with either a strongly or weakly acidic side chain, p*I* is the average of the two *lowest* pK_a values. For the three amino acids with a basic side chain, p*I* is the average of the two *highest* pK_a values.

$pK_a = 3.65$ (side-chain CO₂H), $pK_a = 1.88$ (α-CO₂H), $pK_a = 9.60$ (NH_3^+)

$HOCCH_2CHCOH$ (with C=O on both carbonyl carbons and NH_3^+ on the α carbon)

$$pI = \frac{1.88 + 3.65}{2} = 2.77$$

Acidic amino acid
Aspartic acid

$pK_a = 2.34$ (CO₂H), $pK_a = 9.69$ (NH_3^+)

CH_3CHCOH (with C=O and NH_3^+ on the α carbon)

$$pI = \frac{2.34 + 9.69}{2} = 6.01$$

Neutral amino acid
Alanine

$pK_a = 10.53$ (side-chain NH_3^+), $pK_a = 2.18$ (CO₂H), $pK_a = 8.95$ (α-NH_3^+)

$^+H_3NCH_2CH_2CH_2CH_2CHCOH$ (with C=O and NH_3^+ on the α carbon)

$$pI = \frac{8.95 + 10.53}{2} = 9.74$$

Basic amino acid
Lysine

Just as individual amino acids have isoelectric points, so too do proteins have an overall p*I* because of the numerous acidic or basic residues they may contain. The protein lysozyme, for instance, has a preponderance of basic residues and thus has a high isoelectric point (p*I* = 11.0). Pepsin, however, has a preponderance of acidic residues and a low isoelectric point (p*I* ~ 1.0). Not surprisingly, the solubilities and properties of proteins with different p*I*'s are strongly affected by the pH of the medium.

We can take advantage of the differences in isoelectric points to separate a mixture of amino acids (or a mixture of proteins) into its pure constituents. In a technique known as **electrophoresis**, a solution of amino acids is placed near the center of a strip of paper or gel. The paper or gel is moistened with an aqueous buffer of a given pH, and electrodes are connected to the ends of the strip. When an electric potential is applied, the amino acids with negative charges (those that are deprotonated because the pH of the buffer is above their isoelectric point) migrate slowly toward the positive electrode. At the same time, the amino acids with positive charges (those that are protonated because the pH of the buffer is below their isoelectric point) migrate toward the negative electrode.

Different amino acids migrate at different rates, depending on their isoelectric points and on the pH of the buffer. Thus, the mixture of amino acids can be separated. Figure 15.1 illustrates this separation for a mixture of lysine (basic), glycine (neutral), and aspartic acid (acidic).

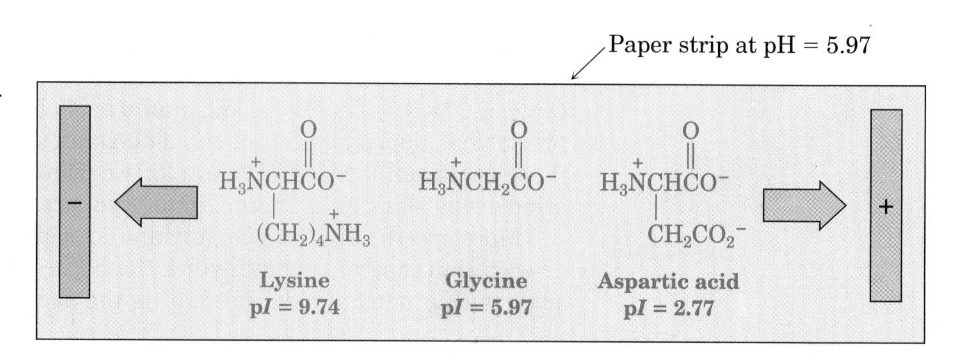

FIGURE 15.1 Separation of an amino acid mixture by electrophoresis. At pH 5.97, glycine molecules are primarily neutral and do not migrate, lysine molecules are protonated and migrate toward the negative electrode, and aspartic acid molecules are deprotonated and migrate toward the positive electrode.

PRACTICE PROBLEM 15.2

Draw structures of the predominant forms of glycine at pH 3.0, pH 6.0, and pH 9.0.

STRATEGY According to Table 15.1, the isoelectric point of glycine is 5.97. At a pH substantially lower than 6.0, glycine is largely protonated; at pH 6.0, glycine is zwitterionic; and at a pH substantially higher than 6.0, glycine is largely deprotonated.

SOLUTION

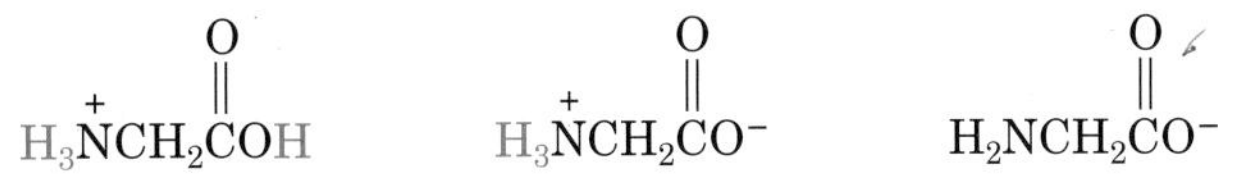

$$\overset{+}{H_3N}CH_2\overset{O}{\overset{\|}{C}}OH \qquad \overset{+}{H_3N}CH_2\overset{O}{\overset{\|}{C}}O^- \qquad H_2NCH_2\overset{O}{\overset{\|}{C}}O^-$$

At pH 3.0 — At pH 6.0 — At pH 9.0

PROBLEM 15.5 Draw the structure of the predominant form of each of the following amino acids:
(a) Lysine at pH 2.0 (b) Aspartic acid at pH 6.0
(c) Lysine at pH 11.0 (d) Alanine at pH 3.0

PROBLEM 15.6 For the mixtures of amino acids indicated, predict the direction of migration of each component (toward the positive or negative electrode) and the relative rate of migration during electrophoresis.
(a) Valine, glutamic acid, and histidine at pH 7.6
(b) Glycine, phenylalanine, and serine at pH 5.7
(c) Glycine, phenylalanine, and serine at pH 6.0

15.3 Peptide Synthesis in the Laboratory

When two amino acids are linked together through formation of an amide bond, we call the product a *dipeptide.* Let's think about the dipeptide that results when serine and alanine react. Depending on which amine and carboxylic acid groups react, two different dipeptides containing serine and alanine are produced. We name peptides from the amine end (or ***N*-terminus**, where *N* denotes the nitrogen of the amine group) to the carboxylic acid end (or ***C*-terminus**, where *C* denotes the carbon of the carboxylic acid group) using the abbreviations listed in Table 15.1.

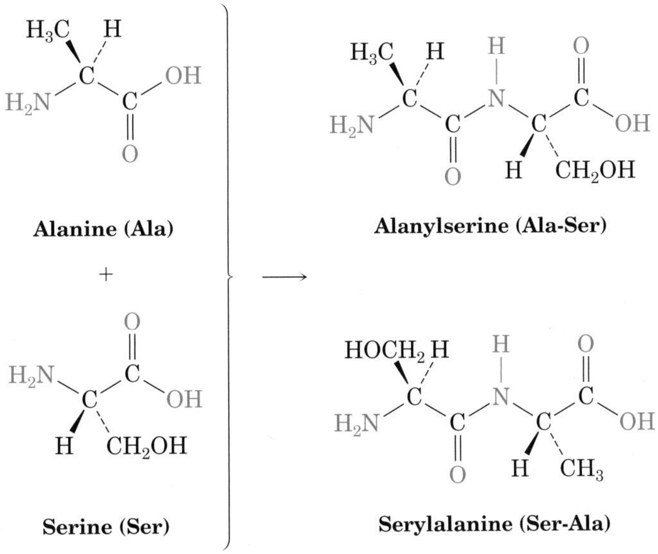

Alanine (Ala): $H_2N-CH(CH_3)-COOH$

+

Serine (Ser): $H_2N-CH(CH_2OH)-COOH$

⟶

Alanylserine (Ala-Ser): $H_2N-CH(CH_3)-C(=O)-NH-CH(CH_2OH)-COOH$

Serylalanine (Ser-Ala): $H_2N-CH(CH_2OH)-C(=O)-NH-CH(CH_3)-COOH$

In fact, by mixing serine and alanine together under conditions that favor amide bond formation, four dipeptides result: Ala-Ser, Ser-Ala, Ser-Ser, and Ala-Ala. In addition, tripeptides, tetrapeptides, and larger structures are formed as well. What do we do if we are interested in only a specific dipeptide? We need to exert control during synthesis, and we do so by employing *protecting groups.* Conceptually, we can think of these protecting groups as "shields" that prevent groups from reacting.

If only one amine group and one carboxylic acid group are available for reaction, then only one dipeptide can form. Therefore, protecting the amine group of alanine and the carboxylic acid group of serine allows only the desired protected dipeptide to form: the shielded groups do not react. A subsequent step (or steps) can be used to remove these protecting groups.

Protected alanine (Ala) + **Protected serine (Ser)** ⟶ **Protected alanylserine (Ala-Ser)**

Protecting groups have three important properties: (1) they are readily installed, (2) they are stable to a range of reaction conditions, and (3) they are readily removed under specific conditions. Esters are commonly used to protect the carboxylic acid group. Two common protecting groups for amines are abbreviated Fmoc (Fluorenylmethyloxycarbonyl) and Boc (*t*-Butyloxycarbonyl). Fmoc groups are removed with base. Boc groups are removed with acid.

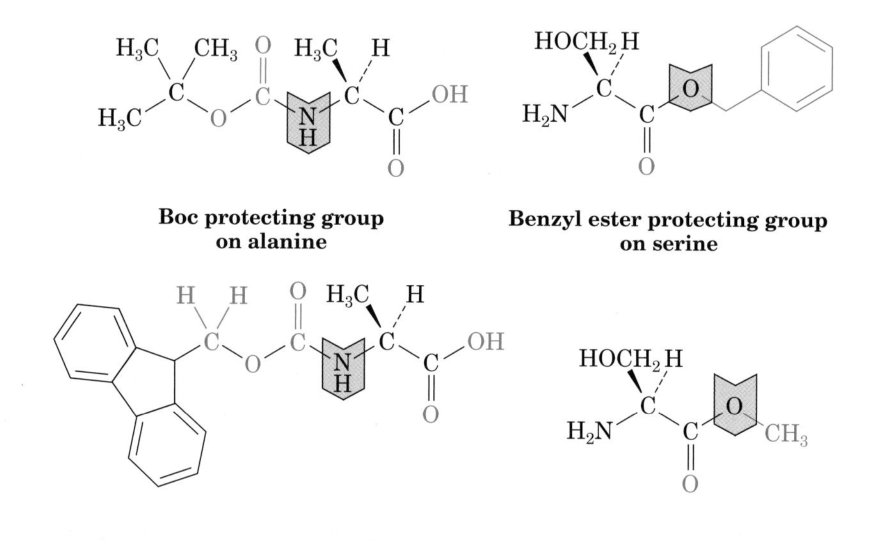

Boc protecting group on alanine

Benzyl ester protecting group on serine

Fmoc protecting group on alanine

Methyl ester protecting group on serine

The reagents used to install these groups, di-*tert*-butyl dicarbonate and Fmoc-Cl, undergo reactions with amines very similar to those discussed in Chapter 10.

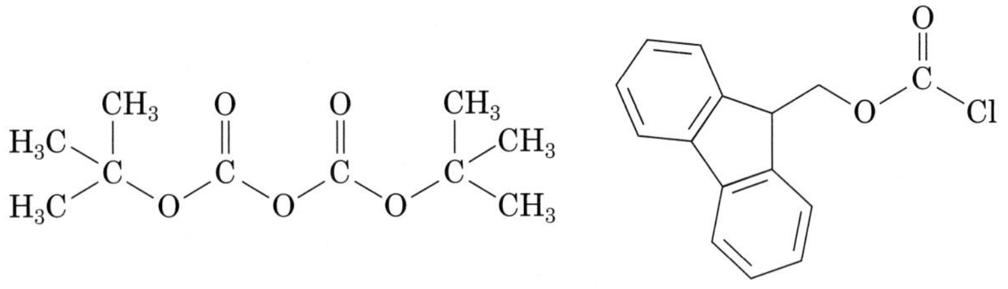

Di-*tert*-butyl dicarbonate
(a source of Boc groups)

Fmoc-Cl
(a source of Fmoc)

What about the amide bond-forming step? Although we have discussed amide-forming reactions between carboxylic acids and amines in Section 10.12, all these conditions are relatively harsh, typified by high temperatures and a strong acid to promote dehydration. Today, *coupling reagents* like DCC (dicyclohexylcarbodiimide) are commonly used to rapidly form the amide under very mild conditions. DCC converts the hydroxyl group of the carboxylic acid into a good leaving group and in the process is ultimately converted to dicyclohexylurea.

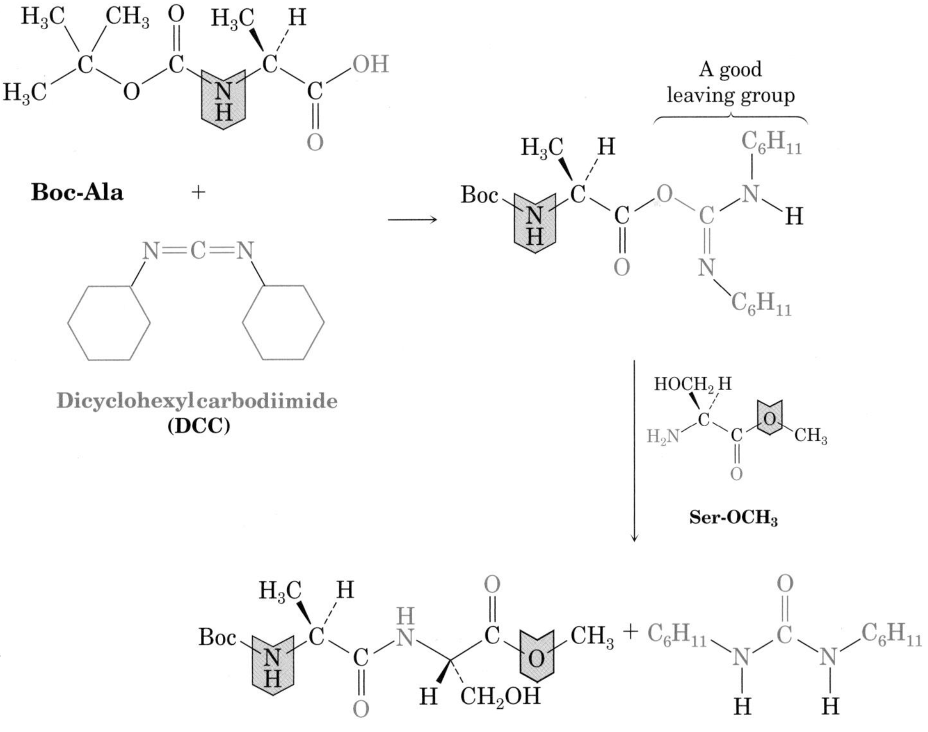

Boc-Ala +

Dicyclohexylcarbodiimide
(DCC)

A good leaving group

Ser-OCH_3

Boc-Ala-Ser-OCH_3

Dicyclohexylurea

Today, instead of doing this chemistry in a flask, it is often performed by *solid-phase synthesis*. We start with one amino acid that is chemically attached to a polymer bead with a cleavable linker, and the reactions that we have discussed are run iteratively. Let's look at the synthesis of a pentapeptide on a solid phase in Figure 15.2. We start with the C-terminal amino acid attached to the solid support. After deprotecting the amine group with base, we react it with the next amino acid using DCC as a coupling reagent. When the support is washed, excess reagents and the dicyclohexylurea that forms during the reaction are washed away. We can deprotect this supported dipeptide and iterate through the process until all the amino acids have been incorporated. Typically, the peptide side chains are deprotected at the same time the peptide is cleaved from the solid support. Filtering away the support provides a solution of the peptide that is normally isolated by precipitation.

FIGURE 15.2 Peptide synthesis on a solid support.

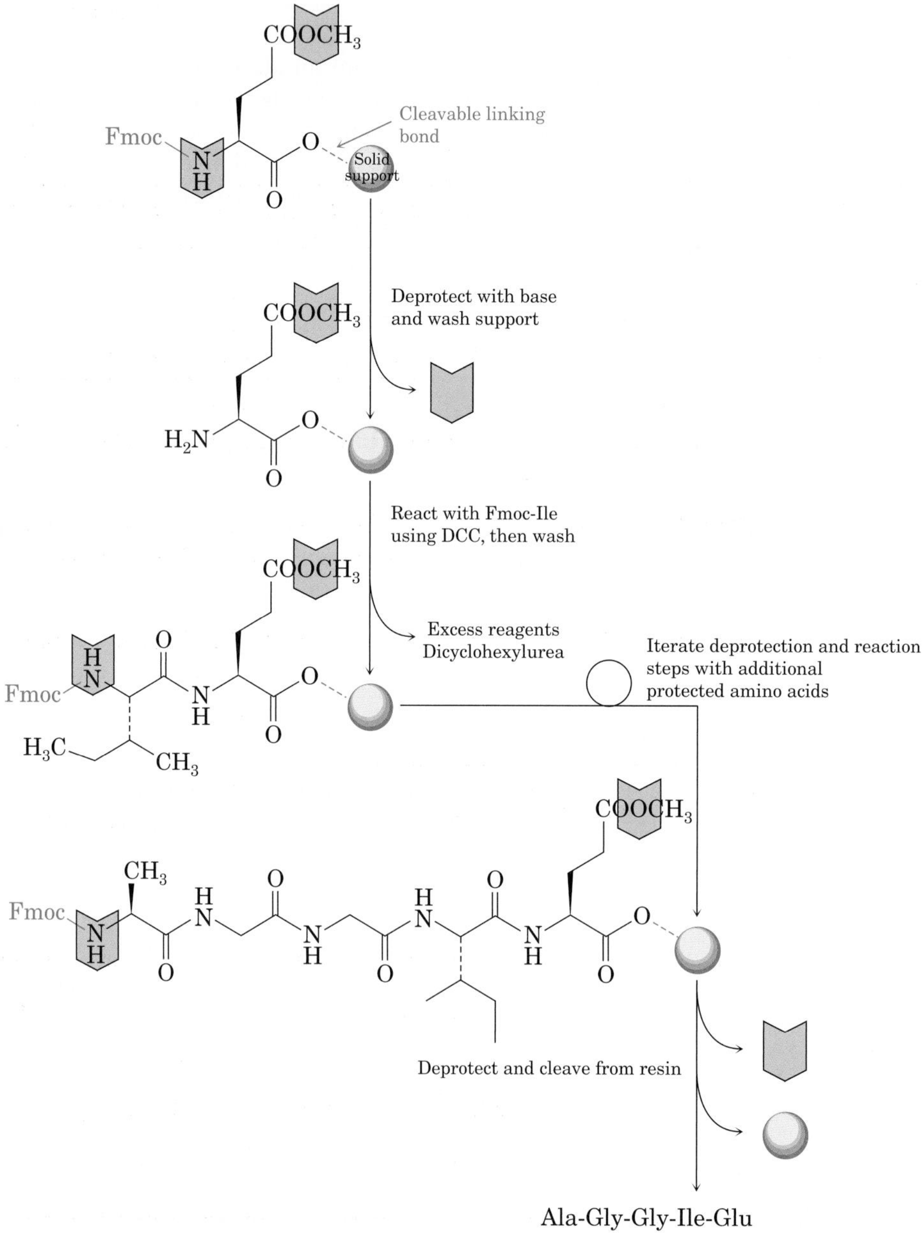

PRACTICE PROBLEM 15.3

Draw the structure of Ala-Val.

STRATEGY By convention, the N-terminal amino acid is written on the left and the C-terminal amino acid on the right. Thus, alanine is N-terminal, valine is C-terminal, and the amide bond is formed between the alanine $-CO_2H$ and the valine $-NH_2$.

SOLUTION

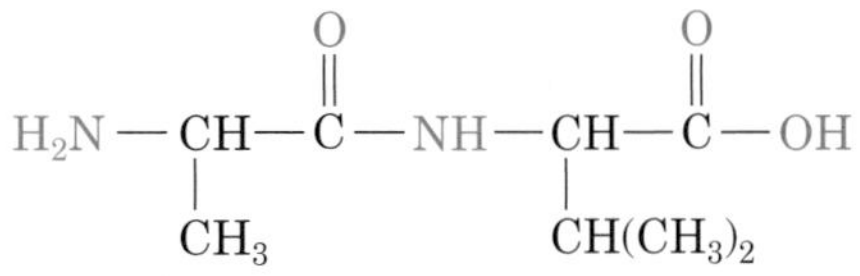

$$H_2N-\underset{\displaystyle CH_3}{\underset{|}{CH}}-\overset{\displaystyle O}{\overset{\|}{C}}-NH-\underset{\displaystyle CH(CH_3)_2}{\underset{|}{CH}}-\overset{\displaystyle O}{\overset{\|}{C}}-OH \quad \textbf{Ala-Val}$$

PRACTICE PROBLEM 15.4

There are six tripeptides that contain methionine, lysine, and isoleucine. Name them using both three- and one-letter abbreviations.

SOLUTION

Met-Lys-Ile (M-K-I)	Lys-Met-Ile (K-M-I)	Ile-Met-Lys (I-M-K)
Met-Ile-Lys (M-I-K)	Lys-Ile-Met (K-I-M)	Ile-Lys-Met (I-K-M)

PROBLEM 15.7 Draw structures of the two dipeptides made from leucine and cysteine.

PROBLEM 15.8 Using both three- and one-letter notations for each amino acid, name the six possible isomeric tripeptides that contain valine, tyrosine, and glycine.

PROBLEM 15.9 Draw the structure of Met-Pro-Val-Gly, and indicate where the amide bonds are.

PRACTICE PROBLEM 15.5

Write equations for the reaction of methionine with the following:
(a) CH_3OH, HCl (b) Di-*tert*-butyl dicarbonate

SOLUTION

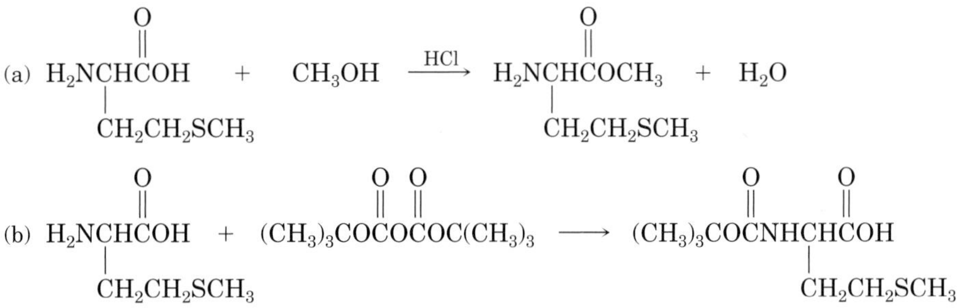

$$\text{(a)}\ H_2N\underset{\displaystyle CH_2CH_2SCH_3}{\underset{|}{C}}H\overset{\displaystyle O}{\overset{\|}{C}}OH + CH_3OH \xrightarrow{HCl} H_2N\underset{\displaystyle CH_2CH_2SCH_3}{\underset{|}{C}}H\overset{\displaystyle O}{\overset{\|}{C}}OCH_3 + H_2O$$

$$\text{(b)}\ H_2N\underset{\displaystyle CH_2CH_2SCH_3}{\underset{|}{C}}H\overset{\displaystyle O}{\overset{\|}{C}}OH + (CH_3)_3CO\overset{\displaystyle O}{\overset{\|}{C}}O\overset{\displaystyle O}{\overset{\|}{C}}OC(CH_3)_3 \longrightarrow (CH_3)_3CO\overset{\displaystyle O}{\overset{\|}{C}}NH\underset{\displaystyle CH_2CH_2SCH_3}{\underset{|}{C}}H\overset{\displaystyle O}{\overset{\|}{C}}OH$$

PROBLEM 15.10 Write the structures of the intermediates in the five-step synthesis of Leu-Ala from alanine and leucine.

PROBLEM 15.11 Show all the steps involved in the solid-phase synthesis of the tripeptide Val-Phe-Gly.

15.4 Peptide Synthesis in Nature

Peptides arise in nature through two different processes: either ribosomal or nonribosomal peptide syntheses. We are most familiar with ribosomal syntheses wherein the ribosome, an aggregate of rRNA (ribosomal RNA) and proteins, catalyzes peptide synthesis with amino acids delivered on tRNA (transfer RNA) as indicated by a mRNA (messenger RNA) sequence. We'll discuss this process in greater depth in the next chapter. Typically, the peptides produced by ribosomal syntheses contain a large number of amino acids.

In microbes, some of the short peptides are obtained through nonribosomal peptide synthesis. These nonribosomal peptides differ markedly from ribosomal peptides in that they are not limited to the 20 common amino acids. In addition, unusual D-amino acids are sometimes incorporated. Sometimes, the nitrogen atom of the amide is methylated. Often these peptides are cyclic. Many of them are biologically active, and some serve as drugs in people and livestock. The antibiotic vancomycin and the iron-binding enterobactin are examples of this diverse class of natural products.

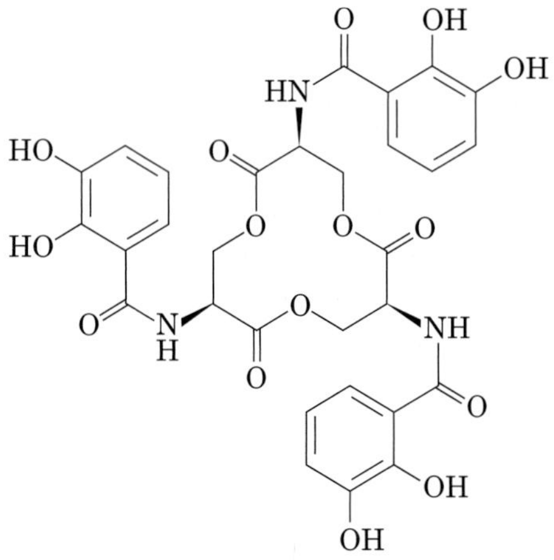

Enterobactin

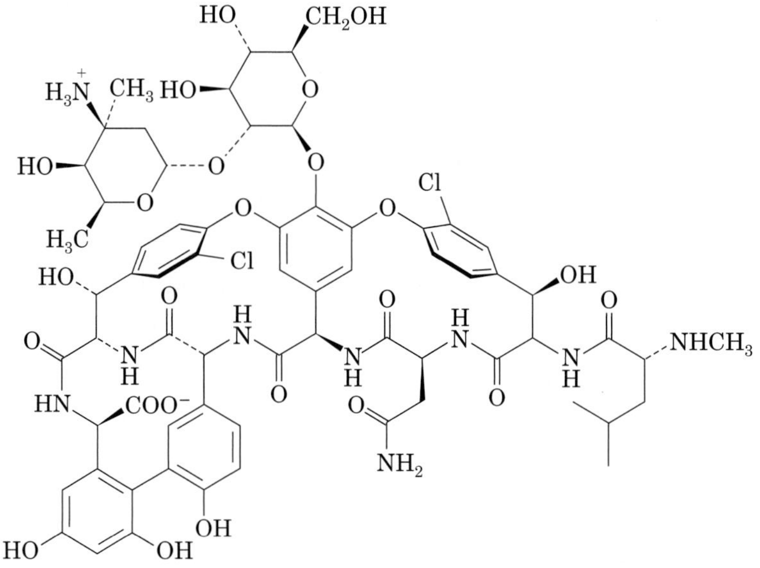

Vancomycin

How are such structures assembled? While the ribosome catalyzes protein synthesis from any mRNA template, nonribosomal peptides are prepared by *sequence-specific* assemblies of multiple enzymes. For the antibiotic tyrocidin, for example, the linear peptide is prepared by enzyme assembly responsible for the sequence and the novel amino acid ornithine (green) and two D-phenylalanine residues (blue). Upon completion of the sequence, another enzyme, thioesterase, converts the linear thioester to a cyclic amide with expulsion of the thiol leaving group (red).

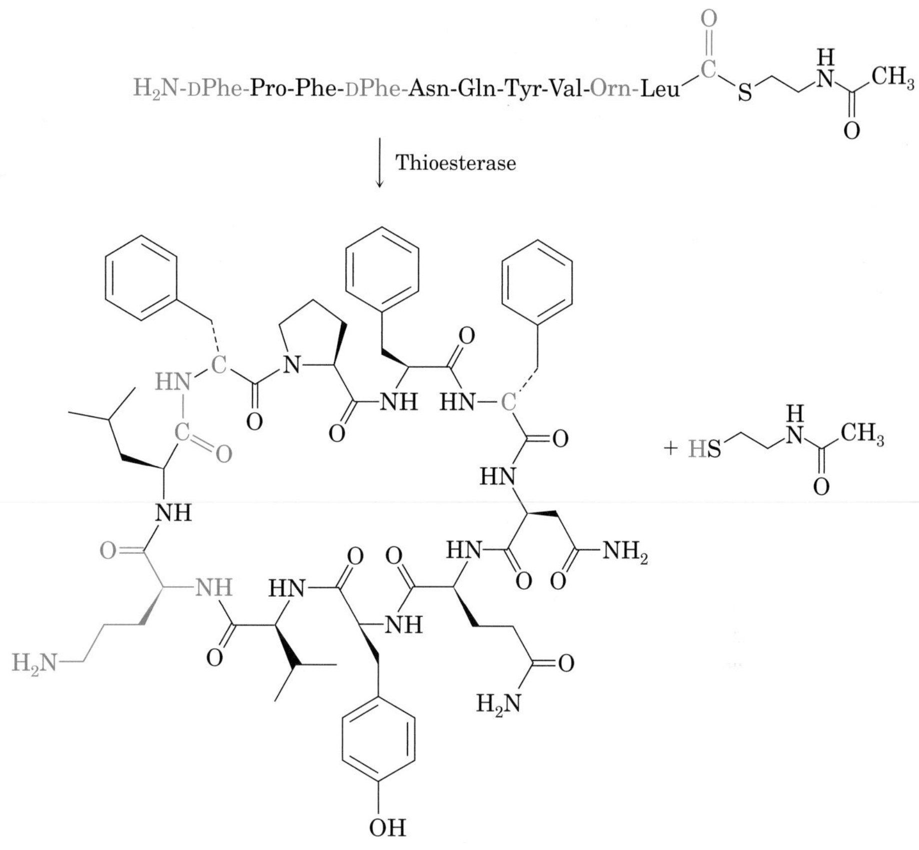

H_2N-DPhe-Pro-Phe-DPhe-Asn-Gln-Tyr-Val-Orn-Leu–C(=O)–S–CH_2CH_2–NH–C(=O)–CH_3

↓ Thioesterase

Tyrocidin

+ $HSCH_2CH_2NHC(=O)CH_3$

15.5 Covalent Bonding in Peptides

The covalent amide bond that links different amino acids together in peptides is the same as any other amide bond. Amide nitrogens are nonbasic (see Section 12.3) because their nonbonding electron pair is shared by interaction with the carbonyl group. This overlap of the nitrogen *p* orbital with the *p* orbitals of the carbonyl group imparts a certain amount of double-bond character to the C–N bond and restricts rotation around it. The amide bond is therefore planar, and the N–H is oriented 180° to the C=O.

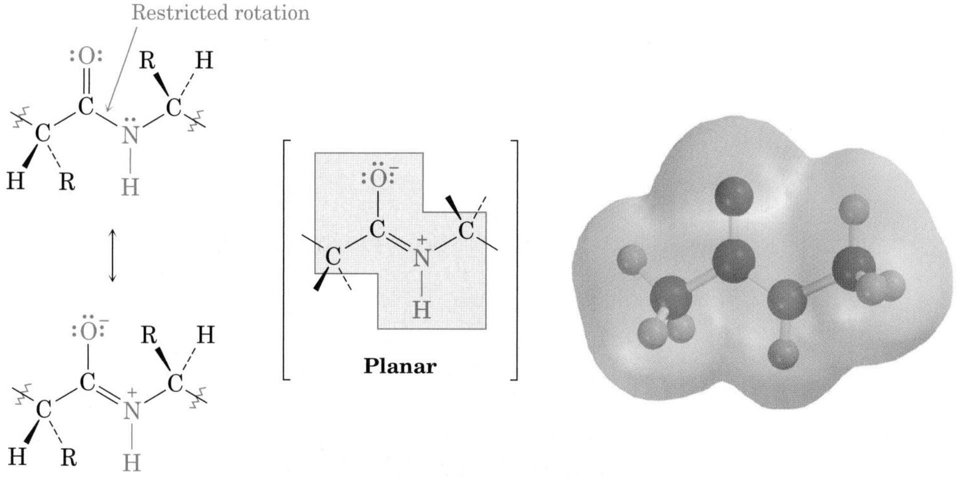

Restricted rotation

Planar

A second kind of covalent bonding in peptides occurs when a disulfide linkage, RS–SR, is formed between two cysteine residues. As we saw in Section 8.9, a disulfide bond is easily formed by mild oxidation of a thiol, RSH, and is easily cleaved by mild reduction.

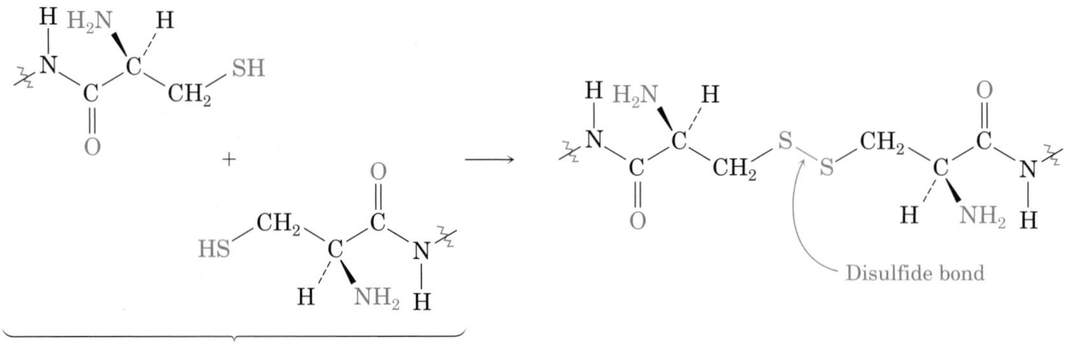

Disulfide bond

Two cysteines

A disulfide bond between cysteine residues in different peptide chains links the otherwise separate chains together, while a disulfide bond between cysteine residues in the same chain forms a loop in the chain. Such is the case with vasopressin, an antidiuretic hormone found in the pituitary gland. Note that the C-terminal end of vasopressin occurs as the primary amide, $-CONH_2$, rather than as the free acid.

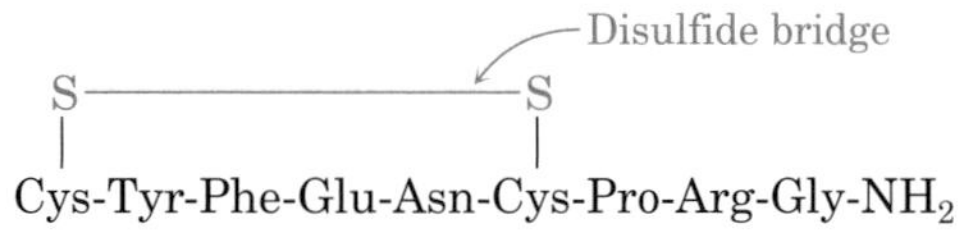

Disulfide bridge

Cys-Tyr-Phe-Glu-Asn-Cys-Pro-Arg-Gly-NH_2

Vasopressin

Another example is insulin. Insulin is composed of two chains totaling 51 amino acids and linked by two cysteine disulfide bridges. Its structure was determined by Frederick Sanger, who received the 1958 Nobel Prize for his work.

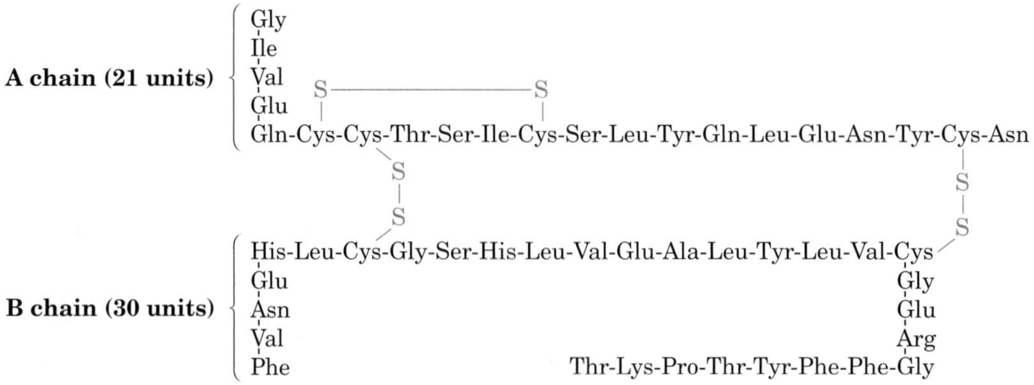

A chain (21 units)

Gly-Ile-Val-Glu-Gln-Cys-Cys-Thr-Ser-Ile-Cys-Ser-Leu-Tyr-Gln-Leu-Glu-Asn-Tyr-Cys-Asn

B chain (30 units)

Phe-Val-Asn-Glu-His-Leu-Cys-Gly-Ser-His-Leu-Val-Glu-Ala-Leu-Tyr-Leu-Val-Cys-Gly-Glu-Arg-Gly-Phe-Phe-Tyr-Thr-Pro-Lys-Thr

Insulin

15.6 Peptide Structure Determination: Amino Acid Analysis

Determining the structure of a peptide or protein requires answering three questions: What amino acids are present? How much of each is present? In what sequence do the amino acids occur in the peptide chain? The answers to the first two questions are provided by an instrument called an *amino acid analyzer.*

In preparation for analysis, the peptide is broken into its constituent amino acids by reducing all disulfide bonds and hydrolyzing all amide bonds with aqueous HCl (see Section 10.12). The resultant amino acid mixture is then analyzed by placing it at the top of a glass column filled with a special adsorbent material and pumping a series of aqueous buffers through the column. The various amino acids migrate down the column at different rates depending on their structures and are thus separated as they exit (*elute* from) the end of the column.

As each amino acid elutes from the glass column, it mixes with a solution of *ninhydrin*, a reagent that forms a purple color on reaction with an α-amino acid. The purple color is detected by a spectrometer, which measures its intensity and plots it as a function of time.

$$\text{Ninhydrin} + H_2NCH(R)COOH \xrightarrow[H_2O]{^-OH} \text{(Purple color)} + RCHO + CO_2$$

Ninhydrin   An α-amino acid   (Purple color)

Because the time required for a given amino acid to elute from a standard column is reproducible, the identities of all amino acids in a peptide are determined simply by noting the various elution times. The amount of each amino acid in the sample is determined by measuring the intensity of the purple color resulting from its reaction with ninhydrin. Figure 15.3 shows the results of amino acid analysis of a standard equimolar mixture of 17 α-amino acids.

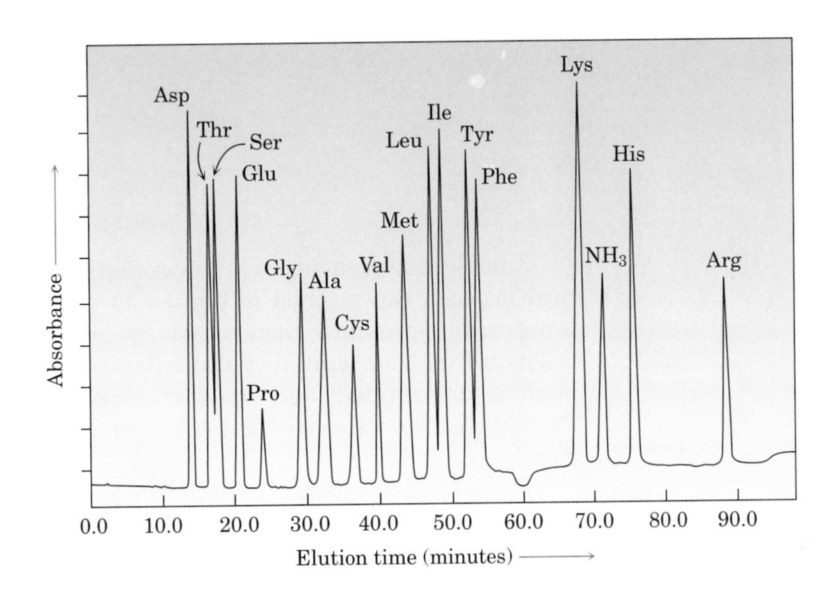

FIGURE 15.3 Amino acid analysis of an equimolar mixture of 17 amino acids.

Typically, amino acid analysis requires about 150 picomoles (4–5 μg) of sample for a protein containing about 200 residues.

PROBLEM 15.12 Write an equation for the reaction of valine with ninhydrin.

15.7 Peptide Sequencing: The Edman Degradation Method

With the identities and amounts of the amino acids known, the peptide is *sequenced* to find the order in which the amino acids are linked. The general idea of peptide sequencing is to cleave one residue at a time from the end of the peptide chain (either C terminus or N terminus). That terminal amino acid is then separated and identified, and the cleavage reaction is repeated on the chain-shortened peptide until the entire sequence is known.

Most peptide sequencing is done by **Edman degradation**, an efficient method of N-terminal analysis. Automated instruments are available that allow as many as 50 repetitive sequencing cycles to be carried out. So efficient are these instruments that sequence information can be obtained from as little as 1 to 5 picomoles of sample—less than 0.1 μg.

Edman degradation involves treatment of a peptide with phenyl isothiocyanate (PITC), $C_6H_5—N=C=S$, followed by mild acid hydrolysis. PITC first attaches to the $–NH_2$ group of the N-terminal amino acid, and the N-terminal residue then splits from the chain giving a *phenylthiohydantoin* derivative (PTH) along with chain-shortened peptide. The PTH is identified by comparison with known derivatives of the common amino acids, and the chain-shortened peptide is automatically resubmitted to another round of Edman degradation.

(PITC)
1. C_6H_5—N=C=S
2. H_3O^+
+ H_2N—Peptide

A phenylthiohydantoin (PTH)

Complete sequencing of large proteins by Edman degradation is impractical because the method is limited to about 50 cycles due to buildup of unwanted by-products. Instead, a large protein chain is first cleaved by partial hydrolysis into a number of smaller fragments, the sequence of each fragment is determined, and the individual pieces are then fitted together. Protein chains with more than 400 amino acids have been sequenced in this way.

Partial hydrolysis of a protein can be carried out either chemically with aqueous acid or enzymatically. Acid hydrolysis is unselective and gives a more or less random mixture of fragments. Enzymatic hydrolysis, however, is quite specific. The enzyme *trypsin*, for instance, catalyzes hydrolysis of peptides only at the carboxyl side of the basic amino acids arginine and lysine; *chymotrypsin* cleaves only at the carboxyl side of the aryl-substituted amino acids phenylalanine, tyrosine, and tryptophan.

Val-Phe-Leu-Met-Tyr-Pro-Gly-Trp-Cys-Glu-Asp-Ile-Lys-Ser-Arg-His

Chymotrypsin cleaves these bonds. Trypsin cleaves these bonds.

Practice Problem 15.6

Angiotensin II is a hormonal octapeptide involved in controlling hypertension by regulating the sodium–potassium salt balance in the body. Amino acid analysis shows the presence of eight different amino acids in equimolar amounts: Arg, Asp, His, Ile, Phe, Pro, Tyr, and Val. Partial hydrolysis of angiotensin II with dilute hydrochloric acid yields the following fragments:

Asp-Arg-Val-Tyr, Ile-His-Pro, Pro-Phe, Val-Tyr-Ile-His

What is the sequence of angiotensin II?

Strategy Line up the fragments to identify the overlapping regions, and then write the sequence.

Asp-Arg-Val-Tyr
Val-Tyr-Ile-His
Ile-His-Pro
Pro-Phe

Solution The sequence is Asp-Arg-Val-Tyr-Ile-His-Pro-Phe.

Problem 15.13 What fragments would result if angiotensin II (Practice Problem 15.5) were cleaved with trypsin? With chymotrypsin?

Problem 15.14 Give the amino acid sequence of a hexapeptide containing Arg, Gly, Ile, Leu, Pro, and Val that produces the following fragments on partial acid hydrolysis: Pro-Leu-Gly, Arg-Pro, Gly-Ile-Val.

Problem 15.15 What is the N-terminal residue on a peptide that gives the following PTH derivative on Edman degradation?

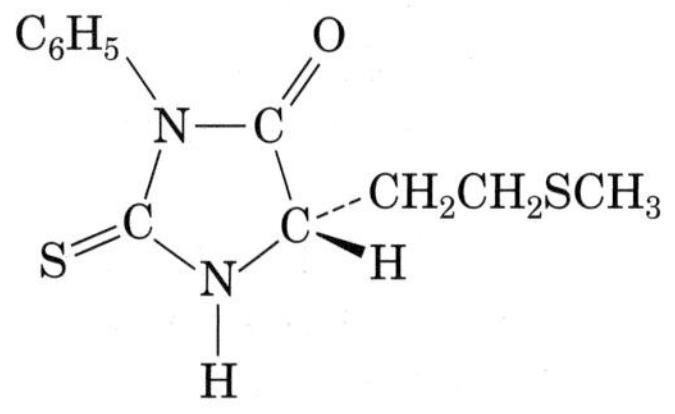

C_6H_5, O, N—C, C, $CH_2CH_2SCH_3$, S=C, N, H, H

15.8 Classification of Proteins

Proteins are classified into two major types according to their composition. **Simple proteins**, such as blood serum albumin, are those that yield only amino acids and no other compounds on hydrolysis. **Conjugated proteins**, which are much more common than simple proteins, yield other compounds such as carbohydrates, fats, or nucleic acids in addition to amino acids on hydrolysis. As shown in Table 15.2, conjugated proteins can be further classified according to the chemical nature of the non–amino acid portion.

Another way to classify proteins is as either *fibrous* or *globular*, according to their three-dimensional shape. **Fibrous proteins**, such as collagen and keratin, consist of polypeptide chains arranged side by side in long filaments. Because these proteins are tough and insoluble in water, they are used in nature for structural materials such as tendons, hooves, horns, and muscles. **Globular proteins**, by contrast, are usually coiled into compact, nearly spherical shapes. These proteins are generally soluble in water and are mobile within cells. Most of the 2000 or so known enzymes are globular. Table 15.3 lists some common examples of both kinds.

TABLE 15.2 Some Conjugated Proteins

Name	Composition
Glycoproteins	Proteins bonded to a carbohydrate; cell membranes have a glycoprotein coating
Lipoproteins	Proteins bonded to fats and oils (lipids); these proteins transport cholesterol and other fats through the body
Metalloproteins	Proteins bonded to a metal ion; the enzyme cytochrome oxidase, necessary for biological energy production, is an example
Nucleoproteins	Proteins bonded to RNA (ribonucleic acid); these are found in cell ribosomes
Phosphoproteins	Proteins bonded to a phosphate group; milk casein, which stores nutrients for growing embryos, is an example

TABLE 15.3 Some Common Fibrous and Globular Proteins

Name	Occurrence and use
Fibrous proteins (insoluble)	
Collagens	Animal hide, tendons, connective tissues
Elastins	Blood vessels, ligaments
Fibrinogen	Necessary for blood clotting
Keratins	Skin, wool, feathers, hooves, silk, fingernails
Myosins	Muscle tissue
Globular proteins (soluble)	
Hemoglobin	Involved in oxygen transport
Immunoglobulins	Involved in immune response
Insulin	Hormone for controlling glucose metabolism
Ribonuclease	Enzyme controlling RNA synthesis

Table 15.4 Some Biological Functions of Proteins

Type	Function and example
Enzymes	Proteins such as chymotrypsin that act as biological catalysts
Hormones	Proteins such as insulin that regulate body processes
Protective proteins	Proteins such as antibodies that fight infection
Storage proteins	Proteins such as casein that store nutrients
Structural proteins	Proteins such as keratin, elastin, and collagen that form the structure of an organism
Transport proteins	Proteins such as hemoglobin that transport oxygen and other substances through the body

Yet a third way to classify proteins is according to function. As shown in Table 15.4, there is an extraordinary diversity to the biological roles of proteins.

15.9 Protein Structure

Proteins are so large that the word *structure* takes on a broader meaning than it does with simpler organic compounds. In fact, chemists speak of four different levels of structure when describing proteins:

- The **primary structure** of a protein is simply the amino acid sequence.
- The **secondary structure** of a protein describes how *segments* of the peptide backbone orient into a regular pattern.
- The **tertiary structure** describes how the *entire* protein molecule coils into an overall three-dimensional shape.
- The **quaternary structure** describes how different protein molecules come together to yield large aggregate structures.

Let's look at three examples—α-keratin (fibrous), fibroin (fibrous), and myoglobin (globular)—to see how higher structure affects a protein's properties.

α-Keratin

α-Keratin is the fibrous structural protein found in wool, hair, nails, and feathers. Studies show that segments of the α-keratin chain are coiled into a right-handed helical secondary structure like that of a telephone cord. Illustrated in Figure 15.4, this so-called **α-helix** is stabilized by hydrogen bonding between amide N–H groups and C=O groups four residues away. Each coil of the helix (the *repeat distance*) contains 3.6 amino acid residues, and the distance between coils is 540 pm, or 5.40 Å.

Evidence also shows that the α-keratins of wool and hair have a quaternary structure. The individual helical strands are themselves coiled about one another in stiff bundles to form a *superhelix* that accounts for the thread-like properties and strength of these proteins. Although α-keratin is the best example of an almost entirely helical protein, most globular proteins also contain α-helical segments.

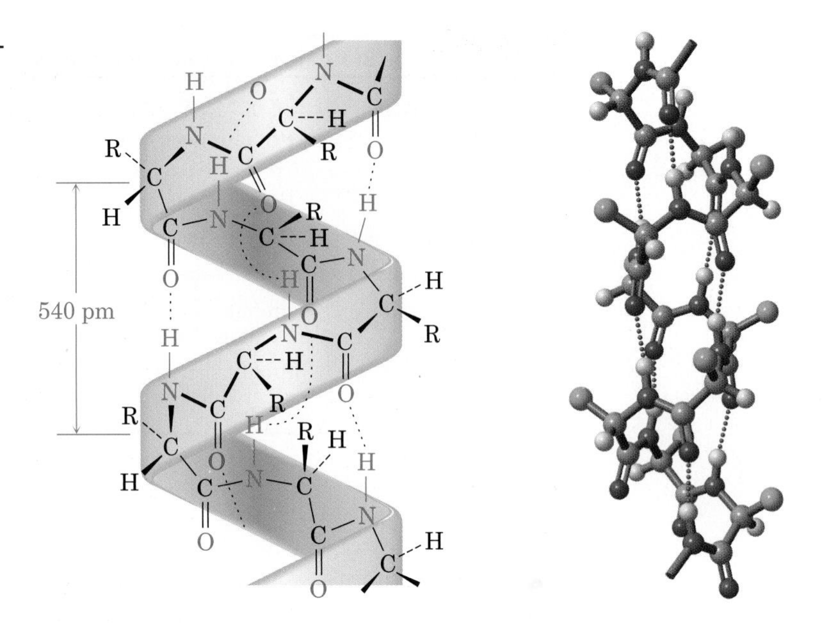

FIGURE 15.4 The helical secondary structure present in α-keratin.

Fibroin

Fibroin, the fibrous protein found in silk, has a secondary structure called a ***β*-pleated sheet** in which neighboring polypeptide chains line up in a parallel arrangement held together by hydrogen bonds between chains (Figure 15.5). The neighboring chains can run either in the same direction (parallel) or in opposite directions (antiparallel), although the antiparallel arrangement is more common and energetically somewhat more favorable. While not as common as the α-helix, small β-pleated-sheet regions occur frequently in globular proteins.

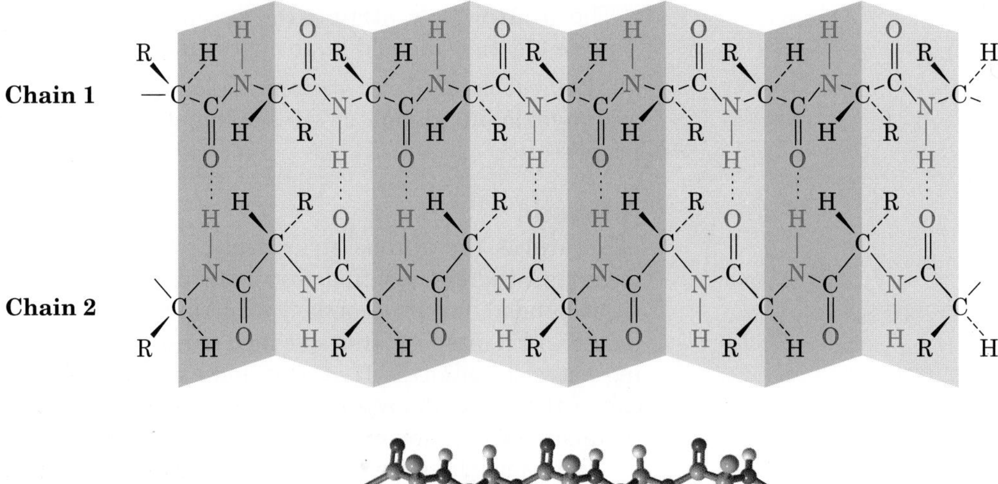

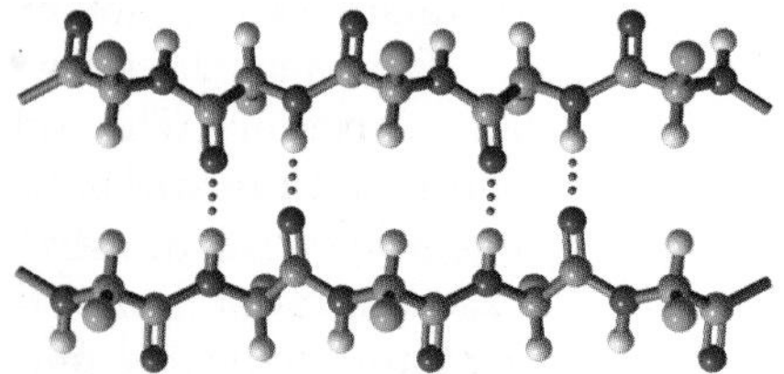

FIGURE 15.5 The β-pleated-sheet structure in silk fibroin.

Myoglobin

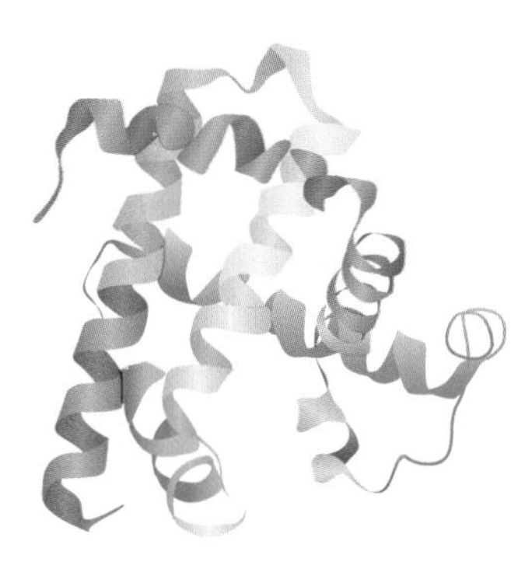

FIGURE 15.6 Secondary and tertiary structure of myoglobin, a small protein with extensive helical sections, shown here as ribbons.

Myoglobin is a small globular protein containing 153 amino acid residues in a single chain. A relative of hemoglobin, myoglobin is found in the skeletal muscles of sea mammals, where it stores oxygen needed to sustain the animals during long dives. Myoglobin consists of eight helical segments connected by bends to form a compact, nearly spherical, tertiary structure (Figure 15.6).

Why does myoglobin adopt the shape it does? The forces that determine the tertiary structure of myoglobin and other globular proteins are the same simple forces that act on all molecules, regardless of size, to provide maximum stability. Particularly important are the hydrophobic (water-repelling) interactions of hydrocarbon side chains on neutral amino acids. Those amino acids with neutral, nonpolar side chains have a strong tendency to congregate on the hydrocarbon-like interior of a protein molecule, away from the aqueous medium. Those acidic or basic amino acids with charged side chains, by contrast, tend to congregate on the exterior of the protein where they can be solvated by water.

Also important for stabilizing a protein's tertiary structure are the formation of disulfide bridges between cysteine residues, the formation of hydrogen bonds between nearby amino acid residues, and the development of ionic attractions, called *salt bridges*, between positively and negatively charged sites on various amino acid side chains within the protein. The various kinds of stabilizing forces are summarized in Figure 15.7.

FIGURE 15.7 Kinds of interactions among amino acid side chains that stabilize a protein's tertiary structure.

Salt bridge

$CH_2S—SCH_2$

Disulfide bridge

Hydrophobic interactions

$CH_2–O–H\cdots{:}O$

Hydrogen bond

Protein backbone

15.10 Enzymes

An **enzyme** is a substance—usually a protein—that acts as a catalyst for a biological reaction. Like all catalysts, enzymes don't affect the equilibrium constant of a reaction and can't bring about a chemical change that is otherwise unfavorable. Enzymes act only to lower the activation energy for a reaction, thereby making the reaction take place more rapidly. Sometimes, in fact, the rate acceleration brought about by enzymes is extraordinary. The glycosidase enzymes that hydrolyze polysaccharides, for example, increase the reaction rate by a factor of more than 10^{17}, changing the time required for the reaction from millions of years to milliseconds.

Unlike many of the catalysts that chemists use in the laboratory, enzymes are usually specific in their action. Often, in fact, an enzyme will catalyze only a single reaction of a single compound, called the enzyme's *substrate.* For example, the enzyme amylase found in the human digestive tract catalyzes only the hydrolysis of starch to yield glucose; cellulose and other polysaccharides are untouched by amylase.

Different enzymes have different specificities. Some, such as amylase, are specific for a single substrate, but others operate on a range of substrates. Papain, for instance, a globular protein of 212 amino acids isolated from papaya fruit, catalyzes the hydrolysis of many kinds of peptide bonds. In fact, it's this ability to hydrolyze peptide bonds that makes papain useful as a meat tenderizer and a cleaner for contact lenses.

$$-\!\!\left(\text{NHCH(R)C(=O)}-\text{NHCH(R')C(=O)}-\text{NHCH(R'')C(=O)}\right)\!\!- \xrightarrow[\text{H}_2\text{O}]{\text{Papain}} \text{H}_2\text{NCH(R)COOH} + \text{H}_2\text{NCH(R')COOH} + \text{H}_2\text{NCH(R'')COOH}$$

Most of the more than 2000 known enzymes are globular proteins. In addition to the protein part, most enzymes also have a small nonprotein part called a **cofactor**. The protein part in such an enzyme is called an **apoenzyme**, and the combination of apoenzyme plus cofactor is called a **holoenzyme**. Only holoenzymes have biological activity; neither cofactor nor apoenzyme can catalyze reactions by themselves.

Holoenzyme = Cofactor + Apoenzyme

A cofactor can be either an inorganic ion, such as Zn^{2+}, or a small organic molecule, called a **coenzyme**. The requirement of many enzymes for inorganic cofactors is the main reason for our dietary need of trace minerals. Iron, zinc, copper, manganese, and numerous other metal ions are all essential minerals that act as enzyme cofactors, although the exact biological role is not known in all cases.

A variety of organic molecules act as coenzymes. Many, though not all, coenzymes are **vitamins**, small organic molecules that must be obtained in the diet and are required in trace amounts for proper growth. Table 15.5 lists the 13 known vitamins required in the human diet and their enzyme functions.

Enzymes are grouped into six classes according to the kind of reaction they catalyze (Table 15.6). *Hydrolases* catalyze hydrolysis reactions, *isomerases* catalyze isomerizations, *ligases* catalyze the bonding together of two molecules with participation of adenosine triphosphate (ATP), *lyases* catalyze the breaking away of a small molecule such as H_2O from a substrate, *oxidoreductases* catalyze oxidations and reductions, and *transferases* catalyze the transfer of a group from one substrate to another.

Although some enzymes, like papain and trypsin, have uninformative common names, the systematic name of an enzyme has two parts, ending with *-ase.* The first part identifies the enzyme's substrate, and the second part identifies its class. For example, *hexose kinase* is an enzyme that catalyzes the transfer of a phosphate group from adenosine triphosphate to glucose.

Table 15.5 Vitamins and Their Enzyme Functions

Vitamin	Enzyme function	Deficiency symptoms
Water-soluble vitamins		
Ascorbic acid (vitamin C)	Hydrolases	Bleeding gums, bruising
Thiamin (vitamin B_1)	Reductases	Fatigue, depression
Riboflavin (vitamin B_2)	Reductases	Cracked lips, scaly skin
Pyridoxine (vitamin B_6)	Transaminases	Anemia, irritability
Niacin	Reductases	Dermatitis, dementia
Folic acid (vitamin M)	Methyltransferases	Megaloblastic anemia
Vitamin B_{12}	Isomerases	Megaloblastic anemia, neurodegeneration
Pantothenic acid	Acyltransferases	Weight loss, irritability
Biotin (vitamin H)	Carboxylases	Dermatitis, anorexia, depression
Fat-soluble vitamins		
Vitamin A	Visual system	Night blindness, dry skin
Vitamin D	Calcium metabolism	Rickets, osteomalacia
Vitamin E	Antioxidant	Hemolysis of red blood cells
Vitamin K	Blood clotting	Hemorrhage, delayed blood clotting

Table 15.6 Classification of Enzymes

Main class	Some subclasses	Type of reaction catalyzed
Hydrolases	Lipases	Hydrolysis of an ester group
	Nucleases	Hydrolysis of a phosphate group
	Proteases	Hydrolysis of an amide group
Isomerases	Epimerases	Isomerization of a stereocenter
Ligases	Carboxylases	Addition of CO_2
	Synthetases	Formation of new bond
Lyases	Decarboxylases	Loss of CO_2
	Dehydrases	Loss of H_2O
Oxidoreductases	Dehydrogenases	Introduction of double bond by removal of H_2
	Oxidases	Oxidation
	Reductases	Reduction
Transferases	Kinases	Transfer of a phosphate group
	Transaminases	Transfer of an amino group

Problem 15.16 To what classes do the following enzymes belong?

(a) Pyruvate decarboxylase (b) Chymotrypsin (c) Alcohol dehydrogenase

15.11 How Do Enzymes Work? Citrate Synthase

Enzymes exert their catalytic activity by bringing reactant molecules together, holding them in the orientation necessary for reaction, and providing any necessary acidic or basic sites to catalyze specific steps. Let's look, for example, at *citrate synthase*, an enzyme that catalyzes the aldol-like addition of acetyl CoA to oxaloacetate to give citrate (see Section 11.12). This reaction is the first step in the so-called citric acid cycle, in which acetyl groups produced by degradation of food molecules are metabolically "burned" to yield CO_2 and H_2O. We'll look at the details of the citric acid cycle in Section 17.4.

$$\underset{\textbf{Oxaloacetate}}{{}^{-}O_2CCH_2\overset{\overset{\displaystyle O}{\|}}{C}CO_2^{-}} + \underset{\textbf{Acetyl CoA}}{CH_3\overset{\overset{\displaystyle O}{\|}}{C}SCoA} \xrightarrow[\text{synthase}]{\text{Citrate}} \underset{\textbf{Citrate}}{{}^{-}O_2CCH_2\overset{\overset{\displaystyle OH}{|}}{\underset{\underset{\displaystyle CO_2^{-}}{|}}{C}}CH_2CO_2^{-}} + HSCoA$$

Citrate synthase is a globular protein with a deep cleft lined by an array of functional groups that can bind to oxaloacetate. Upon binding oxaloacetate, the original cleft closes and another opens up to bind acetyl CoA. This second cleft is also lined by appropriate functional groups, including a histidine at position 274 and an aspartic acid at position 375. The two reactants are now held by the enzyme in close proximity and with a suitable orientation for reaction. Figure 15.8 shows the structure of citrate synthase as determined by X-ray crystallography.

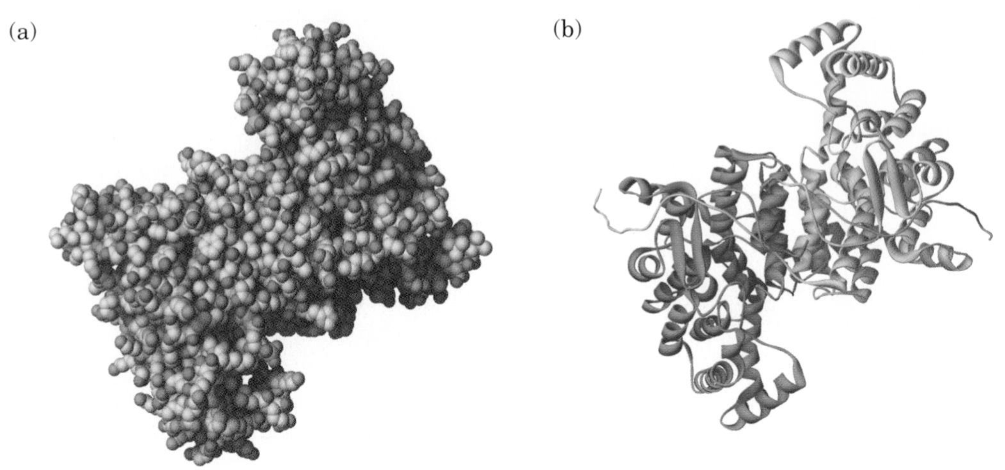

FIGURE 15.8 Models of citrate synthase. Part (a) is a space-filling model, which shows the deep clefts in the enzyme. Part (b) is a ribbon model, which emphasizes the α-helical segments of the protein chain and indicates that the enzyme is *dimeric*; that is, it consists of two identical chains held together by hydrogen bonds and other intermolecular attractions.

The first step in the aldol reaction is generation of the enol of acetyl CoA. The side-chain carboxyl of Asp-375 acts as base to abstract an acidic α proton, while at the same time the side-chain imidazole ring of His-274 donates H^+ to the carbonyl oxygen. The enol thus produced then does a nucleophilic addition to the ketone carbonyl group of oxaloacetate. The His-274 acts as a base to remove the —OH hydrogen from the enol, while another histidine residue at position 320 simultaneously donates a proton to the oxaloacetate carbonyl group, giving citryl CoA. Water then hydrolyzes the thioester group in citryl CoA, releasing citrate and coenzyme A as the final products. The mechanism is shown in Figure 15.9.

FIGURE 15.9 MECHANISM: Mechanism of action of the enzyme citrate synthase.

Acetyl CoA is held in the cleft of the citrate synthase enzyme with His-274 and Asp-375 nearby. The side-chain carboxylate group of Asp-375 acts as a base and removes an acidic α proton, while an N–H group on the side chain of His-274 acts as an acid and donates a proton to the carbonyl oxygen. The net result is formation of an enol.

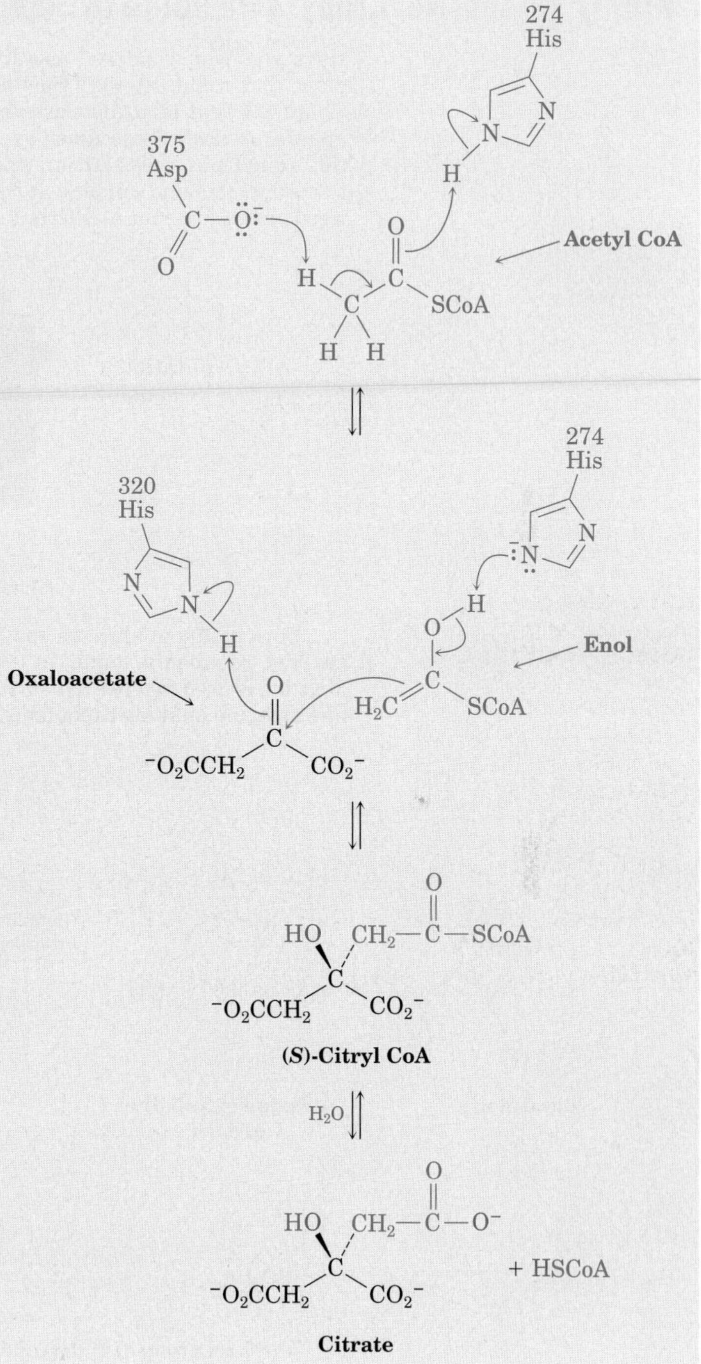

A nitrogen atom on the His-274 side chain acts as a base to deprotonate the acetyl CoA enol, which adds to the ketone carbonyl group of oxaloacetate in an aldol-like reaction. Simultaneously, an acidic N–H proton on the side chain of His-320 protonates the carbonyl oxygen, producing citryl CoA.

The thioester group of citryl CoA is hydrolyzed in a typical nucleophilic acyl substitution step, breaking the C–S bond and producing citrate plus coenzyme A.

INTERLUDE

Killing Weeds, Not Crops: One Route to Selectivity

Have you ever wondered how herbicides know the difference between a beloved plant and noxious weed? How does atrazine kill weeds but not corn? We saw in Chapter 5 that atrazine works by binding to a specific site on the photosynthetic apparatus normally occupied by a quinone. We might guess that the binding site for this quinone in weeds accommodates atrazine, while the site in corn does not. As it turns out, atrazine can bind at this site in corn plants as well, so selectivity against weeds must arise for a different reason.

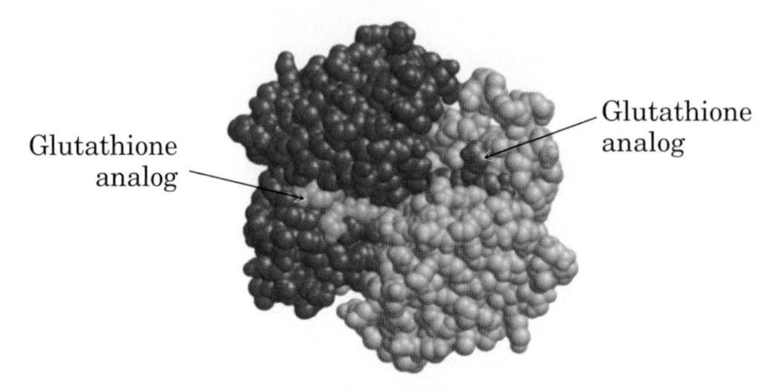

Glutathione *S*-transferase dimer

Corn plants produce an enzyme called glutathione *S*-transferase at levels that are not commonly found in weeds, particularly the broad-leafed weeds that atrazine is used against. This enzyme binds glutathione to small organic molecules like atrazine, making them more water-soluble and more readily excreted.

Glutathione + **Atrazine metabolite in corn** —(Glutathione *S*-transferase)→ ***S*-(4-Amino-6-isopropylamino-*S*-triazinyl-2)glutathione** + HCl

The thiol group of glutathione is sufficiently nucleophilic in the presence of the enzyme that a substitution reaction can occur. This reaction occurs readily on the metabolite of atrazine shown. The resulting glutathione conjugate is larger and more water-soluble than the herbicide by virtue of the two carboxylic acid groups, the amine, and amides. Accordingly, this molecule is less likely to penetrate the

Continued

lipid membrane and bind in the hydrophobic quinone binding site. The larger size also reduces the ability of this molecule to bind.

So, in summary, how does corn avoid the toxic effects of atrazine? It detoxifies the molecule by doing chemistry (*metabolism*) to change its physical and chemical properties. We'll talk more about metabolic pathways in the final chapter of this book. A variety of glutathione *S*-transferases are produced in people too. In addition to detoxifying organic molecules, including those that lead to some cancers, levels of some of these variants have recently been linked to neurodegenerative diseases such as Alzheimer's and Parkinson's. The role that the enzymes play in these diseases is still not understood.

Summary and Key Words

Proteins and **peptides** are biomolecules made of **amino acid residues** linked together by amide bonds. Twenty amino acids are commonly found in proteins; all are **α-amino acids**, and all except glycine have stereochemistry similar to that of L sugars.

Determining the structure of a peptide or protein begins with *amino acid analysis*. The peptide is first hydrolyzed to its constituent α-amino acids, which are then separated and identified. Next, the peptide is *sequenced*. **Edman degradation** by treatment with phenyl isothiocyanate (PITC) cleaves one residue from the **N terminus** of the peptide and forms an easily identifiable derivative phenylthiohydantoin (PTH) of that residue. An automated series of Edman degradations can sequence peptide chains up to 50 residues in length.

Peptide synthesis involves the use of *protecting* groups. An N-protected amino acid with a free $-CO_2H$ group is coupled using DCC to an O-protected amino acid with a free $-NH_2$ group. Amide formation occurs, the protecting groups are removed, and the sequence is repeated. Amines are usually protected as their *tert*-butoxycarbonyl (Boc) derivatives; acids are usually protected as esters.

Proteins are classified as either **globular** or **fibrous**, depending on their **secondary** and **tertiary structures**. Fibrous proteins such as α-keratin are tough and water-insoluble; globular proteins such as myoglobin are water-soluble and mobile within cells. Most of the 2000 or so known enzymes are globular proteins.

Enzymes are globular proteins that act as biological catalysts. They are classified into six groups according to the kind of reaction they catalyze: *oxidoreductases* catalyze oxidations and reductions, *transferases* catalyze transfers of groups, *hydrolases* catalyze hydrolysis, *isomerases* catalyze isomerizations, *lyases* catalyze bond breakages, and *ligases* catalyze bond formations.

In addition to their protein part, many enzymes contain **cofactors**, which can be either metal ions or small organic molecules. If the cofactor is an organic molecule, it is called a **coenzyme**. The combination of protein (**apoenzyme**) plus coenzyme is called a **holoenzyme**. Often, the coenzyme is a **vitamin**, a small molecule that must be obtained in the diet and is required in trace amounts for proper growth and functioning.

EXERCISES

Visualizing Chemistry

15.17 Identify the following amino acids:

(a) (b) (c)

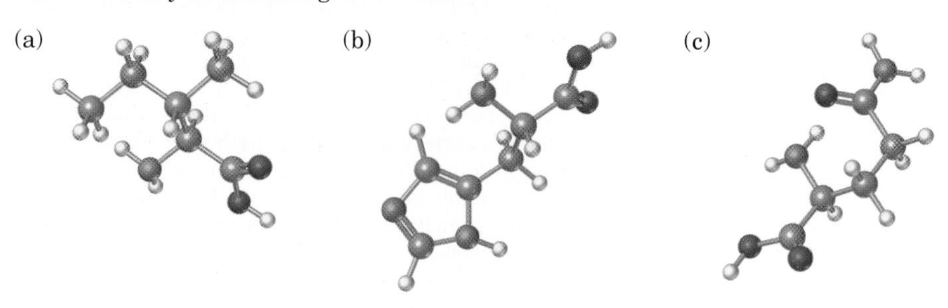

15.18 Give the sequence of the following tetrapeptide (yellow = S):

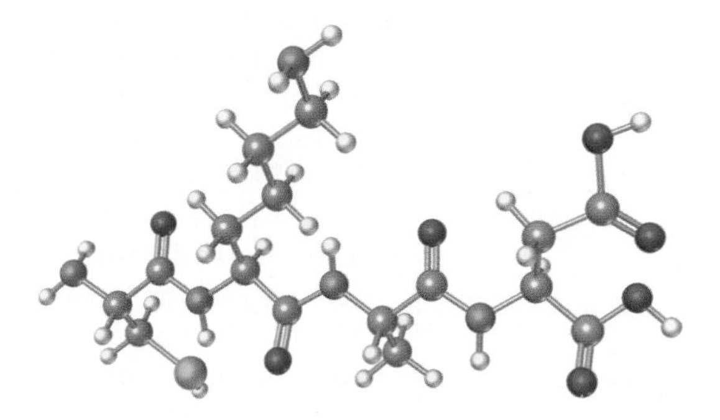

15.19 Isoleucine and threonine are the only two amino acids with two stereocenters. Assign *R* or *S* configuration to the methyl-bearing carbon atom of isoleucine.

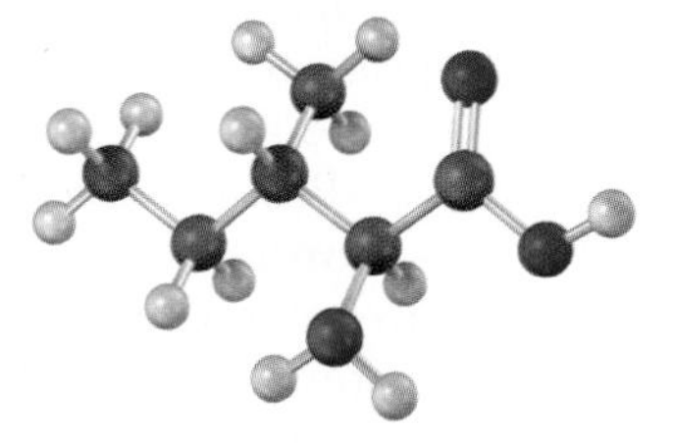

15.20 Is the following molecule a D amino acid or an L amino acid? Identify it.

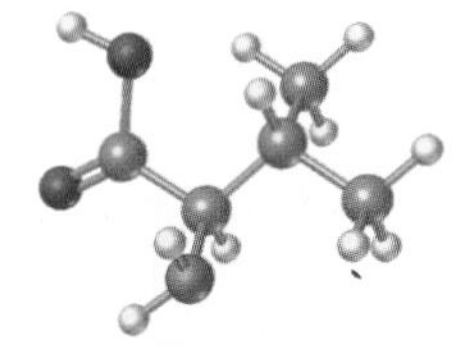

Additional Problems

NOMENCLATURE

15.21 What does the prefix "α" mean when referring to an α-amino acid?

15.22 What amino acids do the following abbreviations stand for?
(a) Ser (b) Thr (c) Pro (d) F (e) Q (f) D

15.23 What kinds of molecules are found in the following conjugated proteins in addition to the protein part?
(a) Nucleoproteins (b) Glycoproteins (c) Lipoproteins

15.24 Why is cysteine such an important amino acid for determining the tertiary structure of a protein?

15.25 The *endorphins* are a group of naturally occurring compounds in the brain that act to control pain. The active part of an endorphin is a pentapeptide called an *enkephalin*, which has the structure Tyr-Gly-Gly-Phe-Met. Draw the structure.

CHIRALITY AND STRUCTURE

15.26 Although only S amino acids occur in proteins, several R amino acids are found elsewhere in nature. For example, (R)-serine is found in earthworms and (R)-alanine is found in insect larvae. Draw Fischer projections of (R)-serine and (R)-alanine.

15.27 Draw a Fischer projection of (S)-proline, the only secondary amino acid.

15.28 Using both one- and three-letter code names for each amino acid, write the structures of all the peptides containing the following amino acids:
(a) Val, Leu, Ser (b) Ser, Leu_2, Pro

15.29 Write full structures for the following peptides, and indicate the positions of the amide bonds:
(a) Val-Phe-Cys (b) Glu-Pro-Ile-Leu

15.30 The amino acid threonine, (2S,3R)-2-amino-3-hydroxybutanoic acid, has two stereocenters and a stereochemistry similar to that of the four-carbon sugar D-threose. Draw a Fischer projection of threonine.

15.31 Draw the Fischer projection of a diastereomer of threonine (see Problem 15.30).

ACID–BASE CHEMISTRY OF AMINO ACIDS

15.32 Draw the following amino acids in their zwitterionic forms:
(a) Serine (b) Tyrosine (c) Threonine

15.33 Draw structures of the predominant forms of lysine and aspartic acid at pH 3.0 and pH 9.7.

15.34 At what pH would you carry out an electrophoresis experiment if you wanted to separate a mixture of histidine, serine, and glutamic acid? Explain.

REAGENTS AND SYNTHESIS

15.35 The amino acid analysis data in Figure 15.2 indicate that proline is not easily detected by reaction with ninhydrin. Suggest a reason.

15.36 Draw the structure of the phenylthiohydantoin product you would expect to obtain from Edman degradation of the following peptides:
(a) Val-Leu-Gly (b) Ala-Pro-Phe

15.37 Predict the product of the reaction of valine with the following reagents:
(a) CH_3CH_2OH, H^+ (b) NaOH, H_2O (c) Di-*tert*-butyl dicarbonate

15.38 Show the steps involved in a synthesis of Phe-Ala-Val by the solid phase method.

15.39 When an unprotected α-amino acid is treated with dicyclohexylcarbodiimide (DCC), a 2,5-diketopiperazine results. Explain.

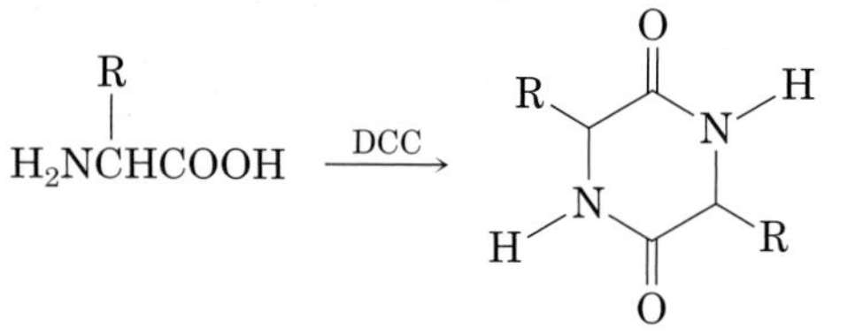

$$\underset{\mathrm{R}}{\mathrm{H_2N\overset{|}{C}HCOOH}} \xrightarrow{\text{DCC}} \text{A 2,5-diketopiperazine}$$

A 2,5-diketopiperazine

Enzymes

15.40 What kinds of reactions do the following classes of enzymes catalyze?
(a) Hydrolases (b) Lyases (c) Transferases

15.41 What kind of reaction does each of the following enzymes catalyze?
(a) A protease (b) A kinase (c) A carboxylase

Sequencing

15.42 Which amide bonds in the following polypeptide are cleaved by trypsin? By chymotrypsin?

Phe-Leu-Met-Lys-Tyr-Asp-Gly-Gly-Arg-Val-Ile-Pro-Tyr

15.43 A heptapeptide shows the composition Asp, Gly, Leu, Phe, Pro_2, Val on amino acid analysis. Edman degradation shows glycine to be the N-terminal group. Acidic hydrolysis gives the following fragments:

Val-Pro-Leu, Gly, Gly-Asp-Phe-Pro, Phe-Pro-Val

Propose a structure for the starting heptapeptide.

15.44 Give the amino acid sequence of hexapeptides that produce the following fragments on partial acid hydrolysis:
(a) Arg, Gly, Ile, Leu, Pro, Val gives Pro-Leu-Gly, Arg-Pro, Gly-Ile-Val
(b) Asp, Leu, Met, Trp, Val_2 gives Val-Leu, Val-Met-Trp, Trp-Asp-Val

15.45 What is the structure of a nonapeptide that gives the following fragments when cleaved by chymotrypsin and by trypsin?

Trypsin cleavage: Val-Val-Pro-Tyr-Leu-Arg, Ser-Ile-Arg
Chymotrypsin cleavage: Leu-Arg, Ser-Ile-Arg-Val-Val-Pro-Tyr

Integrated Problems

15.46 How can you account for the fact that proline is never encountered in a protein α-helix? The α-helical segments of myoglobin and other proteins stop when a proline residue is encountered in the chain.

15.47 Which of the following amino acids are more likely to be found on the outside of a globular protein, and which on the inside? Explain.
(a) Valine (b) Aspartic acid (c) Isoleucine (d) Lysine

15.48 Cysteine is the only amino acid that has L stereochemistry but an *R* configuration. Design another L amino acid that also has an *R* configuration.

15.49 Arginine, which contains a *guanidino* group in its side chain, is the most basic of the 20 common amino acids. How can you account for this basicity? (*Hint:* Use resonance structures to see how the protonated guanidino group is stabilized.)

$$\underbrace{\mathrm{H_2N-\overset{\overset{\displaystyle NH}{\|}}{C}-}}_{\text{Guanidino group}}\mathrm{NHCH_2CH_2CH_2\underset{\underset{\displaystyle NH_2}{|}}{C}HCOOH}$$

Arginine

15.50 Look up the structure of human insulin in Section 15.7, and indicate where in each chain the molecule is cleaved by trypsin and by chymotrypsin.

15.51 Propose two structures for a tripeptide that gives Leu, Ala, and Phe on hydrolysis but does not react with phenyl isothiocyanate.

15.52 Draw as many resonance forms as you can for the purple anion obtained by reaction of ninhydrin with an amino acid:

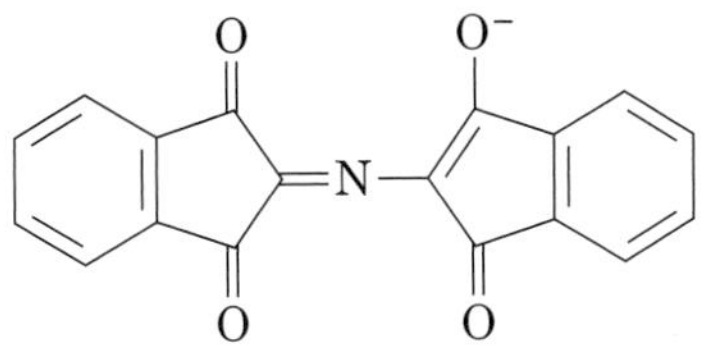

15.53 *Cytochrome c*, an enzyme found in the cells of all aerobic organisms, plays a role in respiration. Elemental analysis of cytochrome *c* reveals it to contain 0.43% iron. What is the minimum molecular weight of this enzyme?

15.54 A hexapeptide with the composition Arg, Gly, Leu, Pro_3 has proline at both C-terminal and N-terminal positions. What is the structure of the hexapeptide if partial hydrolysis gives: Gly-Pro-Arg, Arg-Pro, Pro-Leu-Gly?

15.55 *Aspartame*, a nonnutritive sweetener marketed under the trade name NutraSweet (among others), is the methyl ester of a simple dipeptide, Asp-Phe-OCH_3.

(a) Draw the full structure of aspartame.
(b) The isoelectric point of aspartame is 5.9. Draw the principal structure present in aqueous solution at this pH.
(c) Draw the principal form of aspartame present at physiological pH 7.6.
(d) Show the products of hydrolysis on treatment of aspartame with H_3O^+.

IN THE MEDICINE CABINET

15.56 The most common use for leuprolide acetate (Lupron Depot) in women is for the treatment of endometriosis. This peptide stimulates estrogen production.

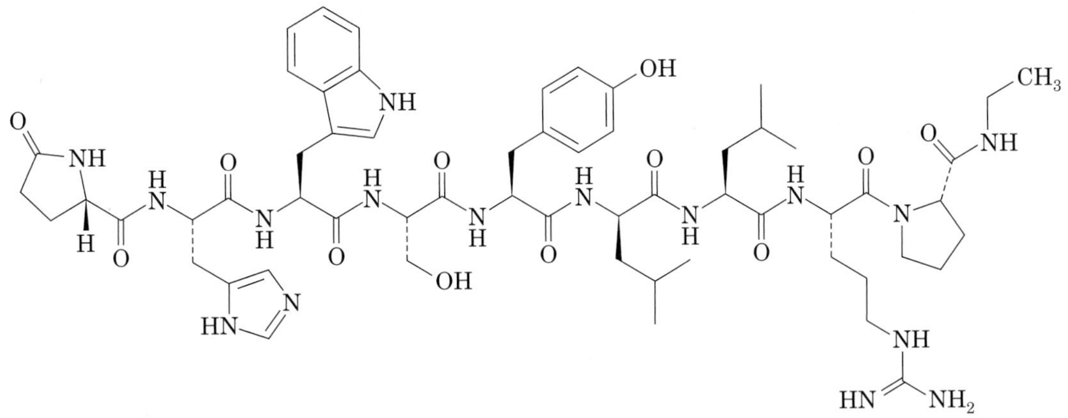

Leuprolide acetate

(a) The first amino acid in this peptide is derived from glutamic acid. How do you reconcile this fact based on the data provided in Table 15.1?
(b) Identify the remaining amino acids by their three-letter codes.
(c) How many of these amino acids are in the common L form?
(d) At neutral pH, what is the charge on this peptide?

15.57 *Oxytocin*, a nonapeptide hormone secreted by the pituitary gland, stimulates uterine contraction and lactation during childbirth. Its sequence was determined from the following evidence:

1. Oxytocin is a cyclic peptide containing a disulfide bridge between two cysteine residues.
2. When the disulfide bridge is reduced, oxytocin has the constitution Asn, Cys_2, Gln, Gly, Ile, Leu, Pro, Tyr.
3. Partial hydrolysis of reduced oxytocin yields seven fragments:

   Asp-Cys, Ile-Glu, Cys-Tyr, Leu-Gly, Tyr-Ile-Glu, Glu-Asp-Cys, Cys-Pro-Leu

4. Gly is the C-terminal group.
5. Both Glu and Asp are present as their side-chain amides (Gln and Asn) rather than as free side-chain acids.

What is the amino acid sequence of reduced oxytocin? What is the structure of oxytocin?

IN THE FIELD WITH AGROCHEMICALS

15.58 The herbicide atrazine binds to the photosynthetic apparatus and in the process is anchored by two critical protein residues: a phenylalanine and a serine.

(a) Describe three interactions that occur between these residues and the herbicide. (*Hint:* See Figure 15.6 for a list of the interaction types.)

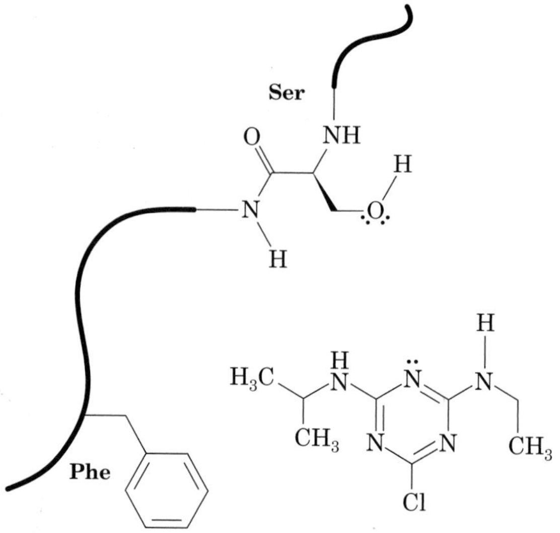

Ser
O NH
H
N O
H
H
H
H_3C N N N
CH_3 N N CH_3
Phe
Cl
Atrazine

(b) Over time, some weeds have become resistant to atrazine by substituting glycine for serine in the photosynthetic apparatus. Similar substitution of the phenylalanine residue has not been observed in resistant weeds. What does this tell you about the relative strengths of hydrogen bonds versus hydrophobic interactions?

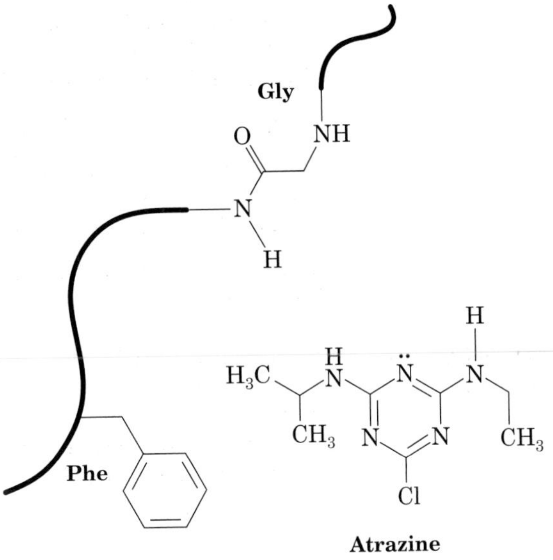

Gly
O NH
N
H
H
H
H_3C N N N
CH_3 N N CH_3
Phe
Cl

Atrazine

15.59 Chemists have countered atrazine-resistant weeds with a derivative of atrazine that has the following structure.

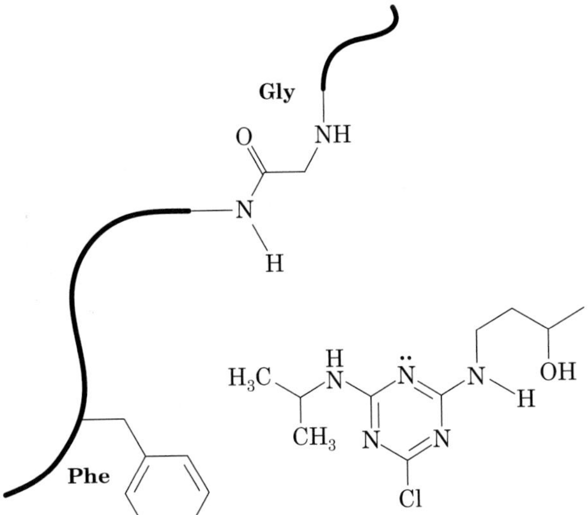

Gly
O NH
N
H
H
H_3C N N N OH
H
CH_3 N N
Phe
Cl

Rationalize the structural basis for the effectiveness of this molecule. (*Hint*: You may need to reorient the new side chain: C–C single bonds enjoy free rotation.)

Sea lions need a heavy layer of fat, a lipid, to survive in cold waters. (Nevada Wier/The Image Bank/Getty Images)

16 CHAPTER

Biomolecules: Lipids and Nucleic Acids

In the previous two chapters, we've discussed the organic chemistry of carbohydrates and proteins, two of the four major classes of biomolecules. Let's now look at the two remaining classes: *lipids* and *nucleic acids.* Although chemically quite different from one another, all four types of biomolecules are essential for life.

16.1 Lipids

Lipids are small naturally occurring molecules that have limited solubility in water and can be isolated from organisms by extraction with a nonpolar organic solvent. Fats, oils, waxes, many vitamins and hormones, and most nonprotein cell-membrane components are examples. Note that this definition differs from the sort used for carbohydrates and proteins in that lipids are defined by a physical property (solubility) rather than by structure.

Lipids are classified into two general types: those like fats and waxes, which contain ester linkages and can be hydrolyzed, and those like cholesterol and other steroids, which don't have ester linkages and can't be hydrolyzed.

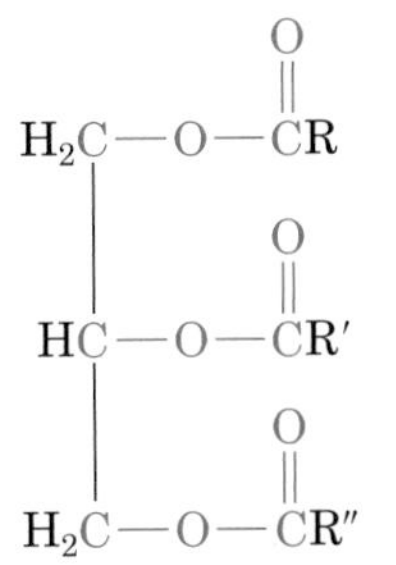

Animal fat, an ester
(R, R′, R″ = C_{11}–C_{19} chains)

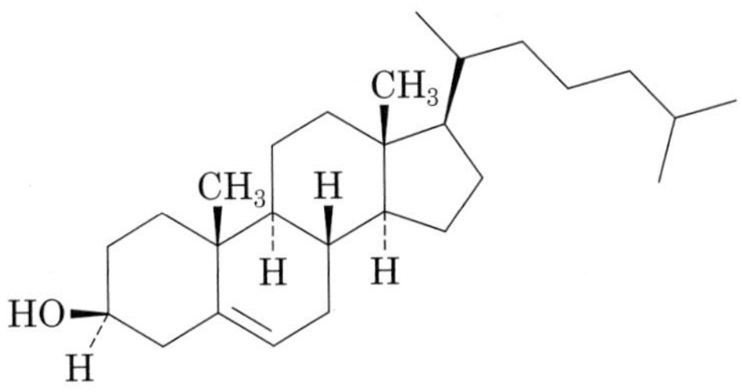

Cholesterol

PROBLEM 16.1 Beeswax contains a lipid with the structure $CH_3(CH_2)_{20}CO_2(CH_2)_{27}CH_3$. What products would you obtain by reaction of this lipid with aqueous NaOH followed by acidification?

16.2 Fats and Oils

Animal fats and vegetable oils are the most widely occurring lipids. Although they appear different—animal fats like butter and lard are solids, whereas vegetable oils like corn oil and peanut oil are liquids—their structures are closely related. Chemically, fats and oils are **triacylglycerols** (also called *triglycerides*), triesters of glycerol with three long-chain carboxylic acids. Hydrolysis of a fat or oil with aqueous NaOH yields glycerol and three long-chain **fatty acids**:

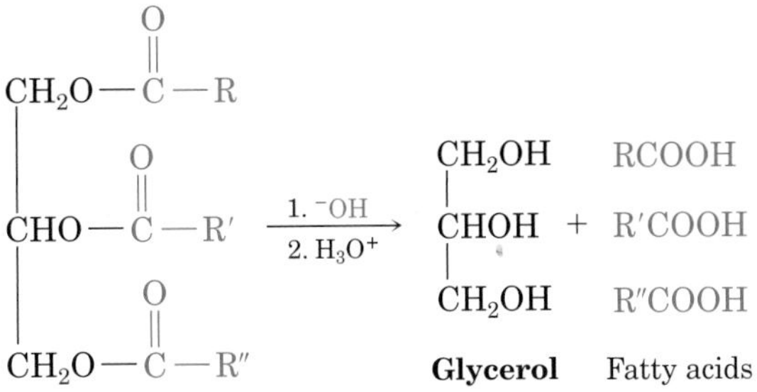

$$\text{A fat} \xrightarrow[\text{2. } H_3O^+]{\text{1. } ^-OH} \text{Glycerol } (CH_2OH{-}CHOH{-}CH_2OH) + \text{Fatty acids } (RCOOH,\ R'COOH,\ R''COOH)$$

A fat

Glycerol

Fatty acids

The fatty acids obtained by hydrolysis of triacylglycerols are generally unbranched and contain an even number of carbon atoms between 12 and 20. If one or more double bonds are present, they usually have *Z* (cis) geometry. The three fatty acids of a specific molecule need not be the same, and a fat or oil from a given source is likely to be a complex mixture of many different triacylglycerols. Table 16.1 lists some commonly occurring fatty acids, and Table 16.2 lists the approximate composition of fats and oils from different sources.

More than 100 different fatty acids are known, and about 40 occur widely. Palmitic acid (C_{16}) and stearic acid (C_{18}) are the most abundant saturated fatty acids; oleic and linoleic acids (both C_{18}) are the most abundant unsaturated ones. Oleic acid is monounsaturated because it has only one double bond, whereas linoleic, linolenic, and arachidonic acids are **polyunsaturated fatty acids** because they have more than one double bond. Linoleic and

TABLE 16.1 Structures of Some Common Fatty Acids

Name	No. of carbons	Melting point (°C)	Structure
Saturated			
Lauric	12	43.2	$CH_3(CH_2)_{10}CO_2H$
Myristic	14	53.9	$CH_3(CH_2)_{12}CO_2H$
Palmitic	16	63.1	$CH_3(CH_2)_{14}CO_2H$
Stearic	18	68.8	$CH_3(CH_2)_{16}CO_2H$
Arachidic	20	76.5	$CH_3(CH_2)_{18}CO_2H$
Unsaturated			
Palmitoleic	16	−0.1	(*Z*)-$CH_3(CH_2)_5CH{=}CH(CH_2)_7CO_2H$
Oleic	18	13.4	(*Z*)-$CH_3(CH_2)_7CH{=}CH(CH_2)_7CO_2H$
Linoleic	18	−12	(*Z,Z*)-$CH_3(CH_2)_4(CH{=}CHCH_2)_2(CH_2)_6CO_2H$
Linolenic	18	−11	(all *Z*)-$CH_3CH_2(CH{=}CHCH_2)_3(CH_2)_6CO_2H$
Arachidonic	20	−49.5	(all *Z*)-$CH_3(CH_2)_4(CH{=}CHCH_2)_4CH_2CH_2CO_2H$

TABLE 16.2 Approximate Fatty Acid Composition of Some Common Fats and Oils

	Saturated fatty acids (%)				Unsaturated fatty acids (%)	
Source	C_{12} lauric	C_{14} myristic	C_{16} palmitic	C_{18} stearic	C_{18} oleic	C_{18} linoleic
Animal fat						
Lard	—	1	25	15	50	6
Butter	2	10	25	10	25	5
Human fat	1	3	25	8	46	10
Whale blubber	—	8	12	3	35	10
Vegetable oil						
Coconut	50	18	8	2	6	1
Corn	—	1	10	4	35	45
Olive	—	1	5	5	80	7
Peanut	—	—	7	5	60	20
Linseed	—	—	5	3	20	20

linolenic acids are essential in the human diet; infants grow poorly and develop skin lesions if fed a diet of nonfat milk for prolonged periods.

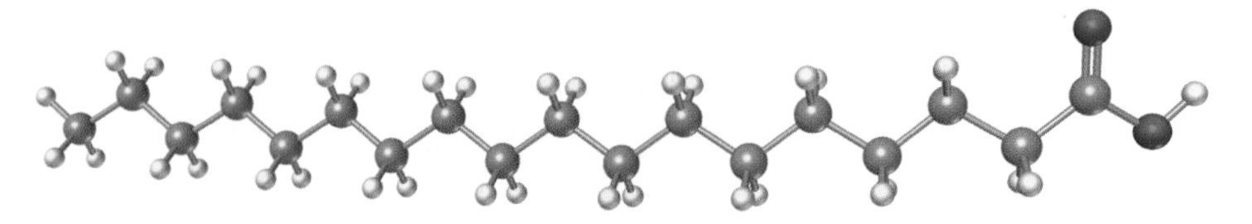

$$CH_3CH_2CH_2CH_2CH_2CH_2CH_2CH_2CH_2CH_2CH_2CH_2CH_2CH_2CH_2CH_2CH_2\overset{O}{\overset{\|}{C}}OH$$

Stearic acid

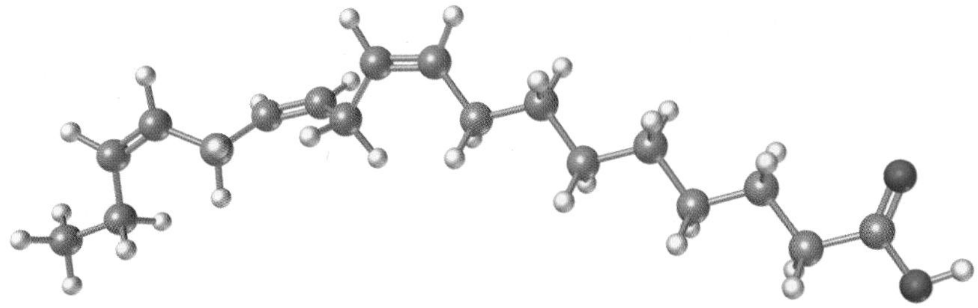

$$CH_3CH_2CH{=}CHCH_2CH{=}CHCH_2CH{=}CHCH_2CH_2CH_2CH_2CH_2CH_2CH_2\overset{\overset{\displaystyle O}{\|}}{C}OH$$

Linolenic acid, a polyunsaturated fatty acid (PUFA)

The data in Table 16.1 show that unsaturated fatty acids generally have lower melting points than their saturated counterparts, a trend that also holds true for triacylglycerols. Since vegetable oils generally have a higher proportion of unsaturated to saturated fatty acids than animal fats (Table 16.2), they have lower melting points. The difference is a consequence of structure. Saturated fats have a uniform shape that allows them to pack together easily in a crystal. Unsaturated vegetable oils, however, have C=C bonds, which introduce bends and kinks into the hydrocarbon chains and make crystal formation more difficult. The more double bonds there are, the harder it is for the molecule to crystallize and the lower the melting point.

The C=C bonds in vegetable oils can be reduced by catalytic hydrogenation (see Section 4.6) to produce saturated solid or semisolid fats. Margarine and solid cooking fats such as Crisco are produced by hydrogenating soybean, peanut, or cottonseed oil until the preferred consistency is obtained.

PRACTICE PROBLEM 16.1

Draw the structure of glyceryl tripalmitate, a typical fat molecule.

STRATEGY As the name implies, glyceryl tripalmitate is the triester of glycerol with three molecules of palmitic acid, $CH_3(CH_2)_{14}CO_2H$.

SOLUTION

$$\begin{array}{l} CH_2O\overset{\overset{\displaystyle O}{\|}}{C}CH_2CH_2CH_2CH_2CH_2CH_2CH_2CH_2CH_2CH_2CH_2CH_2CH_2CH_2CH_3 \\ | \\ CHO\overset{\overset{\displaystyle O}{\|}}{C}CH_2CH_2CH_2CH_2CH_2CH_2CH_2CH_2CH_2CH_2CH_2CH_2CH_2CH_2CH_3 \\ | \\ CH_2O\overset{\overset{\displaystyle O}{\|}}{C}CH_2CH_2CH_2CH_2CH_2CH_2CH_2CH_2CH_2CH_2CH_2CH_2CH_2CH_2CH_3 \end{array}$$

Glyceryl tripalmitate

PROBLEM 16.2 Draw structures of the following compounds. Which would you expect to have a higher melting point?
(a) Glyceryl trioleate (b) Glyceryl monooleate distearate

PROBLEM 16.3 Fats and oils can be either optically active or optically inactive, depending on their structures. Draw the structure of an optically active fat that gives 2 equivalents of palmitic acid and 1 equivalent of stearic acid on hydrolysis. Draw the structure of an optically inactive fat that gives the same products on hydrolysis.

16.3 Soaps

Soap has been known since at least 600 BC, when the Phoenicians prepared a curdy material by boiling goat fat with extracts of wood ash. The cleansing properties of soap weren't generally recognized, however, and the use of soap didn't become widespread until the 18th century. Chemically, soap is a mixture of the sodium or potassium salts of long-chain fatty acids produced by hydrolysis (*saponification*) of animal fat with alkali:

$$\begin{array}{l} CH_2OCOR \\ | \\ CHOCOR \\ | \\ CH_2OCOR \end{array} \xrightarrow[H_2O]{NaOH} 3\ RCO_2^-\ Na^+ + \begin{array}{l} CH_2OH \\ | \\ CHOH \\ | \\ CH_2OH \end{array}$$

A fat
($R = C_{11}$–C_{19} aliphatic chains)

Soap

Glycerol

Crude soap curds contain glycerol and excess alkali as well as soap but can be purified by boiling with water and adding NaCl to precipitate the pure sodium carboxylate salts. The smooth soap that results is dried, perfumed, and pressed into bars. Dyes are added for colored soaps, antiseptics are added for medicated soaps, pumice is added for scouring soaps, and air is blown in for soaps that float.

Soaps act as cleansers because the two ends of a soap molecule are so different. The carboxylate end of the long-chain molecule is ionic and therefore *hydrophilic* (water-loving). As a result, it tries to dissolve in water. The long aliphatic chain portion of the molecule, however, is nonpolar and *hydrophobic* (water-fearing). It tries to avoid water and to dissolve in grease. The net effect of these two opposing tendencies is that soaps are attracted to both grease and water and are therefore useful as cleansers.

When soap molecules are dispersed in water, the long hydrocarbon tails cluster together into a hydrophobic ball, while the ionic heads on the surface of the cluster stick out into the water layer. These spherical clusters, called **micelles**, are shown schematically in Figure 16.1. Grease and oil droplets are solubilized in water when they become coated by the hydrophobic nonpolar tails of soap molecules in the center of micelles. Once solubilized, the grease and dirt can be rinsed away.

FIGURE 16.1 A soap micelle solubilizing a grease particle in water.

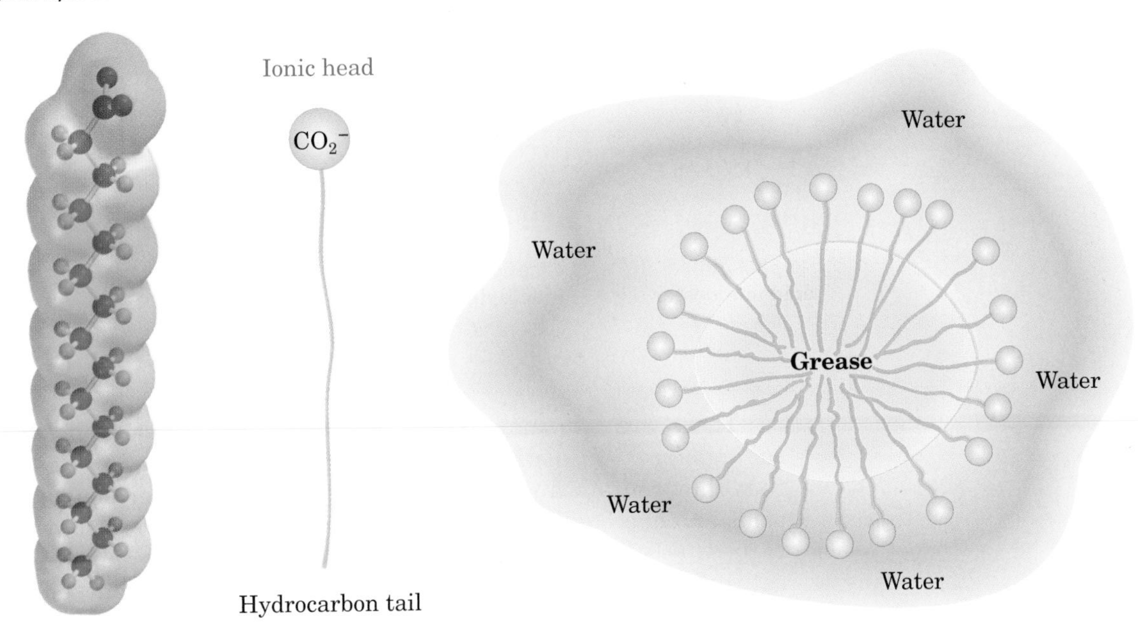

Soaps make life much more pleasant than it would otherwise be, but they also have drawbacks. In hard water, which contains metal ions, soluble sodium carboxylates are converted into insoluble calcium and magnesium salts, leaving the familiar ring of scum around bathtubs and the gray tinge on clothes. Chemists have circumvented these problems by synthesizing a class of synthetic detergents based on salts of long-chain alkylbenzenesulfonic acids. The principle of synthetic detergents is identical to that of soaps: the alkylbenzene end of the molecule is attracted to grease, and the ionic sulfonate end is attracted to water. Unlike soaps, though, sulfonate detergents don't form insoluble metal salts in hard water and don't leave an unpleasant scum.

$$R-C_6H_4-S(=O)_2-O^-\ Na^+$$

A synthetic detergent
(R = a mixture of C_{12} aliphatic chains)

PROBLEM 16.4 Draw the structure of magnesium oleate, one of the components of bathtub scum.

PROBLEM 16.5 Write the saponification reaction of glyceryl monopalmitate dioleate with aqueous NaOH.

16.4 Phospholipids

Just as waxes, fats, and oils are esters of carboxylic acids, **phospholipids** are esters of phosphoric acid, H_3PO_4.

A phosphoric acid monoester **A phosphoric acid diester** **A phosphoric acid triester** **A carboxylic acid ester**

There are two main kinds of phospholipids: *phosphoglycerides* and *sphingolipids.* **Phosphoglycerides** are closely related to fats and oils in that they contain a glycerol backbone linked by ester bonds to two fatty acids and one phosphoric acid. Although the fatty acid residues can be any of the C_{12}–C_{20} units normally present in fats, the acyl group at C1 is usually saturated, and that at C2 is usually unsaturated. The phosphate group at C3 is also bonded by a separate ester link to an amino alcohol such as choline, $[(CH_3)_3NCH_2CH_2OH]^+$, or ethanolamine, $HOCH_2CH_2NH_2$. The most important phosphoglycerides are the *lecithins* and the *cephalins.* Note that these compounds are chiral and that they have the L, or *R*, configuration at C2.

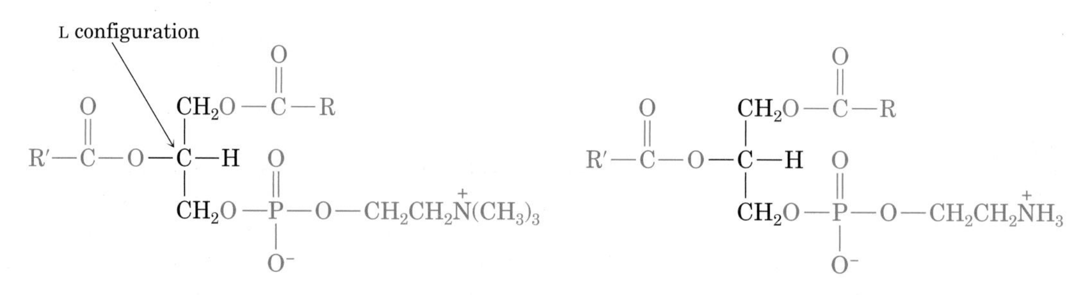

Phosphatidylcholine, a lecithin **Phosphatidylethanolamine, a cephalin**

where R is saturated and R′ is unsaturated

Found widely in plant and animal tissues, phosphoglycerides are the major lipid component of cell membranes (approximately 40%). Like soaps, phosphoglycerides have a long, nonpolar hydrocarbon tail bonded to a polar ionic head (the phosphate group). Cell membranes are composed mostly of phosphoglycerides organized into a **lipid bilayer** about 5.0 nm, or 50 Å, thick. As shown in Figure 16.2, the hydrophobic tails aggregate in the center of the bilayer in much the same way that soap tails aggregate in the center of a micelle (Figure 16.1). The bilayer thus forms an effective barrier to the passage of ions and other polar components into and out of the cell.

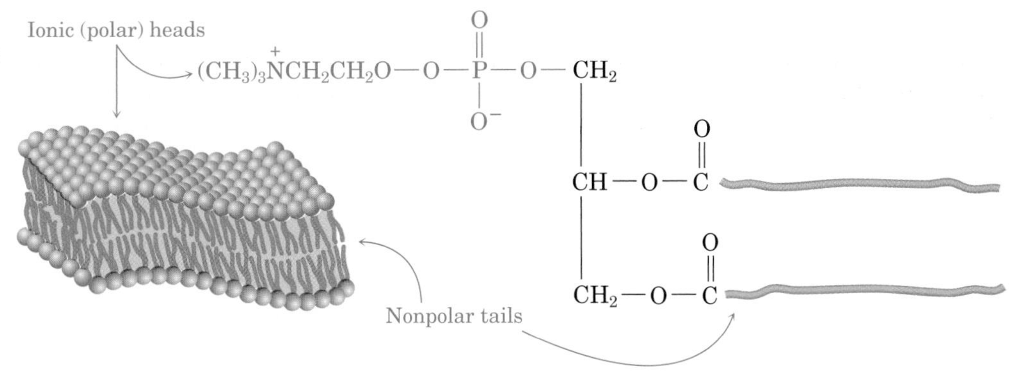

FIGURE 16.2 Aggregation of phosphoglycerides into the lipid bilayer that composes cell membranes.

The second major group of phospholipids is composed of **sphingolipids**. These substances, which have *sphingosine* or a related dihydroxyamine as their backbones, are constituents of plant and animal cell membranes. They are particularly abundant in brain and nerve tissue, where compounds called *sphingomyelins* are a major constituent of the coating around nerve fibers.

CH_2OH
|
$CHNH_2$
|
$CHOH$
|
$CH{=}CH(CH_2)_{12}CH_3$

Sphingosine

$CH_2O—P(=O)(O^-)—OCH_2CH_2\overset{+}{N}(CH_3)_3$
|
$CHNHCO(CH_2)_{16–24}CH_3$
|
$CHOH$
|
$CH{=}CH(CH_2)_{12}CH_3$

Sphingomyelin, a sphingolipid

16.5 Steroids

In addition to fats and phospholipids, the lipid extracts of plants and animals also contain **steroids**, molecules whose structures are based on the tetracyclic ring system that follows. The four rings are designated A, B, C, and D, beginning at the lower left, and the carbon atoms are numbered beginning in the A ring. The three six-membered rings (A, B, and C) adopt minimum-energy chair conformations but are constrained by their rigid conformations from undergoing the usual cyclohexane ring-flip (see Section 2.11).

A steroid
(R = various side chains)

In humans, most steroids function as *hormones*, chemical messengers that are secreted by glands and carried through the bloodstream to target tissues. There are two main classes of steroid hormones: the *sex hormones*, which control maturation and reproduction, and the *adrenocortical hormones*, which regulate a variety of metabolic processes.

Sex Hormones

Testosterone and androsterone are the two most important male sex hormones, or **androgens**. Androgens are responsible for the development of male secondary sex characteristics during puberty and for promoting tissue and muscle growth. Both are synthesized in the testes from cholesterol. Androstenedione is another minor hormone that has received particular attention because of its use by prominent athletes.

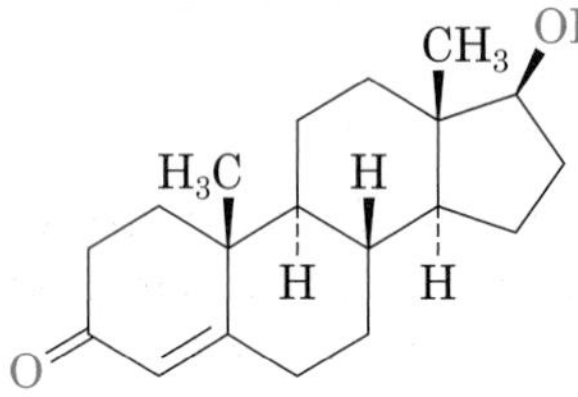

Testosterone 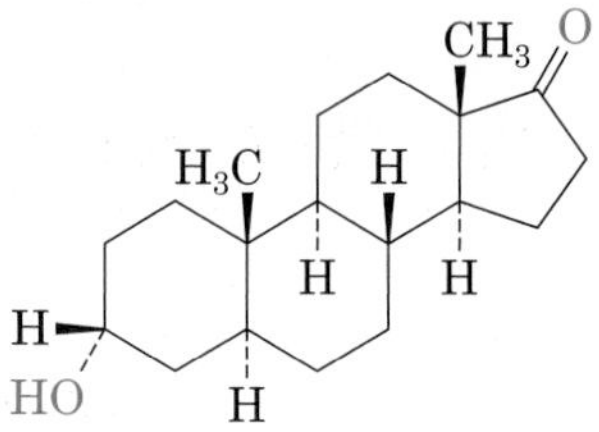 **Androsterone** 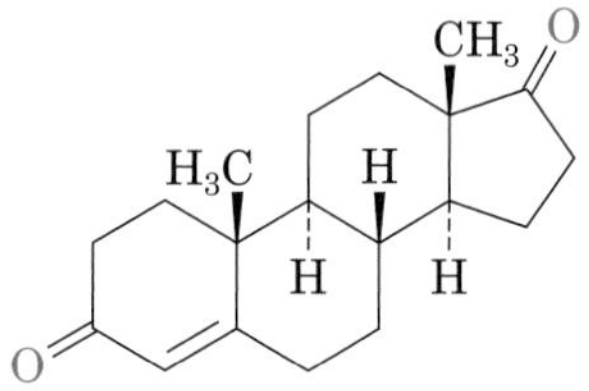 **Androstenedione**

(Androgens)

Estrone and estradiol are the two most important female sex hormones, or **estrogens**. Synthesized in the ovaries from testosterone, estrogenic hormones are responsible for the development of female secondary sex characteristics and for regulation of the menstrual cycle. Note that both have a benzene-like aromatic A ring. In addition, another kind of sex hormone, called a *progestin*, is essential for preparing the uterus for implantation of a fertilized ovum during pregnancy. *Progesterone* is the most important progestin.

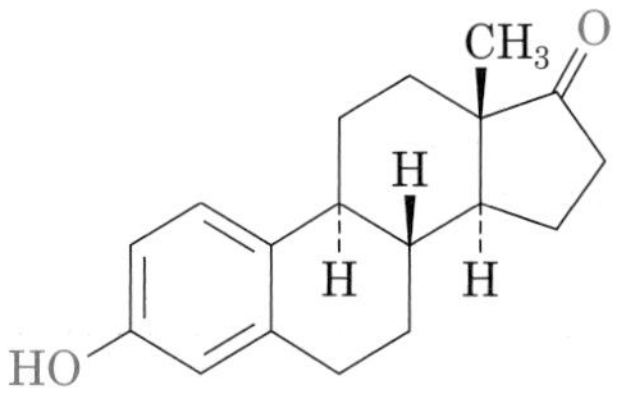

Estrone 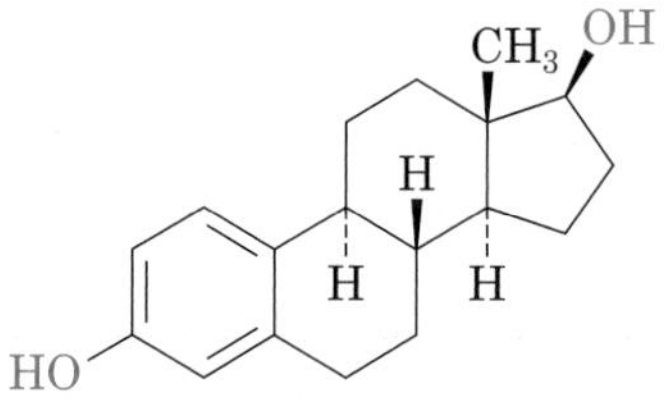 **Estradiol** **Progesterone (a progestin)**

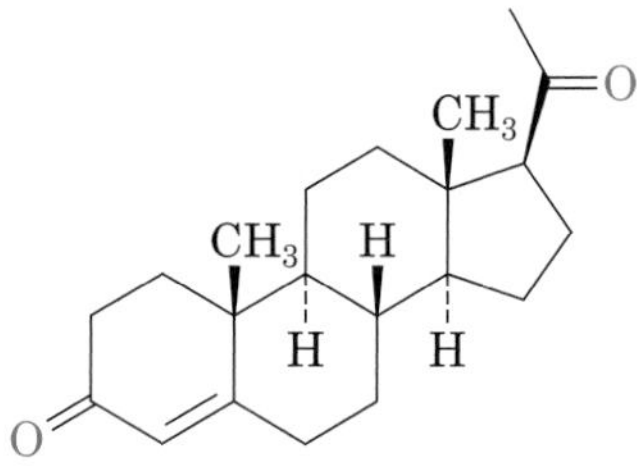

(Estrogens)

Adrenocortical Hormones

Adrenocortical steroids are secreted by the adrenal glands, small organs located near the upper end of each kidney. There are two types of adrenocortical steroids, called *mineralocorticoids* and *glucocorticoids*. Mineralocorticoids, such as aldosterone, control tissue swelling by regulating cellular salt balance between Na^+ and K^+. Glucocorticoids, such as hydrocortisone, are involved in the regulation of glucose metabolism and in the control of inflammation. Glucocorticoid ointments are widely used to bring down the swelling from exposure to poison oak or poison ivy.

Aldosterone
(a mineralocorticoid)

Hydrocortisone
(a glucocorticoid)

Synthetic Steroids

In addition to the many hundreds of steroids isolated from plants and animals, thousands more have been synthesized in pharmaceutical laboratories in a search for new drugs. The idea is to start with a natural hormone, carry out a chemical modification of the structure, and then see what biological properties the modified steroid has.

Among the best-known synthetic steroids are the oral contraceptives and the anabolic agents. Most birth-control pills are a mixture of two compounds, a synthetic estrogen, such as ethynylestradiol, and a synthetic progestin, such as norethindrone. Anabolic steroids, such as methandrostenolone (Dianabol), are synthetic androgens that mimic the tissue-building effects of natural testosterone.

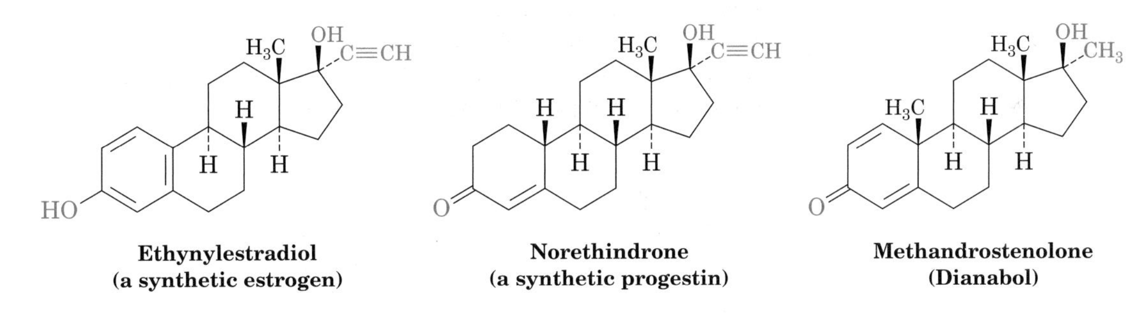

Ethynylestradiol
(a synthetic estrogen)

Norethindrone
(a synthetic progestin)

Methandrostenolone
(Dianabol)

PROBLEM 16.6 Look at the structure of progesterone, and identify all the functional groups in the molecule.

PROBLEM 16.7 Look at the structures of estradiol and ethynylestradiol, and point out the differences. What common structural feature do they share that makes both estrogens?

16.6 The Chemistry of Statin Drugs

Although cholesterol is an essential part of cell membranes, *high blood* cholesterol is an undesirable condition that is implicated in heart disease and artherosclerosis, two leading causes of death in humans. Cholesterol derives from two sources: our body naturally makes cholesterol, and we consume it in our diets. While we can control cholesterol intake with proper nutrition, we usually rely on medications, so-called **statin drugs**, to control naturally occurring high cholesterol. How do statin drugs work? The first step of cholesterol synthesis starts with the conversion of 3-hydroxy-3-methylglutaryl coenzyme A (HMG-CoA) to mevalonate, catalyzed by the enzyme HMG-CoA reductase (Figure 16.3). Statin drugs are structurally similar to HMG-CoA, and they inhibit the action of HMG-CoA reductase, thereby decreasing cholesterol synthesis. The best-selling statin drug generates $10 billion per year, more revenue than generated by any other drug. While most major pharmaceutical companies have one or more statin drugs on the market, the structures of these drugs vary slightly and they all have the same mechanism of action.

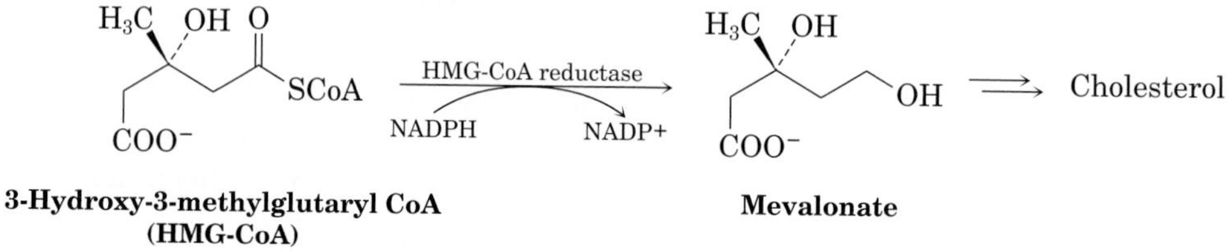

FIGURE 16.3 Cholesterol biosynthesis starts when HMG-CoA reductase converts HMG-CoA to mevalonate, which is ultimately converted to cholesterol.

HMG-CoA reductase; NADPH → NADP+; ⟹ Cholesterol

3-Hydroxy-3-methylglutaryl CoA (HMG-CoA)

Mevalonate

The story of statin drugs began with experiments designed to discover molecules that blocked HMG-CoA reductase. To find the new drug, chemists performed two experiments with samples of unknown chemicals produced by soil microbes. In the first experiment, the unknown chemical was added to growing cells. In the second experiment, the unknown chemicals and mevalonate were added to growing cells. Scientists then looked for differences in cholesterol production between the cells. If cholesterol was *absent* when only the unknown chemical was added but *present* when mevalonate was added, then chemists hypothesized that the unknown chemical interfered with mevalonate synthesis, thereby making it a drug candidate.

Using this strategy, the first naturally occurring statin drug, compactin, was discovered. Since then many other statins have been prepared in the laboratory. These statins fall into two main categories. Type 1 statins are structurally similar to compactin. The **lactone** rings (cyclic ester) mimic HMG-CoA. Type 2 statins have the HMG-CoA mimicked by a hydroxy acid rather than a lactone.

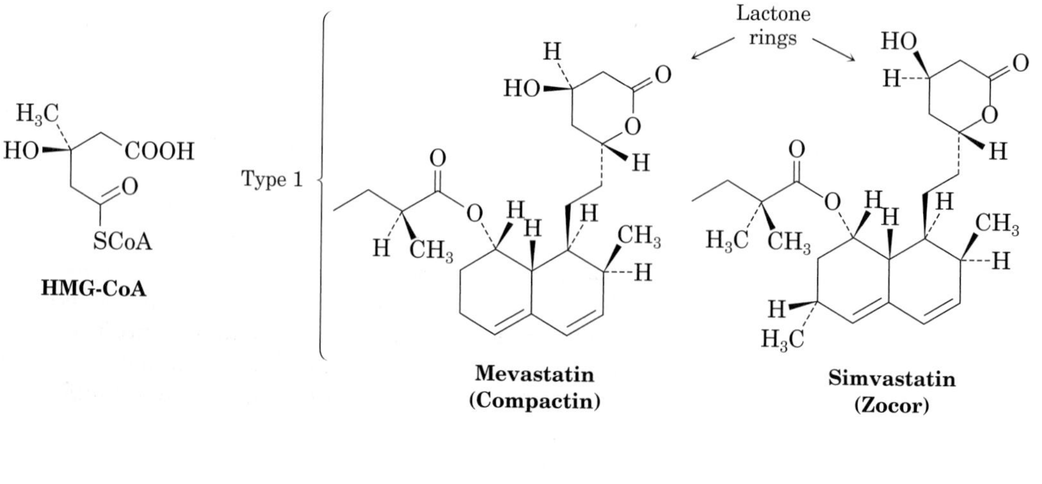

Lactone rings

Type 1

HMG-CoA

Mevastatin (Compactin)

Simvastatin (Zocor)

Type 2

Fluvastatin
(Lescol)

Atorvastatin
(Lipitor)

Rosuvastatin
(Crestor)

The top half of every statin has a region that is very similar to HMG-CoA, a lactone, or a hydroxy acid. The difference between statin drugs lies in the hydrophobic bottom half. Recent crystal structures show that all these statins bind to HMG-CoA reductase in the same way and HMG-CoA is prevented from reaching the active site of the enzyme. The bottom half of each drug, the hydrophobic portion, allows them to bind to the enzyme up to 10,000 times more tightly than HMG-CoA (Figure 16.4).

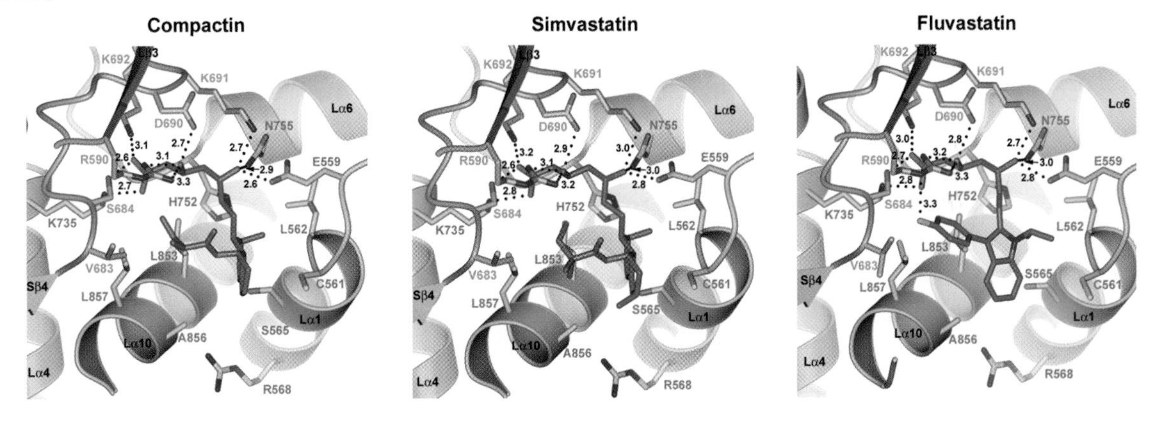

FIGURE 16.4 The statins (blue) bind in the active site of HMG-CoA reductase (green) beneath a flexible loop (gold), where HMG-CoA ordinarily binds. ("Structural Mechanism for Statin Inhibition of HMG-CoA Reductase," by E.S. Istvan and J. Deisenhofer. Copyright 2001 AAAS.)

16.7 Nucleic Acids and Nucleotides

The **nucleic acids**, **deoxyribonucleic acid (DNA)** and **ribonucleic acid (RNA)**, are the carriers and processors of a cell's genetic information. Coded in a cell's DNA is all the information that determines the nature of the cell, controls cell growth and division, and directs biosynthesis of the enzymes and other proteins required for all cellular functions.

Just as proteins are polymers of amino acid units, nucleic acids are polymers of individual building blocks called **nucleotides** linked together to form a long chain. Each nucleotide is composed of a **nucleoside** bonded to a phosphate group, and each nucleoside is composed of an aldopentose sugar joined through its anomeric carbon to the nitrogen atom of a heterocyclic amine base (see Section 12.6).

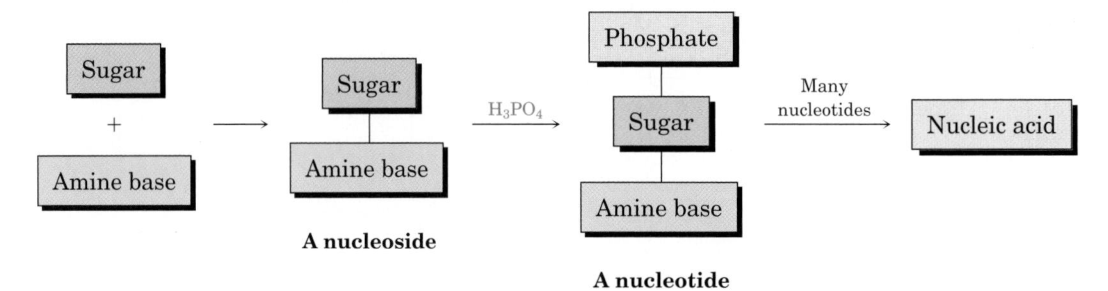

The sugar component in RNA is ribose, and the sugar in DNA is 2-deoxyribose. Recall that the prefix *2-deoxy* means that oxygen is missing from C2 of ribose.

Ribose **2-Deoxyribose**

DNA contains four different heterocyclic amine bases. Two are substituted *purines* (adenine and guanine), and two are substituted *pyrimidines* (cytosine and thymine). Adenine, guanine, and cytosine also occur in RNA, but thymine is replaced in RNA by a different pyrimidine base called uracil.

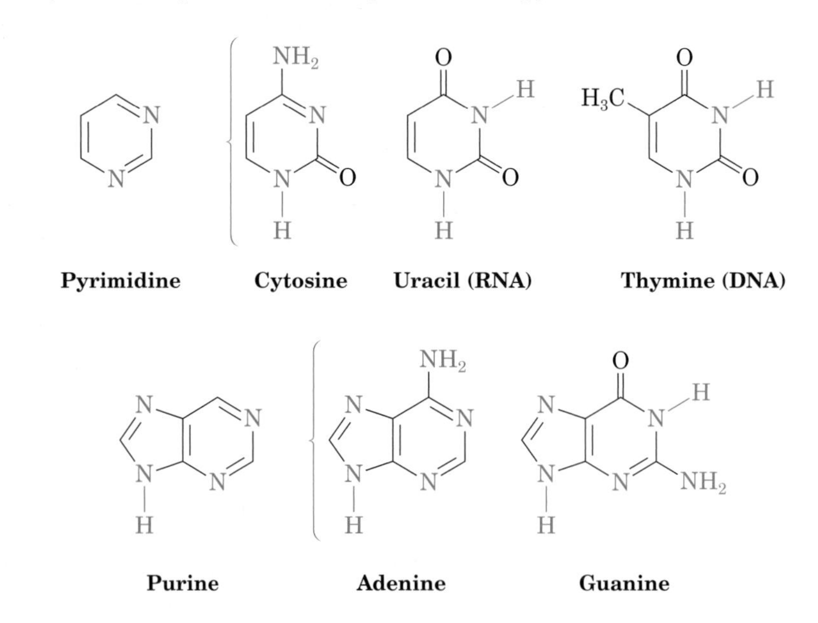

In both DNA and RNA, the heterocyclic amine base is bonded to C1′ of the sugar, and the phosphoric acid is bonded by a phosphate ester linkage to the C5′ sugar position. (In referring to nucleic acids, numbers with a prime superscript refer to positions on the sugar, and numbers without a prime refer to positions on the heterocyclic amine base.) The complete structures of all four deoxyribonucleotides and all four ribonucleotides are shown in Figure 16.5.

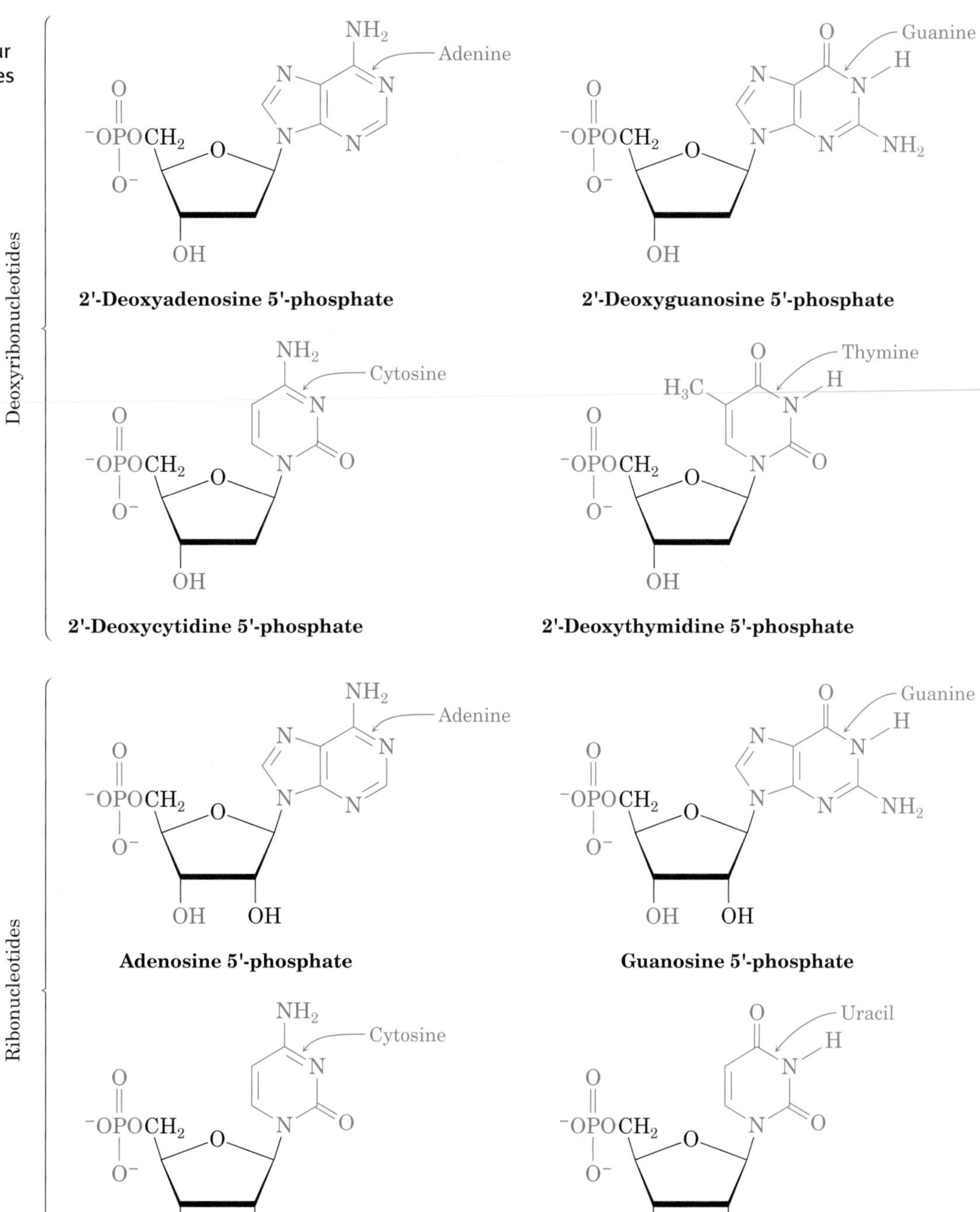

FIGURE 16.5 Structures of the four deoxyribonucleotides and the four ribonucleotides.

Although chemically similar, DNA and RNA differ in size and have different roles within the cell. Molecules of DNA are enormous. They have molecular weights of up to 150 *billion* and lengths of up to 12 cm when stretched out, and they are found mostly in the nucleus of cells. Molecules of RNA, by contrast, are much smaller (as low as 35,000 in molecular weight) and are found mostly outside the cell nucleus. We'll consider the two kinds of nucleic acids separately, beginning with DNA.

16.8 Structure of DNA

Nucleotides join together in DNA by forming a phosphate ester bond between the 5′-phosphate group on one nucleotide and the 3′-hydroxyl group on the sugar of another nucleotide (Figure 16.6). One end of the nucleic acid polymer thus has a free hydroxyl at C3′ (the *3′ end*), and the other end has a phosphate at C5′ (the *5′ end*).

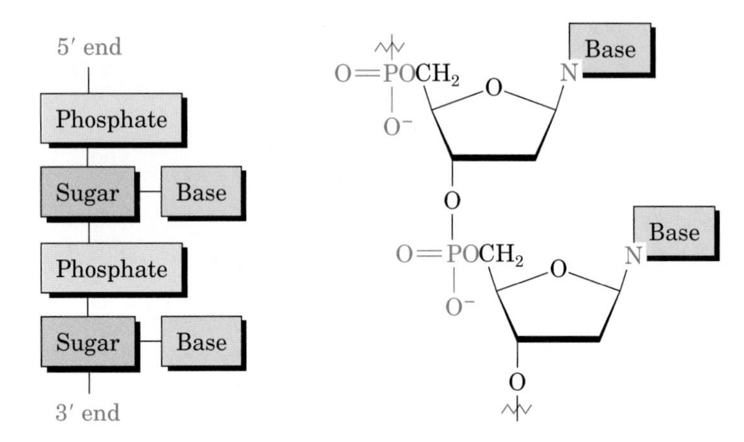

FIGURE 16.6 Generalized structure of DNA.

Just as the structure of a protein depends on the sequence in which individual amino acids are connected, the structure of a nucleic acid depends on the sequence of individual nucleotides. To carry the analogy further, just as a protein has a polyamide backbone with different side chains attached to it, a nucleic acid has an alternating sugar–phosphate backbone with different amine bases attached.

A protein

Different side chains

N terminus — C terminus

R1, R2, R3, R4, R5

NH—CH(R1)—C(=O)—NH—CH(R2)—C(=O)—NH—CH(R3)—C(=O)—NH—CH(R4)—C(=O)—NH—CH(R5)—C(=O)

Amide bonds

A nucleic acid

Different bases

5′ end — 3′ end

Base 1, Base 2, Base 3

Phosphate—Sugar(Base 1)—Phosphate—Sugar(Base 2)—Phosphate—Sugar(Base 3)

Phosphate ester bonds

The sequence of nucleotides in a chain is described by starting at the 5′ end and identifying the bases in order of occurrence, using the abbreviations A for adenosine, G for guanosine, C for cytidine, and T for thymine (or U for uracil in RNA). Thus, a typical sequence might be written as TAGGCT.

PRACTICE PROBLEM 16.2

Draw the full structure of the DNA dinucleotide CT.

SOLUTION

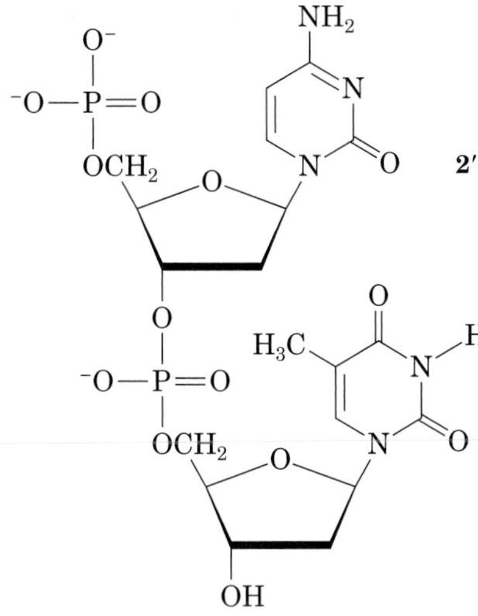

PROBLEM 16.8 Draw the full structure of the DNA dinucleotide AG.

PROBLEM 16.9 Draw the full structure of the RNA dinucleotide UA.

16.9 Base Pairing in DNA: The Watson–Crick Model

Samples of DNA isolated from different tissues of the same species have the same proportions of heterocyclic bases, but samples from different species can have greatly different proportions of bases. Human DNA, for example, contains about 30% each of A and T and about 20% each of G and C. The bacterium *Clostridium perfringens*, however, contains about 37% each of A and T and only 13% each of G and C. Note that in both examples, the bases occur in pairs; A and T are usually present in equal amounts, as are G and C. Why should this be?

In 1953, James Watson and Francis Crick made their now classic proposal for the secondary structure of DNA. According to the Watson–Crick model, DNA consists of two polynucleotide strands, running in opposite directions and coiled around each other in a **double helix** like the handrails on a spiral staircase. The strands run in opposite directions and are held together by hydrogen bonds between specific pairs of bases. Adenine (A) forms strong hydrogen bonds to thymine (T) but not to G or C. Similarly, G and C form strong hydrogen bonds to each other but not to A or T. The nature of this hydrogen bonding is particularly apparent in electrostatic potential maps, which show the alignment of electron-rich and electron-poor regions along the edges of the bases (Figure 16.7).

The two strands of the DNA double helix are not identical; rather, they're complementary because of hydrogen bonding. Whenever a G occurs in one strand, a C occurs opposite it in the other strand. When an A occurs in one strand, a T occurs in the other strand. This complementary pairing of bases explains why A and T are always found in equal amounts, as are G and C. A full turn of the DNA double helix is shown in Figure 16.8. The helix is 2.0 nm (20 Å) wide, there are ten base pairs per turn, and each turn is 3.4 nm (34 Å) in length.

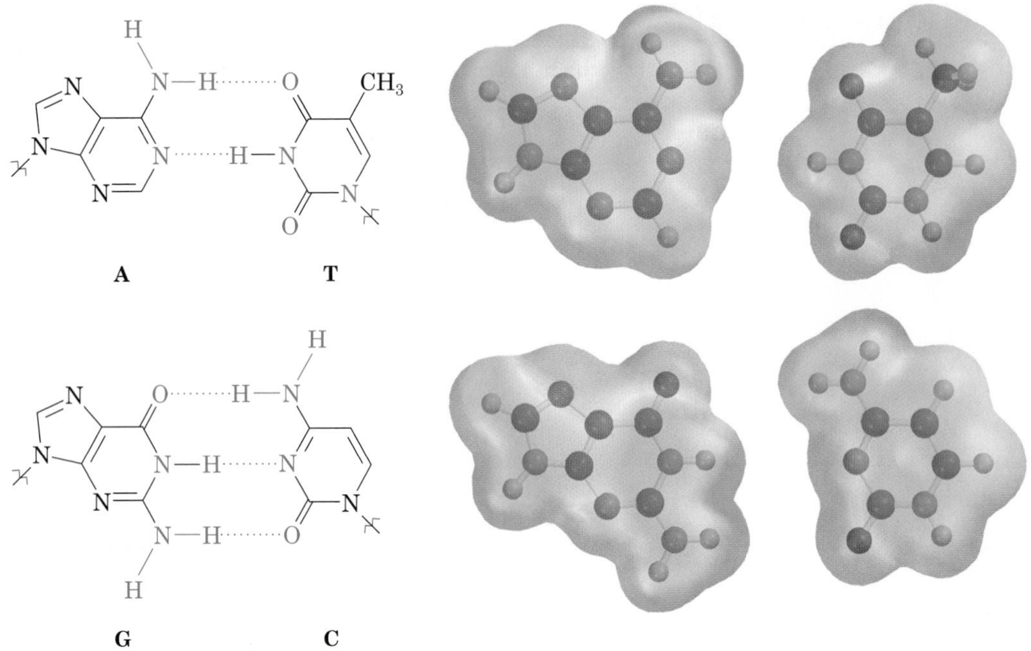

FIGURE 16.7 Hydrogen bonding between complementary base pairs in DNA. The faces of the bases are relatively neutral (green), while the edges have positive (blue) and negative (red) regions. Base A is aligned for hydrogen bonding with T, and G is aligned with C.

The two strands of the double helix coil in such a way that two kinds of "grooves" result, a *major groove* 1.2 nm (12 Å) wide and a *minor groove* 600 pm (6 Å) wide. The major groove is slightly deeper than the minor groove, and both are lined by potential hydrogen bond donors and acceptors. As a result, a variety of flat, polycyclic aromatic molecules are able to insert sideways, or *intercalate*, between the stacked bases. Many cancer-causing and cancer-preventing agents function by interacting with DNA in this way.

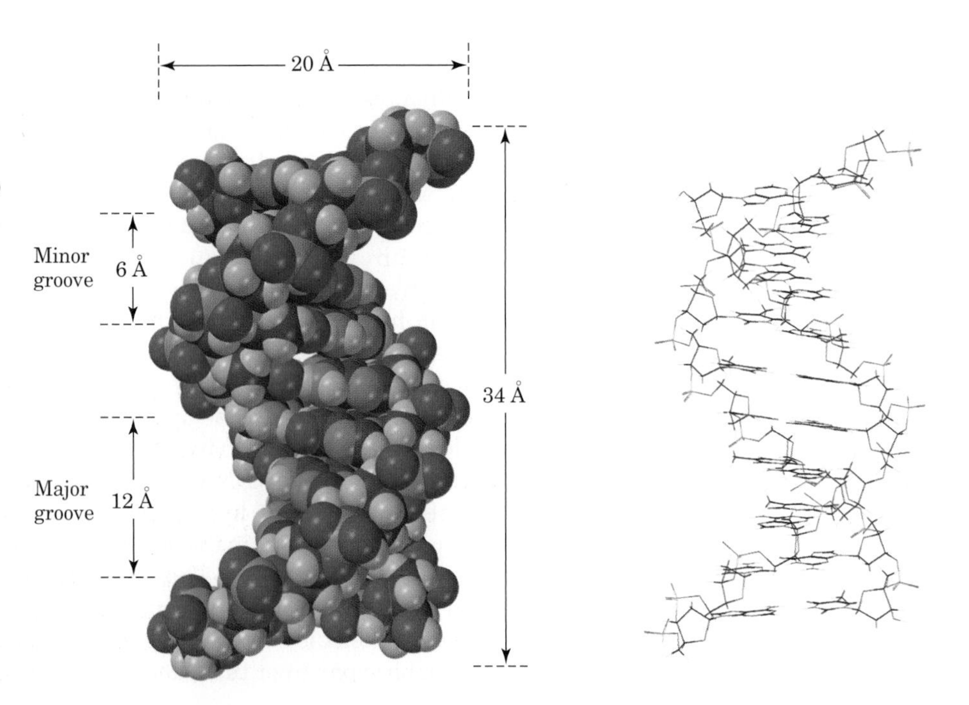

FIGURE 16.8 A turn of the DNA double helix in both space-filling and wire-frame formats. The sugar–phosphate backbone runs along the outside of the helix, and the amine bases hydrogen bond to one another on the inside.

PRACTICE PROBLEM 16.3

What sequence of bases on one strand of DNA is complementary to the sequence TATGCAT on another strand?

STRATEGY Remember that A and G form complementary pairs with T and C, respectively, and then go through the sequence replacing A by T, G by C, T by A, and C by G.

SOLUTION

Original: TATGCAT
Complement: ATACGTA

PROBLEM 16.10 What sequence of bases on one strand of DNA is complementary to the following sequence on another strand?

GGCTAATCCGT

16.10 Nucleic Acids and Heredity

The genetic information of an organism is stored as a sequence of deoxyribonucleotides strung together in the DNA chain. For the information to be preserved and passed on to future generations, a mechanism must exist for copying DNA. For the information to be used, a mechanism must exist for decoding the DNA message and implementing the instructions it contains.

What Crick called the "central dogma of molecular genetics" says that the function of DNA is to store information and pass it on to RNA. The function of RNA, in turn, is to read, decode, and use the information received from DNA to make proteins. By decoding the right bit of DNA at the right time, an organism uses genetic information to synthesize the thousands of proteins necessary for functioning.

Three fundamental processes take place in the transfer of genetic information:

- **Replication** is the process by which identical copies of DNA are made so that genetic information can be preserved and handed down to succeeding generations.
- **Transcription** is the process by which the genetic messages are read and carried out of the cell nucleus to ribosomes, where protein synthesis occurs.
- **Translation** is the process by which the genetic messages are decoded and used to synthesize proteins.

16.11 Replication of DNA

DNA **replication** is an enzyme-catalyzed process that begins by a partial unwinding of the double helix. As the strands separate and bases are exposed, new nucleotides line up on each strand in a complementary manner, A to T and C to G, and two new strands begin to grow. Each new strand is complementary to its old template strand, and two new DNA double helices are produced (Figure 16.9). Since each of the new DNA molecules contains one old strand and one new strand, the process is described as **semiconservative replication**.

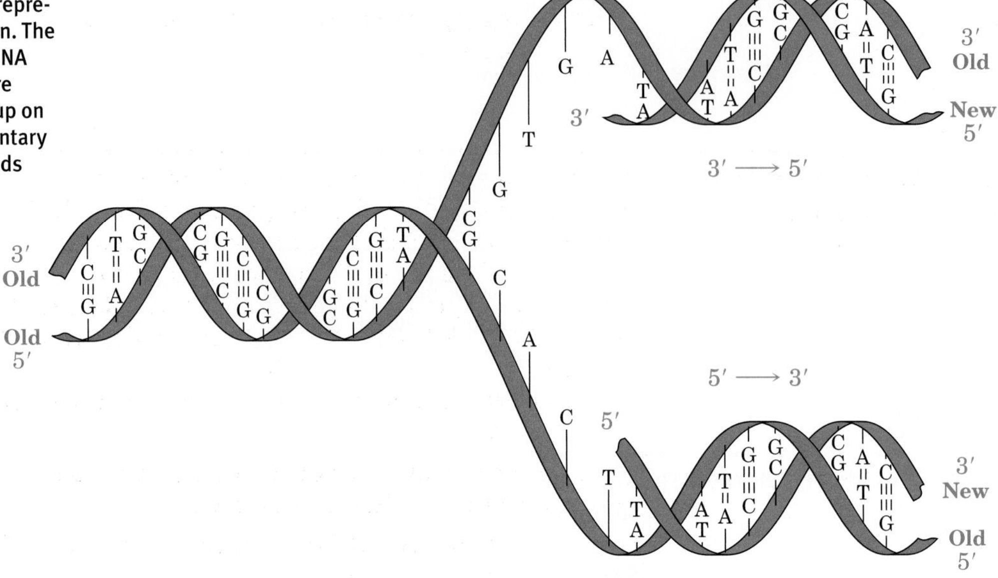

FIGURE 16.9 Schematic representation of DNA replication. The original double-stranded DNA partially unwinds, bases are exposed, nucleotides line up on each strand in a complementary manner, and two new strands begin to grow.

The process by which the individual nucleotides are joined to create new DNA strands involves many steps and many different enzymes. Addition of new nucleotide units to the growing chain takes place in the 5′ → 3′ direction and is catalyzed by the enzyme *DNA polymerase.* The key step is the addition of a 5′-mononucleoside triphosphate to the free 3′-hydroxyl group of the growing chain as the 3′-hydroxyl attacks the triphosphate and expels a diphosphate leaving group.

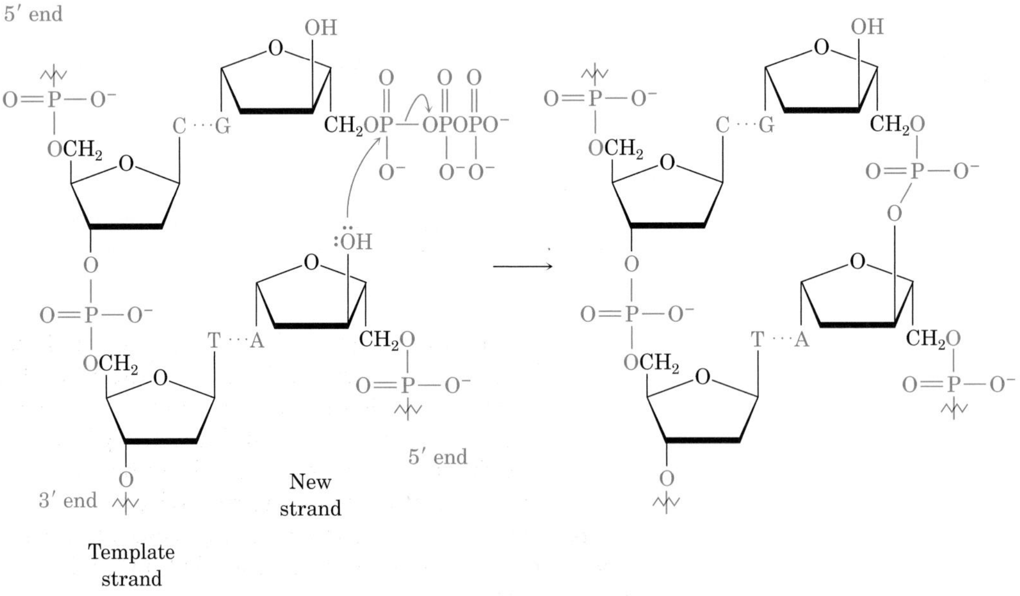

5′ end

3′ end

Template strand

New strand

5′ end

The magnitude of the replication process is staggering. The nucleus of a human cell contains 46 chromosomes (23 pairs), each of which consists of one very large DNA molecule. Each chromosome, in turn, is made up of several thousand DNA segments called *genes*, and the sum of all genes in a human cell (the *genome*) is estimated to be approximately 3 billion base pairs. Despite the size of these massive molecules, the base sequence is faithfully copied during replication, with an error occurring only about once each 10 to 100 billion bases.

16.12 Structure and Synthesis of RNA: Transcription

RNA is structurally similar to DNA. Both are sugar–phosphate polymers, and both have heterocyclic bases attached. The only differences are that RNA contains ribose rather than 2-deoxyribose and uracil rather than thymine. Uracil in RNA forms strong hydrogen bonds to its complementary base, adenine, just as thymine does in DNA. In addition, RNA molecules are much smaller than DNA, and RNA remains single-stranded rather than double-stranded.

There are three major kinds of ribonucleic acid, each of which serves a specific function.

- **Messenger RNA (mRNA)** carries genetic messages from DNA to ribosomes, where protein synthesis occurs.
- **Ribosomal RNA (rRNA)** provides the physical makeup of ribosomes.
- **Transfer RNA (tRNA)** transports specific amino acids to the ribosomes, where they are joined together to make proteins.

The conversion of the information in DNA into proteins begins in the nucleus of cells with the synthesis of mRNA by the process of **transcription**. Several turns of the DNA double helix unwind, forming a "bubble" and exposing the bases of the two strands. Ribonucleotides line up in the proper order by hydrogen bonding to their complementary bases on DNA, bond formation occurs in the $5' \rightarrow 3'$ direction, and the growing RNA molecule unwinds from DNA (Figure 16.10).

```
                 T—C—A—G—C—T—G—G—C—T—G—A—A—C—G—C—G—T—T        DNA coding strand
~—C—A—T                                                G—A—C—~
  :  :  :                                                :  :  :
~—G—T—A                                                C—T—G—~
                 A—G—T—C—G—A—C—C—G—A—C—T—T—G—C—G—C—A—A        DNA template strand
                 :  :  :  :  :  :  :  :  :  :  :  :  :  :  :  :  :  :  :
               —U—C—A—G—C—U—G—G—C—U—G—A—A—C—G—C—G—U—U—        mRNA
```

FIGURE 16.10 Synthesis of RNA using a DNA segment as template.

Unlike what happens in DNA replication, where both strands are copied, only one of the two DNA strands is transcribed into mRNA. The strand that contains the gene is called the **coding strand**, or **sense strand**, and the strand that gets transcribed is called the **template strand**, or **antisense strand**. Since the template strand and the coding strand are complementary, and since the template strand and the RNA molecule are also complementary, *the RNA molecule produced during transcription is a copy of the coding strand.* The only difference is that the RNA molecule has a U everywhere the DNA coding strand has a T.

Transcription of DNA by the process just discussed raises many questions. How does the DNA know where to unwind? Where along the chain does one gene stop and the next one start? How do the ribonucleotides know the right place along the template strand to begin lining up and the right place to stop? The picture that has emerged is that a DNA chain contains specific base sequences called *promoter sites* that lie at positions 10 base pairs and 35 base pairs upstream from the coding region and signal the beginning of a gene. Similarly, there are other base sequences near the end of the gene that signal a stop.

PRACTICE PROBLEM 16.4

What RNA base sequence is complementary to the following DNA base sequence?

(5′) TAAGCCGTG (3′)

STRATEGY Go through the sequence replacing A by U, G by C, T by A, and C by G.

SOLUTION

Original DNA:	(5′) TAAGCCGTG (3′)
Complementary RNA:	(3′) AUUCGGCAC (5′)

PROBLEM 16.11 Show how uracil can form strong hydrogen bonds to adenine, just as thymine can.

PROBLEM 16.12 What RNA base sequence is complementary to the following DNA base sequence?

(5′) GATTACCGTA (3′)

PROBLEM 16.13 From what DNA base sequence was the following RNA sequence transcribed?

(5′) UUCGCAGAGU (3′)

16.13 RNA and Protein Biosynthesis: Translation

The primary cellular function of RNA is to direct biosynthesis of the thousands of diverse peptides and proteins required by an organism. The mechanics of protein biosynthesis are directed by mRNA and take place on *ribosomes*, small granular particles in the cytoplasm of a cell that consist of about 60% rRNA and 40% protein. On the ribosome, mRNA serves as a template to pass on the genetic information it has transcribed from DNA.

The specific ribonucleotide sequence in mRNA forms a message that determines the order in which different amino acid residues are to be joined. Each "word," or **codon**, along the mRNA chain consists of a sequence of three ribonucleotides that is specific for a given amino acid. For example, the series UUC on mRNA is a codon directing incorporation of the amino acid phenylalanine into the growing protein. Of the $4^3 = 64$ possible triplets of the four bases in RNA, 61 code for specific amino acids (most amino acids are specified by more than one codon) and 3 code for chain termination. Table 16.3 shows the meaning of each codon.

The message carried by mRNA is read by tRNA in a process called **translation**. There are 61 different tRNA's, one for each of the 61 codons in Table 16.3 that specifies an amino acid. A typical tRNA is roughly the shape of a cloverleaf, as shown in Figure 16.11. It consists of about 70 to 100 ribonucleotides and is bonded to a specific amino acid by an ester linkage through the 3′-hydroxyl on ribose at the end of the tRNA. Each tRNA also contains in its chain a segment called an **anticodon**, a sequence of three ribonucleotides that is complementary to the codon sequence. For example, the codon sequence UUC present on mRNA is read

TABLE 16.3 Codon Assignments of Base Triplets

First base (5′ end)	Second base	Third base (3′ end) U	C	A	G
U	U	Phe	Phe	Leu	Leu
	C	Ser	Ser	Ser	Ser
	A	Tyr	Tyr	Stop	Stop
	G	Cys	Cys	Stop	Trp
C	U	Leu	Leu	Leu	Leu
	C	Pro	Pro	Pro	Pro
	A	His	His	Gln	Gln
	G	Arg	Arg	Arg	Arg
A	U	Ile	Ile	Ile	Met
	C	Thr	Thr	Thr	Thr
	A	Asn	Asn	Lys	Lys
	G	Ser	Ser	Arg	Arg
G	U	Val	Val	Val	Val
	C	Ala	Ala	Ala	Ala
	A	Asp	Asp	Glu	Glu
	G	Gly	Gly	Gly	Gly

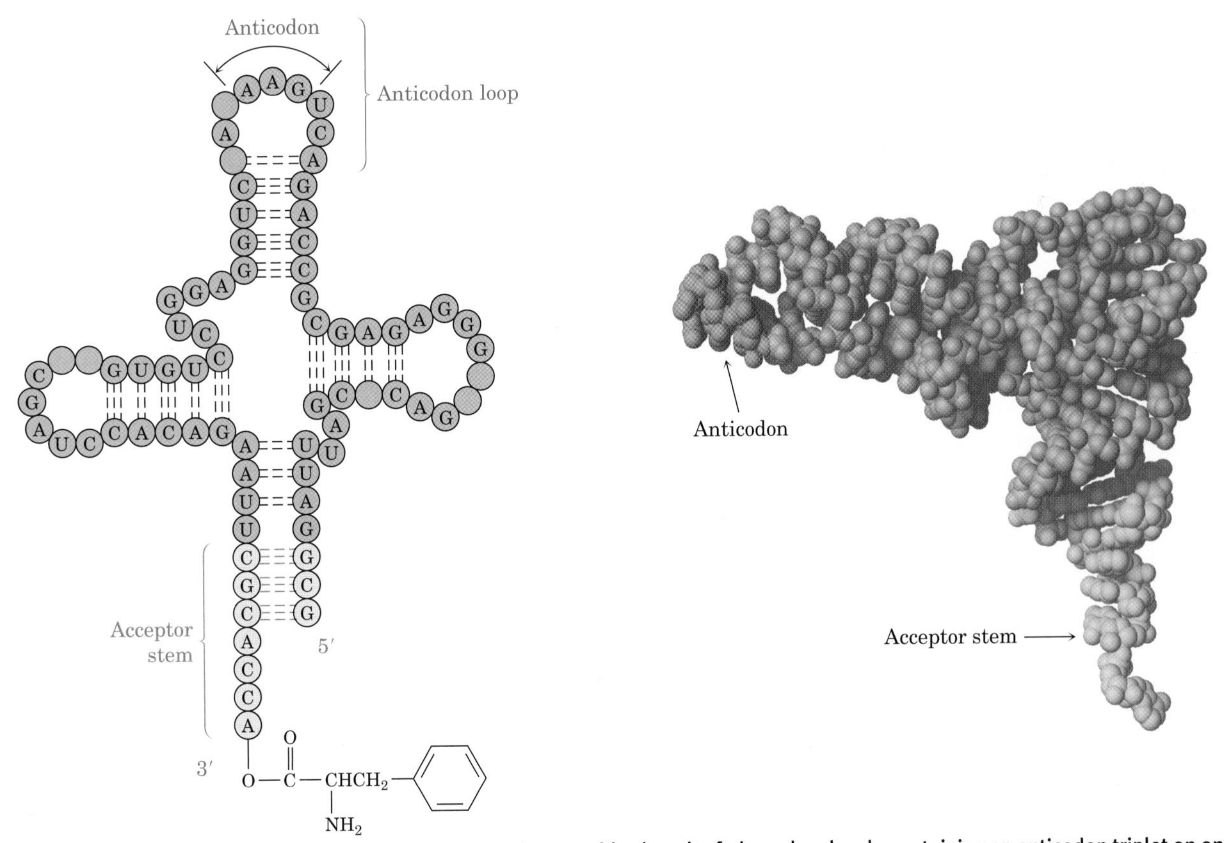

FIGURE 16.11 Structure of a tRNA molecule. The tRNA is a roughly cloverleaf-shaped molecule containing an anticodon triplet on one "leaf" and a covalently attached amino acid residue at its 3' end. The example shown is a yeast tRNA that codes for phenylalanine. The nucleotides not specifically identified are chemically modified analogs of the four usual nucleotides.

by a phenylalanine-bearing tRNA having the complementary anticodon sequence AAG. [Remember that nucleotide sequences are written in the $5' \rightarrow 3'$ direction, so the sequence in an anticodon must be reversed. That is, the complement to (5′)-UUC-(3′) is (3′)-AAG-(5′), which is written as (5′)-GAA-(3′).]

As each successive codon on mRNA is read, appropriate tRNA's bring the correct amino acids into position for enzyme-mediated transfer to the growing peptide. When synthesis of the proper protein is completed, a "stop" codon signals the end and the protein is released from the ribosome. The entire process of protein biosynthesis is illustrated schematically in Figure 16.12.

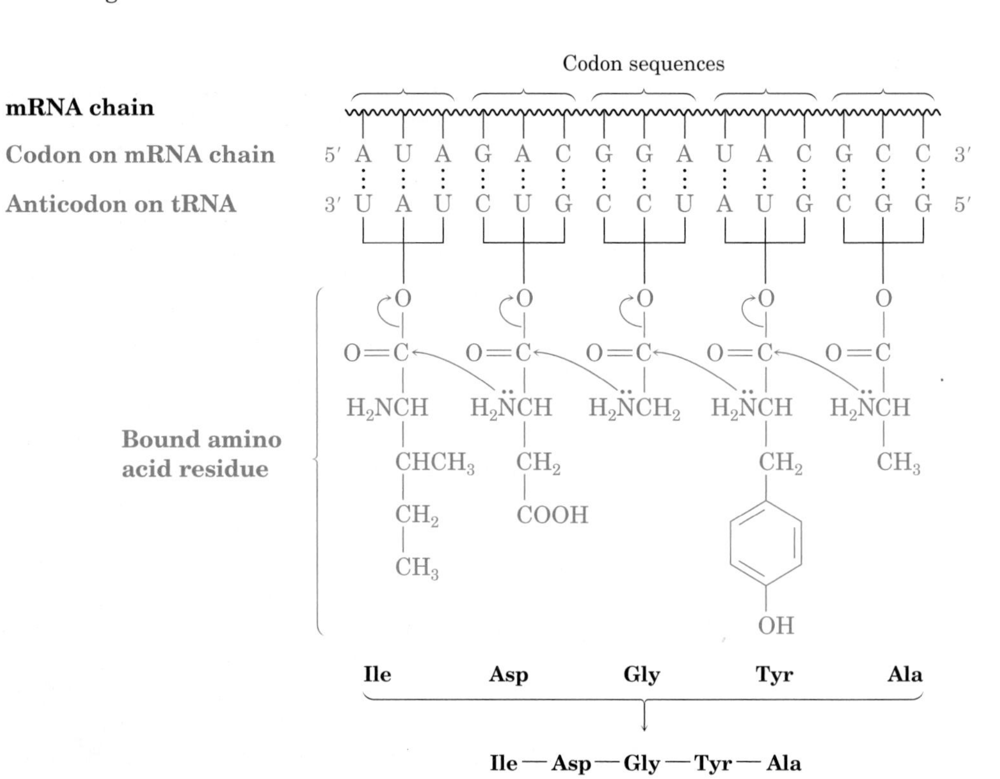

FIGURE 16.12 A schematic representation of protein biosynthesis. The mRNA containing codon base sequences is read by tRNA containing complementary anticodon base sequences. Transfer RNA's assemble the proper amino acids into position for incorporation into the peptide.

PRACTICE PROBLEM 16.5

Give a codon sequence for valine.

SOLUTION According to Table 16.3, there are four codons for valine: GUU, GUC, GUA, GUG.

PRACTICE PROBLEM 16.6

What amino acid sequence is coded by the mRNA base sequence AUC-GGU?

SOLUTION Table 16.3 indicates that AUC codes for isoleucine and GGU codes for glycine. Thus, AUC-GGU codes for Ile-Gly.

PROBLEM 16.14 List codon sequences for the following amino acids:
(a) Ala (b) Phe (c) Leu (d) Tyr

PROBLEM 16.15 What amino acid sequence is coded by the following mRNA base sequence?

CUU-AUG-GCU-UGG-CCC-UAA

PROBLEM 16.16 What anticodon sequences of tRNA's are coded by the mRNA in Problem 16.15?

PROBLEM 16.17 What was the base sequence in the original DNA strand on which the mRNA sequence in Problem 16.15 was made?

16.14 Sequencing DNA

One of the greatest scientific revolutions in history is now occurring in molecular biology as scientists are learning how to manipulate and harness the genetic machinery of organisms. None of the recent extraordinary advances would have been possible, however, were it not for the discovery in 1977 of methods for sequencing immense DNA chains.

The first step in DNA sequencing is to cleave the enormous chain at predictable points to produce smaller, more manageable pieces, a task accomplished by the use of enzymes called *restriction endonucleases.* Each different restriction enzyme, of which more than 200 are available, cleaves a DNA molecule at a well-defined point in the chain wherever a specific base sequence occurs. For example, the restriction enzyme *Alu*I cleaves between G and C in the four-base sequence AG-CT. Note that the sequence is a *palindrome*, meaning that it reads the same from left to right and right to left; that is, the *sequence* (5′)-AG-CT-(3′) is identical to its *complement*, (3′)-TC-GA-(5′). The same is true for other restriction endonucleases.

If the original DNA molecule is cut with another restriction enzyme having a different specificity for cleavage, still other segments are produced whose sequences partially overlap those produced by the first enzyme. Sequencing of all the segments, followed by identification of the overlapping regions, then allows complete DNA sequencing.

Two methods of DNA sequencing are in general use. Both operate along similar lines, but the *Maxam–Gilbert method* uses chemical techniques, while the **Sanger dideoxy method** uses enzymatic reactions. The Maxam–Gilbert method is used in specialized instances, but it is the Sanger method that has allowed the sequencing of the entire human genome of 3 billion base pairs. The dideoxy method used in commercial sequencing instruments begins with a mixture of the following:

- The restriction fragment to be sequenced
- A small piece of DNA called a *primer*, whose sequence is complementary to that on the 3′ end of the restriction fragment
- The four 2′-deoxyribonucleoside triphosphates (dNTPs)
- Very small amounts of the four 2′,3′-dideoxyribonucleoside triphosphates (ddNTPs), each of which is labeled with a fluorescent dye of a different color. (A 2′,3′-dideoxyribonucleoside triphosphate is one in which both 2′ and 3′ –OH groups are missing from ribose.)

A 2′-deoxyribonucleoside triphosphate

A labeled 2′,3′-dideoxyribonucleoside triphosphate

DNA polymerase enzyme is then added to this mix, and a strand of DNA complementary to the restriction fragment begins to grow from the end of the primer. Most of the time, only normal deoxyribonucleotides are incorporated into the growing chain, but every so often, a *dideoxy*ribonucleotide is incorporated. When that happens, DNA synthesis stops because the chain end no longer has a 3′-hydroxyl group for adding further nucleotides.

After reaction is complete, the product consists of a mixture of DNA fragments of all possible lengths, each terminated by one of the four dye-labeled dideoxyribonucleotides. When this product mixture is then submitted to electrophoresis (see Section 15.2), each fragment migrates at a rate that depends on the number of negatively charged phosphate groups (the number of nucleotides) it contains. Smaller pieces move rapidly, and larger pieces move more slowly. The technique is so sensitive that up to 1100 DNA fragments, differing in size by only one nucleotide, can be separated.

After separation by electrophoresis according to size, the identity of the terminal dideoxyribonucleotide in each piece—and thus the sequence of the restriction fragment—is identified simply by noting the color with which it fluoresces. Figure 16.13 shows a typical result.

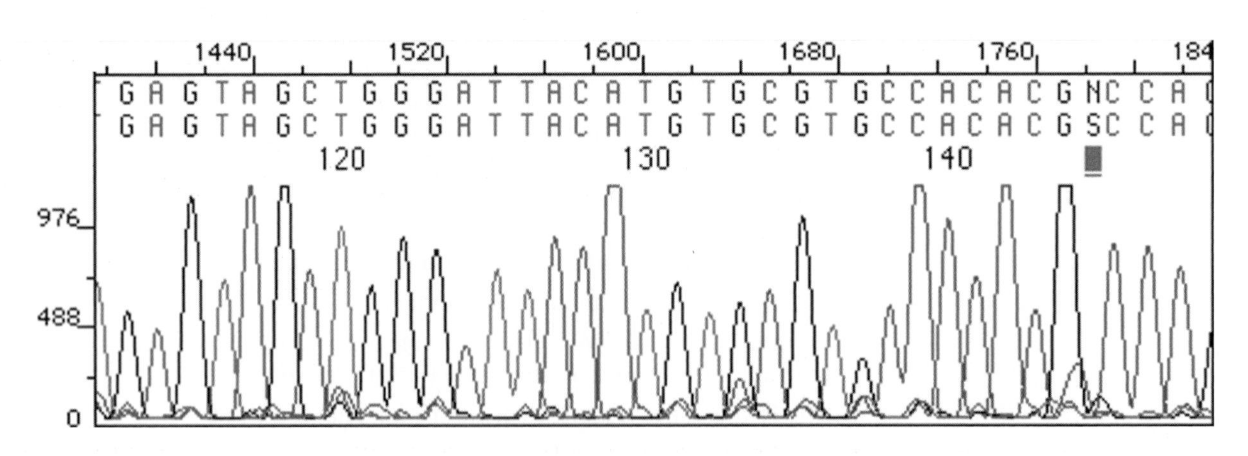

FIGURE 16.13 The sequence of a restriction fragment determined by the Sanger dideoxy method can be read simply by noting the colors of the dye attached to each of the various terminal nucleotides.

So efficient is the automated dideoxy method that sequences up to 1100 nucleotides in length can be rapidly sequenced with 98% accuracy. After a decade of work, preliminary sequence information for the entire human genome of 3 billion base pairs was announced early in 2001. Remarkably, our genome appears to contain only 30,000 to 40,000 genes, about one-third the generally predicted number and only twice the number found in the common roundworm.

16.15 The Polymerase Chain Reaction

Once a gene sequence is known, obtaining an amount of DNA large enough for study is often the next step. The method used is the **polymerase chain reaction (PCR)**, which has been described as being to genes what Gutenberg's invention of the printing press was to the written word. Just as the printing press produces multiple copies of a book, PCR produces multiple copies of a given DNA sequence. Starting from less than 1 *picogram* of DNA with a chain length of 10,000 nucleotides (1 pg = 10^{-12} g; about 100,000 molecules), PCR makes it possible to obtain several micrograms (1 μg = 10^{-6} g; about 10^{11} molecules) in just a few hours.

The key to the polymerase chain reaction is the discovery of *Taq* DNA polymerase, a heat-stable enzyme isolated from the thermophilic bacterium *Thermus aquaticus* found in a hot spring in Yellowstone National Park. *Taq* polymerase is able to take a single strand of DNA and, starting from a short "primer" piece that is complementary to one end of the chain, finish constructing the entire complementary strand. The overall process takes three steps, as shown schematically in Figure 16.14.

STEP 1 The double-stranded DNA to be amplified is heated in the presence of *Taq* polymerase, Mg^{2+} ion, the four deoxyribonucleotide triphosphate monomers (dNTPs), and a large excess of two short DNA primer pieces of about 20 bases each. Each primer is complementary to the sequence at the end of one of the target DNA segments. At a temperature of 95 °C, double-stranded DNA spontaneously breaks apart into two single strands.

STEP 2 The temperature is lowered to between 37 and 50 °C, allowing the primers, because of their relatively high concentration, to anneal to a complementary sequence at the end of each target strand.

STEP 3 The temperature is then raised to 72 °C, and *Taq* polymerase catalyzes the addition of further nucleotides to the two primed DNA strands. When replication of each strand is finished, *two* copies of the original DNA now exist. Repeating the denature–anneal–synthesize cycle a second time yields four DNA copies, repeating a third time yields eight copies, and so on, in an exponential series.

PCR has been automated, and 30 or so cycles can be carried out in an hour, resulting in a theoretical amplification factor of 2^{30} ($\sim 10^9$). In practice, however, the efficiency of each cycle is less than 100%, and an experimental amplification of about 10^6 to 10^8 is routinely achieved for 30 cycles.

16.16 RNA: A Paradigm Breaker

For a long time, it was generally believed that DNA functioned as a permanent repository of information, that RNA functioned primarily as a transient communicator of information, and that proteins did chemistry. This belief,

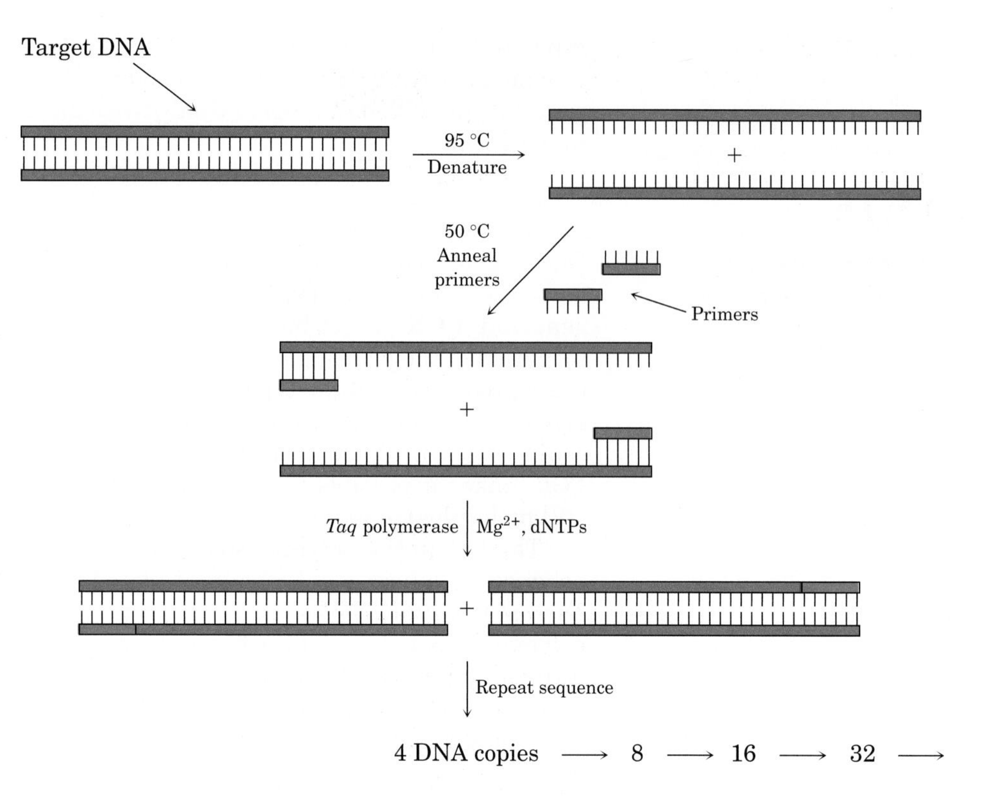

FIGURE 16.14 The polymerase chain reaction. Double-stranded DNA is heated to 95 °C in the presence of two short primer sequences, each of which is complementary to the end of one of the strands. After the DNA denatures, the temperature is lowered and the primer sequences anneal to the strand ends. Raising the temperature in the presence of *Taq* polymerase, Mg^{2+}, and a mixture of the four deoxynucleotide triphosphates (dNTPs) effects strand replication, producing two DNA copies. Each further repetition of the sequence again doubles the number of copies.

often referred to as the *central dogma,* has slowly been refined, and these paradigms have been broken. Viruses, such as the AIDS virus, that use RNA as a repository of genetic information offer one example. The discovery of viral proteins called reverse transcriptases that synthesize DNA from a viral RNA template overturned what was classically believed to be a one-way flow of information from DNA to RNA.

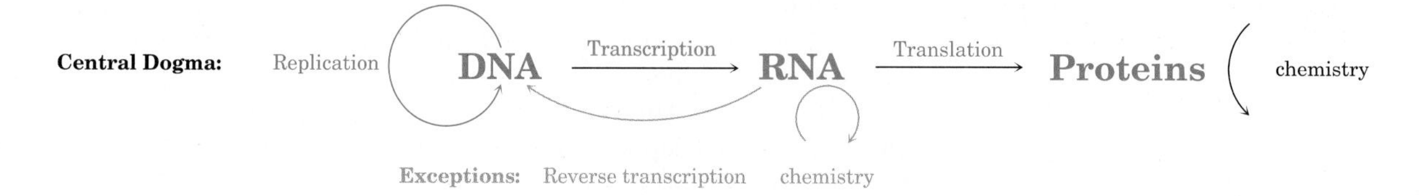

In the last few years, scientists have asked the question, "Can RNA do catalysis like proteins?" The answer is clearly yes: RNA can do chemistry. RNA enzymes, called **ribozymes**, have been found in nature and created in the laboratory. In Figure 16.15, one strand of a ribozyme (shown in red) catalyzes the cleavage of the phosphodiester bond of another RNA (the yellow strand) at the position indicated by the green residue. Groups on the ribozyme act as acids (HA^+) and bases (B:). The ribozyme is certainly a catalyst: it is neither created or destroyed during the reaction and has been shown to greatly accelerate the rate of this cleavage reaction.

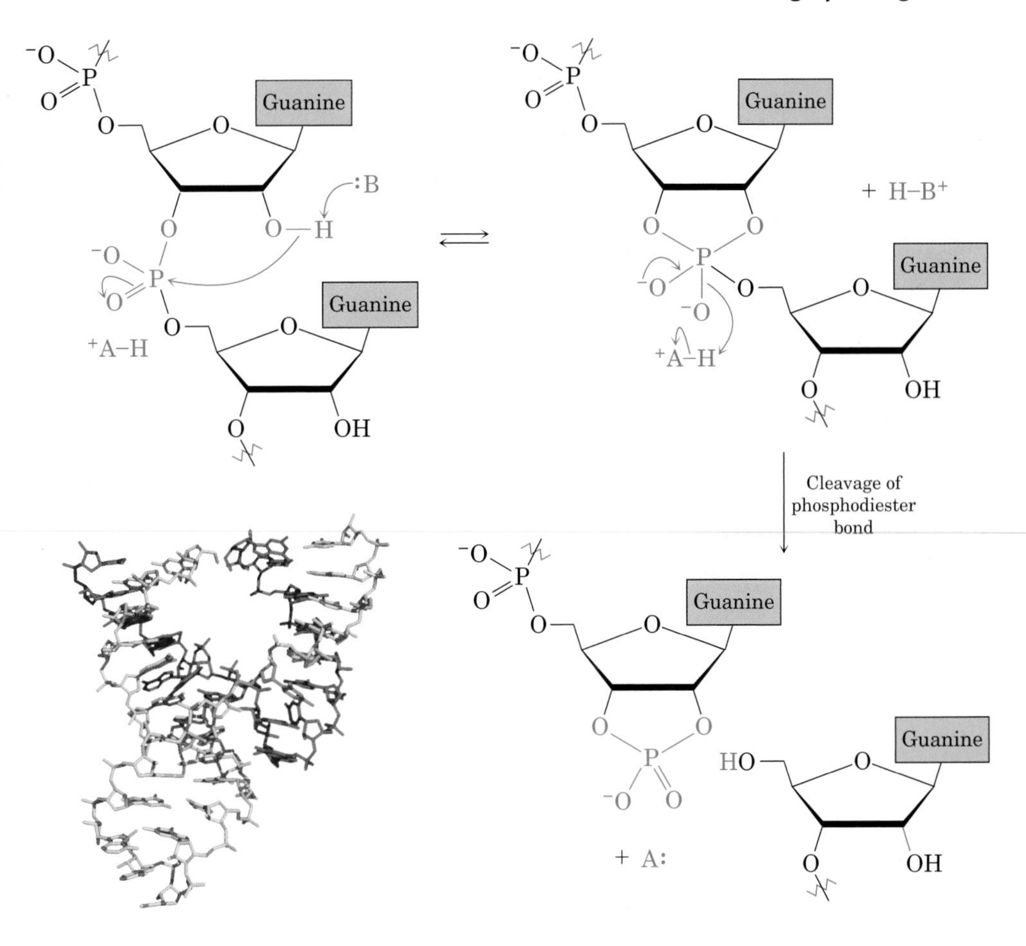

FIGURE 16.15 A ribozyme (red) is an RNA that catalyzes a chemical reaction. In this case, the ribozyme cleaves the phosphodiester bond of another RNA (yellow) at the position denoted in green. (© William G. Scott/Scott Research Group.)

INTERLUDE

DNA Fingerprinting

The invention of DNA sequencing has affected society in many ways, few more dramatic than those stemming from the development of *DNA fingerprinting.* DNA fingerprinting arose from the discovery in 1984 that human genes contain short, repeating sequences of noncoding DNA, called *short tandem repeat* (STR) loci. Furthermore, the STR loci are slightly different for every individual (except identical twins). By sequencing these loci, a pattern unique to each person can be obtained.

Perhaps the most common and well-publicized use of DNA fingerprinting is that carried out by crime laboratories to link suspects to biological evidence—blood, hair, skin, semen, or even items of clothing—found at a crime scene. Thousands of court cases have now been decided based on DNA evidence.

For use in criminal cases, forensic laboratories in the United States have agreed on 13 core STR loci that are most accurate for identification of an

Continued

© Bettmann/CORBIS

Historians have wondered for many years whether Thomas Jefferson fathered a child by Sally Hemings. DNA fingerprinting evidence obtained in 1998 suggests that he may have.

individual. Based on these 13 loci, a Combined DNA Index System (CODIS) has been established to serve as a registry of convicted offenders. When a DNA sample is obtained from a crime scene, the sample is subjected to cleavage with restriction endonucleases to cut out fragments containing the STR loci, the fragments are amplified using the polymerase chain reaction, and the sequences of the fragments are determined.

If the profile of sequences from a known individual and the profile from DNA obtained at a crime scene match, the probability is approximately 82 billion to 1 that the DNA is from the same individual. In paternity cases, where the DNA of father and offspring are related but not fully identical, the identity of the father can be established with a probability of 100,000 to 1. Even after several generations have passed, paternity can still be implied by DNA analysis of the Y chromosome of direct male-line descendants. The most well-known such case is that of Thomas Jefferson, who may have fathered a child by his slave Sally Hemings. Although Jefferson himself has no male-line descendants, DNA analysis of the male-line descendants of Jefferson's paternal uncle contained the same Y chromosome as a male-line descendant of Eston Hemings, the youngest son of Sally Hemings.

Among its many other applications, DNA fingerprinting is widely used for the diagnosis of genetic disorders, both prenatally and in newborns. Cystic fibrosis, hemophilia, Huntington's disease, Tay–Sachs disease, sickle cell anemia, and thalassemia are among the many diseases that can be detected, enabling early treatment of an affected child. Furthermore, by studying the DNA fingerprints of relatives with a history of some particular disorder, it's possible to identify DNA patterns associated with the disease and perhaps obtain clues for eventual cure. In addition, the U.S. Department of Defense now requires blood and saliva samples from all military personnel. The samples are stored, and DNA is extracted should the need for identification of a casualty arise.

Summary and Key Words

Androgen, p. 518
Anticodon, p. 530
Antisense strand, p. 529
Coding strand, p. 529
Codon, p. 530
Deoxyribonucleic acid (DNA), p. 521
Double helix, p. 525
Estrogen, p. 518
Fatty acid, p. 511
Lactone, p. 520
Lipid, p. 510
Lipid bilayer, p. 516
Messenger RNA (mRNA), p. 529
Micelle, p. 514
Nucleic acid, p. 521
Nucleoside, p. 521

Lipids are the naturally occurring substances isolated from plants and animals by extraction with organic solvents. Animal fats and vegetable oils are the most widely occurring lipids. Both fats and oils are **triacylglycerols**—triesters of glycerol with long-chain **fatty acids**. **Phosphoglycerides** such as lecithin and cephalin are closely related to fats. The glycerol backbone in these molecules is esterified to two fatty acids and one phosphate ester. **Sphingolipids** have an amino alcohol such as sphingosine for their backbone.

Steroids are plant and animal lipids with a characteristic four-ring carbon skeleton. Steroids occur widely in body tissue and have many different kinds of physiological activity. Among the more important kinds of steroids are the sex hormones (**androgens** and **estrogens**) and the adrenocortical hormones.

The **nucleic acids**, **DNA** (**deoxyribonucleic acid**) and **RNA** (**ribonucleic acid**), are biological polymers that act as chemical carriers of an organism's genetic information. Nucleic acids are made of **nucleotides**, which consist of a purine or pyrimidine heterocyclic amine base linked to C1′ of a pentose sugar (ribose in RNA and 2-deoxyribose in DNA), with the sugar in turn linked through its C5′ hydroxyl to a phosphate group.

Molecules of DNA consist of two complementary strands held together by hydrogen bonds between heterocyclic bases on the different strands and coiled into a **double helix**. Adenine (A) and thymine (T) form hydrogen bonds to each other, as do cytosine (C) and guanine (G).

Three main processes take place in deciphering the genetic information in DNA. **Replication** of DNA is the process by which identical copies are made. The DNA double helix unwinds, complementary deoxyribonucleotides line up, and two new DNA molecules are produced. **Transcription** is the process by which RNA is produced. This occurs when a segment of the DNA double helix unwinds and complementary ribonucleotides line up to produce **messenger RNA (mRNA)**. **Translation** is the process by which mRNA directs protein synthesis. Each mRNA has a three-base segment called a **codon** along its chain. Codons are recognized by small molecules of **transfer RNA (tRNA)**, which carry the appropriate amino acids needed for protein synthesis.

Sequencing of DNA fragments is done by the **Sanger dideoxy method**. Small amounts of DNA can be amplified using the **polymerase chain reaction (PCR)**.

A **ribozyme** is a ribonucleic acid that catalyzes a chemical reaction.

EXERCISES

Visualizing Chemistry

16.18 Identify the following bases, and tell whether each is found in DNA, RNA, or both:

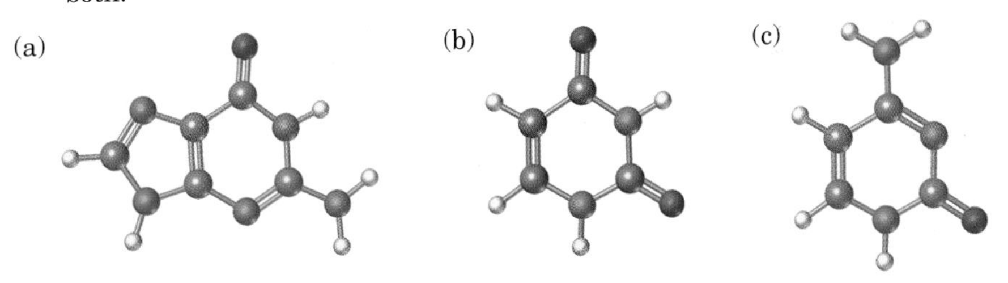

16.19 Identify the following nucleotide, and tell how it is used:

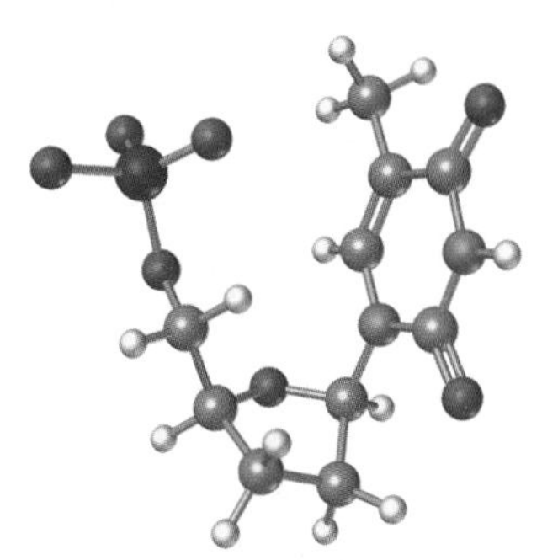

16.20 Cholesterol has the following structure. Tell whether the –OH group is axial or equatorial.

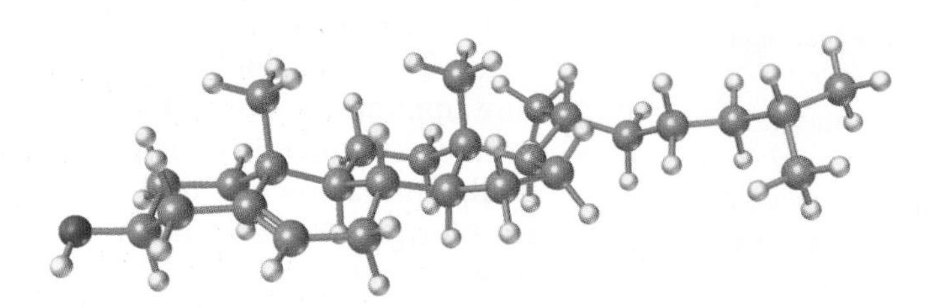

Additional Problems

NOMENCLATURE AND STRUCTURES OF LIPIDS

16.21 Write representative structures for the following:
(a) A fat (b) A vegetable oil (c) A steroid

16.22 Write the structures of the following molecules:
(a) Sodium stearate (b) Ethyl linoleate (c) Glyceryl palmitodioleate

REACTIONS OF LIPIDS

16.23 Show the products you would expect to obtain from the reaction of glyceryl trioleate with the following:
(a) Excess Br_2 in CCl_4 (b) H_2/Pd (c) NaOH, H_2O
(d) $KMnO_4$, H_3O^+ (e) $LiAlH_4$, then H_3O^+

16.24 How would you convert oleic acid into the following substances?
(a) Methyl oleate (b) Methyl stearate (c) Nonanedioic acid

16.25 Eleostearic acid, $C_{18}H_{30}O_2$, is a rare fatty acid found in tung oil. On oxidation with $KMnO_4$, eleostearic acid yields 1 part pentanoic acid, 2 parts oxalic acid ($HO_2C—CO_2H$), and 1 part nonanedioic acid. Propose a structure for eleostearic acid.

16.26 Stearolic acid, $C_{18}H_{32}O_2$, yields oleic acid on catalytic hydrogenation over the Lindlar catalyst. Propose a structure for stearolic acid.

16.27 Draw the products you would obtain from treatment of cholesterol with the following reagents:
(a) Br_2 (b) H_2, Pd catalyst (c) CH_3COCl, pyridine

16.28 If the average molecular weight of soybean oil is 1500, how many grams of NaOH are needed to saponify 5.00 g of the oil?

TRANSCRIPTION AND TRANSLATION

16.29 The DNA from sea urchins contains about 32% A and about 18% G. What percentages of T and C would you expect in sea urchin DNA? Explain.

16.30 What DNA sequence is complementary to the following sequence?

(5′) GAAGTTCATGC (3′)

16.31 Give codons for the following amino acids:
(a) Ile (b) Asp (c) Thr

16.32 Draw the complete structure of the ribonucleotide codon UAC. For what amino acid does this sequence code?

16.33 Draw the complete structure of the deoxyribonucleotide sequence from which the mRNA codon in Problem 16.32 was transcribed.

16.34 What amino acids do the following ribonucleotide codons code for?
(a) AAU (b) GAG (c) UCC (d) CAU (e) ACC

16.35 From what DNA sequences were each of the mRNA codons in Problem 16.34 transcribed?

16.36 What anticodon sequences of tRNA's are coded by each of the codons in Problem 16.34?

16.37 The codon UAA stops protein synthesis. Why does the sequence UAA in the following stretch of mRNA not cause any problems?

-GCA-UUC-GAG-GUA-ACG-CCC-

16.38 If the gene sequence TAACCGGAT on DNA were miscopied during replication and became TGACCGGAT, what effect would the mutation have on the sequence of the protein produced?

16.39 Give an mRNA sequence that codes for synthesis of metenkephalin, a small peptide with morphine-like properties:

Tyr-Gly-Gly-Phe-Met

16.40 Give a DNA gene sequence that will code for metenkephalin (Problem 16.39).

Integrated Problems

16.41 Human and horse insulin both have two polypeptide chains, with one chain containing 21 amino acids and the other containing 30 amino acids. How many nitrogen bases are present in the DNA to code for each chain?

16.42 Human and horse insulin (see Problem 16.41) differ in primary structure at two amino acids: at the 9th position in one chain (human has Ser and horse has Gly) and at the 30th position in the other chain (human has Thr and horse has Ala). How must the DNA differ?

16.43 What amino acid sequence is coded by the following mRNA sequence?

CUA-GAC-CGU-UCC-AAG-UGA

16.44 What anticodon sequences of tRNA's are coded by the mRNA in Problem 16.43? What was the base sequence in the original DNA strand on which this mRNA was made? What was the base sequence in the DNA strand *complementary* to that from which this mRNA was made?

16.45 Look up the structure of angiotensin II in Practice Problem 15.6, and give an mRNA sequence that codes for its synthesis.

16.46 Diethylstilbestrol (DES) exhibits estradiol-like activity even though it is structurally unrelated to steroids. Once used widely as an additive in animal feed, DES has been implicated as a causative agent in several types of cancers. Look up the structure of estradiol (Section 16.5), and show how DES can be drawn so that it is sterically similar to estradiol.

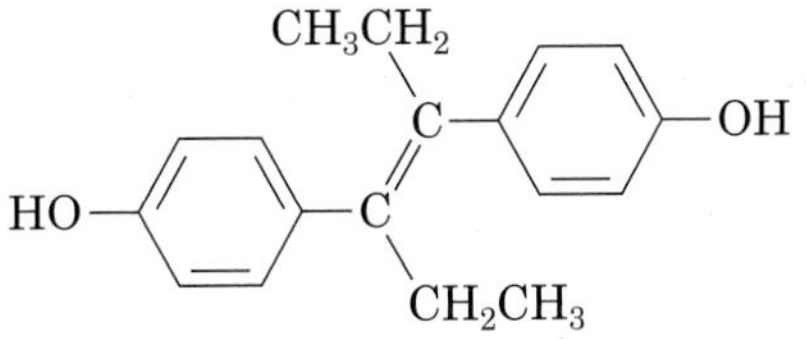

CH_3CH_2
C
OH
HO
C
CH_2CH_3

Diethylstilbestrol

16.47 How many stereocenters are present in estradiol (see Problem 16.46)? Label them.

16.48 What products would you obtain from reaction of estradiol (Problem 16.46) with the following reagents?
(a) NaOH, then CH_3I (b) CH_3COCl, pyridine (c) Br_2 (1 equiv)

16.49 *Nandrolone* is an anabolic steroid sometimes taken by athletes to build muscle mass. Compare the structures of nandrolone and testosterone, and point out their structural similarities.

Nandrolone (an anabolic steroid)

16.50 Draw the structure of cyclic adenosine monophosphate (cAMP), a messenger involved in the regulation of glucose production in the body. Cyclic AMP has a phosphate ring connecting the 3′ and 5′ hydroxyl groups on adenosine.

IN THE MEDICINE CABINET

16.51 Acyclovir is a potent antiviral agent because it mimics a guanine nucleotide.

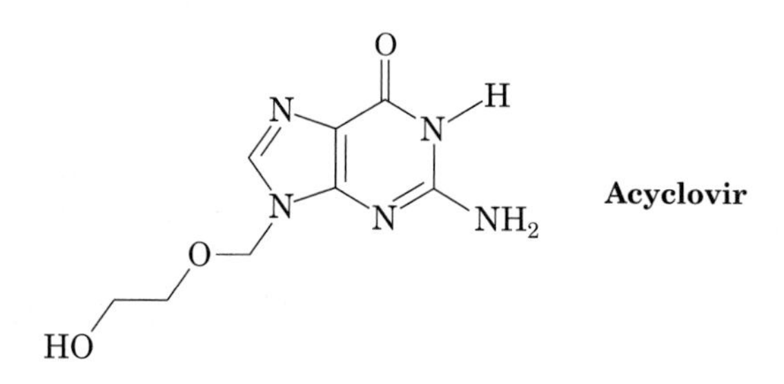

Acyclovir

(a) What role do the atoms missing in acyclovir play in DNA replication?
(b) How might acyclovir interfere with DNA synthesis?

IN THE FIELD WITH AGROCHEMICALS

16.52 Weeds derive resistance to atrazine by a single amino acid substitution wherein a serine is replaced by a glycine residue. Identify the six codons that correspond to serine and the four codons that correspond to glycine.

16.53 Based on your answer to Problem 16.52, what are the corresponding DNA sequences?

16.54 Single nucleotide changes on DNA can have profound effects. Consider your answers to Problems 16.52 and 16.53 as you address these questions:
(a) What is the minimum number of mutations of DNA required for the evolution of resistance?
(b) If we had six weeds, one with each serine codon identified in Problem 16.53, which ones should derive resistance first if the rate of mutation is constant? Which weeds should derived resistance later?
(c) Based on Problems 16.52 and 16.53, if alanine is incorporated instead of serine, would you predict that a partially resistant weed should appear? Why?

Herbicides, like glyphosate, interfere with plant metabolism and produce the clear paths between these growing rows of corn. (Holt Studios/Photo Researchers, Inc.)

17 CHAPTER The Organic Chemistry of Metabolic Pathways

The organic chemical reactions that take place in even the smallest and simplest living organism are more complex than those carried out in any laboratory. Yet the reactions in living organisms, regardless of their complexity, follow the same rules of reactivity and proceed by the same mechanisms we've developed in the preceding chapters.

In this chapter, we'll look at four of the most important reaction pathways by which organisms carry out their chemistry, processes that we refer to as **metabolism**. Pathways that break down larger molecules into small ones are called **catabolic**. Pathways that put small molecules together to make larger molecules are called **anabolic**. Catabolic reactions usually produce energy. Anabolic reactions usually require energy. *All* these reactions are simple organic reactions that we have already discussed. When these reactions occur within a complicated-looking molecule, it is easy to be intimidated, but don't panic. By ignoring everything but the atoms involved in a reaction, you'll quickly realize that *biochemistry is organic chemistry.*

17.1 ATP: The Currency of Biochemical Reactions

Adenosine triphosphate, or ATP, has been called the "energy currency" of the cell. When ATP is hydrolyzed to adenosine diphosphate, or ADP, energy is released. The energy released by this reaction allows other energy-requiring reactions to proceed. Catabolic reactions "pay off" in ATP by synthesizing it from ADP plus hydrogen phosphate ion, HPO_4^{2-}. Anabolic reactions "spend"

ATP by transferring a phosphate group to another molecule, thereby regenerating ADP. Energy production and use thus revolves around the ATP $\rightleftarrows$ ADP interconversion:

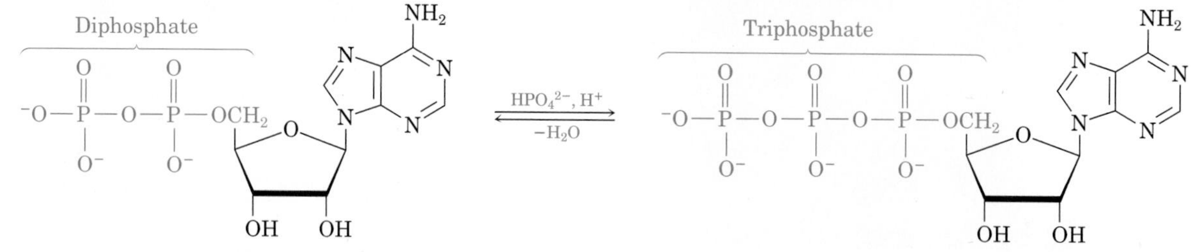

Diphosphate

Triphosphate

HPO_4^{2-}, H^+

$-H_2O$

Adenosine diphosphate (ADP)

Adenosine triphosphate (ATP)

Where does ATP come from? Figure 17.1 summarizes the chemical processes that lead to the production of ATP. In the first stage of catabolism, the foods we eat—fats, carbohydrates, and proteins—are **digested** into simpler organic molecules through chemical and enzymatic hydrolyses in our mouths, stomachs, and intestines. Once absorbed by cells, these organic molecules are further catabolized in the second stage of catabolism into a two-carbon acetyl group linked to a large carrier molecule called *coenzyme A*. The resultant compound, acetyl coenzyme A, or acetyl CoA, is a **thioester**, RCOSR′, a sulfur analog of an ester.

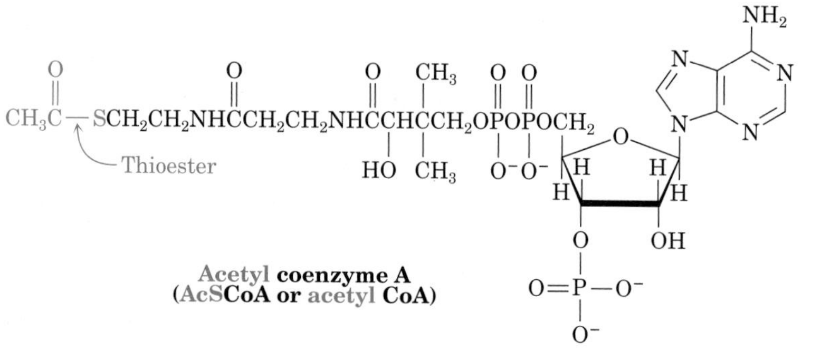

Thioester

Acetyl coenzyme A
(AcSCoA or acetyl CoA)

Acetyl CoA is an intermediate in the breakdown of all main classes of food molecules in human metabolism. It is derived from fatty acids through a process called *β-oxidation*. When glucose is catabolized in the glycolysis pathway (Section 17.3), it also yields acetyl CoA. These acetyl CoA molecules then enter the third stage of catabolism, the *citric acid cycle* (Section 17.4), and are oxidized to CO_2. In the process, NAD^+ (nicotinamide adenine dinucleotide) and FAD (flavin adenine dinucleotide) are reduced to NADH and $FADH_2$. These molecules are used to prepare ATP through the **electron-transport chain**, the fourth stage of catabolism.

How does the body use ATP? For a chemical reaction to have a favorable equilibrium constant and occur spontaneously, energy (actually, *free energy, G*) must be released. This means that the free-energy change for the reaction (ΔG) must be negative. If ΔG is positive, then the reaction is unfavorable and

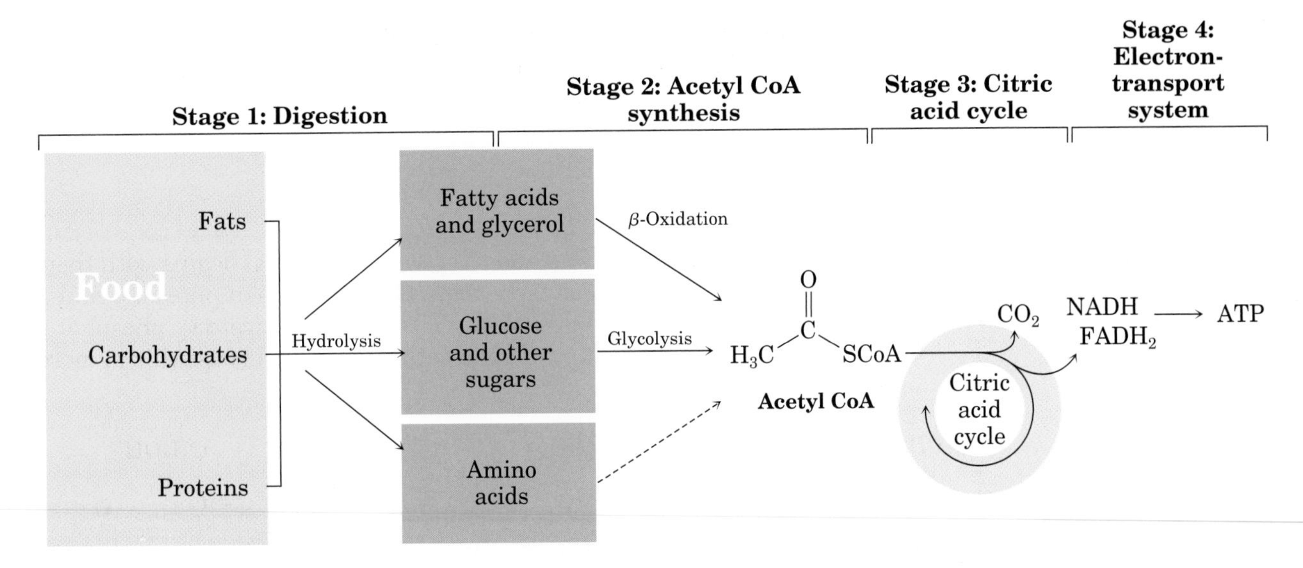

FIGURE 17.1 The production of ATP commences with the hydrolysis of food into organic molecules that travel different catabolic routes (glycolysis, β-oxidation) to form acetyl CoA. Acetyl CoA enters the citric acid cycle and produces CO_2 as well as reduced organic molecules (NADH and $FADH_2$) that are used to manufacture ATP in the electron-transport chain.

the process can't occur spontaneously. What normally happens for an energetically unfavorable reaction to occur is that it is "coupled" to an energetically favorable reaction so that the *overall* free-energy change for the two reactions together is favorable. Take, for example, the phosphorylation reaction of glucose to yield glucose 6-phosphate plus water, an important step in the breakdown of dietary carbohydrates. The reaction of glucose with HPO_4^{2-} does not occur spontaneously because it is energetically unfavorable, with $\Delta G° = +13.8$ kJ/mol (3.3 kcal/mol).

$$\underset{\textbf{Glucose}}{HOCH_2CH(OH)CH(OH)CH(OH)CH(OH)CHO} \xrightarrow{HPO_4^{2-}} \underset{\textbf{Glucose 6-phosphate}}{{}^{-}OP(=O)(O^{-})OCH_2CH(OH)CH(OH)CH(OH)CH(OH)CHO} + H_2O \qquad \Delta G = +13.8 \text{ kJ}$$

With ATP, however, glucose undergoes an energetically favorable reaction to yield glucose 6-phosphate plus ADP. The effect is as if the HPO_4^{2-} reacted with glucose and ATP then reacted with the water by-product, making the *coupled* process favorable by about 16.7 kJ/mol (4.0 kcal/mol). We therefore say that ATP "drives" the phosphorylation reaction of glucose.

	Glucose + HPO_4^{2-} →	Glucose 6-phosphate + H_2O	$\Delta G = +13.8$ kJ
	ATP + H_2O →	ADP + HPO_4^{2-} + H^+	$\Delta G = -30.5$ kJ
Net:	Glucose + ATP →	Glucose 6-phosphate + ADP + H^+	$\Delta G = -16.7$ kJ

It's this ability to drive otherwise unfavorable phosphorylation reactions that makes ATP so useful. The resultant phosphates are much more reactive substances than the corresponding compounds they are derived from.

PROBLEM 17.1 One of the steps in fat metabolism is the reaction of glycerol (propane-1,2,3-triol) with ATP to yield glycerol 1-phosphate. Write the reaction, and draw the structure of glycerol 1-phosphate.

17.2 Catabolism of Fats: β-Oxidation Pathway

The metabolic breakdown of fats and oils (triacylglycerols) begins with their hydrolysis to yield glycerol plus fatty acids. Glycerol is then phosphorylated by reaction with ATP and oxidized to yield dihydroxyacetone phosphate, which enters the carbohydrate catabolic pathway. (We'll discuss this in more detail in Section 17.3.)

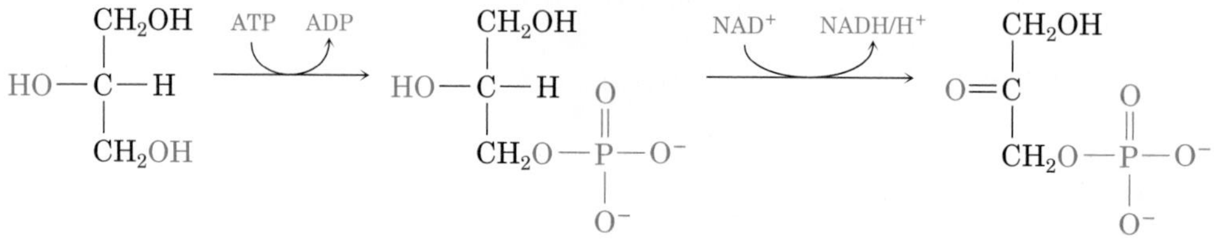

CH₂OH / HO—C—H / CH₂OH —(ATP → ADP)→ CH₂OH / HO—C—H / CH₂O—P(=O)(O⁻)—O⁻ —(NAD⁺ → NADH/H⁺)→ CH₂OH / O=C / CH₂O—P(=O)(O⁻)—O⁻

Glycerol **Glycerol 1-phosphate** **Dihydroxyacetone phosphate**

Note how the preceding reactions are written. It's common practice when writing biochemical transformations to show only the structures of the reactant and product, while abbreviating the structures of coenzymes (see Section 15.10) and other substances. Thus, the curved arrow intersecting the usual straight reaction arrow in the first step shows that ATP is also a reactant and that ADP is also a product. The coenzyme *NAD⁺* is required in the second step, and *NADH* plus a proton are products. We'll see shortly that NAD⁺ is often involved as a biochemical oxidizing agent for converting alcohols to aldehydes or ketones.

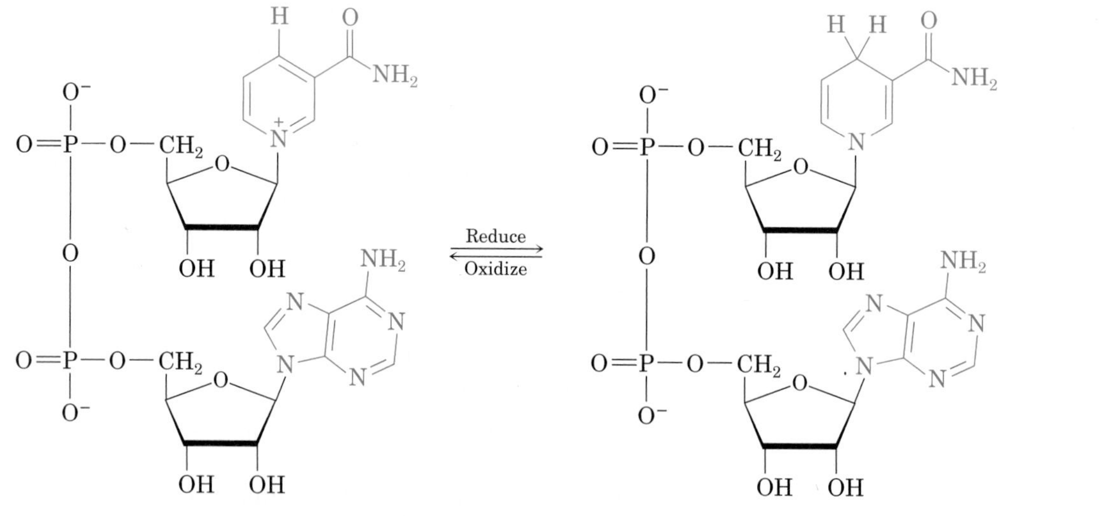

Reduce ⇌ Oxidize

Nicotinamide adenine **dinucleotide (NAD⁺)** Reduced nicotinamide adenine **dinucleotide (NADH)**

Note also that glycerol 1-phosphate and dihydroxyacetone phosphate are written with their phosphate groups dissociated, that is, as $-OPO_3^{2-}$ rather than $-OPO_3H_2$. As previously remarked, it's standard practice in writing biochemical structures to show carboxylic acids and phosphoric acids as their anions, because they exist in this form at the physiological pH of 7.3 found in the cells of organisms.

Fatty acids are catabolized by a repetitive four-step sequence of enzyme-catalyzed reactions called the *fatty-acid spiral*, or **β-oxidation pathway**, shown in Figure 17.2. Each passage through the pathway results in the cleavage of a two-carbon acetyl group from the end of the fatty-acid chain, until the entire molecule is ultimately degraded. As each acetyl group is produced, it enters the citric acid cycle and is further degraded (Section 17.4).

FIGURE 17.2 MECHANISM: The four steps of the β-oxidation pathway, resulting in the cleavage of an acetyl group from the end of the fatty-acid chain. The chain-shortening step is a retro-Claisen reaction of a β-keto ester.

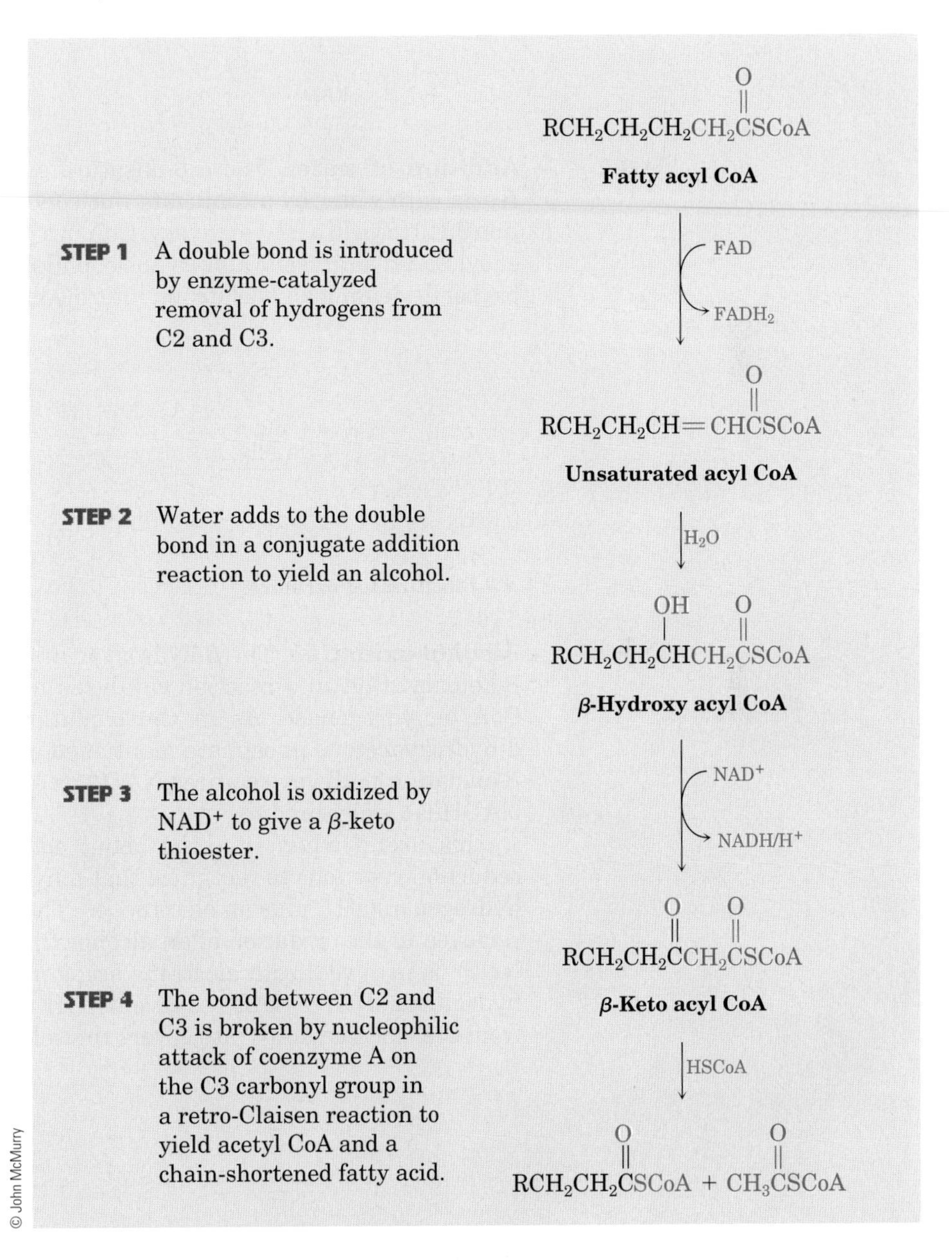
RCH2CH2CH2CH2CSCoA
Fatty acyl CoA
STEP 1 A double bond is introduced by enzyme-catalyzed removal of hydrogens from C2 and C3.
FAD
FADH2
RCH2CH2CH=CHCSCoA
Unsaturated acyl CoA
STEP 2 Water adds to the double bond in a conjugate addition reaction to yield an alcohol.
H2O
RCH2CH2CHCH2CSCoA
β-Hydroxy acyl CoA
STEP 3 The alcohol is oxidized by NAD+ to give a β-keto thioester.
NAD+
NADH/H+
RCH2CH2CCH2CSCoA
β-Keto acyl CoA
STEP 4 The bond between C2 and C3 is broken by nucleophilic attack of coenzyme A on the C3 carbonyl group in a retro-Claisen reaction to yield acetyl CoA and a chain-shortened fatty acid.
HSCoA
RCH2CH2CSCoA + CH3CSCoA

STEP 1 **Introduction of a double bond.** The β-oxidation pathway begins when a fatty acid forms a thioester with coenzyme A to give a fatty acyl CoA. Two hydrogen atoms from C2 and C3 are then removed by an acyl CoA dehydrogenase enzyme to yield an α,β-unsaturated acyl CoA.

This kind of oxidation—the introduction of a conjugated double bond into a molecule—occurs frequently in biochemical pathways and is usually carried out by the coenzyme *FAD*. Reduced $FADH_2$ is the by-product.

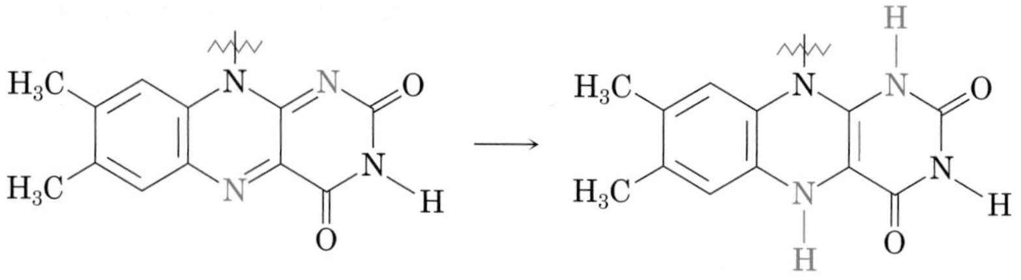

FAD **$FADH_2$**

STEP 2 **Addition of water.** The α,β-unsaturated acyl CoA produced in step 1 reacts with water by a conjugate nucleophilic addition pathway (see Section 9.12) to yield a β-hydroxyacyl CoA in a process catalyzed by the enzyme enoyl CoA hydratase. Water as nucleophile adds to the β carbon of the double bond, yielding an enolate ion intermediate, which is then protonated.

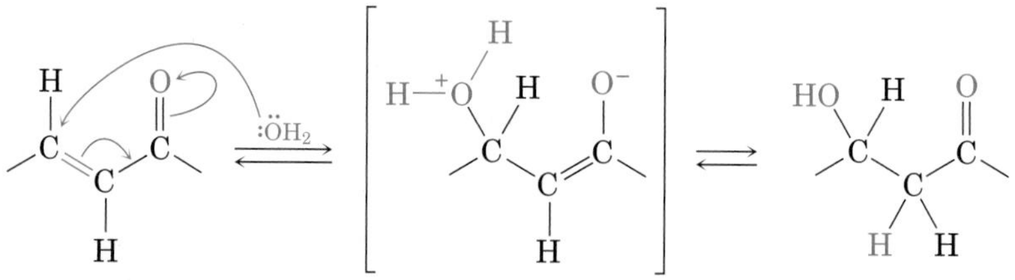

α,β-Unsaturated carbonyl **β-Hydroxy carbonyl**

STEP 3 **Alcohol oxidation.** The β-hydroxyacyl CoA from step 2 is oxidized to a β-ketoacyl CoA in a reaction catalyzed by the enzyme L-3-hydroxyacyl CoA dehydrogenase. As in the oxidation of glycerol 1-phosphate to dihydroxyacetone phosphate mentioned earlier, alcohol oxidation in the β-oxidation pathway requires NAD^+ as a coenzyme and yields reduced $NADH/H^+$ as by-product.

It's useful when thinking about enzyme-catalyzed oxidation and reduction reactions to recognize that a hydrogen *atom* is equivalent to a hydrogen *ion*, H^+, plus an electron, e^-. Thus, for the two hydrogen atoms removed in the oxidation of an alcohol, 2 H atoms = $2\ H^+ + 2\ e^-$. When NAD^+ is involved, both electrons accompany one H^+, in effect adding a hydride ion, $H:^-$, to NAD^+ to give NADH. The second hydrogen removed from the oxidized substrate enters the solution as H^+.

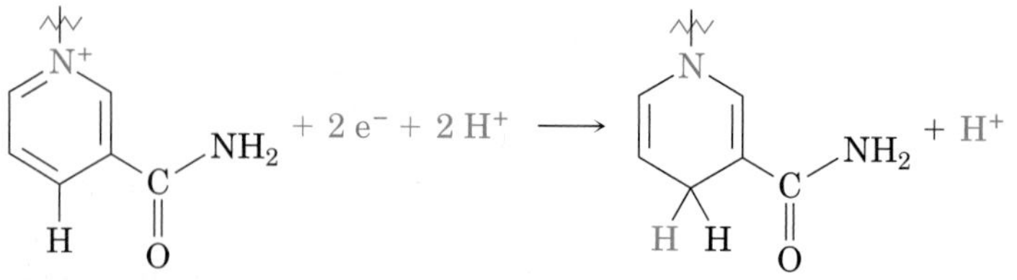

$+ 2\ e^- + 2\ H^+ \longrightarrow \quad + H^+$

NAD^+ **$NADH/H^+$**

The mechanism of oxidation of glycerol 1-phosphate is similar in some respects to that of the conjugate nucleophilic addition reaction in step 2. Thus, a hydride ion expelled from the alcohol acts as a nucleophile and adds to the C=C–C=N$^+$ part of NAD$^+$ in much the same way that water acts as a nucleophile and adds to the C=C–C=O part of the unsaturated acyl CoA.

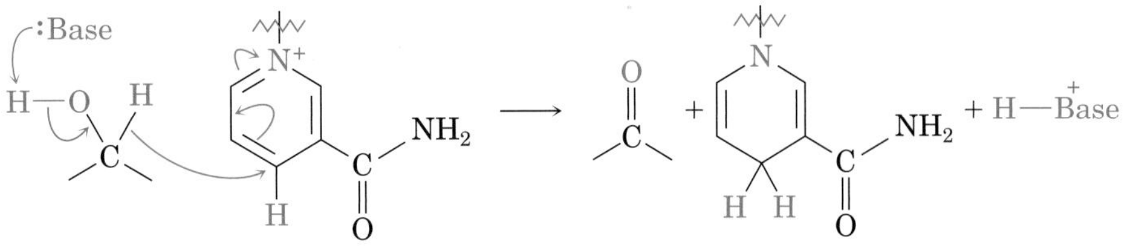

Alcohol NAD$^+$ Ketone NADH

STEP 4 **Chain cleavage.** Acetyl CoA is split off from the acyl chain in the final step of β-oxidation, leaving behind an acyl CoA that is two carbon atoms shorter than the original. The reaction is catalyzed by the enzyme β-ketothiolase and is mechanistically the reverse of the Claisen condensation reaction discussed in Section 11.11. In the *forward* direction, a Claisen condensation joins two esters together to form a β-keto ester product. In the *reverse* direction, a retro-Claisen reaction splits a β-keto ester (or β-keto thioester in this case) apart to form two esters (or two thioesters).

The reaction occurs by nucleophilic addition of coenzyme A to the keto group of the β-keto acyl CoA to yield an alkoxide ion intermediate, followed by cleavage of the C2–C3 bond with expulsion of an acetyl CoA enolate ion. Protonation of the enolate ion gives acetyl CoA, and the chain-shortened acyl CoA enters another round of the β-oxidation pathway for further degradation.

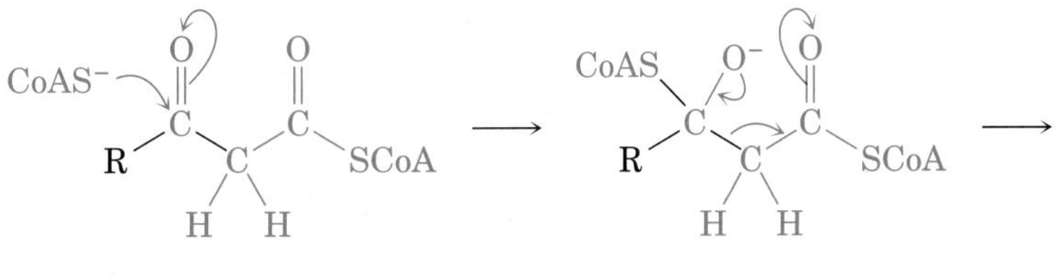

β-Keto acyl CoA

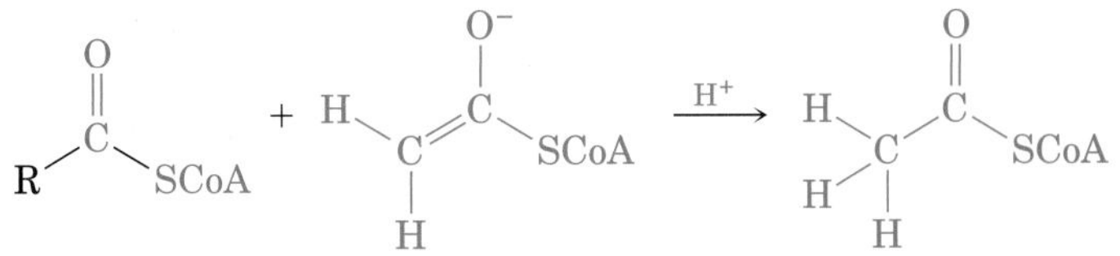

Chain-shortened acyl CoA

Acetyl CoA

Look at the catabolism of myristic acid shown in Figure 17.3 to see the overall results of the β-oxidation pathway. The first passage converts the C_{14} myristyl CoA into the C_{12} lauryl CoA plus acetyl CoA, the second passage converts lauryl CoA into the C_{10} capryl CoA plus acetyl CoA, the third passage converts capryl CoA into the C_8 caprylyl CoA, and so on. Note that the last passage produces *two* molecules of acetyl CoA because the precursor has four carbons.

You can predict how many molecules of acetyl CoA will be obtained from a given fatty acid simply by counting the number of carbon atoms and dividing by 2. For example, the C_{14} myristic acid yields seven molecules of acetyl CoA after six passages along the β-oxidation pathway. The number of passages is always 1 less than the number of acetyl CoA molecules produced because the last passage cleaves a C_4 chain into two acetyl CoA's.

Most fatty acids have an even number of carbon atoms, so none are left over after β-oxidation. Those fatty acids with an odd number of carbon atoms or with double bonds require additional steps for degradation, but all carbon atoms are ultimately released for further oxidation in the citric acid cycle (Section 17.4).

FIGURE 17.3 Catabolism of the C_{14} myristic acid in the β-oxidation pathway yields seven molecules of acetyl CoA after six passages.

$$CH_3CH_2-CH_2CH_2-CH_2CH_2-CH_2CH_2-CH_2CH_2-CH_2CH_2-CH_2\overset{O}{\overset{\|}{C}}SCoA$$

Myristyl CoA

$\downarrow$ β-Oxidation (passage 1)

$$CH_3CH_2-CH_2CH_2-CH_2CH_2-CH_2CH_2-CH_2CH_2-CH_2\overset{O}{\overset{\|}{C}}SCoA + CH_3\overset{O}{\overset{\|}{C}}SCoA$$

Lauryl CoA

$\downarrow$ β-Oxidation (passage 2)

$$CH_3CH_2-CH_2CH_2-CH_2CH_2-CH_2CH_2-CH_2\overset{O}{\overset{\|}{C}}SCoA + CH_3\overset{O}{\overset{\|}{C}}SCoA$$

Capryl CoA

$\downarrow$ β-Oxidation (passage 3)

$$CH_3CH_2-CH_2CH_2-CH_2CH_2-CH_2\overset{O}{\overset{\|}{C}}SCoA + CH_3\overset{O}{\overset{\|}{C}}SCoA$$

Caprylyl CoA $\Downarrow$

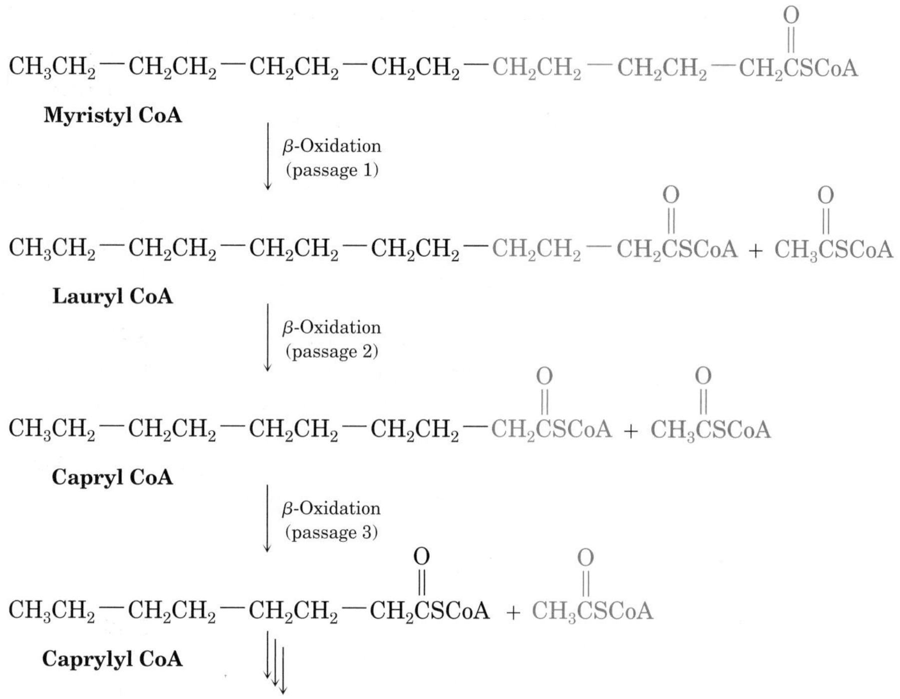

PROBLEM 17.2 Write the equations for the remaining passages of the β-oxidation pathway following those shown in Figure 17.3.

PROBLEM 17.3 How many molecules of acetyl CoA are produced by catabolism of the following fatty acids, and how many passages of the β-oxidation pathway are needed?
(a) Palmitic acid, $CH_3(CH_2)_{14}CO_2H$ (b) Arachidic acid, $CH_3(CH_2)_{18}CO_2H$

17.3 Catabolism of Carbohydrates: Glycolysis

Dietary carbohydrates are catabolized in the **glycolysis** pathway, a series of ten enzyme-catalyzed reactions that break down glucose molecules into 2 equivalents of pyruvate, $CH_3COCO_2^-$. The ten steps of glycolysis, also called the *Embden–Meyerhoff pathway* after its discoverers, are summarized in Figure 17.4.

FIGURE 17.4 The ten-step glycolysis pathway for catabolizing glucose to pyruvate. Individual steps are described in the text.

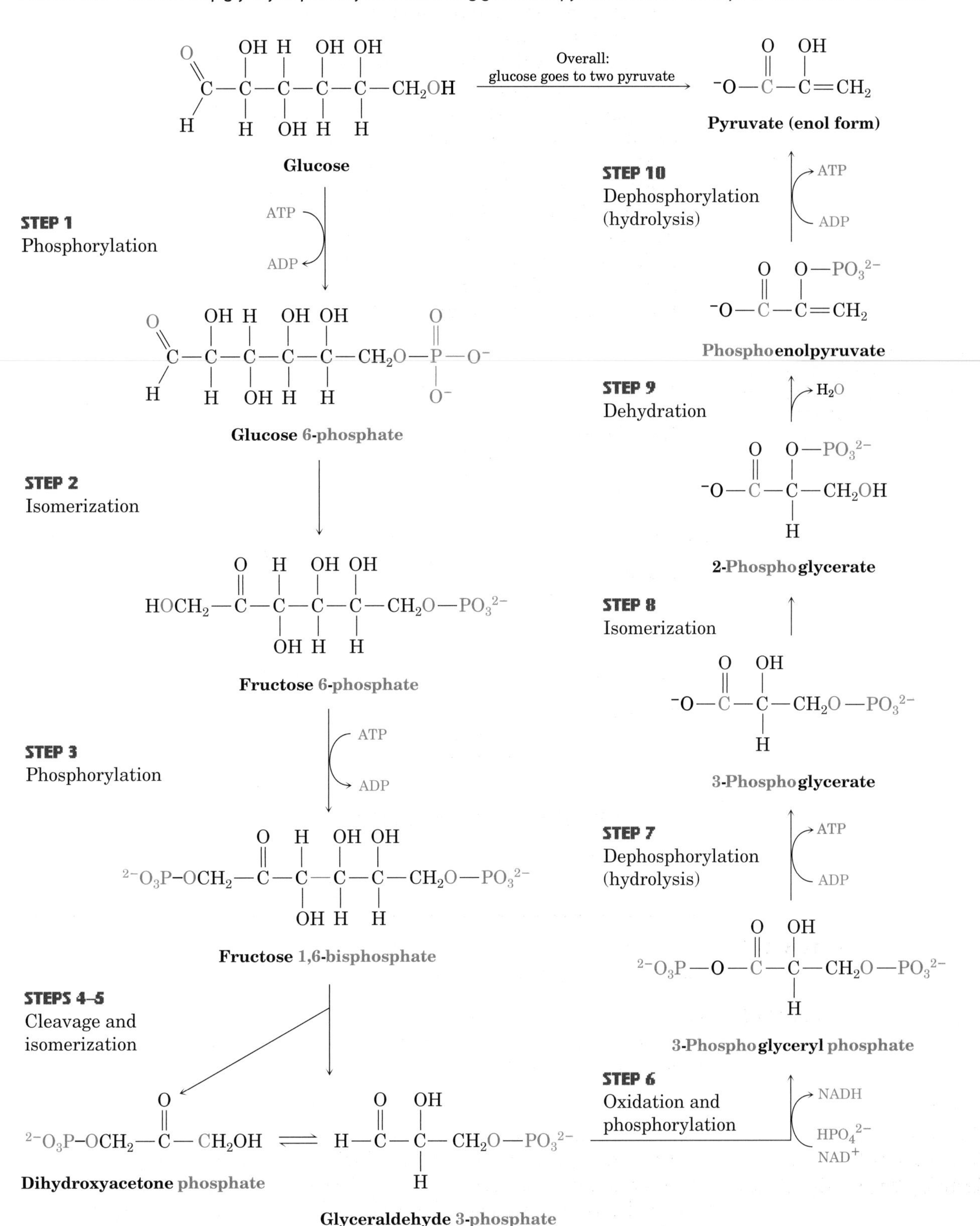

STEPS 1–3

Phosphorylation and isomerization. Glucose, produced by the digestion of dietary carbohydrates, is first phosphorylated by reaction with ATP in a reaction catalyzed by the enzyme hexose kinase. The glucose 6-phosphate that results is then isomerized by phosphoglucose isomerase to fructose 6-phosphate. This isomerization reaction takes place by keto–enol tautomerism (see Section 11.1), since both glucose and fructose share a common enol:

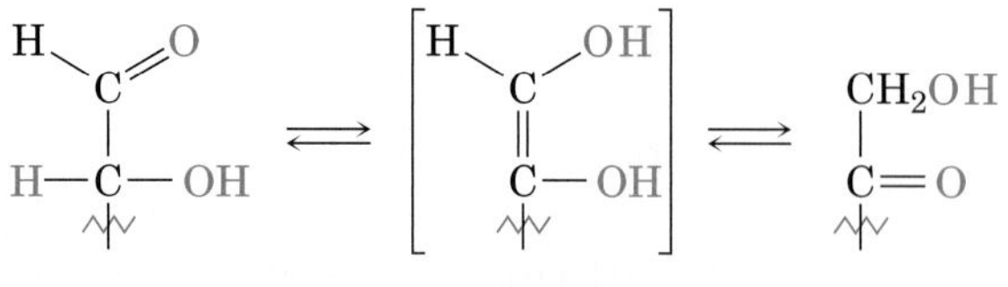

Glucose **Glucose/fructose enol** **Fructose**

Fructose 6-phosphate is then converted to fructose 1,6-bisphosphate (the *bis-* prefix means two) by phosphofructokinase-catalyzed reaction with ATP. The result is a molecule ready to be split into the two three-carbon intermediates that will ultimately become two molecules of pyruvate.

STEPS 4–5

Cleavage and isomerization. Fructose 1,6-bisphosphate is cleaved in step 4 into two 3-carbon monophosphates, one an aldose and one a ketose. The bond between C3 and C4 in fructose 1,6-bisphosphate breaks, and a C=O group is formed. Mechanistically, this cleavage is the reverse of an aldol reaction (see Section 11.9) and is carried out by an aldolase enzyme. (A *forward*-aldol reaction joins two aldehydes or ketones to give a β-hydroxy carbonyl compound; a *retro*-aldol reaction cleaves a β-hydroxy carbonyl compound into two aldehydes or ketones.)

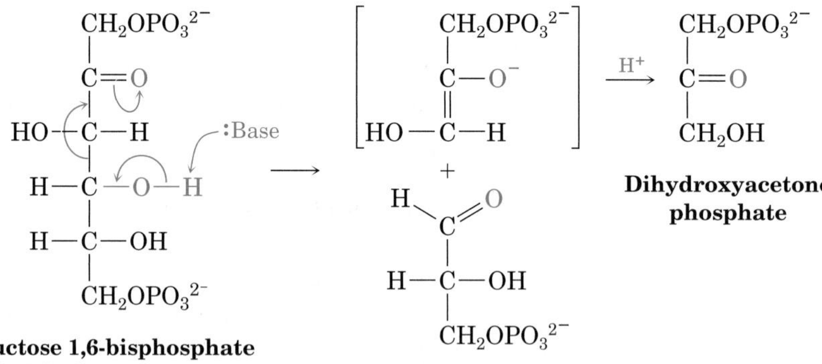

Fructose 1,6-bisphosphate (a β-hydroxy ketone) **Glyceraldehyde 3-phosphate** **Dihydroxyacetone phosphate**

Glyceraldehyde 3-phosphate continues on in the glycolysis pathway, but dihydroxyacetone phosphate must first be isomerized by the enzyme triose phosphate isomerase. As in the glucose-to-fructose conversion of step 2, the isomerization of dihydroxyacetone phosphate to glyceraldehyde 3-phosphate takes place by keto–enol tautomerization through a common enol.

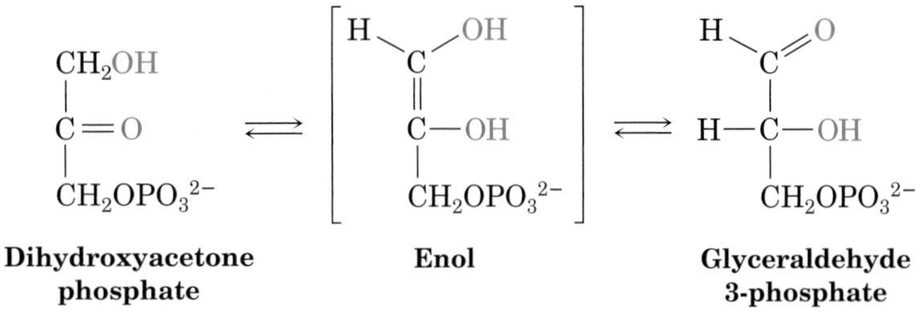

$$CH_2OH-C(=O)-CH_2OPO_3^{2-} \rightleftharpoons [HC(OH)=C(OH)-CH_2OPO_3^{2-}] \rightleftharpoons HC(=O)-CH(OH)-CH_2OPO_3^{2-}$$

Dihydroxyacetone phosphate **Enol** **Glyceraldehyde 3-phosphate**

The net result of steps 4 and 5 is the production of *two* glyceraldehyde 3-phosphate molecules, both of which pass down the rest of the pathway. Thus, each of the remaining five steps of glycolysis takes place twice for every glucose molecule that enters at step 1.

STEPS 6–8 **Oxidation and phosphorylation.** Glyceraldehyde 3-phosphate is oxidized and phosphorylated to give 3-phosphoglyceryl phosphate. The reaction requires the coenzyme NAD^+ in the presence of phosphate ion, HPO_4^{2-} (abbreviated P_i), and is catalyzed by the enzyme glyceraldehyde 3-phosphate dehydrogenase. Transfer of a phosphate group from the carboxyl of 3-phosphoglyceryl phosphate to ADP then yields 3-phosphoglycerate, which is isomerized to 2-phosphoglycerate. The phosphorylation is catalyzed by phosphoglycerate kinase, and the isomerization is catalyzed by phosphoglyceromutase.

STEPS 9–10 **Dehydration and dephosphorylation.** Like the β-hydroxy carbonyl compounds produced in aldol reactions (see Section 11.10), 2-phosphoglycerate undergoes a ready dehydration, yielding phosphoenolpyruvate (PEP). The process is catalyzed by enolase.

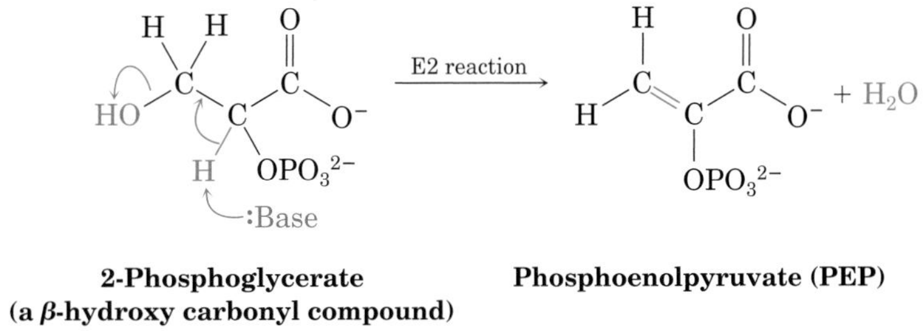

$$HOCH_2-CH(OPO_3^{2-})-COO^- \xrightarrow{\text{E2 reaction}} H_2C=C(OPO_3^{2-})-COO^- + H_2O$$

(:Base removes H)

2-Phosphoglycerate (a β-hydroxy carbonyl compound) **Phosphoenolpyruvate (PEP)**

Transfer of the phosphate group to ADP then generates ATP and gives pyruvate, a reaction catalyzed by pyruvate kinase.

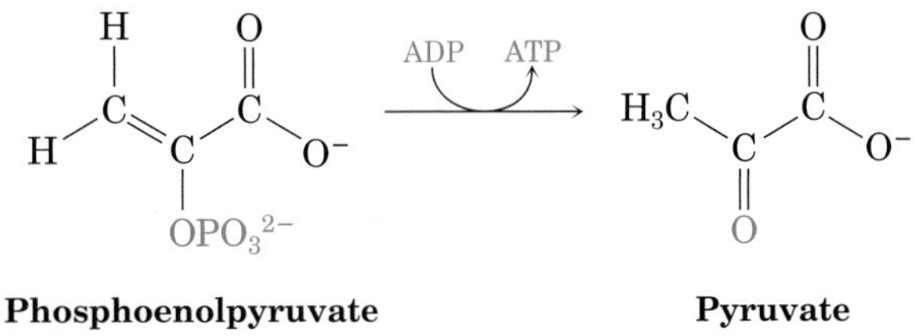

$$H_2C=C(OPO_3^{2-})-COO^- \xrightarrow{ADP \rightarrow ATP} H_3C-C(=O)-COO^-$$

Phosphoenolpyruvate **Pyruvate**

The net result of glycolysis can be summarized as:

$$\underset{\text{Glucose}}{C_6H_{12}O_6} + 2\ NAD^+ + 2\ HPO_4^{2-} + 2\ ADP \longrightarrow 2\ \underset{\text{Pyruvate}}{CH_3\overset{O}{\overset{\|}{C}}-\overset{O}{\overset{\|}{C}}O^-} + 2\ NADH + 2\ ATP + 2\ H_2O + 2\ H^+$$

Pyruvate can undergo several further transformations, depending on the conditions and on the organism. Most commonly, pyruvate is converted to acetyl CoA through a complex, multistep sequence of reactions that requires three different enzymes and four different coenzymes. All the individual steps are well understood and have simple laboratory analogies, although their explanations are a bit outside the scope of this book.

$$\underset{\text{Pyruvate}}{CH_3\overset{O}{\overset{\|}{C}}-\overset{O}{\overset{\|}{C}}O^-} \xrightarrow[\text{HSCoA}]{NAD^+ \quad NADH/H^+} \underset{\text{Acetyl CoA}}{CH_3\overset{O}{\overset{\|}{C}}SCoA} + CO_2$$

PROBLEM 17.4 Identify the steps in glycolysis in which ATP is produced.

PROBLEM 17.5 Look at the entire glycolysis pathway and make a list of the kinds of organic reactions that take place—nucleophilic acyl substitutions, aldol reactions, E2 reactions, and so forth.

17.4 The Citric Acid Cycle

The first two stages of catabolism result in the conversion of fats and carbohydrates into acetyl groups that are bonded through a thioester link to coenzyme A. Acetyl CoA now enters the third stage of catabolism, the **citric acid cycle**, also called the *tricarboxylic acid (TCA) cycle* or *Krebs cycle* after Hans Krebs, who unraveled its complexities in 1937. The eight steps of the citric acid cycle, along with brief descriptions, are given in Figure 17.5.

As its name implies, the citric acid *cycle* is a closed loop of reactions in which the product of the last step is a reactant in the first step. The intermediates are constantly regenerated and flow continuously through the cycle, which operates as long as the oxidizing coenzymes NAD^+ and FAD are available. To meet this condition, the reduced coenzymes NADH and $FADH_2$ must be reoxidized via the electron-transport chain, which in turn relies on oxygen as the final electron acceptor. Thus, the citric acid cycle is also dependent on the availability of oxygen and on the operation of the electron-transport chain.

STEPS 1–2 **Carbonyl condensation and isomerization.** Acetyl CoA enters the citric acid cycle by nucleophilic addition to the ketone carbonyl group of oxaloacetate to give citrate. The addition is an aldol reaction of an enolate ion from acetyl CoA and is catalyzed by the enzyme citrate synthetase, as

FIGURE 17.5 The citric acid cycle is an eight-step series of reactions that results in the conversion of an acetyl group into two molecules of CO_2 plus reduced coenzymes. Individual steps are explained in the text.

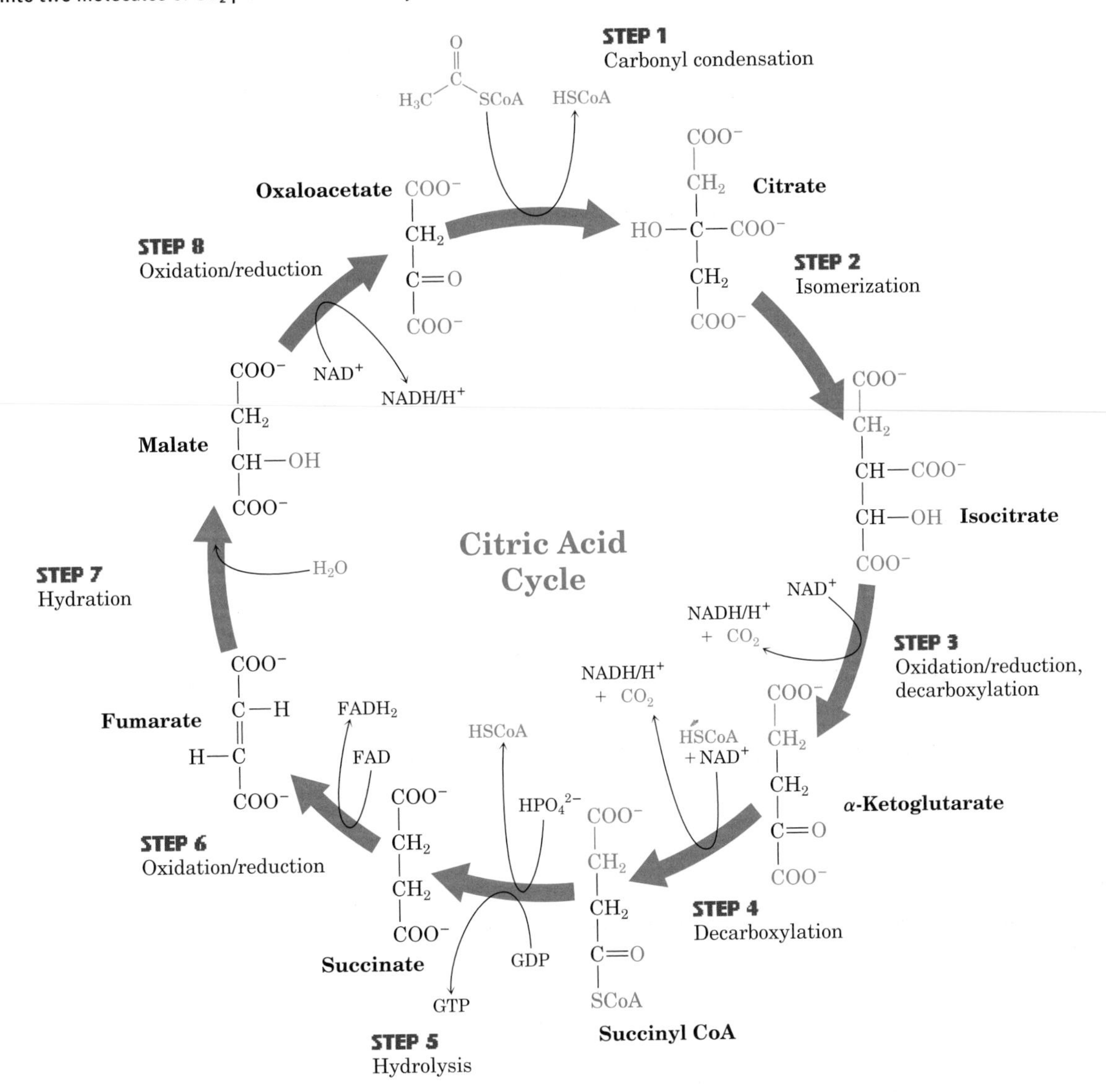

discussed in detail in Section 15.11. Citryl CoA is then hydrolyzed to citrate.

$$\text{Oxaloacetate} + {}^{-}\text{:}CH_2CSCoA \longrightarrow \left[{}^{-}O\text{–}C(COSCoA\text{–}CH_2)(COO^-)(CH_2COO^-) \right] \xrightarrow{H_2O \quad HSCoA} HO\text{–}C(CH_2COO^-)(COO^-)(CH_2COO^-)$$

Oxaloacetate **Citrate**

Citrate, a tertiary alcohol, is next converted into its isomer, isocitrate, a secondary alcohol. The isomerization occurs in two steps, both of which are catalyzed by the same aconitase enzyme. The initial step is an E2 dehydration of the same sort that occurs in step 9 of glycolysis (Figure 17.4). The second step is a conjugate nucleophilic addition of water of the same sort that occurs in step 2 of the β-oxidation pathway (Figure 17.2). Note that the dehydration of citrate takes place specifically *away* from the carbon atoms of the acetyl group that added to oxaloacetate in step 1.

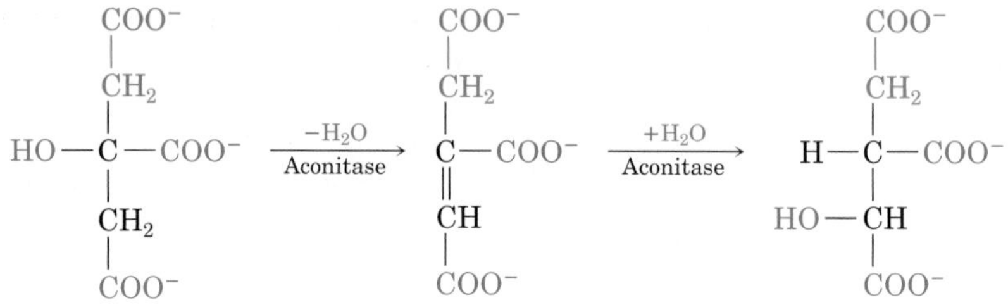

$$\underset{\textbf{Citrate}}{\mathrm{HO{-}C(CH_2COO^-)(COO^-)(CH_2COO^-)}} \xrightarrow[\text{Aconitase}]{-H_2O} \underset{\textbf{Aconitate}}{\mathrm{^-OOCCH_2{-}C(COO^-){=}CH{-}COO^-}} \xrightarrow[\text{Aconitase}]{+H_2O} \underset{\textbf{Isocitrate}}{\mathrm{^-OOCCH_2{-}CH(COO^-){-}CH(OH){-}COO^-}}$$

Steps 3–4 **Oxidation and decarboxylations.** Isocitrate, a secondary alcohol, is oxidized by NAD^+ in step 3 to give a ketone, which loses CO_2 to give α-ketoglutarate. Catalyzed by the enzyme isocitrate dehydrogenase, the decarboxylation is a typical reaction of a carboxylic acid that has a second carbonyl group two atoms away (a β-keto acid). A similar kind of decarboxylation reaction occurs in the malonic ester synthesis (see Section 11.6).

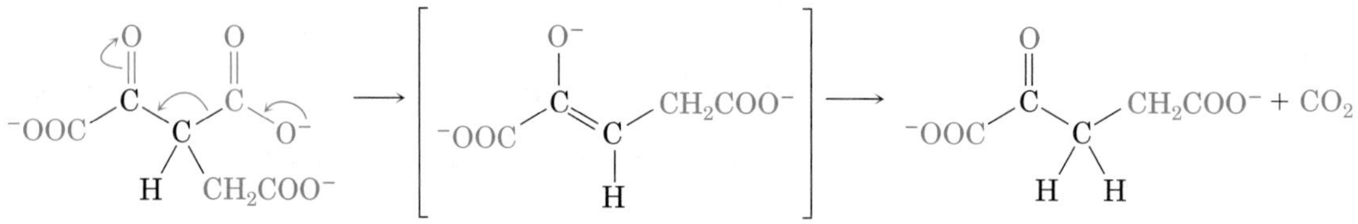

$$\mathrm{^-OOC{-}C({=}O){-}CH(CH_2COO^-){-}C({=}O)O^-} \longrightarrow \left[\mathrm{^-OOC{-}C(O^-){=}CH{-}CH_2COO^-}\right] \longrightarrow \underset{\boldsymbol{\alpha}\textbf{-Ketoglutarate}}{\mathrm{^-OOC{-}C({=}O){-}CH_2{-}CH_2COO^-}} + CO_2$$

The transformation of α-ketoglutarate to succinyl CoA in step 4 is a multistep process analogous to the transformation of pyruvate to acetyl CoA that we saw in the previous section. Like the pyruvate conversion, the α-ketoglutarate conversion requires a number of different enzymes and coenzymes.

Steps 5–6 **Hydrolysis and oxidation of succinyl CoA.** Succinyl CoA is hydrolyzed to succinate in step 5. The reaction is catalyzed by succinyl CoA synthetase and is coupled with phosphorylation of guanosine diphosphate (GDP) to give guanosine triphosphate (GTP). Succinate is then dehydrogenated by FAD and succinate dehydrogenase to give fumarate—a process analogous to that of step 1 in the β-oxidation pathway.

Steps 7–8 **Hydration and oxidation (regeneration of oxaloacetate).** Catalyzed by the enzyme fumarase, conjugate nucleophilic addition of water to fumarate yields malate in a reaction similar to that of step 2 in the β-oxidation pathway. Oxidation with NAD^+ then gives oxaloacetate in a step catalyzed by malate dehydrogenase, and the cycle has returned to its starting point, ready to revolve again.

The net result of the citric acid cycle can be summarized as:

$$\text{Acetyl CoA} + 3\ NAD^+ + FAD + ADP + HPO_4^{2-} + 2\ H_2O \rightarrow$$
$$\text{HSCoA} + 3\ NADH + 3\ H^+ + FADH_2 + ATP + 2\ CO_2$$

PROBLEM 17.6 Which of the substances in the citric acid cycle are tricarboxylic acids, thus giving the cycle its alternate name?

PROBLEM 17.7 Write mechanisms for the reactions in step 2 of the citric acid cycle, the dehydration of citrate and the addition of water to aconitate.

17.5 Catabolism of Proteins: Transamination

The breakdown of dietary proteins is more complex than that of fats and carbohydrates because each of the 20 amino acids is degraded through its own unique pathway. The general idea, however, is that the amino nitrogen atoms are removed and the substances that remain are converted into compounds that enter the citric acid cycle.

Most amino acids lose their nitrogen atom by a **transamination** reaction in which the $-NH_2$ group of the amino acid changes places with the keto group of α-ketoglutarate. The products are a new α-keto acid and glutamate:

$$\underset{\textbf{An amino acid}}{RCH(NH_3^+)COO^-} + \underset{\alpha\textbf{-Ketoglutarate}}{^-OOCCH_2CH_2C(=O)COO^-} \rightleftharpoons \underset{\textbf{An }\alpha\textbf{-keto acid}}{RC(=O)COO^-} + \underset{\textbf{Glutamate}}{^-OOCCH_2CH_2CH(NH_3^+)COO^-}$$

Transaminations use pyridoxal phosphate, a derivative of vitamin B_6, as cofactor. As shown in Figure 17.6 for the reaction of alanine, the key step in transamination is nucleophilic addition of an amino acid $-NH_2$ group to the pyridoxal aldehyde group to yield an imine (see Section 9.10). Loss of a proton from the α position then results in a bond rearrangement to give a different imine, which is hydrolyzed (the exact reverse of imine formation) to yield pyruvate and a nitrogen-containing derivative of pyridoxal phosphate. Pyruvate is converted into acetyl CoA (Section 17.3), which enters the citric acid cycle for further catabolism. The pyridoxal phosphate derivative transfers its nitrogen atom to α-ketoglutarate by the reverse of the steps in Figure 17.6, thereby forming glutamate and regenerating pyridoxal phosphate for further use.

Glutamate, which now contains the nitrogen atom of the former amino acid, next undergoes an *oxidative deamination* to yield ammonium ion and regenerated α-ketoglutarate. The oxidation of the amine to an imine is similar to the oxidation of a secondary alcohol to a ketone and is carried out by NAD^+. The imine is then hydrolyzed (see Section 9.10).

FIGURE 17.6 MECHANISM: Oxidative deamination of alanine requires the cofactor pyridoxal phosphate and yields pyruvate as product.

$^{2-}O_3POCH_2$ H C=O :NH_2 + CH_3CHCOO^- $^+$N H OH CH_3

Pyridoxal phosphate

STEP 1 Nucleophilic attack of the amino acid on the pyridoxal phosphate carbonyl group gives an imine.

$^{2-}O_3POCH_2$ H H CH_3 C N C COO^- + H_2O $^+$N H OH CH_3

STEP 2 Loss of a proton moves the double bonds and gives a second imine intermediate.

$^{2-}O_3POCH_2$ H CH_3 C N C COO^- N H OH CH_3

STEP 3 Hydrolysis of the imine then yields an α-keto acid along with a nitrogen-containing pyridoxal phosphate derivative.

H_2O

$^{2-}O_3POCH_2$ H C NH_2 N H OH CH_3 + CH_3CCOO^- (O)

Pyruvate

STEP 4 Bond tautomerization regenerates an aromatic pyridine ring.

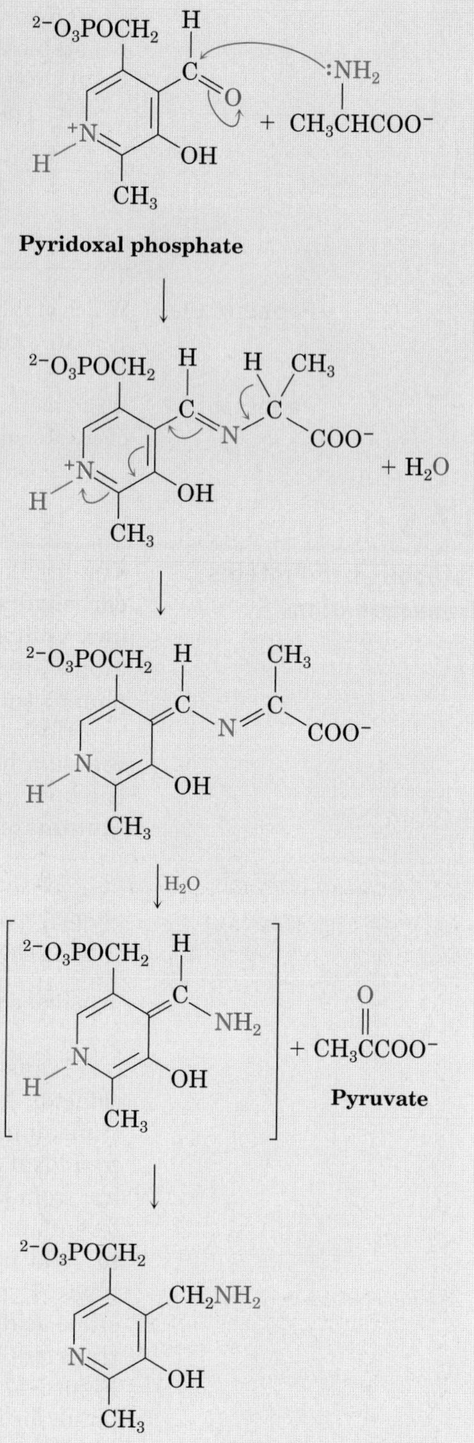

$^{2-}O_3POCH_2$ CH_2NH_2 N OH CH_3

$$\mathrm{^{-}O_2CCH_2CH_2\overset{\overset{\Large NH_3^+}{|}}{C}HCO_2^-}\xrightarrow{\text{NAD}^+\ \ \text{NADH/H}^+}\left[\mathrm{^{-}O_2CCH_2CH_2\overset{\overset{\Large NH}{||}}{C}CO_2^-}\right]\xrightarrow{\text{H}_2\text{O, H}^+}\mathrm{^{-}O_2CCH_2CH_2\overset{\overset{\Large O}{||}}{C}CO_2^-}+\mathrm{NH_4^+}$$

Glutamate **α-Ketoglutarate**

PROBLEM 17.8 Show the structure of the α-keto acid formed by transamination of leucine.

PROBLEM 17.9 From what amino acid was the following α-keto acid derived?

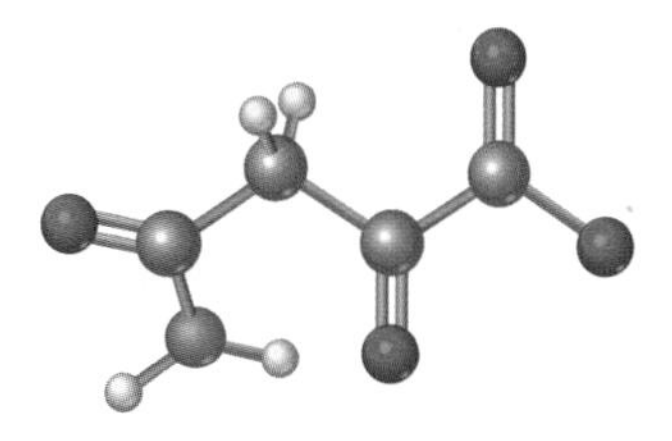

17.6 The Organic Chemistry of Metabolic Pathways: A Summary

As the chapter introduction warned, the past few sections have been a quick-paced tour of a large number of reactions. Following it all undoubtedly required a lot of work and a lot of page turning to look at earlier sections. When you take the time the think about it though, *none* of the many reactions in the four metabolic pathways discussed were new. All were examples of the fundamental processes covered in earlier chapters—eliminations, nucleophilic addition reactions, carbonyl condensation reactions, and so forth.

What's true of β-oxidation, glycolysis, the citric acid cycle, and transamination is also true for all other biological pathways: biochemistry *is* organic chemistry. Understanding how nature works and how living organisms function is a fascinating field of study.

INTERLUDE

Borrowing Metabolic Pathways for Bioremediation

Microorganisms are creative. When exposed to new chemicals, like herbicides, they often adapt their own metabolic pathways to do chemistry on these molecules. In some cases, the chemicals are introduced in concentrations that are toxic, so the microorganisms are under selective pressure to detoxify their environment. In

Continued

other cases, the microorganisms use these chemicals as food. Often, different microorganisms work together. Scientists have recognized that this cooperation can be put to good use, especially when it comes to cleaning up, or *remediating*, the environment. Remediation using microorganisms is called *bioremediation*.

In cases where the pesticide atrazine is present in the environment in too large a concentration, scientists have used microorganisms for the cleanup. The enzymes (labeled in blue) responsible for the reactions shown are not naturally present in a single organism. However, the DNA corresponding to these enzymes has been isolated, and "supermicrobes" capable of converting atrazine into CO_2 and water have been genetically engineered. Turning organic molecules into inorganic precursors is the ultimate form of remediation: it is a process called *mineralization*. Currently, these supermicrobes are being tested in areas where large amounts of atrazine have been spilled.

Summary and Key Words

Metabolism is the sum of all chemical reactions in the body. Reactions that break down larger molecules into smaller ones are called **catabolism**; reactions that build up larger molecules from smaller ones are called **anabolism**. Although the details of specific biochemical pathways are sometimes complex, all the reactions that occur follow the normal rules of organic chemical reactivity.

The catabolism of fats begins with hydrolysis to give glycerol and fatty acids. The fatty acids are degraded in the four-step ***β*-oxidation pathway** by removal of two carbons at a time, yielding acetyl CoA. Catabolism of carbohydrates begins with the hydrolysis of glycoside bonds to give glucose, which is degraded in the ten-step **glycolysis** pathway. Pyruvate, the initial product of glycolysis, is then converted into acetyl CoA. The acetyl groups produced by degradation of fats and carbohydrates next enter the eight-step **citric acid cycle**, where they are further degraded into CO_2.

Protein catabolism is more complex than that of fats or carbohydrates because each of the 20 different amino acids is degraded by its own unique pathway. In general, the amino nitrogen atoms are removed and the substances that remain are converted into compounds that enter the citric acid cycle. Most amino acids lose their nitrogen atom by **transamination**, a reaction in which the $-NH_2$ group of the amino acid changes places with the keto group of an α-keto acid such as α-ketoglutarate. The products are a new α-keto acid and glutamate.

The energy released in all catabolic pathways is used in the **electron-transport chain** to make molecules of *adenosine triphosphate* (ATP), the final result of food catabolism. ATP couples with and drives many otherwise unfavorable reactions.

EXERCISES

Visualizing Chemistry

17.10 Identify the amino acid that is a catabolic precursor of each of the following α-keto acids:

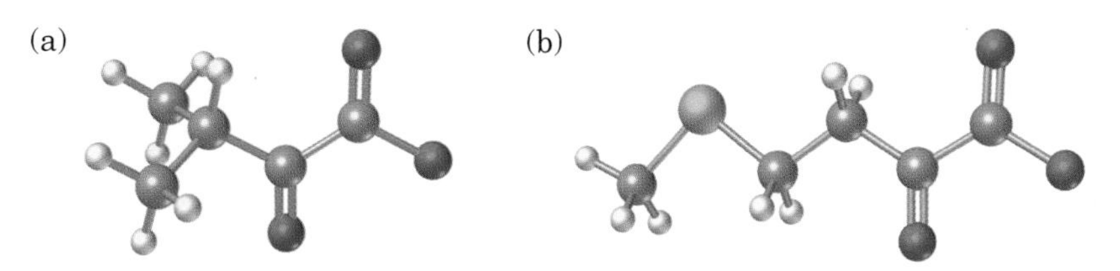

17.11 Identify the following intermediate in the citric acid cycle, and tell whether it has R or S stereochemistry:

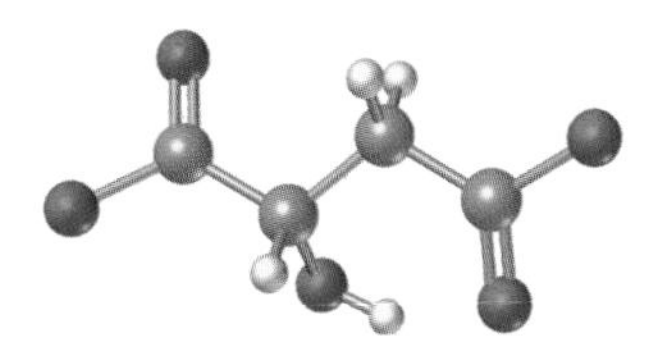

Additional Problems

General Metabolism

17.12 What chemical events occur during the digestion of the following kinds of food molecules?
(a) Fats (b) Complex carbohydrates (c) Proteins

17.13 What is the difference between digestion and metabolism? Between anabolism and catabolism?

17.14 Draw the structure of adenosine monophosphate (AMP), an intermediate in some biochemical pathways.

17.15 What general kind of reaction does ATP carry out?

17.16 What general kind of reaction does NAD^+ carry out?

17.17 What general kind of reaction does FAD carry out?

17.18 Lactate, a product of glucose catabolism in oxygen-starved muscles, can be converted into pyruvate by oxidation. What coenzyme do you think is needed? Write the equation in the normal biochemical format using a curved arrow.

$$\underset{\text{OH}}{CH_3\overset{|}{C}HCOO^-} \quad \textbf{Lactate}$$

Specific Metabolic Pathways

17.19 What substance is the starting point of the citric acid cycle, reacting with acetyl CoA in the first step and being regenerated in the last step?

17.20 How many moles of acetyl CoA are produced by catabolism of the following substances?
(a) 1.0 mol glucose (b) 1.0 mol palmitic acid ($C_{15}H_{31}CO_2H$)
(c) 1.0 mol maltose

17.21 How many grams of acetyl CoA (mol wt = 809.6 amu) are produced by catabolism of the following substances?
(a) 100.0 g glucose (b) 100.0 g palmitic acid (c) 100.0 g maltose

17.22 Which of the substances listed in Problem 17.21 is the most efficient precursor of acetyl CoA on a weight basis?

17.23 List the sequence of intermediates involved in the catabolism of glycerol from hydrolyzed fats to yield acetyl CoA.

17.24 Write the equation for the final step in the β-oxidation pathway of any fatty acid with an even number of carbon atoms.

17.25 Show the products of each of the following reactions:

(a) $CH_3CH_2CH_2CH_2CH_2\overset{O}{\overset{\|}{C}}SCoA \xrightarrow[\text{Acetyl SCoA dehydrogenase}]{\text{FAD} \quad \text{FADH}_2} ?$

(b) Product of (a) + $H_2O \xrightarrow[\text{hydratase}]{\text{Enoyl SCoA}} ?$

(c) Product of (b) $\xrightarrow[\beta\text{-Hydroxyacyl SCoA dehydrogenase}]{\text{NAD}^+ \quad \text{NADH/H}^+} ?$

17.26 What is the structure of the α-keto acid formed by transamination of each of the following amino acids?
(a) Valine (b) Phenylalanine (c) Methionine

17.27 What enzyme cofactor is associated with transamination?

Integrated Problems

17.28 Fatty acids are synthesized in the body by the *lipogenesis* cycle, which begins with acetyl CoA. The first step in lipogenesis is the condensation of two acetyl CoA molecules to yield acetoacetyl CoA, which undergoes three further enzyme-catalyzed steps, yielding butyryl CoA. Based on the kinds of reactions that occur in the β-oxidation pathway, what do you think are the three further steps of lipogenesis?

$$CH_3\overset{O}{\overset{\|}{C}}CH_2\overset{O}{\overset{\|}{C}}SCoA \xrightarrow{\text{3 steps}} CH_3CH_2CH_2\overset{O}{\overset{\|}{C}}SCoA$$

Acetoacetyl CoA **Butyryl CoA**

17.29 In the *pentose phosphate* pathway for degrading sugars, ribulose 5-phosphate is converted to ribose 5-phosphate. Propose a mechanism for the isomerization.

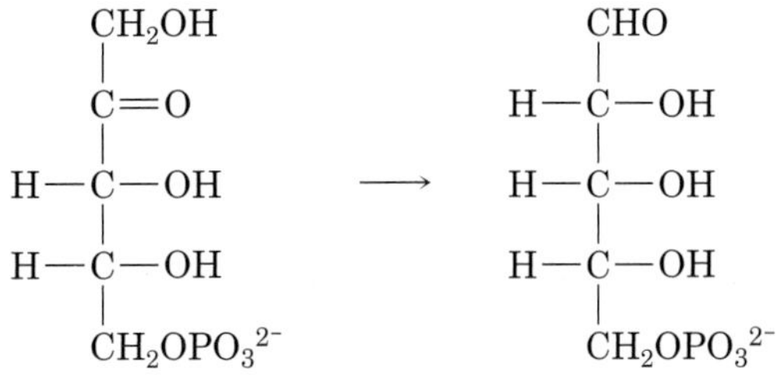

$$\begin{array}{c} CH_2OH \\ | \\ C{=}O \\ | \\ H{-}C{-}OH \\ | \\ H{-}C{-}OH \\ | \\ CH_2OPO_3^{2-} \end{array} \longrightarrow \begin{array}{c} CHO \\ | \\ H{-}C{-}OH \\ | \\ H{-}C{-}OH \\ | \\ H{-}C{-}OH \\ | \\ CH_2OPO_3^{2-} \end{array}$$

Ribulose 5-phosphate **Ribose 5-phosphate**

17.30 Another step in the pentose phosphate pathway for degrading sugars (see Problem 17.29) is the conversion of ribose 5-phosphate to glyceraldehyde 3-phosphate. What kind of organic process is occurring? Propose a mechanism for the conversion.

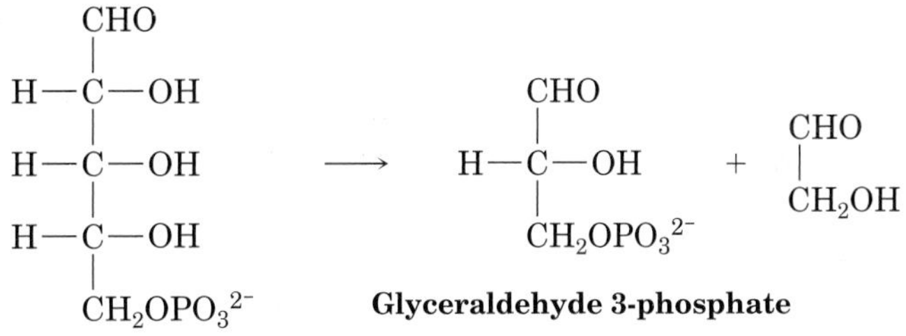

$$\begin{array}{c} CHO \\ | \\ H{-}C{-}OH \\ | \\ H{-}C{-}OH \\ | \\ H{-}C{-}OH \\ | \\ CH_2OPO_3^{2-} \end{array} \longrightarrow \begin{array}{c} CHO \\ | \\ H{-}C{-}OH \\ | \\ CH_2OPO_3^{2-} \end{array} + \begin{array}{c} CHO \\ | \\ CH_2OH \end{array}$$

Ribose 5-phosphate **Glyceraldehyde 3-phosphate**

17.31 One of the steps in the *gluconeogenesis* pathway for synthesizing glucose in the body is the reaction of pyruvate with CO_2 to yield oxaloacetate. Tell what kind of reaction is occurring, and suggest a mechanism.

$$CO_2 + CH_3\overset{O}{\overset{\|}{C}}{-}\overset{O}{\overset{\|}{C}}O^- \longrightarrow {}^-O\overset{O}{\overset{\|}{C}}CH_2\overset{O}{\overset{\|}{C}}{-}\overset{O}{\overset{\|}{C}}O^-$$

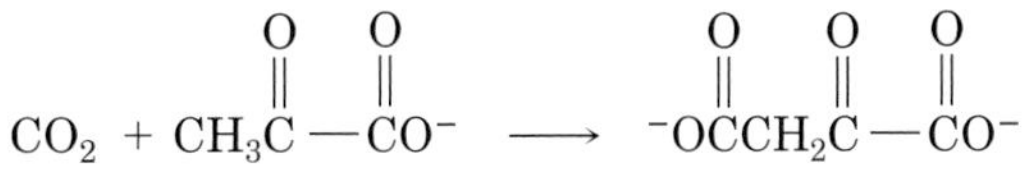

Pyruvate **Oxaloacetate**

17.32 Another step in gluconeogenesis (see Problem 17.31) is the conversion of oxaloacetate to phosphoenolpyruvate by decarboxylation and phosphorylation. Tell what kind of reaction is occurring, and suggest a mechanism.

$${}^-O\overset{O}{\overset{\|}{C}}CH_2\overset{O}{\overset{\|}{C}}{-}\overset{O}{\overset{\|}{C}}O^- \xrightarrow{ATP \;\; ADP} H_2C{=}\overset{^{2-}O_3PO}{\overset{|}{C}}{-}\overset{O}{\overset{\|}{C}}O^- + CO_2$$

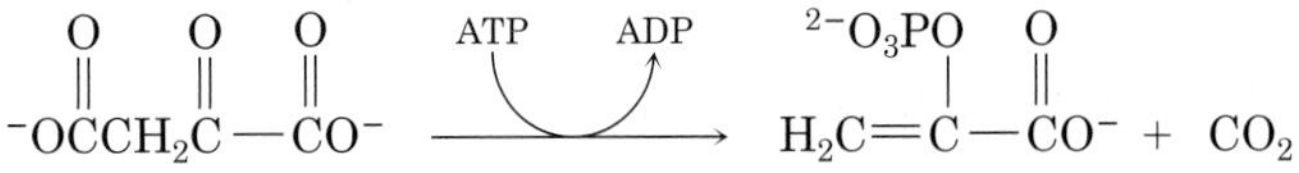

Oxaloacetate **Phosphoenolpyruvate**

17.33 The primary fate of acetyl CoA under normal metabolic conditions is degradation in the citric acid cycle to yield CO_2. When the body is stressed by prolonged starvation, however, acetyl CoA is converted into compounds called *ketone bodies*, which can be used by the brain as a temporary fuel. The biochemical pathway for the synthesis of ketone bodies from acetyl CoA is shown. Fill in the missing information represented by the four question marks.

$2\ CH_3C(=O)SCoA$ (**Acetyl CoA**) $\xrightarrow{HSCoA}$ $CH_3C(=O)CH_2C(=O)SCoA$ (**Acetoacetyl CoA**) $\xrightarrow{H_2O,\ ?}$ $CH_3C(=O)CH_2C(=O)O^-$ (**Acetoacetate**)

Acetoacetate $\xrightarrow{?}$ $CH_3C(=O)CH_3$ (**Acetone**)

Acetoacetate $\xrightarrow{?,\ ?}$ $CH_3CH(OH)CH_2C(=O)O^-$ (**3-Hydroxybutyrate**)

Ketone bodies

17.34 The initial reaction in Problem 17.33, conversion of two molecules of acetyl CoA to one molecule of acetoacetyl CoA, is a Claisen reaction. Assuming that there is a base present, show the mechanism of the reaction.

17.35 The amino acid leucine is biosynthesized from α-ketoisocaproate, which is itself prepared from α-ketoisovalerate by a multistep route that involves: (1) aldol-like reaction with acetyl CoA, (2) hydrolysis, (3) dehydration, (4) hydration, (5) oxidation, and (6) decarboxylation. Show the steps in the transformation.

$(CH_3)_2CHC(=O)COO^-$ (**α-Ketoisovalerate**) $\xrightarrow{\text{Acetyl CoA, } H_2O,\ NAD^+ \ \to\ HSCoA,\ CO_2,\ NADH/H^+}$ $(CH_3)_2CHCH_2C(=O)CO_2^-$ (**α-Ketoisocaproate**)

17.36 The amino acid tyrosine is metabolized by a series of steps that include the following transformations. Propose a mechanism for the conversion of fumaroylacetoacetate into fumarate plus acetoacetate.

Tyrosine $\rightrightarrows$ $^-O_2C\text{–}CH{=}CH\text{–}C(=O)CH_2C(=O)CH_2CO_2^-$ (**Fumaroylacetoacetate**) $\longrightarrow$ $CH_3C(=O)CH_2CO_2^-$ (**Acetoacetate**) $\longrightarrow$ $CH_3C(=O)SCoA$

\+ $^-O_2C\text{–}CH{=}CH\text{–}CO_2^-$ (**Fumarate**)

17.37 Propose a mechanism for the conversion of acetoacetate into acetyl CoA (Problem 17.36).

IN THE MEDICINE CABINET

17.38 Many drugs interfere with metabolic pathways. The anticancer drug methotrexate is an inhibitor of the enzyme dihydrofolate reductase that catalyzes both of the following reactions.

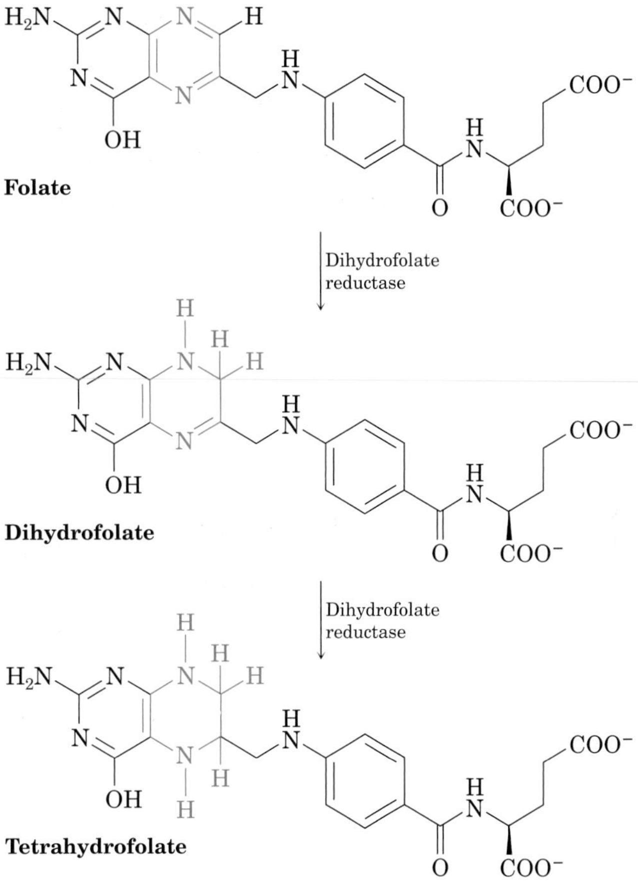

Folate

Dihydrofolate reductase

Dihydrofolate

Dihydrofolate reductase

Tetrahydrofolate

This reduction occurs with oxidation of another molecule. What molecule have we studied in this chapter that could play this role? Write a reaction for this oxidation.

17.39 Methylation of tetrahydrofolate produces a cofactor (see Section 15.10) that is used in the conversion of homocysteine to methionine. Draw a mechanism for this reaction assuming an S_N2 mechanism and that the enzyme acts like a base, "B:".

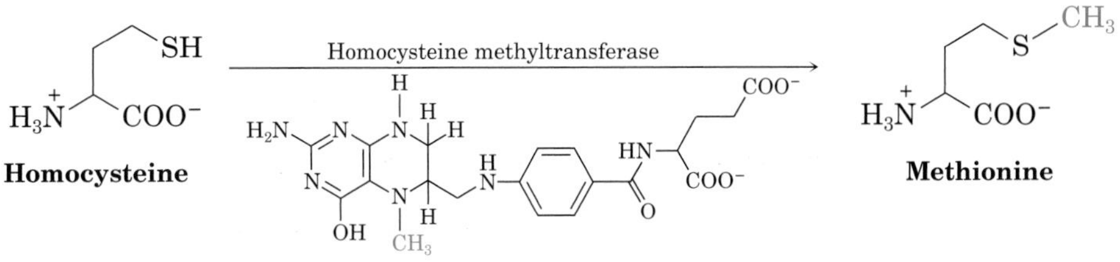

Homocysteine —Homocysteine methyltransferase→ **Methionine**

The cofactor that donates a methyl group

17.40 In addition to methyl groups, tetrahydrofolate (THF) can carry formaldehyde groups that are critical for the biosynthesis of thymidine. Draw a mechanism for incorporation of the formaldehyde into THF to yield the product shown.

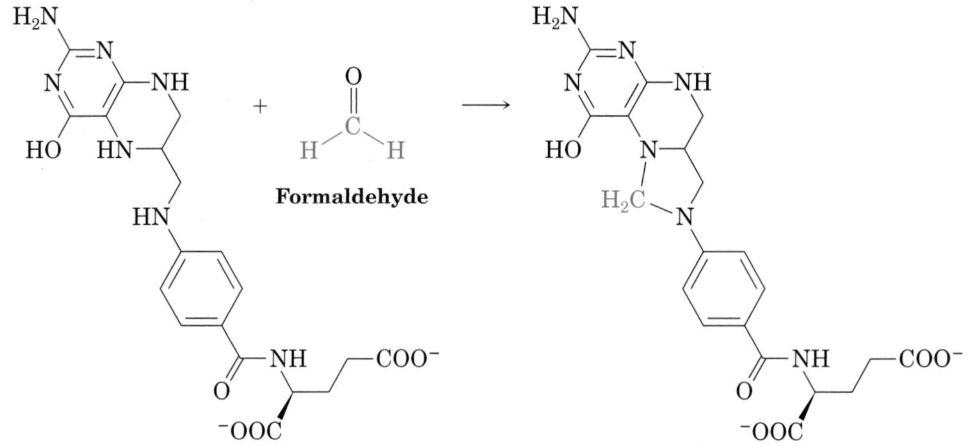

Tetrahydrofolate

Formaldehyde

17.41 Based on Problem 17.40, why is methotrexate an effective anticancer drug?

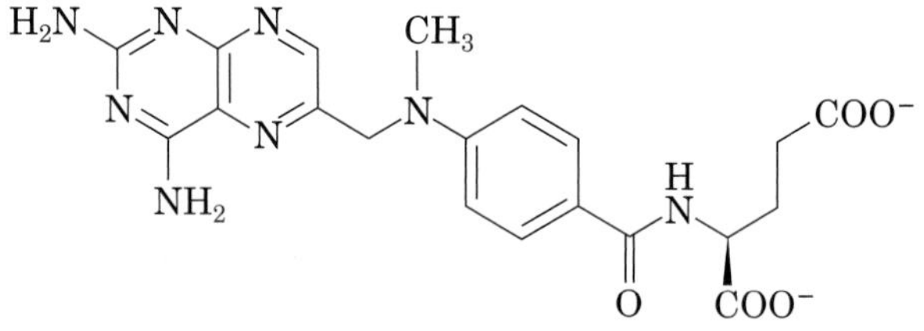

Methotrexate

IN THE FIELD WITH AGROCHEMICALS

17.42 Many herbicides work by interfering with the metabolic pathways, including the shikimic acid pathway, which produces aromatic amino acids such as tyrosine and phenylalanine. A key step in this pathway uses the enzyme EPSP synthase to generate EPSP. What metabolic pathway produces phosphoenopyruvate?

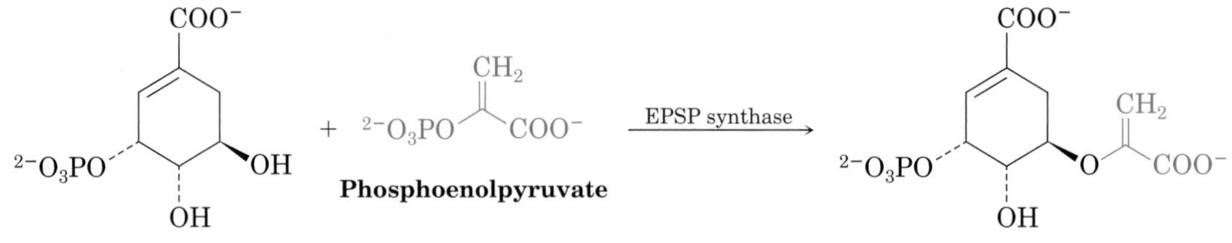

EPSP synthase

Shikimate 3-phosphate

Phosphoenolpyruvate

Enolpyruvylshikimate 3-phosphate (EPSP)

17.43 Glyphosate, marketed under the name Roundup, is a widely used herbicide that blocks EPSP synthase. Hypothesize how glyphosate works.

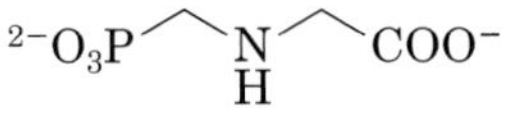

$^{2-}O_3P-CH_2-NH-CH_2-COO^-$

Glyphosate (Roundup)

Nomenclature of Polyfunctional Organic Compounds

With more than 26 million organic compounds now known and several thousand more being created daily, naming them all is a real problem. Part of the problem is due to the sheer complexity of organic structures, but part is also due to the fact that chemical names have more than one purpose. For Chemical Abstracts Service (CAS), which catalogs and indexes the worldwide chemical literature, each compound must have only one correct name. It would be chaos if half the entries for CH_3Br were indexed under "M" for methyl bromide and half under "B" for bromomethane. Furthermore, a CAS name must be strictly systematic so that it can be assigned and interpreted by computers; common names are not allowed.

People, however, have different requirements than computers. For people—which is to say chemists in their spoken and written communications—it's best that a chemical name be pronounceable and that it be as easy as possible to assign and interpret. Furthermore, it's convenient if names follow historical precedents, even if that means a particularly well-known compound might have more than one name. People can readily understand that bromomethane and methyl bromide both refer to CH_3Br.

As noted earlier, chemists overwhelmingly use the nomenclature system devised and maintained by the International Union of Pure and Applied Chemistry, or IUPAC. Rules for naming monofunctional compounds were given throughout the text as each new functional group was introduced, and a list of where these rules can be found is given in Table A.1.

Naming a monofunctional compound is reasonably straightforward, but even experienced chemists often encounter problems when faced with naming a complex polyfunctional compound. Take the following compound, for instance. It has three functional groups, ester, ketone, and C=C, but how

TABLE A.1 Nomenclature Rules for Functional Groups

Functional group	Text section	Functional group	Text section
Acid anhydrides	10.1	Aromatic compounds	5.3
Acid halides	10.1	Carboxylic acids	10.1
Alcohols	8.1	Cycloalkanes	2.7
Aldehydes	9.2	Esters	10.1
Alkanes	2.3	Ethers	8.1
Alkenes	3.1	Ketones	9.2
Alkyl halides	7.1	Nitriles	10.1
Alkynes	4.12	Phenols	8.1
Amides	10.1	Sulfides	8.9
Amines	12.1	Thiols	8.9

should it be named? As an ester with an *-oate* ending, a ketone with an *-one* ending, or an alkene with an *-ene* ending? It's actually named methyl 3-(2-oxocyclohex-6-enyl)propanoate.

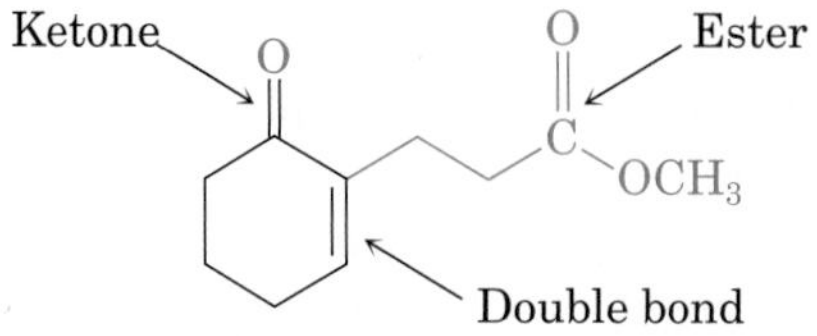

Ketone

Ester

OCH_3

Double bond

Methyl 3-(2-oxocyclohex-6-enyl)propanoate

The name of a polyfunctional organic molecule has four parts—suffix, parent, prefixes, and locants—which must be identified and expressed in the proper order and format. Let's look at each of the four.

Name Part 1. The Suffix: Functional-Group Precedence

Although a polyfunctional organic molecule might contain several different functional groups, we must choose just one suffix for nomenclature purposes. It's not correct to use two suffixes. Thus, keto ester **1** must be named either as a ketone with an *-one* suffix or as an ester with an *-oate* suffix but can't be named as an *-onoate*. Similarly, amino alcohol **2** must be named either as an alcohol (*-ol*) or as an amine (*-amine*) but can't be named as an *-olamine* or *-aminol*.

1. $CH_3C(=O)CH_2CH_2C(=O)OCH_3$ **2.** $CH_3CH(OH)CH_2CH_2CH_2NH_2$

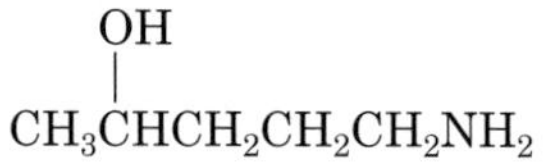

The only exception to the rule requiring a single suffix is when naming compounds that have double or triple bonds. Thus, the unsaturated acid $H_2C{=}CHCH_2CO_2H$ is but-3-enoic acid, and the acetylenic alcohol $HC{\equiv}CCH_2CH_2CH_2OH$ is pent-5-yn-1-ol.

How do we choose which suffix to use? Functional groups are divided into two classes, **principal groups** and **subordinate groups**, as shown in Table A.2. Principal groups can be cited either as prefixes or as suffixes, while subordinate groups are cited only as prefixes. Within the principal groups, an order of priority has been established, with the proper suffix for a given compound determined by choosing the principal group of highest priority. For example, Table A.2 indicates that keto ester **1** should be named as an ester rather than as a ketone because an ester functional group is higher in priority than a ketone. Similarly, amino alcohol **2** should be named as an alcohol rather than as an amine. Thus, the name of **1** is methyl

TABLE A.2 Classification of Functional Groups[a]

Functional group	Name as suffix	Name as prefix
Principal groups		
Carboxylic acids	-oic acid -carboxylic acid	carboxy
Acid anhydrides	-oic anhydride -carboxylic anhydride	—
Esters	-oate -carboxylate	alkoxycarbonyl
Thioesters	-thioate -carbothioate	alkylthiocarbonyl
Acid halides	-oyl halide -carbonyl halide	halocarbonyl
Amides	-amide -carboxamide	carbamoyl
Nitriles	-nitrile -carbonitrile	cyano
Aldehydes	-al -carbaldehyde	oxo
Ketones	-one	oxo
Alcohols	-ol	hydroxy
Phenols	-ol	hydroxy
Thiols	-thiol	mercapto
Amines	-amine	amino
Imines	-imine	imino
Ethers	ether	alkoxy
Sulfides	sulfide	alkylthio
Disulfides	disulfide	—
Alkenes	-ene	—
Alkynes	-yne	—
Alkanes	-ane	—
Subordinate groups		
Azides	—	azido
Halides	—	halo
Nitro compounds	—	nitro

[a]Principal groups are listed in order of decreasing priority; subordinate groups have no priority order.

4-oxopentanoate, and the name of **2** is 5-aminopentan-2-ol. Further examples are shown:

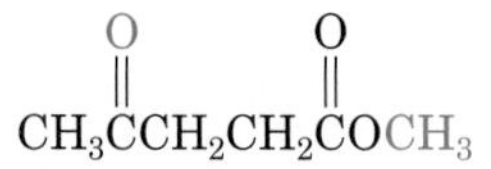

$$CH_3\overset{O}{\overset{\|}{C}}CH_2CH_2\overset{O}{\overset{\|}{C}}OCH_3$$

1. Methyl 4-oxopentanoate
(an ester with a ketone group)

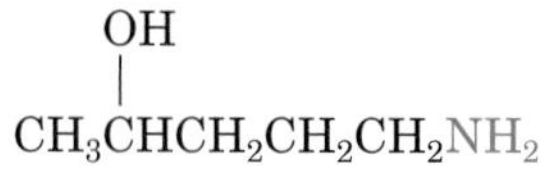

$$CH_3\overset{OH}{\overset{|}{C}}HCH_2CH_2CH_2NH_2$$

2. 5-Aminopentan-2-ol
(an alcohol with an amine group)

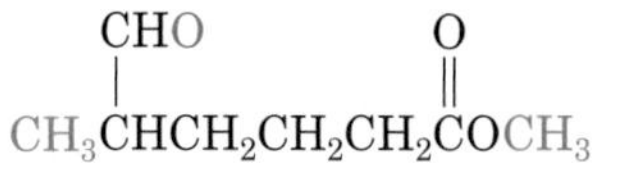

$$CH_3\overset{CHO}{\overset{|}{C}}HCH_2CH_2CH_2\overset{O}{\overset{\|}{C}}OCH_3$$

3. Methyl 5-methyl-6-oxohexanoate
(an ester with an aldehyde group)

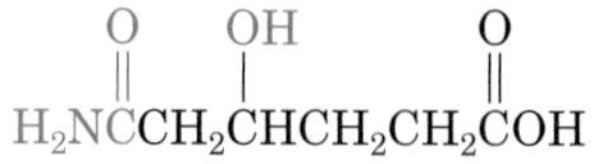

$$H_2N\overset{O}{\overset{\|}{C}}CH_2\overset{OH}{\overset{|}{C}}HCH_2CH_2\overset{O}{\overset{\|}{C}}OH$$

4. 5-Carbamoyl-4-hydroxypentanoic acid
(a carboxylic acid with amide and alcohol group)

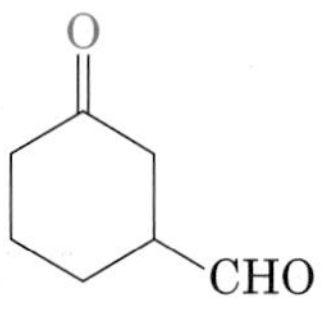

5. 3-Oxocyclohexanecarbaldehyde
(an aldehyde with a ketone group)

Name Part 2. The Parent: Selecting the Main Chain or Ring

The parent, or base, name of a polyfunctional organic compound is usually easy to identify. If the principal group of highest priority is part of an open chain, the parent name is that of the longest chain containing the largest number of principal groups. For example, compounds **6** and **7** are isomeric aldehydo amides, which must be named as amides rather than as aldehydes according to Table A.2. The longest chain in compound **6** has six carbons, and the substance is therefore named 5-methyl-6-oxohexanamide. Compound **7** also has a chain of six carbons, but the longest chain that contains both principal functional groups has only four carbons. The correct name of **7** is 4-oxo-3-propylbutanamide.

$$H\overset{O}{\overset{\|}{C}}\underset{CH_3}{\underset{|}{C}}HCH_2CH_2CH_2\overset{O}{\overset{\|}{C}}NH_2$$

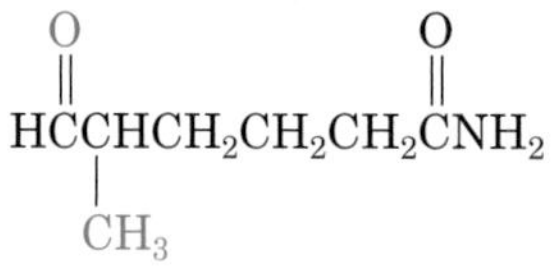

6. 5-Methyl-6-oxohexanamide

$$CH_3CH_2CH_2\overset{CHO}{\overset{|}{C}}HCH_2\overset{O}{\overset{\|}{C}}NH_2$$

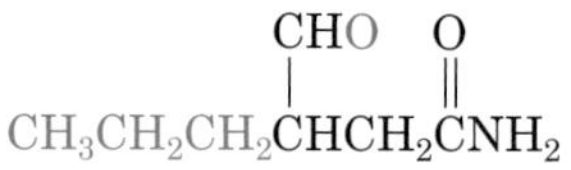

7. 4-Oxo-3-propylbutanamide

If the highest-priority principal group is attached to a ring, the parent name is that of the ring system. Compounds **8** and **9**, for instance, are isomeric keto nitriles and must both be named as nitriles according to Table A.2. Substance **8** is named as a benzonitrile because the –CN functional group is a substituent on the aromatic ring, but substance **9** is named as an acetonitrile because the –CN functional group is on an open chain. The correct names are 2-acetyl-(4-bromomethyl)benzonitrile (**8**) and (2-acetyl-4-bromophenyl)acetonitrile (**9**). As further examples, compounds **10** and **11** are both keto acids and must be named as acids, but the parent name in (**10**) is that of a ring system (cyclohexanecarboxylic acid) and the parent name in (**11**) is that of an open chain (propanoic acid). The full names are *trans*-2-(3-oxopropyl)cyclohexanecarboxylic acid (**10**) and 3-(2-oxocyclohexyl)propanoic acid (**11**).

8. 2-Acetyl-(4-bromomethyl)**benzonitrile**

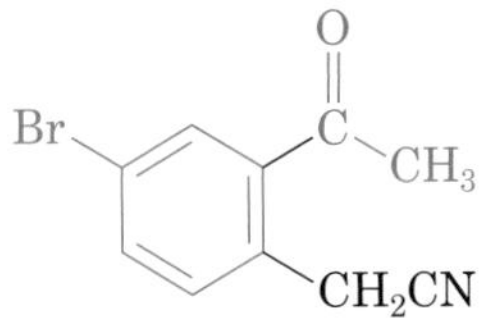

9. (2-Acetyl-4-bromophenyl)**acetonitrile**

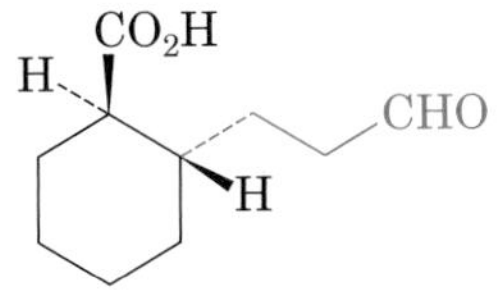

10. ***trans***-2-(3-oxopropyl)**cyclohexanecarboxylic acid**

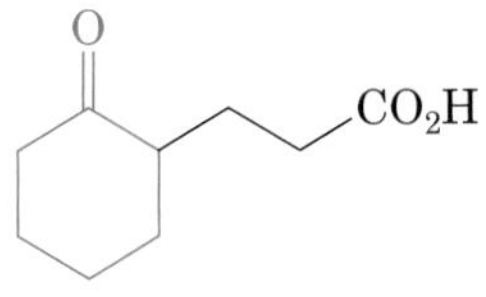

11. 3-(2-Oxocyclohexyl)**propanoic acid**

Name Parts 3 and 4. The Prefixes and Locants

With parent name and suffix established, the next step is to identify and give numbers, or *locants,* to all substituents on the parent chain or ring. These substituents include all alkyl groups and all functional groups other than the one cited in the suffix. For example, compound **12** contains three different functional groups (carboxyl, keto, and double bond). Because the carboxyl group is highest in priority and because the longest chain containing the functional groups has seven carbons, compound **12** is a heptenoic acid. In addition, the main chain has a keto (oxo) substituent and three methyl groups. Numbering from the end nearer the highest-priority functional group, compound **12** is named (*E*)-2,5,5-trimethyl-4-oxohept-2-enoic acid. Look back at some of the other compounds we've named to see other examples of how prefixes and locants are assigned.

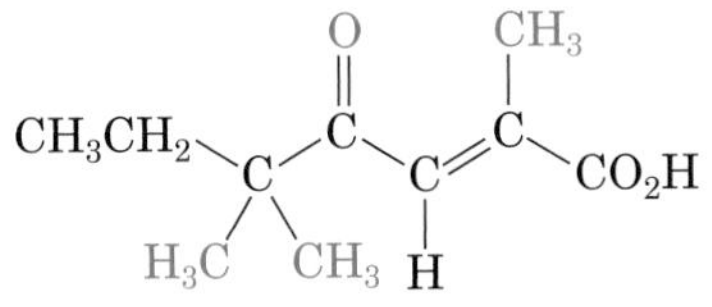

12. (***E***)-2,5,5-Trimethyl-4-oxo**hept-2-enoic acid**

Writing the Name

Once the name parts have been established, the entire name is written out. Several additional rules apply:

1. **Order of prefixes** When the substituents have been identified, the main chain has been numbered, and the proper multipliers such as *di-* and *tri-* have been assigned, the name is written with the substituents listed in alphabetical, rather than numerical, order. Multipliers such as *di-* and *tri-* are not used for alphabetization purposes, but the prefix *iso-* is used.

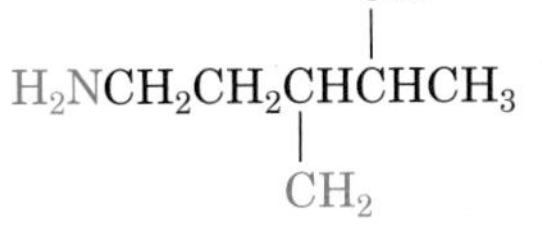

$H_2NCH_2CH_2CH(CH_3)CH(OH)CH_3$

13. 5-Amino-3-methyl**pentan**-2-ol

2. **Use of hyphens; single- and multiple-word names** The general rule is to determine whether the parent is itself an element or compound. If it is, then the name is written as a single word; if it isn't, then the name is written as multiple words. Methylbenzene is written as one word, for instance, because the parent—benzene—is itself a compound. Diethyl ether, however, is written as two words because the parent—ether—is a class name rather than a compound name. Some further examples follow:

$H_3C—Mg—CH_3$

14. Dimethylmagnesium (one word, because magnesium is an element)

$HOCH_2CH_2\overset{O}{\overset{\|}{C}}O\underset{CH_3}{\underset{|}{C}}HCH_3$

15. Isopropyl 3-hydroxypropanoate (two words, because "propanoate" is not a compound)

16. 4-(Dimethylamino)pyridine (one word, because pyridine is a compound)

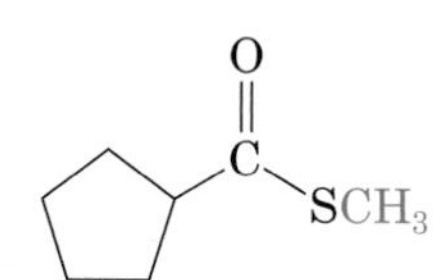

17. Methyl cyclopentanecarbothioate (two words, because "cyclopentanecarbothioate" is not a compound)

3. **Parentheses** Parentheses are used to denote complex substituents when ambiguity would otherwise arise. For example, chloromethylbenzene has two substituents on a benzene ring, but (chloromethyl)benzene has only one complex substituent. Note that the expression in parentheses is not set off by hyphens from the rest of the name.

18. *p*-Chloromethylbenzene

19. (Chloromethyl)benzene

$HO\overset{O}{\overset{\|}{C}}\underset{CH_3CHCH_2CH_3}{\underset{|}{C}}HCH_2CH_2\overset{O}{\overset{\|}{C}}OH$

20. 2-(1-Methylpropyl)pentanedioic acid

Additional Reading

Further explanations of the rules of organic nomenclature can be found online at http://www.acdlabs.com/iupac/nomenclature/ and in the following references:

1. "A Guide to IUPAC Nomenclature of Organic Compounds," CRC Press, Boca Raton, FL, 1993.
2. "Nomenclature of Organic Chemistry, Sections A, B, C, D, E, F, and H," International Union of Pure and Applied Chemistry, Pergamon Press, Oxford, 1979.

APPENDIX B Glossary

Absorption spectrum (Section 13.6): A plot of wavelength of incident light versus amount of light absorbed. Organic molecules show absorption spectra in both the infrared and ultraviolet regions of the electromagnetic spectrum.

Acetal (Section 9.8): A functional group consisting of two ether-type oxygen atoms bonded to the same carbon, $R_2C(OR')_2$.

Acetylide anion (Section 4.12): The anion of a terminal alkyne, $R—C{\equiv}C{:}^-$.

Achiral (Section 6.2): Lacking handedness. A molecule is achiral if it has a plane of symmetry and is thus superimposable on its mirror image.

Acid chloride (Section 10.9): A substance with the general formula RCOCl.

Acidity constant, K_a (Section 1.10): A value that expresses the strength of an acid in water solution. The larger the K_a, the stronger the acid.

Activating group (Section 5.8): An electron-donating group such as hydroxyl (–OH) or amino ($–NH_2$) that increases the reactivity of an aromatic ring toward electrophilic aromatic substitution.

Activation energy, E_{act} (Section 3.9): The difference in energy between ground state and transition state. The amount of activation energy required by a reaction determines the rate at which the reaction proceeds.

Acyl group (Section 5.6): A name for the $—\overset{O}{\overset{\|}{C}}—R$ group.

Acylation (Section 5.6): The introduction of an acyl group, –COR, onto a molecule. For example, acylation of an aromatic ring yields a ketone ($ArH \rightarrow ArCOR$).

1,4-Addition (Section 4.9): The addition of an electrophile to carbons 1 and 4 of a conjugated diene.

Addition reaction (Section 3.5): The reaction that occurs when two reactants combine to form a single new product with no atoms left over.

Alcohol (Section 8.1): A compound with an –OH group bonded to a saturated, sp^3-hybridized carbon atom.

Aldaric acid (Section 14.7): The dicarboxylic acid that results from oxidation of an aldose.

Aldehyde (Section 9.1): A substance with the general formula RCHO.

Alditol (Section 14.7): The polyalcohol that results from reduction of the carbonyl group of a monosaccharide.

Aldol reaction (Section 11.9): A carbonyl condensation reaction between two ketones or aldehydes leading to a β-hydroxy ketone or aldehyde product.

Aldonic acid (Section 14.7): The monocarboxylic acid that results from mild oxidation of an aldose.

Aldose (Section 14.1): A simple sugar with an aldehyde carbonyl group.

Alicyclic (Section 2.7): Referring to an aliphatic cyclic hydrocarbon such as a cycloalkane or cycloalkene.

Aliphatic (Section 2.2): Referring to a nonaromatic hydrocarbon such as a simple alkane, alkene, or alkyne.

Alkaloid (Section 12.7): A naturally occurring compound that contains a basic amine functional group.

Alkane (Section 2.2): A compound that contains only carbon and hydrogen and has only single bonds.

Alkene (Section 3.1): A hydrocarbon with one or more carbon–carbon double bonds.

Alkoxide ion (Section 8.3): The anion formed by loss of H^+ from an alcohol.

Alkyl group (Section 2.2): A part structure, formed by removing a hydrogen from an alkane.

Alkylamine (Section 12.1): An alkyl-substituted amine, RNH_2, as opposed to an aryl-substituted amine, $ArNH_2$.

Alkylation (Sections 5.6 and 11.6): The introduction of an alkyl group onto a molecule. For example, aromatic rings can be alkylated to yield arenes ($ArH \rightarrow ArR$).

Alkyne (Section 4.12): A hydrocarbon that has a carbon–carbon triple bond.

Allylic (Section 4.9): The position next to a double bond.

Alpha amino acid (Section 15.1): A compound with an amino group attached to the carbon atom next to the carboxyl group, $RCH(NH_2)CO_2H$.

Alpha helix (Section 15.10): A common secondary structure in which a protein chain coils into a spiral.

Alpha position (Section 11.1): The position next to a carbonyl carbon.

Alpha substitution reaction (Section 11.2): A reaction that results in substitution of a hydrogen on the α carbon of a carbonyl compound.

Amide (Section 10.12): A substance with the general formula $RCONH_2$.

Amine (Section 12.1): An organic derivative of ammonia, RNH_2, R_2NH, or R_3N.

Amino acid (Section 15.1) See Alpha amino acid.

Amino sugar (Section 14.10): A sugar with an $-NH_2$ group in place of an $-OH$ group on one carbon.

Amplitude (Section 13.1): The height of a wave from midpoint to peak.

Anabolism (Section 17.1): Metabolic reactions that synthesize larger molecules from smaller precursors.

Androgen (Section 16.5): A steroidal male sex hormone such as testosterone.

Angle strain (Section 2.9): The strain introduced into a molecule when a bond angle is deformed from its ideal value.

Anomeric center (Section 14.6): The hemiacetal carbon in a pyranose or furanose sugar.

Anomers (Section 14.6): Cyclic stereoisomers of sugars that differ only in their configurations at the hemiacetal (anomeric) carbon.

Anti stereochemistry (Section 4.5): Referring to opposite sides of a double bond or molecule. For example, an anti addition reaction is one in which the two carbon atoms of the double bond react on different faces.

Anticodon (Section 16.13): A sequence of three bases on tRNA that read the codons on mRNA and bring the correct amino acids into position for protein synthesis.

Antisense strand (Section 16.12): An alternate name for the template strand of DNA.

Apoenzyme (Section 15.10): The protein part of an enzyme that needs a cofactor for biological activity.

Aromatic (Section 5.1): The class of compounds that contain a benzene-like six-membered ring with three double bonds.

Arylamine (Section 12.1): An amine that has its nitrogen atom bonded to an aromatic ring, $Ar{-}NH_2$.

Axial bond (Section 2.10): A bond to chair cyclohexane that lies along the ring axis perpendicular to the rough plane of the ring.

β-Oxidation pathway (Section 17.2): A series of four enzyme-catalyzed reactions that cleave two carbon atoms at a time from the end of a fatty-acid chain.

β-Pleated sheet (Section 15.10): A secondary structure in which a protein chain folds back on itself so that two sections of the chain run parallel.

Basicity constant, K_b (Section 12.3): A value that expresses the strength of a base in water solution. The larger the K_b, the stronger the base.

Benzyl group (Section 5.3): The $C_6H_5CH_2-$ group.

Benzylic (Section 5.9): The position next to an aromatic ring.

Bimolecular reaction (Section 7.5): A reaction that occurs between two molecules.

Bond angle (Section 1.6): The angle formed between two adjacent bonds.

Bond length (Section 1.5): The equilibrium distance between the nuclei of two atoms that are bonded to each other.

Bond strength (Section 1.5): The amount of energy needed to break a bond to produce two radical fragments.

Branched-chain alkane (Section 2.2): An alkane that contains a branching arrangement of carbon atoms in its chain.

Brønsted–Lowry acid (Section 1.10): A substance that donates a hydrogen ion (proton, H^+) to a base.

Brønsted–Lowry base (Section 1.10): A substance that accepts a hydrogen ion, H^+ from an acid.

C-terminus (Section 15.3): The amino acid with a free $-CO_2H$ group at one end of a protein chain.

Carbocation (Section 3.8): A carbon cation, or substance that contains a trivalent, positively charged carbon atom having six electrons in its outer shell (R_3C^+).

Carbocycle (Section 12.6): A cyclic molecule that has only carbon atoms in the ring.

Carbohydrate (Section 14.1): A polyhydroxy aldehyde or polyhydroxy ketone. Carbohydrates can be either simple sugars such as glucose or complex sugars such as cellulose.

Carbonyl condensation reaction (Section 11.8): A reaction between two carbonyl compounds in which the α carbon of one partner bonds to the carbonyl carbon of the other.

Carbonyl group (Section 9.1): The carbon–oxygen double bond functional group, C=O.

Carboxyl group (Section 10.1): The $-CO_2H$ group.

Carboxylic acid (Section 10.1): A substance with the general formula RCO_2H.

Catabolism (Section 17.1): Metabolic reactions that break down large molecules.

Catalyst (Section 3.10): A species that changes the rate of a chemical reaction without being altered or consumed.

Chain-growth polymer (Section 10.14): A polymer produced by chain reaction of a monofunctional monomer.

Chair cyclohexane (Section 2.9): A three-dimensional conformation of cyclohexane that resembles the rough shape of a chair. The chair form of cyclohexane has neither angle strain nor eclipsing strain.

Chemical shift (Section 13.9): The position on the NMR chart where a nucleus absorbs. By convention, the chemical shift of tetramethylsilane is set at zero, and all other absorptions usually occur downfield (to the left on the chart).

Chiral (Section 6.2): Having handedness. A chiral molecule does not have a plane of symmetry, is not superimposable on its mirror image, and thus exists in right- and left-handed forms.

Chiral enviornment (Section 6.1): An environment resulting from the presence of a source of chirality (i.e., enzyme or chiral solid) that allows enantiomers to be distinguished.

Cis–trans isomers (Section 2.8): Stereoisomers that differ in their stereochemistry about a double bond or a ring.

Citric acid cycle (Section 17.4): The third stage of catabolism, in which acetyl groups are degraded to CO_2.

Claisen condensation reaction (Section 11.11): A carbonyl condensation reaction between two esters leading to formation of a β-keto ester product.

Coding strand (Section 16.12): The strand of the DNA double helix that contains genes.

Codon (Section 16.13): A three-base sequence on the mRNA chain that encodes the genetic information necessary to cause specific amino acids to be incorporated into proteins.

Coenzyme (Section 15.10): A small organic molecule that acts as an enzyme cofactor.

Cofactor (Section 15.10): A small nonprotein part of an enzyme necessary for biological activity.

Complex carbohydrate (Section 14.1): A carbohydrate composed of two or more simple sugars linked together by acetal bonds.

Condensed structure (Section 2.2): A shorthand way of drawing structures in which bonds are understood rather than shown.

Configuration (Section 6.6): The three-dimensional arrangement of atoms bonded to a stereocenter.

Conformation (Section 2.5): The exact three-dimensional shape of a molecule at any given instant, assuming that rotation around single bonds is frozen.

Conjugate acid (Section 1.10): The product that results when a base accepts H^+.

Conjugate (1,4) addition reaction (Section 9.12): The addition of a nucleophile to the β carbon atom of an α,β-unsaturated carbonyl compound.

Conjugate base (Section 1.10): The anion that results from dissociation of an acid.

Conjugated protein (Section 15.8): A protein composed of both an amino acid part and a non–amino acid part.

Conjugation (Section 4.9): A series of alternating single and multiple bonds with overlapping *p* orbitals.

Constitutional isomers (Section 2.2): Isomers such as butane and 2-methylpropane, which have their atoms connected in a different order.

Coupling (Section 13.12): The interaction of neighboring nuclear spins that results in spin–spin splitting.

Coupling constant (*J*) (Section 13.12): The magnitude of the spin–spin splitting interaction between nuclei whose spins are coupled.

Covalent bond (Section 1.5): A bond formed by sharing electrons between two nuclei.

Cycloalkane (Section 2.7): An alkane with a ring of carbon atoms.

D Sugar (Section 14.3): A sugar whose hydroxyl group at the stereocenter farthest from the carbonyl group points to the right when the molecule is drawn in Fischer projection.

Deactivating group (Section 5.8): An electron-withdrawing substituent that decreases the reactivity of an aromatic ring toward electrophilic aromatic substitution.

Decarboxylation (Section 11.6): A reaction that involves loss of CO_2. β-Keto acids decarboxylate readily on heating.

Dehydration (Section 8.5): Elimination of water from an alcohol to yield an alkene.

Dehydrohalogenation (Section 7.7): Elimination of HX from an alkyl halide to yield an alkene on treatment with a strong base.

Delta (δ) scale (Section 13.9): The arbitrary scale used for defining the position of NMR absorptions. $1\ \delta = 1$ ppm of spectrometer frequency.

Deoxy sugar (Section 14.10): A sugar with an –OH group missing from one carbon.

Deoxyribonucleic acid (DNA) (Section 16.7): A biopolymer of deoxyribonucleotide units.

Deshielding (Section 13.9): An effect observed in NMR that causes a nucleus to absorb downfield because of a withdrawal of electron density from the nucleus.

Dextrorotatory (Section 6.3): An optically active substance that rotates the plane of polarization of plane-polarized light in a right-handed (clockwise) direction.

Diastereomers (Section 6.7): Stereoisomers that have a non–mirror-image relationship.

1,3-Diaxial interaction (Section 2.11): A spatial interaction between two axial substituents separated by three carbons in a substituted chair cyclohexane.

Digestion (Section 17.1): The first stage of catabolism, in which food molecules are hydrolyzed to yield fatty acids, amino acids, and monosaccharides.

Disaccharide (Section 14.8): A complex carbohydrate having two simple sugars bonded together.

Distillation (Section 2.4): Separation based on differences in boiling points.

Disulfide link (Section 15.5): A sulfur–sulfur link between two cysteine residues in a peptide.

DNA (Section 16.7): See Deoxyribonucleic acid.

Double helix (Section 16.9): The conformation into which double-stranded DNA coils.

Downfield (Section 13.9): The left-hand portion of the NMR chart.

E1 reaction (Section 7.8): An elimination reaction that takes place in two steps through a unimolecular mechanism.

E2 reaction (Section 7.7): An elimination reaction that takes place in a single step through a bimolecular mechanism.

***E,Z* system** (Section 3.4): A series of sequence rules for specifying the cis–trans geometry of double bonds.

Eclipsed conformation (Section 2.5): The geometric arrangement around a carbon–carbon single bond in which the bonds on one carbon are parallel to the bonds on the neighboring carbon as viewed in a Newman projection.

Edman degradation (Section 15.7): A method for selectively cleaving the N-terminal amino acid from a peptide.

Electromagnetic spectrum (Section 13.1): The range of electromagnetic energy, including infrared, ultraviolet, and visible radiation.

Electron shell (Section 1.1): An imaginary layer around the nucleus occupied by electrons.

Electron-transport chain (Section 17.1): The fourth stage of catabolism, in which ATP is synthesized.

Electronegativity (Section 1.9): The ability of an atom to attract electrons and thereby polarize a covalent bond. Electronegativity generally increases from left to right and from bottom to top of the periodic table.

Electrophile (Section 3.7): An "electron-lover," or substance that accepts an electron pair from a nucleophile in a polar bond-forming reaction.

Electrophilic aromatic substitution reaction (Section 5.4): The substitution of an electrophile for a hydrogen atom on an aromatic ring.

Electrophoresis (Section 15.2): A technique for separating charged organic molecules, particularly proteins and amino acids, by placing them in an electric field.

Elimination reaction (Section 3.5): The reaction that occurs when a single reactant splits apart into two products.

Enantiomers (Section 6.1): Stereoisomers that have a mirror-image relationship, with opposite configurations at all stereocenters.

3′ End (Section 16.8): The end of a nucleic acid chain that has a free sugar hydroxyl group.

5′ End (Section 16.8): The end of a nucleic acid chain that has a phosphoric acid unit.

Enol (Section 11.1): A vinylic alcohol, C=C—OH.

Enolate ion (Sections 9.11 and 11.1): The anion of an enol; a resonance-stabilized α-keto carbanion.

Enone (Section 11.10): An unsaturated ketone.

Entgegen (*E*) (Section 3.4): A term used to describe the stereochemistry of a carbon–carbon double bond in which high-priority groups on each carbon are on opposite sides of the double bond.

Enzyme (Sections 10.15 and 15.10): A biological catalyst. Enzymes are large proteins that catalyze specific biochemical reactions.

Epoxide (Section 8.8): A three-membered ring ether functional group.

Equatorial bond (Section 2.10): A bond to cyclohexane that lies along the rough equator of the ring. (See Axial bond.)

Essential amino acid (Section 15.1): An amino acid that must be obtained in the diet.

Ester (Section 10.11): A substance with the general formula RCO_2R'.

Esterase (Section 10.15): An enzyme that catalyzes the hydrolysis of esters.

Estrogen (Section 16.5): A female steroid sex hormone.

Ether (Section 8.1): A compound with two organic groups bonded to the same oxygen atom, R–O–R′.

Fat (Section 16.2): A solid triacylglycerol derived from animal sources.

Fatty acid (Section 16.2): A long straight-chain carboxylic acid found in fats and oils.

Fibrous protein (Section 15.8): A protein that consists of polypeptide chains arranged side by side in long threads.

Fingerprint region (Section 13.6): The complex region of the infrared spectrum from 1500 cm^{-1} to 400 cm^{-1}.

Fischer esterification reaction (Section 10.8): The conversion of a carboxylic acid into an ester by acid-catalyzed reaction with an alcohol.

Fischer projection (Section 14.2): A method for depicting the configuration of a stereocenter using crossed lines. Horizontal bonds come out of the plane of the page, and vertical bonds go back into the plane of the page.

Fractional distillation (Section 2.4): A distillation where multiple fractions are collected to afford separation of complex mixtures.

Frequency (ν) (Section 13.1): The number of electromagnetic wave cycles that travel past a fixed point in a given unit of time, usually expressed in reciprocal seconds, s^{-1}, or Hertz.

Friedel–Crafts reaction (Section 5.6): The introduction of an alkyl or acyl group onto an aromatic ring by an electrophilic substitution reaction.

Functional group (Section 2.1): An atom or group of atoms that is part of a larger molecule and has a characteristic chemical reactivity.

Furanose (Section 14.5): The five-membered ring structure of a simple sugar.

Geminal (Section 9.7): Referring to two groups attached to the same carbon atom.

Globular protein (Section 15.8): A protein that is coiled into a compact, nearly spherical shape.

Glycol (Section 8.8): A diol, such as ethylene glycol, $HOCH_2CH_2OH$.

Glycolysis (Section 17.3): A series of ten enzyme-catalyzed reactions that break down a glucose molecule into two pyruvate molecules.

Glycoside (Section 14.7): A cyclic acetal formed by reaction of a sugar with another alcohol.

Grignard reagent (Section 7.3): An organomagnesium halide, RMgX.

Ground-state electron configuration (Section 1.2): A list of orbitals occupied by the electrons in an atom in its lowest-energy state.

Hemiacetal (Section 9.8): A compound that has one –OR group and one –OH group bonded to the same carbon atom.

Hertz (Hz) (Section 13.9): The standard unit for frequency.

Heterocycle (Section 12.6): A cyclic molecule whose ring contains more than one kind of atom.

Heterogenic (Section 3.6): Electronically unsymmetrical formation of a covalent bond by combination of an anion and a cation.

Heterolytic (Section 3.6): Electronically unsymmetrical breaking of a covalent bond to yield an anion and a cation.

Holoenzyme (Section 15.10): The combination of enzyme and cofactor.

Homogenic (Section 3.6): Electronically symmetrical formation of a covalent bond by combination of two radicals.

Homolytic (Section 3.6): Electronically symmetrical breaking of a covalent bond to yield two radicals.

Hormone (Section 16.5): A chemical messenger secreted by a specific gland and carried through the bloodstream to affect a target tissue.

Hybrid orbital (Section 1.6): An orbital derived from a combination of ground-state atomic orbitals. Hybrid orbitals, such as the sp^3, sp^2, and sp hybrids of carbon, are strongly directed and form strong bonds.

Hydration (Section 4.4): Addition of water to a molecule, such as occurs when alkenes are treated with strong aqueous acid.

Hydrocarbon (Section 2.2): A compound that has only carbon and hydrogen.

Hydrogen bond (Section 8.2): An attraction between a hydrogen atom bonded to an electronegative atom and an electron lone pair on another atom.

Hydrogenation (Section 4.6): Addition of hydrogen to a double or triple bond to yield a saturated product.

Hydroquinone (Section 8.6): A compound that contains a *p*-dihydroxybenzene group.

Hydroxylation (Section 4.7): The addition of one or more –OH groups to a molecule.

Imine (Section 9.10): A compound with a C=N functional group.

Inductive effect (Section 1.9): The electron-attracting or electron-withdrawing effect that is transmitted through σ bonds.

Infrared (IR) spectroscopy (Section 13.6): A kind of optical spectroscopy that uses infrared energy. IR spectroscopy is particularly useful in organic chemistry for determining the kinds of functional groups in molecules.

Integration (Section 13.11): A means of electronically measuring the ratios of the number of nuclei responsible for each peak in an NMR spectrum.

Intermediate (Section 3.9): A species that is formed during the course of a multistep reaction but is not the final product.

Ionic bond (Section 1.4): A bond between two ions due to the electrical attraction of unlike charges.

Isoelectric point (Section 15.2): The pH at which the number of positive charges and the number of negative charges on a protein or amino acid are exactly balanced.

Isomers (Section 2.2): Compounds with the same molecular formula but different structures.

Isotopes (Section 13.3): Atoms of an element that vary in the number of neutrons and as a result, vary in molecular weight.

Kekulé structure (Section 1.4): A representation of a molecule in which a line between atoms represents a covalent bond.

Ketone (Section 9.1): A substance with the general formula $R_2C{=}O$.

Ketose (Section 14.1): A simple sugar with a ketone carbonyl group.

L Sugar (Section 14.3): A sugar whose hydroxyl group at the stereocenter farthest from the carbonyl group points to the left when the molecule is drawn in Fischer projection.

Leaving group (Section 7.5): The group that is replaced in a substitution reaction.

Levorotatory (Section 6.3): An optically active substance that rotates the plane of polarization of plane-polarized light in a left-handed (counterclockwise) direction.

Lewis acid (Section 1.11): A substance with a vacant low-energy orbital that can accept an electron pair from a base.

Lewis base (Section 1.11): A substance that donates an electron lone-pair to an acid.

Lewis structure (Section 1.4): A representation of a molecule showing covalent bonds as a pair of electron dots between atoms.

Line-bond structure (Section 1.4): A representation of a molecule showing covalent bonds as lines between atoms.

1,4′ Link (Section 14.8): A glycosidic link between the C1 carbonyl group of one sugar and the C4 hydroxyl group of another sugar.

Lipid (Section 16.1): A naturally occurring substance that can be isolated from plants or animals by extraction with a nonpolar organic solvent.

Lipid bilayer (Section 16.4): The double layer of phospholipids that makes up cell walls.

Lone-pair electrons (Section 1.4): A nonbonding electron pair that occupies a valence orbital.

Major groove (Section 16.9): The larger of two grooves in double helical DNA.

Malonic ester synthesis (Section 11.6): A method for forming α-substituted acetic acids by reaction of diethyl malonate with an alkyl halide, followed by decarboxylation.

Markovnikov's rule (Section 4.2): A guide for determining the regiochemistry (orientation) of electrophilic addition reactions. In the addition of HX to an alkene, the hydrogen atom becomes bonded to the alkene carbon that has fewer alkyl substituents.

Mass spectrometry (Section 13.3): An analytical method that provides the mass-to-charge ratio (m/z) of an organic molecule. When the molecule has a single charge, m/z corresponds to the molecular weight of the molecule.

Mass-to-charge ratio (Section 13.3): The data provided by mass spectrometry, m/z. When the molecule has a single charge, m/z corresponds to the molecular weight of the molecule.

Mechanism (Section 3.6): A complete description of how a reaction occurs. A mechanism accounts for all reactants and all products and describes the details of each individual step in the overall reaction process.

Mercapto group (Section 8.9): An alternative name for the thiol group, –SH.

Meso (Section 6.8): A compound that contains one or more chirality centers but is nevertheless achiral because it has a symmetry plane.

Messenger RNA (mRNA) (Section 16.12): The kind of RNA transcribed from DNA.

Meta director (Section 5.8): A group on an aromatic ring that directs substitution to the meta position.

Metabolism (Section 17.1): The total of all reactions in living organisms.

Micelle (Section 16.3): A spherical cluster of soaplike molecules that aggregate in aqueous solution. The ionic heads of the molecules lie on the outside, where they are solvated by water, and the organic tails bunch together on the inside of the micelle.

Minor groove (Section 16.9): The smaller groove in double helical DNA.

Molecule (Section 1.5): A group of atoms joined by covalent bonds.

Monomer (Sections 4.8 and 10.11): The starting unit from which a polymer is made.

Monosaccharide (Section 14.1): A simple sugar.

Multiplet (Section 13.12): An NMR splitting pattern caused by coupling with neighboring protons.

Mutarotation (Section 14.6): The spontaneous change in optical rotation observed when a pure anomer of a sugar is dissolved in water and equilibrates to an equilibrium mixture of anomers.

***n* + 1 rule** (Section 13.12): The signal of a proton with n neighboring protons splits into $n + 1$ peaks in the NMR spectrum.

N-terminus (Section 15.3): The amino acid with a free $-NH_2$ group at one end of a protein chain.

Newman projection (Section 2.5): A way of viewing a molecule's spatial arrangement by looking end-on at a carbon–carbon bond.

Nitrile (Section 10.13): A compound with a $-C{\equiv}N$ functional group.

Nonbonding electron (Section 1.4): A valence electron not used for bonding.

Normal (*n*) alkane (Section 2.2): A straight-chain alkane.

Nuclear magnetic resonance (NMR) (Section 13.7): A spectroscopic technique that provides information about the carbon–hydrogen framework of a molecule.

Nucleic acid (Section 16.7): A biopolymer, either DNA or RNA, made of nucleotides joined together.

Nucleophile (Section 3.7): An electron-rich species that can donate an electron pair to an electrophile in a polar reaction.

Nucleophilic acyl substitution reaction (Section 10.5): A substitution reaction that replaces one nucleophile bonded to a carbonyl group by another.

Nucleophilic addition reaction (Section 9.5): A reaction that involves the addition of a nucleophile to a carbonyl group.

Nucleophilic substitution reaction (Section 7.4): A substitution reaction in which one nucleophile replaces another.

Nucleoside (Section 16.7): A nucleic acid constituent, consisting of a sugar residue bonded to a heterocyclic purine or pyrimidine base.

Nucleotide (Section 16.7): A nucleic acid constituent, consisting of a sugar residue bonded both to a heterocyclic purine or pyrimidine base and to phosphoric acid.

Nylon (Section 10.14): A polyamide prepared by reaction between a diacid and a diamine.

Octane number (Section 2.4): A rating applied to gasoline that describes performance of a complex mixture instead of its exact composition.

Optical activity (Section 6.3): The ability of a chiral molecule in solution to rotate plane-polarized light.

Orbital (Section 1.1): A region of space occupied by a given electron or pair of electrons.

Organic chemistry (Section 1.1): The chemistry of carbon compounds.

Ortho director (Section 5.8): A group on an aromatic ring that directs substitution to the ortho-position.

Oxidation (Section 4.7): The addition of oxygen to a molecule or removal of hydrogen from it.

Oxirane (Section 8.8): An alternative name for an epoxide.

Oxonium ion (Section 9.8): An oxygen with three bonds and a positive charge.

Para director (Section 5.8): A group on an aromatic ring that directs substitution to the para-position.

Paraffin (Section 2.4): A common name for an alkane.

Peptide (Section 15.1): A small amino acid polymer in which the individual amino acid residues are linked by amide bonds. (See Proteins.)

Phenol (Section 8.1): A compound with an –OH group bonded to an aromatic ring.

Phenoxide ion (Section 8.3): The anion formed by loss of H^+ from a phenol.

Phenyl group (Section 5.3): The $-C_6H_5$ group.

Phosphoglyceride (Section 16.4): A phospholipid in which glycerol has ester links to two fatty acids and to phosphoric acid.

Phospholipid (Section 16.4): A lipid that contains a phosphate residue.

Phosphorylation (Section 17.1): A reaction that transfers a phosphate group from a phosphoric anhydride to an alcohol.

Pi (π) bond (Section 1.8): A covalent bond formed by sideways overlap of two *p* orbitals.

Plane-polarized light (Section 6.3): Light that has its electric vectors in a single plane rather than in random planes.

Polar covalent bond (Section 1.9): A covalent bond in which the electrons are shared unequally between the atoms.

Polar reaction (Section 3.6): A reaction in which bonds are made when a nucleophile donates two electrons to an electrophile and in which bonds are broken when one fragment leaves with both electrons from the bond.

Polarity (Sections 1.9): The unsymmetrical distribution of electrons in a molecule that results when one atom attracts electrons more strongly than another.

Polycyclic aromatic hydrocarbon (Section 5.10): A molecule that has two or more benzene rings fused together.

Polyester (Section 10.14): A polymer prepared by reaction between a diacid and a dialcohol.

Polymer (Sections 4.8 and 10.14): A large molecule made up of repeating smaller units.

Polymerase chain reaction (PCR) (Section 16.15): A method for amplifying small amounts of DNA to prepare large amounts.

Polysaccharide (Section 14.9): A complex carbohydrate having many simple sugars bonded together by acetal links.

Polyunsaturated fatty acid (Section 16.2): A fatty acid with more than one double bond in its chain.

Primary amine (Section 12.1): An amine with one organic substituent on nitrogen, RNH_2.

Primary structure (Section 15.9): The amino acid sequence of a protein.

Protecting group (Section 9.8): A group that is temporarily introduced into a molecule to protect a functional group from reaction elsewhere in the molecule.

Protein (Section 15.1): A large biological polymer containing 50 or more amino acid residues.

Pyranose (Section 14.5): The six-membered ring structure of a simple sugar.

Quaternary ammonium salt (Section 12.1): A compound with four organic substituents attached to a positively charged nitrogen, $R_4N^+ X^-$.

Quaternary structure (Section 15.9): The way in which several protein molecules aggregate together to yield a larger structure.

Quinone (Section 8.6): A compound that contains the cyclohexadienedione functional group.

R group (Section 2.2): A general symbol used for an organic partial structure.

***R,S* convention** (Section 6.6): A method for defining the absolute configuration around a stereocenter.

Racemic mixture (Section 6.10): A 50:50 mixture of the two enantiomers of a chiral substance.

Radical (Section 3.6): A species that has an odd number of electrons, such as the chlorine radical, Cl·.

Radical reaction (Section 3.6): A reaction in which bonds are made by donation of one electron from each of two reagents and in which bonds are broken when each fragment leaves with one electron.

Reaction energy diagram (Section 3.9): A graph depicting the energy changes that occur during a reaction.

Reaction intermediate (Section 3.9): A substance formed transiently during the course of a multistep reaction.

Reaction mechanism (Section 3.6): A complete description of how a reaction occurs.

Reaction rate (Section 7.5): The exact speed of a reaction under defined conditions.

Rearrangement reaction (Section 3.5): The reaction that occurs when a single reactant undergoes a reorganization of bonds and atoms to give an isomeric product.

Reducing sugar (Section 14.7): A sugar that reduces Ag^+ in the Tollens test or Cu^{2+} in Fehling's or Benedict's tests.

Reduction (Section 4.6): The addition of hydrogen to a molecule or the removal of oxygen from it.

Reductive amination (Section 12.4): A method for synthesizing amines by treatment of an aldehyde or ketone with ammonia or an amine in the presence of a reducing agent.

Replication (Section 16.11): The process by which double-stranded DNA uncoils and is replicated to produce two new copies.

Resolution (Section 6.10): Separation of a racemic mixture into its pure component enantiomers.

Resonance forms (Section 4.10): Representations of a molecule that differ only in where the bonding electrons are placed.

Resonance hybrid (Section 4.10): The composite structure of a molecule described by different resonance forms.

Restriction endonuclease (Section 16.14): An enzyme that is able to cut a DNA strand at a specific base sequence in the chain.

Ribonucleic acid (RNA) (Section 16.7): A biopolymer of ribonucleotide units.

Ribosomal RNA (rRNA) (Section 16.12): A kind of RNA that makes up ribosomes.

Ribozyme (Section 16.16): An RNA that catalyzes a chemical reaction.

Ring-flip (Section 2.11): The molecular motion that converts one chair conformation of cyclohexane into another chair conformation, thereby interconverting axial and equatorial bonds.

RNA (Section 16.8): See Ribonucleic acid.

Saccharide (Section 14.1): A sugar.

Salt bridge (Section 15.9): The ionic attraction between charged amino acid side chains that helps stabilize a protein's tertiary structure.

Sanger dideoxy method (Section 16.14): A method for sequencing DNA strands.

Saponification (Sections 10.11 and 16.3): An old term for the base-induced hydrolysis of an ester to yield a carboxylic acid salt.

Saturated (Section 2.2): A compound that has only single bonds.

Sawhorse structure (Section 2.5): A perspective view of the conformation around single bonds.

Secondary amine (Section 12.1): An amine with two organic substituents on nitrogen, R_2NH.

Secondary structure (Section 15.9): The specific way in which segments of a protein chain are oriented into a regular pattern.

Semiconservative replication (Section 16.11): A description of DNA replication in which each new DNA molecule contains one old strand and one new strand.

Sense strand (Section 16.12): An alternative name for the coding strand of DNA.

Sequence rules (Sections 3.4 and 6.6): A series of rules for assigning relative priorities to substituent groups on a double-bond carbon atom or on a stereocenter.

Shielding (Section 13.8): An effect observed in NMR that causes a nucleus to absorb toward the right (upfield) side of the chart. Shielding is caused by donation of electron density to the nucleus. (See Deshielding.)

Side chain (Section 15.1): The organic substituent bonded to the α carbon of an α amino acid.

Sigma (σ) bond (Section 1.6): A covalent bond formed by head-on overlap of atomic orbitals.

Simple protein (Section 15.8): A protein composed entirely of amino acids.

Simple sugar (Section 14.1): A carbohydrate like glucose that can't be hydrolyzed to smaller molecules.

Skeletal structure (Section 2.6): A shorthand way of drawing structures that shows only bonds, not atoms.

S_N1 reaction (Section 7.6): A nucleophilic substitution reaction that takes place in two steps through a carbocation intermediate.

S_N2 reaction (Section 7.5): A nucleophilic substitution reaction that takes place in a single step by backside displacement of the leaving group.

***sp* Hybrid orbital** (Section 1.8): An atomic orbital formed by combination of one *s* and one *p* atomic orbital.

***sp*^{2} Hybrid orbital** (Section 1.8): An atomic orbital formed by combination of one *s* and two *p* atomic orbitals.

***sp*^{3} Hybrid orbital** (Section 1.6): An atomic orbital formed by combination of one *s* and three *p* atomic orbitals.

Specific rotation, $[\alpha]_D$ (Section 6.4): The amount by which an optically active compound rotates plane-polarized light under standard conditions.

Sphingolipid (Section 16.4): A phospholipid based on the sphingosine backbone rather than on glycerol.

Spin–spin splitting (Section 13.12): The splitting of an NMR signal into a multiplet caused by an interaction between nearby magnetic nuclei whose spins are coupled.

Staggered conformation (Section 2.5): The three-dimensional arrangement of atoms around a carbon–carbon single bond in which the bonds on one carbon bisect the bond angles on the second carbon as viewed end-on.

Statin drug (Section 16.6): A drug that is used to control naturally occurring high cholesterol.

Step-growth polymer (Section 10.14): A polymer produced by a series of polar reactions between two difunctional monomers.

Stereocenter (Section 6.2): An atom in a molecule that is a cause of chirality.

Stereochemistry (Section 6.1): The branch of chemistry concerned with the three-dimensional arrangement of atoms in molecules.

Stereoisomers (Section 2.8): Isomers that have their atoms connected in the same order but with a different three-dimensional arrangement.

Steric strain (Section 2.11): The strain imposed on a molecule when two groups are too close together and try to occupy the same space.

Steroid (Section 16.5): A lipid whose structure is based on a characteristic tetracyclic carbon skeleton.

Straight-chain alkane (Section 2.2): An alkane whose carbon atoms are connected in a row.

Substitution reaction (Section 3.5): The reaction that occurs when two reactants exchange parts to give two products.

Sulfide (Section 8.9): A compound that has two organic groups bonded to a sulfur atom, R–S–R′.

Syn stereochemistry (Section 4.6): A syn addition reaction is one in which the two ends of the double bond are attacked from the same face.

Tautomers (Section 11.1): Isomers that are rapidly interconverted.

Template strand (Section 16.12): The strand of the DNA double helix that is used for transcription.

Tertiary amine (Section 12.1): An amine with three organic substituents on nitrogen, R_3N.

Tertiary structure (Section 15.9): The way in which a protein molecule is oriented into an overall three-dimensional shape.

Thioester (Section 17.1): The sulfur analog of an ester, RCOSR′.

Thiol (Section 8.9): A compound with the –SH functional group.

Transamination (Section 17.5): A reaction in which the NH_2 group of an amine changes places with the keto group of an α-keto acid.

Transcription (Section 16.12): The process by which RNA is synthesized from DNA.

Transition state (Section 3.9): An activated complex between reactants, representing the highest energy point on a reaction curve.

Translation (Section 16.13): The process by which the genetic information transcribed from DNA onto mRNA is read by tRNA and used to direct protein synthesis.

Triacylglycerol (Section 16.2): A lipid such as animal fat or vegetable oil; a triester of glycerol with long-chain fatty acids.

Ultraviolet (UV) spectroscopy (Section 13.4): An optical spectroscopy employing ultraviolet irradiation. UV spectroscopy provides structural information about the extent of electron conjugation in organic molecules.

Unimolecular (Section 7.6): A reaction step that involves only one molecule.

Unsaturated (Section 3.1): A molecule that has one or more double or triple bonds and thus has fewer hydrogens than the corresponding alkane.

α,β-Unsaturated carbonyl compound (Section 9.12): A compound containing the C=C–C=O functional group.

Upfield (Section 13.9): The right-hand portion of the NMR chart.

Valence bond theory (Section 1.5): A theory of chemical bonding that describes bonds as resulting from overlap of atomic orbitals.

Valence shell (Section 1.4): The outermost electron shell of an atom.

Vinylic (Section 4.12): Referring to a substituent attached to a double-bond carbon atom.

Vitamin (Section 15.10): A small organic molecule that must be obtained in the diet and that is required for proper growth.

Wavelength (λ) (Section 13.1): The length of a wave from peak to peak.

Wavenumber ($\tilde{\nu}$) (Section 13.6): A unit of frequency measurement equal to the reciprocal of the wavelength in centimeters, cm^{-1}.

Williamson ether synthesis (Section 8.5): The reaction of an alkoxide ion with an alkyl halide to yield an ether.

X-ray crystallography (Section 13.2): A method that provides the three-dimensional connectivity of atoms in a molecule.

Zaitsev's rule (Section 7.7): A rule stating that E2 elimination reactions normally yield the more highly substituted alkene as major product.

Zusammen (*Z*) (Section 3.4): A term used to describe the stereochemistry of a carbon–carbon double bond in which the two high priority groups on each carbon are on the same side of the double bond.

Zwitterion (Section 15.2): A neutral dipolar molecule whose positive and negative charges are not adjacent.

Answers to Selected In-Chapter Problems

The following answers to in-chapter problems are meant only as a quick check. Full answers and explanations for all problems, both in-chapter and end-of-chapter, are provided in the accompanying *Study Guide and Solutions Manual.*

Chapter 1

1.1 (a) 1 (b) 2 (c) 3

1.2 (a) B: $1s^2\ 2s^2\ 2p$ (b) P: $1s^2\ 2s^2\ 2p^6\ 3s^2\ 3p^3$
(c) O: $1s^2\ 2s^2\ 2p^4$ (d) Ar: $1s^2\ 2s^2\ 2p^6\ 3s^2\ 3p^6$

1.3

```
     Cl
     |
     C
   .' /\
  H  /  H
    H
```

1.4

```
  H        H
   \      /..H
    C — C
  H/ '     \
   H        H
```

1.5 (a) CCl_4 (b) AlH_3 (c) CH_2Cl_2 (d) SiF_4

1.6

```
       ..
      :Cl:                 Cl                H          H                  H           H
       ..  ..              |                 ..         |                  ..  ..      |
(a) H:C:Cl:            H—C—Cl        (b) :S:H      S—H       (c) H:C:N:H    H—C—N—H
       ..  ..              |                 ..                            ..  ..      |   |
      :Cl:                 Cl                                              H   H       H   H
       ..
```

1.7 C_2H_7 has too many hydrogens for a compound with two carbons.

1.8

```
     Cl
     |
     C
   .' /\
 Cl  /  Cl
    Cl
```

1.9 A carbon atom is larger than a hydrogen atom.

1.10

```
    H   H   H                     H  H
    |   |   |                      \ /
H — C — C — C — H             H    C    H
    |   |   |                  \  / \  /
    H   H   H                   C     C
                               / \   / \
                              H   H H   H
```

1.11

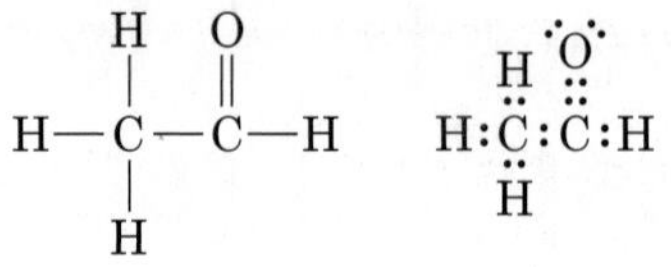

```
    H  O             ··
    |  ||          H :O:
H — C — C — H     H:C:C:H
    |              H
    H
```

1.12 The CH_3 carbon is sp^3, the double-bond carbons are sp^2, and the C=C–C bond angle is approximately 120°.

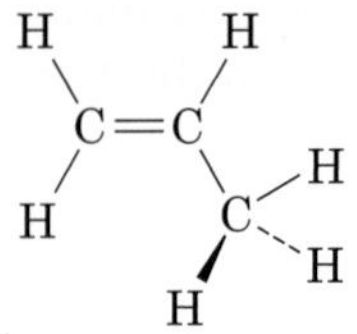

```
H       H
 \     /
  C = C
 /     \    H
H       C <
       /    H
      H
```

1.13 All carbons are sp^2, and all bond angles are near 120°.

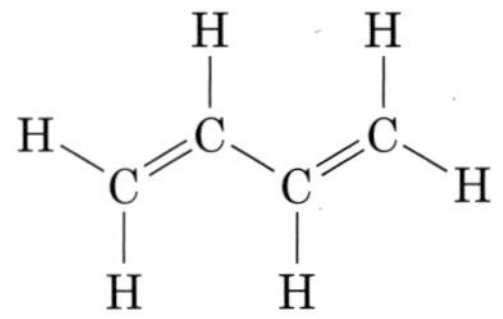

```
       H       H
       |       |
H     C       C
  \ //  \   //  \
   C     C        H
   |     |
   H     H
```

1.14

```
       H     :O:
       |      ||
H     C       C     ··    H
  \ //  \   /   \  O  /
   C     C          ··
   |     ||
   C     C     ··
H /  \\ / \   O:
      C       |       H
      |       C      /
      H     //  \   C  — H
         :O:      \ |
                    H
```

sp^3, all other C are sp^2

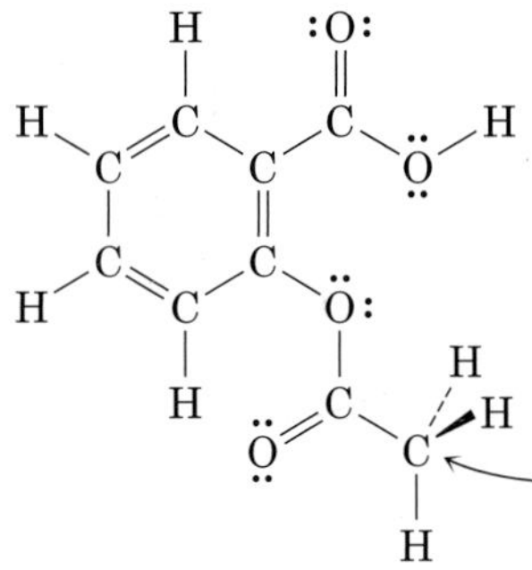

1.15 The CH_3 carbon is sp^3, the triple-bond carbons are sp, and the C≡C–C bond angle is 180°.

1.16 (a) H (b) Br (c) Cl

1.17 (a) C is δ+, Br is δ− (b) C is δ+, N is δ− (c) H is δ+, N is δ−
(d) C is δ+, O is δ− (e) Mg is δ+, C is δ− (f) C is δ+, F is δ−

1.18 CCl_4 and ClO_2 < $TiCl_3$ < $MgCl_2$

1.19

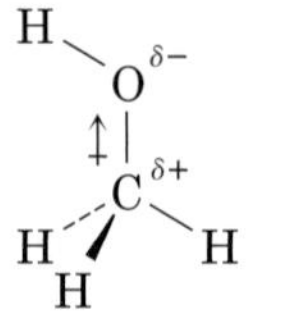

```
H
  \  δ−
   O
 ↕ |
   C δ+
 /   \
H  H   H
```

1.20 (a) Formic acid: $K_a = 1.8 \times 10^{-4}$; picric acid: $K_a = 0.42$
(b) Picric acid is stronger.

1.21 Water is stronger than ammonia because its conjugate base is weaker than NH_2^-.

1.22 (a) No (b) No

1.23 Lewis acids: (c), (d), (e); Lewis bases: (b), (f); both: (a)

1.24 (a) $CH_3CH_2OH + HCl \rightarrow CH_3CH_2OH_2^+\ Cl^-$;
$(CH_3)_2NH + HCl \rightarrow (CH_3)_2NH_2^+\ Cl^-$; $(CH_3)_3P + HCl \rightarrow (CH_3)_3PH^+\ Cl^-$
(b) $HO^- + CH_3^+ \rightarrow HO{-}CH_3$; $HO^- + B(CH_3)_3 \rightarrow HO{-}B(CH_3)_3^-$;
$HO^- + MgBr_2 \rightarrow HO{-}MgBr_2^-$

1.25

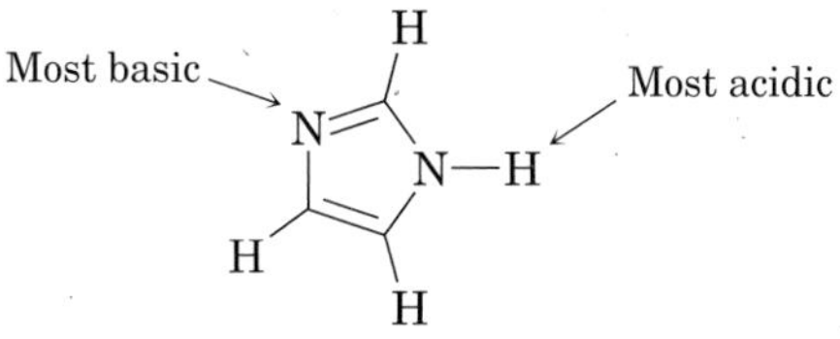

Most basic → N (ring nitrogen of imidazole, N=C)
Most acidic → N–H

(Imidazole ring: N=CH–N(H)–CH=CH, with H atoms on each ring carbon)

Chapter 2

2.1 (a) Carboxylic acid, double bond (b) Carboxylic acid, aromatic ring, ester
(c) Aldehyde, alcohols

2.2 (a) CH_3OH (b) benzene ring (c) $CH_3\overset{O}{\overset{\|}{C}}OH$ (d) CH_3NH_2

(e) $CH_3\overset{O}{\overset{\|}{C}}CH_2CH_2NH_2$ (f) $H_2C{=}CHCH{=}CH_2$

2.3

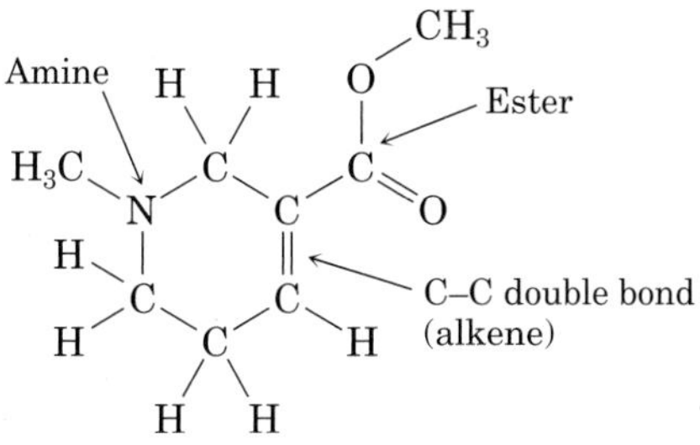

Ring structure: H_3C–N (Amine), N–CH_2 (C with H, H), C–C(=O)–O–CH_3 (Ester), C=C (C–C double bond (alkene)), ring C atoms bearing H atoms.

Labels: Amine; Ester; C–C double bond (alkene)

2.4

$CH_3CH_2CH_2CH_2CH_2CH_3$ $\quad CH_3\overset{CH_3}{\overset{|}{C}}HCH_2CH_2CH_3$ $\quad CH_3CH_2\overset{CH_3}{\overset{|}{C}}HCH_2CH_3$

$CH_3\underset{CH_3}{\underset{|}{\overset{CH_3}{\overset{|}{C}}}}CH_2CH_3$ $\quad CH_3\overset{CH_3}{\overset{|}{C}}H\underset{CH_3}{\underset{|}{C}}HCH_3$

2.5 (a) $CH_3CH_2CH_2CH_2CH_2CH_2CH_2CH_3$
$(CH_3)_2CHCH_2CH_2CH_2CH_2CH_3$
$(CH_3)_3CCH_2CH_2CH_2CH_3$

(b) $H{-}\underset{H}{\underset{|}{\overset{H}{\overset{|}{C}}}}{-}\underset{H}{\underset{|}{\overset{H}{\overset{|}{C}}}}{-}\underset{H}{\underset{|}{\overset{H}{\overset{|}{C}}}}{-}\overset{O}{\overset{\|}{C}}{-}OH$

$H{-}\underset{H}{\underset{|}{\overset{H}{\overset{|}{C}}}}{-}\underset{H}{\underset{|}{\overset{H}{\overset{|}{C}}}}{-}O{-}\overset{O}{\overset{\|}{C}}{-}\underset{H}{\underset{|}{\overset{H}{\overset{|}{C}}}}{-}H$

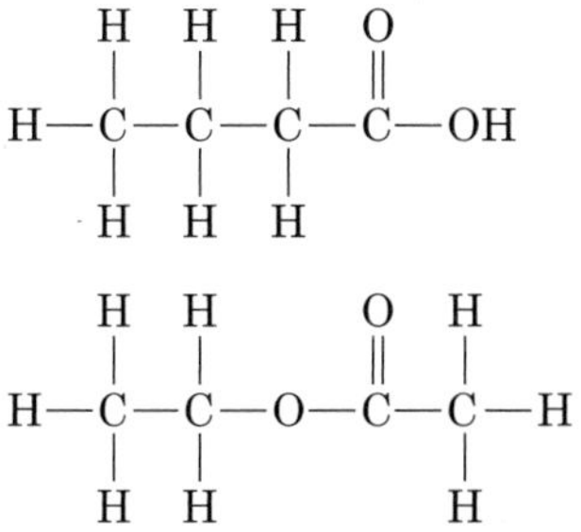

2.6

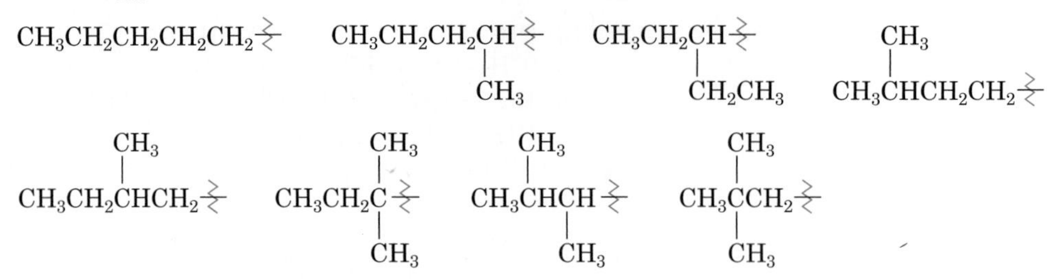

2.7 (a)

$$\begin{array}{c} CH_3 \\ | \\ CH_3CHCHCH_3 \\ | \\ CH_3 \end{array}$$

(b)

$$\begin{array}{c} CH_3CHCH_3 \\ | \\ CH_3CH_2CHCH_2CH_3 \end{array}$$

(c)

$$\begin{array}{c} CH_3 \\ | \\ CH_3CCH_2CH_3 \\ | \\ CH_3 \end{array}$$

2.8

(a)

$$\begin{array}{c} \overset{p}{CH_3} \\ | \\ \underset{p}{CH_3}\underset{t}{CH}\underset{s}{CH_2}\underset{s}{CH_2}\underset{p}{CH_3} \end{array}$$

(b)

$$\begin{array}{c} \overset{p}{CH_3}\overset{t}{CH}\overset{p}{CH_3} \\ | \\ \underset{p}{CH_3}\underset{s}{CH_2}\underset{t}{CH}\underset{s}{CH_2}\underset{p}{CH_3} \end{array}$$

(c)

$$\begin{array}{c} \overset{p}{CH_3} \qquad \overset{p}{CH_3} \\ | \qquad\qquad | \\ \underset{p}{CH_3}\underset{t}{CH}\underset{s}{CH_2}-\underset{q}{C}-\underset{p}{CH_3} \\ | \\ \underset{p}{CH_3} \end{array}$$

2.9 (a) Pentane, 2-methylbutane, 2,2-dimethylpropane
(b) 3,4-Dimethylhexane
(c) 2,4-Dimethylpentane
(d) 2,2,5-Trimethylheptane

2.10 (a)

$$\begin{array}{c} CH_3 \\ | \\ CH_3CH_2CHCHCH_2CH_2CH_2CH_2CH_3 \\ | \\ CH_3 \end{array}$$

(b)

$$\begin{array}{c} H_3C \quad CH_2CH_3 \\ | \qquad | \\ CH_3CH_2CH_2C-CHCH_2CH_3 \\ | \\ H_3C \end{array}$$

(c)

$$\begin{array}{c} CH_3 \quad CH_2CH_2CH_3 \\ | \qquad | \\ CH_3CCH_2CHCH_2CH_2CH_2CH_3 \\ | \\ CH_3 \end{array}$$

(d)

$$\begin{array}{c} CH_3 \quad CH_3 \\ | \qquad | \\ CH_3CCH_2CHCH_3 \\ | \\ CH_3 \end{array}$$

2.11 3,3,4,5-Tetramethylheptane

2.12

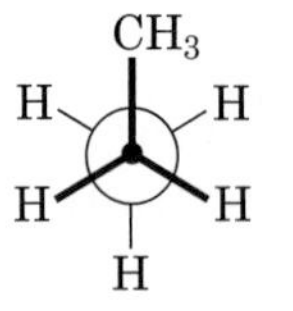

Most stable conformation (staggered): front carbon CH₃, H, H; back carbon H, H, H

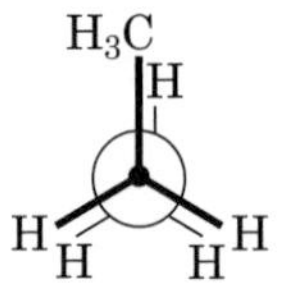

Least stable conformation (eclipsed): front carbon H₃C, H, H; back carbon H, H, H

Most stable conformation
(staggered)

Least stable conformation
(eclipsed)

2.13

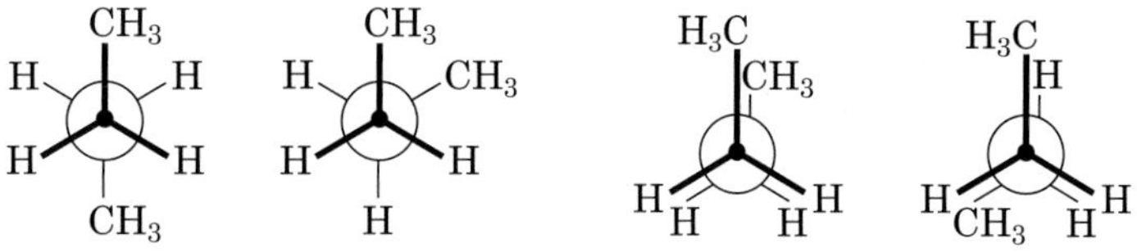

CH_3 H H H H CH_3 — CH_3 H CH_3 H H H

Staggered butane

H_3C CH_3 H H H H — H_3C H H CH_3 H H

Eclipsed butane

2.14 The first staggered conformation of butane is the most stable.

2.15 (a) C_5H_5N (b) $C_6H_{10}O$ (c) C_8H_7N

2.16 (a) $CH_3CH_2CH{=}CH_2$ (b) 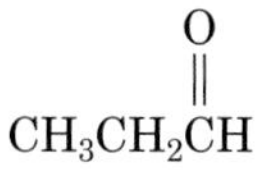 $CH_3CH_2CH{=}O$ (c) 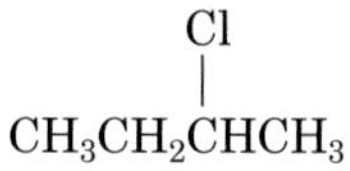 $CH_3CH_2CH(Cl)CH_3$

2.17

H_2N–C_6H_4–C(=O)OH

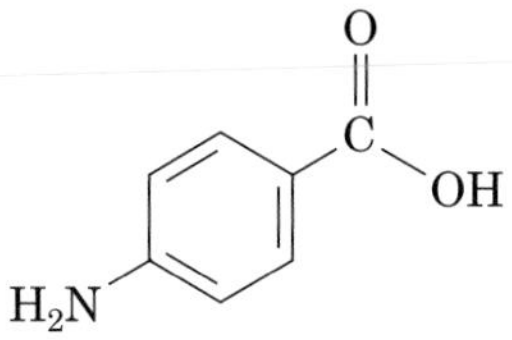

2.18 (a) 1,4-Dimethylcyclohexane (b) 1-Ethyl-3-methylcyclopentane
(c) Isopropylcyclobutane

2.19 (a) 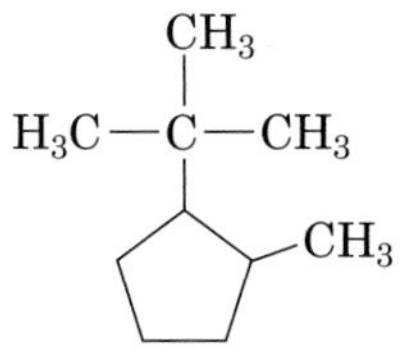 H_3C—C(CH_3)(CH_3)— (2-methylcyclopentyl) (b) 1,1-dimethylcyclobutane (CH_3, CH_3) (c) CH_3CH_2–cyclohexane–$CHCH_3$ with CH_3

2.20 H_3C Cl H H

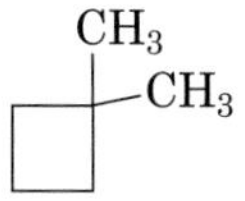

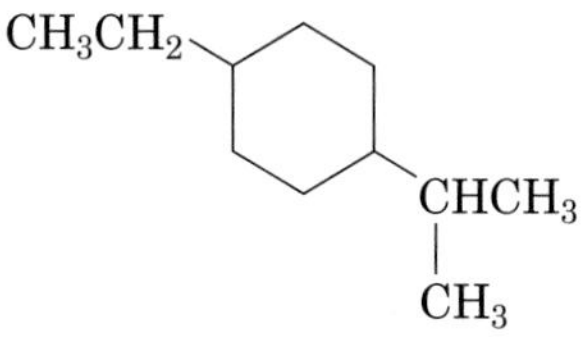

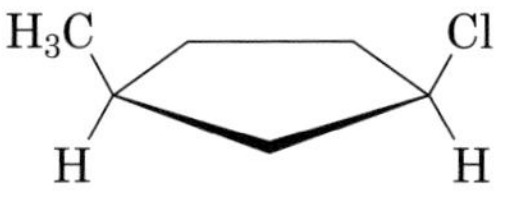

2.21 Br Br H H — Br H H Br

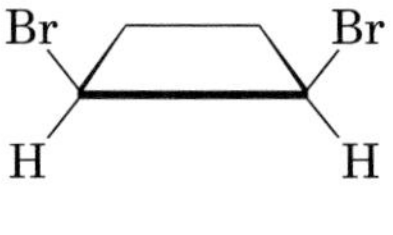

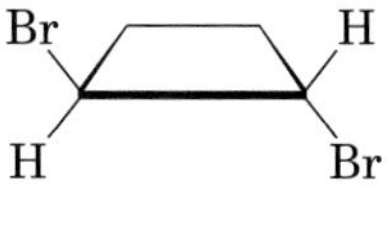

Cis Trans

2.22 CH_3 H — H CH_3

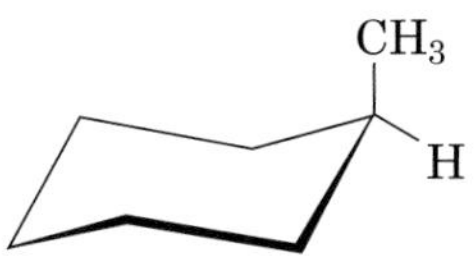

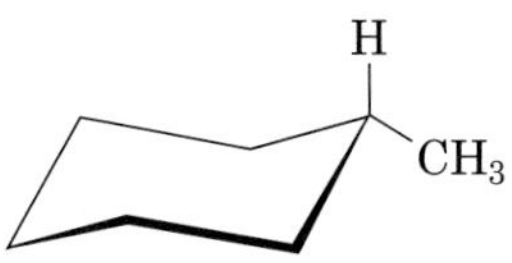

Axial Equatorial

2.23

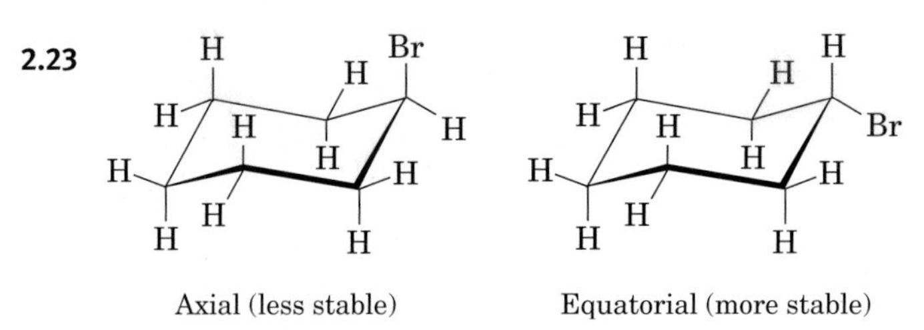

Axial (less stable) Equatorial (more stable)

2.24 Axial and equatorial positions alternate on each side of a ring.

Ring-flip

2.25 Axial and equatorial positions alternate on each side of a ring.

Ring-flip

2.26 1-Chloro-2,4-dimethylcyclohexane; less stable

Chapter 3

3.1 (a) 4-Methylpent-1-ene (b) Hept-3-ene (c) Hepta-1,5-diene (d) 2-Methylhex-3-ene

3.2 (a) 1,2-Dimethylcyclohexene (b) 4,4-Dimethylcycloheptene (c) 3-Isopropylcyclopentene

3.3

(a) $CH_3CH_2CH_2CH_2C(CH_3){=}CH_2$ (b) $(CH_3)_3CCH{=}CHCH_3$

(c) $H_2C{=}CHCH_2CH_2C(CH_3){=}CH_2$ (d) $CH_3CH_2CH_2CH{=}C(CH_2CH_3)C(CH_3)_3$

3.4 Compounds (c), (d), (e), and (f) can exist as pairs of isomers.

3.5 (a) *cis*-3,4-Dimethylhex-2-ene (b) *trans*-6-Methylhept-3-ene

3.6 (a) Br (b) Br (c) CH_2CH_3 (d) OH (e) CH_2OH (f) $CH{=}O$

3.7 CO_2CH_3 is higher.

3.8 (a) *Z* (b) *E*

3.9 *Z*

3.10 (a) Substitution (b) Elimination (c) Addition

3.11 (a) (b) (c) (d)

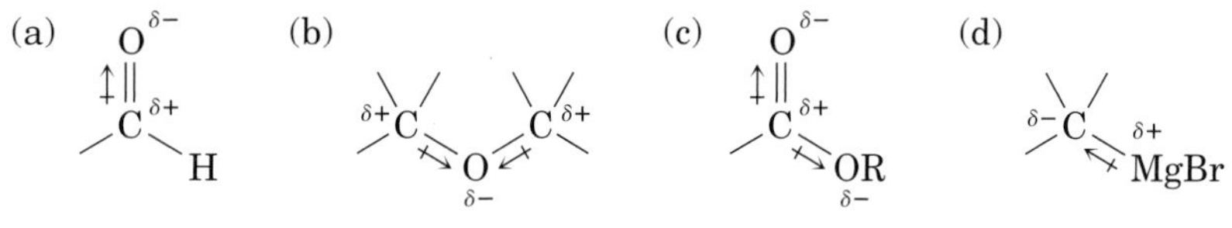

3.12 Electrophile: (a), (c); nucleophile: (b), (d), (e)

3.13 Boron is a Lewis acid/electrophile because it has only six outer-shell electrons.

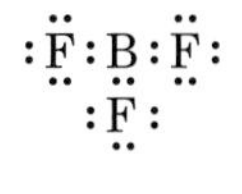

$:\ddot{F}:B:\ddot{F}:$
$:\ddot{F}:$

3.14 $(CH_3)_3C^+$

3.15 2-Chloropentane and 3-chloropentane

3.16 $E_{act} = 60$ kJ/mol is faster.

Chapter 4

4.1 (a) 2-Chlorobutane (b) 2-Iodo-2-methylpentane (c) Chlorocyclohexane

4.2 (a) Cyclopentene (b) Hex-3-ene (c) 1-Isopropylcyclohexene
(d) Cyclohexylethylene (vinylcyclohexane)

4.3 (a) $CH_3CH_2\overset{+}{C}(CH_3)CH_2CH(CH_3)CH_3$ (b) CH_2CH_3 (on cyclopentyl cation carbon, +)

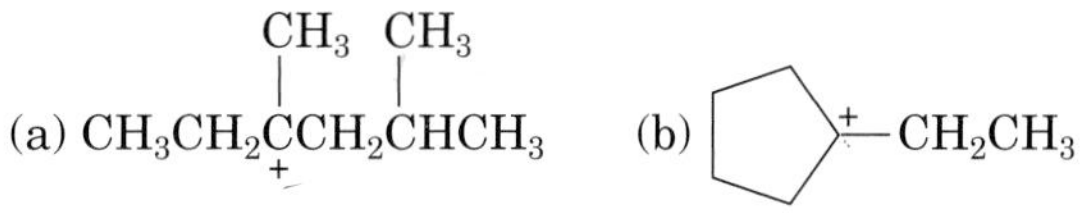

4.4 (a) $CH_3CH_2C(OH)(CH_3)CH_2CH_2CH_3$ (b) CH_3, OH (1-methylcyclopentanol) (c) $CH_3CH_2CH(CH_3)CH_2CH_2C(CH_3)(OH)-CH_3$

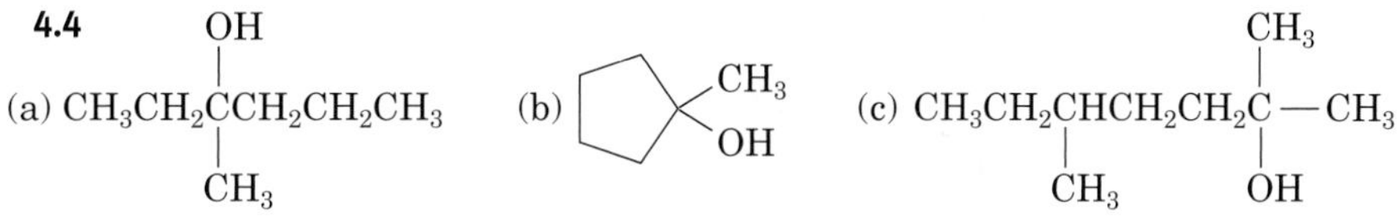

4.5 (a) But-1-ene or but-2-ene (b) 3-Methylpent-2-ene or 2-ethylbut-1-ene
(c) 1,2-Dimethylcyclohexene or 2,3-dimethylcyclohexene

4.6 *trans*-1,2-Dibromo-1,2-dimethylcyclohexane

4.7 $+:\ddot{Br}$ bridging ring carbons bearing H_3C and CH_3

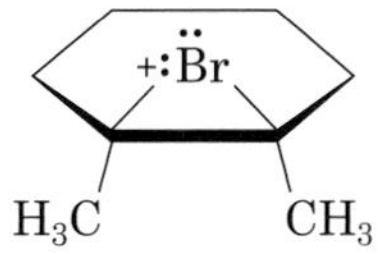

4.8 (a) 2-Methylpentane (b) 1,1-Dimethylcyclopentane

4.9 (a) O, O (octane-2,7-dione) (b) CH_3, OH, OH, CH_3 (1,2-dimethylcyclohexane-1,2-diol)

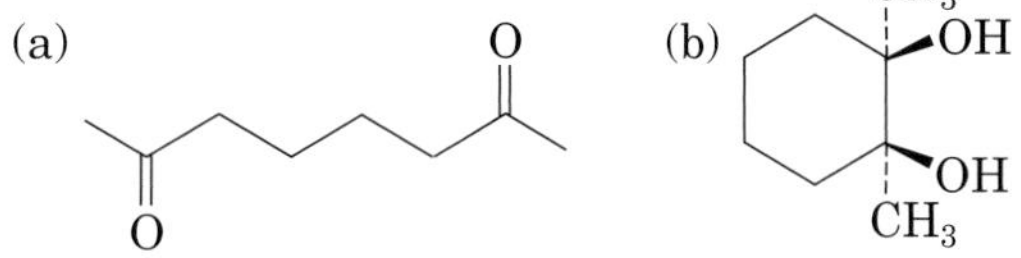

4.10 (a) 2-Methylpropene (b) Hex-3-ene

4.11

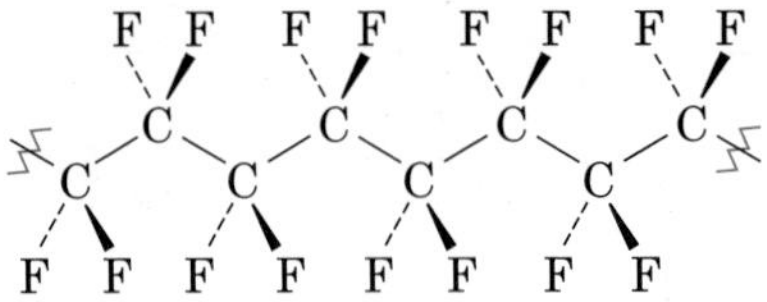

F F F F F F F F
C C C C
C C C C
F F F F F F F F

4.12 1,4-Dibromobut-2-ene and 3,4-dibromobut-1-ene

4.13 4-Chloropent-2-ene, 3-chloropent-1-ene, 1-chloropent-2-ene

4.14 $CH_3CH_2\overset{\delta+}{C}H{=\!=}CH{=\!=}\overset{\delta+}{C}H_2$ and $CH_3\overset{\delta+}{C}H{=\!=}CH{=\!=}\overset{\delta+}{C}HCH_3$

4.15

(a) $\overset{+}{C}H_2$ ⟷ CH_2 +

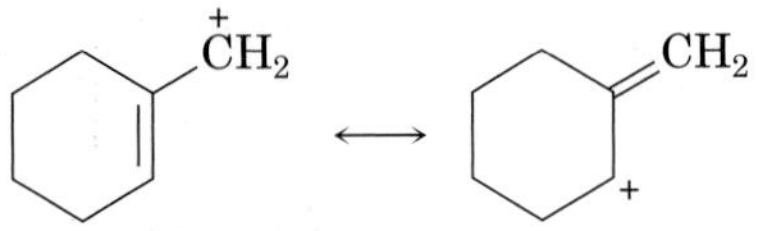

(b) $CH_3-C(=\ddot{O}:)-\ddot{C}H_2^{-} \longleftrightarrow CH_3-C(-\ddot{O}:^{-})=CH_2$

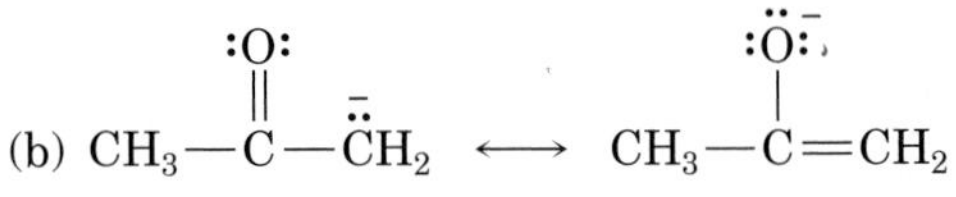

(c) + ⟷ + ⟷ +

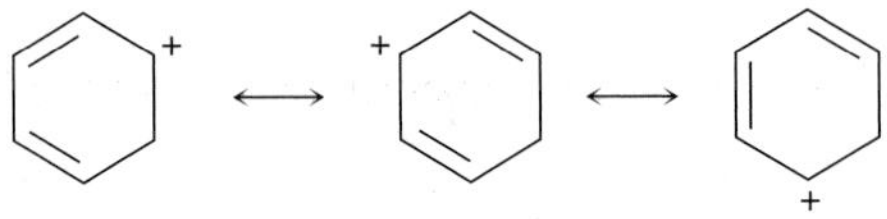

4.16 (a) 6-Methylhept-3-yne (b) 3,3-Dimethylbut-1-yne
(c) 5-Methylhex-2-yne (d) Hept-2-en-5-yne

4.17 (a) 1,2-Dichloropent-1-ene (b) 4-Bromohept-3-ene and 3-bromohept-3-ene
(c) *cis*-6-Methylhept-3-ene

4.18 Octan-4-one

4.19 (a) Pent-1-yne (b) Hex-3-yne

4.20 (a) 1-Bromo-3-methylbutane + acetylene
(b) 1-Bromopropane + prop-1-yne, or bromomethane + pent-1-yne
(c) Bromomethane + 3-methylbut-1-yne

Chapter 5

5.1

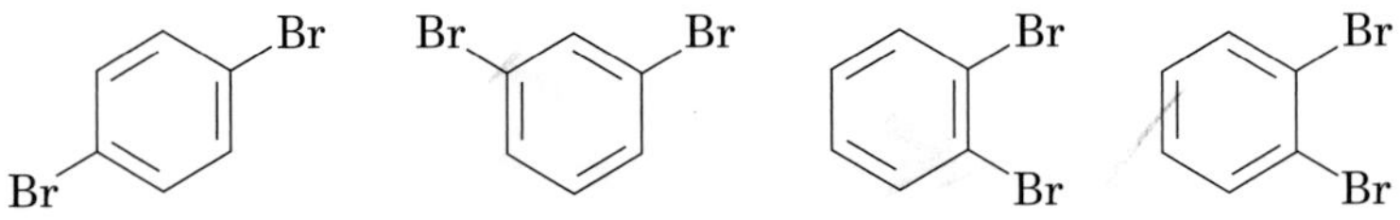

Br Br | Br Br | Br Br | Br Br

5.2 Two Kekulé structures are resonance forms.

5.3 (a) Meta (b) Para (c) Ortho

5.4 (a) *m*-Bromochlorobenzene (b) Isobutylbenzene (c) *p*-Bromoaniline

5.5 (a) 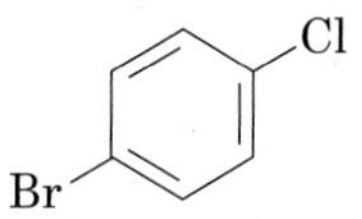 Br–C₆H₄–Cl (b) 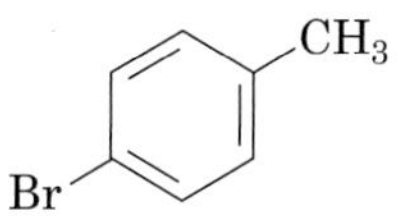 Br–C₆H₄–CH_3 (c) 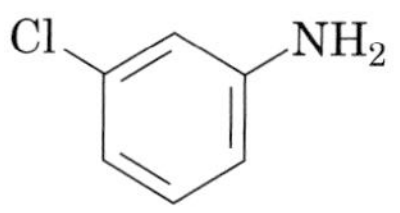 Cl–C₆H₄–NH_2

(d) 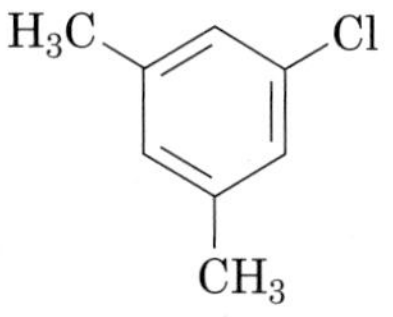 H_3C, Cl, CH_3 substituted benzene

5.6 *o*-, *m*-, and *p*-Bromotoluene

5.7 H, $\overset{+}{N}O_2$ → [NO_2, H, +] → NO_2 + H^+

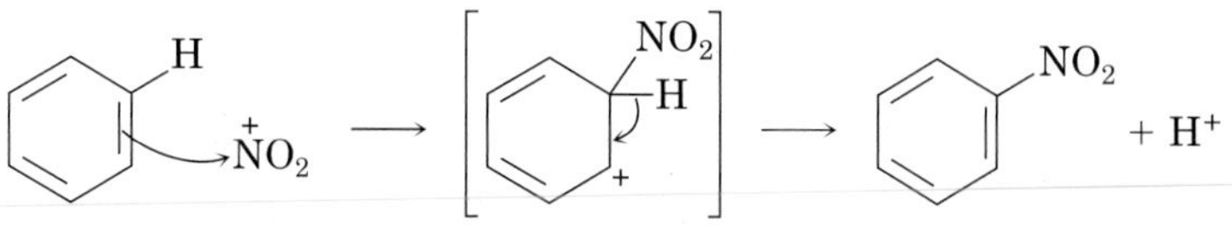

Carbocation intermediate

5.8 *p*-Xylene has one kind of ring position; *o*-xylene has two.

5.9 Three

5.10 (a) Ethylbenzene (b) 2-Ethyl-1,4-dimethylbenzene

5.11 (a) *tert*-Butylbenzene (b) Propanoylbenzene, $C_6H_5COCH_2CH_3$

5.12 (a) Nitrobenzene < toluene < phenol
(b) Benzoic acid < chlorobenzene < benzene < phenol
(c) Benzaldehyde < bromobenzene < benzene < aniline

5.13 (a) *m*-Chlorobenzonitrile (b) *o*- and *p*-Bromochlorobenzene

5.14 (a) *m*-Nitrobenzenesulfonic acid
(b) *o*- and *p*-Bromobenzenesulfonic acid
(c) *o*- and *p*-Methylbenzenesulfonic acid
(d) *m*-Carboxybenzenesulfonic acid
(e) *m*-Cyanobenzenesulfonic acid

5.15 Ortho

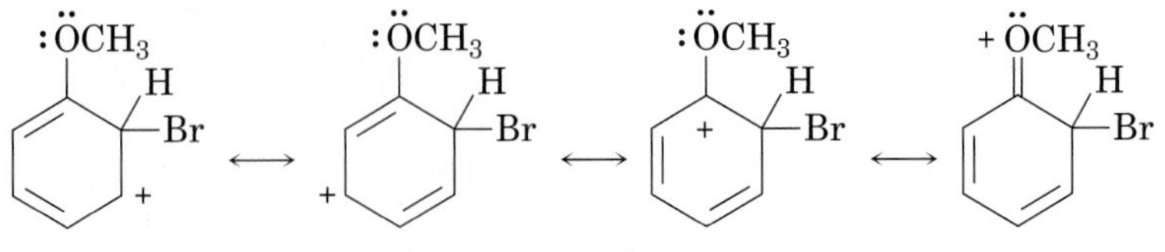

:ÖCH_3 H Br ⟷ :ÖCH_3 H Br ⟷ :ÖCH_3 H Br ⟷ +ÖCH_3 H Br

Meta

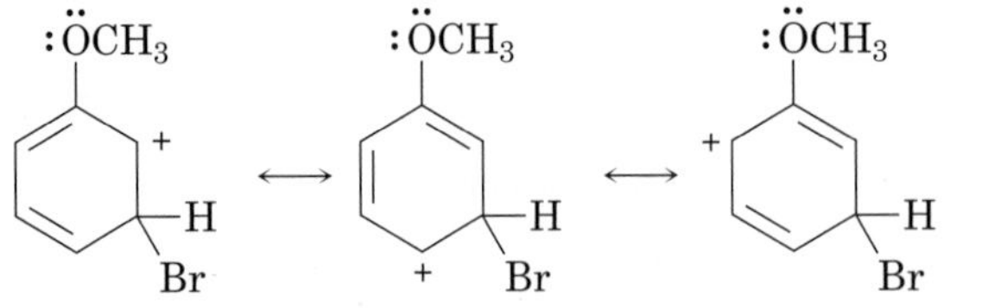

:ÖCH_3 H Br ⟷ :ÖCH_3 H Br ⟷ :ÖCH_3 H Br

Para

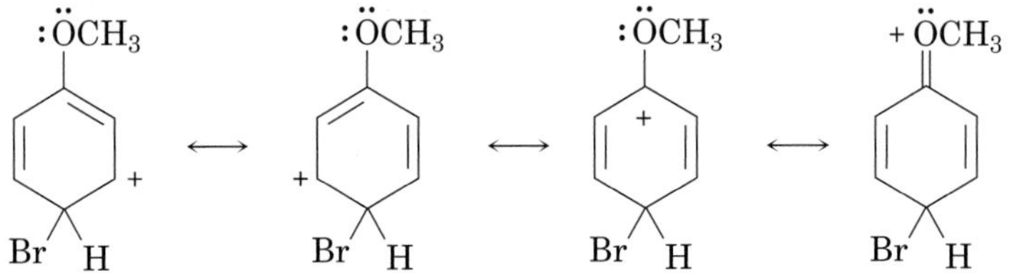

:ÖCH_3 Br H ⟷ :ÖCH_3 Br H ⟷ :ÖCH_3 Br H ⟷ +ÖCH_3 Br H

5.16 Ortho

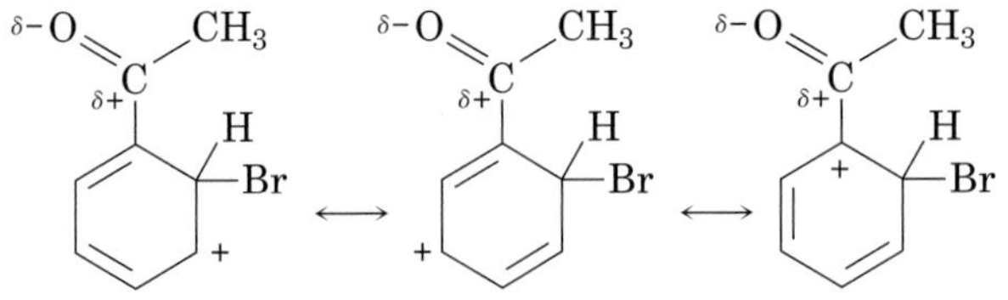

$\delta-$O C CH_3 $\delta+$ H Br ⟷ $\delta-$O C CH_3 $\delta+$ H Br ⟷ $\delta-$O C CH_3 $\delta+$ H Br

Meta

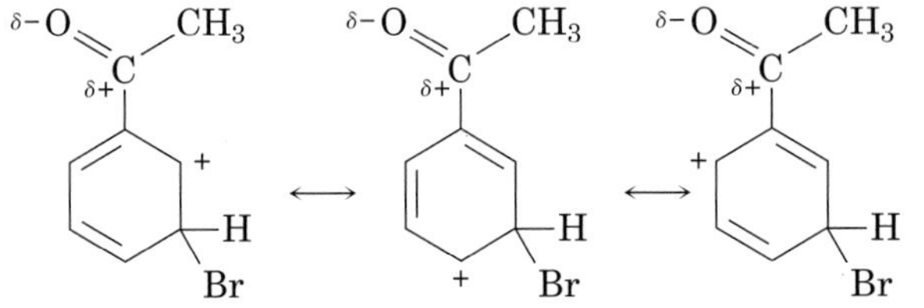

$\delta-$O C CH_3 $\delta+$ H Br ⟷ $\delta-$O C CH_3 $\delta+$ H Br ⟷ $\delta-$O C CH_3 $\delta+$ H Br

Para

$\delta-$O C CH_3 $\delta+$ Br H ⟷ $\delta-$O C CH_3 $\delta+$ Br H ⟷ $\delta-$O C CH_3 $\delta+$ Br H

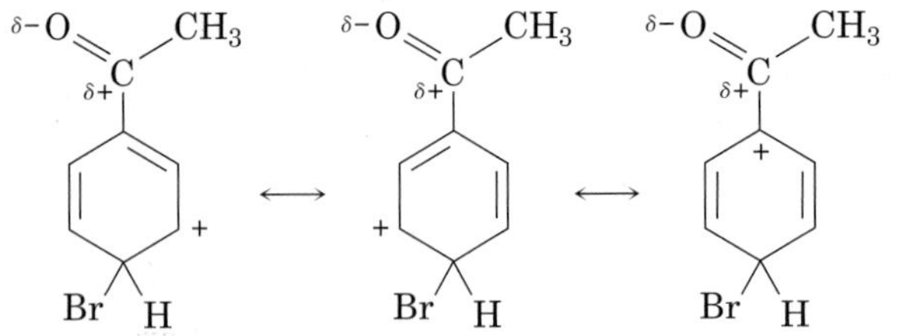

5.17 (a) *m*-Chlorobenzoic acid (b) *o*-Benzenedicarboxylic acid

5.18 (a), (c), and (d) are aromatic.

5.19

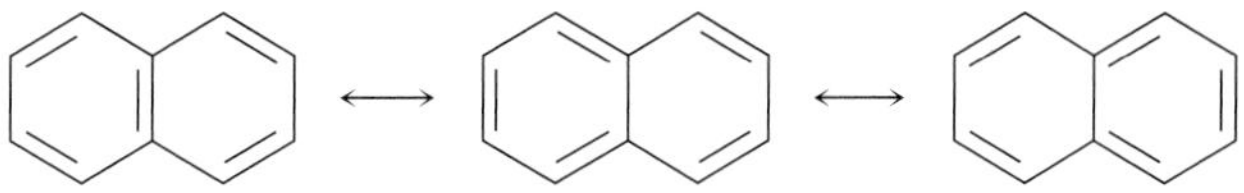

5.20 (a) 1. CH_3Cl, $AlCl_3$; 2. CH_3COCl, $AlCl_3$
(b) 1. Cl_2, $FeCl_3$; 2. HNO_3, H_2SO_4

5.21 (a) 1. Br_2, $FeBr_3$; 2. CH_3Cl, $AlCl_3$
(b) 1. 2 CH_3Cl, $AlCl_3$; 2. Br_2, $FeBr_3$

5.22 1. CH_3Cl, $AlCl_3$; 2. $KMnO_4$, H_2O; 3. Cl_2, $FeCl_3$

Chapter 6

6.1 Chiral: screw, beanstalk, shoe

6.2 Chiral: (b), (c)

6.3 Chiral: (b)

6.4

(a) 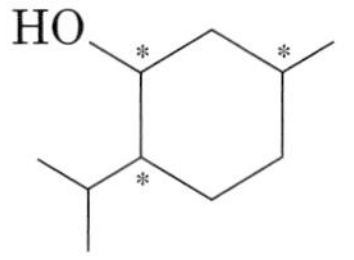 HO (b) 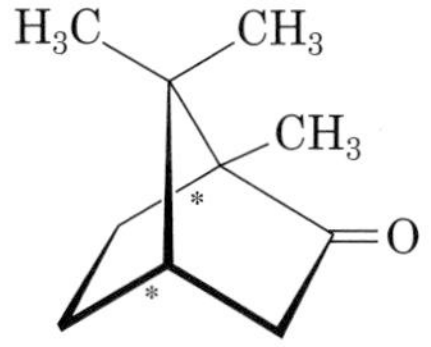 H_3C CH_3 CH_3 O (c) 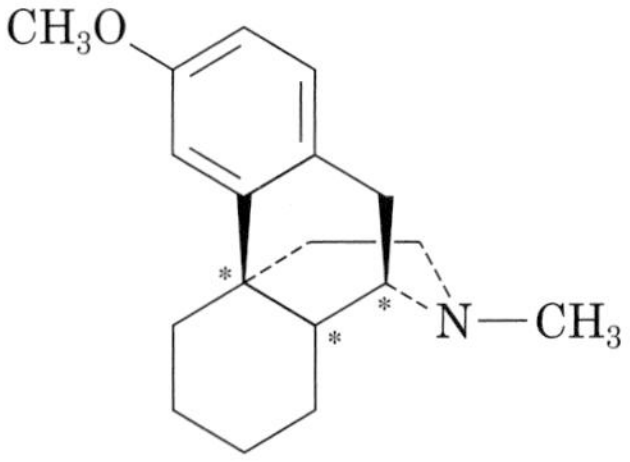 CH_3O N—CH_3

6.5 COOH C H CH_3 NH_2 and COOH C H_3C H NH_2

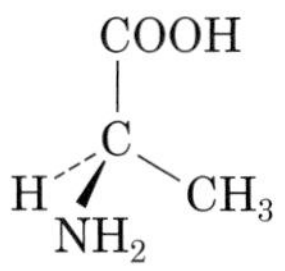

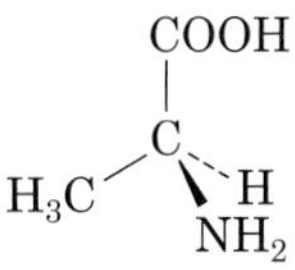

6.6 Levorotatory

6.7 +16.1°

6.8 (a) —Br, —CH_2CH_2OH, —CH_2CH_3, —H
(b) —OH, —CO_2CH_3, —CO_2H, —CH_2OH
(c) —Br, —Cl, —CH_2Br, —CH_2Cl

6.9 (a) *S* (b) *S* (c) *R*

6.10 H C HO $CH_2CH_2CH_3$ CH_3

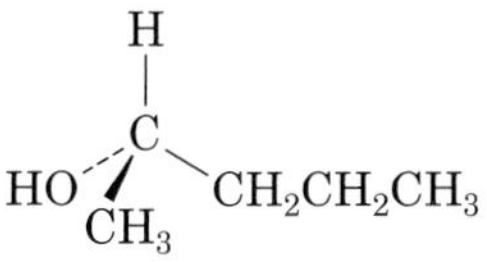

6.11 *S*

6.12 (a) *R*,*R* (b) *S*,*R* (c) *R*,*S*

6.13 Molecules (b) and (c) are enantiomers (mirror images). Molecule (a) is the diastereomer of (b) and (c).

6.14 *R*,*R*

6.15 *S*,*S*

6.16 Meso: (a) and (c)

6.17 Meso: (a) and (c)

6.18 6 stereocenters; 64 stereoisomers

6.19 The product is the pure *S* ester.

6.20 (a) Constitutional isomers (b) Diastereomers

Chapter 7

7.1 (a) 2-Bromobutane (b) 3-Chloro-2-methylpentane
(c) 1-Chloro-3-methylbutane (d) 1,3-Dichloro-3-methylbutane
(e) 1-Bromo-4-chlorobutane (f) 4-Bromo-1-chloropentane

7.2 (a) $CH_3CH_2CH_2C(CH_3)_2CH(Cl)CH_3$ (b) $CH_3CH_2CH_2C(Cl)_2CH(CH_3)_2$
(c) $CH_3CH_2C(Br)(CH_2CH_3)_2$ (d) $CH_3CH(Cl)CH_2CH(CH_3)CH(Br)CH_3$

7.3 1-Chloro-3-methylpentane, 2-chloro-3-methylpentane, 3-chloro-3-methylpentane, 3-(chloromethyl)pentane. The first two are chiral.

7.4 (a) 2-Methylpropan-2-ol + HCl (b) 4-Methylpentan-2-ol + PBr_3
(c) 5-Methylhexan-1-ol + PBr_3 (d) 2,4-Dimethylhexan-2-ol + HCl

7.5 (a) 4-Bromo-2-methylhexane (b) 1-Chloro-1-methylcyclohexane
(c) 1-Chloro-3,3-dimethylcyclopentane

7.6 React the halide with Mg, and then treat the Grignard reagent with D_2O.

7.7 1. PBr_3; 2. Mg, ether; 3. H_2O

7.8 (a) $CH_3CH_2CH(I)CH_3$ (b) $(CH_3)_2CHCH_2SH$ (c) $C_6H_5CH_2CN$

7.9 (a) 1-Bromobutane + NaOH (b) 1-Bromo-3-methylbutane + NaN_3

7.10 (a) Rate is tripled. (b) Rate is quadrupled.

7.11 (*R*) $CH_3CO_2CH(CH_3)CH_2CH_2CH_2CH_3$

7.12

H CH_3 Br H → H CH_3 H SH

H_3C–C–C–C(*S*)–CH_3 → H_3C–C–C–C(*R*)–CH_3

H H   H H

7.13 (a) Reaction with $CH_3CH_2CH_2Br$ is faster.
(b) Reaction with $(CH_3)_2CHCH_2Cl$ is faster.

7.14 $CH_3I > CH_3Br > CH_3F$

7.15 (a) Rate is unchanged. (b) Rate is doubled.

7.16 Racemic 3-bromo-3-methyloctane

7.17 The *S* substrate gives a racemic mixture of alcohols.

7.18 (a) 2-Methylpent-2-ene (b) 2,3,5-Trimethylhex-2-ene
(c) $=CHCH_3$

7.19 (a) 1-Bromo-3,6-dimethylheptane (b) 1,2-Dimethyl-4-bromocyclopentane

7.20 The rate is tripled.

7.21 (a) S_N2 (b) E2 (c) S_N1

Chapter 8

8.1 (a) 5-Methylhexane-2,4-diol (b) 2-Methyl-4-phenylbutan-2-ol
(c) 4,4-Dimethylcyclohexanol (d) *cis*-2-Bromocyclopentanol

8.2 Secondary: (a), (c), (d); tertiary: (b)

8.3

(a) OH
$CH_3CH_2CH_2CH_2C(CH_3)_2$

(b) OH
$CH_3CHCH_2CH_2CH_2CH_2OH$

(c) CH_2CH_3
$CH_3CH{=}CCH_2OH$

(d) OH

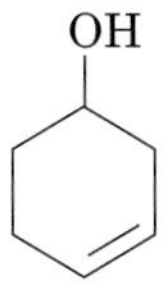

(e) OH
Br

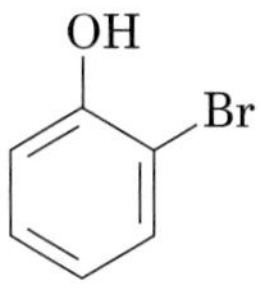

(f) OH
O_2N NO_2
NO_2

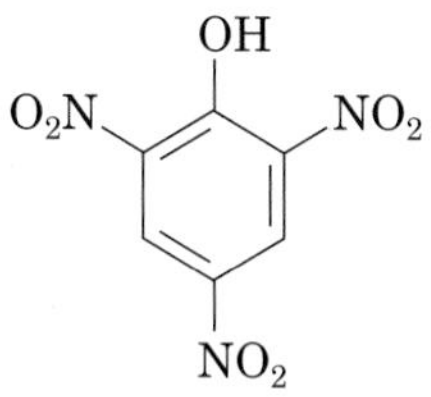

8.4 (a) Diisopropyl ether
(b) Cyclopentyl propyl ether
(c) *p*-Bromoanisole or 1-bromo-4-methoxybenzene
(d) Ethyl isobutyl ether

8.5 (a) $NaBH_4$ (b) $LiAlH_4$

8.6 (a) C_6H_5CHO, $C_6H_5CO_2H$, $C_6H_5CO_2R$ (b) $C_6H_5COCH_3$ (c) Cyclohexanone

8.7 (a) 2,3-Dimethylpent-2-ene (b) 2-Methylpent-2-ene

8.8 (a) 2,3-Dimethylcyclohexanol (b) Heptan-4-ol

8.9 (a) 1-Phenylethanol (b) 2-Methylpropan-1-ol (c) Cyclopentanol

8.10 (a) Cyclohexanone (b) Hexanoic acid (c) Hexan-2-one

8.11 (a) Cyclohexanone (b) Hexanal (c) Hexan-2-one

8.12 $C_6H_{11}O^- + CH_3CH_2Br \rightarrow C_6H_{11}OCH_2CH_3$

8.13 (a) $CH_3CH_2CH_2O^- + CH_3Br$ (b) $C_6H_5O^- + CH_3Br$
(c) $(CH_3)_2CHO^- + C_6H_5CH_2Br$

8.14 (a) Bromoethane > chloroethane > 2-bromopropane > 2-chloro-2-methylpropane

8.15 $CH_3CH_2COCH_2CH(CH_3)_2$; (i) $CH_3CH_2CH(OCH_3)CH_2CH(CH_3)_2$
(ii) $CH_3CH_2CH(Cl)CH_2CH(CH_3)_2$ (iii) $CH_3CH_2COCH_2CH(CH_3)_2$

8.16 (a) 1. CH_3Cl, $AlCl_3$; 2. SO_3, H_2SO_4; 3. NaOH, heat

8.17 (a) Ethanol + iodoethane (b) Cyclohexanol + iodoethane
(c) 2-Iodo-2-methylpropane + ethanol

8.18 The product is a racemic mixture of *R*,*R* and *S*,*S* butane-1,2-diols.

8.19 (a) Butane-2-thiol (b) 2,2,6-Trimethylheptane-4-thiol
(c) Cyclopent-3-ene-1-thiol

8.20 (a) Ethyl methyl sulfide (b) *tert*-Butyl ethyl sulfide
(c) *o*-Di(methylthio)benzene

8.21 (a) 1. PBr_3; 2. $Na^+ {}^-SH$ (b) 1. $LiAlH_4$; 2. PBr_3; 3. $Na^+ {}^-SH$

Chapter 9

9.1 (a) Pentan-2-one (b) $CH_3CH_2CH_2CH{=}CHCHO$
(c) $CH_3CH_2COCH_2CH_2CHO$ (d) Cyclopentanone

9.2 (a) 2-Methylpentan-3-one (b) 3-Phenylpropanal
(c) Octane-2,6-dione (d) *trans*-2-Methylcyclohexanecarbaldehyde
(e) Pentanedial (f) *cis*-2,5-Dimethylcyclohexanone

9.3 (a) $CH_3CH(CH_3)CH_2CHO$ (b) $H_2C{=}C(CH_3)CH_2CHO$ (c) $CH_3CH(Cl)CH_2C(O)CH_3$

(d) $C_6H_5CH_2CHO$ (e) 2,2-dimethylcyclohexane-CHO (f) cyclohexane-1,3-dione

9.4 (a) PCC (b) 1. $LiAlH_4$; 2. PCC (c) 1. $KMnO_4$; 2. $LiAlH_4$; 3. PCC

9.5 (a) PCC (b) H_3O^+, $HgSO_4$ (c) $KMnO_4$, H_3O^+

9.6 (a) 1. H_3O^+; 2. PCC (b) 1. CH_3COCl, $AlCl_3$; 2. $NaBH_4$

9.7 (a) Pentanoic acid (b) 2,2-Dimethylhexanoic acid (c) No reaction

9.8 NC OH (on a $C(CH_3)_2$ carbon)

9.9 $C_6H_5CH(OCH_3)OH$

9.10 $Cl_3CCH(OH)OH$

9.11 Labeled water adds reversibly to the carbonyl group.

9.12 1,1-diethoxycyclohexane (OCH_2CH_3, OCH_2CH_3)

9.13 1,3-dioxolane with H and C_6H_5 on C2

9.14 1. $HOCH_2CH_2OH$, H^+ catalyst; 2. $LiAlH_4$; 3. H_3O^+

9.15 (a) cyclohexylidene$=N-CH_3$ (b) cyclohexane with OCH_2CH_3, OCH_2CH_3 (c) cyclohexane with OH, H

9.16 $(CH_3)_2COCH_2CH_3 + CH_3NH_2$

9.17 (a) 1-Methylcyclopentanol (b) 1,1-Diphenylethanol (c) 3-Methylhexan-3-ol

9.18 (a) Acetone + CH_3MgBr (b) Cyclohexanone + CH_3MgBr
(c) Pentan-3-one + CH_3MgBr, or butan-2-one + CH_3CH_2MgBr

9.19 $C_5H_9COCH_3 + CH_3MgBr$, or $CH_3COCH_3 + C_5H_{11}MgBr$

9.20 6-Methylcyclohex-2-enone + $(CH_3)_2CHOH$

Chapter 10

10.1 (a) 3-Methylbutanoic acid (b) 4-Bromopentanoic acid
(c) Hex-4-enoic acid (d) 2-Ethylpentanoic acid
(e) *trans*-2-Methylcyclohexanecarboxylic acid

10.2

(a)

H_3C CH_3

$CH_3CH_2CH_2CHCHCOOH$

(b)

CH_3

$CH_3CHCH_2CH_2COOH$

(c)

COOH

OH

(d)

H

COOH

H

COOH

10.3 (a) 4-Methylpentanoyl chloride
(b) 2-Methylbutanenitrile
(c) Pent-4-enamide
(d) 2-Ethylbutanenitrile
(e) Cyclopentyl 2,2-dimethylpropanoate
(f) 2,3-Dimethylbut-2-enoyl chloride
(g) Benzoic anhydride
(h) Isopropyl cyclopentanecarboxylate

10.5 (a) $C_6H_5CO_2^- \, Na^+$ (b) $(CH_3)_3CCO_2^- \, K^+$

10.6 Methanol < phenol < *p*-nitrophenol < acetic acid < sulfuric acid

10.7 (a) $CH_3CH_2CO_2H < BrCH_2CH_2CO_2H < BrCH_2CO_2H$
(b) Ethanol < benzoic acid < *p*-cyanobenzoic acid

10.8 1. NaCN; 2. NaOH, H_2O. Iodobenzene cannot be converted to benzoic acid by this method.

10.9 (a) CH_3COCl (b) $CH_3CH_2CO_2CH_3$ (c) $CH_3CO_2COCH_3$ (d) $CH_3CO_2CH_3$

10.10 The electron-withdrawing trifluoromethyl group polarizes the carbonyl carbon.

10.11 (a) C_6H_5COCl (b) $C_6H_5CO_2CH_3$ (c) $C_6H_5CH_2OH$ (d) $C_6H_5CO_2^- \, Na^+$

10.12 (a) Acetic acid + butan-1-ol (b) Butanoic acid + methanol
(c) Benzoic acid + propan-2-ol

10.13 (a) Propanoyl chloride + methanol (b) Acetyl chloride + ethanol
(c) Acetyl chloride + cyclohexanol

10.15 (a) Propanoyl chloride + NH_3 (b) 3-Methylbutanoyl chloride + CH_3NH_2
(c) Propanoyl chloride + $(CH_3)_2NH$
(d) Benzoyl chloride + diethylamine

10.17

O

$COCH_3$

CO_2H

10.18 (a) Isopropyl alcohol + acetic acid
(b) Methanol + cyclohexanecarboxylic acid

10.19 Reaction of an acid with an alkoxide ion gives the unreactive carboxylate ion.

10.20 (a) $CH_3CH_2CH_2CH(CH_3)CH_2OH + CH_3OH$ (b) $C_6H_5OH + C_6H_5CH_2OH$

10.21 $HOCH_2CH_2CH_2CH_2OH$

10.22 (a) Ethyl benzoate + 2 CH_3MgBr (b) Ethyl acetate + 2 C_6H_5MgBr
(c) Ethyl pentanoate + 2 CH_3CH_2MgBr

10.23 (a) H_2O, NaOH (b) 1. H_2O, NaOH; 2. $LiAlH_4$ (c) $LiAlH_4$

10.24 [structure: 2,2-dimethylpyrrolidine]

10.25 (a) $CH_3CH_2CN + CH_3CH_2MgBr$
(b) $CH_3CH_2CN + (CH_3)_2CHMgBr$, or $(CH_3)_2CHCN + CH_3CH_2MgBr$
(c) $C_6H_5CN + CH_3MgBr$, or $CH_3CN + C_6H_5MgBr$
(d) $C_6H_{11}CN + C_6H_{11}MgBr$

10.26 $\left(-\overset{O}{\overset{\|}{C}}-C_6H_4-\overset{O}{\overset{\|}{C}}-NH-C_6H_4-NH- \right)_n$

Chapter 11

11.1

(a) 1-cyclopentenol (OH on cyclopentene ring)
(b) $H_2C{=}C(OH)Cl$
(c) $H_2C{=}C(OH)OCH_2CH_3$
(d) $H_2C{=}C(OH)OH$
(e) $C_6H_5C(OH){=}CH_2$

11.2 (a) 4 (b) 3 (c) 3 (d) 4 (e) 3

11.3 6-methyl-1-cyclohexenol (OH, CH_3) and 2-methyl-1-cyclohexenol (OH, CH_3)

11.4 (a) $CH_3C(H)(CH_3)-\overset{O}{\overset{\|}{C}}-C(Cl)(CH_3)CH_3$ (b) 2-bromo-6,6-dimethylcyclohexanone (H, Br, O, CH_3, CH_3)

11.5 1. Br_2; 2. Pyridine, heat

11.6 (a) CH_3CH_2CHO (b) $(CH_3)_3CCOCH_3$
(c) CH_3CO_2H (d) $CH_3CH_2CH_2C{\equiv}N$
(e) 1,3-cyclohexanedione with acidic hydrogens marked (H, H, H, H, H H)

11.7

(a) CH_3CH_2—CH=C(H)—$\ddot{O}$:$^-$

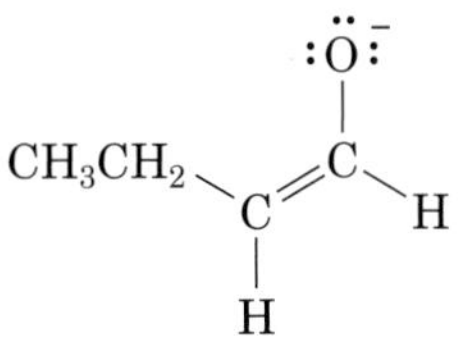

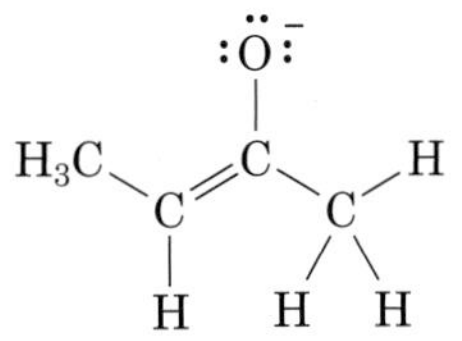

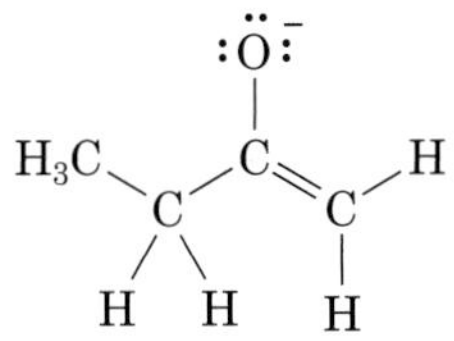

(b) H_3C—CH=C($\ddot{O}$:$^-$)—CH_3 and H_3C—CH_2—C($\ddot{O}$:$^-$)=CH_2

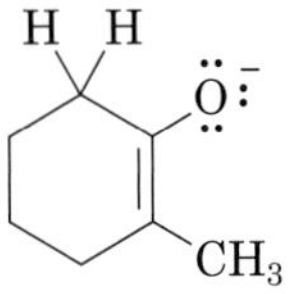

(c) H H, $\ddot{O}$:$^-$, CH_3 and H, $\ddot{O}$:$^-$, CH_3, H

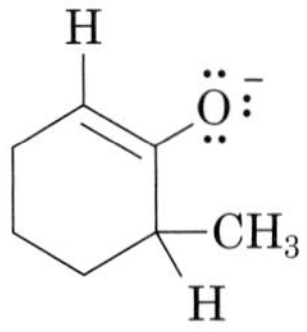

11.8

H_3C—C(=O)—$\ddot{C}$H—C(=O)—OCH_3 ⟷ H_3C—C($\ddot{O}$:$^-$)=CH—C(=O)—OCH_3 ⟷ H_3C—C(=O)—CH=C($\ddot{O}$:$^-$)—OCH_3

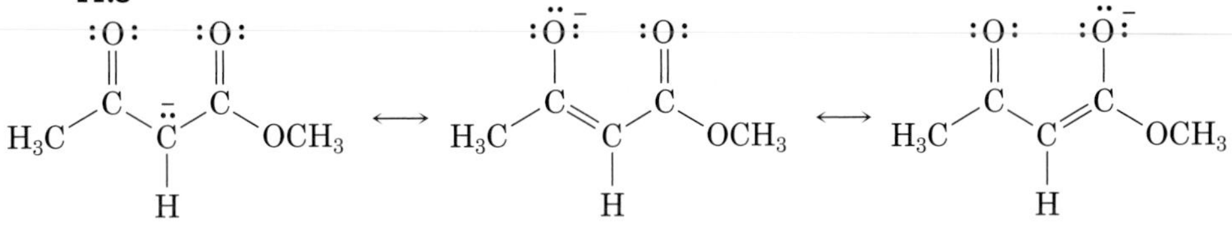

11.9 (a) CH_3CH_2Br (b) $C_6H_5CH_2Br$ (c) $(CH_3)_2CHCH_2CH_2Br$

11.10 (a) Alkylate with $(CH_3)_2CHCH_2Br$
(b) Alkylate first with $CH_3CH_2CH_2Br$ and then with CH_3Br

11.11 Alkylate first with $(CH_3)_2CHCH_2Br$ and then with CH_3Br

11.12 Only (a) undergoes an aldol reaction.

11.13 (a) $CH_3CH_2CH_2CH(OH)$—$CH(CH_2CH_3)CHO$ (b) OH, O (c) OH, O, C—CH_2C, CH_3

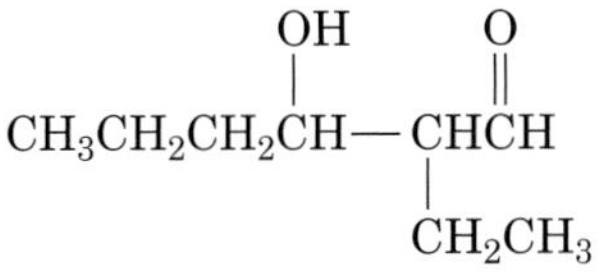

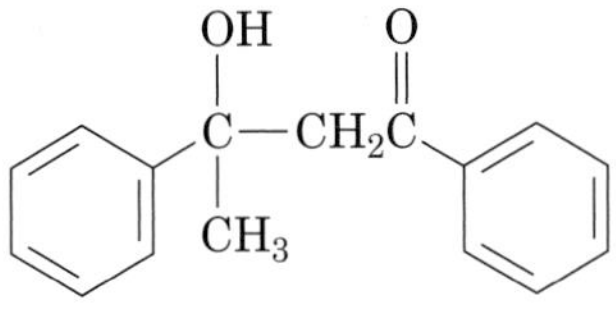

11.14 (a) $CH_3C(CH_3)$=$CHCCH_3$ (C=O) (b) O (c) CH_3CH_2CH=$C(CH_3)CH$ (C=O)

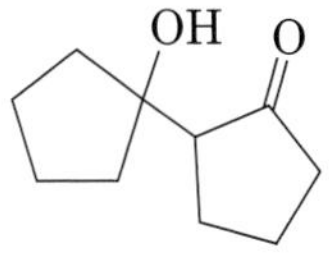

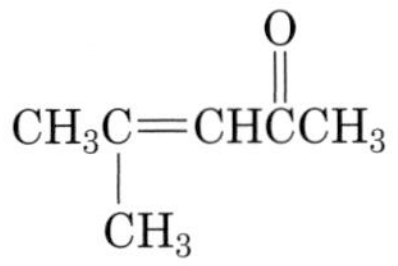

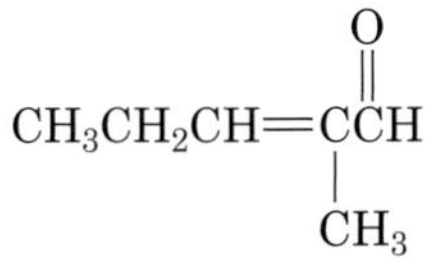

11.15

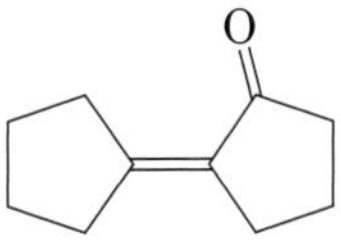

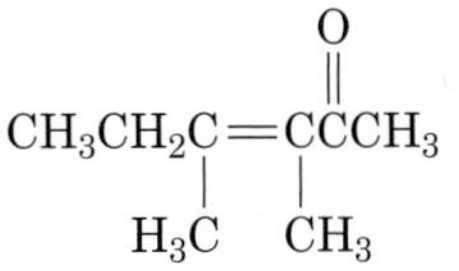

$CH_3CH_2C(H_3C)$=$C(CH_3)CCH_3$ (C=O) and $CH_3CH_2C(CH_3)$=$CHCCH_2CH_3$ (C=O)

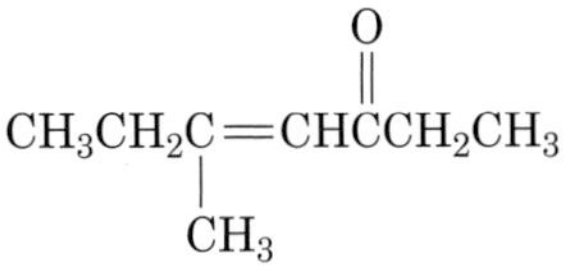

11.16 $C_6H_5COCH_3$

11.17 Only (c) undergoes a Claisen reaction.

11.18

(a) $(CH_3)_2CHCH_2\overset{O}{\overset{\|}{C}}\underset{CH(CH_3)_2}{CH}\overset{O}{\overset{\|}{C}}OCH_3$

(b) $C_6H_5CH_2\overset{O}{\overset{\|}{C}}\underset{C_6H_5}{CH}\overset{O}{\overset{\|}{C}}OCH_3$

(c) $C_6H_{11}CH_2\overset{O}{\overset{\|}{C}}\underset{C_6H_{11}}{CH}\overset{O}{\overset{\|}{C}}OCH_3$

Chapter 12

12.1 (a) Primary (b) Secondary (c) Tertiary (d) Quaternary

12.2

(a) $CH_3\underset{}{\overset{CH_3}{\overset{|}{C}}}HNHCH_3$ (b) $C_6H_5\overset{CH_3}{\overset{|}{N}}CH_2CH_3$ (c) $C_6H_{11}\overset{+}{N}(CH_3)(CH_2CH_3)(CH_2CH_2CH_3)\ Br^-$

12.3 (a) *N*-Methylethylamine (b) Tricyclohexylamine (c) *N*-Methylpyrrole
(d) *N*-Methyl-*N*-propylcyclohexylamine (e) Butane-1,3-diamine

12.4

(a) $(CH_3CH_2)_3N$ (b) $C_6H_5NHCH_3$ (c) $(CH_3CH_2)_4N^+\ Br^-$

(d) $Br-C_6H_4-NH_2$ (para) (e) $C_5H_9\overset{CH_3}{\overset{|}{N}}CH_2CH_3$

12.5 *N*-Methylcyclopentylammonium bromide

12.6 (a) $CH_3CH_2NH_2$ (b) NaOH (c) CH_3NHCH_3 (d) $(CH_3)_3N$

12.7 (a) Propanamide (b) *N*-Propylpropanamide (c) Benzamide

12.8 (a) 3-Methylbutanenitrile (b) Benzonitrile

12.9 (a) 3 $CH_3CH_2Br + NH_3$ (b) 4 $CH_3Br + NH_3$

12.10 $CH_3NH_2 + BrCH_2CH_2CH_2CH_2CH_2Br$

12.11 (a) $CH_3CH_2NH_2 + CH_3COCH_3$, or $(CH_3)_2CHNH_2 + CH_3CHO$
(b) $C_6H_5NH_2 + CH_3CHO$
(c) $C_5H_{11}NH_2 + CH_2O$, or CH_3NH_2 + cyclopentanone

12.12 $(CH_3)_2NH$ + *m*-methylbenzaldehyde

12.13 (a) 1. CH_3Cl, $AlCl_3$; 2. $KMnO_4$, H_2O; 3. HNO_3, H_2SO_4; 4. H_2, Pt catalyst
(b) 1. HNO_3, H_2SO_4; 2. H_2/Pt catalyst; 3. 3 Br_2

12.14 (a) *N*-Methyl-2-bromopyrrole (b) *N*-Methyl-2-methylpyrrole
(c) *N*-Methyl-2-acetylpyrrole

12.15 Pyrrole (N–H) $\xrightarrow{+\ NO_2}$ cationic intermediate (H, NO_2 at C-2) $\longrightarrow$ 2-nitropyrrole (N–H, NO_2)

12.16

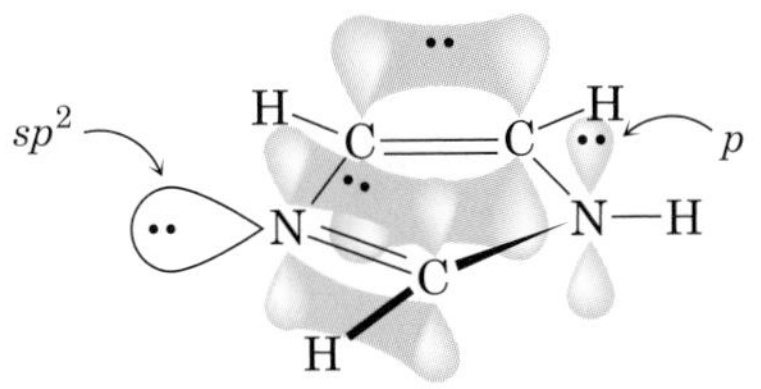

12.17 The pyridine-like doubly bonded nitrogen is more basic.

Chapter 13

13.1 IR: $\epsilon = 2.0 \times 10^{-19}$ J; X ray: $\epsilon = 6.6 \times 10^{-17}$ J

13.2 $\lambda = 9.0 \times 10^{-6}$ m is higher in energy.

13.3 We can estimate the energies associated with a single wave of light in the X-ray region (10^{-16} J), UV (10^{-18} J), cell phone (10^{-25} J), and radio (10^{-23} J) ranges using $E = h\nu$ and a frequency chosen from Figure 13.1. Given that a mole of C–C bonds has an energy of 300 kJ, one C–C bond should have an energy of $\sim 10^{-17}$ J. This energy is similar to that of X rays or UV light but greater than that of cell phone and radio waves.

13.4 Two lines of identical height at *m/z* 84 and *m/z* 86.

13.5 I_2

13.6 Butanoic acid

13.7 (a), (c), (d), and (f) have UV absorptions.

13.8 Hexa-1,3,5-triene absorbs at a longer wavelength.

13.9 3×10^{-5} M

13.10 3.5×10^{7}

13.11 (a) Ketone or aldehyde (b) Nitro (c) Nitrile or alkyne
(d) Carboxylic acid (e) Alcohol and ester

13.12 (a) CH_3CH_2OH has an –OH absorption.
(b) Hex-1-ene has a double-bond absorption.
(c) Propanoic acid has a very broad –OH absorption.

13.13 Nitrile: 2210–2260 cm^{-1}; ketone: 1690 cm^{-1}; double bond: 1640 cm^{-1}

13.14 The energy used by NMR spectroscopy is less than that used by IR spectroscopy.

13.15 (a) ^{1}H, 1; ^{13}C, 1 (b) ^{1}H, 1; ^{13}C, 1 (c) ^{1}H, 2; ^{13}C, 2 (d) ^{1}H, 1; ^{13}C, 1
(e) ^{1}H, 1; ^{13}C, 1 (f) ^{1}H, 1; ^{13}C, 1 (g) ^{1}H, 2; ^{13}C, 2 (h) ^{1}H, 2; ^{13}C, 2
(i) ^{1}H, 1; ^{13}C, 2

13.16 The vinylic C–H protons are nonequivalent.

13.17 ^{1}H, 5; ^{13}C, 7

13.18 (a) 210 Hz (b) 2.1 δ (c) 460 Hz

13.19 (a) 7.27 δ (b) 3.05 δ (c) 3.47 δ (d) 5.30 δ

13.20 (a) 0.88 δ (b) 2.17 δ (c) 7.17 δ (d) 2.22 δ

13.21 Two peaks; 3:2 ratio

13.22 (a) Doublet and ten-line multiplet (b) Doublet and quartet
(c) Singlet and two triplets (d) Singlet, triplet, and quartet
(e) Triplet and quintet (f) Singlet, doublet, and septet

13.23 (a) CH_3OCH_3 (b) $CH_3CO_2CH_3$ (c) $(CH_3)_2CHCl$

13.24

O, H (4)

H (2) — C — C(=O) — C = C(H) — CH_3 (5)

H_3C (1), CH_3 (1), CH_3 (3)

1 doublet
2 septet
3 singlet
4 quartet
5 doublet

13.25 (a) 1 (b) 5 (c) 4 (d) 7

13.26 (a) Hept-1-ene (b) 2-Methylpentane (c) 1-Chloro-2-methylpropane

Chapter 14

14.1 (a) Aldotetrose (b) Ketopentose (c) Ketohexose (d) Aldopentose

14.2

CHO
HO — H
CH_2OH

14.3

H
H_3C — Cl
CH_2CH_3
R

H
Cl — CH_3
CH_2CH_3
S

14.4 (a) *S* (b) *R* (c) *S*

14.5

H
$HOCH_2$ — CH_3 *R*
Cl

14.6 (a) L (b) D (c) D

14.7 (a)

CHO
H — OH
H — OH
CH_2OH
D

(b)

CHO
HO — H
H — OH
HO — H
CH_2OH
L

(c)

CH_2OH
C=O
H — OH
HO — H
CH_2OH
L

14.8 (a)

CHO
H — OH
HO — H
HO — H
CH_2OH

(b)

CHO
H — OH
HO — H
CH_2OH

(c)

CHO
HO — H
H — OH
H — OH
HO — H
CH_2OH

14.9 There are 16 D and 16 L aldoheptoses.

14.10

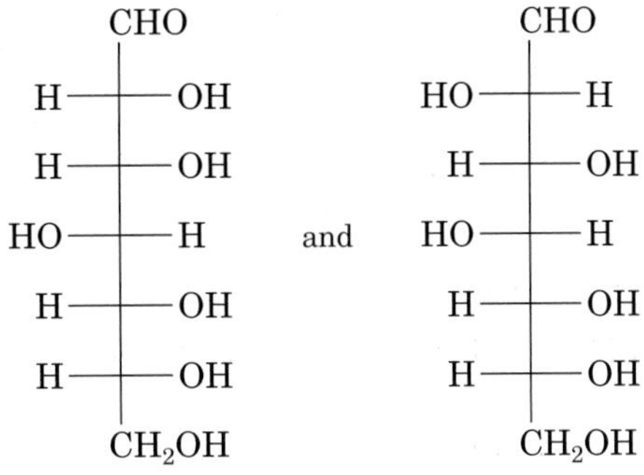

14.11

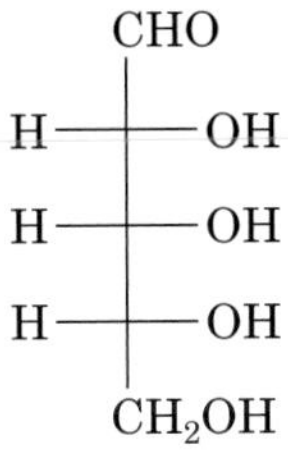

D-Ribose

14.12

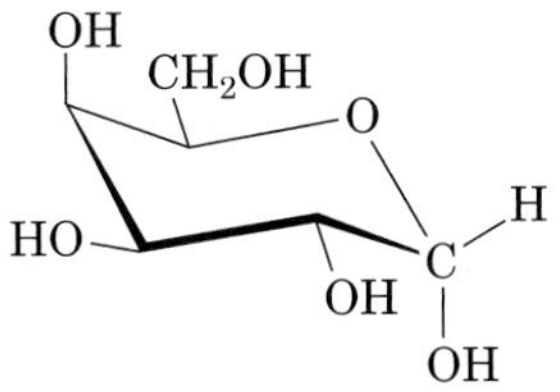

14.13

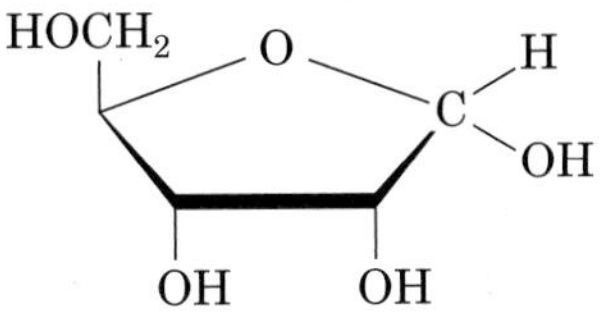

14.14

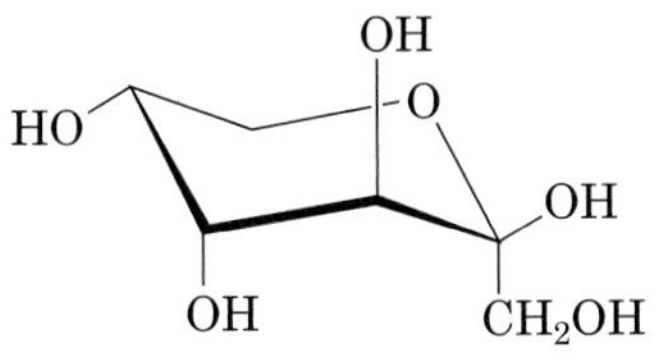

β-D-Fructopyranose

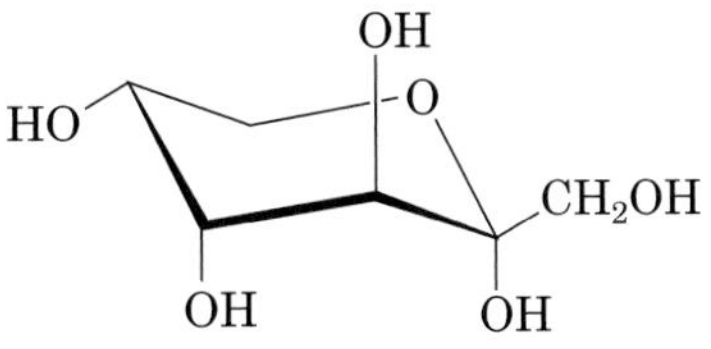

α-D-Fructopyranose

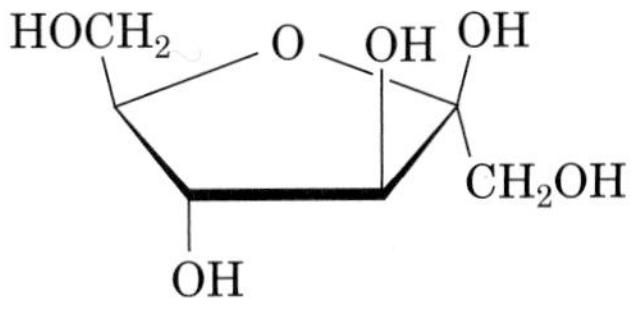

β-D-Fructofuranose

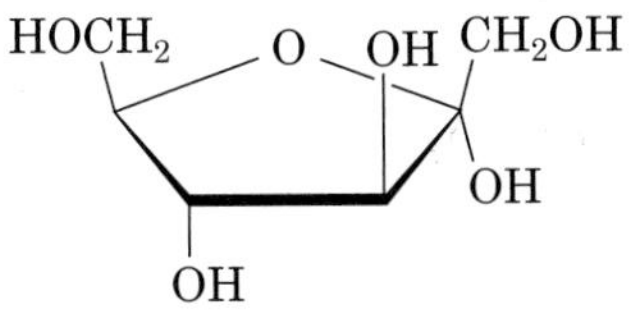

α-D-Fructofuranose

14.15 Equal stability

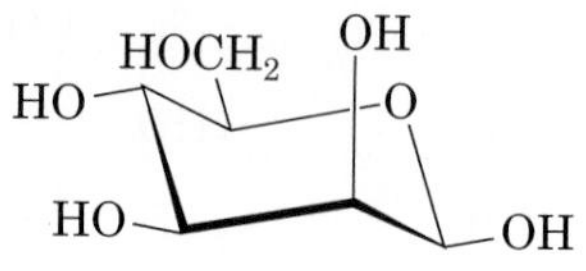

HOCH$_2$ OH
HO -O
HO OH

14.16

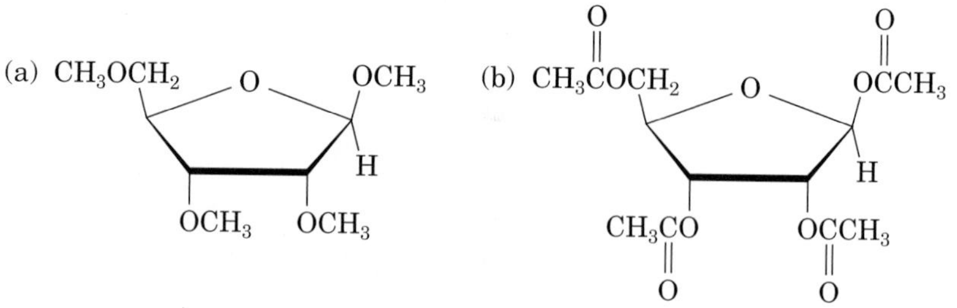

(a) CH_3OCH_2 O OCH_3 H OCH_3 OCH_3

(b) $CH_3\overset{O}{\overset{\|}{C}}OCH_2$ O $O\overset{O}{\overset{\|}{C}}CH_3$ H $CH_3\underset{O}{\underset{\|}{C}}O$ $O\underset{O}{\underset{\|}{C}}CH_3$

14.17

OH HOCH$_2$ H -O H HO OCH$_2$CH$_3$ H H OH H

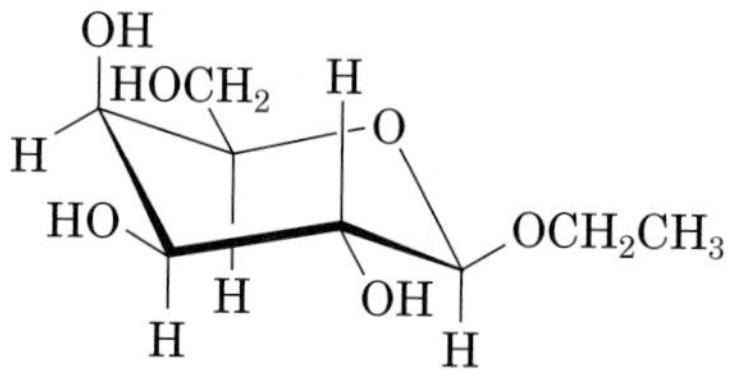

14.18 D-Galactitol is a meso compound.

14.19 An alditol has a –CH_2OH group at both ends; either could have been a –CHO group in the parent sugar.

14.20 D-Allaric acid is a meso compound; D-glucaric acid is not.

14.21 D-Allose and D-galactose yield meso aldaric acids; the other six D-aldohexoses yield optically active aldaric acids.

Chapter 15

15.1 Aromatic: Phe, Tyr, Trp, His; sulfur-containing: Cys, Met; alcohols: Ser, Thr, Tyr; hydrocarbon side chains: Ala, Ile, Leu, Val.

15.2 The sulfur atom in the –CH_2SH group of cysteine makes the side chain higher in priority than the –CO_2H group.

15.3

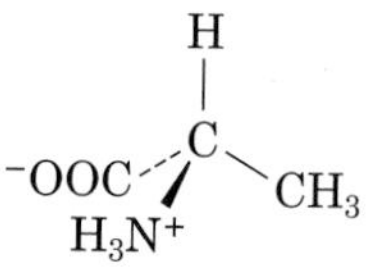

H
C
^-OOC CH_3
H_3N^+

15.4

(a) C_6H_5–$CH_2CH(NH_2)\overset{O}{\overset{\|}{C}}O^-$

(b) C_6H_5–$CH_2CH(NH_3^+)\overset{O}{\overset{\|}{C}}O^-$

(c) C_6H_5–$CH_2CH(NH_3^+)\overset{O}{\overset{\|}{C}}OH$

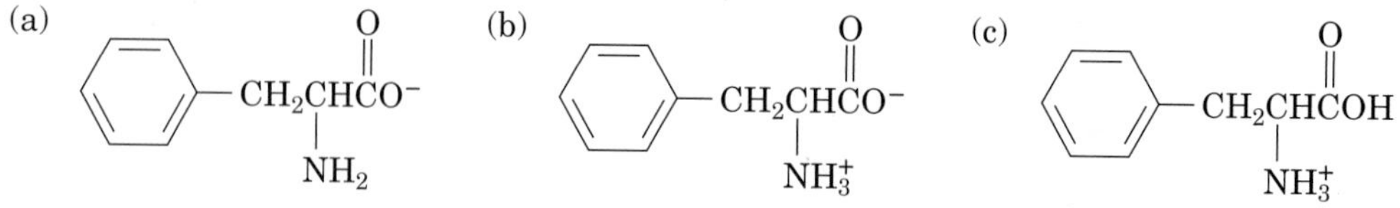

15.5 (a) $H_3\overset{+}{N}CH_2CH_2CH_2CH_2CH(^{+}NH_3)COOH$ (b) $^{-}OOCCH_2CH(^{+}NH_3)COO^{-}$

(c) $H_2NCH_2CH_2CH_2CH_2CH(NH_2)COO^{-}$ (d) $CH_3CH(^{+}NH_3)COO^{-}$

15.6 (a) Toward (+): Glu > Val; toward (−): none
(b) Toward (+): Phe; toward (−): Gly
(c) Toward (+): Phe > Ser; toward (−): none

15.8 Val-Tyr-Gly (VYG), Tyr-Gly-Val (YGV), Gly-Val-Tyr (GVY), Val-Gly-Tyr (VGY), Tyr-Val-Gly (YVG), Gly-Tyr-Val (GYV)

15.9

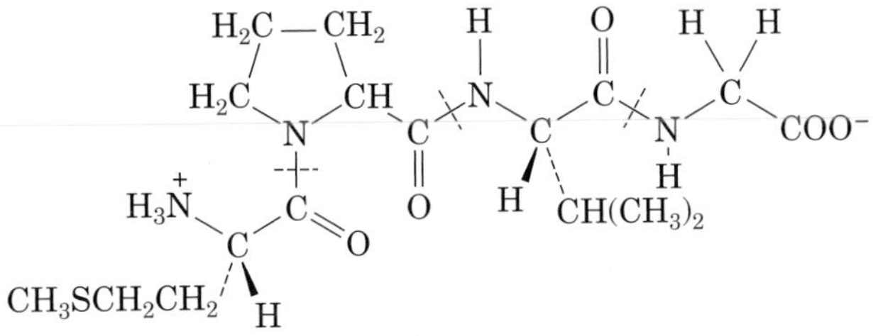

H_2C—CH_2, H, O, H, H
H_2C, CH, N, C, C, N, C, COO^-
$H_3\overset{+}{N}$, C, C=O, O, H, $CH(CH_3)_2$, H
$CH_3SCH_2CH_2$, H

15.10

$$(CH_3)_2CHCH(^{+}NH_3)COO^{-} \xrightarrow{\text{Ninhydrin}} (CH_3)CHCH{=}O + CO_2 + 3\,H_2O$$

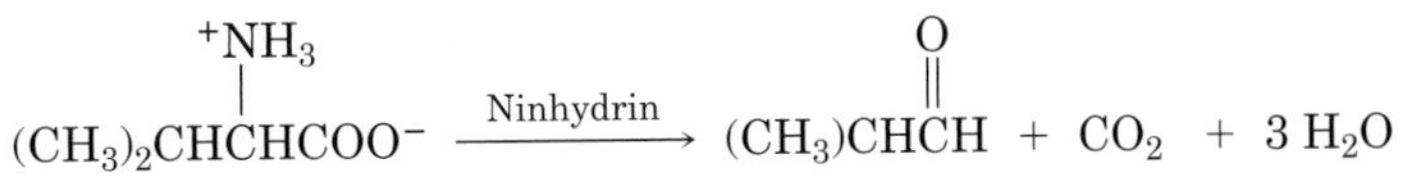

15.11 Trypsin: Asp-Arg + Val-Tyr-Ile-His-Pro-Phe
Chymotrypsin: Asp-Arg-Val-Tyr + Ile-His-Pro-Phe

15.12 Arg-Pro-Leu-Gly-Ile-Val

15.13 Methionine

15.14 Steps 1 and 2: Protect the N-terminus of leucine. Protect the C-terminus of alanine. Step 3: Couple protected amino acids together using a coupling reagent like DCC. Steps 4 and 5: Deprotect N-terminus, deprotect C-terminus.

15.15 Starting with a protected glycine on the solid phase, remove the protective group, treat with protected phenylalanine and coupling reagent such as DCC, deprotect the dipeptide and treat with protected valine and DCC, deprotect the tripeptide, and cleave it from the solid phase.

15.16 (a) Lyase (b) Hydrolase (c) Oxidoreductase

Chapter 16

16.1 $CH_3(CH_2)_{20}CO_2H + HO(CH_2)_{27}CH_3$

16.2 Glyceryl monooleate distearate is higher melting.

16.3 The fat molecule with stearic acid esterified to the central –OH group of glycerol is optically inactive.

16.4 $[CH_3(CH_2)_7CH{=}CH(CH_2)_7CO_2^{-}]_2\ Mg^{2+}$

16.6 Two ketones, double bond

16.7 Both have an aromatic ring.

16.10 ACGGATTAGCC

16.11

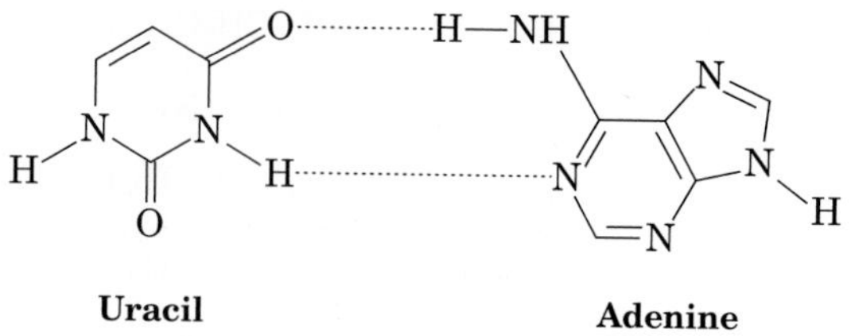

Uracil — Adenine

16.12 UACGGUAAUC

16.13 ACTCTGCGAA

16.14 (a) GCU, GCC, GCA, GCG (b) UUU, UUC
(c) UUA, UUG, CUU, CUC, CUA, CUG (d) UAU, UAC

16.15 Leu-Met-Ala-Trp-Pro-Stop

16.16 AAG, CAU, AGC, CCA, GGG, UUA

16.17 TTA-GGG-CCA-AGC-CAT-AAG

Chapter 17

17.1 $HOCH_2CH(OH)CH_2OH + ATP \rightarrow HOCH_2CH(OH)CH_2OPO_3^{2-} + ADP$

17.3 (a) 8 acetyl CoA; 7 passages (b) 10 acetyl CoA; 9 passages

17.4 Steps 7 and 10

17.5 Step 1: nucleophilic acyl substitution at phosphorus; Step 2: isomerization by keto–enol tautomerization; Step 3: like step 1; Step 4: retro aldol condensation; Step 5: like step 2; Step 6: oxidation; Step 7: like step 2; Step 8: isomerization; Step 9: E2 reaction; Step 10: substitution at phosphorus, followed by tautomerization

17.6 Citrate and isocitrate

17.7

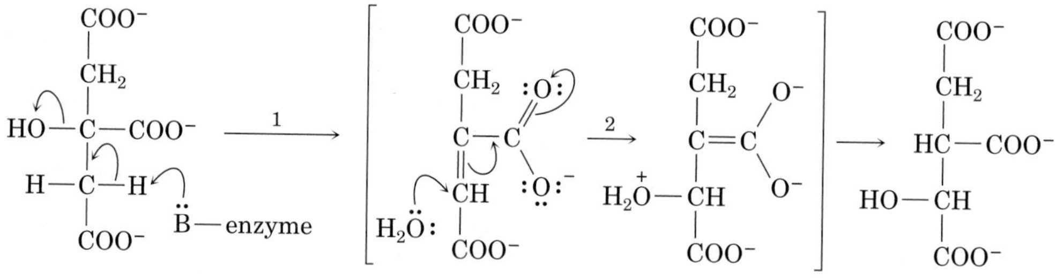

17.8 $(CH_3)_2CHCH_2COCO_2^-$

17.9 Asparagine

Index

Boldfaced references refer to pages where terms are defined.